THE ATMOSPHERE

12TH EDITION

THE ATMOSPHERE

AN INTRODUCTION TO METEOROLOGY

Frederick K. Lutgens

Edward J. Tarbuck

Illustrated by

Dennis Tasa

PEARSON

Boston   Columbus   Indianapolis   New York   San Francisco   Upper Saddle River
Amsterdam   Cape Town   Dubai   London   Madrid   Milan   Munich   Paris   Montréal   Toronto
Delhi   Mexico City   São Paulo   Sydney   Hong Kong   Seoul   Singapore   Taipei   Tokyo

Geography Editor: *Christian Botting*
Marketing Manager: *Maureen McLaughlin*
Senior Project Editor: *Crissy Dudonis*
VP/Executive Director, Development: *Carol Trueheart*
Development Editor: *Jonathan Cheney*
Media Producer: *Tim Hainley*
Assistant Editor: *Sean Hale*
Editorial Assistant: *Bethany Sexton*
Marketing Assistant: *Nicola Houston*
Managing Editor, Geosciences and Chemistry: *Gina M. Cheselka*
Project Manager, Production: *Edward Thomas*
Full Service/Composition: *Element-Thomson North America*
Full Service Project Manager: *Heidi Allgair*
Senior Art Specialist: *Connie Long*
Cartography: *Kevin Lear, Spatial Graphics*
Interior and Cover Design: *Tamara Newman*
Photo Manager: *Maya Melenchuk*
Photo Researcher: *Kristin Piljay*
Text Permissions Manager: *Beth Wollar*
Text Permissions Researcher: *Jenny Bevington*
Operations Specialist: *Michael Penne*
Front Cover and Title Page Photo Credit: *Wing of a jet aircraft. Dreamstime image #11158189, photo by Adisa.*

Credits and acknowledgments borrowed from other sources and reproduced, with permission, in this textbook appear on the appropriate page within text.

Library of Congress Cataloging-in-Publication Data

Lutgens, Frederick K.
  The atmosphere : an introduction to meteorology / Frederick K. Lutgens, Edward J. Tarbuck ; illustrated by Dennis Tasa. — 12th ed.
    p. cm.
  Includes index.
  ISBN-13: 978-0-321-75631-2
  ISBN-10: 0-321-75631-2
1. Atmosphere. 2. Meteorology. 3. Weather. I. Tarbuck, Edward J.
  II. Title.
  QC861.2.L87 2013
  551.5—dc23

                                        2011037045

1 2 3 4 5 6 7 8 9 10—DOW—15 14 13 12 11

ISBN-10: 0-321-75631-2; ISBN-13: 978-0-321-75631-2 (Student Edition)
ISBN-10: 0-321-78035-3; ISBN-13: 978-0-321-78035-5 (Instructor's Review Copy)

To Our Grandchildren
Allison and Lauren
Shannon, Amy, Andy, Ali, and Michael
Each is a bright promise for the future

About Our Sustainability Initiatives

Pearson recognizes the environmental challenges facing this planet, as well as acknowledges our responsibility in making a difference. This book is carefully crafted to minimize environmental impact. The binding, cover, and paper come from facilities that minimize waste, energy consumption, and the use of harmful chemicals. Pearson closes the loop by recycling every out-of-date text returned to our warehouse.

Along with developing and exploring digital solutions to our market's needs, Pearson has a strong commitment to achieving carbon-neutrality. As of 2009, Pearson became the first carbon- and climate-neutral publishing company. Since then, Pearson remains strongly committed to measuring, reducing, and offsetting our carbon footprint.

The future holds great promise for reducing our impact on Earth's environment, and Pearson is proud to be leading the way. We strive to publish the best books with the most up-to-date and accurate content, and to do so in ways that minimize our impact on Earth. To learn more about our initiatives, please visit www.pearson.com/responsibility.

Brief Contents

GEODe: ATMOSPHERE

The *GEODe: Atmosphere* interactive learning aid is accessed from the book's Website (www.MyMeteorologyLab.com). This dynamic instructional tool reinforces atmospheric science concepts by using tutorials, animations, and interactive exercises. The *GEODe* chapter numbers relate to equivalent chapters in the text. The *GEODe: Atmosphere* icon appears throughout the book wherever a text discussion has a corresponding activity in *GEODe*.

Contents

WAVE RF/PHOTOLIBRARY

MARTIN WOIKE/AGEFOTOSTOCK

PAT & CHUCK BLACKLEY/ALAMY

PEDRO2009/DREAMSTIME

5 Forms of Condensation and Precipitation   128

6 Air Pressure and Winds   160

ED PRITCHARD/GETTY IMAGES

SAM PELLISSIER/SUPERSTOCK

7 Circulation of the Atmosphere 188

8 Air Masses 220

MICHAEL COLLIER

MIKE HILLINGSHEAD/PHOTO RESEARCHERS, INC.

9 Midlatitude Cyclones   238

10 Thunderstorms and Tornadoes   270

ALEXEY STIOP/ALAMY

PUBLIC DOMAIN

11 Hurricanes 302

12 Weather Analysis and Forecasting 326

SMILEY N. POOL/RAPPORT PRESS/NEWCOM

DOABLE/AMANA IMAGES/GLOW IMAGES

13 Air Pollution   356

14 The Changing Climate   378

DAVID VAUGHAN/PHOTO RESEARCHERS, INC.

MICHAEL GIANNECHINI/PHOTO RESEARCHERS, INC.

Preface

There are few aspects of the physical environment that influence our daily lives more than the phenomena we collectively call *weather*. The media regularly report a wide range of weather events as major news stories—an obvious reflection of people's interest in and curiosity about the atmosphere.

Not only does the atmosphere impact the lives of people, but people have a significant impact on the atmosphere as well. By altering the composition of Earth's atmosphere, people have diminished the stratosphere's ozone layer and created serious air-quality problems in urban and rural areas around the world. Moreover, human-generated emissions likely play an important role in global climate change, one of the most serious environmental issues facing humankind in the twenty-first century.

In order to understand the weather phenomena that affect our daily lives and the serious environmental problems related to the atmosphere, it is important to develop an understanding of meteorological principles. A basic meteorology course takes advantage of our interest in and curiosity about the weather as well as our desire to understand the impact that people have on the atmospheric environment.

The Atmosphere: An Introduction to Meteorology, 12th edition, is designed to meet the needs of students who enroll in such a course. It is our hope that the knowledge gained by taking a class and using this book will encourage many to actively participate in bettering the environment; some may even be sufficiently stimulated to continue their study of meteorology. Equally important, however, is our belief that a basic understanding of the atmosphere and its processes will greatly enhance appreciation of our planet and thereby enrich the reader's life.

In addition to being informative and up-to-date, *The Atmosphere* meets the need of beginning students for a readable and user-friendly text, a book that is a highly usable tool for learning basic meteorological principles and concepts.

New to the 12th Edition

- **New Active Learning Path.** Each chapter begins with *Focus on Concepts,* which identifies the knowledge and skills that students should master by the end of the chapter, helping students prioritize key concepts. Within the chapter, each major section concludes with a *Concept Check* that allows students to check their understanding and comprehension of important topics before moving on to the next section. Each chapter concludes with a new section called *Give It Some Thought.* These questions and problems challenge learners by involving them in activities that require higher-order thinking skills that include application, analysis, and synthesis of material in the chapter.
- **Eye on the Atmosphere.** Within every chapter are two or three images, often aerial or satellite views, that challenge students to apply their understanding of basic facts and principles. A brief explanation of each image is followed by questions that serve to focus students on visual analysis and critical thinking tasks.
- **Severe and Hazardous Weather.** Atmospheric hazards adversely affect millions of people worldwide every day. Severe weather events have a significance and fascination that go beyond ordinary weather phenomena. Two entire chapters (Chapter 10, "Thunderstorms and Tornadoes," and Chapter 11, "Hurricanes") focus entirely on such topics. Moreover, the text contains 15 *Severe and Hazardous Weather* essays that are devoted to a broad variety of topics—heat waves, winter storms, floods, air pollution episodes, drought, wildfires, cold waves, and more.
- **Professional Profiles.** These essays present profiles of professionals who use meteorology in the real world, giving students a sense of professional applications and careers in the science. Included are profiles on research meteorologists, a military meteorologist, a climate scientist, a broadcast meteorologist, a storm chaser, and a senior hurricane specialist and warning coordination meteorologist.
- **An Unparalleled Visual Program That Teaches.** In addition to more than 150 new high-quality photos and satellite images, dozens of figures are new or redrawn by renowned geoscience illustrator Dennis Tasa. Numerous diagrams and maps are paired with photographs for greater effectiveness. Many new and revised art pieces also have additional labels that narrate the process being illustrated and guide students as they examine the figures. The result is a visual program that is clear and easy to understand. In addition, a new removable Cloud Guide appears at the back of the book, to help students with observation and forecasting out in the field.
- **Significant Updating and Revision of Content.** With the goal of keeping the text current and highly readable for beginning students, many discussions, case studies, and examples have been revised. This 12th edition represents perhaps the *most extensive and thorough revision* in the long history of this textbook. See the section "Revised and Updated Content" for more particulars.
- **MyMeteorologyLab with Pearson eText.** www.MyMeteorologyLab.com is a new resource both for student self-study and for instructors to manage their courses online and provide customizable assessments to students. MyMeteorologyLab's assignable content includes Geoscience Animations, GEODe: Atmosphere tutorials, MapMaster™ interactive maps, a variety of chapter quizzes, and more. Students can also access the Pearson eText for *The Atmosphere*, 12th edition, Carbone's *Exercises for Weather & Climate* interactive media, "*In the News*" RSS feeds, glossary flashcards, social networking features, and additional references and resources to extend learning beyond the text.

Distinguishing Features

Readability

The language of this book is straightforward and easy to understand. Clear, readable discussions with a minimum of technical language are the rule. The frequent headings and subheadings help students follow discussions and identify the important ideas presented in each chapter.

Visual Program

Meteorology is highly visual, so art and photographs play a critical role in an introductory textbook. Our aim is to get *maximum effectiveness* from the visual component of the book. As in previous editions, Dennis Tasa, a gifted artist and respected science illustrator, has worked closely with the authors to plan and produce the diagrams, maps, graphs, and sketches that are so basic to student understanding. The result is art that is clear and easy to understand.

Focus on Basic Principles and Instructor Flexibility

Although many topical issues are treated in the 12th edition of *The Atmosphere,* it should be emphasized that the main focus of this new edition remains the same as the focus of each of its predecessors—to promote student understanding of the basic principles of meteorology. Student use of the text is a primary concern, and the book's adaptability to the needs and desires of the instructor is equally important. In keeping with this aim, the organization of the text remains intentionally traditional, allowing for maximum instructor flexibility in terms of the sequence and emphasis of topics.

Focus on Learning

In addition to the new active learning features described earlier, *The Atmosphere,* 12th edition includes other important learning aids. Each chapter includes a number of *Students Sometimes Ask* features that address common student questions, a *Chapter in Review* that recaps all the major points, and a *Vocabulary Review* that provides a checklist of key terms with page references. In most chapters, *Problems,* many with quantitative orientation, are included. Most problems require only basic math skills and allow students to enhance their understanding by applying concepts and principles explained in the chapter. Each chapter ends by reminding students to go to the text's outstanding premium website, www.MyMeteorologyLab.com, to access many useful learning tools.

Revised and Updated Content

This 12th edition of *The Atmosphere* represents a thorough revision. In fact, it is likely the most thorough revision of text and figures in the long history of this text. Those familiar with the previous edition will see much that is new. The following are some examples:*

- In Chapter 1 the section "Earth as a System" is simplified and reduced in size, and the box on "The Origin and Evolution of Earth's Atmosphere" is revised, updated, and rewritten.
- In Chapter 2 the basic discussion of seasons is revised to provide a more understandable introduction to this important topic. "The Role of Gases in the Atmosphere" is revised and expanded, and "Earth's Heat Budget" is reorganized and revised to improve clarity.
- In Chapter 3 significant changes include three new world temperature maps and the all-new section "Heat Stress and Windchill."
- In Chapter 4 the section "Vapor Pressure and Saturation" is substantially revised, and a new figure is added to support this important topic. "How Is Humidity Measured?" is simplified to improve clarity.
- In Chapter 5 the discussion of hail is revised and updated, the section "Precipitation Measurement" is simplified and reduced, and the section "Intentional Weather Modification" is reorganized and streamlined.
- Chapter 6 begins with a revised and expanded introductory section. A new section titled "Why Does Air Pressure Vary?" replaces the section "Horizontal Variation in Air Pressure."
- Significant portions of Chapter 7 are completely rewritten and extensively reorganized. The sections "The Westerlies" and "Waves in the Westerlies" are combined and completely rewritten. The discussions of El Niño and La Niña are completely revised and include all-new figures.
- Chapter 8 includes an all-new case study of a classic nor'easter that hit New England in January 2011.
- Chapter 9 (formerly "Weather Patterns") has a new title, "Midlatitude Cyclones," and has been completely revamped, with several new discussions and others that have been rewritten for greater clarity.
- The material in Chapter 10, "Thunderstorms and Tornadoes," includes new statistics and examples from spring 2011. It includes a new box on downbursts, new material on lightning and tornado destruction, and a Professional Profile that highlights a storm chaser.
- Chapter 11, the second of two chapters that focus on severe weather, includes a revised introduction, updated coverage, a Professional Profile, and two revised case study boxes.
- In Chapter 12 the sections "Weather Forecasting" and "Forecast Accuracy" are reorganized and significantly revised for greater clarity. The section "Satellites in Weather Forecasting" is updated and expanded. The chapter includes a Professional Profile of a research meteorologist.
- Chapter 13 begins with a new heading, "The Threat of Air Pollution." The revised discussion "Secondary

*For a complete and detailed list of changes, contact your local Pearson representative.

Pollutants" is now divided into two parts. The new section "Air Quality Index" is accompanied by new art. A new box, "Viewing an Air Pollution Episode from Space," presents a brief case study.

- Chapter 14, "The Changing Climate," which surveys both natural causes and the human impact on global climate, is thoroughly updated to include the most recent data available. The chapter includes material on trace gases, aerosols, computer models, sea-level rise, and ocean acidity.
- Chapter 15 includes a new, more accurate, and more readable map of world climates and a new chart showing the Köppen classification.
- In Chapter 16 the section "Interactions of Light and Matter" is completely rewritten and includes new supporting illustrations. The discussion of rainbows is reorganized and rewritten for greater clarity, while other sections are reduced and are more readable.

The Atmosphere Teaching and Learning Package

The authors and publisher are pleased to present unparalleled media and supplements resources for students and instructors.

For You, the Student

- **MyMeteorologyLab with Pearson eText**
  www.MyMeteorologyLab.com is a new resource both for student self-study and for instructors to manage their courses online and assign customizable and automatically graded assessments to students. MyMeteorologyLab's assignable content includes Geoscience Animations, GEODe: Atmosphere tutorials, MapMaster™ interactive maps, videos, a variety of chapter quizzes, and more. Students can also access the Pearson eText for *The Atmosphere*, 12th edition, Carbone's *Exercises for Weather & Climate* interactive media, "*In the News*" RSS feeds, glossary flashcards, social networking features, and additional references and resources to extend learning beyond the text.
- **Pearson eText for** *The Atmosphere*, **12th edition** Pearson eText for *The Atmosphere*, 12th edition gives you access to the text whenever and wherever you can access the Internet and includes powerful interactive and customization functions.
- *Exercises for Weather and Climate*, 8th edition by Greg Carbone [0321769651] This bestselling exercise manual's 17 exercises encourage students to review important ideas and concepts through problem solving, simulations, and guided thinking. The graphics program and computer-based simulations and tutorials help students grasp key concepts. This manual is designed to compliment any introductory meteorology or weather and climate text.

- *Encounter Meteorology: Interactive Explorations of Earth Using Google Earth* [0321815912] This workbook and premium website provides rich, interactive explorations of meteorology concepts through Google Earth™ explorations. All chapter explorations are available in print format as well as via online quizzes and downloadable PDFs, accommodating different classroom needs. Each worksheet is accompanied by corresponding Google Earth KMZ media files containing the placemarks, overlays, and annotations referred to in the worksheets, available for download from www.mygeoscienceplace.com.
- *Encounter Geosystems: Interactive Explorations of Earth Using Google Earth* [0321636996] This workbook and premium website provides rich, interactive explorations of physical geography concepts through Google Earth explorations. All chapter explorations are available in print format as well as via online quizzes and downloadable PDFs, accommodating different classroom needs. Each worksheet is accompanied by corresponding Google Earth KMZ media files containing the placemarks, overlays, and annotations referred to in the worksheets, available for download from www.mygeoscienceplace.com.
- *Goode's World Atlas,* 22nd edition [0321652002] *Goode's World Atlas* has been the world's premiere educational atlas since 1923—and for good reason. It features over 250 pages of maps, from definitive physical and political maps to important thematic maps that illustrate the spatial aspects of many important topics. The 22nd edition includes 160 pages of new, digitally produced reference maps, as well as new thematic maps on global climate change, sea-level rise, CO_2 emissions, polar ice fluctuations, deforestation, extreme weather events, infectious diseases, water resources, and energy production.
- *Dire Predictions: Understanding Global Warming* [0136044352] Periodic reports from the Intergovernmental Panel on Climate Change (IPCC) evaluate the risk of climate change brought on by humans. But the sheer volume of scientific data remains inscrutable to the general public, particularly to those who may still question the validity of climate change. In just over 200 pages, this practical text presents and expands upon the essential findings in a visually stunning and undeniably powerful way to the lay reader. Scientific findings that provide validity to the implications of climate change are presented in clear-cut graphic elements, striking images, and understandable analogies.

For You, the Instructor

- *Geoscience Animation Library* on DVD, 5th edition [0321716841] For engaging lectures, the Geoscience Animation Library includes more than 110 animations illuminating many difficult-to-visualize topics in physical geology, physical geography, oceanography, meteorology, and Earth science—created through a unique collaboration among Pearson's leading geoscience authors.

- **MyMeteorologyLab**   www.MyMeteorologyLab.com helps instructors manage their courses online, with robust course management, gradebook, and diagnostic tools. Instructors can assign customizable and automatically graded assessments to students. Assignable content includes Geoscience Animations, GEODe: Atmosphere tutorials, MapMaster™ interactive maps, videos, a variety of chapter quizzes, and more.

- *Instructor Resource* **DVD** [0321780337] The *Instructor Resource* DVD provides high-quality electronic versions of photos and illustrations from the book, as well as customizable PowerPoint™ lecture presentations, Classroom Response System questions in PowerPoint, and the *Instructor Resource Manual* and *Test Bank* in Microsoft Word and TestGen formats. The DVD also includes all the illustrations and photos from the text, in presentation-ready JPEG files, as well as digital transparencies. For easy reference and identification, all resources are organized by chapter. All of the elements on the DVD are also available online to professors at www.pearsonhighered.com/irc.

- *Instructor Resource Manual* **by Neve Duncan Tabb (download only)** [0321780329] The *Instructor Resource Manual* is intended as a resource for both new and experienced instructors. It includes a variety of lecture outlines, additional source materials, teaching tips, advice about how to integrate visual supplements (including the web-based resources), and various other ideas for the classroom. See www.pearsonhighered.com/irc.

- **TestGen® Computerized Test Bank by Jennifer Johnson (download only)** [0321780299] TestGen® is a computerized test generator that lets instructors view and edit *Test Bank* questions, transfer questions to tests, and print tests in a variety of customized formats. This *Test Bank* includes more than 2000 multiple-choice, fill-in-the-blank, and short-answer/essay questions. Questions are correlated to the revised U.S. National Geography Standards and Bloom's Taxonomy to help instructors better map the assessments against both broad and specific teaching and learning objectives. The *Test Bank* is also available in Microsoft Word™ and is importable into Blackboard. See www.pearsonhighered.com/irc.

- *Earth Report Geography Videos* **on DVD** [0321662989] This three-DVD set is designed to help students visualize how human decisions and behavior have affected the environment and how individuals are taking steps toward recovery. With topics ranging from poor land management promoting the devastation of river systems in Central America to the struggles for electricity in China and Africa, these 13 videos from Television for the Environment's global *Earth Report* series recognize the efforts of individuals around the world to unite and protect the planet.

Acknowledgments

Writing a college textbook requires the talents and cooperation of many individuals. Working with Dennis Tasa, who is responsible for all of the text's outstanding illustrations and much of the developmental work on *GEODe: Atmosphere*, is always special for us. We not only value his outstanding artistic talents and imagination but his friendship as well.

Great thanks go to those colleagues who prepared in-depth reviews or suggestions for new Give It Some Thought questions. Their critical comments and thoughtful input helped guide our work and clearly strengthened the 12th edition. We wish to thank:

Jason Allard, Valdosta State University
Deanna Bergondo, U.S. Coast Guard Academy
William Conant, University of Arizona
Ron Dowey, Harrisburg Area Community College–Harrisburg
Douglas Gamble, University of North Carolina, Wilmington
Mark Hildebrand, Southern Illinois University–Edwardsville
Helenmary Hotz, University of Massachusetts–Boston
Timothy and Jennifer Klingler, Delta College
Mark Lemmon, Texas A&M University
Jason Ortegren, University of West Florida
Robert S. Rose, Tidewater Community College–Virginia Beach
Roger D. Shew, University of North Carolina–Wilmington
Steve Simpson, Highland Community College
Eric Snodgrass, University of Illinois–Urbana-Champaign
Andrew Van Tuyl, Galivan College

We also want to acknowledge the team of professionals at Pearson Education. We sincerely appreciate the company's continuing strong support for excellence and innovation. Our special thanks to the outstanding geography and meteorology team—Christian Botting, Crissy Dudonis, and Maureen McLaughlin. In addition to being great people to work with, all are committed to producing the best textbooks possible. The production team, led by Heidi Allgair at Element LLC and Ed Thomas at Pearson, has once again done an outstanding job. All are true professionals with whom we are very fortunate to be associated.

Fred Lutgens

Ed Tarbuck

Structured Learning Path

Powerful pedagogy equips students with the skills to master the science.

Focus On Concepts

After completing this chapter, you should be able to:

- Distinguish between weather and climate and name the basic elements of weather and climate.
- List several important atmospheric hazards and identify those that are storm related.
- Construct a hypothesis and distinguish between a scientific hypothesis and a scientific theory.
- List and describe Earth's four major spheres.
- Define system and explain why Earth can be thought of as a system.

- List the major gases composing Earth's atmosphere and identify those components that are most important meteorologically.
- Explain why ozone depletion is a significant global issue.
- Interpret a graph that shows changes in air pressure from Earth's surface to the top of the atmosphere.
- Sketch and label a graph showing the thermal structure of the atmosphere.
- Distinguish between homosphere and heterosphere.

NEW! Focus on Concepts

Clear, testable learning outcomes at the beginning of each chapter focus students on key concepts and skills.

Concept Check 10.6

1 How is thunder produced?

2 Which is more common: sheet lightning or cloud-to-ground lightning?

3 What is heat lightning?

Concept Check 11.1

1 Define *hurricane*. What other names are used for this storm?

2 In what latitude zone do hurricanes develop?

3 Distinguish between the eye and the eye wall of a hurricane. How do conditions differ in these zones?

NEW!
Concept Checks

These questions appear at the end of each major section, giving students a chance to stop, check, and practice their understanding of key chapter concepts before moving on.

Give It Some Thought

1. If you were asked to identify the coldest city in the United States (or any other designated region), what statistics could you use? Can you list at least three different ways of selecting the coldest city?

2. The accompanying graph shows monthly high temperatures for Urbana, Illinois, and San Francisco, California. Although both cities are located at about the same latitude, the temperatures they experience are quite different. Which line on the graph represents Urbana and which represents San Francisco? How did you figure this out?

3. On which summer day would you expect the greatest temperature range? Which would have the smallest range in temperature? Explain your choices.
   a. Cloudy skies during the day and clear skies at night
   b. Clear skies during the day and cloudy skies at night
   c. Clear skies during the day and clear skies at night
   d. Cloudy skies during the day and cloudy skies at night

4. The accompanying scene shows an island near the equator in the Indian Ocean. Describe how latitude, altitude,  and the differential heating of land and water influence the climate of this place.

5. The accompanying sketch map represents a hypothetical continent in the Northern Hemisphere. One isotherm has been placed on the map.

   a. Is the temperature higher at city A or city B? Explain.
   b. Is the season winter or summer? How are you able to determine this?
   c. Describe (or sketch) the position of this isotherm six months later.

6. The data below are mean monthly temperatures in °C for an inland location that lacks any significant ocean influence. *Based on the annual temperature range,* what is the approximate latitude of this place? Are these temperatures what you would normally expect for this latitude? If not, what control would explain these temperatures?

J	F	M	A	M	J	J	A	S	O	N	D
6.1	6.6	6.6	6.6	6.1	6.1	6.1	6.1	6.1	6.1	6.6	6.6

7. Refer to Figure 3–18. What causes the bend or kink in the isotherms in the North Atlantic?

NEW! Give It Some Thought

GIST questions at the end of each chapter give students an opportunity to synthesize chapter concepts and practice higher-order thinking.

PROBLEMS

1. Figure 8-12 shows the distribution of air temperatures (top number) and dew-point temperatures (lower number) for a December morning. Two well-developed air masses are influencing North America at this time. The air masses are separated by a broad zone that is not affected by either air mass. Draw lines on the map to show the boundaries of each air mass. Label each air mass with the proper classification.

2. Refer to Figure 8-5. Notice the narrow, north-south-oriented zone of relatively heavy snowfall east of Pittsburgh and Charleston. This region is too far from the Great Lakes to receive lake-effect snows. Speculate on a likely reason for the higher snowfalls here. Does your explanation explain the shape of this snowy zone?

3. Albuquerque, New Mexico, is situated in the desert Southwest. Its annual precipitation is just 21.2 centimeters (8.3 inches). Month-by-month data (in centimeters) are as follows:

Figure 8-12 Map to accompany Problem 1.

Jan.	Feb.	Mar.	Apr.	May	Jun.	Jul.	Aug.	Sep.	Oct.	Nov.	Dec.
1.0	1.0	1.3	1.3	1.3	1.3	3.3	3.8	2.3	2.3	1.0	1.3

What are the two rainiest months? The pattern here is similar to the pattern in other southwestern cities, including Tuscon, Arizona. Briefly explain why the rainiest months occur when they do.

Problems

"Problems" extend learning with deeper, quantitative treatments of chapter concepts.

Observe and Apply

Tools to refine students' observation skills and emphasize the relevance of meteorology today.

NEW!
Eye on the Atmosphere

These features ask students to inspect visualizations and data, practicing their critical thinking and visual analysis skills.

NEW!
Professional Profiles

These essays profile a variety of professionals who use meteorology every day, emphasizing opportunities in the field and the relevance of the science.

NEW!
Severe and Hazardous Weather Essays

These essays focus on the dramatic severe and hazardous weather phenomena that increasingly impact our world.

NEW! Cloud Guide

A fold out cloud guide at the back of the book provides students with a tool and reference for real world observation.

Cloud Guide

High Clouds: cloud bases above 6km (20,000 ft) | **Middle Clouds: cloud bases 2–6km (6,500–20,000 ft)**

Cirrus These clouds are made exclusively of ice crystals. They are not as horizontally extensive as cirrostratus clouds.

Cirrocumulus These high clouds can produce striking skies. Composed of ice crystals, they often contain linear bands, numerous patches of greater vertical development, or both.

Altocumulus These midlevel clouds are horizontally layered but exhibit varying thicknesses across their bases. Thicker areas can be arranged as parallel linear bands or as a series of individual puffs.

Altocumulus (Lenticular) These clouds are marked by their lens-shaped appearance. They usually form downwind of mountain barriers as horizontal airflow is disrupted into a sequence of waves.

Cirrostratus These are thin layered clouds composed of ice crystals. They are relatively indistinct and give the sky a whitish appearance.

Contrails A contrail is a long, narrow cloud that is formed as exhaust from a jet aircraft condenses in cold air at high altitude. Upper level winds may gradually cause contrails to spread out.

Altostratus These are midlevel, layered clouds that produce gray skies and obscure the Sun or Moon enough to make them appear as poorly defined bright spots. In this example, the setting sun brightens the clouds near the horizon but the gray appearance remains elsewhere.

Altostratus (Multilayer) These are midlevel layered clouds that are dense enough to completely hide the Sun or Moon.

MyMeteorologyLab
Visualization and Practice

An assortment of assignable, assessable media illustrate complex processes and bring concepts to life.

www.MyMeteorologyLab.com

NEW! Geoscience Animations

These animations help students visualize difficult physical processes over space and time.

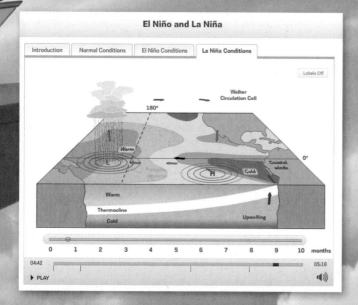

NEW! MapMaster™

These interactive maps act as a mini-GIS tool, allowing students to layer various thematic maps to analyze spatial patterns and data at regional and global scales.

Exercises for Weather and Climate

These interactive simulations and exercises give students practice with problem solving and guided critical thinking.

GEODe: Atmosphere

A dynamic program that reinforces key concepts through animations, tutorials, interactive exercises, and review quizzes.

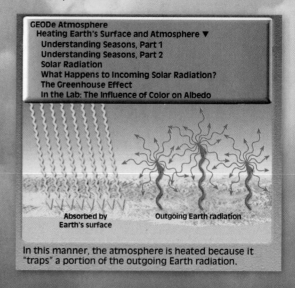

In this manner, the atmosphere is heated because it "traps" a portion of the outgoing Earth radiation.

MyMeteorologyLab
A Full Course Solution

Assignable media, intuitive gradebook, and powerful diagnostics help you focus on teaching.

www.MyMeteorologyLab.com

Course Management

MyMeteorologyLab is a full-featured online course management and homework system.

Gradebook

An easy-to-use and flexible gradebook helps identify struggling students.

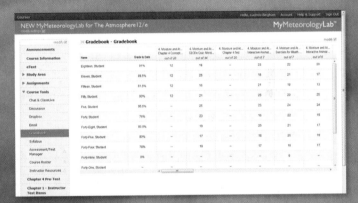

Study Area (media, RSS feeds, flashcards, quizzes)
Students have access to a rich set of study tools to help extend learning, master concepts, and prepare for exams.

eText

MyMeteorologyLab includes the option of a Pearson eText version of *The Atmosphere*, with full search and annotation capability.

Introduction to the Atmosphere

Earth's atmosphere is unique. No other planet in our solar system has an atmosphere with the exact mixture of gases or the heat and moisture conditions necessary to sustain life as we know it. The gases that make up Earth's atmosphere and the controls to which they are subject are vital to our existence. In this chapter we begin our examination of the ocean of air in which we all must live.

Hundreds of cars stranded on Chicago's Lake Shore Drive on February 2, 2011, following a winter blizzard of historic proportions. (AP Photo/ Kiichiro Sato)

Focus On Concepts

After completing this chapter, you should be able to:

- Distinguish between weather and climate and name the basic elements of weather and climate.
- List several important atmospheric hazards and identify those that are storm related.
- Construct a hypothesis and distinguish between a scientific hypothesis and a scientific theory.
- List and describe Earth's four major spheres.
- Define *system* and explain why Earth can be thought of as a system.

- List the major gases composing Earth's atmosphere and identify those components that are most important meteorologically.
- Explain why ozone depletion is a significant global issue.
- Interpret a graph that shows changes in air pressure from Earth's surface to the top of the atmosphere.
- Sketch and label a graph showing the thermal structure of the atmosphere.
- Distinguish between homosphere and heterosphere.

Focus on the Atmosphere

GE●De   Introduction to the Atmosphere
ATMOSPHERE   ▶ Weather and Climate

Weather influences our everyday activities, our jobs, and our health and comfort. Many of us pay little attention to the weather unless we are inconvenienced by it or when it adds to our enjoyment of outdoor activities. Nevertheless, there are few other aspects of our physical environment that affect our lives more than the phenomena we collectively call the weather.

Weather in the United States

The United States occupies an area that stretches from the tropics to the Arctic Circle. It has thousands of miles of coastline and extensive regions that are far from the influence of the ocean. Some landscapes are mountainous, and others are dominated by plains. It is a place where Pacific storms strike the West Coast, while the East is sometimes influenced by events in the Atlantic and the Gulf of Mexico. For those in the center of the country, it is common to experience weather events triggered when frigid southward-bound Canadian air masses clash with northward-moving tropical ones from the Gulf of Mexico.

Stories about weather are a routine part of the daily news. Articles and items about the effects of heat, cold, floods, drought, fog, snow, ice, and strong winds are commonplace (Figure 1–1). Memorable weather events occur everywhere on our planet. The United States likely has the greatest variety of weather of any country in the world. Severe weather events, such as tornadoes, flash floods, and intense thunderstorms, as well as hurricanes and blizzards, are collectively more frequent and more damaging in the United States than in any other nation. Beyond its direct impact on the lives of individuals, the weather has a strong effect on the world economy, by influencing agriculture, energy use, water resources, transportation, and industry.

Weather clearly influences our lives a great deal. Yet it is also important to realize that people influence the atmosphere and its behavior as well (Figure 1–2). There are, and will continue to be, significant political and scientific decisions to make involving these impacts. Answers to questions regarding air pollution and its control and the effects of various emissions on global climate are important examples. So there is a need for increased awareness and understanding of our atmosphere and its behavior.

Meteorology, Weather, and Climate

The subtitle of this book includes the word *meteorology*. **Meteorology** is the scientific study of the atmosphere and the phenomena that we usually refer to as *weather*. Along with geology, oceanography, and astronomy, meteorology is considered one of the *Earth sciences*—the sciences that seek to understand our planet. It is important to point out that there are not strict boundaries among the Earth sciences; in many situations, these sciences overlap. Moreover, all of the Earth sciences involve an understanding and application of knowledge and principles from physics, chemistry, and biology. You will see many examples of this fact in your study of meteorology.

Acted on by the combined effects of Earth's motions and energy from the Sun, our planet's formless and invisible envelope of air reacts by producing an infinite variety of weather, which in turn creates the basic pattern of global climates. Although not identical, weather and climate have much in common.

Weather is constantly changing, sometimes from hour to hour and at other times from day to day. It is a term that refers to the state of the atmosphere at a given time and place. Whereas changes in the weather are continuous and sometimes seemingly erratic, it is nevertheless possible to arrive at a generalization of these variations. Such a description of aggregate weather conditions is termed **climate.** It is based on observations that have been accumulated over many decades. Climate is often defined simply as "average weather," but this is an inadequate definition. In order to accurately portray the character of an area, variations and extremes must also be included, as well as the probabilities that such departures will take place. For example, it is necessary for farmers to know the average rainfall during the growing season, and it is also important to know the frequency of extremely wet and extremely dry years. Thus, climate is the sum of all statistical weather information that helps describe a place or region.

Figure 1–1 Few aspects of our physical environment influence our daily lives more than the weather. Tornadoes are intense and destructive local storms of short duration that cause an average of about 55 deaths each year. (Photo by Wave RF/Photolibrary)

(a)

(b)

Figure 1–2 These examples remind us that people influence the atmosphere and its behavior. (a) Motor vehicles are a significant contributor to air pollution. This traffic jam was in Kuala Lumpur, Malaysia. (Photo by Ron Yue/Alamy) (b) Smoke bellows from a coal-fired electricity-generating plant in New Delhi, India, in June 2008. (AP Photo/Gurindes Osan)

Maps similar to the one in Figure 1–3 are familiar to everyone who checks the weather report in the morning newspaper or on a television station. In addition to showing predicted high temperatures for the day, this map shows other basic weather information about cloud cover, precipitation, and fronts.

Suppose you were planning a vacation trip to an unfamiliar place. You would probably want to know what kind of weather to expect. Such information would help as you selected clothes to pack and could influence decisions regarding activities you might engage in during your stay. Unfortunately, weather forecasts that go beyond a few days are not very dependable. Thus, it would not be possible to get a reliable weather report about the conditions you are likely to encounter during your vacation.

Instead, you might ask someone who is familiar with the area about what kind of weather to expect. "Are

thunderstorms common?" "Does it get cold at night?" "Are the afternoons sunny?" What you are seeking is information about the climate, the conditions that are typical for that

Students Sometimes Ask ...

Does meteorology have anything to do with meteors?

Yes, there is a connection. Most people use the word *meteor* when referring to solid particles (meteoroids) that enter Earth's atmosphere from space and "burn up" due to friction ("shooting stars"). The term *meteorology* was coined in 340 BC, when the Greek philosopher Aristotle wrote a book titled *Meteorlogica*, which included explanations of atmospheric and astronomical phenomena. In Aristotle's day *anything* that fell from or was seen in the sky was called a meteor. Today we distinguish between particles of ice or water in the atmosphere (called *hydrometeors*) and extraterrestrial objects called meteoroids, or *meteors*.

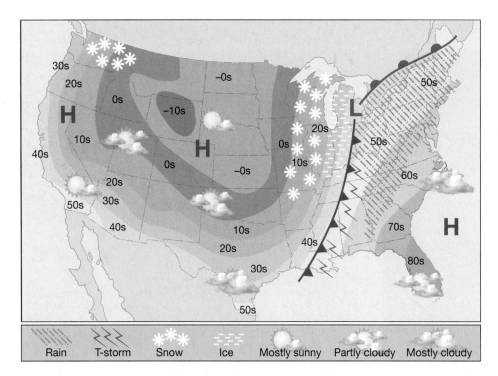

Figure 1–3 A typical newspaper weather map for a day in late December. The color bands show the high temperatures forecast for the day.

place. Another useful source of such information is the great variety of climate tables, maps, and graphs that are available. For example, the map in Figure 1–4 shows the average percentage of possible sunshine in the United States for the month of November, and the graph in Figure 1–5 shows average daily high and low temperatures for each month, as well as extremes, for New York City.

Such information could, no doubt, help as you planned your trip. But it is important to realize that *climate data* *cannot predict the weather.* Although the place may usually (climatically) be warm, sunny, and dry during the time of your planned vacation, you may actually experience cool, overcast, and rainy weather. There is a well-known saying that summarizes this idea: "Climate is what you expect, but weather is what you get."

The nature of both weather and climate is expressed in terms of the same basic **elements**—those quantities or properties that are measured regularly. The most important are (1) the temperature of the air, (2) the humidity of the air, (3) the type and amount of cloudiness, (4) the type and amount of precipitation, (5) the pressure exerted by the air, and (6) the speed and direction of the wind. These elements constitute the variables by which weather patterns and climate types are depicted. Although you will study these elements separately at first, keep in mind that they are very much interrelated. A change in one of the elements often produces changes in the others.

Concept Check 1.1

1 Distinguish among meteorology, weather, and climate.

2 List the basic elements of weather and climate.

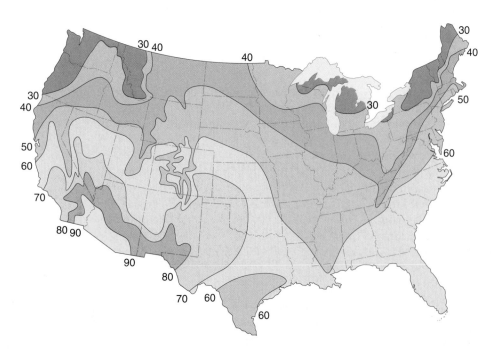

Figure 1–4 Mean percentage of possible sunshine for November. Southern Arizona is clearly the sunniest area. By contrast, parts of the Pacific Northwest receive a much smaller percentage of the possible sunshine. Climate maps such as this one are based on many years of data.

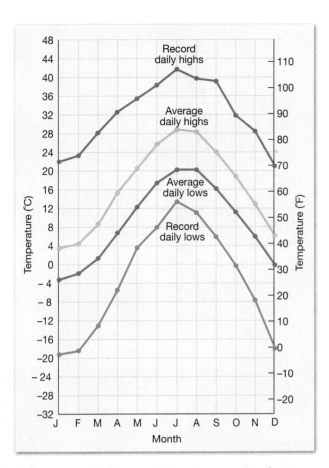

Figure 1-5 Graph showing daily temperature data for New York City. In addition to the average daily maximum and minimum temperatures for each month, extremes are also shown. As this graph shows, there can be significant departures from the average.

Atmospheric Hazards: Assault by the Elements

Natural hazards are a part of living on Earth. Every day they adversely affect literally millions of people worldwide and are responsible for staggering damages. Some, such as earthquakes and volcanic eruptions, are geological. Many others are related to the atmosphere.

Occurrences of severe weather are far more fascinating than ordinary weather phenomena. A spectacular lightning display generated by a severe thunderstorm can elicit both awe and fear (Figure 1–6a). Of course, hurricanes and tornadoes attract a great deal of much-deserved attention. A single tornado outbreak or hurricane can cause billions of dollars in property damage, much human suffering, and many deaths.

Of course, other atmospheric hazards adversely affect us. Some are storm related, such as blizzards, hail, and freezing rain. Others are not direct results of storms. Heat waves, cold waves, fog, wildfires, and drought are important examples (Figure 1–6b). In some years the loss of human life due to excessive heat or bitter cold exceeds that caused

by all other weather events combined. Moreover, although severe storms and floods usually generate more attention, droughts can be just as devastating and carry an even bigger price tag.

Between 1980 and 2010 the United States experienced 99 weather-related disasters in which overall damages and costs reached or exceeded $1 billion (Figure 1–7). The combined costs of these events exceeded $725 billion (normalized to 2007 dollars)! During the decade 1999–2008, an average of 629 direct weather fatalities occurred per year in the United States. During this span, the annual economic impacts of adverse weather on the national highway system alone exceeded $40 billion, and weather-related air traffic delays caused $4.2 billion in annual losses.

At appropriate places throughout this book, you will have an opportunity to learn about atmospheric hazards. Two entire chapters (Chapter 10 and Chapter 11) focus

This is a scene on a day in late April in southern Arizona's Organ Pipe Cactus National Monument. (Photo by Michael Collier)

Question 1 Write two brief statements about the locale in this image—one that relates to weather and one that relates to climate.

(a)

(b)

Figure 1–6 (a) Many people have incorrect perceptions of weather dangers and are unaware of the relative differences of weather threats to human life. For example, they are awed by the threat of hurricanes and tornadoes and plan accordingly on how to respond (for example, "Tornado Awareness Week" each spring) but fail to realize that lightning and winter storms can be greater threats. (Photo by Mark Newman/Superstock) (b) During the summer, dry weather coupled with lightning and strong winds contribute to wildfire danger. Millions of acres are burned each year, especially in the West. The loss of anchoring vegetation sets the stage for accelerated erosion when heavy rains subsequently occur. Near Boulder, Colorado, October 10, 2010. (AP Photo/The Daily Camera, Paul Aiken)

almost entirely on hazardous weather. In addition, a number of the book's special-interest boxes are devoted to a broad variety of severe and hazardous weather, including heat waves, winter storms, floods, dust storms, drought, mudflows, and lightning.

Every day our planet experiences an incredible assault by the atmosphere, so it is important to develop an awareness and understanding of these significant weather events.

Concept Check 1.2

1 List at least five storm-related atmospheric hazards.

2 What are three atmospheric hazards that are not directly storm related?

The Nature of Scientific Inquiry

As members of a modern society, we are constantly reminded of the benefits derived from science. But what exactly is the nature of scientific inquiry? Developing an understanding of how science is done and how scientists work is an important theme in this book. You will explore the difficulties of gathering data and some of the ingenious methods that have been developed to overcome these difficulties. You will also

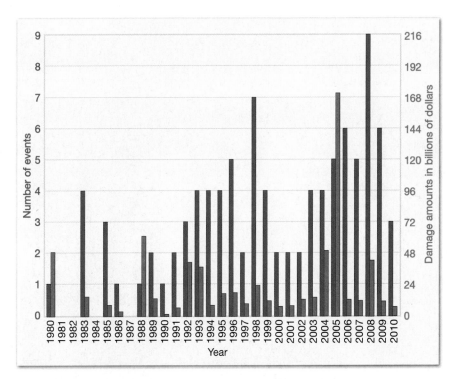

Figure 1–7 Between 1980 and 2010 the United States experienced 99 weather-related disasters in which overall damages and costs reached or exceeded $1 billion. This bar graph shows the number of events that occurred each year and the damage amounts in billions of dollars (normalized to 2007 dollars). The total losses for the 99 events exceeded $725 billion! For more about these extraordinary events see www.ncdc.noaa.gov/oa/reports/billionz.html. (After NOAA)

see examples of how hypotheses are formulated and tested, as well as learn about the development of some significant scientific theories.

All science is based on the assumption that the natural world behaves in a consistent and predictable manner that is comprehensible through careful, systematic study. The overall goal of science is to discover the underlying patterns in nature and then to use this knowledge to make predictions about what should or should not be expected, given certain facts or circumstances. For example, by understanding the processes and conditions that produce certain cloud types, meteorologists are often able to predict the approximate time and place of their formation.

The development of new scientific knowledge involves some basic logical processes that are universally accepted. To determine what is occurring in the natural world, scientists collect scientific *facts* through observation and measurement. The types of facts that are collected often seek to answer a well-defined question about the natural world, such as "Why does fog frequently develop in this place?" or "What causes rain to form in this cloud type?" Because some error is inevitable, the accuracy of a particular measurement or observation is always open to question. Nevertheless, these data are essential to science and serve as a springboard for the development of scientific theories (Box 1–1).

Hypothesis

Once facts have been gathered and principles have been formulated to describe a natural phenomenon, investigators try to explain how or why things happen in the manner observed. They often do this by constructing a tentative (or untested) explanation, which is called a scientific **hypothesis**. It is best if an investigator can formulate more than one hypothesis to explain a given set of observations. If an individual scientist is unable to devise multiple hypotheses, others in the scientific community will almost always develop alternative explanations. A spirited debate frequently ensues. As a result, extensive research is conducted by proponents of opposing hypotheses, and the results are made available to the wider scientific community in scientific journals.

Before a hypothesis can become an accepted part of scientific knowledge, it must pass objective testing and analysis. If a hypothesis cannot be tested, it is not scientifically useful, no matter how interesting it might seem. The verification process requires that *predictions* be made based on the hypothesis being considered and the predictions be tested by being compared against objective observations of nature. Put another way, hypotheses must fit observations other than those used to formulate them in the first place. Hypotheses that fail rigorous testing are ultimately discarded. The history of science is littered with discarded hypotheses. One of the best known is the Earth-centered model of the universe—a proposal that was supported by the apparent daily motion of the Sun, Moon, and stars around Earth.

Theory

When a hypothesis has survived extensive scrutiny and when competing ones have been eliminated, a hypothesis may be elevated to the status of a scientific **theory.** In everyday language we may say, "That's only a theory." But a scientific theory is a well-tested and widely accepted view that the scientific community agrees best explains certain observable facts.

Some theories that are extensively documented and extremely well supported are comprehensive in scope. An example from the Earth sciences is the theory of plate tectonics, which provides the framework for understanding the origin of mountains, earthquakes, and volcanic activity. It also explains the evolution of continents and ocean basins through time. As you will see in Chapter 14, this theory also helps us understand some important aspects of climate change through long spans of geologic time.

Box 1–1 Monitoring Earth from Space

Scientific facts are gathered in many ways, including through laboratory experiments and field observations and measurements. Satellites provide another very important source of data. Satellite images give us perspectives that are difficult to gain from more traditional sources (**Figure 1–A**). Moreover, the high-tech instruments aboard many satellites enable scientists to gather information from remote regions where data are otherwise scarce.

The image in **Figure 1–B** is from NASA's *Tropical Rainfall Measuring Mission* (*TRMM*). *TRMM* is a research satellite designed to expand our understanding of Earth's water (hydrologic) cycle and its role in our climate system. By covering the region between the latitudes 35° north and 35° south, it provides much-needed data on rainfall and the heat release associated with rainfall. Many types of measurements and images are possible. Instruments aboard the *TRMM* satellite have greatly expanded our ability to collect precipitation data. In addition to recording data for land areas, this satellite provides extremely precise measurements of rainfall over the oceans where conventional land-based instruments cannot see. This is especially important because much of Earth's rain falls in ocean-covered tropical areas, and a great deal of the globe's weather-producing energy comes from heat exchanges involved in the rainfall process. Until the *TRMM*, information on the intensity and amount of rainfall over the tropics was scanty. Such data are crucial to understanding and predicting global climate change.

FIGURE 1–A Satellite image of a massive winter storm on February 1, 2011. During a winter marked by several crippling storms, this one stands out. Heavy snow, ice, freezing rain, and frigid winds battered nearly two-thirds of the contiguous United States. In this image, the storm measures about 2000 kilometers (1240 miles) across. Satellites allow us to monitor the development and movement of major weather systems. (NASA)

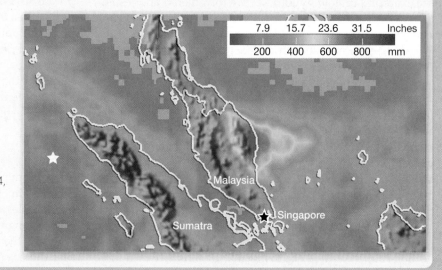

FIGURE 1–B This map of rainfall for December 7–13, 2004, in Malaysia was constructed using *TRMM* data. Over 800 millimeters (32 inches) of rain fell along the east coast of the peninsula (darkest red area). The extraordinary rains caused extensive flooding and triggered many mudflows. (NASA/*TRMM* image)

Scientific Methods

The processes just described, in which scientists gather facts through observations and formulate scientific hypotheses and theories, is called the *scientific method*. Contrary to popular belief, the scientific method is not a standard recipe that scientists apply in a routine manner to unravel the secrets of our natural world. Rather, it is an endeavor that involves creativity and insight. Rutherford and Ahlgren put it this way: "Inventing hypotheses or theories to imagine how the world works and then figuring out how they can be put to the test of reality is as creative as writing poetry, composing music, or designing skyscrapers."*

---

*F. James Rutherford and Andrew Ahlgren, *Science for All Americans* (New York: Oxford University Press, 1990), p. 7.

There is not a fixed path that scientists always follow that leads unerringly to scientific knowledge. Nevertheless, many scientific investigations involve the following steps: (1) A question is raised about the natural world; (2) scientific data are collected that relate to the question (Figure 1–8); (3) questions are posed that relate to the data, and one or more working hypotheses are developed that may answer these questions; (4) observations and experiments are developed to test the hypotheses; (5) the hypotheses are accepted, modified, or rejected, based on extensive testing; (6) data and results are shared with the scientific community for critical and further testing.

Other scientific discoveries may result from purely theoretical ideas that stand up to extensive examination. Some researchers use high-speed computers to simulate what is happening in the "real" world. These models are useful when dealing with natural processes that occur on very long time scales or take place in extreme or inaccessible locations. Still other scientific advancements have been made when a totally unexpected happening occurred during an experiment. These serendipitous discoveries are more than pure luck; as Louis Pasteur stated, "In the field of observation, chance favors only the prepared mind."

Scientific knowledge is acquired through several avenues, so it might be best to describe the nature of scientific

Students Sometimes Ask...

In class you compared a hypothesis to a theory. How is each one different from a scientific law?

A *scientific law* is a basic principle that describes a particular behavior of nature that is generally narrow in scope and can be stated briefly—often as a simple mathematical equation. Because scientific laws have been shown time and time again to be consistent with observations and measurements, they are rarely discarded. Laws may, however, require modifications to fit new findings. For example, Newton's laws of motion are still useful for everyday applications (NASA uses them to calculate satellite trajectories), but they do not work at velocities approaching the speed of light. For these circumstances, they have been supplanted by Einstein's theory of relativity.

inquiry as the *methods* of science rather than *the* scientific method. In addition, it should always be remembered that even the most compelling scientific theories are still simplified explanations of the natural world.

Concept Check 1.3

1 How is a scientific hypothesis different from a scientific theory?

2 List the basic steps followed in many scientific investigations.

Figure 1–8 Gathering data and making careful observations are a basic part of scientific inquiry. (a) This *Automated Surface Observing System* (ASOS) installation is one of nearly 900 in use for data gathering as part of the U.S. primary surface observing network. (Photo by Bobbé Christopherson) (b) These scientists are working with a sediment core recovered from the ocean floor. Such cores often contain useful data about Earth's climate history. (Photo by Science Source/Photo Researchers, Inc.)

(a)

(b)

Earth's Spheres

The images in Figure 1–9 are considered to be classics because they let humanity see Earth differently than ever before. Figure 1–9a, known as "Earthrise," was taken when
the *Apollo 8* astronauts orbited the Moon for the first time
in December 1968. As the spacecraft rounded the Moon,
Earth appeared to rise above the lunar surface. Figure 1–9b,
referred to as "The Blue Marble," is perhaps the most widely
reproduced image of Earth; it was taken in December 1972
by the crew of *Apollo 17* during the last manned lunar mission. These early views profoundly altered our conceptual-
izations of Earth and remain powerful images decades after
they were first viewed. Seen from space, Earth is breathtak-
ing in its beauty and startling in its solitude. The photos
remind us that our home is, after all, a planet—small, self-
contained, and in some ways even fragile. Bill Anders, the

Apollo 8 astronaut who took the "Earthrise" photo, expressed
it this way: "We came all this way to explore the Moon, and
the most important thing is that we discovered the Earth."

As we look closely at our planet from space, it becomes
apparent that Earth is much more than rock and soil. In
fact, the most conspicuous features in Figure 1–9a are not
continents but swirling clouds suspended above the surface of the vast global ocean. These features emphasize the
importance of water on our planet.

The closer view of Earth from space shown in Figure 1–9b
helps us appreciate why the physical environment is tradi-
tionally divided into three major parts: the solid Earth, the
water portion of our planet, and Earth's gaseous envelope.

It should be emphasized that our environment is highly
integrated and is not dominated by rock, water, or air alone.
It is instead characterized by continuous interactions as air
comes in contact with rock, rock with water, and water with
air. Moreover, the biosphere, the totality of life forms on our
planet, extends into each of the three physical realms and is
an equally integral part of the planet.

The interactions among Earth's four spheres are incalculable. Figure 1–10 provides us with one easy-to-visualize
example. The shoreline is an obvious meeting place for
rock, water, and air. In this scene, ocean waves that were
created by the drag of air moving across the water are
breaking against the rocky shore. The force of the water can
be powerful, and the erosional work that is accomplished
can be great.

On a human scale Earth is huge. Its surface area occupies 500,000,000 square kilometers (193 million square
miles). We divide this vast planet into four independent
parts. Because each part loosely occupies a shell around
Earth, we call them *spheres*. The four spheres include the
geosphere (solid Earth), the *atmosphere* (gaseous envelope),
the *hydrosphere* (water portion), and the *biosphere* (life).

(a)

(b)

Figure 1–9 (a) View, called
"Earthrise," that greeted the *Apollo 8*
astronauts as their spacecraft
emerged from behind the Moon.
(NASA) (b) Africa and Arabia are
prominent in this classic image
called "The Blue Marble" taken from
Apollo 17. The tan cloud-free zones
over the land coincide with major
desert regions. The band of clouds
across central Africa is associated
with a much wetter climate that in
places sustains tropical rain forests.
The dark blue of the oceans and the
swirling cloud patterns remind us of
the importance of the oceans and the
atmosphere. Antarctica, a continent
covered by glacial ice, is visible at
the South Pole. (NASA)

Figure 1–10 The shoreline is one obvious example of an *interface*—a common boundary where different parts of a system interact. In this scene, ocean waves (*hydrosphere*) that were created by the force of moving air (*atmosphere*) break against a rocky shore (*geosphere*). (Photo by Radius Images/photolibrary.com)

It is important to remember that these spheres are not separated by well-defined boundaries; rather, each sphere is intertwined with all of the others. In addition, each of Earth's four major spheres can be thought of as being composed of numerous interrelated parts.

The Geosphere

Beneath the atmosphere and the ocean is the solid Earth, or **geosphere.** The geosphere extends from the surface to the center of the planet, a depth of about 6400 kilometers (nearly 4000 miles), making it by far the largest of Earth's four spheres.

Based on compositional differences, the geosphere is divided into three principal regions: the dense inner sphere, called the *core*; the less dense *mantle*; and the *crust*, which is the light and very thin outer skin of Earth.

Soil, the thin veneer of material at Earth's surface that supports the growth of plants, may be thought of as part of all four spheres. The solid portion is a mixture of weathered rock debris (geosphere) and organic matter from decayed plant and animal life (biosphere). The decomposed and disintegrated rock debris is the product of weathering processes that require air (atmosphere) and water (hydrosphere). Air and water also occupy the open spaces between the solid particles.

The Atmosphere

Earth is surrounded by a life-giving gaseous envelope called the **atmosphere** (Figure 1–11). When we watch a high-flying jet plane cross the sky, it seems that the atmosphere extends upward for a great distance. However, when compared to the thickness (radius) of the solid Earth (about 6400 kilometers [4000 miles]), the atmosphere is a very shallow layer. More than 99 percent of the atmosphere is within 30 kilometers (20 miles) of Earth's surface. This thin blanket of air is nevertheless an integral part of the planet. It not only provides the air that we breathe but also acts to protect us from the dangerous radiation emitted by the Sun. The energy exchanges that continually occur between the atmosphere and Earth's surface and between the atmosphere and space produce the effects we call *weather*. If, like the Moon, Earth had no atmosphere, our planet would not only be lifeless, but many of the processes and interactions that make the surface such a dynamic place could not operate.

The Hydrosphere

Earth is sometimes called the *blue planet*. More than anything else, water makes Earth unique. The **hydrosphere** is a dynamic mass that is continually on the move, evaporating from the oceans to the atmosphere, precipitating to

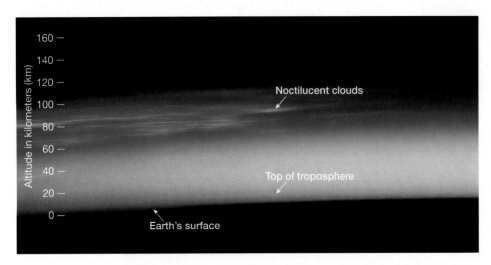

Figure 1-11 This unique image of Earth's atmosphere merging with the emptiness of space resembles an abstract painting. It was taken in June 2007 by a Space Shuttle crew member. The silvery streaks (called *noctilucent clouds*) high in the blue area are at a height of about 80 kilometers (50 miles). Air pressure at this height is less than one-thousandth of that at sea level. The reddish zone in the lower portion of the image is the densest part of the atmosphere. It is here, in a layer called the *troposphere*, that practically all weather phenomena occur. Ninety percent of Earth's atmosphere occurs within just 16 kilometers (10 miles) of the surface. (NASA)

the land, and running back to the ocean again. The global ocean is certainly the most prominent feature of the hydrosphere, blanketing nearly 71 percent of Earth's surface to an average depth of about 3800 meters (12,500 feet). It accounts for about 97 percent of Earth's water (Figure 1–12). However, the hydrosphere also includes the fresh water found in clouds, streams, lakes, and glaciers, as well as that found underground.

Although these latter sources constitute just a tiny fraction of the total, they are much more important than their meager percentage indicates. Clouds, of course, play a vital role in many weather and climate processes. In addition to providing the fresh water that is so vital to life on land, streams, glaciers, and groundwater are responsible for sculpting and creating many of our planet's varied landforms.

The Biosphere

The **biosphere** includes all life on Earth (Figure 1–13). Ocean life is concentrated in the sunlit surface waters of the sea. Most life on land is also concentrated near the surface, with tree roots and burrowing animals reaching a few meters underground and flying insects and birds reaching a kilometer or so above the surface. A surprising variety of life forms are also adapted to extreme environments. For example, on the ocean floor, where pressures are extreme and no light penetrates, there are places where vents spew hot, mineral-rich fluids that support communities of exotic life-forms. On land, some bacteria thrive in rocks as deep as 4 kilometers

(2.5 miles) and in boiling hot springs. Moreover, air currents can carry microorganisms many kilometers into the atmosphere. But even when we consider these extremes, life still must be thought of as being confined to a narrow band very near Earth's surface.

Plants and animals depend on the physical environment for the basics of life. However, organisms do more than just respond to their physical environment. Through

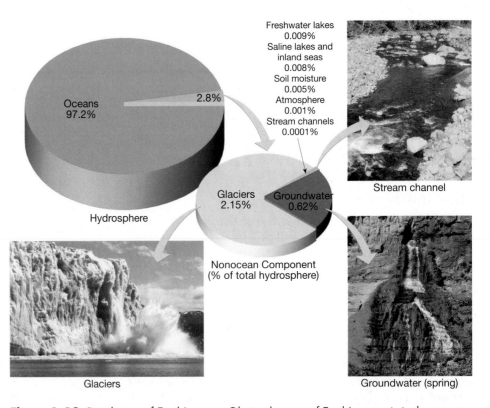

Figure 1-12 Distribution of Earth's water. Obviously, most of Earth's water is in the oceans. Glacial ice represents about 85 percent of all the water *outside* the oceans. When only *liquid freshwater* is considered, more than 90 percent is groundwater. (Glacier photo by Bernhard Edmaier/Photo Researchers, Inc.; stream photo by E.J. Tarbuck; and groundwater photo by Michael Collier)

(a)

(b)

Figure 1–13 (a) The ocean contains a significant portion of Earth's biosphere. Modern coral reefs are unique and complex examples and are home to about 25 percent of all marine species. Because of this diversity, they are sometimes referred to as the ocean equivalent of tropical rain forests. (Photo by Darryl Leniuk/agefotostock) (b) Tropical rain forests are characterized by hundreds of different species per square kilometer. Climate has a strong influence on the nature of the biosphere. Life, in turn, influences the atmosphere. (Photo by agefotostock/SuperStock)

countless interactions, life forms help maintain and alter their physical environment. Without life, the makeup and nature of the geosphere, hydrosphere, and atmosphere would be very different.

> ## Concept Check 1.4
>
> **1** Compare the height of the atmosphere to the thickness of the geosphere.
>
> **2** How much of Earth's surface do oceans cover?
>
> **3** How much of the planet's total water supply do the oceans represent?
>
> **4** List and briefly define the four spheres that constitute our environment.

Earth as a System

Anyone who studies Earth soon learns that our planet is a dynamic body with many separate but highly interactive parts, or *spheres*. The atmosphere, hydrosphere, biosphere, and geosphere and all of their components can be studied separately. However, the parts are *not* isolated. Each is related in many ways to the others, producing a complex and continuously interacting whole that we call the *Earth system.*

Earth System Science

A simple example of the interactions among different parts of the Earth system occurs every winter as moisture evaporates from the Pacific Ocean and subsequently falls as rain in the hills of southern California, triggering destructive debris flows (Figure 1–14). The processes that move water from the hydrosphere to the atmosphere and then to the geosphere have a profound impact on the physical environment and on the plants and animals (including humans) that inhabit the affected regions.

Scientists have recognized that in order to more fully understand our planet, they must learn how its individual components (land, water, air, and life-forms) are interconnected. This endeavor, called *Earth system science,* aims to study Earth as a *system* composed of numerous interacting parts, or *subsystems.* Using an interdisciplinary approach, those who practice Earth system science attempt to achieve the level of understanding necessary to comprehend and solve many of our global environmental problems.

A **system** is a group of interacting, or interdependent, parts that form a complex whole. Most of us hear and use the term *system* frequently. We may service our car's cooling *system*, make use of the city's transportation *system,*

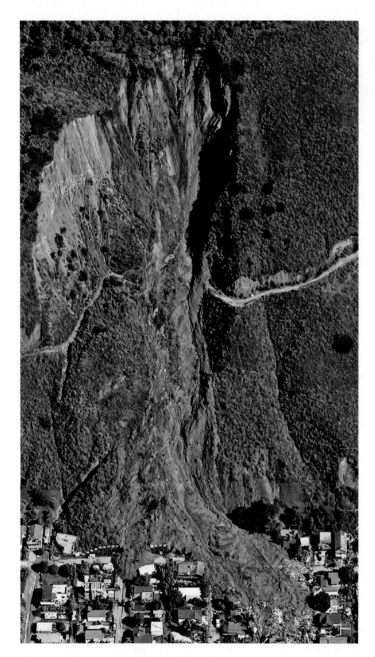

Figure 1–14 This image provides an example of interactions among different parts of the Earth system. On January 10, 2005, extraordinary rains triggered this debris flow (popularly called a mudslide) in the coastal community of La Conchita, California. (AP Wideworld Photo)

and be a participant in the political *system*. A news report might inform us of an approaching weather *system*. Further, we know that Earth is just a small part of a larger system known as the *solar system*, which in turn is a subsystem of an even larger system called the Milky Way Galaxy.

The Earth System

The Earth system has a nearly endless array of subsystems in which matter is recycled over and over again. One example that is described in Box 1–2 traces the movements of carbon among Earth's four spheres. It shows us, for example, that

the carbon dioxide in the air and the carbon in living things and in certain rocks is all part of a subsystem described by the *carbon cycle*.

The parts of the Earth system are linked so that a change in one part can produce changes in any or all of the other parts. For example, when a volcano erupts, lava from Earth's interior may flow out at the surface and block a nearby valley. This new obstruction influences the region's drainage system by creating a lake or causing streams to change course. The large quantities of volcanic ash and gases that can be emitted during an eruption might be blown high into the atmosphere and influence the amount of solar energy that can reach Earth's surface. The result could be a drop in air temperatures over the entire hemisphere.

Where the surface is covered by lava flows or a thick layer of volcanic ash, existing soils are buried. This causes the soil-forming processes to begin anew to transform the new surface material into soil (Figure 1–15). The soil that eventually forms will reflect the interactions among many parts of the Earth system—the volcanic parent material, the climate, and the impact of biological activity. Of course, there would also be significant changes in the biosphere. Some organisms and their habitats would be eliminated by the lava and ash, whereas new settings for life, such as the lake, would be created. The potential climate change could also impact sensitive life-forms.

The Earth system is characterized by processes that vary on spatial scales from fractions of millimeters to thousands of kilometers. Time scales for Earth's processes range from milliseconds to billions of years. As we learn about Earth, it becomes increasingly clear that despite significant separations in distance or time, many processes are connected, and a change in one component can influence the entire system.

The Earth system is powered by energy from two sources. The Sun drives external processes that occur in the atmosphere, hydrosphere, and at Earth's surface. Weather and climate, ocean circulation, and erosional processes are driven by energy from the Sun. Earth's interior is the second source of energy. Heat remaining from when our planet formed and heat that is continuously generated by radioactive decay power the internal processes that produce volcanoes, earthquakes, and mountains.

Humans are *part of* the Earth system, a system in which the living and nonliving components are entwined and interconnected. Therefore, our actions produce changes in all the other parts. When we burn gasoline and coal, dispose of our wastes, and clear the land, we cause other parts of the system to respond, often in unforeseen ways. Throughout this book you will learn about some of Earth's subsystems, including the hydrologic system and the climate system. Remember that these components *and we humans* are all part of the complex interacting whole we call the Earth system.

Concept Check 1.5

1 What is a system? List three examples.

2 What are the two sources of energy for the Earth system?

Figure 1–15 When Mount St. Helens erupted in May 1980, the area shown here was buried by a volcanic mudflow. Now, plants are reestablished and new soil is forming. (Photo by Terry Donnelly/ Alamy)

Composition of the Atmosphere

 GE●De
ATMOSPHERE

Introduction to the Atmosphere
▶ Composition of the Atmosphere

In the days of Aristotle, air was thought to be one of four fundamental substances that could not be further divided into constituent components. The other three substances were fire, earth (soil), and water. Even today the term **air** is sometimes used as if it were a specific gas, which of course it is not. The envelope of air that surrounds our planet is a *mixture* of many discrete gases, each with its own physical properties, in which varying quantities of tiny solid and liquid particles are suspended.

Major Components

The composition of air is not constant; it varies from time to time and from place to place (see Box 1–3). If the water vapor, dust, and other variable components were removed from the atmosphere, we would find that its makeup is very stable up to an altitude of about 80 kilometers (50 miles).

As you can see in Figure 1–16, two gases—nitrogen and oxygen—make up 99 percent of the volume of clean, dry air. Although these gases are the most plentiful components of the atmosphere and are of great significance to life on Earth, they are of little or no importance in affecting weather phenomena. The remaining 1 percent of dry air is mostly the inert gas argon (0.93 percent) plus tiny quantities of a number of other gases.

Carbon Dioxide

Carbon dioxide, although present in only minute amounts (0.0391 percent, or 391 parts per million), is nevertheless a meteorologically important constituent of air. Carbon dioxide is of great interest to meteorologists because it is an efficient absorber of energy emitted by Earth and thus influences the heating of the atmosphere. Although the proportion of carbon dioxide in the atmosphere is relatively

Box 1–2   The Carbon Cycle: One of Earth's Subsystems

To illustrate the movement of material and energy in the Earth system, let us take a brief look at the *carbon cycle* (Figure 1–C). Pure carbon is relatively rare in nature. It is found predominantly in two minerals: diamond and graphite. Most carbon is bonded chemically to other elements to form compounds such as carbon dioxide, calcium carbonate, and the hydrocarbons found in coal and petroleum. Carbon is also the basic building block of life as it readily combines with hydrogen and oxygen to form the fundamental organic compounds that compose living things.

In the atmosphere, carbon is found mainly as carbon dioxide (CO_2). Atmospheric carbon dioxide is significant because it is a greenhouse gas, which means it is an efficient absorber of energy emitted by Earth and thus influences the heating of the atmosphere. Because many of the processes that operate on Earth involve carbon dioxide, this gas is constantly moving into and out of the atmosphere. For example, through the process of photosynthesis, plants absorb carbon dioxide from the atmosphere to produce the essential organic compounds needed for growth. Animals that consume these plants (or consume other animals that eat plants) use these organic compounds as a source of energy and, through the process

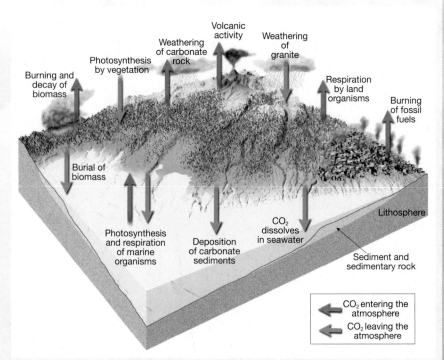

FIGURE 1–C Simplified diagram of the carbon cycle, with emphasis on the flow of carbon between the atmosphere and the hydrosphere, geosphere, and biosphere. The colored arrows show whether the flow of carbon is into or out of the atmosphere.

of respiration, return carbon dioxide to the atmosphere. (Plants also return some CO_2 to the atmosphere via respiration.) Further,

when plants die and decay or are burned, this biomass is oxidized, and carbon dioxide is returned to the atmosphere.

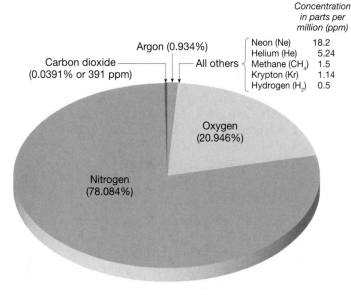

Figure 1–16 Proportional volume of gases composing dry air. Nitrogen and oxygen obviously dominate.

uniform, its percentage has been rising steadily for more than a century. Figure 1–17 is a graph showing the growth in atmospheric CO_2 since 1958. Much of this rise is attributed to the burning of ever-increasing quantities of fossil fuels, such as coal and oil. Some of this additional carbon dioxide is absorbed by the waters of the ocean or is used by plants, but more than 40 percent remains in the air. Estimates project that by sometime in the second half of the twenty-first century, carbon dioxide levels will be twice as high as pre-industrial levels.

Most atmospheric scientists agree that increased carbon dioxide concentrations have contributed to a warming of Earth's atmosphere over the past several decades and will continue to do so in the decades to come. The magnitude of such temperature changes is uncertain and depends partly on the quantities of CO_2 contributed by human activities in the years ahead. The role of carbon dioxide in the atmosphere and its possible effects on climate are examined in more detail in Chapters 2 and 14.

Not all dead plant material decays immediately back to carbon dioxide. A small percentage is deposited as sediment. Over long spans of geologic time, considerable biomass is buried with sediment. Under the right conditions, some of these carbon-rich deposits are converted to fossil fuels—coal, petroleum, or natural gas. Eventually some of the fuels are recovered (mined or pumped from a well) and burned to run factories and fuel our transportation system. One result of fossil-fuel combustion is the release of huge quantities of CO_2 into the atmosphere. Certainly one of the most active parts of the carbon cycle is the movement of CO_2 from the atmosphere to the biosphere and back again.

Carbon also moves from the geosphere and hydrosphere to the atmosphere and back again. For example, volcanic activity early in Earth's history is thought to be the source of much of the carbon dioxide found in the atmosphere. One way that carbon dioxide makes its way back to the hydrosphere and then to the solid Earth is by first combining with water to form carbonic acid (H_2CO_3), which then attacks the rocks that compose the geosphere. One product of this chemical weathering of solid rock is the soluble bicarbonate ion ($2HCO_3^-$), which is carried by groundwater and streams to the ocean. Here water-dwelling organisms extract this dissolved material to produce hard parts (shells) of calcium carbonate ($CaCO_3$). When the organisms die, these skeletal remains settle to the ocean floor as biochemical sediment and become sedimentary rock. In fact, the geosphere is by far Earth's largest depository of carbon, where it is a constituent of a variety of rocks, the most abundant being limestone (Figure 1–D). Eventually the limestone may be exposed at Earth's surface, where chemical weathering will cause the carbon stored in the rock to be released to the atmosphere as CO_2.

In summary, carbon moves among all four of Earth's major spheres. It is essential to every living thing in the biosphere. In the atmosphere carbon dioxide is an important greenhouse gas. In the hydrosphere, carbon dioxide is dissolved in lakes, rivers, and the ocean. In the geosphere, carbon is contained in carbonate-rich sediments and sedimentary rocks and is stored as organic matter dispersed through sedimentary rocks and as deposits of coal and petroleum.

FIGURE 1–D A great deal of carbon is locked up in Earth's geosphere. England's White Chalk Cliffs are an example. Chalk is a soft, porous type of limestone ($CaCO_3$) consisting mainly of the hard parts of microscopic organisms called coccoliths (inset). (Photo by Prisma/SuperStock; inset by Steve Gschmeissner/Photo Researchers, Inc.)

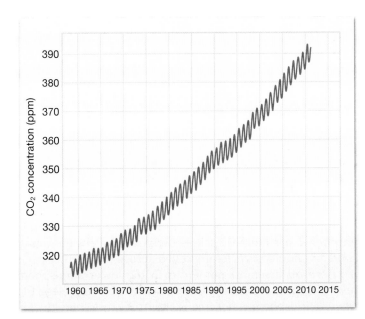

Figure 1–17 Changes in the atmosphere's carbon dioxide (CO_2) as measured at Hawaii's Mauna Loa Observatory. The oscillations reflect the seasonal variations in plant growth and decay in the Northern Hemisphere. During the first 10 years of this record (1958–1967), the average yearly CO_2 increase was 0.81 ppm. During the last 10 years (2001–2010) the average yearly increase was 2.04 ppm. (Data from NOAA)

Variable Components

Air includes many gases and particles that vary significantly from time to time and place to place. Important examples include water vapor, aerosols, and ozone. Although usually present in small percentages, they can have significant effects on weather and climate.

Water Vapor The amount of water vapor in the air varies considerably, from practically none at all up to about 4 percent by volume. Why is such a small fraction of the

Box 1–3   Origin and Evolution of Earth's Atmosphere

The air we breathe is a stable mixture of 78 percent nitrogen, 21 percent oxygen, nearly 1 percent argon, and small amounts of gases such as carbon dioxide and water vapor. However, our planet's original atmosphere 4.6 billion years ago was substantially different.

Earth's Primitive Atmosphere

Early in Earth's formation, its atmosphere likely consisted of gases most common in the early solar system: hydrogen, helium, methane, ammonia, carbon dioxide, and water vapor. The lightest of these gases, hydrogen and helium, escaped into space because Earth's gravity was too weak to hold them. Most of the remaining gases were probably scattered into space by strong *solar winds* (vast streams of particles) from a young active Sun. (All stars, including the Sun, apparently experience a highly active stage early in their evolution, during which solar winds are very intense.)

Earth's first enduring atmosphere was generated by a process called *outgassing*, through which gases trapped in the planet's interior are released. Outgassing from hundreds of active volcanoes still remains an important planetary function worldwide (Figure 1–E). However, early in Earth's history, when massive heating and fluid-like motion occurred in the planet's interior, the gas output must have been immense. Based on our understanding of modern volcanic eruptions, Earth's primitive atmosphere probably consisted of mostly water vapor, carbon dioxide, and sulfur dioxide, with minor amounts of other gases and minimal nitrogen. Most importantly, free oxygen was not present.

FIGURE 1–E Earth's first enduring atmosphere was formed by a process called *outgassing*, which continues today, from hundreds of active volcanoes worldwide. (Photo by Greg Vaughn/Alamy)

---

atmosphere so significant? The fact that water vapor is the source of all clouds and precipitation would be enough to explain its importance. However, water vapor has other roles. Like carbon dioxide, it has the ability to absorb heat given off by Earth, as well as some solar energy. It is therefore important when we examine the heating of the atmosphere.

When water changes from one state to another, such as from a gas to a liquid or a liquid to a solid (see Figure 4–3, p. 99), it absorbs or releases heat. This energy is termed *latent heat*, which means hidden heat. As you will see in later chapters, water vapor in the atmosphere transports this latent heat from one region to another, and it is the energy source that drives many storms.

Aerosols The movements of the atmosphere are sufficient to keep a large quantity of solid and liquid particles suspended within it. Although visible dust sometimes clouds the sky, these relatively large particles are too heavy to stay in the air very long. Still, many particles are microscopic and remain suspended for considerable periods of time. They may originate from many sources, both natural and human made, and include sea salts from breaking waves, fine soil blown into the air, smoke and soot from fires, pollen and microorganisms lifted by the wind, ash and dust from volcanic eruptions, and more (Figure 1–18a). Collectively, these tiny solid and liquid particles are called **aerosols.**

Aerosols are most numerous in the lower atmosphere near their primary source, Earth's surface. Nevertheless, the upper atmosphere is not free of them, because some dust is

Students Sometimes Ask...

Could you explain a little more about why the graph in Figure 1–17 has so many ups and downs?

Sure. Carbon dioxide is removed from the air by photosynthesis, the process by which green plants convert sunlight into chemical energy. In spring and summer, vigorous plant growth in the extensive land areas of the Northern Hemisphere removes carbon dioxide from the atmosphere, so the graph takes a dip. As winter approaches, many plants die or shed leaves. The decay of organic matter returns carbon dioxide to the air, causing the graph to spike upward.

Oxygen in the Atmosphere

As Earth cooled, water vapor condensed to form clouds, and torrential rains began to fill low-lying areas, which became the oceans. In those oceans, nearly 3.5 billion years ago, photosynthesizing bacteria began to release oxygen into the water. During photosynthesis, organisms use the Sun's energy to produce organic material (energetic molecules of sugar containing hydrogen and carbon) from carbon dioxide (CO_2) and water (H_2O). The first bacteria probably used hydrogen sulfide (H_2S) as the source of hydrogen rather than water. One of the earliest bacteria, *cyanobacteria* (once called blue-green algae), began to produce oxygen as a by-product of photosynthesis.

Initially, the newly released oxygen was readily consumed by chemical reactions with other atoms and molecules (particularly iron) in the ocean (Figure 1–F). Once the available iron satisfied its need for oxygen and as the number of oxygen-generating organisms increased, oxygen began to build in the atmosphere. Chemical analyses of rocks suggest that a significant amount of oxygen appeared in the atmosphere as early as 2.2 billion years ago and increased steadily until it reached stable levels about 1.5 billion years ago. Obviously, the availability of free oxygen had a major impact on the development of life and vice versa. Earth's atmosphere evolved together with its life-forms from an oxygen-free envelope to an oxygen-rich environment.

FIGURE 1–F These ancient layered, iron-rich rocks, called banded iron formations, were deposited during a geologic span known as the Precambrian. Much of the oxygen generated as a by-product of photosynthesis was readily consumed by chemical reactions with iron to produce these rocks. (Photo by John Cancalosi/Photolibrary)

Another significant benefit of the "oxygen explosion" is that oxygen molecules (O_2) readily absorb ultraviolet radiation and rearrange themselves to form *ozone* (O_3). Today, ozone is concentrated above the surface in a layer called the *stratosphere*, where it absorbs much of the ultraviolet radiation that strikes the upper atmosphere.

For the first time, Earth's surface was protected from this type of solar radiation, which is particularly harmful to DNA. Marine organisms had always been shielded from ultraviolet radiation by the oceans, but the development of the atmosphere's protective ozone layer made the continents more hospitable.

---

carried to great heights by rising currents of air, and other particles are contributed by meteoroids that disintegrate as they pass through the atmosphere.

From a meteorological standpoint, these tiny, often invisible particles can be significant. First, many act as surfaces on which water vapor may condense, an important function in the formation of clouds and fog. Second, aerosols can absorb or reflect incoming solar radiation. Thus, when an air-pollution episode is occurring or when ash fills the sky following a volcanic eruption, the amount of sunlight reaching Earth's surface can be measurably reduced. Finally, aerosols contribute to an optical phenomenon we have all observed—the varied hues of red and orange at sunrise and sunset (Figure 1–18b).

Ozone Another important component of the atmosphere is **ozone.** It is a form of oxygen that combines three oxygen atoms into each molecule (O_3). Ozone is not the same as the oxygen we breathe, which has two atoms per molecule (O_2). There is very little ozone in the atmosphere. Overall, it represents just 3 out of every 10 million molecules. Moreover,

its distribution is not uniform. In the lowest portion of the atmosphere, ozone represents less than 1 part in 100 million. It is concentrated well above the surface in a layer called the *stratosphere,* between 10 and 50 kilometers (6 and 31 miles).

In this altitude range, oxygen molecules (O_2) are split into single atoms of oxygen (O) when they absorb ultraviolet radiation emitted by the Sun. Ozone is then created when a single atom of oxygen (O) and a molecule of oxygen (O_2) collide. This must happen in the presence of a third, neutral molecule that acts as a *catalyst* by allowing the reaction to take place without itself being consumed in the process. Ozone is concentrated in the 10- to 50-kilometer height range because a crucial balance exists there: The ultraviolet radiation from the Sun is sufficient to produce single atoms of oxygen, and there are enough gas molecules to bring about the required collisions.

The presence of the ozone layer in our atmosphere is crucial to those of us who are land dwellers. The reason is that ozone absorbs the potentially harmful ultraviolet (UV) radiation from the Sun. If ozone did not filter a great deal of the ultraviolet radiation, and if the Sun's UV rays reached

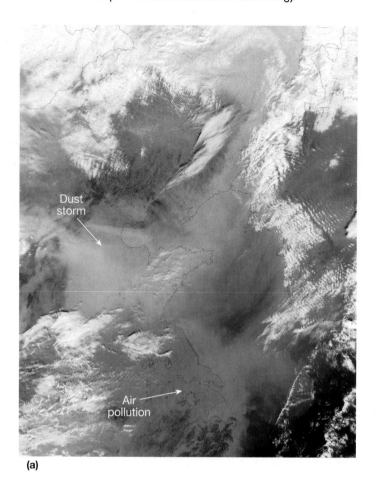

(a)

(b)

Figure 1–18 (a) This satellite image from November 11, 2002, shows two examples of aerosols. First, a large dust storm is blowing across northeastern China toward the Korean Peninsula. Second, a dense haze toward the south (bottom center) is human-generated air pollution. (b) Dust in the air can cause sunsets to be especially colorful. (Satellite image courtesy of NASA; photo by elwynn/ Shutterstock)

the surface of Earth undiminished, land areas on our planet would be uninhabitable for most life as we know it. Thus, anything that reduces the amount of ozone in the atmosphere could affect the well-being of life on Earth. Just such a problem is described in the next section.

Concept Check 1.6

1 Is *air* a specific gas? Explain.

2 What are the two major components of clean, dry air? What proportion does each represent?

3 Why are water vapor and aerosols important constituents of Earth's atmosphere?

4 What is ozone? Why is ozone important to life on Earth?

Ozone Depletion— A Global Issue

The loss of ozone high in the atmosphere as a consequence of human activities is a serious global-scale environmental problem. For nearly a billion years Earth's ozone layer has

protected life on the planet. However, over the past half century, people have unintentionally placed the ozone layer in jeopardy by polluting the atmosphere. The most significant of the offending chemicals are known as chlorofluorocarbons (CFCs). They are versatile compounds that are chemically stable, odorless, nontoxic, noncorrosive, and inexpensive to produce. Over several decades many uses were developed for CFCs, including as coolants for air-conditioning and refrigeration equipment, as cleaning solvents for electronic components, as propellants for aerosol sprays, and in the production of certain plastic foams.

Students Sometimes Ask…

Isn't ozone some sort of pollutant?

Yes, you're right. Although the naturally occurring ozone in the stratosphere is critical to life on Earth, it is regarded as a pollutant when produced at ground level because it can damage vegetation and be harmful to human health. Ozone is a major component in a noxious mixture of gases and particles called *photochemical smog*. It forms as a result of reactions triggered by sunlight that occur among pollutants emitted by motor vehicles and industries. Chapter 13 provides more information about this.

No one worried about how CFCs might affect the atmosphere until three scientists, Paul Crutzen, F. Sherwood Rowland, and Mario Molina, studied the relationship. In 1974 they alerted the world when they reported that CFCs were probably reducing the average concentration of ozone in the stratosphere. In 1995 these scientists were awarded the Nobel Prize in chemistry for their pioneering work.

They discovered that because CFCs are practically inert (that is, not chemically active) in the lower atmosphere, a portion of these gases gradually makes its way to the ozone layer, where sunlight separates the chemicals into their constituent atoms. The chlorine atoms released this way, through a complicated series of reactions, have the net effect of removing some of the ozone.

The Antarctic Ozone Hole

Although ozone depletion by CFCs occurs worldwide, measurements have shown that ozone concentrations take an especially sharp drop over Antarctica during the Southern Hemisphere spring (September and October). Later, during November and December, the ozone concentration recovers to more normal levels (Figure 1–19). Between 1980, when it was discovered, and the early 2000s, this well-publicized *ozone hole* intensified and grew larger until it covered an area roughly the size of North America (Figure 1–20).

The hole is caused in part by the relatively abundant ice particles in the south polar stratosphere. The ice boosts the effectiveness of CFCs in destroying ozone, thus causing a greater decline than would otherwise occur. The zone

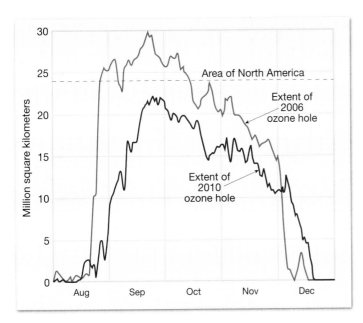

Figure 1-19 Changes in the size of the Antarctic ozone hole during 2006 and 2010. The ozone hole in both years began to form in August and was well developed in September and October. As is typical, each year the ozone hole persisted through November and disappeared in December. At its maximum, the area of the ozone hole was about 22 million square kilometers in 2010, an area nearly as large as all of North America.

of maximum depletion is confined to the Antarctic region by a swirling upper-level wind pattern. When this vortex weakens during the late spring, the ozone-depleted air is no longer restricted and mixes freely with air from other latitudes where ozone levels are higher.

A few years after the Antarctic ozone hole was discovered, scientists detected a similar but smaller ozone thinning in the vicinity of the North Pole during spring and early summer. When this pool breaks up, parcels of ozone-depleted air move southward over North America, Europe, and Asia.

Effects of Ozone Depletion

Because ozone filters out most of the damaging UV radiation in sunlight, a decrease in its concentration permits more of these harmful wavelengths to reach Earth's surface. What are the effects of the increased ultraviolet radiation? Each 1 percent decrease in the concentration of stratospheric ozone increases the amount of UV radiation that reaches Earth's surface by about 2 percent. Therefore, because ultraviolet radiation is known to induce skin cancer, ozone depletion seriously affects human health, especially among fair-skinned people and those who spend considerable time in the sun.

The fact that up to a half million cases of these cancers occur in the United States annually means that ozone depletion could ultimately lead to many thousands more cases each year.* In addition to raising the risk of skin cancer, an increase in damaging UV radiation can negatively impact the human immune system, as well as promote cataracts, a clouding of the eye lens that reduces vision and may cause blindness if not treated.

The effects of additional UV radiation on animal and plant life are also important. There is serious concern that crop yields and quality will be adversely affected. Some scientists also fear that increased UV radiation in the Antarctic will penetrate the waters surrounding the continent and impair or destroy the microscopic plants, called phytoplankton, that represent the base of the food chain. A decrease in phytoplankton, in turn, could reduce the population of copepods and krill that sustain fish, whales, penguins, and other marine life in the high latitudes of the Southern Hemisphere.

Montreal Protocol

What has been done to protect the atmosphere's ozone layer? Realizing that the risks of not curbing CFC emissions were difficult to ignore, an international agreement known as the *Montreal Protocol on Substances That Deplete the Ozone Layer* was concluded under the auspices of the United Nations in late 1987. The protocol established legally binding controls

---

* For more on this, see Severe and Hazardous Weather: "The Ultraviolet Index," p. 49.

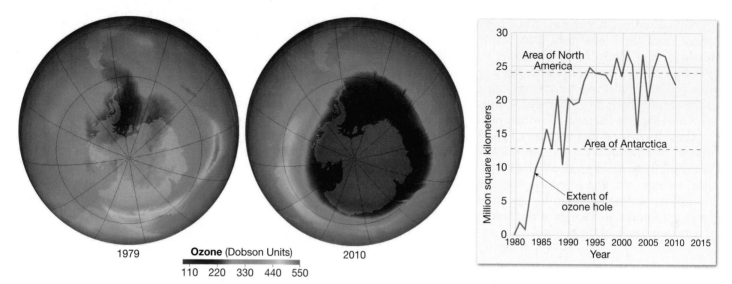

Figure 1–20 The two satellite images show ozone distribution in the Southern Hemisphere on the days in September 1979 and 2010 when the ozone hole was largest. The dark blue shades over Antarctica correspond to the region with the sparsest ozone. The ozone hole is not technically a "hole" where no ozone is present but is actually a region of exceptionally depleted ozone in the stratosphere over the Antarctic that occurs in the spring. The small graph traces changes in the maximum size of the ozone hole, 1980–2010. (NOAA)

on the production and consumption of gases known to cause ozone depletion. As the scientific understanding of ozone depletion improved after 1987 and substitutes and alternatives became available for the offending chemicals, the Montreal Protocol was strengthened several times. More than 190 nations eventually ratified the treaty.

The Montreal Protocol represents a positive international response to a global environment problem. As a result of the action, the total abundance of ozone-depleting gases in the atmosphere has started to decrease in recent years. According to the U.S. Environmental Protection Agency (U.S. EPA), the ozone layer has not grown thinner since 1998 over most of the world.* If the nations of the world continue to follow the provisions of the protocol, the decreases are expected to continue throughout the twenty-first century. Some offending chemicals are still increasing but will begin to decrease in coming decades. Between 2060 and 2075, the abundance of ozone-depleting gases is projected to fall to values that existed before the Antarctic ozone hole began to form in the 1980s.

Concept Check 1.7

1 What are CFCs, and what is their connection to the ozone problem?

2 During what time of year is the Antarctic ozone hole well developed?

3 Describe three effects of ozone depletion.

4 What is the Montreal Protocol?

---

* U.S. EPA, *Achievements in Stratospheric Ozone Protection, Progress Report.* EPA-430-R-07-001, April 2007, p. 5.

Vertical Structure of the Atmosphere

GE⬤De
ATMOSPHERE
Introduction to the Atmosphere
▶ Extent of the Atmosphere/Thermal Structure of the Atmosphere

To say that the atmosphere begins at Earth's surface and extends upward is obvious. However, where does the atmosphere end and where does outer space begin? There is no sharp boundary; the atmosphere rapidly thins as you travel away from Earth, until there are too few gas molecules to detect.

Pressure Changes

To understand the vertical extent of the atmosphere, let us examine the changes in atmospheric pressure with height. Atmospheric pressure is simply the weight of the air above. At sea level the average pressure is slightly more than 1000 millibars. This corresponds to a weight of slightly more than 1 kilogram per square centimeter (14.7 pounds per square inch). Obviously, the pressure at higher altitudes is less (Figure 1–21).

One-half of the atmosphere lies below an altitude of 5.6 kilometers (3.5 miles). At about 16 kilometers (10 miles), 90 percent of the atmosphere has been traversed, and above 100 kilometers (62 miles) only 0.00003 percent of all the gases composing the atmosphere remain.

At an altitude of 100 kilometers the atmosphere is so thin that the density of air is less than could be found in the most perfect artificial vacuum at the surface. Nevertheless, the atmosphere continues to even greater heights. The truly

PROFESSIONAL PROFILE

Kathy Orr, Broadcast Meteorologist

KATHY ORR is an award-winning broadcast meteorologist in Philadelphia. (Photo courtesy of Kathy Orr)

Kathy Orr is a trusted and familiar face on the airwaves of Philadelphia. As chief meteorologist for CBS3, Orr has kept the City of Brotherly Love abreast of the weather for 18 years and earned 10 regional Emmy awards in the process.

Orr calls being a television weathercaster a dream come true.

Orr calls being a television weathercaster a dream come true. Growing up in Syracuse, New York, Orr operated her own miniature weather station and marveled at the snow squalls that howled across Lake Ontario. "It could be a sunny afternoon, then the wind would blow over the lake. All of a sudden there was a blinding blizzard," she says.

When not watching the skies, Orr stayed glued to her family's TV set. At the time, she couldn't see how to combine her two major interests. "There weren't any women doing the weather on television back then. There were also not a lot of meteorologists on TV; it was less about the science and more for comic relief," she says.

She majored in broadcasting at Syracuse University and went on to earn a second degree in meteorology at the State University of New York at Oswego. There she learned the basis for the snow squalls that transfixed her as a girl. "These kinds of phenomena are associated with being on the downwind side of a Great Lake. When wind comes along, the lake acts like a snowmaking machine." While still in school, Orr landed a job as the weathercaster on a Syracuse station's brand-new morning show. She's remained a television meteorologist ever since.

Today, Orr says, being a trained meteorologist "is definitely a competitive advantage. It's not the 'rip and read' of years gone by. We take data from the supercomputers in Washington or models by the Navy and make our own forecasts. There are some services that provide forecasts locally and nationally, but they're not located where we are. I can look out the window and tell whether those forecasts are going to be accurate or not."

As a weathercaster, Orr has worked to promote education in science and math. For three years, she led a community program called *Kidcasters*. By offering children a chance to present the weather on TV, Orr hoped to interest elementary school children in science and math. For the past nine summers, she has conducted a similar program called *Orr at the Shore*. Each program highlights environmental issues along the New Jersey coast.

My job is to explain complicated ideas to people in an uncomplicated way.

Orr continues to promote science literacy by volunteering for the American Meteorological Society's DataStreme Atmosphere Project. As a DataStreme mentor, she has visited dozens of schools to train teachers in the science of meteorology. The teachers then promote the use of weather lessons in their districts to pique student interest in science, mathematics, and technology. Orr considers her forecasts educational as well. "My job is to explain complicated ideas to people in an uncomplicated way."

Being a weathercaster, Orr says, is demanding but also exhilarating. "In TV, the hours are crazy. If you work mornings, you're up at 2 AM; if nights, you're up until midnight. So you really have to love it. But if you do, you'll find a way. And I feel blessed to have done this for so long."

---

rarefied nature of the outer atmosphere is described very well by Richard Craig:

> The earth's outermost atmosphere, the part above a few hundred kilometers, is a region of extremely low density. Near sea level, the number of atoms and molecules in a cubic centimeter of air is about 2×10^{19}; near 600 km, it is only about 2×10^7, which is the sea-level value divided by a million million. At sea level, an atom or molecule can be expected, on the average, to move about 7×10^{-6} cm before colliding with another particle; at the 600-km level, this distance, called the "mean free path," is about 10 km. Near sea level, an atom or molecule, on the average, undergoes about 7×10^9 such collisions each second; near 600 km, this number is reduced to about 1 each minute.*

The graphic portrayal of pressure data (Figure 1–21) shows that the rate of pressure decrease is not constant. Rather, pressure decreases at a decreasing rate with an increase in altitude until, beyond an altitude of about 35 kilometers (22 miles), the decrease is negligible.

---

*Richard Craig, *The Edge of Space: Exploring the Upper Atmosphere* (New York: Doubleday & Company, Inc., 1968), p. 130.

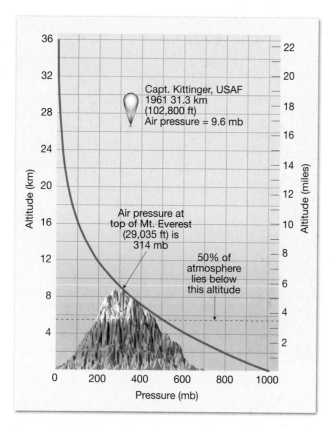

Figure 1–21 Atmospheric pressure changes with altitude. The rate of pressure decrease with an increase in altitude is not constant. Rather, pressure decreases rapidly near Earth's surface and more gradually at greater heights.

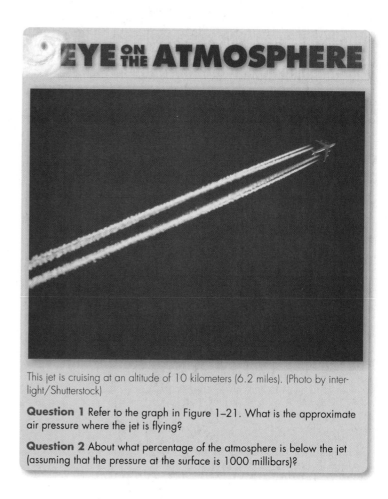

EYE ON THE ATMOSPHERE

This jet is cruising at an altitude of 10 kilometers (6.2 miles). (Photo by inter-light/Shutterstock)

Question 1 Refer to the graph in Figure 1–21. What is the approximate air pressure where the jet is flying?

Question 2 About what percentage of the atmosphere is below the jet (assuming that the pressure at the surface is 1000 millibars)?

Put another way, data illustrate that air is highly compressible—that is, it expands with decreasing pressure and becomes compressed with increasing pressure. Consequently, traces of our atmosphere extend for thousands of kilometers beyond Earth's surface. Thus, to say where the atmosphere ends and outer space begins is arbitrary and, to a large extent, depends on what phenomenon one is studying. It is apparent that there is no sharp boundary.

In summary, data on vertical pressure changes show that the vast bulk of the gases making up the atmosphere is very near Earth's surface and that the gases gradually merge with the emptiness of space. When compared with the size of the solid Earth, the envelope of air surrounding our planet is indeed very shallow.

Temperature Changes

By the early twentieth century much had been learned about the lower atmosphere. The upper atmosphere was partly known from indirect methods. Data from balloons and kites had revealed that the air temperature dropped with increasing height above Earth's surface. This phenomenon is felt by anyone who has climbed a high mountain and is obvious in pictures of snow-capped mountaintops rising above snow-free lowlands (Figure 1–22).

Although measurements had not been taken above a height of about 10 kilometers (6 miles), scientists believed that the temperature continued to decline with height to a value of absolute zero (–273°C) at the outer edge of the atmosphere. In 1902, however, the French scientist Leon Philippe Teisserenc de Bort refuted the notion that temperature

Figure 1–22 Temperatures drop with an increase in altitude in the troposphere. Therefore, it is possible to have snow on a mountaintop and warmer, snow-free lowlands below. (Photo by David Wall/Alamy)

decreases continuously with an increase in altitude. In studying the results of more than 200 balloon launchings, Teisserenc de Bort found that the temperature stopped decreasing and leveled off at an altitude between 8 and 12 kilometers (5 and 7.5 miles). This surprising discovery was at first doubted, but subsequent data-gathering confirmed his findings. Later, through the use of balloons and rocket-sounding techniques, the temperature structure of the atmosphere up to great heights became clear. Today the atmosphere is divided vertically into four layers on the basis of temperature (Figure 1–23).

Troposphere The bottom layer in which we live, where temperature decreases with an increase in altitude, is the **troposphere.** The term was coined in 1908 by Teisserrenc de Bort and literally means the region where air "turns over," a reference to the appreciable vertical mixing of air in this lowermost zone.

The temperature decrease in the troposphere is called the **environmental lapse rate.** Its average value is 6.5°C per kilometer (3.5°F per 1000 feet), a figure known as the *normal lapse rate.* It should be emphasized, however, that the environmental lapse rate is not a constant but rather can be highly variable and must be regularly measured. To determine the actual environmental lapse rate as well as gather information about vertical changes in air pressure, wind, and humidity, radiosondes are used. A **radiosonde** is an instrument package that is attached to a balloon and transmits data by radio as it ascends through the atmosphere (Figure 1–24). The environmental lapse rate can vary during the course of a day with fluctuations of the weather, as well as seasonally and from place to place. Sometimes shallow layers where temperatures actually increase with height are observed in the troposphere. When such a reversal occurs, a *temperature inversion* is said to exist.*

The temperature decrease continues to an average height of about 12 kilometers (7.5 miles). Yet the thickness of the troposphere is not the same everywhere. It reaches heights in excess of 16 kilometers (10 miles) in the tropics, but in polar regions it is more subdued, extending to 9 kilometers (5.5 miles) or less (Figure 1–25). Warm surface temperatures and highly developed thermal mixing are responsible for the greater vertical extent of the troposphere near the equator. As a result, the environmental lapse rate extends to great heights; and despite relatively high surface temperatures below, the lowest tropospheric temperatures are found aloft in the tropics and not at the poles.

*Temperature inversions are described in greater detail in Chapter 13.

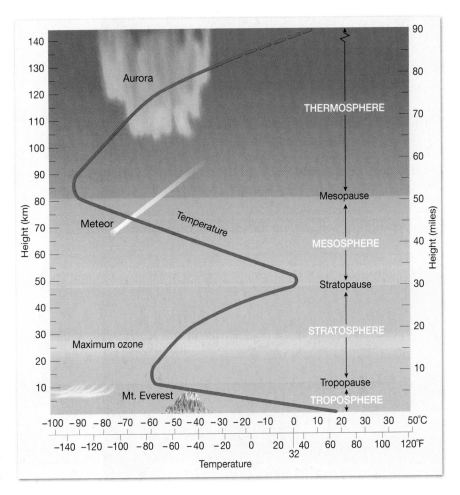

Figure 1–23 Thermal structure of the atmosphere.

The troposphere is the chief focus of meteorologists because it is in this layer that essentially all important weather phenomena occur. Almost all clouds and certainly all precipitation, as well as all our violent storms, are born in this lowermost layer of the atmosphere. This is why the troposphere is often called the "weather sphere."

Stratosphere Beyond the troposphere lies the **stratosphere;** the boundary between the troposphere and the stratosphere is known as the **tropopause.** Below the tropopause, atmospheric properties are readily transferred by large-scale turbulence and mixing, but above it, in the stratosphere, they are not. In the stratosphere, the temperature at first remains nearly constant to a height of about 20 kilometers (12 miles) before it begins a sharp increase that continues until the **stratopause** is encountered at a height of about 50 kilometers (30 miles) above Earth's surface. Higher temperatures occur in the stratosphere because it is in this layer that the atmosphere's ozone is concentrated. Recall that ozone absorbs ultraviolet radiation from the Sun. Consequently, the stratosphere is heated by the Sun. Although the maximum ozone concentration exists between 15 and 30 kilometers (9 and 19 miles), the smaller amounts of ozone above this height range absorb enough UV energy to cause the higher observed temperatures.

Figure 1–24 A lightweight instrument package, the *radiosonde,* is suspended below a 2-meter-wide weather balloon. As the radiosonde is carried aloft, sensors measure pressure, temperature, and relative humidity. A radio transmitter sends the measurements to a ground receiver. By tracking the radiosonde in flight, information on wind speed and direction aloft is also obtained. Observations where winds aloft are obtained are called "rawinsonde" observations. Worldwide, there are about 900 upper-air observation stations. Through international agreements, data are exchanged among countries. (Photo by Mark Burnett/ Photo Researchers, Inc.)

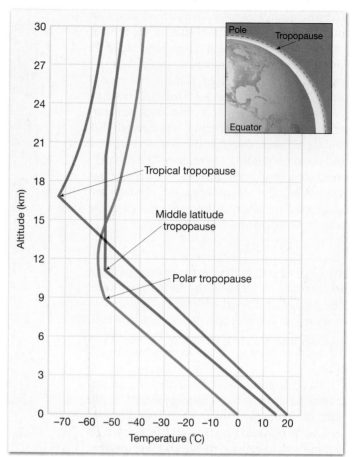

Figure 1–25 Differences in the height of the tropopause. The variation in the height of the tropopause, as shown on the small inset diagram, is greatly exaggerated.

Mesosphere In the third layer, the **mesosphere,** temperatures again decrease with height until at the **mesopause,** some 80 kilometers (50 miles) above the surface, the average temperature approaches −90°C (−130°F). The coldest temperatures anywhere in the atmosphere occur at the mesopause. The pressure at the base of the mesosphere is only about one-thousandth that at sea level. At the mesopause, the atmospheric pressure drops to just one-millionth that at sea level. Because accessibility is difficult, the mesosphere is one of the least explored regions of the atmosphere. The reason is that it cannot be reached by the highest-flying airplanes and research balloons, nor is it accessible to the lowest-orbiting satellites. Recent technical developments are just beginning to fill this knowledge gap.

Thermosphere The fourth layer extends outward from the mesopause and has no well-defined upper limit. It is the **thermosphere,** a layer that contains only a tiny fraction of the atmosphere's mass. In the extremely rarified air of this outermost layer, temperatures again increase, due to the absorption of very shortwave, high-energy solar radiation by atoms of oxygen and nitrogen.

Temperatures rise to extremely high values of more than 1000°C (1800°F) in the thermosphere. But such temperatures are not comparable to those experienced near Earth's surface. Temperature is defined in terms of the average speed at which molecules move. Because the gases of the thermosphere are moving at very high speeds, the temperature is very high. But the gases are so sparse that collectively they possess only an insignificant quantity of heat. For this reason, the temperature of a satellite orbiting Earth in the thermosphere is determined chiefly by the amount of solar radiation it absorbs and not by the high temperature of the almost nonexistent surrounding air. If an astronaut inside were to expose his or her hand, the air in this layer would not feel hot.

Concept Check 1.8

1 Does air pressure increase or decrease with an increase in altitude? Is the rate of change constant or variable? Explain.

2 Is the outer edge of the atmosphere clearly defined? Explain.

3 The atmosphere is divided vertically into four layers on the basis of temperature. List these layers in order from lowest to highest. In which layer does practically all of our weather occur?

4 Why does temperature increase in the stratosphere?

5 Why are temperatures in the thermosphere not strictly comparable to those experienced near Earth's surface?

Vertical Variations in Composition

In addition to the layers defined by vertical variations in temperature, other layers, or zones, are also recognized in the atmosphere. Based on composition, the atmosphere is often divided into two layers: the homosphere and the heterosphere. From Earth's surface to an altitude of about 80 kilometers (50 miles), the makeup of the air is uniform in terms of the proportions of its component gases. That is, the composition is the same as that shown earlier, in Figure 1–16. This lower uniform layer is termed the *homosphere*, the zone of homogeneous composition.

In contrast, the very thin atmosphere above 80 kilometers is not uniform. Because it has a heterogeneous composition, the term *heterosphere* is used. Here the gases are arranged into four roughly spherical shells, each with a distinctive composition. The lowermost layer is dominated by molecular nitrogen (N_2), next, a layer of atomic oxygen (O) is encountered, followed by a layer dominated by helium (He) atoms, and finally a region consisting largely of hydrogen (H) atoms. The stratified nature of the gases making up the heterosphere varies according to their weights. Molecular nitrogen is the heaviest, and so it is lowest. The lightest gas, hydrogen, is outermost.

Ionosphere

Located in the altitude range between 80 to 400 kilometers (50 to 250 miles), and thus coinciding with the lower portions of the thermosphere and heterosphere, is an electrically charged layer known as the **ionosphere.** Here molecules of nitrogen and atoms of oxygen are readily ionized as they

When this weather balloon was launched, the surface temperature was 17°C. It is now at an altitude of 1 kilometer. (Photo by David R. Frazier/Photo Researchers, Inc.)

Question 1 What term is applied to the instrument package being carried aloft by the balloon?

Question 2 In what layer of the atmosphere is the balloon?

Question 3 If average conditions prevail, what air temperature is the instrument package recording? How did you figure this out?

Question 4 How will the size of the balloon change, if at all, as it rises through the atmosphere? Explain.

absorb high-energy shortwave solar energy. In this process, each affected molecule or atom loses one or more electrons and becomes a positively charged ion, and the electrons are set free to travel as electric currents.

Although ionization occurs at heights as great as 1000 kilometers (620 miles) and extends as low as perhaps 50 kilometers (30 miles), positively charged ions and negative electrons are most dense in the range of 80 to 400 kilometers (50 to 250 miles). The concentration of ions is not great below this zone because much of the short-wavelength radiation needed for ionization has already been depleted.

In addition, the atmospheric density at this level results in a large percentage of free electrons being swiftly captured by positively charged ions. Beyond the 400-kilometer (250-mile) upward limit of the ionosphere, the concentration of ions is low because of the extremely low density of the air. Because so few molecules and atoms are present, relatively few ions and free electrons can be produced.

The electrical structure of the ionosphere is not uniform. It consists of three layers of varying ion density. From bottom to top, these layers are called the D, E, and F layers, respectively. Because the production of ions requires direct solar radiation, the concentration of charged particles changes from day to night, particularly in the D and E zones. That is, these layers weaken and disappear at night and reappear during the day. The uppermost layer, or F layer, on the other hand, is present both day and night. The density of the atmosphere in this layer is very low, and positive ions and electrons do not meet and recombine as rapidly as they do at lesser heights, where density is higher. Consequently, the concentration of ions and electrons in the F layer does not change rapidly, and the layer, although weak, remains through the night.

The Auroras

As best we can tell, the ionosphere has little impact on our daily weather. But this layer of the atmosphere is the site of one of nature's most interesting spectacles, the auroras (Figure 1–26). The **aurora borealis** (northern lights) and its Southern Hemisphere counterpart, the **aurora australis** (southern lights), appear in a wide variety of forms. Sometimes the displays consist of vertical streamers in which there can be considerable movement. At other times the auroras appear as a series of luminous expanding arcs or as a quiet glow that has an almost foglike quality.

The occurrence of auroral displays is closely correlated in time with solar-flare activity and, in geographic location, with Earth's magnetic poles. Solar flares are massive magnetic storms on the Sun that emit enormous amounts of energy and great quantities of fast-moving atomic particles. As the clouds of protons and electrons from the solar storm approach Earth, they are captured by its magnetic field, which in turn guides them toward the magnetic poles. Then, as the ions impinge on the ionosphere, they energize the atoms of oxygen and molecules of nitrogen and cause them to emit light—the glow of the auroras. Because the occurrence of solar flares is closely correlated with sunspot activity, auroral displays increase conspicuously at times when sunspots are most numerous.

Concept Check 1.9

1 Distinguish between the homosphere and the heterosphere.

2 What is the ionosphere? Where in the atmosphere is it located?

3 What is the primary cause of the auroras?

Figure 1–26 Aurora borealis (northern lights) as seen from Alaska. The same phenomenon occurs toward the South Pole, where it is called the aurora australis (southern lights). (Photo by agefotostock/SuperStock)

Give It Some Thought

1. Determine which statements refer to weather and which refer to climate. (*Note*: One statement includes aspects of both weather and climate.)
   a. The baseball game was rained out today.
   b. January is Omaha's coldest month.
   c. North Africa is a desert.
   d. The high this afternoon was 25°C.
   e. Last evening a tornado ripped through central Oklahoma.
   f. I am moving to southern Arizona because it is warm and sunny.
   g. Thursday's low of –20°C is the coldest temperature ever recorded for that city.
   h. It is partly cloudy.

2. After entering a dark room, you turn on a wall switch, but the light does not come on. Suggest at least three hypotheses that might explain this observation.

3. Making accurate measurements and observations is a basic part of scientific inquiry. The accompanying radar image, showing the distribution and intensity of precipitation associated with a storm, provides one example. Identify three additional images in this chapter that illustrate ways in which scientific data are gathered. Suggest advantages that might be associated with each example.

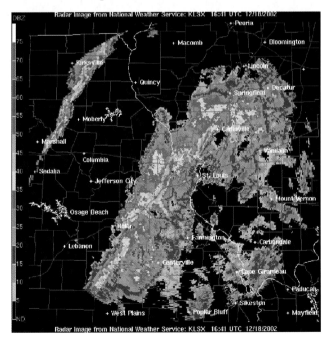

(Image by National Weather Service)

4. During a conversation with your meteorology professor, she makes the two statements listed below. Which can be considered a hypothesis? Which is more likely a theory?
   a. After several decades, the science community has determined that human-generated greenhouse gases have increased global average temperatures.
   b. One or two studies suggest that hurricane intensity is increasing.

5. Refer to Figure 1–21 to answer the following questions.
   a. If you were to climb to the top of Mount Everest, how many breaths of air would you have to take at that altitude to equal one breath at sea level?
   b. If you are flying in a commercial jet at an altitude of 12 kilometers, about what percentage of the atmosphere's mass is below you?

6. If you were ascending from the surface of Earth to the top of the atmosphere, which one of the following would be most useful for determining the layer of the atmosphere you were in? Explain.
   a. Doppler radar
   b. Hygrometer (humidity)
   c. Weather satellite
   d. Barometer (air pressure)
   e. Thermometer (temperature)

7. The accompanying photo provides an example of interactions among different parts of the Earth system. It is a view of a mudflow that was triggered by extraordinary rains. Which of Earth's four "spheres" were involved in this natural disaster that buried a small town on the Philippine island of Leyte? Describe how each contributed to the mudflow.

(Photo by AP Photo/Pat Roque)

8. Where would you expect the thickness of the troposphere (that is, the distance between Earth's surface and the tropopause) to be greater: over Hawaii or Alaska? Why? Do you think it is likely that the thickness of the troposphere over Alaska is different in January than in July? If so, why?

INTRODUCTION TO THE ATMOSPHERE IN REVIEW

- *Meteorology* is the scientific study of the atmosphere. *Weather* refers to the state of the atmosphere at a given time and place. It is constantly changing, sometimes from hour to hour and other times from day to day. *Climate* is an aggregate of weather conditions, the sum of all statistical weather information that helps describe a place or region. The nature of both weather and climate is expressed in terms of the same basic *elements,* those quantities or properties measured regularly. The most important elements are (1) air temperature, (2) humidity, (3) type and amount of cloudiness, (4) type and amount of precipitation, (5) air pressure, and (6) the speed and direction of the wind.

- All science is based on the assumption that the natural world behaves in a consistent and predictable manner. The process by which scientists gather facts through observation and careful measurement and formulate scientific *hypotheses* and *theories* is often referred to as the *scientific method.*

- Earth's four spheres include the *atmosphere* (gaseous envelope), the *geosphere* (solid Earth), the *hydrosphere* (water portion), and the *biosphere* (life). Each sphere is composed of many interrelated parts and is intertwined with all the other spheres.

- Although each of Earth's four spheres can be studied separately, they are all related in a complex and continuously interacting whole that we call the *Earth system. Earth system science* uses an interdisciplinary approach to integrate the knowledge of several academic fields in the study of our planet and its global environmental problems.

- A *system* is a group of interacting parts that form a complex whole. The two sources of energy that power the Earth system are (1) the Sun, which drives the external processes that occur in the atmosphere, hydrosphere, and at Earth's surface, and (2) heat from Earth's interior that powers the internal processes that produce volcanoes, earthquakes, and mountains.

- Air is a mixture of many discrete gases, and its composition varies from time to time and place to place. After water vapor, dust, and other variable components are removed, two gases, *nitrogen* and *oxygen,* make up 99 percent of the volume of the remaining clean, dry air. *Carbon dioxide,* although present in only minute amounts (0.0391 percent, or 391 ppm), is an efficient absorber of energy emitted by Earth and thus influences the heating of the atmosphere.

- The variable components of air include *water vapor, dust particles,* and *ozone.* Like carbon dioxide, water vapor can absorb heat given off by Earth as well as some solar energy. When water vapor changes from one state to another, it absorbs or releases heat. In the atmosphere, water vapor transports this *latent* ("hidden") *heat* from one place to another, and it is the energy source that helps drive many storms. *Aerosols* (tiny solid and liquid particles) are meteorologically important because these often-invisible particles act as surfaces on which water can condense and are also absorbers and reflectors of incoming solar radiation. *Ozone,* a form of oxygen that combines three oxygen atoms into each molecule (O_3), is a gas concentrated in the 10- to 50-kilometer height range in the atmosphere that absorbs the potentially harmful ultraviolet (UV) radiation from the Sun.

- Over the past half century, people have placed Earth's ozone layer in jeopardy by polluting the atmosphere with chlorofluorocarbons (CFCs), which remove some of the gas. Ozone concentrations take an especially sharp drop over Antarctica during the Southern Hemisphere spring (September and October). Ozone depletion seriously affects human health, especially among fair-skinned people and those who spend considerable time in the Sun. The *Montreal Protocol,* concluded under the auspices of the United Nations, represents a positive international response to the ozone problem.

- No sharp boundary to the upper atmosphere exists. The atmosphere simply thins as you travel away from Earth, until there are too few gas molecules to detect. Traces of atmosphere extend for thousands of kilometers beyond Earth's surface.

- Using temperature as the basis, the atmosphere is divided into four layers. The temperature decrease in the *troposphere,* the bottom layer in which we live, is called the *environmental lapse rate.* Its average value is 6.5°C per kilometer, a figure known as the *normal lapse rate.* The environmental lapse rate is not a constant and must be regularly measured using *radiosondes.* The thickness of the troposphere is generally greater in the tropics than in polar regions. Essentially all important weather phenomena occur in the troposphere. Beyond the troposphere lies the *stratosphere;* the boundary between the troposphere and stratosphere is known as the *tropopause.* In the stratosphere, the temperature at first remains constant to a height of about 20 kilometers (12 miles) before it begins a sharp increase due to the absorption of ultraviolet radiation from the Sun by ozone. The temperatures continue to increase until the *stratopause* is encountered at a height of about 50 kilometers (30 miles). In the *mesosphere,* the third layer, temperatures again decrease with height until the *mesopause,* some 80 kilometers (50 miles) above the surface. The fourth layer, the *thermosphere,* with no well-defined upper limit, consists of extremely rarefied air. Temperatures here increase with an increase in altitude.

- The atmosphere is often divided into two layers, based on composition. The *homosphere* (zone of homogeneous composition), from Earth's surface to an altitude of about 80 kilometers (50 miles), consists of air that is uniform in terms of the proportions of its component gases. Above 80 kilometers, the *heterosphere* (zone of heterogenous composition) consists of gases arranged into four roughly spherical shells, each with a distinctive composition. The stratified nature of the gases in the heterosphere varies according to their weights.

- Occurring in the altitude range between 80 and 400 kilometers (50 and 250 miles) is an electrically charged layer known as the *ionosphere.* Here molecules of nitrogen and atoms of oxygen are readily ionized as they absorb high-energy, shortwave solar energy. Three layers of varying ion density make up the ionosphere. Auroras (the *aurora borealis,* northern lights, and its Southern Hemisphere counterpart the *aurora australis,* southern lights) occur within the ionosphere. Auroras form as clouds of protons and electrons ejected from the Sun during solar-flare activity enter the atmosphere near Earth's magnetic poles and energize the atoms of oxygen and molecules of nitrogen, causing them to emit light—the glow of the auroras.

VOCABULARY REVIEW

aerosols (p. 21)
air (p. 17)
atmosphere (p. 13)
aurora australis (p. 29)
aurora borealis (p. 29)
biosphere (p. 15)
climate (p. 5)
elements of weather and
    climate (p. 7)

environmental lapse rate (p. 26)
geosphere (p. 13)
hydrosphere (p. 14)
hypothesis (p. 9)
ionosphere (p. 29)
mesopause (p. 27)
mesosphere (p. 27)
meteorology (p. 4)
ozone (p. 21)

radiosonde (p. 26)
stratopause (p. 27)
stratosphere (p. 27)
system (p. 16)
theory (p. 10)
thermosphere (p. 27)
tropopause (p. 27)
troposphere (p. 26)
weather (p. 5)

PROBLEMS

1. Refer to the newspaper-type weather map in Figure 1–3 to answer the following:

   a. Estimate the predicted high temperatures in central New York State and the northwest corner of Arizona.

   b. Where is the coldest area on the weather map? Where is the warmest?

   c. On this weather map, H stands for the center of a region of high pressure. Does it appear as though high pressure is associated with precipitation or fair weather?

   d. Which is warmer—central Texas or central Maine? Would you normally expect this to be the case?

2. Refer to the graph in Figure 1–5 to answer the following questions about temperatures in New York City:

   a. What is the approximate average daily high temperature in January? In July?

   b. Approximately what are the highest and lowest temperatures ever recorded?

3. Refer to the graph in Figure 1–7. Which year had the greatest number of billion-dollar weather disasters? How many events occurred that year? In which year was the damage amount greatest?

4. Refer to the graph in Figure 1–21 to answer the following:

   a. Approximately how much does the air pressure drop (in millibars) between the surface and 4 kilometers? (Use a surface pressure of 1000 millibars.)

   b. How much does the pressure drop between 4 and 8 kilometers?

   c. Based on your answers to parts a and b, answer the following: With an increase in altitude, air pressure decreases at a(n) (constant, increasing, decreasing) rate. Underline the correct answer.

5. If the temperature at sea level were 23°C, what would the air temperature be at a height of 2 kilometers, under average conditions?

6. Use the graph of the atmosphere's thermal structure (Figure 1–23) to answer the following:

   a. What are the approximate height and temperature of the stratopause?

   b. At what altitude is the temperature lowest? What is the temperature at that height?

7. Answer the following questions by examining the graph in Figure 1–25:

   a. In which one of the three regions (tropics, middle latitudes, poles) is the *surface* temperature lowest?

   b. In which region is the tropopause encountered at the lowest altitude? The highest? What are the altitudes and temperatures of the tropopause in those regions?

8. a. On a spring day a middle-latitude city (about 40°N latitude) has a surface (sea-level) temperature of 10°C. If vertical soundings reveal a nearly constant environmental lapse rate of 6.5°C per kilometer and a temperature at the tropopause of –55°C, what is the height of the tropopause?

   b. On the same spring day a station near the equator has a surface temperature of 25°C, 15°C higher than the middle-latitude city mentioned in part a. Vertical soundings reveal an environmental lapse rate of 6.5°C per kilometer and indicate that the tropopause is encountered at 16 kilometers. What is the air temperature at the tropopause?

MyMeteorologyLab™

Log in to www.mymeteorologylab.com for animations, videos, MapMaster interactive maps, GEODe media, *In the News* RSS feeds, web links, glossary flashcards, self-study quizzes and a Pearson eText version of this book to enhance your study of *Introduction to the Atmosphere*.

Heating Earth's Surface and Atmosphere

From our everyday experiences, we know that the Sun's rays feel warmer and paved roads become much hotter on clear, sunny days compared to overcast days. Pictures of snowcapped mountains remind us that temperatures decrease with altitude. And we know that the fury of winter is always replaced by the newness of spring. What you may not know is that these are manifestations of the same phenomenon that causes the blue color of the sky and the red color of a brilliant sunset. All these are a result of the interaction of solar radiation with Earth's atmosphere and its land–sea surface.

This power plant in Andalucia, Spain, produces clean thermoelectric power from the Sun. (Photo by Kevin Foy/Alamy)

Focus On Concepts

After completing this chapter, you should be able to:

- Explain what causes the Sun angle and length of daylight to change during the year and describe how these changes produce the seasons.

- Calculate the noon Sun angle for any latitude on the equinoxes and the solstices.

- Define *temperature* and explain how it is different from the total kinetic energy contained in a substance.

- Contrast the concepts of latent heat and sensible heat.

- List and describe the three mechanisms of heat transfer.

- Sketch and label a diagram that shows the fate of incoming solar radiation.

- Explain what causes blue skies and red sunsets.

- Explain what is meant by the statement "The atmosphere is heated from the ground up."

- Describe the role of water vapor and carbon dioxide in producing the greenhouse effect.

- Sketch and label a diagram illustrating Earth's heat budget.

35

Earth–Sun Relationships

Heating Earth's Surface and Atmosphere
▶ Understanding Seasons

The amount of solar energy received at any location varies with latitude, time of day, and season of the year. Contrasting images of polar bears and perpetual ice and palm trees along a tropical beach serve to illustrate the extremes (Figure 2–1). The unequal heating of Earth's land–sea surface creates

Figure 2–1 An understanding of Earth-Sun relationships is basic to an understanding of weather and climate. (a) In tropical latitudes, temperature contrasts during the year are modest. (Photo by Maria Skaldina/Shutterstock) (b) In polar regions, seasonal temperature contrasts can be dramatic. (Photo by Michael Collier)

winds and drives the ocean's currents, which in turn transport heat from the tropics toward the poles in an unending attempt to balance energy inequalities.

The consequences of these processes are the phenomena we call *weather*. If the Sun were "turned off," global winds and ocean currents would quickly cease. Yet as long as the Sun shines, winds *will* blow and weather *will* persist. So to understand how the dynamic weather machine works, we must know why different latitudes receive different quantities of solar energy and why the amount of solar energy received changes during the course of a year to produce the seasons.

Earth's Motions

Earth has two principal motions—rotation and revolution. **Rotation** is the spinning of Earth on its axis that produces the daily cycle of day and night.

The other motion, **revolution,** refers to Earth's movement in a slightly elliptical orbit around the Sun. The distance between Earth and Sun averages about 150 million kilometers (93 million miles). Because Earth's orbit is not perfectly circular, however, the distance varies during the course of a year. Each year, on about January 3, our planet is about 147.3 million kilometers (91.5 million miles) from the Sun, closer than at any other time—a position called **perihelion.** About six months later, on July 4, Earth is about 152.1 million kilometers (94.5 million miles) from the Sun, farther away than at any other time—a position called **aphelion.** Although Earth is closest to the Sun and receives up to 7 percent more energy in January than in July, this difference plays only a minor role in producing seasonal temperature variations, as evidenced by the fact that Earth is closest to the Sun during the Northern Hemisphere winter.

What Causes the Seasons?

If variations in the distance between the Sun and Earth do not cause seasonal temperature changes, what does? The gradual but significant *change in day length* certainly accounts for some of the difference we notice between summer and winter. Furthermore, a gradual change in the angle (altitude) of the Sun above the horizon is also a major contributing factor (Figure 2–2). For example, someone living in Chicago, Illinois, experiences the noon Sun highest in the sky in late June. But as summer gives way to autumn, the noon Sun appears lower in the sky, and sunset occurs earlier each evening.

The seasonal variation in the angle of the Sun above the horizon affects the amount of energy received at Earth's surface in two ways. First, when the Sun is directly overhead (at a 90° angle), the solar rays are most concentrated and thus most intense. At lower Sun angles, the rays become more spread out and less intense (Figure 2–3). You have probably experienced this when using a flashlight. If the beam strikes a surface perpendicularly, a small intense spot is produced. By contrast, if the flashlight beam strikes at any other angle, the area illuminated is larger—but noticeably dimmer.

Second, but of less significance, the angle of the Sun determines the path solar rays take as they pass through the

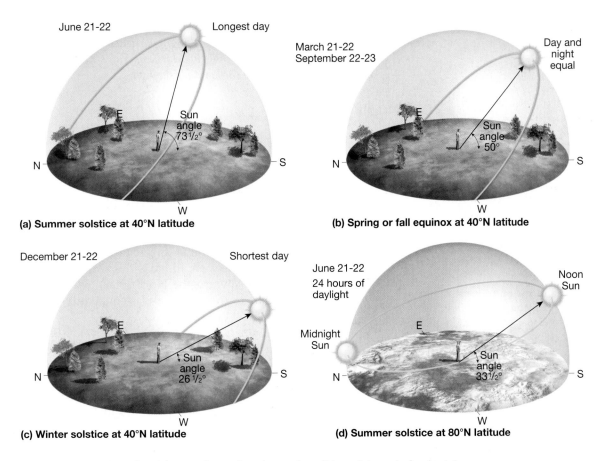

Figure 2–2 Daily paths of the Sun for a place located at 40° north latitude for the (a) summer solstice; (b) spring or fall equinox, and (c) winter solstice and for a place located at 80° north latitude at the (d) summer solstice.

atmosphere (Figure 2–4). When the Sun is directly overhead, the rays strike the atmosphere at a 90° angle and travel the shortest possible route to the surface. This distance is referred to as *1 atmosphere*. However, rays entering the atmosphere at a 30° angle must travel twice this distance before reaching the surface, while rays at a 5° angle travel through a distance roughly equal to the thickness of 11 atmospheres (Table 2–1). The longer the path, the greater the chance that sunlight will be dispersed by the atmosphere, which reduces the intensity at the surface. These conditions account for the fact that we cannot look directly at the midday Sun but we enjoy gazing at a sunset.

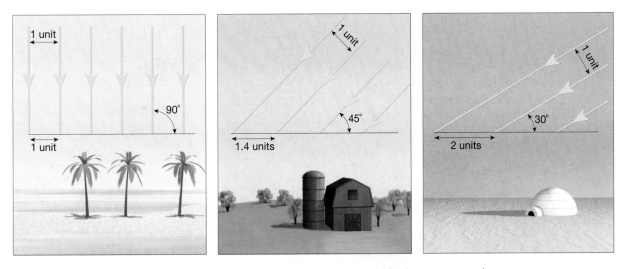

Figure 2–3 Changes in the Sun's angle cause variations in the amount of solar energy reaching Earth's surface. The higher the angle, the more intense the solar radiation.

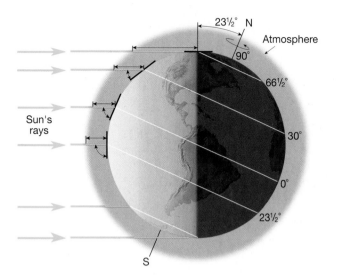

Figure 2–4 Rays striking Earth at a low angle (near the poles) must traverse more of the atmosphere than rays striking at a high angle (around the equator) and thus are subject to greater depletion by reflection and absorption.

It is important to remember that Earth's shape is spherical. Hence, on any given day, the only places that will receive vertical (90°) rays from the Sun are located along one particular line of latitude. As we move either north or south of this location, the Sun's rays strike at decreasing angles. Thus, the nearer a place is situated to the latitude receiving the vertical rays of the Sun, the higher will be its noon Sun, and the more concentrated will be the radiation it receives.

In summary, the most important reasons for variations in the amount of solar energy reaching a particular location are the seasonal changes in the angle at which the Sun's rays strike the surface and changes in the length of daylight.

Earth's Orientation

What causes fluctuations in Sun angle and length of daylight during the course of a year? Variations occur *because Earth's orientation to the Sun continually changes.* Earth's axis (the imaginary line through the poles around which Earth rotates) is not perpendicular to the plane of its orbit around the Sun—called the **plane of the ecliptic.** Instead, it is tilted 23 1/2° from the perpendicular, called the **inclination of the axis.** If the axis were not inclined, Earth would lack seasons. Because the axis remains pointed in the same direction (toward the North Star), the orientation of Earth's axis to the Sun's rays is constantly changing (Figure 2–5).

For example, on one day in June each year, Earth's position in orbit is such that the Northern Hemisphere is "leaning" 23 1/2° *toward* the Sun (left in Figure 2–5). Six months later, in December, when Earth has moved to the opposite side of its orbit, the Northern Hemisphere "leans" 23 1/2° *away* from the Sun (Figure 2–5, right). On days between these extremes, Earth's axis is leaning at amounts less than 23 1/2° to the rays of the Sun. This change in orientation causes the spot where the Sun's rays are vertical to make an annual migration from 23 1/2° north of the equator to 23 1/2° south of the equator. In turn, this migration causes the angle of the noon Sun to vary by up

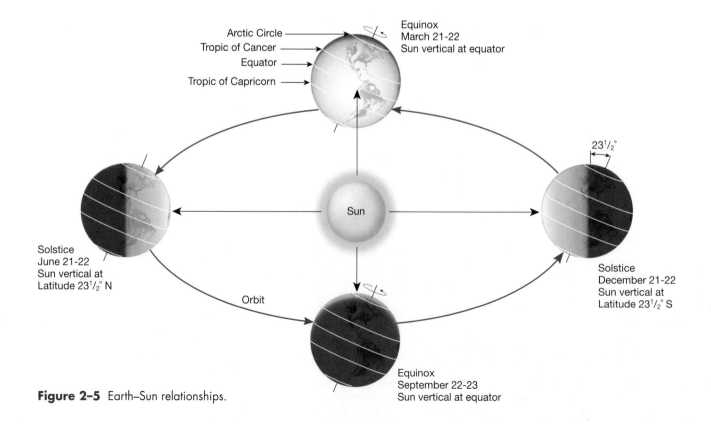

Figure 2–5 Earth–Sun relationships.

TABLE 2-1 Distance Radiation Must Travel Through the Atmosphere

Angle of Sun Above Horizon	Equivalent Number of Atmospheres Sunlight Must Pass Through
90° (directly overhead)	1.00
80°	1.02
70°	1.06
60°	1.15
50°	1.31
40°	1.56
30°	2.00
20°	2.92
10°	5.70
5°	10.80
0° (at horizon)	45.00

to 47° (23 1/2 + 23 1/2) for many locations during a year. A midlatitude city such as New York, for instance, has a maximum noon Sun angle of 73 1/2° when the Sun's vertical rays have reached their farthest northward location in June and a minimum noon Sun angle of 26 1/2° six months later (Figure 2–6).

Solstices and Equinoxes

Based on the annual migration of the direct rays of the Sun, four days each year are especially significant. On June 21 or 22, the vertical rays of the Sun strike 23 1/2° north latitude (23 1/2° north of the equator), a line of latitude known as the **Tropic of Cancer** (Figure 2–5). For people living in the Northern Hemisphere, June 21 or 22 is known as the **summer solstice,** the first "official" day of summer (see Box 2–1).

Six months later, on December 21 or 22, Earth is in an opposite position, with the Sun's vertical rays striking at 23 1/2° south latitude. This line is known as the **Tropic of Capricorn.** For those in the Northern Hemisphere, December 21 or 22 is the **winter solstice,** the first day of winter. However,

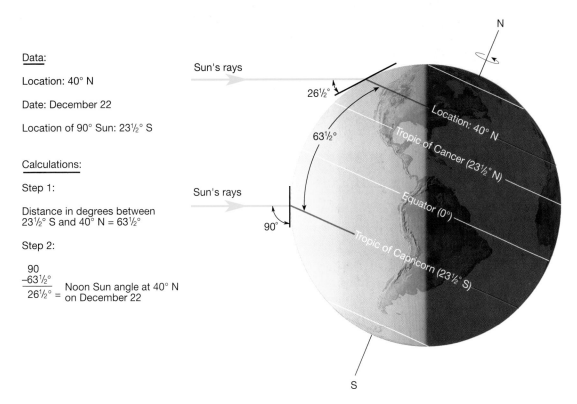

Data:

Location: 40° N

Date: December 22

Location of 90° Sun: 23½° S

Calculations:

Step 1:

Distance in degrees between 23½° S and 40° N = 63½°

Step 2:

$$\begin{array}{r} 90 \\ -63\frac{1}{2}° \\ \hline 26\frac{1}{2}° \end{array} = \text{Noon Sun angle at 40° N on December 22}$$

Figure 2–6 Calculating the noon Sun angle. Recall that on any given day, only one latitude receives vertical (90°) rays of the Sun. A place located 1° away (either north or south) receives an 89° angle; a place 2° away, an 88° angle; and so forth. To calculate the noon Sun angle, simply find the number of degrees of latitude separating the location you want to know about from the latitude that is receiving the vertical rays of the Sun. Then subtract that value from 90°. The example in this figure illustrates how to calculate the noon Sun angle for a city located at 40° north latitude on December 22 (winter solstice).

Box 2-1 When Are the Seasons?

Have you ever been caught in a snowstorm around Thanksgiving, even though winter does not begin until December 21? Or perhaps you have endured several consecutive days of 100° temperatures although summer has not "officially" started? The idea of dividing the year into four seasons originated from the Earth–Sun relationships discussed in this chapter (**Table 2–A**). This astronomical definition of the seasons defines winter (Northern Hemisphere) as the period from the winter solstice (December 21–22) to the spring equinox (March 21–22) and so forth. This is also the definition used most widely by the news media, yet it is not unusual for portions of the United States and Canada to have significant snowfalls weeks before the "official" start of winter (Table 2–A).

Because the weather phenomena we normally associate with each season do not coincide well with the astronomical seasons, meteorologists prefer to divide the year into four three-month periods based primarily on temperature. Thus, winter is defined as December, January, and February, the three coldest months of the year in the Northern Hemisphere. Summer is defined as the three warmest months, June, July, and August. Spring and autumn are the transition periods between these two seasons (**Figure 2–A**). Inasmuch as these four three-month periods better reflect the temperatures and weather that we associate with the respective seasons, this definition of the seasons is more useful for meteorological discussions.

TABLE 2–A Occurrence of the Seasons in the Northern Hemisphere

Season	Astronomical Season	Climatological Season
Spring	March 21 or 22 to June 21 or 22	March, April, May
Summer	June 21 or 22 to September 22 or 23	June, July, August
Autumn	September 22 or 23 to December 21 or 22	September, October, November
Winter	December 21 or 22 to March 21 or 22	December, January, February

FIGURE 2–A Trees turning bright colors before losing their leaves is a common fall scene in the middle latitudes. (Photo by Corbis/SuperStock).

on this same day people in the Southern Hemisphere are experiencing their summer solstice.

The *equinoxes* occur midway between the solstices. September 22 or 23 is the date of the **autumnal (fall) equinox** in the Northern Hemisphere, and March 21 or 22 is the date of the **spring equinox** (also called the **vernal equinox**). On these dates the vertical rays of the Sun strike the equator (0° latitude) because Earth's position is such that its axis is tilted neither toward nor away from the Sun.

The length of daylight versus darkness is also determined by the position of Earth relative to the Sun's rays. The length of daylight on June 21, the summer solstice in the Northern Hemisphere, is greater than the length of night.

This fact can be established by examining Figure 2–7, which illustrates the **circle of illumination**—that is, the boundary separating the dark half of Earth from the lighted half. The length of daylight is established by comparing the fraction of a line of latitude that is on the "day" side of the circle of illumination with the fraction on the "night" side. Notice that on June 21 all locations in the Northern Hemisphere experience longer periods of daylight than darkness (Figure 2–7). By contrast, during the December solstice the length of darkness exceeds the length of daylight at all locations in the Northern Hemisphere. For example, consider New York City (about 40° north latitude), which has about 15 hours of daylight on June 21 and only 9 hours on December 21.

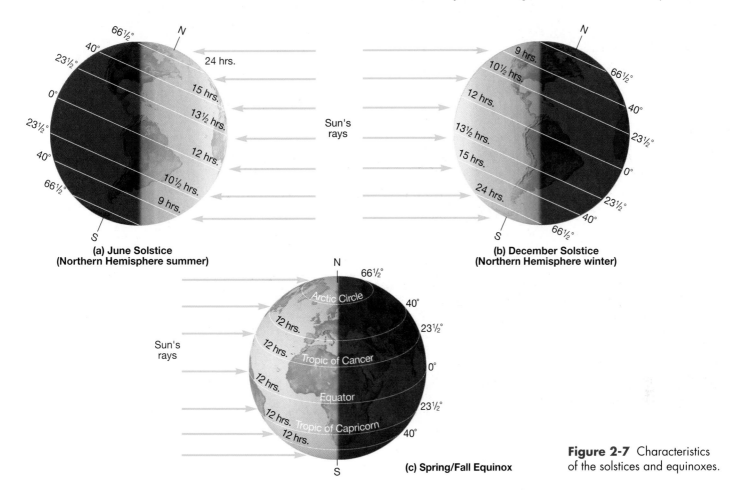

(a) June Solstice
(Northern Hemisphere summer)

(b) December Solstice
(Northern Hemisphere winter)

(c) Spring/Fall Equinox

Figure 2-7 Characteristics of the solstices and equinoxes.

Also note from Table 2–2 that the farther north a location is from the equator on June 21, the longer the period of daylight. When you reach the Arctic Circle (66 1/2° north latitude), the length of daylight is 24 hours. Places located at or north of the Arctic Circle experience the "midnight Sun," which does not set for a period that ranges from one day to about six months (Figure 2–8).

As a review of the characteristics of the summer solstice for the Northern Hemisphere, examine Figure 2–7 and Table 2–2 and consider the following:

1. The date of occurrence is June 21 or 22.

2. The vertical rays of the Sun are striking the Tropic of Cancer (23 1/2° north latitude).

3. Locations in the Northern Hemisphere are experiencing their longest length of daylight and highest Sun angle. (The opposite is true for the Southern Hemisphere.)

4. The farther north a location is from the equator, the longer the period of daylight, until the Arctic Circle is reached, where the length of daylight becomes 24 hours. (The opposite is true for the Southern Hemisphere.)

The winter solstice facts are the reverse. It should now be apparent why a midlatitude location is warmest in the summer—when the days are longest and the angle of the Sun is highest.

During an equinox (meaning "equal night"), the length of daylight is 12 hours everywhere on Earth because the cir-

cle of illumination passes directly through the poles, thus dividing the lines of latitude in half.

These seasonal changes, in turn, cause the month-to-month variations in temperature observed at most locations outside the tropics. Figure 2–9 shows mean monthly temperatures for selected cities at different latitudes. Notice that the cities located at more poleward latitudes experience larger temperature differences from summer to winter than do cities located nearer the equator. Also notice that temperature minimums for Southern Hemisphere locations occur in July,

TABLE 2–2 Length of Daylight

Latitude (degrees)	Summer Solstice	Winter Solstice	Equinoxes
0	12 hr	12 hr	12 hr
10	12 hr 35 min	11 hr 25 min	12 hr
20	13 hr 12 min	10 hr 48 min	12 hr
30	13 hr 56 min	10 hr 04 min	12 hr
40	14 hr 52 min	9 hr 08 min	12 hr
50	16 hr 18 min	7 hr 42 min	12 hr
60	18 hr 27 min	5 hr 33 min	12 hr
70	2 mo	0 hr 00 min	12 hr
80	4 mo	0 hr 00 min	12 hr
90	6 mo	0 hr 00 min	12 hr

whereas they occur in January for most places in the Northern Hemisphere.

In summary, seasonal fluctuations in the amount of solar energy reaching various places on Earth's surface are caused by the migrating vertical rays of the Sun and the resulting variations in Sun angle and length of daylight.

All locations situated at the same latitude have identical Sun angles and lengths of daylight. If the Earth–Sun relationships previously described were the only controls of temperature, we would expect these places to have identical temperatures as well. Obviously, such is not the case. Although the angle of the Sun above the horizon and the length of daylight are the primary controls of temperature, other factors must be considered—a topic that will be addressed in Chapter 3.

Concept Check 2.1

1 Do the annual variations in Earth–Sun distance adequately account for seasonal temperature changes? Explain.

2 Use a simple sketch to show why the intensity of solar radiation striking Earth's surface changes when the Sun angle changes.

3 Briefly explain the primary cause of the seasons.

4 What is the significance of the Tropic of Cancer and the Tropic of Capricorn?

5 After examining Table 2–2, write a general statement that relates the season, latitude, and the length of daylight.

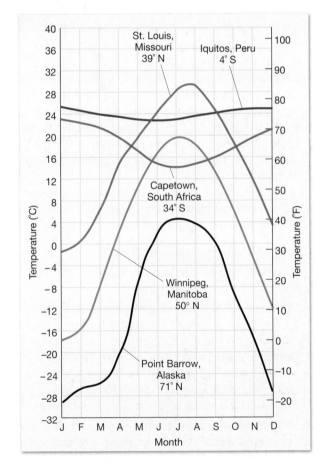

Figure 2–9 Mean monthly temperatures for five cities located at different latitudes. Note that Cape Town, South Africa, experiences winter in June, July, and August.

EYE ON THE ATMOSPHERE

This image, which shows the first sunrise of 2008 at the South Pole, was taken at the U.S. Amundsen–Scott Station. At the moment the Sun cleared the horizon, the weathered American flag was seen whipping in the wind above a sign marking the location of the geographic South Pole. (NASA)

Question 1 What was the approximate date that this photograph was taken?

Question 2 How long after this photo was taken did the Sun set at the South Pole?

Question 3 Over the course of one year, what is the highest position the Sun can reach (measured in degrees) at the South Pole? On what date does this occur?

Students Sometimes Ask . . .

Where Is the Land of the Midnight Sun?

Any place located north of the Arctic Circle (66 1/2° north latitude) or south of the Antarctic Circle (66 1/2° south latitude) experiences 24 hours of continuous daylight (or darkness) at least one day each year. The closer a place is to a pole, the longer the period of continuous daylight (or darkness). Someone stationed at either pole will experience six months of continuous daylight followed by six months of darkness. Thus, the Land of the Midnight Sun refers to any location where the latitude is greater than 66 1/2°, such as the northern portions of Alaska, Canada, Russia, and Scandinavia, as well as most of Antarctica.

Energy, Temperature, and Heat

The universe is made up of a combination of matter and energy. The concept of matter is easy to grasp because it is the "stuff" we can see, smell, and touch. Energy, on the other hand, is abstract and therefore more difficult to describe and understand. Energy comes to Earth from the Sun in the form of electromagnetic radiation, which we see as light and feel as heat. There are countless places and situations where energy is present—in the food we eat, the water located at the top of a waterfall, and the waves that break along the shore.

Forms of Energy

Energy can be defined simply as *the capacity to do work.* Work is done whenever matter moves. Common examples include the chemical energy from gasoline that powers automobiles, the heat energy from stoves that excites water molecules (boils water), and the gravitational energy that has the capacity to move snow down a mountain slope in the form of an avalanche. These examples illustrate that energy takes many forms and can also change from one form to another. For example, the chemical energy in gasoline is first converted to heat in the engine of an automobile, which is then converted to mechanical energy that moves the automobile along.

You are undoubtedly familiar with some of the common forms of energy, such as heat, chemical, nuclear, radiant (light), and gravitational energy. Energy is also placed into one of two major categories: *kinetic energy* and *potential energy.*

Kinetic Energy *Energy associated with an object by virtue of its motion* is described as **kinetic energy.** A simple example of kinetic energy is the motion of a hammer when driving a nail. Because of its motion, the hammer is able to move another object (do work). The faster the hammer is swung, the greater its kinetic energy (energy of motion). Similarly, a larger (more massive) hammer possesses more kinetic energy than a smaller one, provided that both are swung at the same velocity. Likewise, the winds associated with a hurricane possess much more kinetic energy than do light, localized breezes because they are both larger in scale (cover a larger area) and travel at higher velocities.

Kinetic energy is also significant at the atomic level. All matter is composed of atoms and molecules that are continually vibrating, and by virtue of this motion have kinetic energy. For example, when a pan of water is placed over a fire, the water becomes warmer because the fire causes the water molecules to vibrate faster. Thus, when a solid, liquid, or gas is heated, its atoms or molecules move faster and possess more kinetic energy.

Potential Energy   As the term implies, **potential energy** has the capability to do work. For example, large hailstones suspended by an updraft in a towering cloud have potential energy. Should the updraft subside, these hailstones may fall to Earth and do destructive work on roofs and vehicles. Many substances, including wood, gasoline, and the food you eat, contain potential energy, which is capable of doing work given the right circumstances.

Temperature

Humans think of temperature as how warm or cold an object is with respect to some standard measure. However, our senses are often poor judges of warm and cold (see Students Sometimes Ask, p. 46). **Temperature** is a *measure of the average kinetic energy of the atoms or molecules in a substance.* When a substance gains energy, its particles move faster and its temperature rises. By contrast, when energy is lost, the atoms and molecules vibrate more slowly and its temperature drops. In the United States the Fahrenheit scale is used most often for everyday expressions of temperature. However, scientists and the majority of other countries use the Celsius and Kelvin temperature scales. A discussion of these scales is provided in Chapter 3.

It is important to note that temperature is *not* a measure of the total kinetic energy of an object. For example, a cup of boiling water has a much higher temperature than a bathtub of lukewarm water. However, the quantity of water in the cup is small, so it contains far less kinetic energy than the water in the tub. Much more ice would melt in the tub of lukewarm water than in the cup of boiling water. The temperature of the water in the cup is higher because the atoms and molecules are vibrating faster, but the total amount of kinetic energy (also called *thermal energy*) is much smaller because there are fewer particles.

Students Sometimes Ask . . .

What would the seasons be like if Earth were not tilted on its axis?

The most obvious change would be that all locations on the globe would experience 12 hours of daylight every day of the year. Moreover, for any latitude, the Sun would always follow the path it does during an equinox. There would be no seasonal temperature changes, and daily temperatures would be roughly equivalent to the "average" for that location.

Heat

We define **heat** as *energy transferred into or out of an object because of temperature differences between that object and its surroundings.* If you hold a hot mug of coffee, your hand will begin to feel warm or even hot. By contrast, when you hold an ice cube, heat is transferred from your hand to the ice cube. Heat flows from a region of higher temperature to one of lower temperature. Once the temperatures become equal, heat flow stops.

Meteorologists further subdivide heat into two categories, *latent heat* and *sensible heat.* Latent heat is the energy involved when water changes from one state of matter to another—when liquid water evaporates and becomes water vapor, for example. During evaporation heat is required to break the hydrogen bonds between water molecules that occurs when water vapor escapes a water body. Because the most energetic water molecules escape, the average kinetic energy (temperature) of the water body drops. Therefore, evaporation is a cooling process, something you may have experienced upon stepping out, dripping wet, from a swimming pool or bathtub. The energy absorbed by the escaping water vapor molecules is termed **latent heat** (meaning "hidden") because it does not result in a temperature increase. The latent heat stored in water vapor is eventually released in the atmosphere during condensation—when water vapor returns to the liquid state during cloud formation. Therefore, latent heat is responsible for transporting considerable amounts of energy from Earth's land–sea surface to the atmosphere.

By contrast, **sensible heat** is the heat we can feel and measure with a thermometer. It is called sensible heat because it can be "sensed." Warm air that originates over the Gulf of Mexico and flows into the Great Plains in the winter is an example of the transfer of sensible heat.

Concept Check 2.2

1 Distinguish between heat and temperature.

2 Describe how latent heat is transferred from Earth's land–sea surface to the atmosphere.

3 Compare latent heat and sensible heat.

Mechanisms of Heat Transfer

 Heating Earth's Surface and Atmosphere
▶ Solar Radiation

The flow of energy can occur in three ways: *conduction, convection,* and *radiation* (Figure 2–10). Although they are presented separately, all three mechanisms of heat transfer can operate simultaneously. In addition, these processes may transfer heat between the Sun and Earth and between Earth's surface, the atmosphere, and outer space.

Conduction

Anyone who attempts to pick up a metal spoon left in a boiling pot of soup realizes that heat is transmitted along the entire length of the spoon. The transfer of heat in this manner is called *conduction.* The hot soup causes the molecules at the lower end of the spoon to vibrate more rapidly.

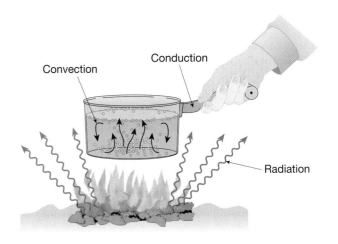

Figure 2–10 The three mechanisms of heat transfer: conduction, convection, and radiation.

These molecules and free electrons collide more vigorously with their neighbors and so on up the handle of the spoon. Thus, **conduction** is the transfer of heat through electron and molecular collisions from one molecule to another. The ability of substances to conduct heat varies considerably. Metals are good *conductors*, as those of us who have touched a hot spoon have quickly learned. Air, in contrast, is a very poor conductor of heat. Consequently, conduction is important only between Earth's surface and the air immediately in contact with the surface. As a means of heat transfer for the atmosphere as a whole, conduction is the least significant and can be disregarded when considering most meteorological phenomena.

Objects that are poor conductors, such as air, are called *insulators*. Most objects that are good insulators, such as cork, plastic foam, or goose down, contain many small air spaces. The poor conductivity of the trapped air gives these materials their insulating value. Snow is also a poor conductor (good insulator), and like other insulators, it con-

tains numerous air spaces that impair the flow of heat. This is why wild animals may burrow into a snowbank to escape the "cold." The snow, like a down-filled comforter, does not supply heat; it simply retards the loss of the animal's own body heat.

Convection

Much of the heat transport in Earth's atmosphere and oceans occurs by convection. **Convection** is heat transfer that involves the actual movement or circulation of a substance. It takes place in fluids (liquids such as water and gases such as air) where the material is able to flow.

The pan of water being heated over a campfire in Figure 2–10 illustrates the nature of a simple convective circulation. The fire warms the bottom of the pan, which conducts heat to the water inside. Because water is a relatively poor conductor, only the water in close proximity to the bottom of the pan is heated by conduction. Heating causes water to expand and become less dense. Thus, the hot, buoyant water near the bottom of the pan rises, while the cooler, denser water above sinks. As long as the water is heated from the bottom and cools near the top, it will continue to "turn over," producing a *convective circulation*.

In a similar manner, some of the air in the lowest layer of the atmosphere that was heated by radiation and conduction is transported by convection to higher layers of the atmosphere. For example, on a hot, sunny day the air above a plowed field will be heated more than the air above the surrounding woodlands. As warm, less dense air above the plowed field buoys upward, it is replaced by the cooler air above the woodlands (Figure 2–11). In this way a convective flow is established. The warm parcels of rising air are called **thermals** and are what hang-glider pilots use to keep their crafts soaring. Convection of this type not only transfers heat but also transports moisture aloft. The result is an increase in cloudiness that frequently can be observed on warm summer afternoons.

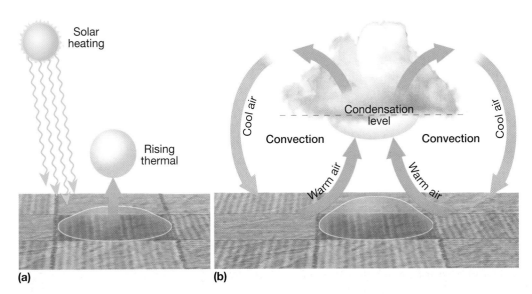

(a) (b)

Figure 2–11 (a) Heating of Earth's surface produces thermals of rising air that transport heat and moisture aloft. (b) The rising air cools, and if it reaches the condensation level, clouds form. Rising warmer air and descending cooler air are an example of convective circulation.

On a much larger scale is the global convective circulation of the atmosphere, which is driven by the unequal heating of Earth's surface. These complex movements are responsible for the redistribution of heat between hot equatorial regions and frigid polar latitudes and will be discussed in detail in Chapter 7.

Students Sometimes Ask . . .

In the morning when I get out of bed, why does the tile flooring in the bathroom feel much colder than the carpeted area, even though both materials are the same temperature?

The difference you feel is due mainly to the fact that floor tile is a much better conductor of heat than carpet. Hence, heat is more rapidly conducted from your bare feet to the tile floor than from your bare feet to the carpeted floor. Even at room temperature (20°C [68°F]), objects that are good conductors can feel chilly to the touch. (Remember, body temperature is about 37°C [98.6°F].)

Atmospheric circulation consists of vertical as well as horizontal components, so both vertical and horizontal heat transfer occurs. Meteorologists often use the term *convection* to describe the part of the atmospheric circulation that involves *upward and downward* heat transfer. By contrast, the term **advection** is used to denote the primarily horizontal component of convective flow. (The common term for advection is "wind," a phenomenon we will examine closely in later chapters.) Residents of the midlatitudes often experience the effects of heat transfer by advection. For example, when frigid Canadian air invades the Midwest in January, it brings bitterly cold winter weather.

Radiation

The third mechanism of heat transfer is radiation. Unlike conduction and convection, radiation is the only mechanism of heat transfer that can travel through the vacuum of space and thus is responsible for solar energy reaching Earth.

Solar Radiation   The Sun is the ultimate source of energy that drives the weather machine. We know that the Sun emits light and heat energy as well as the rays that darken skin pigmentation. Although these forms of energy constitute a major portion of the total energy that radiates from the Sun, they are only a part of a large array of energy called **radiation, or electromagnetic radiation.** This array or spectrum of electromagnetic energy is shown in Figure 2–12.

All types of radiation, whether X-rays, radio waves, or heat waves, travel through the vacuum of space at 300,000 kilometers (186,000 miles) per second, a value known as the *speed of light.* To help visualize radiant energy, imagine ripples made in a calm pond when a pebble is tossed in. Like the waves produced in the pond, electromagnetic waves come in various sizes, or **wavelengths**—the distance from one crest to the next (Figure 2–12). Radio waves have the longest wavelengths, up to tens of kilometers in length. Gamma waves are the shortest, at less than one-billionth of a centimeter long. Visible light is roughly in the middle of this range.

Radiation is often identified by the effect that it produces when it interacts with an object. The retinas of our eyes, for instance, are sensitive to a range of wavelengths called **visible light.** We often refer to visible light as white light because it appears "white" in color. It is easy to show, however, that white light is really an array of colors, each color corresponding to a specific range of wavelengths. By using a prism, white light can be divided into the colors

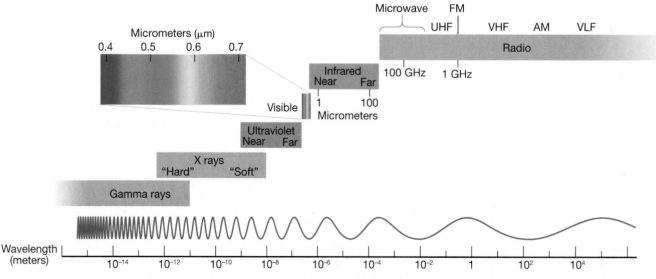

Figure 2–12 The electromagnetic spectrum, illustrating the wavelengths and names of various types of radiation.

Figure 2–13 Visible light consists of an array of colors we commonly call the "colors of the rainbow." Rainbows are relatively common phenomena produced by the bending and reflection of light by drops of water. There is more about rainbows in Chapter 16. (Photo by David Robertson/Alamy)

of the rainbow, from violet with the shortest wavelength, 0.4 micrometer (μm) (1 micrometer is one-millionth of a meter), to red with the longest wavelength, 0.7 micrometer (Figure 2–13).

Located adjacent to the color red, and having a longer wavelength, is **infrared radiation,** which cannot be seen by the human eye but is detected as heat. Only the infrared energy that is nearest the visible part of the spectrum is intense enough to be felt as heat and is referred to as *near infrared.* On the opposite side of the visible range, located next to violet, the energy emitted is called **ultraviolet radiation** and consists of wavelengths that may cause sunburned skin.

Although we divide radiant energy into categories based on our ability to perceive them, all wavelengths of radiation behave similarly. When an object absorbs any form of electromagnetic energy, the waves excite subatomic particles (electrons). This results in an increase in molecular motion and a corresponding increase in temperature. Thus, electromagnetic waves from the Sun travel through space and, upon being absorbed, increase the molecular motion of other molecules—including those that make up the atmosphere, Earth's land–sea surfaces, and human bodies.

One important difference among the various wavelengths of radiant energy is that *shorter wavelengths are more energetic.* This accounts for the fact that relatively short (high-energy) ultraviolet waves can damage human tissue more readily than can similar exposures to longer-wavelength radiation. The damage can result in skin cancer and cataracts.

It is important to note that the Sun emits all forms of radiation, as shown in Figure 2–12, but in varying quantities. Over 95 percent of all solar radiation is emitted in wavelengths between 0.1 and 2.5 micrometers, with much of this energy concentrated in the visible and near-infrared parts of the electromagnetic spectrum (Figure 2–14). The narrow band of visible light, between 0.4 and 0.7 micrometer, represents over 43 percent of the total energy emitted. The bulk of the remainder lies in the infrared zone (49 percent) and ultraviolet (UV) section (7 percent). Less than 1 percent of solar radiation is emitted as X-rays, gamma rays, and radio waves.

Laws of Radiation

To obtain a better appreciation of how the Sun's radiant energy interacts with Earth's atmosphere and land–sea surface, it is helpful to have a general understanding of the basic radiation laws. Although the mathematics of these laws is beyond the scope of this text, the concepts are fundamental to understanding radiation:

1. **All objects continually emit radiant energy over a range of wavelengths.*** Thus, not only do hot objects such as the Sun continually emit energy, but Earth does as well, even the polar ice caps.

---

*The temperature of the object must be above a theoretical value called *absolute zero* (–273°C) in order to emit radiant energy. The letter K is used for values on the Kelvin temperature scale. For more explanation, see the section on "Temperature Scales" in Chapter 3.

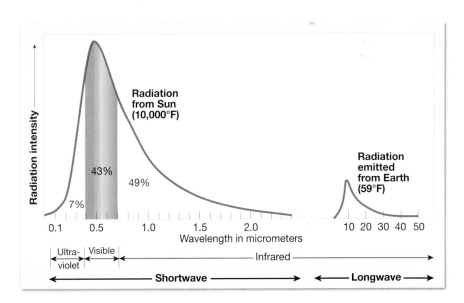

Figure 2–14 Comparison of the intensity of solar radiation and radiation emitted by Earth. Because of the Sun's high surface temperature, most of its energy is radiated at wavelengths shorter than 4 micrometers, with the greatest intensity in the visible range of the electromagnetic spectrum. Earth, in contrast, radiates most of its energy in wavelengths longer than 4 micrometers, primarily in the infrared band. Thus, we call the Sun's radiation *shortwave* and Earth's radiation *longwave.* (From PHYSICAL GEOGRAPHY: A LANDSCAPE APPRECIATION, 9th Edition, by Tom L. McKnight and Darrell Hess, © 2008. Reprinted by permission of Pearson Education, Inc., Upper Saddle River, NJ.)

Box 2-2 Radiation Laws

Gregory J. Carbone[*]

All bodies radiate energy. Both the rate and the wavelength of radiation emitted depend on the temperature of the radiating body.

Stefan–Boltzmann Law

This law mathematically expresses the rate of radiation emitted per unit area:

$$E = \sigma T^4$$

E, the rate of radiation emitted by a body, is proportional to the fourth power of the body's temperature (T). The Stefan–Boltzmann constant (σ) is equal to $5.67 \times 10^{-8}\,W/m^2\,K^4$. Compare the difference between the radiation emission from the Sun and Earth. The Sun, with an average temperature of 6000 K, emits 73,483,200 watts per square meter (Wm^{-2}):

$$E = (5.67 \times 10^{-8}\,W/m^2\,K^4)(6000\,K)^4$$
$$= 73,483,200\,W/m^2$$

By contrast, Earth has an average temperature of only 288 K. If we round the value to 300 K, we have:

$$E = (5.67 \times 10^{-8}\,W/m^2\,K^4)(300\,K)^4$$
$$= 459\,W/m^2$$

The Sun has a temperature that is approximately 20 times higher than that of Earth and thus it emits approximately 160,000 times more radiation per unit area. This makes sense because $20^4 = 160,000$.

Wien's Displacement Law

Wien's displacement law describes mathematically the relationship between the temperature (T) of a radiating body and its wavelength of maximum emission (λ_{max}):

$$\lambda_{max} = C/T$$

Wien's constant (C) is equal to $2898\,\mu mK$. If we use the Sun and Earth as examples, we find:

$$\lambda_{max}(Sun) = \frac{2898\,\mu mK}{6000\,K} = 0.483\,\mu m$$

and:

$$\lambda_{max}(Earth) = \frac{2898\,\mu mK}{300\,K} = 9.66\,\mu m$$

Note that the Sun radiates its maximum energy within the visible portion of the electromagnetic spectrum. The cooler Earth radiates its maximum energy in the infrared portion of the electromagnetic spectrum.

[*]Professor Carbone is a faculty member in the Department of Geography at the University of South Carolina.

2. **Hotter objects radiate more total energy per unit area than do colder objects.** The Sun, which has a surface temperature of 6000 K (10,000°F), emits about 160,000 times more energy per unit area than does Earth, which has an average surface temperature of 288 K (59°F). (This concept is called the *Stefan–Boltzmann law* and is expressed mathematically in Box 2–2.)

3. **Hotter objects radiate more energy in the form of short-wavelength radiation than do cooler objects.** We can visualize this law by imagining a piece of metal that, when heated sufficiently (as occurs in a blacksmith's shop), produces a white glow. As it cools, the metal emits more of its energy in longer wavelengths and glows a reddish color. Eventually, no light is given off, but if you place your hand near the metal, the still longer infrared radiation will be detectable as heat. The Sun radiates maximum energy at 0.5 micrometer, which is in the visible range (Figure 2–14). The maximum radiation emitted from Earth occurs at a wavelength of 10 micrometers, well within the infrared (heat) range. Because the maximum Earth radiation is roughly 20 times longer than the maximum solar radiation, it is often referred to as **longwave radiation,** whereas solar radiation is called **shortwave radiation.** (This concept, known as *Wien's displacement law*, is expressed mathematically in Box 2–2.)

4. **Objects that are good absorbers of radiation are also good emitters.** Earth's surface and the Sun are nearly perfect radiators because they absorb and radiate with nearly 100 percent efficiency. By contrast, the gases that compose our atmosphere are *selective* absorbers and emitters of radiation. For some wavelengths the atmosphere is nearly transparent (little radiation absorbed). For others, however, it is nearly opaque (absorbs most of the radiation that strikes it). Experience tells us that the atmosphere is quite transparent to visible light emitted by the Sun because it readily reaches Earth's surface.

To summarize, although the Sun is the ultimate source of radiant energy, all objects continually radiate energy over a range of wavelengths. Hot objects, such as the Sun, emit mostly shortwave (high-energy) radiation. By contrast, most objects at everyday temperatures (Earth's surface and atmosphere) emit longwave (low-energy) radiation. Objects that are good absorbers of radiation, such as Earth's surface, are also good emitters. By contrast, most gases are good absorbers (emitters) only in certain wavelengths but poor absorbers in other wavelengths.

Concept Check 2.3

1 Describe the three basic mechanisms of energy transfer. Which mechanism is least important meteorologically?

2 What is the difference between convection and advection?

3 Compare visible, infrared, and ultraviolet radiation. For each, indicate whether it is considered shortwave or longwave radiation.

4 In what part of the electromagnetic spectrum does the Sun radiate maximum energy? How does this compare to Earth?

5 Describe the relationship between the temperature of a radiating body and the wavelengths it emits.

SEVERE AND HAZARDOUS WEATHER

The Ultraviolet Index*

Most people welcome sunny weather. On warm days, when the sky is cloudless and bright, many spend a great deal of time outdoors "soaking up" the sunshine (Figure 2–B). For many, the goal is to develop a dark tan, one that sunbathers often describe as "healthy looking." Ironically, there is strong evidence that too much sunshine (specifically, too much ultraviolet radiation) can lead to serious health problems, mainly skin cancer and cataracts of the eyes.

Since June 1994 the National Weather Service (NWS) has issued the next-day ultraviolet index (UVI) for the United States to warn the public of potential health risks of exposure to sunlight (Figure 2–C). The UV index is determined by taking into account the predicted cloud cover and reflectivity of the surface, as well as the Sun angle and atmospheric depth for each forecast location. Because atmospheric ozone strongly absorbs ultraviolet radiation, the extent of the ozone layer is also considered. The UVI values lie on a scale from 0 to 15, with larger values representing greatest risk.

The U.S. Environmental Protection Agency has established five exposure categories based on UVI values—Low, Moderate, High, Very High, and Extreme (Table 2–B). Precautionary measures have been developed for each category. The public is advised to minimize outdoor activities when the UVI is Very High or Extreme. Sunscreen with a sun-protection factor (SPF) of 15 or higher is recommended for all exposed skin. This is especially important after swimming or while sunbathing, even on cloudy days with the UVI in the Low category.

UV Index

0 1 2 3 4 5 6 7 8 9 10 11 12 13 14 15

June 8, 2008

FIGURE 2–C UV index forecast for June 8, 2008. To view the current UVI forecast, go to www.epa.gov/sunwise/uvindex.html.

> The public is advised to minimize outdoor activities when the UVI is Very High or Extreme.

Table 2–B shows the range of minutes to burn for the most susceptible skin type (pale or milky white) for each exposure category. Note that the exposure that results in sunburn varies from over 60 minutes for the Low category to less than 10 minutes for the Extreme category. It takes approximately five times longer to cause sunburn of the least susceptible skin type, brown to dark. The most susceptible skin type develops red sunburn, painful swelling, and skin peeling when exposed to excessive sunlight. By contrast, the least susceptible skin type rarely burns and shows very rapid tanning response.

*Based on material prepared by Professor Gong-Yuh Lin, a faculty member in the Department of Geography at California State University, Northridge.

FIGURE 2–B Exposing sensitive skin to too much solar ultraviolet radiation has potential health risks. (Photo by Eddie Gerald/Rough Guides)

TABLE 2–B The UV Index: Minutes to Burn for the Most Susceptible Skin Type

UVI Value	Exposure Category	Description	Minutes to Burn
0–2	Low	Low danger from the Sun's UV rays for the average person.	> 60
3–5	Moderate	Moderate risk from unprotected Sun exposure. Take precautions during the midday, when Sun is strongest.	40–60
6–7	High	Protection against sunburn is needed. Cover up, wear a hat and sunglasses, and use sunscreen.	25–40
8–10	Very High	Try to avoid the Sun between 11 AM and 4 PM. Otherwise, cover up and use sunscreen.	10–25
11–15	Extreme	Take all precautions. Unprotected skin will burn in minutes. Do not pursue outdoor activities if possible. If outdoors, apply sunscreen liberally every 2 hours.	< 10

What Happens to Incoming Solar Radiation?

 Heating Earth's Surface and Atmosphere
▶ What Happens to Incoming Solar Radiation?

When radiation strikes an object, three different things may occur simultaneously. First, some of the energy may be *absorbed*. Recall that when radiant energy is absorbed, the molecules begin to vibrate faster, which causes an increase in temperature. The amount of energy absorbed by an object depends on the intensity of the radiation and the object's **absorptivity.** In the visible range, the degree of absorptivity is largely responsible for the brightness of an object. Surfaces that are good absorbers of all wavelengths of visible light appear black in color, whereas light-colored surfaces have a much lower absorptivity. That is why wearing light-colored clothing on a sunny summer day may help keep you cooler. Second, substances such as water and air, which are transparent to certain wavelengths of radiation, may simply *transmit* energy—allowing it to pass through without being absorbed. Third, some radiation may "bounce off" the object without being absorbed or transmitted. In summary, *radiation may be absorbed, transmitted, or redirected (reflected or scattered).*

Figure 2–15 shows the fate of incoming solar radiation averaged for the entire globe. On average, about 50 percent of incoming solar energy is absorbed at Earth's surface. Another 30 percent is reflected and scattered back to space by the atmosphere, clouds, and reflective surfaces such as snow and water, and about 20 percent is absorbed by clouds and atmospheric gases.

What determines whether solar radiation will be transmitted to the surface, scattered, reflected back to space, or absorbed by the gases and particles in the atmosphere? As you will see, it depends greatly upon the *wavelength* of the radiation, as well as the size and nature of the intervening material.

Reflection and Scattering

Reflection is the process whereby light bounces back from an object at the same angle and intensity (Figure 2–16a). By contrast, **scattering** produces a larger number of weaker rays, traveling in different directions. Scattering disperses light both forward and backward (**backscattering**)

(Figure 2–16b). Whether solar radiation is reflected or scattered depends largely on the size of the intervening particles and the wavelength of the light.

Reflection and Earth's Albedo About 30 percent of the solar energy that reaches our planet is reflected back to space (Figure 2–15). Included in this figure is the amount sent skyward by backscattering. This energy is lost to Earth and does not play a role in heating the atmosphere or Earth's surface.

The fraction of radiation that is reflected by an object is called its **albedo.** The albedo for Earth as a whole (planetary albedo) is 30 percent. The amount of light reflected from Earth's land–sea surface represents only about 5 percent of the total planetary albedo (Figure 2–15). Not surprisingly, clouds are largely responsible for most of Earth's "brightness" as seen from space. The high reflectivity of clouds is experienced when looking down on a cloud during an airline flight.

In comparison to Earth, the Moon, which has neither clouds nor an atmosphere, has an average albedo of only 7 percent. Although a full Moon appears bright, the much brighter and larger Earth would, by comparison, provide far more light for an astronaut's "Earth-lit" Moon walk at night.

Figure 2–17 gives the albedos for various surfaces. Fresh snow and thick clouds have high albedos (good reflectors). By contrast, dark soils and parking lots have low albedos and thus absorb much of the radiation they receive. In the case of a lake or the ocean, note that the angle at which the Sun's rays strike the water surface greatly affects its albedo.

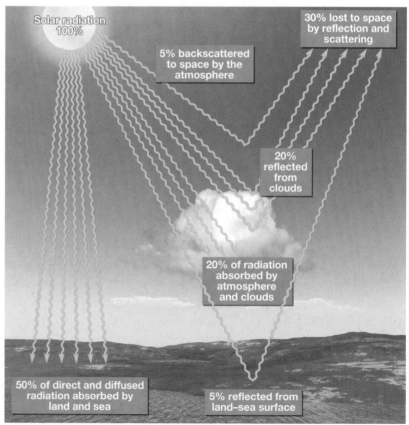

Solar radiation 100%

30% lost to space by reflection and scattering

5% backscattered to space by the atmosphere

20% reflected from clouds

20% of radiation absorbed by atmosphere and clouds

50% of direct and diffused radiation absorbed by land and sea

5% reflected from land–sea surface

Figure 2–15 Average distribution of incoming solar radiation, by percentage. More solar energy is absorbed by Earth's surface than by the atmosphere.

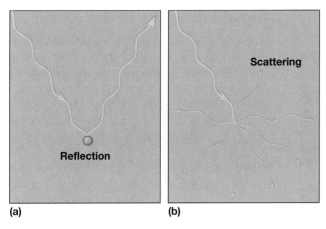

(a)   **(b)**

Figure 2-16 Reflection and scattering. (a) Reflected light bounces back from a surface at the same angle at which it strikes that surface and with the same intensity. (b) When a beam of light is scattered, it results in a larger number of weaker rays, traveling in different directions. Usually more energy is scattered in the forward direction than is backscattered.

Scattering and Diffused Light Although incoming solar radiation travels in a straight line, small dust particles and gas molecules in the atmosphere scatter some of this energy in different directions. The result, called **diffused light,** explains how light reaches the area under the limbs of a tree and how a room is lit in the absence of direct sunlight. Further, on clear days scattering accounts for the brightness and the blue color of the daytime sky. In contrast, bodies like the Moon and Mercury, which are without atmospheres, have dark skies and "pitch-black" shadows, even during daylight hours. Overall, about one-half of solar radiation that is absorbed at Earth's surface arrives as diffused (scattered) light.

Blue Skies and Red Sunsets Recall that sunlight appears white, but it is composed of all colors. Gas molecules more effectively scatter blue and violet light that have shorter wavelengths than they scatter red and orange. This accounts for the blue color of the sky and the orange and red colors seen at sunrise and sunset (Figure 2–18). On clear days, you can look in any direction away from the direct Sun and see a blue sky, the wavelength of light more readily scattered by the atmosphere.

Conversely, the Sun appears to have an orange-to-reddish hue when viewed near the horizon (Figure 2–19). This is the result of the great distance solar radiation must travel through the atmosphere before it reaches your eyes (see Table 2–1). During its travel, most of the blue and violet

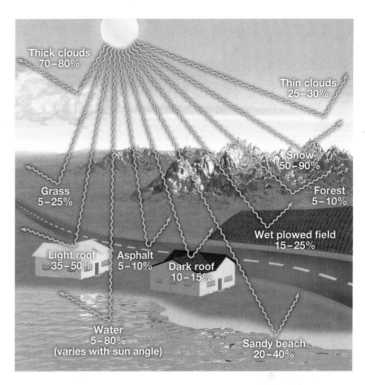

Figure 2-17 Albedo (reflectivity) of various surfaces. In general, light-colored surfaces tend to be more reflective than dark-colored surfaces and thus have higher albedos.

wavelengths are scattered out. Consequently, the light that reaches your eyes consists mostly of reds and oranges. The reddish appearance of clouds during sunrise and sunset also results because the clouds are illuminated by light from which the blue color has been removed by scattering.

Figure 2-18 At sunset, clouds often appear red because they are illuminated by sunlight in which most of the blue light has been lost due to scattering. (Photo by Michael Collier)

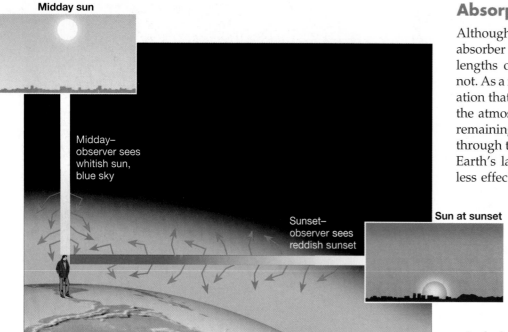

Midday sun

Midday–
observer sees
whitish sun,
blue sky

Sunset–
observer sees
reddish sunset

Sun at sunset

Figure 2–19 Short wavelengths (blue and violet) of visible light are scattered more effectively than are longer wavelengths (red and orange). Therefore, when the Sun is overhead, an observer can look in any direction and see predominantly blue light that was selectively scattered by the gases in the atmosphere. By contrast, at sunset, the path that light must take through the atmosphere is much longer. Consequently, most of the blue light is scattered before it reaches an observer. Thus, the Sun appears reddish in color.

The most spectacular sunsets occur when large quantities of fine dust or smoke particles penetrate the stratosphere. For three years after the great eruption of the Indonesian volcano Krakatau in 1883, brilliant sunsets occurred worldwide. In addition, the European summer that followed this colossal explosion was cooler than normal, which has been attributed to the loss of incoming solar radiation due to an increase in backscattering.

Large particles associated with haze, fog, and smog scatter light more equally in all wavelengths. Because no color is predominant over any other, the sky appears white or gray on days when large particles are abundant. Scattering of sunlight by haze, water droplets, or dust particles makes it possible for us to observe bands (or rays) of sunlight called *crepuscular rays*. These bright fan-shaped bands are most commonly seen when the Sun shines through a break in the clouds, as shown in Figure 2–20. Crepuscular rays can also be observed around twilight, when towering clouds cause alternating lighter and darker bands (light rays and shadows) to streak across the sky.

In summary, the color of the sky gives an indication of the number of large or small particles present. Numerous small particles produce red sunsets, whereas large particles produce white (gray) skies. Thus, the bluer the sky, the less polluted, or dryer, the air.

Absorption of Solar Radiation

Although Earth's surface is a relatively good absorber (effectively absorbing most wavelengths of solar radiation), the atmosphere is not. As a result, only 20 percent of the solar radiation that reaches Earth is absorbed by gases in the atmosphere (see Figure 2–15). Much of the remaining incoming solar radiation transmits through the atmosphere and is absorbed by the Earth's land–sea surface. The atmosphere is a less effective absorber because gases are selective absorbers (and emitters) of radiation.

Freshly fallen snow is another example of a selective absorber. Snow is a poor absorber of visible light (reflecting up to 85 percent) and, as a result, the temperature directly above a snow-covered surface is colder than it would otherwise be because much of the incoming radiation is reflected away. By contrast, snow is a very good absorber (absorbing up to 95 percent) of the infrared (heat) radiation that is emitted from Earth's surface. As the ground radiates heat upward, the lowest layer of snow absorbs this energy and, in turn, radiates most of the energy downward. Thus, the depth at which a winter's frost can penetrate into the ground is much less when the ground is snow covered than in an equally cold region without snow—giving credence to the statement "The ground is blanketed with snow." Farmers who plant winter wheat desire a deep snow cover because it insulates their crops from bitter winter temperatures.

Figure 2–20 Crepuscular rays are produced when haze scatters light. Crepuscular rays are most commonly seen when the Sun shines through a break in the clouds. (Photo by Tetra Images/Jupiter Images)

Earth's surface is emitted at wavelengths between 2.5 and 30 micrometers, placing it in the long end of the infrared band of the electromagnetic spectrum. Heating of the atmosphere requires an understanding of how gases interact with the short wavelength *incoming* solar radiation and the long wavelength *outgoing* radiation emitted by Earth (Figure 2–21, top).

Heating the Atmosphere

When a gas molecule absorbs radiation, this energy is transformed into internal molecular motion, which is detectable as a rise in temperature (sensible heat). The lower part of Figure 2–21 gives the absorptivity of the principal atmospheric gases. Note that nitrogen, the most

This image was taken by astronauts aboard the International Space Station as it traveled over the west coast of South America. On average, these astronauts experience 16 sunrises and sunsets during a 24-hour orbital period. The separation between day and night is marked by a line called the *terminator* (NASA).

Question 1 Locate the *terminator* in this image. Would you describe it as a sharp line? Explain.

Question 2 Are the astronauts looking at a sunrise or a sunset?

The Role of Gases in the Atmosphere

 Heating Earth's Surface and Atmosphere
▶ The Greenhouse Effect

Figure 2–21 shows that the majority of solar radiation is emitted in wavelengths shorter than 2.5 micrometers—shortwave radiation. By contrast, most radiation from

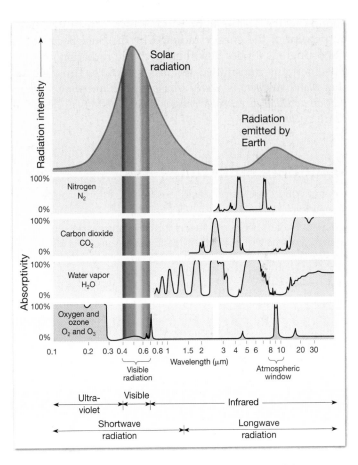

Figure 2–21 The effectiveness of selected gases of the atmosphere in absorbing incoming shortwave radiation (left side of graph) and outgoing longwave terrestrial radiation (right side). The blue areas represent the percentage of radiation absorbed by the various gases. The atmosphere as a whole is quite transparent to solar radiation between 0.3 and 0.7 micrometer, which includes the band of visible light. Most solar radiation falls in this range, explaining why a large amount of solar radiation penetrates the atmosphere and reaches Earth's surface. Also, note that longwave infrared radiation in the zone between 8 and 12 micrometers can escape the atmosphere most readily. This zone is called the atmospheric window.

abundant constituent in the atmosphere (78 percent), is a relatively poor absorber of incoming solar radiation. The only significant absorbers of incoming solar radiation are water vapor, oxygen, and ozone, which account for most of the solar energy absorbed directly by the atmosphere. Oxygen and ozone are efficient absorbers of high-energy, shortwave radiation. Oxygen removes most of the shorter-wavelength UV radiation high in the atmosphere, and ozone absorbs UV rays in the stratosphere between 10 and 50 kilometers (6 and 30 miles). The absorption of UV energy in the stratosphere accounts for the high temperatures experienced there. More importantly, without the removal of most UV radiation, human life would not be possible because UV energy disrupts our genetic code.

Looking at the bottom of Figure 2–21, you can see that for the atmosphere as a whole, none of the gases are effective absorbers of radiation with wavelengths between 0.3 and 0.7 micrometer. This region of the spectrum corresponds to the visible light band, which constitutes about 43 percent of the energy radiated by the Sun. Because the atmosphere is a poor absorber of visible radiation, most of this energy is transmitted to Earth's surface. Thus, we say that *the atmosphere is nearly transparent to incoming solar radiation and that direct solar energy is not an effective "heater" of Earth's atmosphere.*

Students Sometimes Ask . . .

What causes leaves on deciduous trees to change color each fall?

The leaves of all deciduous trees contain the pigment chlorophyll, which gives them a green color. The leaves of some trees also contain the pigment carotene, which is yellow, and still others produce a class of pigments that appear red in color. During summer, leaves are factories that generate sugar from carbon dioxide and water by the action of light on chlorophyll. As the dominant pigment, chlorophyll causes the leaves of most trees to appear green. The shortening days and cool nights of autumn trigger changes in deciduous trees. With a drop in chlorophyll production, the green color of the leaves fades, allowing other pigments to be seen. If the leaves contain carotene, as do birch and hickory, they will change to bright yellow. Other trees, such as red maple and sumac, display the brightest reds and purples in the autumn landscape.

We can also see in Figure 2–21 that the atmosphere is generally a relatively efficient absorber of longwave (infrared) radiation emitted by Earth (see the bottom right of Figure 2–21). Water vapor and carbon dioxide are the principal absorbing gases, with water vapor absorbing about 60 percent of the radiation emitted by Earth's surface. Therefore, water vapor, more than any other gas, accounts for the warm temperatures of the lower troposphere, where it is most highly concentrated.

Although the atmosphere is an effective absorber of radiation emitted by Earth's surface, it is nevertheless quite transparent to the band of radiation between 8 and 12 micrometers. Notice in Figure 2–21 (lower right) that the gases in the atmosphere (N_2, CO_2, H_2O) absorb minimal energy in these wavelengths. Because the atmosphere is transparent to radiation between 8 and 12 micrometers, much as window glass is transparent to visible light, this band is called the **atmospheric window.** Although other "atmospheric windows" exist, the one located between 8 and 12 micrometers is the most significant because it is located where Earth's radiation is most intense.

By contrast, clouds that are composed of tiny liquid droplets (not water vapor) are excellent absorbers of the energy in the atmospheric window. Clouds absorb outgoing radiation and radiate much of this energy back to Earth's surface. Thus, clouds serve a purpose similar to window blinds because they effectively block the atmospheric window and lower the rate at which Earth's surface cools. This explains why nighttime temperatures remain higher on cloudy nights than on clear nights.

Because the atmosphere is largely transparent to solar (shortwave) radiation but more absorptive of the longwave radiation emitted by Earth, the atmosphere is heated from the ground up. This explains the general drop in temperature with increased altitude in the troposphere. The farther from the "radiator" (Earth's surface), the colder it gets. On average, the temperature drops 6.5°C for each kilometer (3.5°F per 1000 feet) increase in altitude, a figure known as the *normal lapse rate.* The fact that the atmosphere does not acquire the bulk of its energy directly from the Sun but is heated by Earth's surface is of utmost importance to the dynamics of the weather machine.

The Greenhouse Effect

Research of "airless" planetary bodies such as the Moon have led scientists to determine that if Earth had no atmosphere, it would have an average surface temperature below freezing. However, Earth's atmosphere "traps" some of the outgoing radiation, which makes our planet habitable. The extremely important role the atmosphere plays in heating Earth's surface has been named the **greenhouse effect.**

As discussed earlier, cloudless air is largely transparent to incoming shortwave solar radiation and, hence, transmits it to Earth's surface. By contrast, a significant fraction of the longwave radiation emitted by Earth's land–sea surface is absorbed by water vapor, carbon dioxide, and other trace gases in the atmosphere. This energy heats the air and increases the rate at which it radiates energy, both out to space and back toward Earth's surface. Without this complicated game of "pass the hot potato," Earth's average temperature would be –18°C (0°F) rather than the current temperature of 15°C (59°F) (Figure 2–22). These absorptive gases in our atmosphere make Earth habitable for humans and other life forms.

This natural phenomenon was named the greenhouse effect because greenhouses are heated in a similar manner

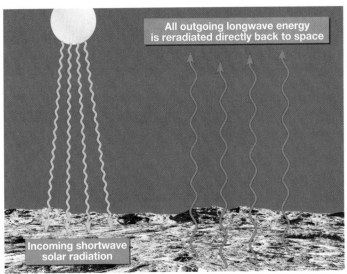

(a) Airless bodies like the Moon

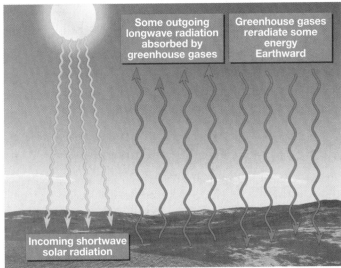

(b) Bodies with modest amounts of greenhouse gases like Earth

Figure 2-22 The greenhouse effect. (a) All incoming solar radiation reaches the surface of airless bodies such as the Moon. However, all of the energy that is absorbed by the surface is radiated directly back to space. This causes the lunar surface to have a much lower average surface temperature than Earth. Because the Moon experiences days and nights that are about 2 weeks long, the lunar days are hot and the nights are frigid. (b) On bodies with modest amounts of greenhouse gases, such as Earth, much of the short-wavelength radiation from the Sun passes through the atmosphere and is absorbed by the surface. This energy is then emitted from the surface as longer-wavelength radiation, which is absorbed by greenhouse gases in the atmosphere. Some of the energy absorbed will be radiated back to the surface and is responsible for keeping Earth's surface 33°C (59°F) warmer than it would be otherwise. (c) Bodies with abundant greenhouse gases, such as Venus, experience extraordinary greenhouse warming, which is estimated to raise its surface temperature by 523°C (941°F).

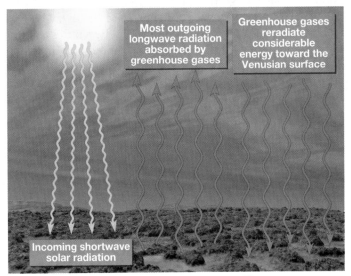

(c) Bodies with abundant greenhouse gases like Venus

(Figure 2–22). The glass in a greenhouse allows shortwave solar radiation to enter and be absorbed by the objects inside. These objects, in turn, radiate energy but at longer wavelengths, to which glass is nearly opaque. The heat, therefore, is "trapped" in the greenhouse. Although this analogy is widely used, it has been shown that air inside greenhouses attains higher temperatures than outside air in part due to the restricted exchange of warmer air inside and cooler air outside. Nevertheless, the term "greenhouse effect" is still used to describe atmospheric heating.

Media reports frequently and erroneously identify the greenhouse effect as the "villain" in the global warming problem. However, the greenhouse effect and global warming *are different concepts.* Without the greenhouse effect, Earth would be uninhabitable. Scientists have mounting evidence that human activities (particularly the release of carbon dioxide into the atmosphere) are responsible for a rise in global temperatures (see Chapter 14). Thus, hu-

mans are compounding the effects of an otherwise natural process (the greenhouse effect). It is incorrect to equate the greenhouse phenomenon, which makes life possible, with global warming—which involves undesirable changes to our atmosphere caused by human activities.

Concept Check 2.5

1 Explain why the atmosphere is heated chiefly by radiation from Earth's surface rather than by direct solar radiation.

2 Which gases are the primary heat absorbers in the lower atmosphere? Which gas is most influential in weather?

3 What is the atmospheric window? How is it "closed"?

4 How does Earth's atmosphere act as a greenhouse?

5 What is the "villain" in the global warming problem?

Students Sometimes Ask . . .

Is Venus so much hotter than Earth because it is closer to the Sun?

Proximity to the Sun is actually not the primary factor. On Earth, water vapor and carbon dioxide are the primary greenhouse gases and are responsible for elevating Earth's average surface temperature by 33°C (59°F). However, greenhouse gases make up less than 1 percent of Earth's atmosphere. By contrast, the Venusian atmosphere is much denser and consists of 97 percent carbon dioxide. Thus, the Venusian atmosphere experiences extraordinary greenhouse warming, which is estimated to raise its surface temperature by 523°C (941°F)—hot enough to melt lead.

Earth's Heat Budget

Globally, Earth's average temperature remains relatively constant, despite seasonal cold spells and heat waves. This stability indicates that a balance exists between the amount of incoming solar radiation and the amount of radiation emitted back to space; otherwise, Earth would be getting progressively colder or progressively warmer. The annual balance of incoming and outgoing radiation is called Earth's **heat budget.**

Annual Energy Balance

Figure 2–23 illustrates Earth's annual energy budget. For simplicity we will use 100 units to represent the solar radiation intercepted at the outer edge of the atmosphere. You have already seen that, of the total radiation that reaches Earth, roughly 30 units (30 percent) are reflected and scattered back to space. The remaining 70 units are absorbed, 20 units within the atmosphere and 50 units by Earth's land–sea surface. How does Earth transfer this energy back to space?

If all of the energy absorbed by our planet were radiated directly and immediately back to space, Earth's heat budget would be simple—100 units of radiation received and 100 units returned to space. In fact, this does happen *over time* (minus small quantities of energy that become locked up in biomass that may eventually become fossil fuel). What complicates the heat budget is the behavior of certain greenhouse gases, particularly water vapor and carbon dioxide. As you learned, these greenhouse gases absorb a large share of outward-directed infrared radiation and radiate much of that energy back to Earth. This "recycled" energy significantly increases the radiation received by Earth's surface. In addition to the 50 units received directly from the Sun, Earth's surface receives longwave radiation emitted downward by the atmosphere (94 units).

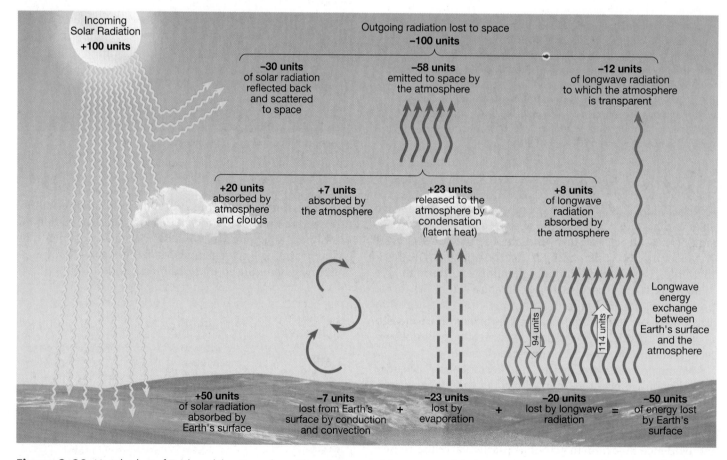

Figure 2–23 Heat budget of Earth and the atmosphere. These estimates of the average global energy budget come from satellite observations and radiation studies. As more data are accumulated, these numbers will be modified. (Data from Kiehl, Trenberth, Liou, and others)

EYE ON THE ATMOSPHERE

North America

Inter Tropical Convergence Zone

South America

(NASA)

This infrared (IR) image produced by the GOES-14 satellite displays cold objects as bright white and hot objects as black. The hottest (blackest) features shown are land surfaces, and the coldest (whitest) features are the tops of towering storm clouds. Recall that we cannot see infrared (thermal) radiation, but we have developed artificial detectors capable of extending our vision into the long-wavelength portion of the electromagnetic spectrum.

Question 1 Several areas of cloud development and potential storms are shown on this IR image. One is a well-developed tropical storm named Hurricane Bill. Can you locate Hurricane Bill?

Question 2 What is an advantage of IR images over visible images?

A balance is maintained because all the energy absorbed by Earth's surface is returned to the atmosphere and eventually radiated back to space. Earth's surface loses energy through a variety of processes: the emission of longwave radiation; conduction and convection; and energy loss to Earth's surface through the process of evaporation—latent heat (Figure 2–23). Most of the longwave radiation emitted skyward is re-absorbed by the atmosphere. Conduction results in the transfer of energy between Earth's surface to the air directly above, while convection carries the warm air located near the surface upward as thermals (7 units).

Earth's surface also loses a substantial amount of energy (23 units) through evaporation. This occurs because energy is required for liquid water molecules to leave the surface of a body of water and change to its gaseous form, water vapor. The energy lost by a water body is carried into the atmosphere by molecules of water vapor. Recall that the heat used to evaporate water does not produce a temperature change and is referred to as *latent heat* (hidden heat). If the water vapor condenses to form cloud droplets, the energy will be detectable as *sensible heat* (heat we can feel and measure with a thermometer). Thus through the process of evaporation, water molecules carry latent heat into the atmosphere, where it is eventually released.

In summary, a careful examination of Figure 2–23 confirms that the quantity of incoming solar radiation is, over time, balanced by the quantity of longwave radiation that is radiated back to space.

Box 2-3   Solar Power

Nearly 95 percent of the world's energy needs are derived from fossil fuels, primarily oil, coal, and natural gas. Current estimates indicate that, at the present rate of consumption, we have enough fossil fuels to last about 150 years. However, with the rapid increase in demand by developing countries, the rate of consumption continues to climb. Thus, reserves will be in short supply sooner rather than later. How can a growing demand for energy be met without radically altering the planet we inhabit? Although no clear answer has yet emerged, we must consider greater use of alternate energy sources, such as solar and wind power.

The term *solar energy* generally refers to the direct use of the Sun's rays to supply energy for the needs of people. The simplest, and perhaps most widely used, *passive solar collectors* are south-facing windows. As shortwave sunlight passes through the glass, its energy is absorbed by objects in the room that, in turn, radiate longwave heat that warms the air in the room.

More elaborate systems used for home heating involve *active solar collectors*. These roof-mounted devices are normally large blackened boxes that are covered with glass. The heat they collect can be transferred to where it is needed by circulating air or fluids through pipes. Solar collectors are also used successfully to heat water for domestic and commercial needs. In Israel, for example, about 80 percent of all homes are equipped with solar collectors that provide hot water.

Research is currently under way to improve the technologies for concentrating sunlight. One method uses parabolic troughs as solar energy collectors. Each collector resembles a large tube that has been cut in half. Their highly polished surfaces reflect sunlight onto a collection pipe. A fluid (usually oil) runs through the pipe and is heated by the concentrated sunlight. The fluid can reach temperatures of over 400°F and is typically used to make steam that drives a turbine to produce electricity.

Another type of collector uses photovoltaic (solar) cells that convert the Sun's energy

FIGURE 2–D Nearly cloudless deserts, such as California's Mojave Desert, are prime sites for photovoltaic cells that convert solar radiation directly into electricity. (Photo by Jim West/Alamy)

directly into electricity. Photovoltaic cells are usually connected together to create solar panels in which sunlight knocks electrons into higher-energy states, to produce electricity (**Figure 2–D**). For many years, solar cells were used mainly to power calculators and novelty devices. Today, however, large photovoltaic power stations are connected to electric grids to supplement other power-generating facilities. The leading countries in photovoltaic capacity are Germany, Japan, Spain, and the United States. The high cost of solar cells makes generating solar electricity more expensive than electricity created by other sources. As the cost of fossil fuels continues to increase, advances in photovoltaic technology should narrow the price differential.

A new technology being developed, called the *Stirling dish*, converts thermal energy to electricity by using a mirror array to focus the Sun's rays on the receiver end of a Stirling engine (**Figure 2–E**). The internal side of a receiver then heats hydrogen gas, causing it to expand. The pressure created by the expanding gas drives a piston, which turns a small electric generator.

FIGURE 2–E This Stirling dish is located near Phoenix, Arizona. (Photo by Brian Green/Alamy)

Latitudinal Heat Balance

Because incoming solar radiation is roughly equal to the amount of outgoing radiation, on average, worldwide temperatures remain constant. However, the balance of

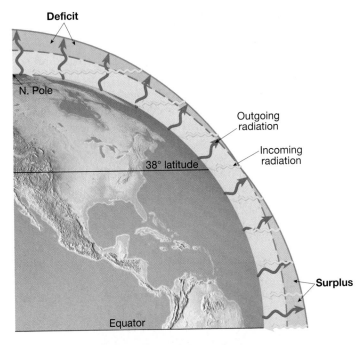

Figure 2–24 Latitudinal heat balance averaged over the entire year. In the zone extending 38° on both sides of the equator, the amount of incoming solar radiation exceeds the loss from outgoing Earth radiation. The reverse is true for the middle and high polar latitudes, where losses from outgoing Earth radiation exceed gains from incoming solar radiation.

incoming and outgoing radiation that is applicable for the entire planet is not maintained at each latitude. Averaged over the entire year, a zone around Earth that lies within 38° latitude of the equator *receives more solar radiation than is lost to space* (Figure 2–24). The opposite is true for higher latitudes, where *more heat is lost through radiation emitted by Earth than is received from the Sun.*

We might conclude that the tropics are getting hotter and the poles are getting colder. But that is not the case. Instead, the global wind systems and, to a lesser extent, the oceans act as giant thermal engines, transferring surplus heat from the tropics poleward. In effect, the energy imbalance drives the winds and the ocean currents.

It should be of interest to those who live in the middle latitudes—in the Northern Hemisphere, from the latitude of New Orleans at 30° north to the latitude of Winnipeg, Manitoba, at 50° north—that most heat transfer takes place across this region. Consequently, much of the stormy weather experienced in the middle latitudes can be attributed to this unending transfer of heat from the tropics toward the poles. These processes are discussed in more detail in later chapters.

Concept Check 2.6

1 The tropics receive more solar radiation than is lost. Why then don't the tropics keep getting hotter?

2 What two phenomena are driven by the imbalance of heating that exists between the tropics and poles?

Give It Some Thought

1. How would our seasons be affected if Earth's axis were not inclined $23\frac{1}{2}°$ to the plane of its orbit but were instead perpendicular?

2. Describe the seasons if Earth's axis were inclined 40°. Where would the Tropics of Cancer and Capricorn be located? How about the Arctic and Antarctic Circles?

3. The accompanying four diagrams (labeled a–d) are intended to illustrate the Earth–Sun relationships that produce the seasons.
   a. Which one of these diagrams most accurately shows this relationship?
   b. Identify what is *inaccurately* shown in each of the other three diagrams.

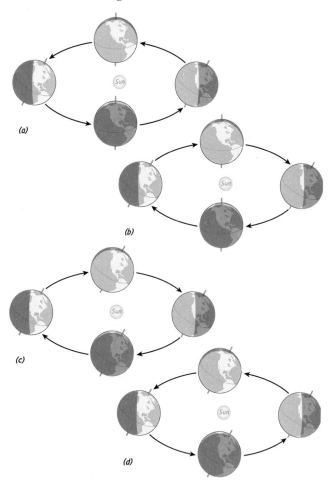

4. On what date is Earth closest to the Sun? On that date, what season is it in the Northern Hemisphere? Explain this apparent contradiction.

5. When a person in the United States watches the Sun move through the sky during a typical day, he or she sees it travel from left to right. However, on the same day, a person in Australia will see the Sun travel across the sky from right to left. Explain or make a sketch that illustrates why this difference occurs.

6. Which of the three mechanisms of heat transfer is most significant in each of the following situations:
   a. Driving a car with the seat heater turned on
   b. Sitting in an outdoor hot tub
   c. Lying inside a tanning bed
   d. Driving a car with the air conditioning turned on

7. During a "shore lunch" on a fishing trip to a remote location, the fishing guide will sometimes place a pail of lake water next to the cooking fire, as shown in the accompanying illustration. When the water in the pail begins to boil, the guide will lift the pail from the fire and "impress" the guests by placing their other hand on the bottom. Use what you have learned about the three mechanisms of heat transfer to explain why the guide's hands aren't burned by touching the bottom of the pail.

8. The Sun shines continually at the North Pole for six months, from the spring equinox until the fall equinox, yet temperatures never get very warm. Explain why this is the case.

9. The accompanying image shows an area of our galaxy where stars having surface temperatures much hotter than the Sun have recently formed. Imagine that an Earth-like planet formed around one of these stars, at a distance where it receives the same intensity of light as Earth. Use the laws of radiation to explain why this planet may not provide a habitable environment for humans.

(NASA)

10. Rank the following according to the wavelength of radiant energy each emits, from the shortest wavelength to the longest:
    a. A light bulb with a filament glowing at 4000°C
    b. A rock at room temperature
    c. A car engine at 140°C
11. Figure 2–15 shows that about 30 percent of the Sun's energy is reflected or scattered back to space. If Earth's albedo were to increase to 50 percent, how would you expect Earth's average surface temperature to change?
12. Explain why Earth's equatorial regions are not becoming warmer, despite the fact that they receive more incoming solar radiation than they radiate back to space.
13. The accompanying photo shows the explosive 1991 eruption of Mount Pinatubo in the Philippines. How would you expect global temperatures to respond to the ash and debris this volcano spewed high into the atmosphere?

(U.S. Geological Survey)

HEATING EARTH'S SURFACE AND ATMOSPHERE IN REVIEW

- The *seasons* are caused by changes in the angle at which the Sun's rays strike the surface and the changes in the length of daylight at each latitude. These seasonal changes are the result of the tilt of Earth's axis as it revolves around the Sun.
- *Energy* is the ability to do work. The two major categories of energy are (1) *kinetic energy,* which can be thought of as energy of motion, and (2) *potential energy,* energy that has the capability to do work.
- *Temperature* is a measure of the average kinetic energy of the atoms or molecules in a substance.
- *Heat* is the transfer of energy into or out of an object because of temperature differences between that object and its surroundings.
- *Latent heat* (hidden heat) is the energy involved when water changes from one state of matter to another. During evaporation, for example, energy is stored by the escaping water vapor molecules, and this energy is eventually released when water vapor condenses to form water droplets in clouds.
- *Conduction* is the transfer of heat through matter by electron and molecular collisions between molecules. Because air is a poor conductor, conduction is significant only between Earth's surface and the air immediately in contact with the surface.
- *Convection* is heat transfer that involves the actual movement or circulation of a substance. Convection is an important mechanism of heat transfer in the atmosphere, where warm air rises and cooler air descends.
- *Radiation* or *electromagnetic radiation* consists of a large array of energy that includes X-rays, visible light, heat waves, and radio waves that travel as waves of various sizes. Shorter wavelengths of radiation have greater energy.
- These are four basic *laws of radiation:* (1) All objects emit radiant energy; (2) hotter objects radiate more total energy per unit area than colder objects; (3) the hotter the radiating body, the shorter is the wavelength of maximum radiation; and (4) objects that are good absorbers of radiation are also good emitters.
- Approximately 50 percent of the solar radiation that strikes the atmosphere reaches Earth's surface. About 30 percent is reflected back to space. The fraction of radiation reflected by a surface is called its *albedo.* The remaining 20 percent of the energy is absorbed by clouds and the atmosphere's gases.
- Radiant energy absorbed at Earth's surface is eventually radiated skyward. Because Earth has a much lower surface temperature than the Sun, its radiation is in the form of longwave infrared radiation. Because the atmospheric gases, primarily water vapor and carbon dioxide, are more efficient absorbers of terrestrial (longwave) radiation, the atmosphere is heated from the ground up.
- The selective absorption of Earth radiation by atmospheric gases, mainly water vapor and carbon dioxide, that results in Earth's average temperature being warmer than it would be otherwise is referred to as the *greenhouse effect.*
- The greenhouse effect is a natural phenomenon that makes Earth habitable. Human activities that release greenhouse gases (primarily carbon dioxide) are the "villains" of global warming—not the "greenhouse effect" as it is often, but inaccurately, portrayed.

- Because of the annual balance that exists between incoming and outgoing radiation, called the *heat budget,* Earth's average temperature remains relatively constant.
- Averaged over the entire year, a zone around Earth between 38° north and 38° south receives more solar radiation than is lost to space. The opposite is true for higher latitudes, where more heat is lost through outgoing longwave radiation than is received. It is this energy imbalance between the low and high latitudes that drives the weather system and in turn transfers surplus heat from the tropics poleward.

VOCABULARY REVIEW

absorptivity (p. 50)
advection (p. 46)
albedo (p. 50)
aphelion (p. 36)
atmospheric window (p. 54)
autumnal (fall) equinox (p. 40)
backscattering (p. 50)
circle of illumination (p. 40)
conduction (p. 45)
convection (p. 45)
diffused light (p. 51)
energy (p. 43)
greenhouse effect (p. 54)
heat (p. 44)

heat budget (p. 56)
inclination of the axis (p. 38)
infrared radiation (p. 47)
kinetic energy (p. 43)
latent heat (p. 44)
longwave radiation (p. 48)
perihelion (p. 36)
plane of the ecliptic (p. 38)
potential energy (p. 44)
radiation, or electromagnetic radiation (p. 46)
reflection (p. 50)
revolution (p. 36)
rotation (p. 36)
scattering (p. 50)

sensible heat (p. 44)
shortwave radiation (p. 48)
spring (vernal) equinox (p. 40)
summer solstice (p. 39)
temperature (p. 44)
thermal (p. 45)
Tropic of Cancer (p. 39)
Tropic of Capricorn (p. 39)
ultraviolet radiation (p. 47)
visible light (p. 46)
wavelength (p. 46)
winter solstice (p. 39)

PROBLEMS

1. Refer to Figure 2–6 and calculate the noon Sun angle on June 21 and December 21 at 50° north latitude, 0° latitude (the equator), and 20° south latitude. Which of these latitudes has the greatest variation in noon Sun angle between summer and winter?

2. For the latitudes listed in Problem 1, determine the length of daylight and darkness on June 21 and December 21 (refer to Table 2–2). Which of these latitudes has the largest seasonal variation in length of daylight? Which latitude has the smallest variation?

3. Calculate the noon Sun angle at your location for the equinoxes and solstices.

4. If Earth had no atmosphere, its longwave radiation emission would be lost quickly to space, making the planet approximately 33 K cooler. Calculate the rate of radiation emitted (E), and the wavelength of maximum radiation emission (λ_{max}) for Earth at 255 K. (Hint: See Box 2–2.)

5. The intensity of solar radiation can be calculated using trigonometry, as shown in Figure 2–25. For simplicity, consider a solar beam of 1 unit width. The surface area over which the beam would be spread changes with Sun angle, such that:

$$\text{Surface area} = \frac{1 \text{ unit}}{\sin(\text{Sun angle})}$$

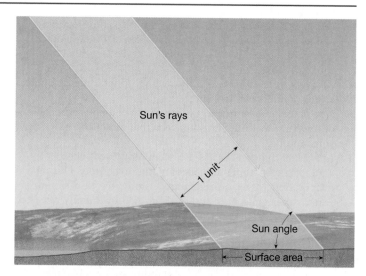

Figure 2–25 Calculating solar intensity.

Therefore, if the Sun angle at solar noon is 56°:

$$\text{Surface area} = \frac{1 \text{ unit}}{\sin 56°} = \frac{1 \text{ unit}}{0.829} = 1.206 \text{ units}$$

Using this method and your answers to Problem 3, calculate the intensity of solar radiation (surface area) for your location at noon during the summer and winter solstices.

6. Figure 2–26 is an analemma—a graph used to determine the latitude where the overhead noon Sun is located for any date. To determine the latitude of the overhead noon Sun from the analemma, find the desired date on the graph and read the coinciding latitude along the left axis. Determine the latitude of the overhead noon Sun for the following dates. Remember to indicate *north* (N) or *south* (S).

   a. March 21

   b. June 5

   c. December 10

7. Use Figure 2–6 and the analemma in Figure 2–26 to calculate the noon Sun angle at your location (latitude) on the following dates:

   a. September 7

   b. July 5

   c. January 1

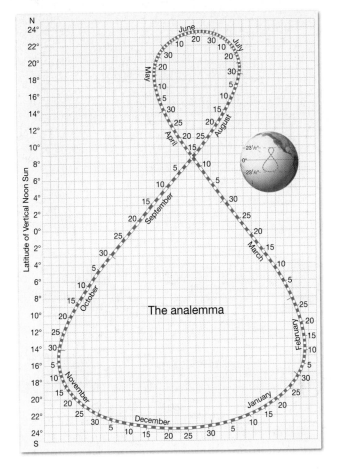

Figure 2–26

MyMeteorologyLab™

Log in to www.mymeteorologylab.com for animations, videos, MapMaster interactive maps, GEODe media, *In the News* RSS feeds, web links, glossary flashcards, self-study quizzes and a Pearson eText version of this book to enhance your study of *Heating Earth's Surface and Atmosphere*.

Temperature

Temperature is one of the basic elements of weather and climate. When someone asks what the weather is like outside, air temperature is often the first element we mention. From everyday experience, we know that temperatures vary on different time scales: seasonally, daily, and even hourly. Moreover, we all realize that substantial temperature differences exist from one place to another. In Chapter 2 you learned how air is heated and examined the role of Earth–Sun relationships in causing temperature variations from season to season and from latitude to latitude. In this chapter you will focus on several other aspects of this very important atmospheric property, including factors other than Earth–Sun relationships, that act as temperature controls. You will also look at how temperature is measured and expressed and see that temperature data can be of very practical value to us all. Applications include calculations that are useful in evaluating energy consumption, crop maturity, and human comfort.

Focus On Concepts

After completing this chapter you should be able to:

- Calculate five commonly used types of temperature data and interpret a map that depicts temperature data using isotherms.

- List the principal controls of temperature and use examples to describe their effects.

- Explain why water and land heat and cool differently.

- Interpret the patterns depicted on world maps of January and July temperatures and on a world map of annual temperature ranges.

- Discuss the basic daily and annual cycles of air temperature.

- Explain how different types of thermometers work and why the placement of thermometers is an important factor in obtaining accurate readings.

- Distinguish among Fahrenheit, Celsius, and Kelvin temperature scales.

- Discuss the concept of apparent temperature and compare two basic indices that are expressions of this idea.

For the Record: Air-Temperature Data

GE○De ATMOSPHERE — Temperature Data and the Controls of Temperature
▶ Basic Temperature Data

Temperatures recorded daily at thousands of weather stations worldwide provide much of the temperature data compiled by meteorologists and climatologists (Figure 3–1). Hourly temperatures may be recorded by an observer or obtained from automated observing systems that continually monitor the atmosphere. At many locations only the maximum and minimum temperatures are obtained see (Box 3–1).

Basic Calculations

The **daily mean temperature** is determined by averaging the 24 hourly readings or by adding the maximum and minimum temperatures for a 24-hour period and dividing by 2. From the maximum and minimum, the **daily temperature range** is computed by finding the difference between these figures. Other data involving longer periods are also compiled:

- The **monthly mean temperature** is calculated by adding together the daily means for each day of the month and dividing by the number of days in the month.
- The **annual mean temperature** is an average of the 12 monthly means.
- The **annual temperature range** is computed by finding the difference between the warmest and coldest monthly mean temperatures.

Mean temperatures are especially useful for making daily, monthly, and annual comparisons. It is common to hear a weather reporter state, "Last month was the warmest February on record" or "Today Omaha was 10° warmer than Chicago." Temperature ranges are also useful statistics because they give an indication of extremes, a necessary part of understanding the weather and climate of a place or an area.

Isotherms

To examine the distribution of air temperatures over large areas, isotherms are commonly used. An **isotherm** is a line that connects points on a map that have the same temperature (*iso* = "equal," *therm* = "temperature"). Therefore, all points through which an isotherm passes have identical temperatures for the time period indicated. Generally, isotherms representing 5° or 10° temperature differences are used, but *any* interval may be chosen. Figure 3–2 illustrates how isotherms are drawn on a map. Notice that most isotherms do not pass directly through the observing stations because the station readings may not coincide with the values chosen for the isotherms. Only an occasional station temperature will be exactly the same as the value of the isotherm, so it is usually necessary to draw the lines by estimating the proper position between stations.

Isothermal maps are valuable tools because they clearly make temperature distribution visible at a glance. Areas of low and high temperatures are easy to pick out. In addition, the amount of temperature change per unit of distance, called the **temperature gradient,** is easy to visualize. Closely spaced isotherms indicate a rapid rate of temperature change, whereas more widely spaced lines indicate a more gradual rate of change. For example, notice in Figure 3–2 that the isotherms are more closely spaced in Colorado and Utah (steeper temperature gradient), whereas the isotherms are spread farther apart in Texas (gentler temperature gradient). Without isotherms a map would be covered with numbers representing temperatures at tens or hundreds of places, which would make patterns difficult to see.

Concept Check 3.1

1 How are the following temperature data calculated: daily mean, daily range, monthly mean, annual mean, and annual range?

2 What are isotherms, and what is their purpose?

(a)

(b)

Figure 3–1 People living in the middle latitudes can experience a wide range of temperatures during a year. (a) Pedestrian navigating through a Chicago neighborhood during a February 2011 blizzard that dropped more than 50 centimeters (nearly 20 inches) of snow on the city. (Photo by Zuma Press/Newscom) (b) Beating the heat on a hot summer day. (Photo by mylife photos/Alamy).

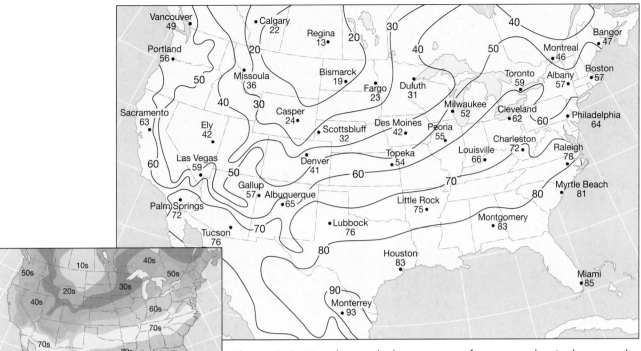

Figure 3–2 Map showing high temperatures for a spring day. Isotherms are lines that connect points of equal temperature. Showing temperature distribution in this way makes patterns easier to see. Notice that most isotherms do not pass directly through the observing stations. It is usually necessary to draw isotherms by estimating their proper position between stations. On television and in many newspapers, temperature maps are in color. Rather than labeling isotherms, the area *between* isotherms is labeled. For example, the zone between the 60° and 70° isotherms is labeled "60s."

Students Sometimes Ask...

What's the hottest city in the United States?

It depends on how you define "hottest." If average annual temperature is used, then Key West, Florida, is the hottest, with an annual mean of 25.6°C (78°F) for the 30-year span 1971–2000. However, if we look at cities with the highest July maximums during the 1971–2000 span, then the desert community of Palm Springs, California, has the distinction of being hottest. Its average daily high in July is a blistering 42.4°C (108.3°F)! Yuma, Arizona (41.7°C/107°F), Phoenix, Arizona (41.4°C/106°F), and Las Vegas, Nevada (40°C/104.1°F), aren't far behind.

Why Temperatures Vary: The Controls of Temperature

 Temperature Data and the Controls of Temperature
▶ Controls of Temperature

The **controls of temperature** are factors that cause temperatures to vary from place to place and from time to time. Chapter 2 examined the most important cause for temperature variation—differences in the receipt of solar radiation. Because variations in Sun angle and length of daylight depend on latitude, they are responsible for warm temperatures in the tropics and colder temperatures poleward. Of course, seasonal temperature changes at a given latitude occur as the Sun's vertical rays migrate toward and away from a place during the year. Figure 3–3 reminds us of the importance of latitude as a control of temperature.

But latitude is not the only control of temperature. If it were, we would expect all places along the same parallel to have identical temperatures. This is clearly not the case. For instance, Eureka, California, and New York City are both coastal cities at about the same latitude, and both places have an annual mean temperature of 11°C (51.8°F). Yet New York City is 9.4°C (16.9°F) warmer than Eureka in July and 9.4°C (16.9°F) colder than Eureka in January. In another example, two cities in Ecuador—Quito and Guayaquil—are relatively close to one another, but the mean annual temperatures at these two cities differ by 12.2°C (22°F). To explain these situations and countless others, we must realize that factors other than latitude also exert a strong influence on temperature. In the next sections we examine these other controls, which include:

- Differential heating of land and water
- Ocean currents
- Altitude
- Geographic position
- Cloud cover and albedo

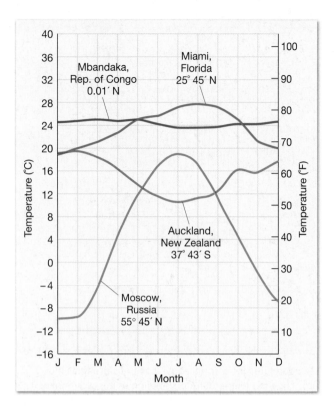

Figure 3–3 The data for these four cities reminds us that latitude (Earth–Sun relationship) is a major control of temperature.

Concept Check 3.2

1 List five controls of temperature other than latitude.

2 Provide two examples that illustrate that latitude is *not* the only temperature control.

Land and Water

In Chapter 2 you saw that the heating of Earth's surface controls the heating of the air above it. Therefore, to understand variations in air temperature, we must understand the variations in heating properties of the different surfaces that Earth presents to the Sun—soil, water, trees, ice, and so on. Different land surfaces reflect and absorb varying amounts of incoming solar energy, which in turn cause variations in the temperature of the air above. The greatest contrast, however, is not between different land surfaces but between land and water. Figure 3–4 illustrates this idea nicely. This satellite image shows surface temperatures in portions of Nevada, California, and the adjacent Pacific Ocean on the afternoon of May 2, 2004, during a spring heat wave. Land-surface temperatures are clearly much higher than water-surface temperatures. The image shows the extreme high surface temperatures in southern California and Nevada in dark red.*

_____

*Realize that air temperatures are cooler than surface temperatures. For example, the surface of a sandy beach can be painfully hot even though the temperature of the air above the sand is comfortable.

Surface temperatures in the Pacific Ocean are much lower. The peaks of the Sierra Nevada, still capped with snow, form a cool blue line down the eastern side of California.

In side-by-side bodies of land and water, such as is shown in Figure 3–4, *land heats more rapidly and to higher temperatures than water, and it cools more rapidly and to lower temperatures than water*. Variations in air temperatures, therefore, are much greater over land than over water. Why do land and water heat and cool differently? Several factors are responsible.

1. An important reason that the surface temperature of water rises and falls much more slowly than the surface temperature of land is that *water is highly mobile*. As water is heated, convection distributes the heat through a considerably larger mass. Daily temperature changes occur to depths of 6 meters (20 feet) or more below the surface, and yearly, oceans and deep lakes experience temperature variations through a layer between 200 and 600 meters (650 and 2000 feet) thick.

   In contrast, heat does not penetrate deeply into soil or rock; it remains near the surface. Obviously, no mixing can occur on land because it is not fluid. Instead, heat must be transferred by the slow process of

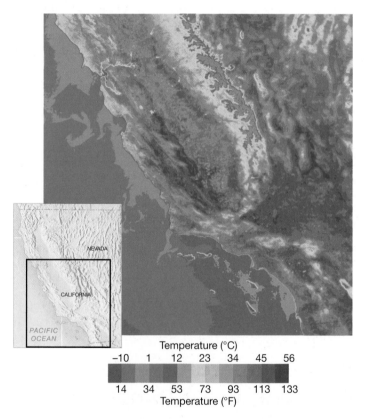

Figure 3–4 The differential heating of land and water is an important control of air temperatures. In this satellite image from the afternoon of May 2, 2004, water-surface temperatures in the Pacific Ocean are much lower than land-surface temperatures in California and Nevada. The narrow band of cool temperatures in the center of the image is associated with mountains (the Sierra Nevada). The cooler water temperatures immediately offshore are associated with the California Current (see Figure 3–9). (NASA)

Box 3–1   North America's Hottest and Coldest Places

Most people living in the United States have experienced temperatures of 38°C (100°F) or more. When statistics for the 50 states are examined for the past century or longer, we find that every state has a maximum temperature record of 38°C or higher. Even Alaska has recorded a temperature this high. Its record was set June 27, 1915, at Fort Yukon, a town along the Arctic Circle in the interior of the state.

Maximum Temperature Records

Surprisingly, the state that ties Alaska for the "lowest high" is Hawaii. Panala, on the south coast of the big island, recorded 38°C on April 27, 1931. Although humid tropical

FIGURE 3–A At 1886 meters (6288 feet), Mount Washington is the highest peak in the White Mountains and in fact the entire Northeast. Some consider it to be "the home of the world's worst weather" due to its extreme cold, heavy snows, high winds, frequent icing, and dense fog. The observatory at its summit keeps detailed weather records. (Photo by Jim Salge/Mount Washington Observatory)

and subtropical places such as Hawaii are known for being warm throughout the year, they seldom experience maximum temperatures that surpass the low to mid-30s Celsius (90s Fahrenheit).

The highest accepted temperature record for the United States as well as the entire Western Hemisphere is 57°C (134°F). This long-standing record was set at Death Valley, California, on July 10, 1913. Summer temperatures at Death Valley are consistently among the highest in the Western Hemisphere. During June, July, and August, temperatures exceeding 49°C (120°F) are to be expected. Fortunately, Death Valley has few human summertime residents.

Why are summer temperatures at Death Valley so high? In addition to having the lowest elevation in the Western Hemisphere (53 meters (174 feet) below sea level), Death Valley is a desert. Although it is only about 300 kilometers (less than 200 miles) from the Pacific Ocean, mountains cut off the valley from the ocean's moderating influence and moisture. Clear skies allow a maximum of sunshine to strike the dry, barren surface. Because no energy is used to evaporate moisture as occurs in humid regions, all the energy is available to

heat the ground. In addition, subsiding air that warms by compression as it descends is also common to the region and contributes to its high maximum temperatures.

Minimum Temperature Records

The temperature controls that produce truly frigid temperatures are predictable, and they should come as no surprise. We should expect extremely cold temperatures during winter in high-latitude places that lack the moderating influence of the ocean (Figure 3–A). Moreover, stations located on ice sheets and glaciers should be especially cold, as should stations positioned high in the mountains. All of these criteria apply to Greenland's Northice Station (elevation 2307 meters (7567 feet)). Here on January 9, 1954, the temperature plunged to −66°C (−87°F). If we exclude Greenland from consideration, Snag, in Canada's Yukon Territory, holds the record for North America. This remote outpost experienced a temperature of −63°C (−81° F) on February 3, 1947. When only locations in the United States are considered, Prospect Creek, located north of the Arctic Circle in the Endicott Mountains of Alaska, came close to the North American record on January 23, 1971, when the temperature plunged to −62°C (−80°F). In the lower 48 states the record of −57°C (−70°F) was set in the mountains at Rogers Pass, Montana, on January 20, 1954. Remember that many other places have no doubt experienced equally low or even lower temperatures; they just were not recorded.

---

conduction. Consequently, daily temperature changes are small below a depth of 10 centimeters (4 inches), although some change can occur to a depth of perhaps 1 meter (3 feet). Annual temperature variations usually reach depths of 15 meters (50 feet) or less. Thus, as a result of the mobility of water and the lack of mobility in the solid Earth, a relatively thick layer of water is heated to moderate temperatures during the summer. On land only a thin layer is heated but to much higher temperatures.

During winter the shallow layer of rock and soil that was heated in summer cools rapidly. Water bodies, in contrast, cool slowly as they draw on the reserve of heat stored within. As the water surface cools, vertical motions are established. The chilled surface water, which

is dense, sinks and is replaced by warmer water from below, which is less dense. Consequently, a larger mass of water must cool before the temperature at the surface will drop appreciably.

2. Because land surfaces are opaque, heat is absorbed only at the surface. This fact is easily demonstrated at a beach on a hot summer afternoon by comparing the surface temperature of the sand to the temperature just a few centimeters beneath the surface. Water, being more transparent, allows some solar radiation to penetrate to a depth of several meters.

3. The **specific heat** (the amount of heat needed to raise the temperature of 1 gram of a substance 1°C) is more than three times greater for water than for land. Thus, water

requires considerably more heat to raise its temperature the same amount as an equal volume of land.

4. Evaporation (a cooling process) from water bodies is greater than from land surfaces. Energy is required to evaporate water. When energy is used for evaporation, it is not available for heating.*

All these factors collectively cause water to warm more slowly, store greater quantities of heat, and cool more slowly than land.

Monthly temperature data for two cities will demonstrate the moderating influence of a large water body and the extremes associated with land (Figure 3–5). Vancouver, British Columbia, is located along the windward Pacific coast, whereas Winnipeg, Manitoba, is in a continental position far from the influence of water. Both cities are at about the same latitude and thus experience similar sun angles and lengths of daylight. Winnipeg, however, has a mean January temperature that is 20°C (36°F) lower than Vancouver's. Conversely, Winnipeg's July mean is 2.6°C (4.7°F) higher than Vancouver's. Although their latitudes are nearly the same, Winnipeg, which has no water influence, experiences much greater temperature extremes than does Vancouver. The key to Vancouver's moderate year-round climate is the Pacific Ocean.

On a different scale, the moderating influence of water may also be demonstrated when temperature variations in the Northern and Southern Hemispheres are compared. The views of Earth in Figure 3–6 show the uneven distribution of land and water over the globe. Water covers 61 percent of the Northern Hemisphere; land represents the remaining 39 percent. However, the figures for the Southern Hemisphere (81 percent water and 19 percent land) reveal why it is correctly called the *water hemisphere*. Between 45° north and 79° north latitude there is actually more land than water, whereas between 40° south and 65° south latitude there is almost no land to interrupt the oceanic and atmospheric circulation. Figure 3–7 portrays the considerably smaller annual temperature ranges in the water-dominated Southern Hemisphere compared with the Northern Hemisphere.

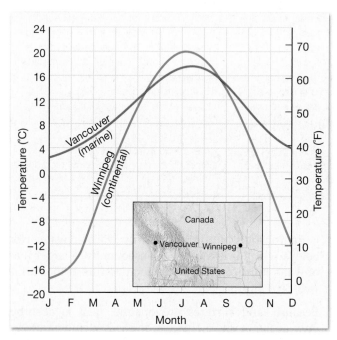

Figure 3–5 Mean monthly temperatures for Vancouver, British Columbia, and Winnipeg, Manitoba. Vancouver has a much smaller annual temperature range, due to the strong marine influence of the Pacific Ocean. Winnipeg illustrates the greater extremes associated with an interior location.

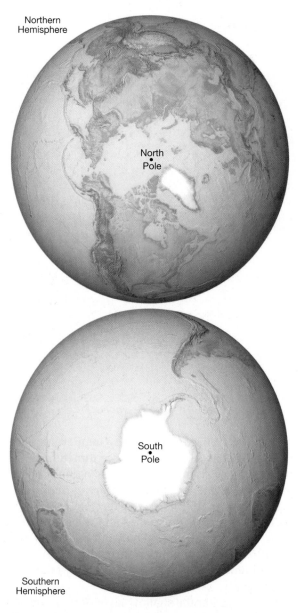

Figure 3–6 The uneven distribution of land and water between the Northern and Southern Hemispheres. Almost 81 percent of the Southern Hemisphere is covered by the oceans—20 percent more than the Northern Hemisphere.

---

*Evaporation is an important process that is discussed more thoroughly in the section "Water's Changes of State" in Chapter 4.

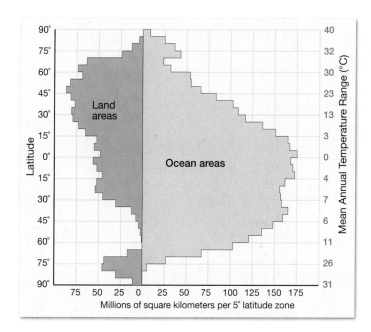

Figure 3–7 The graph shows the amount of land and water in each 5° latitude belt. Mean annual temperature ranges are noted on the right. The moderating influence of water is demonstrated when the Northern and Southern Hemispheres are compared. Annual temperature ranges in the water-dominated midlatitudes (30°–60°) of the Southern Hemisphere are considerably smaller than in the land-dominated midlatitudes of the Northern Hemisphere.

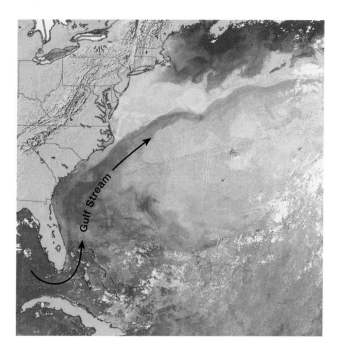

Figure 3–8 Satellite image of the Gulf Stream off the East Coast of the United States. Red represents higher water temperatures and blue cooler water temperatures. The current transports heat from the tropics far into the North Atlantic. (Johns Hopkins University Applied Physics Laboratory)

Concept Check 3.3

1 State the relationship between the heating and cooling of land and water.

2 List and briefly describe the factors that cause land and water to heat and cool differently.

Ocean Currents

You probably have heard of the Gulf Stream, an important surface current in the Atlantic Ocean that flows northward along the East Coast of the United States (Figure 3–8). Surface currents like this one are set in motion by the wind. At the water surface, where the atmosphere and ocean meet, energy is passed from moving air to the water through friction. As a consequence, the drag exerted by winds blowing steadily across the ocean causes the surface layer of water to move. Thus, major horizontal movements of surface waters are closely related to the circulation of the atmosphere, which in turn is driven by the unequal heating of Earth by the Sun (Figure 3–9).*

Surface ocean currents have an important effect on climate. It is known that for Earth as a whole, the gains in solar energy equal the losses to space of heat radiated from the surface. When most latitudes are considered individually, however, this is not the case. There is a net gain of energy in lower latitudes, and there is a net loss at higher latitudes. Because the tropics are not becoming progressively warmer, nor are the polar regions becoming colder, there must be a large-scale transfer of heat from areas of excess to areas of deficit. This is indeed the case. *The transfer of heat by winds and ocean currents equalizes these latitudinal energy imbalances.* Ocean water movements account for about one-quarter of this total heat transport, and winds account for the remaining three-quarters.

The moderating effect of poleward-moving warm ocean currents is well known. The North Atlantic Drift, an extension of the warm Gulf Stream, keeps wintertime temperatures in Great Britain and much of Western Europe warmer than would be expected for their latitudes (London is farther north than St. John's, Newfoundland.) Because of the prevailing westerly winds, the moderating effects are carried far inland. For example, Berlin (52° north latitude) has a mean January temperature similar to that experienced in New York City, which lies 12° latitude farther south. The January mean in London (51° north latitude) is 4.5°C (8.1°F) higher than in New York City.

In contrast to warm ocean currents such as the Gulf Stream, the effects of which are felt most during the winter, cold currents exert their greatest influence in the tropics or during the summer months in the middle latitudes. For example, the cool Benguela Current off the western coast of southern Africa moderates the tropical heat along this coast. Walvis Bay (23° south latitude), a

*The relationship between global winds and surface ocean currents is examined in Chapter 7.

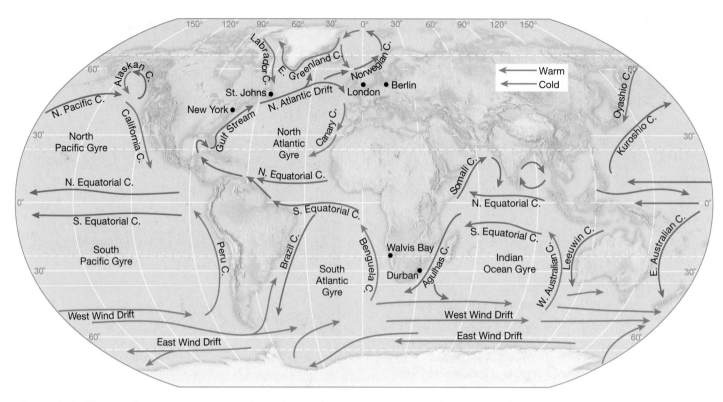

Figure 3–9 Major surface ocean currents. Poleward-moving currents are warm, and equatorward-moving currents are cold. Surface ocean currents are driven by global winds and play an important role in redistributing heat around the globe. Note that cities mentioned in the text discussion are shown on this map.

town adjacent to the Benguela Current, is 5°C (9°F) cooler in summer than Durban, which is 6° latitude farther poleward but on the eastern side of South Africa, away from the influence of the current (Figure 3–9). The east and west coasts of South America provide another example. Figure 3–10 shows monthly mean temperatures for Rio de Janeiro, Brazil, which is influenced by the warm Brazil Current and Arica, Chile, which is adjacent to the cold Peru Current. Closer to home, because of the cold California current, summer temperatures in subtropical coastal southern California are lower by 6°C (10.8°F) or more compared to East Coast stations.

Concept Check 3.4

1 What force drives ocean currents?

2 Are poleward-moving ocean currents warm or cold?

3 How do ocean currents that move toward the equator influence the temperatures of adjacent land areas?

Altitude

Recall from Chapter 1 that temperatures decrease with an increase in altitude in the troposphere. As a result, some mountaintops are snow-covered year round. This can even

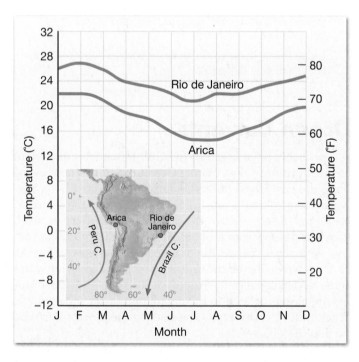

Figure 3–10 Monthly mean temperatures for Rio de Janeiro, Brazil, and Arica, Chile. Both are coastal cities near sea level. Even though Arica is closer to the equator than Rio de Janeiro, its temperatures are cooler. Arica is influenced by the cold Peru Current, whereas Rio de Janeiro is adjacent to the warm Brazil Current.

Figure 3–11 Recall from Chapter 2 that temperatures decrease with an increase in altitude in the troposphere. As a result, some mountaintops are snow-covered all year. This can even occur in the tropics. Africa's Mt. Kilimanjaro actually has a small glacier at its summit. (Photo by Corbis/SuperStock)

occur in the tropics if the mountains are high enough (Figure 3–11).

The two cities in Ecuador mentioned earlier, Quito and Guayaquil, demonstrate the influence of altitude on mean temperature. Both cities are near the equator and relatively close to one another, but the annual mean temperature at Guayaquil is 25.5°C (77.9°F), compared with Quito's mean of 13.3°C (55.9°F). The difference may be understood when the cities' elevations are noted. Guayaquil is only 12 meters (39 feet) above sea level, whereas Quito is high in the Andes Mountains at 2800 meters (9200 feet). Figure 3–12 provides another example.

In Chapter 1 you learned that temperatures drop an average of 6.5°C per kilometer (3.5°F per 1000 feet) in the troposphere. However, if this figure is applied, we would expect Quito to be about 18.2°C (32.7°F) cooler than Guayaquil, but the difference is only 12.2°C (22°F). The fact that high-altitude places, such as Quito, are warmer than the value calculated using the normal lapse rate results from the absorption and reradiation of solar energy by the ground surface.

In addition to the effect of altitude on mean temperatures, the daily temperature range also changes with variations in height. Not only do temperatures drop with an increase in altitude but atmospheric pressure and density also diminish. Because of the reduced density at high altitudes, the overlying atmosphere absorbs and reflects a smaller portion of the incoming solar radiation. Consequently, with an increase in altitude, the intensity of solar radiation increases, resulting in relatively rapid and intense daytime heating. Conversely, rapid nighttime cooling is also the rule in high mountain locations. Therefore, stations located high in the mountains generally have a greater daily temperature range than do stations at lower elevations.

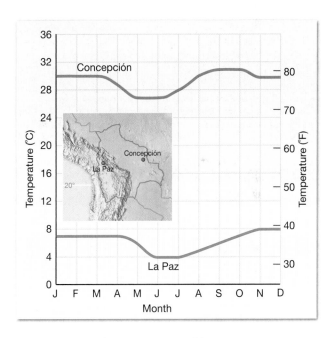

Figure 3–12 Graph comparing monthly mean temperatures at Concepción and La Paz, Bolivia. Both cities have nearly the same latitude (about 16° South). However, because La Paz is high in the Andes, at 4103 meters (13,461 feet), it experiences much cooler temperatures than Concepcieón, which is at an elevation of 490 meters (1608 feet).

Concept Check 3.5

1 How does altitude influence average temperatures? Provide an example.

2 Where is daily temperature range usually greater: at the base or the top of a mountain? Explain.

EYE ON THE ATMOSPHERE

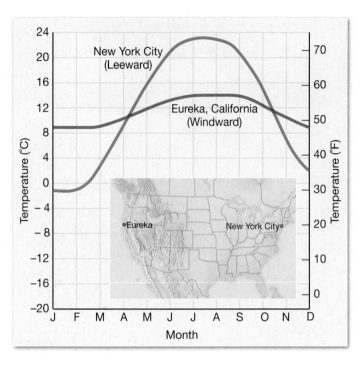

Imagine being at this beach on a warm, sunny summer afternoon. (Photo by Tequilab/Shutterstock)

Question 1 Describe the temperatures you would expect if you measured the surface of the beach and at a depth of 12 inches.

Question 2 If you stood waist deep in the water and measured the water's surface temperature and the temperature at a depth of 12 inches, how would these measurements compare to those taken at the beach?

Figure 3–13 Monthly mean temperatures for Eureka, California, and New York City. Both cities are coastal and located at about the same latitude. Because Eureka is strongly influenced by prevailing winds from the ocean and New York City is not, the annual temperature range at Eureka is much smaller.

Geographic Position

The geographic setting can greatly influence the temperatures experienced at a specific location. A coastal location where prevailing winds blow from the ocean onto the shore (a *windward* coast) experiences considerably different temperatures than does a coastal location where prevailing winds blow from the land toward the ocean (a *leeward* coast). In the first situation the windward coast will experience the full moderating influence of the ocean—cool summers and mild winters—compared to an inland station at the same latitude.

A leeward coastal situation, however, will have a more continental temperature regime because the winds do not carry the ocean's influence onshore. Eureka, California, and New York City, the two cities mentioned earlier, illustrate this aspect of geographic position (Figure 3–13). The annual temperature range in New York City is 19°C (34°F) greater than Eureka's.

Seattle and Spokane, both in the state of Washington, illustrate a second aspect of geographic position: mountains acting as barriers. Although Spokane is only about 360 kilometers (225 miles) east of Seattle, the towering Cascade Range separates the cities. Consequently, Seattle's temperatures show a marked marine influence, but Spokane's are more typically continental (Figure 3–14). Spokane is

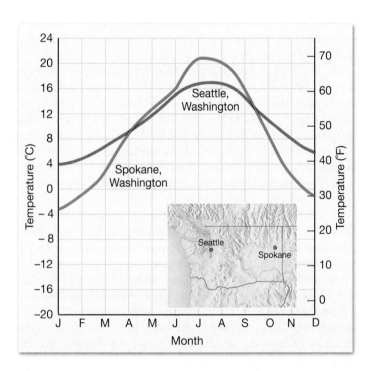

Figure 3–14 Monthly mean temperatures for Seattle and Spokane, Washington. Because the Cascade Mountains cut off Spokane from the moderating influence of the Pacific Ocean, its annual temperature range is greater than Seattle's.

7°C (12.6°F) cooler than Seattle in January and 4°C (7.2°F) warmer than Seattle in July. The annual range at Spokane is 11°C (nearly 20°F) greater than in Seattle. The Cascade Range effectively cuts off Spokane from the moderating influence of the Pacific Ocean.

Concept Check 3.6

1 Three cities are at the same latitude (about 45° North). One city is located along a windward coast, another in the center of the continent, and the third along a leeward coast. Compare the annual temperature ranges of these cities.

2 Why does Spokane, Washington, have a higher annual temperature range than Seattle, Washington?

Cloud Cover and Albedo

You may have noticed that clear days are often warmer than cloudy ones and that clear nights usually are cooler than cloudy ones. This demonstrates that cloud cover is another factor that influences temperature in the lower atmosphere. Studies using satellite images show that at any particular time, about half of our planet is covered by clouds. Cloud cover is important because many clouds have a high albedo and therefore reflect a significant proportion of the sunlight that strikes them back to space. By reducing the amount of incoming solar radiation, daytime temperatures will be lower than if the clouds were absent and the sky were clear (Figure 3–15). As was noted in Chapter 2, the albedo of clouds depends on the thickness of the cloud cover and can vary from 25 to 80 percent (see Figure 2–17, page 51).

At night, clouds have the opposite effect as during daylight. They absorb outgoing Earth radiation and emit a portion of it toward the surface. Consequently, some of the heat that otherwise would have been lost remains near the ground. Thus, nighttime air temperatures do not drop as low as they would on a clear night. The effect of cloud cover is to reduce the daily temperature range by lowering the daytime maximum and raising the nighttime minimum. This is illustrated nicely by the graph in Figure 3–15.

The effect of cloud cover on reducing maximum temperatures can also be detected when monthly mean temperatures are examined for some stations. For example, each year much of southern Asia experiences an extended period of relative drought during the cooler low-Sun period; it is then followed by heavy monsoon rains.* The graph for Yangon, Myanmar (also known as Rangoon, Burma), illustrates this pattern (Figure 3–16). Notice that the highest monthly mean temperatures occur in April and May, before the summer solstice, rather than in July and August, as normally occurs at most stations in the Northern Hemisphere. Why?

*This pattern is associated with the monsoon circulation and is discussed in Chapter 7.

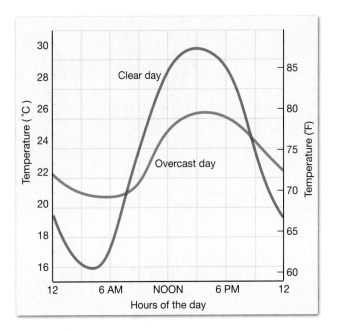

Figure 3–15 The graph shows the daily temperature cycle at Peoria, Illinois, for two July days—one clear and the other overcast. On a clear day the maximum temperature is higher, and the minimum temperature is lower than if the day had been overcast. On an overcast day, the daily temperature range is lower than if it had been cloud free.

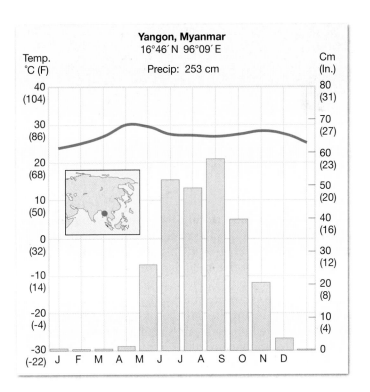

Figure 3–16 Monthly mean temperatures (curve) and monthly mean precipitation (bar graph) for Yangon, Myanmar. The highest mean temperature occurs in April, just before the onset of heavy summer rains. The abundant cloud cover associated with the rainy period reflects back into space the solar energy that otherwise would strike the ground and raise summer temperatures.

SEVERE AND HAZARDOUS WEATHER

Heat Waves*

A *heat wave* is a prolonged period of abnormally hot and usually humid weather that typically lasts from a few days to several weeks. The impact of heat waves on individuals varies greatly. The elderly are the most vulnerable because heat puts more stress on weak hearts and bodies. The poor, who often cannot afford air-conditioning, also suffer disproportionately. Studies also show that the temperature at which death rates increase varies from city to city. In Dallas, Texas, a temperature of 39°C (103°F) is required before the death rate climbs. In San Francisco the key temperature is just 29°C (84°F).

The 1936 North American heat wave is likely the continent's most severe in modern history

Heat waves do not elicit the same sense of fear or urgency as tornadoes, hurricanes, and flash floods. One reason is that it may take many days of oppressive temperatures for a heat wave to exact its toll rather than just a few minutes or a few hours. Another reason is that they cause far less property damage than other extreme weather events. Nevertheless, heat waves can be deadly and costly.

The 1936 North American heat wave is likely the continent's most severe in modern history. It took place during the economic hard times of the Great Depression and coincided with a significant drought in many parts of the Great Plains and Midwest. The prolonged heat wave began in late June and did not end until early September. The estimated death toll exceeded 5000, and agricultural losses were catastrophic in many areas.

Many records from this summer of exceptional temperatures still stand. In fact, current record high temperatures for 13 states are from July and August 1936 (Table 3–A). There are other extraordinary records as well. For example, Mt. Vernon, Illinois, experienced 18 straight days (August 12–29) when temperatures surpassed 38 °C (100 °F).

A recent example occurred in the summer of 2003, when much of Europe experienced perhaps its worst heat wave in more than a century. The image in Figure 3–B relates to this deadly event. An estimate based on government records puts the death toll at between 20,000 and 35,000. The majority died during the first two weeks of August—the hottest span. France suffered the greatest number of heat-related fatalities—about 14,000.

The severity of heat waves is usually greatest in cities

The dangerous impact of summer heat is also reinforced when examining Figure 3–C, which shows average annual weather-related deaths in the United States for the 10-year period 2001–2010. A comparison of values reveals that the number of heat deaths is among the highest.

The severity of heat waves is usually greatest in cities because of the *urban heat island* (see Box 3–3). Large cities do not cool off as much at night during heat waves as rural areas do, and this can be a critical difference in the amount of heat stress within

---

*For a related discussion, see the section on heat stress later in this chapter.

TABLE 3–A State Temperature Records Remaining from 1936

State	Temperature (°F)	Date
Arkansas	120	August 10
Indiana	116	July 14
Kansas	121	July 24
Louisiana	114	August 10
Maryland	109	July 10
Michigan	112	July 13
Minnesota	114	July 13
Nebraska	118	July 24
New Jersey	110	July 10
North Dakota	121	July 6
Pennsylvania	111	July 10
West Virginia	112	July 10
Wisconsin	114	July 13

The reason is that during the summer months, when we would usually expect temperatures to climb, the extensive cloud cover increases the albedo of the region, which reduces incoming solar radiation at the surface. As a result, the highest monthly mean temperatures occur in late spring, when the skies are still relatively clear.

Cloudiness is not the only phenomenon that increases albedo and thereby reduces air temperature. We also recognize that snow- and ice-covered surfaces have high albedos (Figure 3–17). This is one reason mountain glaciers do not melt away in the summer and why snow may still be present on a mild spring day. In addition, during the winter when snow covers the ground, daytime maximums on a sunny day are cooler than they otherwise would be because energy that the land would have absorbed and used to heat the air is reflected and lost.

Concept Check 3.7

1 Contrast the daily temperature range on an overcast day with that on a cloudless sunny day.

2 Examine Figure 3–16 and explain why Yangon's monthly mean temperature for April is higher than the July monthly mean.

World Distribution of Temperatures

Take a moment to study the two world isothermal maps in (Figures 3–18 and 3–19). From hot colors near the equator to cool colors toward the poles, these maps portray sea-level

Land Surface Temperature difference (°C)

-10   -5   0   5   10

FIGURE 3–B This image is derived from satellite data and shows the difference in daytime land surface temperatures during the 2003 European heat wave (July 20–August 20) as compared to the years 2000, 2001, 2002, and 2004. The zone of deep red shows where temperatures were 10°C (18°F) hotter than in the other years. France was hardest hit. (NASA)

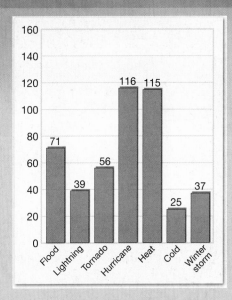

FIGURE 3–C Average annual weather-related fatalities for the 10-year period 2001–2010. The figure for the hurricane category was dramatically affected by a single storm—Hurricane Katrina in 2005. Katrina was responsible for 1016 of the 1154 hurricane-related fatalities that occurred during this 10-year span. (Data from NOAA)

the inner city. In addition, the stagnant atmospheric conditions usually associated with heat waves trap pollutants in urban areas and add the stresses of severe air pollution to the already dangerous stresses caused by the high temperatures.

In July 1995, a brief but intense heat wave developed in the central United States. A total of 830 deaths were attributed to this severe 5-day event, the worst in 50 years in the northern Midwest. The greatest loss of life occurred in Chicago, where there were 525 fatalities. This event provided a sobering lesson by focusing attention on the need for more effective warning and response plans, especially in major urban areas where heat stress is greatest.

Figure 3–17 Ice- and snow-covered surfaces have high albedos, thus keeping air temperatures lower than if the surface were not highly reflective. This view shows sea ice (frozen seawater) near Barrow, Alaska. When sea ice melts, as has occurred on the left side of the image, a bright reflective surface is replaced by a darker surface that absorbs a higher percentage of incoming solar radiation. (Photo by Michael Collier)

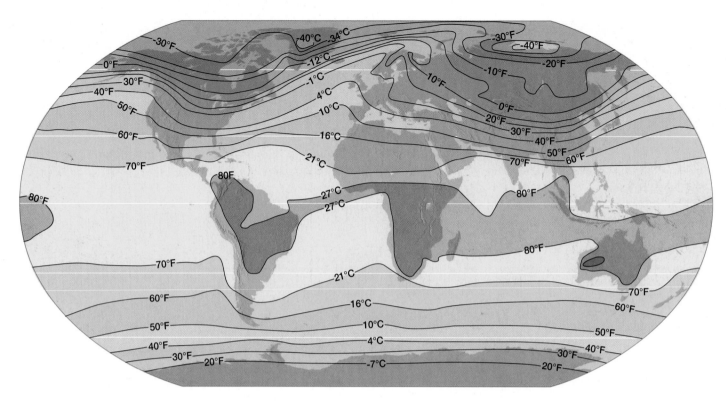

Figure 3–18 World mean sea-level temperatures in January, in Celsius (°C) and Fahrenheit (°F).

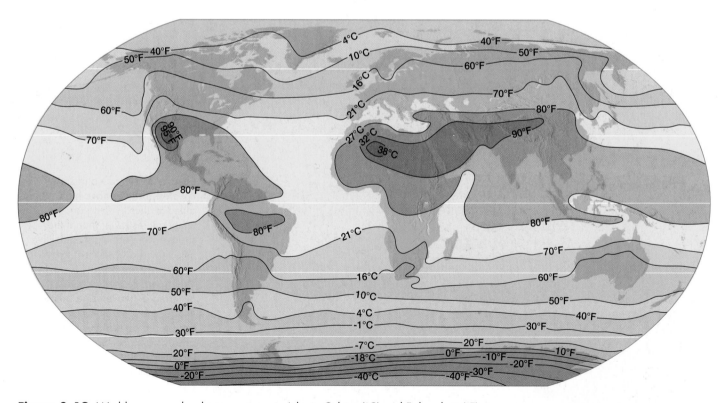

Figure 3–19 World mean sea-level temperatures in July, in Celsius (°C) and Fahrenheit (°F).

temperatures in the seasonally extreme months of January and July. On these maps you can study global temperature patterns and the effects of the controls of temperature, especially latitude, the distribution of land and water, and ocean currents. Like most other isothermal maps of large regions, all temperatures on these world maps have been reduced to sea level to eliminate the complications caused by differences in altitude.

On both maps the isotherms generally trend east and west and show a decrease in temperatures poleward from the tropics. They illustrate one of the most fundamental aspects of world temperature distribution: that the effectiveness of incoming solar radiation in heating Earth's surface and the atmosphere above it is largely a function of latitude. Moreover, there is a latitudinal shifting of temperatures caused by the seasonal migration of the Sun's vertical rays. To see this, compare the color bands by latitude on the two maps.

If latitude were the only control of temperature distribution, our analysis could end here, but this is not the case. The added effect of the differential heating of land and water is clearly reflected on the January and July temperature maps. The warmest and coldest temperatures are found over land—note the coldest area, a purple oval in Siberia, and the hottest areas, the deep orange ovals—all over land. Consequently, because temperatures do not fluctuate as much over water as over land, the north–south migration of isotherms is greater over the continents than over the oceans. In addition, it is clear that the isotherms in the Southern Hemisphere, where there is little land and where

the oceans predominate, are much more regular than in the Northern Hemisphere, where they bend sharply northward in July and southward in January over the continents.

Isotherms also reveal the presence of ocean currents. Warm currents cause isotherms to be deflected toward the poles, whereas cold currents cause an equatorward bending. The horizontal transport of water poleward warms the overlying air and results in air temperatures that are higher than would otherwise be expected for the latitude. Conversely, currents moving toward the equator produce cooler-than-expected air temperatures.

Figures 3–18 and 3–19 show the seasonal extremes of temperature, and comparing them enables us to see the annual range of temperature from place to place. Comparing the two maps shows that a station near the equator has a very small annual range because it experiences little variation in the length of daylight, and it always has a relatively high Sun angle. A station in the middle latitudes, however, experiences wide variations in Sun angle and length of daylight and hence large variations in temperature. Therefore, we can state that the annual temperature range increases with an increase in latitude (see Box 3–2).

Moreover, land and water also affect seasonal temperature variations, especially outside the tropics. A continental location must endure hotter summers and colder winters than a coastal location. Consequently, outside the tropics the annual range will increase with an increase in continentality.

Figure 3–20, which shows the global distribution of annual temperature ranges, serves to summarize the pre-

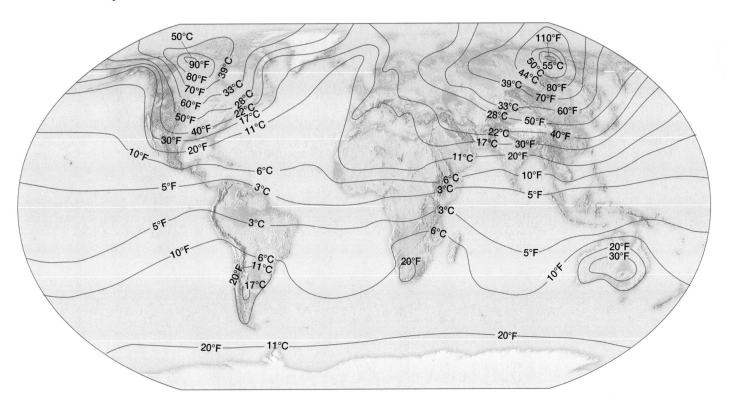

Figure 3-20 Global annual temperature ranges in Celsius (°C) and Fahrenheit (°F). Annual ranges are small near the equator and increase toward the poles. Outside the tropics, annual temperature ranges increase as we move away from the ocean and toward the interior of large landmasses. (After Robert W. Christopherson, *Geosystems: An Introduction to Physical Geography*, 8th ed., Pearson Prentice Hall, 2012)

EYE ON THE ATMOSPHERE

This image shows a snow-covered area in the middle latitudes on a sunny day in late winter. Assume that one week after this photo was taken, conditions were essentially identical except that the snow was gone. (Photo by CoolR/Shutterstock)

Question 1 Would you expect the air temperatures to be different on the two days? If so, which day would likely be warmer?

Question 2 Suggest a reason for the different temperatures.

much smaller than in the Northern Hemisphere with its large continents.

Concept Check 3.8

1 Why do isotherms on the January and July temperature maps generally trend east–west?

2 On the January map (Figure 3–18), why do the isotherms bend in the North Atlantic?

3 Explain why the isotherms bend northward over North America on the July map (Figure 3–19).

4 Refer to Figure 3–20 to determine which area on Earth experiences the highest annual temperature range. Why is the annual range so high in this region?

Students Sometimes Ask . . .

Where in the world would I experience the greatest contrast between summer and winter temperatures?

Among places for which records exist, it appears as though Yakutsk, a station in the heart of Siberia, is the best candidate. The latitude of Yakutsk is 62° north, just a few degrees south of the Arctic Circle. Moreover, it is far from the influence of water. The January mean at Yakutsk is a frigid −43°C (−45°F), whereas its July mean is a pleasant 20°C (68°F). The result is an average annual temperature range of 63°C (113°F), among the highest ranges anywhere on the globe.

ceding two paragraphs. By examining this map, it is easy to see the influence of latitude and continentality on this temperature statistic. The tropics clearly experience small annual temperature variations. As expected, the highest values occur in the middle of large landmasses in the subpolar latitudes. It is also obvious that annual temperature ranges in the ocean-dominated Southern Hemisphere are

Cycles of Air Temperature

You know from experience that a rhythmic rise and fall of air temperature occurs almost every day. Your experience is confirmed by thermograph records like the one in Figure 3–21. (A *thermograph* is an instrument that continuously records temperature.) The temperature curve reaches a mini-

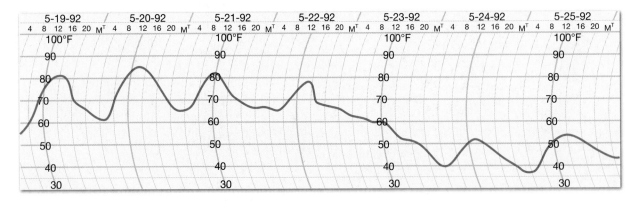

Figure 3–21 Thermograph of temperatures in Peoria, Illinois, during a seven-day span in May 1992. The typical daily rhythm, with minimums around sunrise and maximums in mid- to late afternoon, occurred on most days. The obvious exception occurred on May 23, when the maximum was reached at midnight and temperatures dropped throughout the day.

Box 3–2 Latitude and Temperature Range

Gregory J. Carbone*

Latitude, because of its influence on Sun angle, is the most important temperature control. Figures 3–18 and 3–19 clearly show higher temperatures in tropical locations and lower temperatures in polar regions. The maps also show that higher latitudes experience a greater range of temperatures during the year than do lower latitudes. Notice also that the temperature gradient between the subtropics and the poles is greatest during the winter season. A look at two cities—San Antonio, Texas, and Winnipeg, Manitoba—illustrates how seasonal differences in Sun angle and day length account for these temperature patterns. Figure 3–D shows the annual march of temperatures for the two cities, and Figure 3–E illustrates the Sun angles for the June and December solstices.

San Antonio and Winnipeg are a fixed distance apart (approximately 20.5° latitude), so the difference in Sun angles between the two cities is the same throughout the year. However, in December, when the Sun's rays are least direct, this difference more strongly affects the intensity of solar radiation received at Earth's surface. Therefore, we expect a greater difference in temperatures between the two stations in winter than in summer. Moreover, the seasonal difference in intensity (spreading out of the light beam) at Winnipeg is considerably greater than at San Antonio. This helps to explain the greater annual temperature range at the more northerly station. Table 2–2, Chapter 2, shows that seasonal contrasts in day length also contribute to different temperature patterns at the two cities.

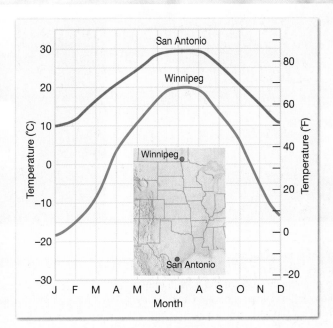

FIGURE 3–D The annual temperature range at Winnipeg is much greater than at San Antonio.

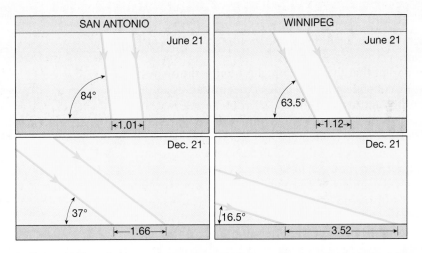

FIGURE 3–E A comparison of Sun angles (solar noon) for summer and winter solstices at San Antonio and Winnipeg. The space covered by a 90° angle is 1.00.

*Professor Carbone is a faculty member in the Department of Geography at the University of South Carolina.

mum around sunrise (Figure 3–22). It then climbs steadily to a maximum between 2 PM and 5 PM. The temperature then declines until sunrise the following day.

Daily Temperature Variations

The primary control of the daily cycle of air temperature is as obvious as the cycle itself: It is Earth's daily rotation, which causes a location to move into daylight for part of each day and then into darkness. As the Sun's angle in-

creases throughout the morning, the intensity of sunlight also rises, reaching a peak at local noon and gradually diminishing in the afternoon.

Figure 3–23 shows the daily variation of incoming solar energy versus outgoing Earth radiation and the resulting temperature curve for a typical middle-latitude location at the time of an equinox. During the night the atmosphere and the surface of Earth cool as they radiate away heat that is not replaced by incoming solar energy. The minimum temperature, therefore, occurs about the time of sunrise,

(a)

(b)

Figure 3–22 The minimum daily temperature usually occurs near the time of sunrise. As the ground and air cool during the nighttime hours, familiar early-morning phenomena such as the frost in photo (a) and the ground fog in photo (b) may form. ((a): AP Photo/Daily Telegram/Jed Carlson; (b): Michael Collier)

after which the Sun again heats the ground, which in turn heats the air.

It is apparent that the time of highest temperature does not generally coincide with the time of maximum radiation. By comparing Figures 3–21 and 3–23, you can see that the curve for incoming solar energy is symmetrical with respect to noon, but the daily air temperature curves are not. The delay in the occurrence of the maximum until mid- to late afternoon is termed the *lag of the maximum*.

Although the intensity of solar radiation drops in the afternoon, it still exceeds outgoing energy from Earth's surface for a period of time. This produces an energy surplus for up to several hours in the afternoon and contributes substantially to the lag of the maximum. In other words, as long as the solar energy gained exceeds the rate of Earth radiation lost, the air temperature continues to rise. When the input of solar energy no longer exceeds the rate of energy lost by Earth, the temperature falls.

The lag of the daily maximum is also a result of the process by which the atmosphere is heated. Recall that air is a poor absorber of most solar radiation; consequently, it is heated primarily by energy reradiated from Earth's surface. The rate at which Earth supplies heat to the atmosphere through radiation, conduction, and other means, however, is not in balance with the rate at which the atmosphere radiates away heat. Generally, for a few hours after the period of maximum solar radiation, more heat is supplied to the atmosphere by Earth's surface than is emitted by the atmosphere to space. As a result, most locations experience an increase in air temperature during the afternoon.

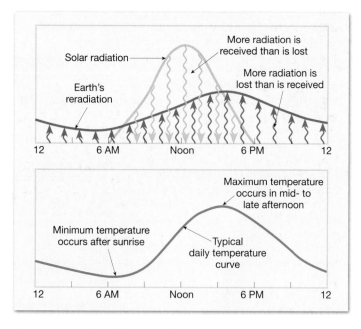

Figure 3–23 The daily cycle of incoming solar radiation, Earth's radiation, and the resulting temperature cycle. This example is for a midlatitude site around the time of an equinox. As long as solar energy gained exceeds outgoing energy emitted by Earth, the temperature rises. When outgoing energy from Earth exceeds the input of solar energy, temperature falls. Note that the daily temperature cycle *lags* behind the solar radiation input by a couple hours.

In dry regions, particularly on cloud-free days, the amount of radiation absorbed by the surface will generally be high. Therefore, the time of the maximum temperature at these locales will often occur quite late in the afternoon. Humid locations, in contrast, will frequently experience a shorter time lag in the occurrence of their temperature maximum.

EYE ON THE ATMOSPHERE

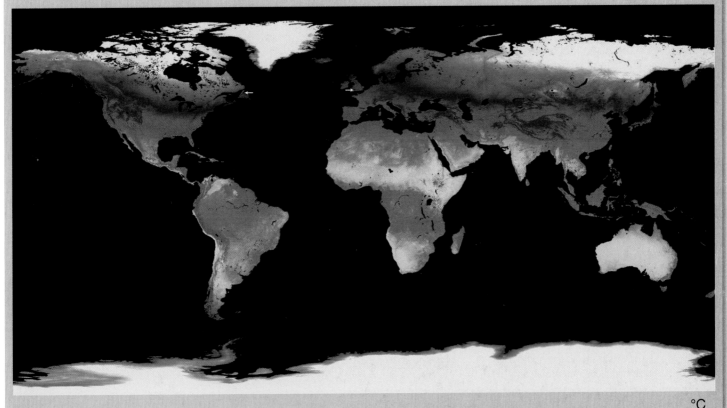

°C

−25   10   45

For more than a decade scientists have used the Moderate Resolution Imaging Spectroradiometer (MODIS, for short) aboard NASA's Aqua and Terra satellites to gather surface temperature data from around the globe. This image shows average land-surface temperatures for the month of February over a 10-year span (2001-2010). (NASA)

Question 1 What are the approximate temperatures for southern Great Britain and northern Newfoundland (white arrows)?

Question 2 Both are coastal areas at the same latitude yet average February temperatures are quite different. Suggest a cause for this disparity.

Magnitude of Daily Temperature Changes

The magnitude of daily temperature changes is variable and may be influenced by locational factors or local weather conditions or both (see Box 3–3). Four common examples illustrate this point. The first two relate to location, and the second two pertain to local weather conditions:

1. Variations in Sun angle are relatively great during the day in the middle and low latitudes. However, points near the poles experience a low Sun angle all day. Consequently, the temperature change experienced during a day in the high latitudes is small.

2. A windward coast is likely to experience only modest variations in the daily cycle. During a typical 24-hour period the ocean warms less than 1°C. As a result, the air above it shows a correspondingly slight change in temperature. For example, Eureka, California, a windward coastal station, consistently has a lower daily temperature range than Des Moines, Iowa, an inland city at about the same latitude. Annually the daily

range at Des Moines averages 10.9°C (19.6°F) compared with 6.1°C (11°F) at Eureka, a difference of 4.8°C (8.6°F).

3. As mentioned earlier, an overcast day is responsible for a flattened daily temperature curve (see Figure 3–15). By day, clouds block incoming solar radiation and so reduce daytime heating. At night the clouds retard the loss of radiation by the ground and air. Therefore, nighttime temperatures are warmer than they otherwise would have been.

4. The amount of water vapor in the air influences daily temperature range because water vapor is one of the atmosphere's important heat-absorbing gases. When the air is clear and dry, heat readily escapes at night, and the temperature falls rapidly. When the air is humid, absorption of outgoing long-wavelength radiation by water vapor slows nighttime cooling, and the temperature does not fall to as low a value. Thus, dry conditions are associated with a higher daily temperature range because of greater nighttime cooling.

Box 3–3 How Cities Influence Temperature: The Urban Heat Island

One of the most apparent human impacts on climate is the modification of the atmospheric environment by the building of cities (Figure 3–F). The construction of every factory, road, office building, and house destroys microclimates and creates new ones of great complexity.

The most studied and well-documented urban climatic effect is the *urban heat island*. The term refers to the fact that temperatures within cities are generally higher than in rural areas. The heat island is evident when temperature data such as those that appear in Table 3–B are examined. As is typical, the data for Philadelphia show the heat island is most pronounced when minimum temperatures are examined. The magnitude of the temperature differences shown by Table 3–B is probably even greater than the figures indicate because temperatures observed at suburban airports are usually higher than those in truly rural environments.

Figure 3–G, which shows the distribution of average minimum temperatures in the Washington, D.C., metropolitan area for the three-month winter period (December through February) over a five-year span, also illustrates a well-developed heat island. The warmest winter temperatures occurred in the heart of the city, whereas the suburbs and surrounding countryside experienced average minimum temperatures that were as much as 3.3°C (6°F) lower. Remember that these temperatures are averages. On many clear, calm nights the temperature difference between the city center and the countryside was considerably greater, often 11°C (20°F) or more. Conversely, on many overcast or windy nights the temperature differential approached 0°.

One effect of an urban heat island is to influence the biosphere by extending

FIGURE 3–F The radical change in the surface that occurs when rural areas are transformed into cities leads to the creation of an urban heat island. (Photo by Grin Maria/Shutterstock)

the plant cycle.[*] After examining data for 70 cities in eastern North America, researchers found that the growing cycle in cities was about 15 days longer than in surrounding rural areas. Plants began the growth cycle an average of about 7 days earlier in spring and continued growing an average of 8 days longer in fall.

Why are cities warmer than rural areas? The radical change in the surface that results when rural areas are transformed into cities is a significant cause of an urban heat island (Figure 3–H). First, the tall buildings and the concrete and asphalt of the city absorb and store greater quantities of solar radiation than do the vegetation and soil typical

of rural areas. In addition, because the city surface is impermeable, the runoff of water following a rain is rapid, resulting in a significant reduction in the evaporation rate. Hence, heat that once would have been used to convert liquid water to a gas now goes to increase further the surface temperature. At night, as both the city and countryside cool by radiative losses, the stonelike surface of the city gradually releases the additional heat accumulated during the day, keeping the urban air warmer than that of the outlying areas.

A portion of the urban temperature rise is also attributable to waste heat from sources such as home heating and air-conditioning, power generation, industry, and transportation. In addition, the "blanket" of pollutants over a city contributes to a heat island by absorbing a portion of the upward-directed longwave radiation emitted by the surface and reemitting some of it back to the ground.

TABLE 3–B Average Temperatures (°C) for Suburban Philadelphia Airport and Downtown Philadelphia (10-year averages)

	Airport	Downtown
Annual mean	12.8	13.6
Mean June max	27.8	28.2
Mean December max	6.4	6.7
Mean June min	16.5	17.7
Mean December min	−2.1	−0.4

Source: After H. Neuberger and J. Cahir, *Principles of Climatology* (New York: Holt, Rinehart and Winston, 1969), 128.

[*]Reported in *Weatherwise*, 57, no. 6 (November/December 2004), 20.

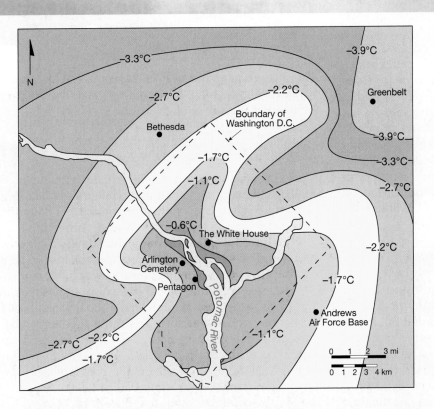

FIGURE 3–G The heat island of Washington, D.C., as shown by the average minimum temperatures (°C) during the winter season (December through February). The city center had an average minimum that was nearly 4°C higher than some outlying areas. (After Clarence A. Woolum, "Notes from the Study of the Microclimatology of the Washington, D.C., area for the Winter and Spring Seasons," *Weatherwise,* 17, no. 6 (1964), 264, 267.)

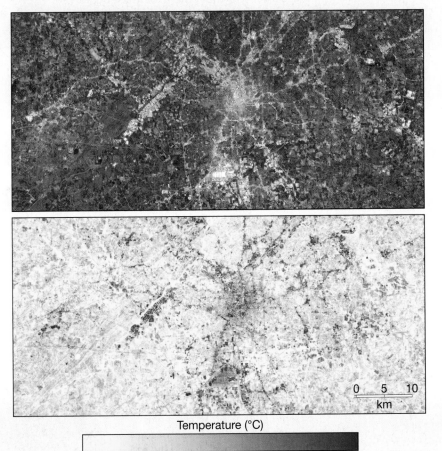

FIGURE 3–H This pair of satellite images provides two views of Atlanta, Georgia, on September 28, 2000. The urban core is in the center of the images. The top image is a photo-like view in which vegetation is green, roads and dense development appear gray, and bare ground is tan or brown. The bottom image is a land-surface temperature map in which cooler temperatures are yellow and hotter temperatures are red. It is clear that where development is densest, the land-surface temperatures are highest. (NASA)

Although the rise and fall of daily temperatures usually reflects the general rise and fall of incoming solar radiation, such is not always the case. For example, a glance back at Figure 3–21 shows that on May 23 the maximum temperature occurred at midnight, after which temperatures fell throughout the day. If records for a station are examined for a period of several weeks, apparently random variations are seen. Obviously these are not Sun controlled. Such irregularities are caused primarily by the passage of atmospheric disturbances (weather systems) that are often accompanied by variable cloudiness and winds that bring air having contrasting temperatures. Under these circumstances, the maximum and minimum temperatures may occur at any time of the day or night.

Annual Temperature Variations

In most years the months with the highest and lowest mean temperatures do not coincide with the periods of maximum and minimum incoming solar radiation. North of the tropics the greatest intensity of solar radiation occurs at the time of the summer solstice in June, yet the months of July and August are generally the warmest of the year in the Northern Hemisphere. Conversely, a minimum of solar energy is received in December at the time of the winter solstice, but January and February are usually colder.

The fact that the occurrence of annual maximum and minimum radiation does not coincide with the times of temperature maximums and minimums indicates that the amount of solar radiation received is not the only factor determining the temperature at a particular location. Recall from Chapter 2 that places equatorward of about 38° north and 38° south receive more solar radiation than is lost to space and that the opposite is true of more poleward regions. Based on this imbalance between incoming and outgoing radiation, any location in the southern United States, for example, should continue to get warmer late into autumn.

But this does not occur because more poleward locations begin experiencing a negative radiation balance shortly after the summer solstice. As the temperature contrasts become greater, the atmosphere and ocean currents "work harder" to transport heat from lower latitudes poleward.

Concept Check 3.9

1 Although the intensity of incoming solar radiation is greatest at local noon, the warmest part of the day is most often midafternoon. Why? Use Figure 3–23 to explain.

2 List at least three factors that might cause the daily temperature range to vary significantly from place to place and from time to time.

Temperature Measurement

 GEODe Temperature Data and the Controls of Temperature
ATMOSPHERE   ▶ Basic Temperature Data

Thermometers are "meters of therms": They measure temperature (Figure 3–24). Thermometers measure temperature either mechanically or electrically.

Mechanical Thermometers

Most substances expand when heated and contract when cooled, and many common thermometers operate using this property. More precisely, they rely on the fact that different substances react to temperature changes differently.

The **liquid-in-glass thermometer** shown in Figure 3–25 is a simple instrument that provides relatively accurate readings over a wide temperature range. Its design has remained essentially unchanged since it was developed in the late 1600s. When temperature rises, the molecules of fluid grow more active and spread out (the fluid expands). Expansion of the fluid in the bulb is much greater than the expansion of the enclosing glass. As a consequence, a thin "thread" of fluid is forced up the capillary tube. Conversely, when temperature falls, the liquid contracts, and the thread of fluid moves back down the tube toward the bulb. The movement of the end of this thread (known as the *meniscus*) is calibrated against an established scale to indicate the temperature.

The highest and lowest temperatures that occur each day are of considerable importance and are often obtained by using specially designed liquid-in-glass thermometers. Mercury is the liquid used in the **maximum thermometer,** which has a narrowed passage called a *constriction* in the bore of the glass tube just above the bulb (Figure 3–26a). As the temperature rises, the mercury expands

Figure 3–24 The design of this thermometer is based on an instrument called a *thermoscope* that was invented by Galileo in the late 1500s. Today such devices, which are fairly accurate, are used mostly for decoration. The instrument is made of a sealed glass cylinder containing a clear liquid and a series of glass bulbs of different densities that can "float" up and down. (Photo by Frank Labua/Pearson Education)

and is forced through the constriction. When the temperature falls, the constriction prevents a return of mercury to the bulb. As a result, the top of the mercury column remains at the highest point (maximum temperature attained during the measurement period). The instrument is reset by shaking or by whirling it to force the mercury through the constriction back into the bulb. Once the thermometer is reset, it indicates the current air temperature.

In contrast to a maximum thermometer that contains mercury, a **minimum thermometer** contains a liquid of low density, such as alcohol. Within the alcohol, and resting at the top of the column, is a small dumbbell-shaped index (Figure 3–26b). As the air temperature drops, the column shortens, and the index is pulled toward the bulb by the effect of surface tension with the meniscus. When the temperature subsequently rises, the alcohol flows past the index, leaving it at the lowest temperature reached. To return the index to the top of the alcohol column, the thermometer is simply tilted. Because the index is free to move, the minimum thermometer must be

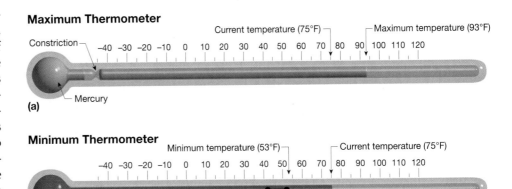

Maximum Thermometer

Figure 3–26 (a) Maximum thermometer and (b) minimum thermometer.

mounted horizontally; otherwise, the index will fall to the bottom.

Another commonly used mechanical thermometer is the **bimetal strip.** As the name indicates, this thermometer consists of two thin strips of metal that are bonded together and have widely different expansion properties. When the temperature changes, both metals expand or contract, but they do so unequally, causing the strips to curl. This change corresponds to the change in temperature.

The primary meteorological use of the bimetal strip is in the construction of a **thermograph,** an instrument that continuously records temperature. The changes in the curvature of the strip can be used to move a pen arm that records the temperature on a calibrated chart that is attached to a clock-driven, rotating drum (Figure 3–27). Although very convenient, thermograph records are generally less accurate than readings obtained from a mercury-in-glass thermometer. To obtain the most reliable values, it is necessary to check and correct the thermograph periodically by comparing it with an accurate, similarly exposed thermometer.

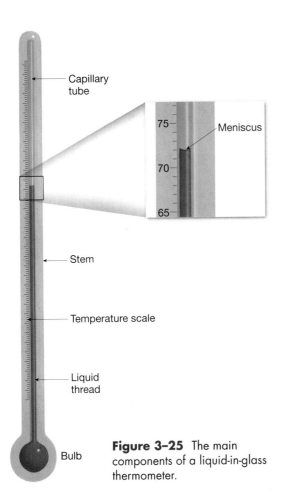

Figure 3–25 The main components of a liquid-in-glass thermometer.

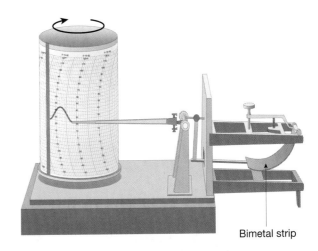

Figure 3–27 A common use of the bimetal strip is in the construction of a thermograph, an instrument that continuously records temperatures.

Electrical Thermometers

Some thermometers do not rely on differential expansion but instead measure temperature electrically.

A resistor is a small electronic part that resists the flow of electrical current. A **thermistor** (thermal resistor) is similar, but its resistance to current flow varies with temperature. As temperature increases, so does the resistance of the thermistor, reducing the flow of current. As temperature drops, so does the resistance of the thermistor, allowing more current to flow. The current operates a meter or digital display that is calibrated in degrees of temperature. The thermistor thus is used as a temperature sensor—an electrical thermometer.

Thermistors are rapid-response instruments that quickly register temperature changes. Therefore, they are commonly used in radiosondes where rapid temperature changes are often encountered. The National Weather Service also uses a thermistor system for ground-level readings. The sensor is mounted inside a shield made of louvered plastic rings, and a digital readout is placed indoors (Figure 3–28).

> ## Students Sometimes Ask...
>
> What are the highest and lowest temperatures ever recorded at Earth's surface?
>
> The world's record-high temperature is nearly 58°C (136°F)! It was recorded on September 13, 1922, at Azizia, Libya, in North Africa's Sahara Desert. The lowest recorded temperature is −89°C (−129°F). This incredibly frigid temperature was recorded in Antarctica, at the Russian Vostok Station, on July 21, 1983.

Instrument Shelters

How accurate are thermometer readings? Accuracy depends not only on the design and quality of the instruments but also on where they are placed. Placing a thermometer in direct sunlight will give a grossly excessive reading because the instrument itself absorbs solar energy much more efficiently than does the air. Placing a thermometer near a heat-radiating surface, such as a building or the ground, also yields inaccurate readings. Another way to assure false readings is to prevent air from moving freely around the thermometer.

So where should a thermometer be placed to read air temperature accurately? The ideal location is an instrument shelter (Figure 3–29). An instrument shelter is a white box that has louvered sides to permit the free movement of air through it, while shielding the instruments from direct sunshine, heat from the ground, and precipitation. Furthermore, the shelter is placed over grass whenever possible and as far away from buildings as circumstances permit. Finally, the shelter must conform to a standardized height so that the thermometers will be mounted at 1.5 meters (5 feet) above the ground (Box 3–4).

Figure 3–29 This traditional standard instrument shelter is white (for high albedo) and louvered (for ventilation). It protects instruments from direct sunlight and allows for the free flow of air. (Courtesy of Qualimetrics, Inc.)

Figure 3–28 This modern shelter contains an electrical thermometer called a *thermistor*. (Photo by Bobbé Christopherson)

Temperature Scales

In the United States, TV weather reporters give temperatures in degrees Fahrenheit. But scientists as well as most people outside the United States use degrees Celsius. Scientists sometimes also use the Kelvin, or absolute, scale. What are the differences among these three temperature scales? To make quantitative measurement of temperature possible, it was necessary to establish scales. Such temperature scales are based on the use of reference points, sometimes called **fixed points.** In 1714, Gabriel Daniel Fahrenheit, a German physicist, devised the **Fahrenheit** scale.

He constructed a mercury-in-glass thermometer in which the zero point was the lowest temperature he could attain with a mixture of ice, water, and common salt. For his second fixed point he chose human body temperature, which he arbitrarily set at 96°.

On this scale, he determined that the melting point of ice (the **ice point**) was 32° and the boiling point of water (the **steam point**) was 212°. Because Fahrenheit's original reference points were difficult to reproduce accurately, his scale is now defined by using the ice point and the steam point. As thermometers improved, average human body temperature was later changed to 98.6°F.*

In 1742, 28 years after Fahrenheit invented his scale, Anders Celsius, a Swedish astronomer, devised a decimal scale on which the melting point of ice was set at 0° and the boiling point of water at 100°.† For many years it was called the *centigrade scale,* but it is now known as the **Celsius scale,** after its inventor.

Because the interval between the melting point of ice and the boiling point of water is 100° on the Celsius scale and 180° on the Fahrenheit scale, a Celsius degree (°C) is larger than a Fahrenheit degree (°F) by a factor of 180/100, or 1.8. So, to convert from one system to the other, allowance must be made for this difference in the size of the degrees. Also, conversions must be adjusted because the ice point on the Celsius scale is at 0° rather than at 32°. This relationship is shown graphically in Figure 3–30.

---

*This traditional value for "normal body temperature" was established in 1868. A recent assessment places the value at 98.2°F, with a range of 4.8°F.

†The boiling point referred to in the Celsius and Fahrenheit scales pertains to pure water at standard sea-level pressure. It is necessary to remember this fact because the boiling point of water gradually decreases with altitude.

The Celsius–Fahrenheit relationship also is shown by the following formulas:

$$°F = (1.8 \times °C) + 32$$

and:

$$°C = \frac{°F - 32}{1.8}$$

You can see that the formulas adjust for degree size with the 1.8 factor and adjust for the different 0° points with the ±32 factor.

For some scientific purposes a third temperature scale is used, the **Kelvin,** or **absolute, scale.** On this scale, degrees Kelvin are called *Kelvins* (abbreviated K). This scale is similar to the Celsius scale because its divisions are exactly the same; there are 100° separating the melting point of ice and the boiling point of water. However, on the Kelvin scale, the ice point is set at 273 K, and the steam point is at 373 K (Figure 3–30). The reason is that the zero point represents the temperature at which all molecular motion is presumed to cease (called **absolute zero**). Thus, unlike with the Celsius and Fahrenheit scales, it is not possible to have a negative

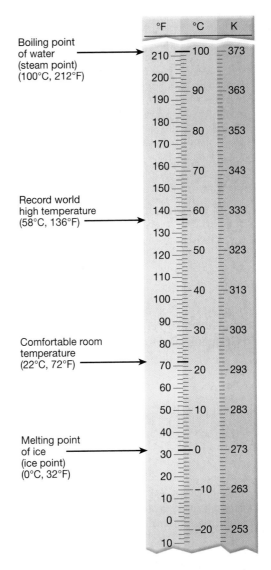

Figure 3–30 The three temperature scales compared.

Box 3–4 Applying Temperature Data

To make weather data more useful to people, many different applications have been developed. This box focuses on three indices that all have the term *degree-days* as part of their name: heating degree-days, cooling degree-days, and growing degree-days. The first two are relative measures that allow us to evaluate the weather-produced needs and costs of heating and cooling. The third is a simple index used by farmers to estimate the maturity of crops.

Heating Degree-Days

Heating degree-days represent a practical method for evaluating energy demand and consumption. This index starts from the assumption that heating is not required in a building when the daily mean temperature is 65°F (18.3°C) or higher.* Simply, each degree of temperature below 65°F is counted as 1 heating degree-day. Therefore, heating degree-days are determined each day by subtracting the daily mean below 65°F from 65°F. Thus, a day with a mean temperature of 50°F has 15 heating degree-days (65 − 50 = 15), and one with an average temperature of 65°F or higher has none.

The amount of heat required to maintain a certain temperature in a building is pro-

portional to the total heating degree-days. This linear relationship means that doubling the heating degree-days usually doubles the fuel consumption. Consequently, a fuel bill will generally be twice as high for a month with 1000 heating degree-days as for a month with 500. When seasonal totals are compared for different places, we can estimate differences in seasonal fuel consumption (Table 3–C). For example, more than five times as much fuel is required to

heat a building in Chicago (nearly 6500 total heating degree-days) than to heat a similar building in Los Angeles (almost 1300 heating degree-days). This statement is true, however, only if we assume that building construction and living habits in these areas are the same.

Each day, the previous day's accumulation is reported along with the total thus far in the season. For reporting purposes the heating season is defined as the period

---

*Because the National Weather Service and the news media in the United States still compute and report degree-day information in Fahrenheit degrees, we will use the Fahrenheit scale throughout this discussion.

TABLE 3–C Average Annual Heating and Cooling Degree-Days for Selected Cities*

City	Heating Degree-Days	Cooling degree-days
Anchorage, AK	10,470	3
Baltimore, MD	3807	1774
Boston, MA	5630	777
Chicago, IL	6498	830
Denver, CO	6128	695
Detroit, MI	6422	736
Great Falls, MT	7828	288
International Falls, MN	10,269	233
Las Vegas, NV	2239	3214
Los Angeles, CA	1274	679
Miami, FL	149	4361
New York City, NY	4754	1151
Phoenix. AZ	1125	4189
San Antonio, TX	1573	3038
Seattle, WA	4797	173

*Source: NOAA, National Climatic Data Center.

---

value when using the Kelvin scale, for there is no temperature lower than absolute zero. The relationship between the Kelvin and Celsius scales is easily written as follows:

$$°C = K - 273$$

or:

$$K = °C + 273$$

Concept Check 3.11

1 What is meant by the terms *steam point* and *ice point*? What values are given to these points on each of the three temperature scales presented here?

2 Why is it not possible to have a negative value when using the Kelvin scale?

Heat Stress and Windchill: Indices of Human Discomfort

Summertime weather reports often try to make people aware of potential harmful effects of high humidity coupled with high temperatures. In winter we are reminded about the effect of strong winds combined with low temperatures.

FIGURE 3–I Growing degree-days are used to determine the approximate date when crops will be ready for harvest. (Photo by Kletr/Shutterstock)

from July I through June 30. These reports often include a comparison with the total up to this date last year or with the long-term average for this date or both, and so it is a relatively simple matter to judge whether the season thus far is above, below, or near normal.

Cooling Degree-Days

Just as fuel needs for heating can be estimated and compared by using heating degree-days, the amount of power required to cool a building can be estimated by using a similar index called the *cooling degree-day.* Because the 65°F base temperature is also used in calculating this index, cooling degree-days are determined each day by subtracting 65°F from the daily mean. Thus, if the mean temperature for a given day is 80°F, 15 cooling degree-days would be accumulated. Mean annual totals of cooling degree-days for selected cities are shown in Table 3–C. By comparing the totals for Baltimore and Miami, we can see that the fuel requirements for cooling a building in Miami are almost 2½ times as great as for a similar building in Baltimore. The "cooling season" is conventionally measured from January 1 through December 31. Therefore, when cooling degree-day totals are reported, the number represents the accumulation since January 1 of that year.

Although indices that are more sophisticated than heating and cooling degree-days have been proposed to take into account the effects of wind speed, solar radiation, and humidity, degree-days continue to be widely used.

Growing Degree-Days

Another practical application of temperature data is used in agriculture to determine the approximate date when crops will be ready for harvest. This simple index is called the *growing degree-day.* The number of growing degree-days for a particular crop on any day is the difference between the daily mean temperature and the base temperature of the crop, which is the minimum temperature required for it to grow. For example, the base temperature for sweet corn is 50°F, and for peas it is 40°F. Thus, on a day when the mean temperature is 75°F, the number of growing degree-days for sweet corn is 25 and the number for peas is 35.

Starting with the onset of the growth season, the daily growing degree-day values are added. Thus, if 2000 growing degree-days are needed for a crop to mature, it should be ready to harvest when the accumulation reaches 2000 (Figure 3–I). Although many factors important to plant growth are not included in the index, such as moisture conditions and sunlight, this system nevertheless serves as a simple and widely used tool in determining approximate dates of crop maturity.

In the first instance, we are cautioned about heat stress and the possibility of heat stroke, and in the second case we are warned about windchill and the potential dangers of frostbite. These indices are expressions of **apparent temperature**—the temperature that a person perceives. Heat stress and windchill are based on the fact that our sensation of temperature is often different from the actual air temperature recorded by a thermometer.

The human body is a heat generator that continually releases energy. Anything that influences the rate of heat loss from the body also influences our sensation of temperature, thereby affecting our feeling of comfort. Several factors control the thermal comfort of the human body, and certainly air temperature is a major one. Other environmental conditions are also significant, such as relative humidity, wind, and sunshine.

Heat Stress—High Temperatures Plus High Humidities

High humidity contributes significantly to the discomfort people feel during a heat wave. Why are hot, muggy days so uncomfortable? Humans, like other mammals, are warm-blooded creatures who maintain a constant body temperature, regardless of the temperature of the environment. One of the ways the body prevents overheating is by perspiring. However, this process does little to cool the body unless the perspiration can evaporate. It is the cooling created by the evaporation of perspiration that reduces body temperature. Because high humidities retard evaporation, people are more uncomfortable on a hot and humid day than on a hot and dry day.

Generally, temperature and humidity are the most important elements influencing summertime human comfort.

Heat Index

Air Temperature (°F)	Relative Humidity (%) 40	45	50	55	60	65	70	75	80	85	90	95	100
110	136												
108	130	137											
106	124	130	137										
104	119	124	131	137									
102	114	119	124	130	137								
100	109	114	118	124	129	136							
98	105	109	113	117	123	128	134						
96	101	104	108	112	116	121	126	132					
94	97	100	102	106	110	114	119	124	129	135			
92	94	96	99	101	105	108	112	116	121	126	131		
90	91	93	95	97	100	103	106	109	113	117	122	127	132
88	88	89	91	93	95	98	100	103	106	110	113	117	121
86	85	87	88	89	91	93	95	97	100	102	105	108	112
84	83	84	85	86	88	89	90	92	94	96	98	100	103
82	81	82	83	84	84	85	86	88	89	90	91	93	95
80	80	80	81	81	82	82	83	84	84	85	86	86	87

With prolonged exposure and/or physical activity

Extreme danger Heat stroke or sunstroke highly likely

Danger Sunstroke, muscle cramps, and/or heat exhaustion likely

Extreme caution Sunstroke, muscle cramps, and/or heat exhaustion possible

Caution Fatigue possible

Figure 3–31 Heat index is a measure of apparent temperature. For example, if the air temperature is 90°F and the relative humidity is 65%, it would "feel like" 103°F. As relative humidity increases, apparent temperature increases as well.

as Minneapolis would not be well tolerated. This occurs because hot and humid weather is more taxing on people who live where these conditions are relatively rare than it is on people who live where prolonged periods of heat and high humidity are the rule.

Windchill—The Cooling Power of Moving Air

Most everyone is familiar with the wintertime cooling power of moving air. When the wind blows on a cold day, we realize that comfort would improve if the wind would stop. A stiff breeze penetrates ordinary clothing and reduces its capacity to retain body heat while causing exposed parts of the body to chill rapidly. Not only is cooling by evaporation heightened in this situation but the wind is also acting to carry heat away from the body by constantly replacing warmer air next to the body with colder air.

The U.S. National Weather Service and the Meteorological Services of Canada use a Wind Chill Temperature (WCT) index that is designed to calculate how the wind and cold feel on human skin (Figure 3–32). The index accounts for wind effects at face level and takes into account body heat-loss estimates. It was tested on human subjects in a chilled wind tunnel. The results of those trials were used to validate and improve the accuracy of the formula. The wind

One index widely used by the National Weather Service that combines these factors to establish the degree of comfort or discomfort is called the *heat stress index*, or simply the heat index. If you examine Figure 3–31, you will see that as relative humidity increases, the apparent temperature, and thus heat stress, increases as well. Further, when the relative humidity is low, the apparent temperature can have a value that is less than the actual air temperature. It is important to note that factors such as the length of exposure to direct sunlight, the wind speed, and the general health of the individual greatly affect the amount of stress a person will experience. In addition, while a period of hot, humid weather in New Orleans might be reasonably well tolerated by its residents, a similar event in a northern city such

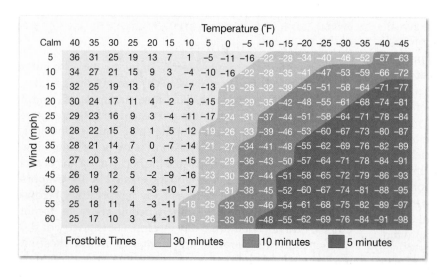

Wind (mph)	Temperature (°F) 40	35	30	25	20	15	10	5	0	−5	−10	−15	−20	−25	−30	−35	−40	−45
Calm																		
5	36	31	25	19	13	7	1	−5	−11	−16	−22	−28	−34	−40	−46	−52	−57	−63
10	34	27	21	15	9	3	−4	−10	−16	−22	−28	−35	−41	−47	−53	−59	−66	−72
15	32	25	19	13	6	0	−7	−13	−19	−26	−32	−39	−45	−51	−58	−64	−71	−77
20	30	24	17	11	4	−2	−9	−15	−22	−29	−35	−42	−48	−55	−61	−68	−74	−81
25	29	23	16	9	3	−4	−11	−17	−24	−31	−37	−44	−51	−58	−64	−71	−78	−84
30	28	22	15	8	1	−5	−12	−19	−26	−33	−39	−46	−53	−60	−67	−73	−80	−87
35	28	21	14	7	0	−7	−14	−21	−27	−34	−41	−48	−55	−62	−69	−76	−82	−89
40	27	20	13	6	−1	−8	−15	−22	−29	−36	−43	−50	−57	−64	−71	−78	−84	−91
45	26	19	12	5	−2	−9	−16	−23	−30	−37	−44	−51	−58	−65	−72	−79	−86	−93
50	26	19	12	4	−3	−10	−17	−24	−31	−38	−45	−52	−60	−67	−74	−81	−88	−95
55	25	18	11	4	−3	−11	−18	−25	−32	−39	−46	−54	−61	−68	−75	−82	−89	−97
60	25	17	10	3	−4	−11	−19	−26	−33	−40	−48	−55	−62	−69	−76	−84	−91	−98

Frostbite Times: ■ 30 minutes ■ 10 minutes ■ 5 minutes

Figure 3–32 Wind chill chart. Fahrenheit temperatures are used because this is how the National Weather Service and the news media in the United States commonly report wind chill information. The shaded areas on the chart indicate frostbite danger. Each shaded zone shows how long a person can be exposed before frostbite develops. For example, a temperature of 0°F and a wind speed of 15 miles per hour will produce a windchill temperature of −19°F. Under these conditions, exposed skin can freeze in 30 minutes. (After NOAA, National Weather Service)

PROFESSIONAL PROFILE
Captain Ryan J. Harris, Military Meteorologist

Before any United States military mission gets a green light, commanders always check the weather. Weather is a crucial element in virtually every task, from flying a bombing mission, to delivering supplies, to conducting covert reconnaissance. The approach of a storm, the percentage of cloud cover, and even magnetic storms on the Sun can influence whether deployed troops will return safe and sound.

As a military meteorologist, "my role is to know exactly what the weather is and how it's going to provide battlefield commanders with an advantage over our adversaries," says Captain Ryan J. Harris. As a Weather Flight Commander in the United States Air Force, Captain Harris leads a unit of 45 Airmen devoted to weather forecasting. He works out of Offutt Air Force Base near Omaha, Nebraska.

"We're not your typical weatherman on TV. We're providing similar information, but tailored to military operations," Captain Harris says. Commanders want to know, "Can I move a battalion through this pass, or will it be too muddy because of heavy snowmelt? Is there enough nasty weather to shroud the actions of my special ops mission? Can I fly this helicopter in these winds?"

To serve the needs of both the Army and the Air Force, weather teams like Captain Harris's operate 24 hours a day. Different weather units, at bases around the world, track atmospheric models on local, regional, and global scales. Their predictions have shaped events ranging from the invasion of Normandy during World War II, to the Navy SEALS mission that tracked down Osama bin Laden.

Captain Harris has watched the skies for almost as long as he can remember. As a kindergartner, he often preferred the Weather Channel to cartoons on TV. He wrote so often about weather in his first grade journal that his teacher suggested he become a weatherman. His plans still hadn't changed by high school. There, a counselor mentioned that the Air Force always needs meteorologists and may even help pay for a college degree. He applied and was accepted to the Reserve Officer Training Corps (ROTC), which supported his bachelor's degree in meteorology, and earned him an officer commission in the Air Force.

> ### Captain Harris's Air Force assignments have taken him around the world.

Captain Harris's Air Force assignments have taken him around the world. In Illinois, he learned to make global weather forecasts for aerial refueling and aeromedical evacuation missions. In Germany, he learned to tailor forecasts to meet fighter aircraft commanders' specific needs. He recently returned to school to earn a master's degree at the Naval Postgraduate School in Monterey, California. There, he specialized in satellite meteorology.

The military, Captain Harris says, "has niches for everyone in meteorology. Those who want to do research and understand the theory, and those who want to take the science and apply it." Captain Harris puts himself in the latter category. "The idea of applying meteorology to aviation and military operations certainly intrigued me and drew me in. It's been a great experience."

> ### "The idea of applying meteorology to aviation and military operations certainly intrigued me and drew me in."

Captain Harris now has a unique job in the Air Force. "The best way to describe it is, anything that doesn't fit in the scope of normal terrestrial weather falls under us," he says. His purview includes tracking airborne ash from volcanoes, delivering global cloud analyses and forecasts, and analyzing meteorological satellite imagery. He also works with scientists to improve forecasting models, and provides snow analyses for Army and Special Operations missions, among other duties.

As sophisticated as this work can be, Captain Harris still applies the meteorology basics he learned in college. "I don't analyze weather charts so often anymore. But I still have to retain the knowledge from my college days to see that this part of this model is not right, and use theoretical physics and meteorology as a critical foundation when my commander asks why we need this model or instrument over that," he says.

The art and science of forecasting weather has improved dramatically over the last century since the first numerical models were developed. Nowhere are these advances more important to get the weather right than in the military. "For us in military weather, the lives of our fellow Airmen, Soldiers, Sailors, and Marines depend on an accurate forecast. All of us take our jobs very seriously," Captain Harris says.

CAPTAIN RYAN J. HARRIS leads a 45-person forecasting team. (Photo courtesy of USAF)

chill chart includes a frostbite indicator that shows where temperature, wind speed, and exposure time will produce frostbite (Figure 3–32).

It is worth pointing out that in contrast to a cold and windy day, a calm and sunny day in winter feels warmer than the thermometer reading. In this situation the warm feeling is caused by the absorption of direct solar radiation by the body. The index does not take into account any offsetting effect for windchill due to solar radiation. Such a factor may be added in the future.

It is important to remember that the windchill temperature is only an estimate of human discomfort. The degree of discomfort felt by different people will vary because it is influenced by many factors. Even if clothing is assumed to be the same, individuals vary widely in their responses because of such factors as age, physical condition, state of health, and level of activity. Nevertheless, as a relative measure, the WCT index is useful because it allows people to make more informed judgments regarding the potential harmful effects of wind and cold.

Concept Check 3.12

1 What is apparent temperature?

2 Why does high humidity contribute to summertime discomfort?

3 Why do strong winds make apparent temperatures in winter feel lower than the thermometer reading?

4 List several reasons heat stress and wind chill do not affect everyone the same.

Give It Some Thought

1. If you were asked to identify the coldest city in the United States (or any other designated region), what statistics could you use? Can you list at least three different ways of selecting the coldest city?

2. The accompanying graph shows monthly high temperatures for Urbana, Illinois, and San Francisco, California. Although both cities are located at about the same latitude, the temperatures they experience are quite different. Which line on the graph represents Urbana and which represents San Francisco? How did you figure this out?

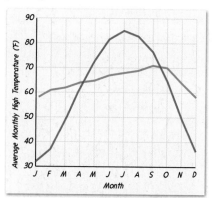

3. On which summer day would you expect the greatest temperature range? Which would have the smallest range in temperature? Explain your choices.

   a. Cloudy skies during the day and clear skies at night
   b. Clear skies during the day and cloudy skies at night
   c. Clear skies during the day and clear skies at night
   d. Cloudy skies during the day and cloudy skies at night

4. The accompanying scene shows an island near the (Photo by Chad Ehlers/ Photolibrary)
equator in the Indian Ocean. Describe how latitude, altitude, and the differential heating of land and water influence the climate of this place.

5. The accompanying sketch map represents a hypothetical continent in the Northern Hemisphere. One isotherm has been placed on the map.

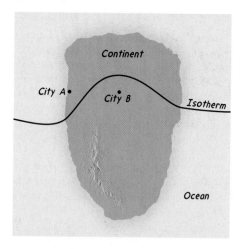

   a. Is the temperature higher at city A or city B? Explain.
   b. Is the season winter or summer? How are you able to determine this?
   c. Describe (or sketch) the position of this isotherm six months later.

6. The data below are mean monthly temperatures in °C for an inland location that lacks any significant ocean influence. *Based on the annual temperature range,* what is the approximate latitude of this place? Are these temperatures what you would normally expect for this latitude? If not, what control would explain these temperatures?

J	F	M	A	M	J	J	A	S	O	N	D
6.1	6.6	6.6	6.6	6.6	6.1	6.1	6.1	6.1	6.1	6.6	6.6

7. Refer to Figure 3–18. What causes the bend or kink in the isotherms in the North Atlantic?

TEMPERATURE IN REVIEW

- Temperature is one of the basic elements of weather and climate. Commonly used temperature data include *daily mean temperature, daily temperature range, monthly mean temperature, annual mean temperature,* and *annual temperature range*. Such data are commonly depicted on maps using *isotherms*, lines of equal temperature.
- The *controls of temperature*—factors that cause temperature to vary from place to place and from time to time—include latitude, differential heating of land and water, ocean currents, altitude, geographic position, and cloud cover and albedo.
- On world maps showing January and July mean temperatures, isotherms generally trend east–west and show a decrease in temperature moving poleward from the equator. When the two maps are compared, a latitudinal shifting of temperatures is easily seen. Bending isotherms reveal the locations of ocean currents.
- Annual temperature range is small near the equator and increases with an increase in latitude. Outside the tropics, annual temperature range also increases as marine influence diminishes.

- The primary control of the daily cycle of air temperature is Earth's rotation. The magnitude of daily changes is variable and influenced by locational factors, local weather conditions, or both.
- As a consequence of the mechanism by which Earth's atmosphere is heated, the months of the highest and lowest temperatures do not coincide with the periods of maximum and minimum incoming solar radiation.
- *Thermometers* measure temperature either mechanically or electrically. Most *mechanical thermometers* are based on the ability of a material to expand when heated and contract when cooled. *Electrical thermometers* use a thermistor (a thermal resistor) to measure temperature.
- Temperature scales are established using reference points, called fixed points. Three common scales are the *Fahrenheit scale*, the *Celsius scale*, and the *Kelvin,* or *absolute, scale*.
- *Heat stress* and *wind chill* are two familiar uses of temperature data that relate to *apparent temperature*—the temperature people perceive.

VOCABULARY REVIEW

absolute zero (p. 89)
apparent temperature (p. 91)
annual mean temperature (p. 66)
annual temperature range (p. 66)
bimetal strip (p. 87)
Celsius scale (p. 89)
controls of temperature (p. 67)
daily mean temperature (p. 66)

daily temperature range (p. 66)
Fahrenheit scale (p. 89)
fixed points (p. 89)
ice point (p. 89)
isotherm (p. 66)
Kelvin, or absolute, scale (p. 89)
liquid-in-glass thermometer (p. 86)
maximum thermometer (p. 86)

minimum thermometer (p. 87)
monthly mean temperature (p. 66)
specific heat (p. 69)
steam point (p. 89)
temperature gradient (p. 66)
thermistor (p. 88)
thermograph (p. 87)
thermometer (p. 86)

PROBLEMS

1. Examine the number of hurricane-related deaths in Figure 3–C (p. 77). Be sure to read the entire caption. If Hurricane Katrina had *not* occurred, what number would have been plotted on the graph for hurricane fatalities?

2. Refer to the thermograph record in Figure 3–21. Determine the maximum and minimum temperature for each day of the week. Use these data to calculate the daily mean and daily range for each day.

3. By referring to the world maps of temperature distribution for January and July (Figures 3–18 and 3–19), determine the approximate January mean, July mean, and annual temperature range for a place located at 60° north latitude, 80° east longitude, and a place located at 60° south latitude, 80° east longitude.

4. Calculate the annual temperature range for three cities in Appendix G. Try to choose cities with different ranges and explain these differences in terms of the controls of temperature.

5. Referring to Figure 3–32, determine wind chill temperatures under the following circumstances:
   a. Temperature = 5°F, wind speed = 15 mph.
   b. Temperature = 5°F, wind speed = 30 mph.

6. The mean temperature is 55°F on a particular day. The following day the mean drops to 45°F. Calculate the number of heating degree-days for each day. How much more fuel would be needed to heat a building on the second day compared with the first day?

7. Use the appropriate formula to convert the following temperatures:

   20°C = ____°F
   −25°C = ____K
   59°F = ____°C

MyMeteorologyLab™

Moisture and Atmospheric Stability

Water vapor is an odorless, colorless gas that mixes freely with the other gases of the atmosphere. Unlike oxygen and nitrogen—the two most abundant components of the atmosphere—water can change from one state of matter to another (solid, liquid, or gas) at the temperatures and pressures experienced on Earth. (By comparison, nitrogen in the atmosphere condenses to a liquid at a temperature of –196°C [–371°F].) Because of this unique property, water leaves the oceans as a gas and returns again as a liquid.

Focus On Concepts

After completing this chapter, you should be able to:

- Sketch and describe the movement of water through the hydrologic cycle.

- List and describe water's unique properties.

- Summarize the processes by which water changes from one state of matter to another and indicate whether heat is absorbed or liberated.

- Write a generalization relating air temperature and the amount of water vapor needed to saturate air.

- Define *vapor pressure* and describe the relationship between vapor pressure and saturation.

- List and describe the ways relative humidity changes in nature.

- Compare and contrast relative humidity and dew-point temperature.

- Describe adiabatic temperature changes and explain why the wet adiabatic rate of cooling is less than the dry adiabatic rate.

- List and describe four mechanisms that cause air to rise.

- Write a statement relating the environmental lapse rate to stability.

97

Movement of Water Through the Atmosphere

Water is found everywhere on Earth—in the oceans, glaciers, rivers, lakes, air, soil, and living tissue (Figure 4-1). The vast majority of the water on or close to Earth's surface (over 97 percent) is saltwater found in the oceans. Much of the remaining 3 percent is stored in the ice sheets of Antarctica and Greenland. Only a meager 0.001 percent is found in the atmosphere, and most of this is in the form of water vapor.

The continuous exchange of water among the oceans, the atmosphere, and the continents is called the **hydrologic cycle** (Figure 4-2). Water from the oceans and, to a lesser extent, from land areas, evaporates into the atmosphere. Winds transport this moisture-laden air, often over great distances, until conditions cause the moisture to condense into cloud droplets.

The process of cloud formation may result in precipitation. The precipitation that falls into the ocean has ended its cycle and is ready to begin another. A portion of the water that falls on the land soaks into the ground, some of it moving downward and then laterally, where it eventually seeps into lakes and streams. Much of the water that soaks in or runs off returns to the atmosphere through evaporation. In addition, some of the water that infiltrates the ground is absorbed by plants through their roots. They then release it into the atmosphere, a process called **transpiration.** Because we cannot clearly distinguish between the amount of water that evaporates from that which is transpired by plants, the term *evapotranspiration* is often used to describe the combined process.

The total amount of water vapor in the atmosphere remains fairly constant. Therefore, the average annual precipitation over Earth must be roughly equal to the quantity of water lost through evaporation. However, over the continents, precipitation exceeds evaporation. Evidence for the roughly balanced hydrological cycle is found in the fact that the level of the world's ocean is not dropping.

In summary, the hydrologic cycle depicts the continuous movement of Earth's water from the oceans to the atmosphere, from the atmosphere to the land, and from the land back to the sea. The movement of water through the cycle holds the key to the distribution of moisture over the surface of our planet and is intricately related to all atmospheric phenomena.

Concept Check 4.1

1 Sketch and describe the movement of water through the hydrologic cycle.

2 The quantity of water lost to evaporation over the oceans is not equaled by precipitation. Why, then, does sea level not drop?

3 Name and describe the two processes that occur during evapotranspiration.

Water: A Unique Substance

Water has several unique properties that set it apart from most other substances. For instance, (1) water is the only *liquid* found at Earth's surface in large quantities; (2) water is readily converted from one state of matter to another (solid, liquid, gas); (3) water's solid phase, ice, is less dense than liquid water; and (4) water has a high heat capacity—meaning it requires considerable energy to change its temperature. All these properties influence Earth's weather and climate and are favorable to life as we know it.

These unique properties are largely a consequence of water's ability to form *hydrogen bonds*. To better grasp the nature of hydrogen bonds, let's examine a water molecule (Figure 4-3a). A water molecule (H_2O) consists of two hydrogen

Figure 4-1 This lake in Alabama is experiencing steam fog on a cool autumn morning. (Photo by Pat & Chuck Blackley/ Alamy)

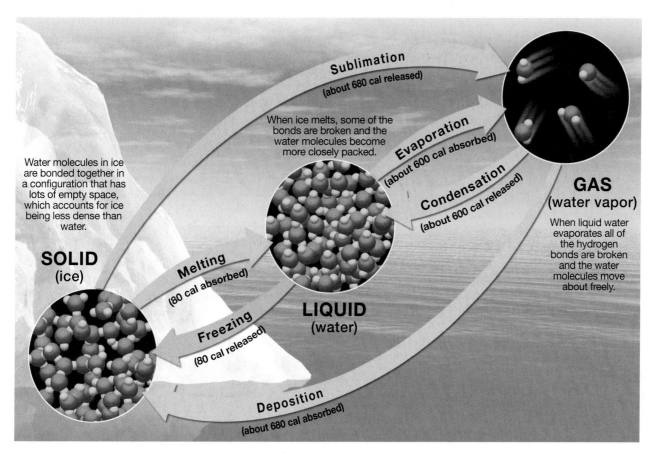

Figure 4-5 Change of state always involves an exchange of heat. The amounts shown here are the approximate numbers of calories absorbed or released when 1 gram of water changes from one state of matter to another.

compared to liquid water and exhibit very energetic and random motion.

To summarize, when water changes state, hydrogen bonds either form or are broken.

Latent Heat

Whenever water changes state, heat is exchanged between water and its surroundings. For example, heat is required to evaporate water. Meteorologists measure the heat involved when water changes state in units called *calories.* One **calorie** is the amount of heat required to raise the temperature of 1 gram of water 1°C (1.8°F). Thus, when 10 calories of heat are absorbed by 1 gram of water, the molecules vibrate faster, and a 10°C (18°F) temperature increase occurs.

Under certain conditions, heat may be added to a substance without an accompanying rise in temperature. For example, when the ice in a glass of ice water melts, the temperature of the mixture remains a constant 0°C (32°F) until all the ice has melted. If the added energy does not raise the temperature of ice water, where does this energy go? In this case, the added energy breaks the hydrogen bonds that once bound the water molecules into a crystalline structure.

Because the heat used to melt ice does not produce a temperature change, it is referred to as **latent heat.** (Latent

means *hidden,* like the latent fingerprints hidden at a crime scene.) This energy is essentially stored in liquid water and released to its surroundings as heat when the liquid returns to the solid state.

It requires 80 calories to melt 1 gram of ice, an amount referred to as *latent heat of melting. Freezing,* the reverse process, releases these 80 calories per gram to the environment as the *latent heat of fusion.* We will consider the importance of latent heat of fusion in Chapter 5, in the section on frost prevention.

Evaporation and Condensation Latent heat is also involved during **evaporation,** the process of converting a liquid to a gas (vapor). The energy absorbed by water molecules during evaporation is used to give them the motion needed to escape the surface of the liquid and become a gas. This energy is referred to as the *latent heat of vaporization* and varies from about 600 calories per gram for water at 0°C to 540 calories per gram at 100°C. (Notice from Figure 4-5 that it takes much more energy to evaporate 1 gram of water than it does to melt the same amount of ice.) During the process of evaporation, the higher-temperature (faster-moving) molecules escape the surface. As a result, the average molecular motion (temperature) of the remaining water is lowered—hence the expression "Evaporation is a cooling process." You have undoubtedly experienced this cooling effect if you have stepped, dripping wet, out of a swimming pool or bathtub full of water. In this situation the energy used to evaporate water comes from your skin—hence, you feel cool.

Condensation, the reverse process, occurs when water vapor changes to the liquid state. During condensation, water-vapor molecules release energy (*latent heat of condensation*) in an amount equivalent to what was absorbed during evaporation. When condensation occurs in the atmosphere, it results in the formation of fog and clouds (Figure 4-6).

Latent heat plays an important role in many atmospheric processes. In particular, when water vapor condenses to form cloud droplets, latent heat is released, which warms the surrounding air and gives it buoyancy. When the moisture content of air is high, this process can spur the growth of towering storm clouds. In addition, the evaporation of water over the tropical oceans and the subsequent condensation at higher latitudes results in significant energy transfer from equatorial to more poleward locations.

Sublimation and Deposition   You are probably least familiar with the last two processes illustrated in Figure 4-5—sublimation and deposition. **Sublimation** is the conversion of a solid directly to a gas, without passing through the liquid state. Examples you may have observed include the gradual shrinkage of unused ice cubes in a freezer and the rapid conversion of dry ice (frozen carbon dioxide) to wispy clouds that quickly disappear.

Deposition refers to the reverse process: the conversion of a vapor directly to a solid. This change occurs, for example, when water vapor is deposited as ice on solid objects

such as grass or windows (Figure 4-7). These deposits are called *white frost* or simply *frost*. A household example of the process of deposition is the "frost" that accumulates in a freezer. As shown in Figure 4-5, deposition releases an amount of energy equal to that which is released by condensation and freezing.

Concept Check 4.3

1 Summarize the processes by which water changes from one state to another. Indicate whether heat is absorbed or liberated in each case.

2 Describe latent heat and explain the role that latent heat of condensation plays in the growth of towering clouds.

3 Explain why evaporation is called a cooling process.

4 Why does the process of sublimation (or deposition) involve the exchange of more latent heat than any of the other processes that cause changes of state?

Students Sometimes Ask...

What is "freezer burn"?

Freezer burn is what happens to poorly wrapped food stored in the freezer of a frost-free refrigerator for an extended time. Because modern refrigerators are designed to remove moisture from the freezer compartment, the air within them is relatively dry. As a result, the moisture in food sublimates—turns from ice to water vapor—and escapes. Thus, the food is not actually burned; it is simply dried out.

Figure 4-6
Condensation of water vapor generates phenomena such as dew, clouds and fog. (Photo by NaturePL/Shutterstock)

Figure 4-7 White frost on a window pane. (Photo by Stockxpert/Thinkstock)

Humidity: Water Vapor in the Air

 Moisture and Cloud Formation
▶ Humidity: Water Vapor in the Air

Water vapor constitutes only a small fraction of the atmosphere, varying from as little as. 0.1 percent up to about 4 percent by volume. But the importance of water in the air is far greater than these small percentages indicate. In fact, *water vapor* is the most important gas in the atmosphere when it comes to atmospheric processes.

Humidity is the general term used to describe the amount of water vapor in the air (Figure 4-8). Meteorologists employ several methods to express the water-vapor content of the air, including (1) absolute humidity, (2) mixing ratio, (3) vapor pressure, (4) relative humidity, and (5) dew point. Two of these methods, *absolute humidity* and *mixing ratio*, are

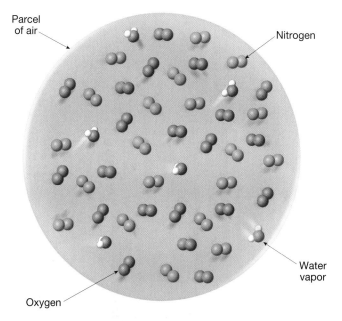

Figure 4-8 Meteorologists use several methods to express the water-vapor content of air.

similar in that both are expressed as the quantity of water vapor contained in a specific amount of air.

Absolute humidity is *the mass of water vapor in a given volume of air* (usually as grams per cubic meter):

$$\text{Absolute humidity} = \frac{\text{Mass of water vapor (grams)}}{\text{Volume of air (cubic meters)}}$$

As air moves from one place to another, changes in pressure and temperature cause changes in its volume. When such volume changes occur, the absolute humidity also changes, even if no water vapor is added or removed. Consequently, it is difficult to monitor the water-vapor content of a moving mass of air when using the absolute humidity index. Therefore, meteorologists generally prefer to use mixing ratio to express the water-vapor content of air.

The **mixing ratio** is *the mass of water vapor in a unit of air compared to the remaining mass of dry air:*

$$\text{Mixing ratio} = \frac{\text{Mass of water vapor (grams)}}{\text{Mass of dry air (kilograms)}}$$

Because it is measured in units of mass (usually grams per kilogram), the mixing ratio is not affected by changes in pressure or temperature (Figure 4-9).*

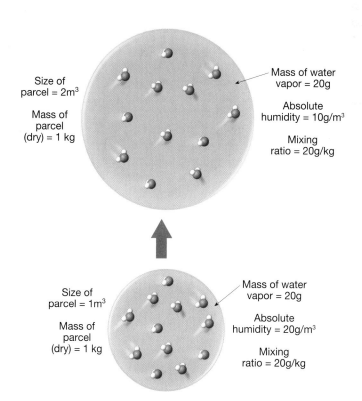

Figure 4-9 Comparison of absolute humidity and mixing ratio for a rising parcel of air. Note that the mixing ratio is not affected by changes in pressure as the parcel of air rises and expands.

---

*Another commonly used expression is "specific humidity," which is the mass of water vapor in a unit mass of air, including the water vapor. Because the amount of water vapor in the air rarely exceeds a few percent of the total mass of the air, the specific humidity of air is equivalent to its mixing ratio for all practical purposes.

Neither the absolute humidity nor the mixing ratio, however, can be easily determined by direct sampling. Therefore, other methods are also used to express the moisture content of the air.

Vapor Pressure and Saturation

The moisture content of the air can be obtained from the pressure exerted by water vapor. To understand how water vapor exerts pressure, imagine a closed flask containing pure water and overlain by dry air, as shown in Figure 4-10a. Almost immediately some of the water molecules begin to leave the water surface and evaporate into the dry air above. The addition of water vapor into the air can be detected by a small increase in pressure (Figure 4-10b). This increase in pressure is the result of the motion of the water-vapor molecules that were added to the air through evaporation. In the atmosphere, this pressure is called **vapor pressure** and is defined as *the part of the total atmospheric pressure attributable to its water-vapor content.*

Initially, many more molecules will leave the water surface (evaporate) than will return (condense). However, as more and more molecules evaporate from the water surface, the steadily increasing vapor pressure in the air above forces more and more water molecules to return to the liquid. Eventually a balance is reached in which the number of water molecules returning to the surface equals the number leaving. At that point the air is said to have reached an equilibrium called **saturation** (Figure 4-10c). When air is

saturated, the pressure exerted by the motion of the water-vapor molecules is called the **saturation vapor pressure.**

Now suppose we were to disrupt the equilibrium by heating the water in our closed container, as illustrated in Figure 4-10d. The added energy would increase the rate at which the water molecules would evaporate from the surface. This in turn would cause the vapor pressure in the air above to increase until a new equilibrium was reached between evaporation and condensation. Thus, we conclude that the saturation vapor pressure is temperature dependent,

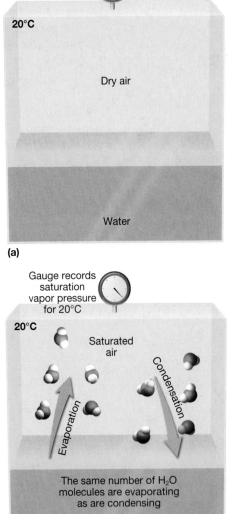
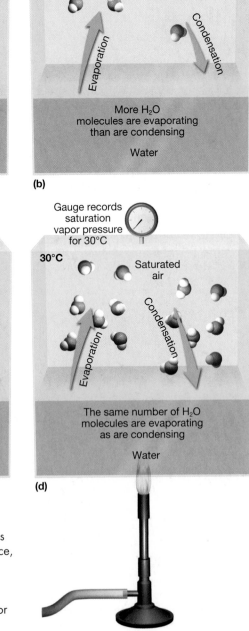
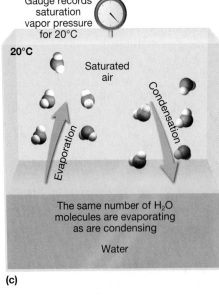

Figure 4-10 The relationship between vapor pressure and saturation. (a) Initial conditions—dry air at 20°C with no observable vapor pressure. (b) Evaporation generates measurable vapor pressure. (c) As more and more molecules escape from the water surface, the steadily increasing vapor pressure forces more and more of these molecules to return to the liquid. Eventually, the number of water-vapor molecules returning to the surface will balance the number leaving. At that point the air is said to be saturated. (d) When the container is heated from 20° to 30°C, the rate of evaporation increases, causing the vapor pressure to increase until a new balance is reached.

EYE ON THE ATMOSPHERE

Water is everywhere on Earth—in the oceans, glaciers, rivers, lakes, air, and living tissue. In addition, water can change from one state of matter to another at the temperatures and pressures experienced on Earth. Refer to this image, taken near Vega Island, Antarctica, to answer the following questions. (Photo by Michael Collier)

Question 1 What two features in this photo are composed of water in the solid state?

Question 2 What three different features are composed of liquid water?

Question 3 Where is water vapor found in this image?

such that at higher temperatures it takes more water vapor to saturate air (Figure 4-11). The amount of water vapor required for the saturation of 1 kilogram (2.2 pounds) of dry air at various temperatures is shown in Table 4-1. Note that for every 10°C (18°F) increase in temperature, the amount

TABLE 4-1 Saturation Mixing Ratio (at Sea-Level Pressure)

Temperature, °C (°F)	Saturation Mixing Ratio, g/kg
−40 (−40)	0.1
−30 (−22)	0.3
−20 (−4)	0.75
−10 (14)	2
0 (32)	3.5
5 (41)	5
10 (50)	7
15 (59)	10
20 (68)	14
25 (77)	20
30 (86)	26.5
35 (95)	35
40 (104)	47

of water vapor needed for saturation almost doubles. Thus, roughly four times more water vapor is needed to saturate 30°C (86°F) air than 10°C (50°F) air.

The atmosphere behaves in much the same manner as our closed container. In nature, gravity, rather than a lid, prevents water vapor (and other gases) from escaping into space. Also as with our container, water molecules are constantly evaporating from liquid surfaces (such as lakes or cloud droplets), and other water vapor molecules are condensing. However, in nature a balance is not always achieved. More often than not, more water molecules are leaving the surface of a water puddle than are arriving, causing what meteorologists call *net evaporation*. By contrast, during the formation of fog, more water molecules are condensing than are evaporating from the tiny fog droplets, resulting in *net condensation*.

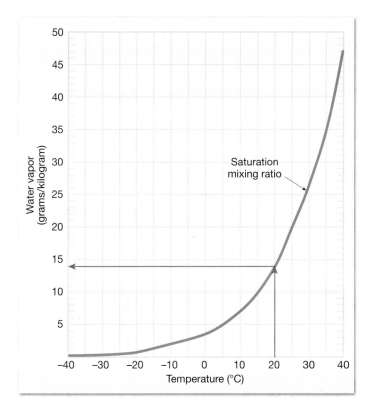

Figure 4-11 This graph shows the amount of water vapor required to saturate 1 kilogram of dry air at various temperatures. For example, the red arrows show that saturated air at 20°C contains 14 grams of water vapor per kilogram of dry air.

What determines whether the rate of evaporation exceeds the rate of condensation (net evaporation) or vice versa? One of the major factors is the temperature of the surface water, which in turn determines how much motion (kinetic energy) the water molecules possess. At higher temperatures the molecules have more energy and can more readily escape. Thus, under otherwise similar conditions, because hot water has more energy, it evaporates faster than cold water.

The other major factor determining whether evaporation or condensation will dominate is the vapor pressure in the air around the liquid. Recall from our example of a closed container that vapor pressure determines the rate at which the water molecules return to the surface (condense). When the air is dry (low vapor pressure), the rate at which water molecules return to the liquid phase is low. However, when the air around a liquid reaches the saturation vapor pressure, the rate of condensation will be equal to the rate of evaporation. Thus, at saturation there is neither a net condensation nor a net evaporation. Therefore, all else being equal, net evaporation is greater when the air is dry (low vapor pressure) than when the air is humid (high vapor pressure).

Students Sometimes Ask...

Why do the sizes of snow piles seem to shrink a few days after a snowfall, even when the temperatures remain below freezing?

On clear, cold days following a snowfall, the air can be very dry. This fact, plus solar heating, causes the ice crystals to sublimate—turn from a solid to a gas. Thus, even without any appreciable melting, these accumulations of snow gradually get smaller.

Concept Check 4.5

1 Define *vapor pressure* and describe the relationship between vapor pressure and saturation. (*Hint:* See Figure 4-10.)

2 After reviewing Table 4-1, write a generalization relating air temperature and the amount of water vapor needed to saturate air.

Relative Humidity

 Moisture and Cloud Formation
▶ Humidity: Water Vapor in the Air

The most familiar and, unfortunately, the most misunderstood term used to describe the moisture content of air is relative humidity. **Relative humidity** *is a ratio of the air's actual water-vapor content compared with the amount of water vapor required for saturation at that temperature (and pressure).* Thus, relative humidity indicates how near the air is to saturation rather than the actual quantity of water vapor in the air (Box 4-1).

To illustrate, we see from Table 4-1 that at 25°C, air is saturated when it contains 20 grams of water vapor per kilogram of air. Thus, if the air contains 10 grams per kilogram on a 25°C day, the relative humidity is expressed as 10/20, or 50 percent. Further, if air with a temperature of 25°C had a water-vapor content of 20 grams per kilogram, the relative humidity would be expressed as 20/20 or 100 percent. On occasions when the relative humidity reaches 100 percent, the air is said to be *saturated.*

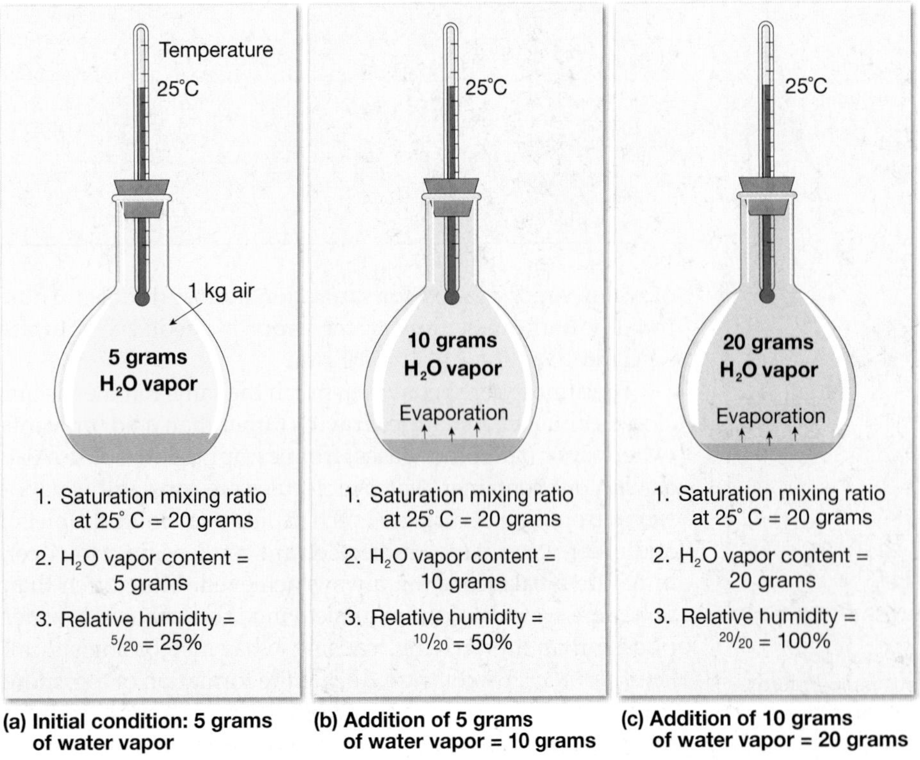

How Relative Humidity Changes

Because relative humidity is based on the air's water-vapor content, as well as the amount of moisture required for saturation, it can change in one of two ways. First, relative humidity changes when water vapor is added to or removed from the atmosphere. Second, because the amount of moisture required for saturation is a function of air temperature, relative humidity varies with temperature.

Adding or Subtracting Moisture Notice in Figure 4-12 that when water vapor is added to a parcel of air through

Figure 4-12 At a constant temperature (in this example it is 25°C), the relative humidity will increase as water vapor is added to the air. The saturation mixing ratio for air at 25°C is 20 g/kg (see Table 4-1). As the water-vapor content in the flask increases, the relative humidity rises from 25 percent in (a) to 100 percent in (c).

Temperature
25°C

25°C

25°C

1 kg air

**5 grams
H₂O vapor**

**10 grams
H₂O vapor**

Evaporation

**20 grams
H₂O vapor**

Evaporation

1. Saturation mixing ratio at 25° C = 20 grams

2. H₂O vapor content = 5 grams

3. Relative humidity = 5/20 = 25%

1. Saturation mixing ratio at 25° C = 20 grams

2. H₂O vapor content = 10 grams

3. Relative humidity = 10/20 = 50%

1. Saturation mixing ratio at 25° C = 20 grams

2. H₂O vapor content = 20 grams

3. Relative humidity = 20/20 = 100%

(a) Initial condition: 5 grams of water vapor

(b) Addition of 5 grams of water vapor = 10 grams

(c) Addition of 10 grams of water vapor = 20 grams

Box 4-1 Dry Air at 100 Percent Relative Humidity?

A common misconception relating to meteorology is the notion that air with a high relative humidity must have a greater water-vapor content than air with a lower relative humidity. This is not always the case (**Figure 4-A**). To illustrate, let us compare a typical January day at International Falls, Minnesota, to one in the desert near Phoenix, Arizona. On this hypothetical day the temperature in International Falls is a cold −10°C (14°F) and the relative humidity is 100 percent. By referring to Table 4-1, we can see that saturated −10°C (14°F) air has a water-vapor content (mixing ratio) of 2 grams per kilogram (g/kg). By contrast, the desert air at Phoenix on this January day is a warm 25°C (77°F), and the relative humidity is just 20 percent. A look at Table 4-1 reveals that 25°C (77°F) air has a saturation mixing ratio of 20 g/kg. Therefore, with a relative humidity of 20 percent, the air at Phoenix has a water-vapor content of 4 g/kg (20 grams × 20 percent). Consequently, the "dry" air at Phoenix actually contains twice the water vapor as the air at International Falls, with a relative humidity of 100 percent.

This example illustrates that places that are very cold are also very dry. The low water-vapor content of frigid air (even when saturated) helps to explain why many arctic areas receive only meager amounts of precipitation and are referred to as "polar deserts." This also helps us understand why people frequently experience dry skin and chapped lips during the winter months. The water-vapor content of cold air is low, even compared to some hot, arid regions.

FIGURE 4-A *Moisture content of hot air versus frigid air. Hot desert air with a low relative humidity generally has a higher water-vapor content than frigid air with a high relative humidity. (Top photo by Matt Duvall; bottom photo by E. J. Tarbuck)*

evaporation, the relative humidity of the air increases until saturation occurs (100 percent relative humidity). What if even more moisture is added to this parcel of saturated air? Does the relative humidity exceed 100 percent? Normally, this situation does not occur. Instead, the excess water vapor condenses to form liquid water.

You may have experienced such a situation while taking a hot shower. The water leaving the shower is composed of very energetic (hot) molecules, which means that the rate of evaporation is high. As long as you run the shower, the process of evaporation continually adds water vapor to the unsaturated air in the bathroom. If you stay in a hot shower long enough, the air eventually becomes saturated, and the excess water vapor begins to condense on the mirror, window, tile, and other surfaces in the room.

In nature, moisture is added to the air mainly via evaporation from the oceans. However, plants, soil, and smaller bodies of water also make substantial contributions. Unlike with your shower, however, the natural processes that add water vapor to the air generally do not operate at rates fast enough to cause saturation to occur directly. One exception is what happens when you exhale on a cold winter day and "see

your breath"—the warm, moist air from your lungs mixes with the cold outside air. Your breath has enough moisture to saturate a small quantity of cold outside air, which results in a miniature "cloud." Almost as fast as the "cloud" forms, it mixes with the surrounding dry air and evaporates.

Changes with Temperature The second condition that affects relative humidity is air temperature (Box 4-2). Examine Figure 4-13a carefully and note that when air at 20°C contains 7 grams of water vapor per kilogram, it has a relative humidity of 50 percent. When the flask in Figure 4-13a is cooled from 20° to 10°C, as shown in Figure 4-13b, the relative humidity increases from 50 to 100 percent. We can conclude that when the water-vapor content remains constant, *a decrease in temperature results in an increase in relative humidity.*

But there is no reason to assume that cooling would cease the moment the air reached saturation. What happens when the air is cooled below the temperature at which saturation occurs? Figure 4-13c illustrates this situation. Notice from Table 4-1 that when the flask is cooled to 0°C, the air is saturated, at 3.5 grams of water vapor per kilogram of

Box 4-2   Humidifiers and Dehumidifiers

In summer, stores sell *dehumidifiers*. As winter rolls around, these same retailers feature *humidifiers*. Why do you suppose so many homes are equipped with both appliances? The answer lies in the relationship between temperature and relative humidity. Recall that if the water-vapor content of air remains at a constant level, an increase in temperature lowers the relative humidity and a lowering of temperature increases the relative humidity.

During the summer months, warm, moist air frequently dominates the weather of the central

and eastern United States. When hot and humid air enters a home, some of it circulates into the cool basement. As a result, the temperature of this air drops, and its relative humidity increases. The result is a damp, musty-smelling basement. The homeowner may install a dehumidifier to alleviate the problem. As air is drawn over the cold coils of the dehumidifier, water vapor condenses and collects in a bucket or flows down a drain. This process reduces the relative humidity and makes for a drier, more comfortable basement.

By contrast, during the winter months, outside air is cool and dry. When this air is drawn into the home, it is heated to room temperature. This process in turn causes the relative humidity to plunge, often to uncomfortable levels of 25 percent or lower. Living with dry air can mean static electrical shocks, dry skin, sinus headaches, and even nosebleeds. Consequently, the homeowner may install a humidifier, which adds water to the air and increases the relative humidity to a more comfortable level.

air. Because this flask originally contained 7 grams of water vapor, 3.5 grams of water vapor will condense to form liquid droplets that collect on the walls of the container. In the meantime, the relative humidity of the air inside remains at 100 percent. This raises an important concept. When air aloft is cooled below its saturation level, some of the water vapor condenses to form clouds. Since clouds are made of liquid droplets, this moisture is no longer part of the water-vapor content of the air.

Review Figure 4-13 and see what would happen if the flask in Figure 4-13a were heated to 35°C. From Table 4-1 we see that at 35°C, saturation occurs at 35 grams of water vapor per kilogram of air. Consequently, by heating the air from 20° to 35°C, the relative humidity would drop from 7/14 (50 percent) to 7/35 (20 percent).

We can summarize the effects of temperature on relative humidity as follows: *When the water-vapor content of air remains constant, a decrease in air temperature results in an increase in relative humidity, and, conversely, an increase in temperature causes a decrease in relative humidity.*

Natural Changes in Relative Humidity

In nature there are three major ways that air temperatures change (over relatively short time spans) to cause corresponding changes in relative humidity:

1. Daily changes in temperatures (daylight versus nighttime temperatures)

2. Temperature changes that result as air moves horizontally from one location to another

3. Temperature changes caused as air moves vertically in the atmosphere

The effect of the first of these three processes (daily changes) is shown in Figure 4-14. Notice that during the midafternoon, relative humidity reaches

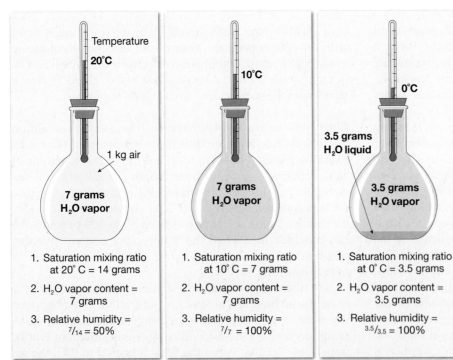

(a) Initial condition: 20°C   (b) Cooled to 10°C   (c) Cooled to 0°C

Figure 4-13 Relative humidity varies with temperature. When the water-vapor content (mixing ratio) remains constant, the relative humidity can be changed by increasing or decreasing the air temperature. In this example, when the temperature of the air in the flask was lowered from 20°C in (a) to 10°C in (b), the relative humidity increased from 50 to 100 percent. Further cooling from 10°C in (b) to 0°C in (c) causes one-half of the water vapor to condense. In nature, when saturated air cools, it causes condensation in the form of clouds, dew, or fog.

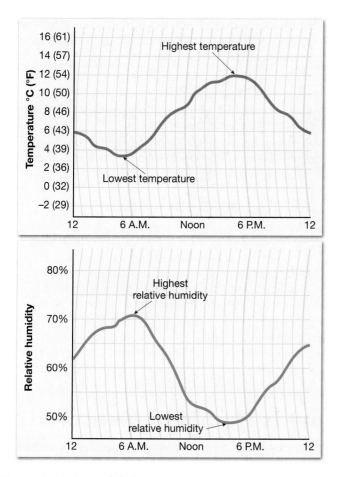

Figure 4-14 Typical daily variation in temperature and relative humidity during a spring day at Washington, D.C.

Dew-Point Temperature

 Moisture and Cloud Formation
▶ Humidity: Water Vapor in the Air

Another important measure of humidity is dew-point temperature. The **dew-point temperature,** or simply the **dew point,** is the temperature to which air needs to be cooled to reach saturation. For example, in Figure 4-13 the unsaturated air in the flask had to be cooled to 10°C before saturation occurred. Therefore, 10°C is the dew-point temperature for this air. In nature, cooling below the dew point causes water vapor to condense, typically as dew, fog, or clouds (Figure 4-15). The term *dew point* stems from the fact that during nighttime hours, objects near the ground often cool below the dew-point temperature and become coated with dew.

Unlike relative humidity, which is a measure of how near the air is to being saturated, dew-point temperature is a measure of the *actual moisture content* of a parcel of air. Recall that

its lowest level, whereas the cooler evening hours are associated with higher relative humidity. In this example, the actual water-vapor content (mixing ratio) of the air remains unchanged; only the relative humidity varies.

In summary, relative humidity indicates how near the air is to being saturated, whereas the air's mixing ratio denotes the actual quantity of water vapor contained in that air.

Concept Check 4.6

1 How is relative humidity different from absolute humidity and the mixing ratio?

2 Refer to Figure 4-14 and then answer the following questions:
  a When is relative humidity highest during a typical day? When is it lowest?
  b At what time of day would dew most likely form?
  c Write a generalization relating changes in air temperature to changes in relative humidity.

3 If the temperature remains unchanged, and if the mixing ratio decreases, how will relative humidity change?

4 List three ways relative humidity changes in nature.

Figure 4-15 Condensation, or "dew," occurs when a cold drinking glass chills the surrounding layer of air below the dew-point temperature. (Photo by Elena Elisseeva/Shutterstock)

TABLE 4-2 Dew-Point Thresholds

Dew-Point Temperature	Threshold
≤ 10°F	Significant snowfall is inhibited
≥ 55°F	Minimum for severe thunderstorms to form
≥ 65°F	Considered humid by most people
≥ 70°F	Typical of the rainy tropics
≥ 75°F	Considered oppressive by most

the saturation vapor pressure is temperature dependent and that for every 10°C (18°F) increase in temperature, the amount of water vapor needed for saturation doubles. Therefore, cold *saturated air* (0°C [32°F]) contains about half the water vapor of *saturated air* having a temperature of 10°C (50°F) and roughly one-fourth that of *saturated air* with a temperature of 20°C (68°F). Because the dew point is the temperature at which saturation occurs, we can conclude that high dew-point temperatures equate to moist air and, conversely, low dew-point temperatures indicate dry air (Table 4-2). More precisely, based on what we have learned about vapor pressure and saturation, we can state that for every 10°C (18°F) increase in the dew-point temperature, air contains about twice as much water vapor. Therefore, we know that when the dew-point temperature is 25°C (77°F), air contains about twice the water vapor as when the dew point is 15°C (59°F) and four times that of air with a dew point of 5°C (41°F).

Because the dew-point temperature is a good measure of the amount of water vapor in the air, it commonly appears on weather maps. When the dew point exceeds 65°F (18°C), most people consider the air to feel humid; air with a dew point of 75°F (24°C) or higher is considered oppressive. Notice on the map in Figure 4-16 that much of the area near the warm Gulf of Mexico has dew-point temperatures that exceed 70°F (21°C). Also notice in Figure 4-16 that although

the Southeast is dominated by humid conditions (dew points above 65°F), most of the remainder of the country is experiencing comparatively drier air.

Concept Check 4.7

1 Define *dew-point temperature*.

2 Which measure of humidity, relative humidity or dew point, best describes the actual quantity of water vapor in a mass of air?

Students Sometimes Ask...

Is frost just frozen dew?

Contrary to popular belief, frost is not frozen dew. Rather, white frost (*hoar frost*) forms when saturation occurs at a temperature of 0°C (32°F) or below (a temperature called the *frost point*). Thus, frost forms when water vapor changes directly from a gas into a solid (ice), without entering the liquid state. This process, called *deposition*, produces delicate patterns of ice crystals that often decorate windows during winter.

How Is Humidity Measured?

Instruments called **hygrometers** are used to measure the moisture content of the air. One of the simplest hygrometers, a *psychrometer*, consists of two identical thermometers mounted side by side (Figure 4-17a). One thermometer is dry and measures air temperature, and the other, called the *wet bulb*, has a thin cloth wick tied at the bottom. To use the psychrometer, the cloth wick is saturated with water, and a continuous current of air is passed over the wick, either by swinging the instrument or by using an electric fan to move air past the instruments (Figure 4-17b,c). As a result, water evaporates from the wick, absorbing heat energy from the wet-bulb thermometer, which causes its temperature to drop. The amount of cooling that takes place is directly proportional to the dryness of the air. The drier the air, the greater the cooling. Therefore, the larger the difference between the wet- and dry-bulb temperatures, the lower the relative humidity. By contrast, if the air is saturated, no evaporation will occur, and the two thermometers will have identical readings. By using a psychrometer and the tables provided in Appendix C, the relative humidity and the dew-point temperature can be easily determined.

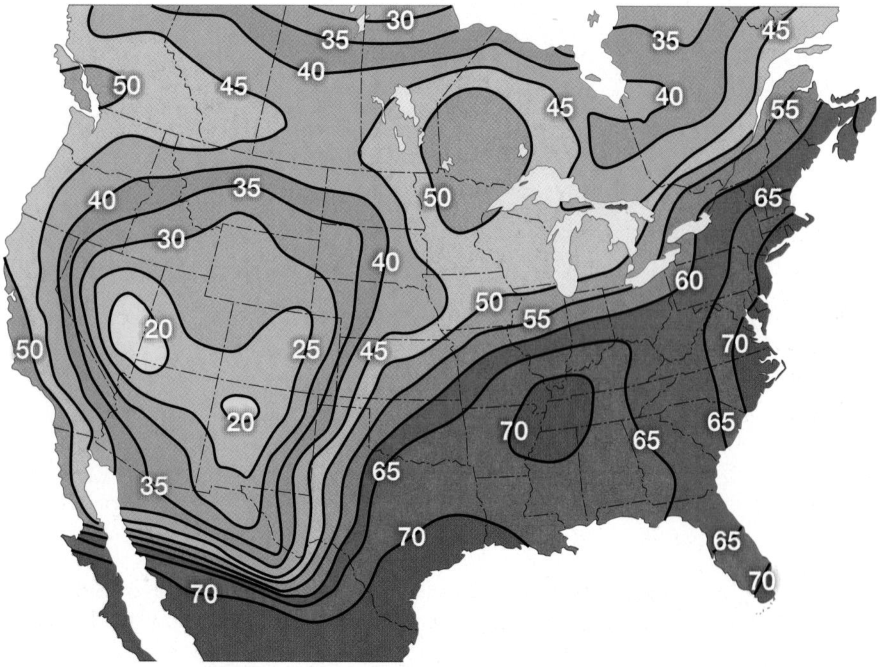

Figure 4-16 Surface map showing dew-point temperatures for September 15, 2005. Dew-point temperatures above 60°F dominate the southeastern United States, indicating that this region is blanketed with humid air.

(a)

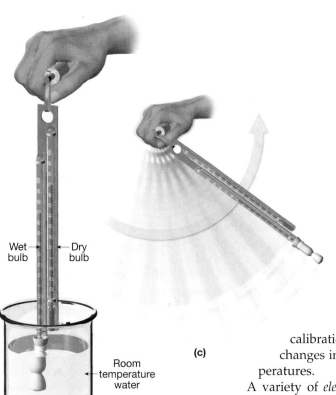

Wet — → ← — Dry
bulb   bulb

Room
temperature
water

(b)

(c)

Figure 4-17 Sling psychrometer.
(a) Sling psychrometers use dry-bulb and wet-bulb thermometers to calculate both relative humidity and dew point. (b) The dry-bulb thermometer measures the current air temperature. The wet-bulb thermometer is covered with a cloth wick dipped in water. (c) As the instrument is spun, evaporation cooling causes the temperature of the wet-bulb thermometer to decrease. The amount of cooling that occurs is directly proportional to the dryness of the air. The temperature difference between the dry- and wet-bulb thermometers is used in conjunction with the tables in Appendix C to determine relative humidity and dew-point temperature. (Photo by E. J. Tarbuck)

calibration. It is also slow in responding to changes in humidity, especially at low temperatures.

A variety of *electric hygrometers* are also used to measure humidity. One relatively accurate type of electric hygrometer uses a chilled mirror and a mechanism that detects condensation on the mirror. Thus, a *chilled mirror hygrometer* accurately measures the dew-point temperature of the air. From the dew point and air temperature, the relative humidity can be easily calculated.

In addition to being used in meteorology, hygrometers are used in greenhouses, humidors, museums, and numerous industrial settings that are sensitive to humidity, such as places where paint is applied to vehicles.

Concept Check 4.8

1 Describe the principle of a psychrometer, the tool used to measure relative humidity.

2 What does a chilled mirror hygrometer measure?

Adiabatic Temperature Changes

 Moisture and Cloud Formation
▶ The Basics of Cloud Formation: Adiabatic Cooling

Recall that condensation occurs when sufficient water vapor is added to the air or, more commonly, when the air is cooled to its dew-point temperature. Condensation may produce dew, fog, or clouds. Heat near Earth's surface is readily exchanged between the ground and the air directly above. As the ground loses heat in the evening (radiation cooling), dew may condense on the grass, while fog may form slightly above Earth's surface. Thus, surface cooling

Students Sometimes Ask...

What are the most humid cities in the United States?

As you might expect, the most humid cities in the United States are located near the ocean, in regions that experience frequent onshore breezes. The record belongs to Quillayute, Washington, with an average relative humidity of 83 percent. However, many coastal cities in Oregon, Texas, Louisiana, and Florida also have average relative humidities that exceed 75 percent. Coastal cities in the Northeast tend to be somewhat less humid because they often experience air masses that originate over the drier continental interior.

Another instrument used for measuring relative humidity, the *hair hygrometer,* can be read directly without using tables. The hair hygrometer operates on the principle that hair changes length in proportion to changes in relative humidity. Hair lengthens as relative humidity increases and shrinks as relative humidity drops. People with naturally curly hair experience this phenomenon, for in humid weather their hair lengthens and hence becomes curlier. The hair hygrometer uses a bundle of hairs linked mechanically to an indicator that is calibrated between 0 and 100 percent. Thus, we need only glance at the dial to read directly the relative humidity. As you might expect, the hair hygrometer is less accurate than a psychrometer and requires frequent

that occurs after sunset produces some condensation. Cloud formation, however, often takes place during the warmest part of the day—an indication that another mechanism must operate aloft that cools air sufficiently to generate clouds.

The process that generates most clouds is easily visualized. Have you ever pumped up a bicycle tire with a hand pump and noticed that the pump barrel became very warm? When you applied energy to *compress* the air, the motion of the gas molecules increased, and the temperature of the air rose. Conversely, if you allow air to escape from a bicycle tire, the air *expands*; the gas molecules move less rapidly, and the air cools. You have probably felt the cooling effect of the propellant gas expanding as you applied hair spray or spray deodorant. The temperature changes just described, in which heat was neither added nor subtracted, are called **adiabatic temperature changes.**

In summary, when air is compressed, it warms and when air is allowed to expand, it cools.

Adiabatic Cooling and Condensation

To simplify the discussion of adiabatic cooling, imagine a volume of air enclosed in a thin balloon-like bubble. Meteorologists call this imaginary volume of air a **parcel.** Typically, we consider a parcel to be a few hundred cubic meters in volume, and we assume that it acts independently of the surrounding air. It is also assumed that no heat is transferred into, or out of, the parcel. Although highly idealized, over short time spans, a parcel of air behaves much like an actual volume of air moving up or down in the atmosphere. In nature, sometimes the surrounding air infiltrates a rising or descending column of air, a process called **entrainment.**

For the following discussion, however, we assume that no mixing of this type occurs.

Any time a parcel of air moves upward, it passes through regions of successively lower pressure. As a result, ascending air expands and cools adiabatically. Unsaturated air cools at a constant rate of 10°C for every 1000 meters of ascent (5.5°F per 1000 feet). Conversely, descending air comes under increasing pressure and is compressed and heated 10°C for every 1000 meters of descent (Figure 4-18). This rate of cooling or healing applies only to *unsaturated air* and is known as the **dry adiabatic rate** ("dry" because the air is unsaturated).

If an air parcel rises high enough, it will eventually cool to its dew point and trigger the process of condensation. The altitude at which a parcel reaches saturation and cloud formation begins is called the **lifting condensation level.** At the lifting condensation level an important change occurs: The *latent heat* that was absorbed by the water vapor when it evaporated is released as *sensible heat*—heat we can measure with a thermometer. Although the parcel will continue to cool adiabatically, the release of latent heat slows the rate of cooling. In other words, when a parcel of air ascends above the lifting condensation level, the rate at which it cools is reduced. This slower rate of cooling is called the **wet adiabatic rate** ("wet" because the air is saturated).

Because the amount of latent heat released depends on the quantity of moisture present in the air (generally between 0 and 4 percent), the wet adiabatic rate varies from 5°C per 1000 meters for air with a high moisture content to 9°C per 1000 meters for air with a low moisture content. Figure 4-19 illustrates the role of adiabatic cooling in the formation of clouds.

To summarize, rising air cools at the faster dry adiabatic rate from the surface up to the lifting condensation level, at which point the slower wet adiabatic rate commences.

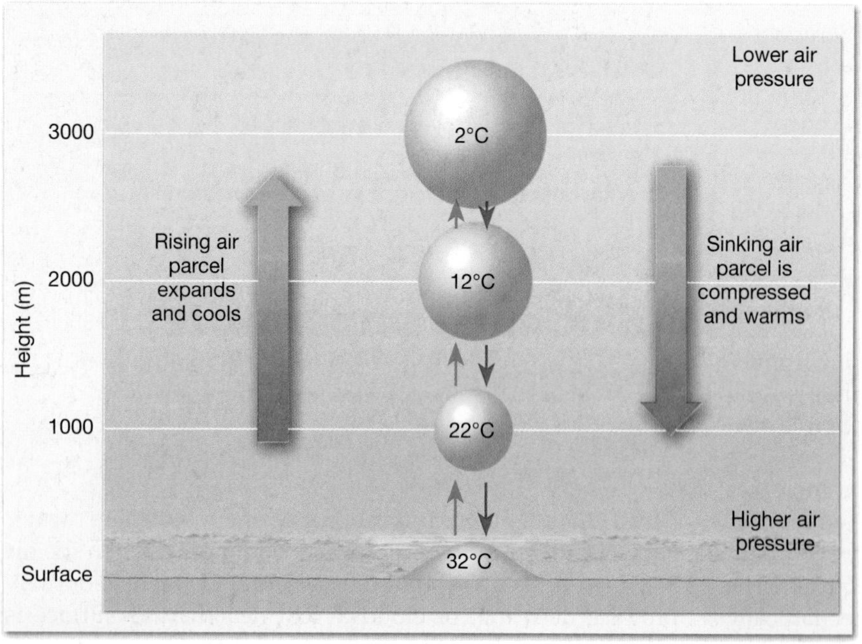

Figure 4-18 Whenever an unsaturated parcel of air is lifted, it expands and cools at the *dry adiabatic rate* of 10°C per 1000 meters. Conversely, when air sinks, it is compressed and heats at the same rate.

Concept Check 4.9

1 What name is given to the processes whereby the temperature of air changes without the addition or subtraction of heat?

2 Why does air expand as it moves upward through the atmosphere?

3 At what rate does unsaturated air cool when it rises through the atmosphere?

4 Why does the adiabatic rate of cooling change when condensation begins?

5 Why is the wet adiabatic rate not a constant figure?

6 The contents of an aerosol can are under very high pressure. When you push the nozzle on such a can, the spray feels cold. Explain.

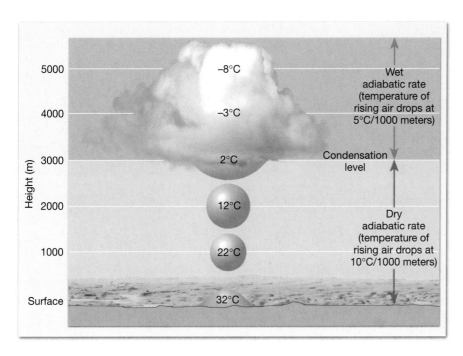

Figure 4-19 Rising air expands and cools at the dry adiabatic rate of 10°C per 1000 meters until the air reaches the dew point and condensation (cloud formation) begins. As air continues to rise, the latent heat released by condensation reduces the rate of cooling. The wet adiabatic rate is therefore always less than the dry adiabatic rate.

Four mechanisms cause air to rise:

1. *Orographic lifting,* in which air is forced to rise over a mountainous barrier
2. *Frontal wedging,* in which warmer, less dense air is forced over cooler, denser air
3. *Convergence,* which is a pileup of horizontal air flow that results in upward movement
4. *Localized convective lifting,* in which unequal surface heating causes localized pockets of air to rise because of their buoyancy

In later chapters, we will consider other mechanisms that cause air to rise.

Processes That Lift Air

 Moisture and Cloud Formation
▶ Processes That Lift Air

Why does air rise on some occasions but not on others? Generally, the tendency is for air to resist vertical movement. Therefore, air located near the surface "wants" to stay near the surface, and air aloft tends to remain aloft. Because clouds are a common occurrence, we are left to conclude that some processes must be at work that cause the air to rise.

Orographic Lifting

Orographic lifting occurs when elevated terrains, such as mountains, act as barriers to the flow of air (Figure 4-20). As air ascends a mountain slope, adiabatic cooling often generates clouds and copious precipitation. In fact, many of the rainiest places in the world are located on windward mountain slopes (Box 4-3).

By the time air reaches the leeward side of a mountain, much of its moisture has been lost. If the air descends, it warms adiabatically, making condensation and precipitation even less likely. As shown in Figure 4-20, the result can be a **rain shadow desert** (Box 4-4). The Great Basin Desert of the western United States lies only a few hundred kilometers from the Pacific Ocean, but it is effectively cut off from the ocean's moisture by the imposing Sierra Nevada (Figure 4-20). The Gobi Desert

Figure 4-20 Orographic lifting leads to precipitation on windward slopes. By the time air reaches the leeward side of the mountains, much of the moisture has been lost. The Great Basin desert is a rain shadow desert that covers nearly all of Nevada and portions of adjacent states. (Photo on left by Dean Pennala/Shutterstock, photo on right by Dennis Tasa)

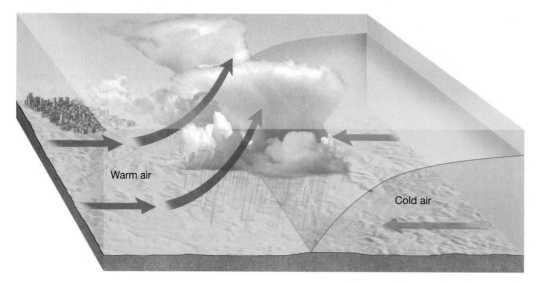

Figure 4-21 Frontal wedging. Colder, denser air acts as a barrier over which warmer, less dense air rises.

of Mongolia, the Takla Makan of China, and the Patagonia Desert of Argentina are other examples of deserts that exist because they are on the leeward sides of large mountain systems.

Frontal Wedging

If orographic lifting were the only mechanism that forced air aloft, the relatively flat central portion of North America would be an expansive desert rather than the area known as "the nation's breadbasket." Fortunately, this is not the case.

In central North America, masses of warm and cold air collide, producing **fronts.** Along a front the cooler, denser air acts as a barrier over which the warmer, less dense air rises. This process, called **frontal wedging,** is illustrated in Figure 4-21.

It should be noted that weather-producing fronts are associated with storm systems called *middle-latitude cyclones.* Because these storms are responsible for producing a high percentage of the precipitation in the middle latitudes, we will examine them in detail in Chapter 9.

Convergence

We saw that the collision of contrasting air masses forces air to rise. In a more general sense, whenever air in the lower troposphere flows together, lifting results—a phenomenon called **convergence** (Figure 4-22).

Convergence can also occur when an obstacle slows or restricts horizontal air flow (wind). For example, when air moves from a relatively smooth surface, such as the ocean, onto an irregular landscape, increased friction reduces its speed. The result is a pileup of air (convergence). Imagine what happens when large numbers of people leave a concert or sporting event—pileups occur at the exits. When air converges, there is a net upward flow of air molecules

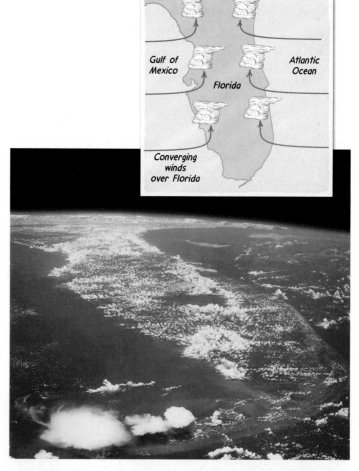

Figure 4-22 Convergence. When surface air converges, the column of air increases in height to allow for the decreased area it occupies. Florida provides a good example. On warm days, airflow from the Atlantic Ocean and Gulf of Mexico onto the Florida peninsula generates many midafternoon thunderstorms. (Photo by NASA/Media Services)

Box 4-3 Precipitation Records and Mountainous Terrain

Many of the rainiest places in the world are located on windward mountain slopes. Typically, these areas are rainy because the mountains act as barriers to Earth's natural circulation. The prevailing winds are forced to ascend the sloping terrain, thereby generating clouds and often abundant precipitation. Mount Waialeale, Hawaii, for example, records the highest average annual rainfall in the world, some 1234 centimeters

(486 inches). The station is located on the windward (northeast) coast of the island of Kauai, at an elevation of 1569 meters (5148 feet). By contrast, only 31 kilometers (19 miles) away lies sunny Barking Sands, with annual precipitation that averages less than 50 centimeters (20 inches).

The largest recorded rainfall for a 12-month period occurred at Cherrapunji, India, where an astounding 2647 centimeters

(1042 inches), over 86 feet, fell from August 1860 through July 1861. Most of this rainfall occurred in the summer, particularly during the month of July, when a record 930 centimeters (366 inches) fell. By comparison, 10 times more rain fell in a *month* at Cherrapunji, India, than falls in Chicago in an average *year*. Cherrapunji's elevation of 1293 meters (4309 feet) lies just north of the Bay of Bengal, making it an ideal location to receive the full effect of India's wet summer monsoon.

Because mountains can be sites of abundant precipitation, they are typically very important sources of water, especially for many dry locations in the western United States. The snow pack, which accumulates high in the mountains during the winter, is a major source of water for the summer season, when precipitation is light and demand for water is great (**Figure 4-B**). The record for greatest annual snowfall in the United States goes to the Mount Baker ski area north of Seattle, Washington, where 2896 centimeters (1140 inches) of snow fell during the winter of 1998–1999.

In addition to providing lift, mountains remove additional moisture in other ways. By slowing the horizontal flow of air, they cause convergence and impede the passage of storm systems. Moreover, the irregular topography of mountains enhances the differential heating that causes some localized convective lifting. These combined effects account for the generally higher precipitation associated with mountainous regions as compared to surrounding lowlands.

FIGURE 4-B This heavy snowpack is at Gotthard Pass in the Swiss Alps. (Photo by agefotostock/SuperStock)

rather than a simple squeezing together of molecules (as happens with people exiting a crowded building).

The Florida peninsula provides an excellent example of the role that convergence can play in initiating cloud development and precipitation. On warm days, the airflow is from the ocean to the land along both coasts of Florida. This leads to a pileup of air along the coasts and general convergence over the peninsula. This pattern of air movement and the uplift that results are aided by intense solar heating of the land. As a result, Florida's peninsula experiences the greatest frequency of midafternoon thunderstorms in the United States (Figure 4-22b).

Convergence as a mechanism of forceful lifting is also a major contributor to the severe weather associated with middle-latitude cyclones and hurricanes. These important weather producers will be covered in more detail later, but for now remember that convergence near the surface results in a general upward flow.

Localized Convective Lifting

On warm summer days, unequal heating of Earth's surface may cause pockets of air to be warmed more than the surrounding air (Figure 4-23). For instance, air above a plowed field will be warmed more than the air above adjacent fields of crops. Consequently, the parcel of air above the field, which is warmer (less dense) than the surrounding air, will be buoyed upward. These rising parcels of warmer air are called *thermals*. Birds such as hawks and eagles use thermals to carry them to great heights, where they can gaze down on unsuspecting prey (Figure 4-24). People have learned to employ these rising parcels to use hang gliders as a way to "fly."

The phenomenon that produces rising thermals is called **localized convective lifting.** When these warm parcels of air rise above the lifting condensation level, clouds form and on occasion produce midafternoon rain

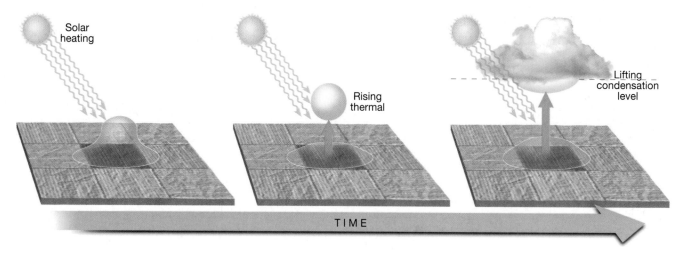

Figure 4-23 Localized convective lifting. Unequal heating of Earth's surface causes pockets of air to be warmed more than the surrounding air. These buoyant parcels of hot air rise, producing thermals, and if they reach the condensation level, clouds form.

EYE ON THE ATMOSPHERE

(Michael Collier)

The Navajo Generating Station, located on the Navajo Indian Reservation near Page, Arizona, has three 236-meter (560-feet) stacks.

Question 1 What fuel is this plant burning to generate electricity? (*Hint:* Look directly behind the facility.)

Question 2 Why do power-generating facilities such as this one have such tall stacks?

Question 3 Can you explain why the "smoke" changes color from bright white to pale yellow when it reaches a height of about 200 feet above the stacks?

showers. The height of clouds produced in this fashion is somewhat limited, because the buoyancy caused solely by unequal surface heating is confined to, at most, the first few kilometers of the atmosphere. Also, the accompanying rains, although occasionally heavy, are of short duration and widely scattered, a phenomenon described as *sun showers.*

Concept Check 4.10

1 How do orographic lifting and frontal wedging act to force air to rise?

2 Explain why the Great Basin of the western United States is dry. What term is applied to such a situation?

3 Explain why Florida has abundant midafternoon thunderstorms.

4 Describe localized convective lifting.

The Critical Weathermaker: Atmospheric Stability

GE**O**De   Moisture and Cloud Formation
ATMOSPHERE   ▶ The Critical Weathermaker: Atmospheric Stability

Why do clouds vary so much in size, and why does the resulting precipitation vary so widely? The answer is closely tied to the *stability* of the air.

Recall that when a parcel is forced to rise, its temperature will decrease because of expansion (adiabatic cooling). By comparing the parcel's temperature to that of the surrounding air, we can determine its stability. If the parcel is *cooler* than the surrounding environment, it will be more dense; and if allowed to do so, it will sink to its original

Figure 4-24 Birds, such as eagles and hawks, glide across the sky using thermals. (Photo by Teekaygee/Dreamstime)

Figure 4-25 As long as air is warmer than its surroundings, it will rise. Hot-air balloons rise up through the atmosphere for this reason. (Photo by Steve Vidler/SuperStock)

position. Air of this type, called **stable air,** resists vertical movement.

If, however, our imaginary rising parcel is *warmer* and hence less dense than the surrounding air, it will continue to rise until it reaches an altitude where its temperature equals that of its surroundings. This type of air is classified as **unstable air.** Unstable air is like a hot-air balloon—it will rise as long as the air in the balloon is sufficiently warmer and less dense than the surrounding air (Figure 4-25).

In summary, stability is a property of air that describes its tendency to remain in its original position (stable) or to rise (unstable).

Types of Stability

The stability of the atmosphere is determined by measuring the air temperature at various heights. This measure, called the *environmental lapse rate,* should not be confused with adiabatic temperature changes. The environmental lapse rate is the *actual temperature* of the atmosphere, as determined from observations made by radiosondes and aircraft. Adiabatic temperature changes, on the other hand, are the *changes in temperature* that a parcel of air experiences as it moves vertically through the atmosphere.

Figure 4-26 illustrates how the stability of the atmosphere is determined. In the example, the air at 1000 meters is 5°C cooler than the air at the surface, whereas the air at 2000 meters is 10°C cooler, and so forth. Thus, the prevailing environmental lapse rate is 5°C per 1000 meters. The air at the surface appears to be less dense than the air at 1000 meters because it is 5°C warmer. However, if the surface air were forced to rise to 1000 meters, it would expand

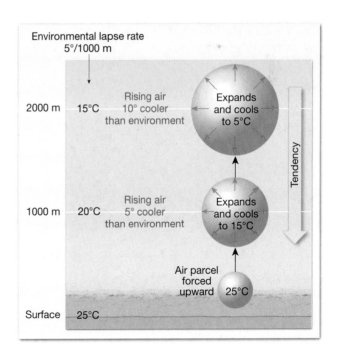

Figure 4-26 In a stable atmosphere, as an unsaturated parcel of air is lifted, it expands and cools at the dry adiabatic rate of 10°C per 1000 meters. Because the temperature of the rising parcel of air is lower than that of the surrounding environment, it will be heavier and, if allowed to do so, will sink to its original position.

and cool at the dry adiabatic rate of 10°C per 1000 meters. Therefore, on reaching 1000 meters, the temperature of the rising parcel would have dropped from 25°C to 15°C. Because the rising air would be 5°C cooler than its environment, it would be more dense and, if allowed to do so, it would sink to its original position. Thus, we say that the air near the surface is *potentially cooler* than the air aloft, and therefore, it will not rise unless forced to do so. (For example, air can be forced to rise over a mountainous terrain.) The air just described is referred to as *stable* and resists vertical movement.

We will now look at three fundamental conditions of the atmosphere: absolute stability, absolute instability, and conditional instability.

Absolute Stability Stated quantitatively, **absolute stability** prevails when *the environmental lapse rate is less than the wet adiabatic rate.* Figure 4-27 depicts this situation by using an environmental lapse rate of 5°C per 1000 meters and a wet adiabatic rate of 6°C per 1000 meters. Note that at 1000 meters the temperature of the rising parcel is 5°C cooler than its environment, which makes it denser. Even if this stable air were forced above the lifting condensation level, it would remain cooler and denser than its environment and would have a tendency to return to the surface.

The most stable conditions occur when the temperature in a layer of air increases with altitude, rather than decreases. When this type of environmental lapse rate occurs, it is called a *temperature inversion.* Many processes, such as radiation cooling at Earth's surface on clear nights, can create temperature inversions. Under these conditions an inversion is created because the surface and the air directly above it cool more rapidly than the air aloft.

Absolute Instability At the other extreme, a layer of air is said to exhibit **absolute instability** when *the environmental lapse rate is greater than the dry adiabatic rate.* As shown in Figure 4-28, the ascending parcel of air is always warmer than its environment and will continue to rise because of its own buoyancy. Absolute instability occurs most often during the warmest months and on clear days, when solar heating is intense. Under these conditions the lowermost layer of the atmosphere is heated to a much higher temperature than the air aloft. This results in a steep environmental lapse rate and a very unstable atmosphere.

Conditional Instability The most common type of atmospheric instability is called **conditional instability.** This situation prevails when *moist air has an environmental lapse rate between the dry and wet adiabatic rates* (between about 5° and 10°C per 1000 meters). Simply stated, the

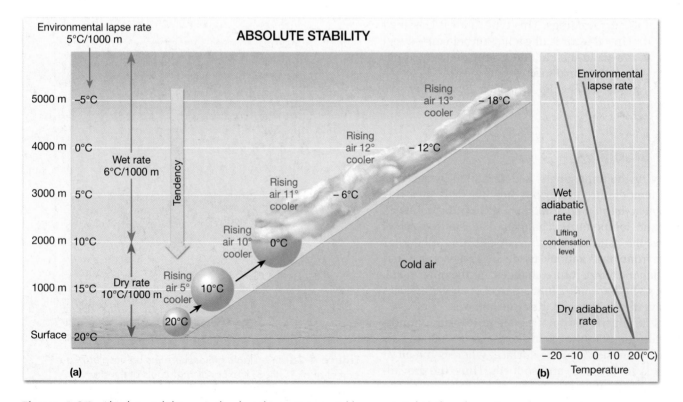

Figure 4-27 *Absolute stability* prevails when the environmental lapse rate is less than the wet adiabatic rate. (a) The rising parcel of air is always cooler and heavier than the surrounding air, producing stability. (b) Graphic representation of the conditions shown in part (a).

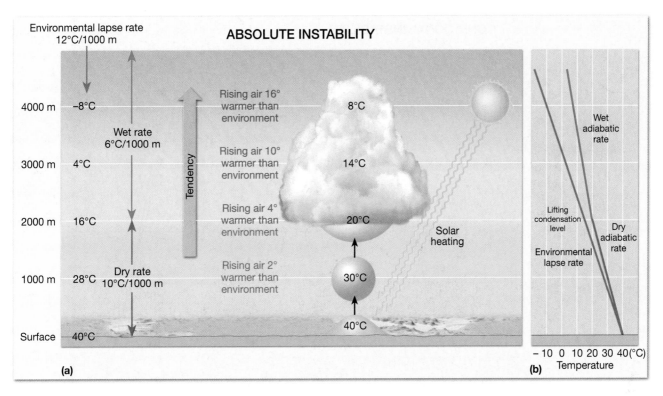

Figure 4-28 Illustration of *absolute instability* that develops when solar heating causes the lowermost layer of the atmosphere to be warmed to a higher temperature than the air aloft. The result is a steep environmental lapse rate that renders the atmosphere unstable. (b) Graphic representation of the conditions shown in part (a).

atmosphere is said to be conditionally unstable when it is *stable* with respect to an *unsaturated* parcel of air but *unstable* with respect to a *saturated* parcel of air. Notice in Figure 4-29 that the rising parcel of air is cooler than the surrounding air for nearly 3000 meters. However, because of the release of latent heat that occurs above the lifting condensation level, the parcel eventually becomes warmer than the surrounding air. From this point along its ascent, the parcel will continue to rise without an outside force. The word *conditional* is used because the air must be forced upward before it reaches the level where it becomes unstable and rises on its own.

In summary, the stability of air is determined by measuring the temperature of the atmosphere at various heights (environmental lapse rate). A column of air is deemed unstable when the air near the bottom of this layer is significantly warmer (less dense) than the air aloft. Conversely, the air is considered to be stable when the temperature decreases gradually with increasing altitude. The most stable conditions occur during a temperature inversion, when the temperature actually increases with height. Under these conditions, there is little vertical air movement.

Stability and Daily Weather

How does stability manifest itself in our daily weather? When stable air is forced aloft, the clouds that form are widespread and have little vertical thickness compared with their

---

Concept Check 4.11

1 How does stable air differ from unstable air?

2 Explain the difference between the environmental lapse rate and adiabatic cooling.

3 How is the stability of air determined?

4 Write a statement relating the environmental lapse rate to stability.

5 Describe conditional instability.

---

horizontal dimension. Precipitation, if any, is light to moderate. In contrast, clouds associated with unstable air are towering and are usually accompanied by heavy precipitation. Thus, we can conclude that on a dreary and overcast day with light drizzle, stable air is forced aloft. Conversely, on a day when towering clouds are forming, we can be relatively certain that the atmosphere is unstable (Figure 4-30).

As noted earlier, the most stable conditions occur during a temperature inversion, when temperature increases with height. In this situation, the air near the surface is cooler and heavier than the air aloft, and therefore little vertical mixing occurs between the layers. Because pollutants are generally added to the air from below, a temperature inversion confines them to the lowermost layer, where their concentration will continue to increase until the temperature inversion

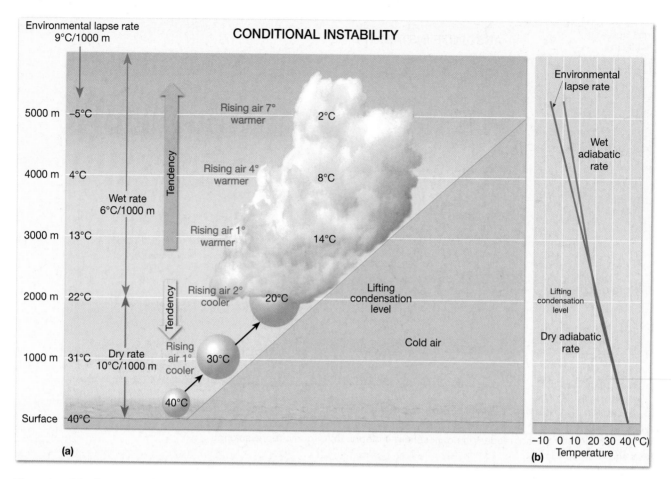

(a)

Figure 4-29 Illustration of *conditional instability,* where warm air is forced to rise along a frontal boundary. Note that the environmental lapse rate of 9°C per 1000 meters lies between the dry and wet adiabatic rates. (a) The parcel of air is cooler than the surrounding air up to nearly 3000 meters, where its tendency is to sink toward the surface (stable). Above this level, however, the parcel is warmer than its environment and will rise because of its own buoyancy (unstable). Thus, when conditionally unstable air is forced to rise, the result can be towering cumulus clouds. (b) Graphic representation of the conditions shown in part (a).

Figure 4-30 Towering clouds provide evidence of unstable conditions in the atmosphere. (Photo by Pedro2009/ Dreamstime)

dissipates. Widespread fog is another sign of stability. Fog generally forms when there is a lack of mixing between the air near the surface and the air aloft.

How Stability Changes

Any factor that causes air near the surface to become warmed in relation to the air aloft increases instability. By contrast, any factor that causes the surface air to be chilled results in the air becoming more stable.

Instability is enhanced by the following:

1. Intense solar heating warming the lowermost layer of the atmosphere
2. The heating of an air mass from below as it passes over a warm surface
3. General upward movement of air caused by processes such as orographic lifting, frontal wedging, and convergence
4. Radiation cooling from cloud tops

Stability is enhanced by the following:

1. Radiation cooling of Earth's surface after sunset
2. The cooling of an air mass from below as it traverses a cold surface
3. General subsidence within an air column

Note that most processes that alter stability result from temperature changes caused by horizontal or vertical air movement, although daily temperature changes are important too. In general, any factor that increases the environmental lapse rate renders the air more unstable, whereas any factor that reduces the environmental lapse rate increases the air's stability.

Temperature Changes and Stability

On clear days when there is abundant surface heating, the lower atmosphere often becomes warmed sufficiently to cause parcels of air to rise. After the Sun sets, surface cooling generally renders the air stable again.

Similar changes in stability occur as air moves horizontally over a surface having markedly different temperatures. In the winter, warm air from the Gulf of Mexico moves northward over the cold, snow-covered Midwest. Because the air is cooled from below, it becomes more stable, often producing widespread fog.

The opposite occurs in the winter, when polar air moves southward over the open waters of the Great Lakes. Although we would die of hypothermia in a few minutes if we fell into these cold waters (perhaps 5°C [41°F]), these waters are warm compared to a polar air mass. Because the water of the Great Lakes is comparatively warm (high vapor pressure) and because the polar air is dry (low vapor pressure), the rate of evaporation will be high. The moisture and heat added to the frigid polar air from the water below render the polar air sufficiently unstable to produce heavy snow-

EYE ON THE ATMOSPHERE

(NASA)

When Earth is viewed from space, the most striking feature of the planet is *water*. It is found as a liquid in the global oceans, as a solid in the polar ice caps, and as clouds and water vapor in the atmosphere. Although only one-thousandth of 1 percent of the water on Earth exists as water vapor, it has a huge influence on our planet's weather and climate.

Question 1 What role does water vapor play in heating Earth's surface?

Question 2 How does water vapor act to transfer heat from Earth's land–sea surface to the atmosphere?

falls on the downwind shores of the Great Lakes—called "lake-effect snows" (see Chapter 8).

Radiation Cooling from Clouds   On a smaller scale, the loss of heat by radiation from cloud tops during evening hours adds to their instability and growth. Unlike air, which is a poor radiator of heat, cloud droplets emit considerable energy to space. Towering clouds that owe their growth to surface heating lose that source of energy at sunset. After sunset, however, radiation cooling at their tops steepens the lapse rate near the tops of these clouds and can lead to additional upward flow of warmer air from below. This process is responsible for producing nocturnal thunderstorms from clouds whose growth temporarily ceased at sunset.

Vertical Air Movement and Stability

Vertical movements of air influence stability. When there is a general downward airflow, called **subsidence,** the upper portion of the subsiding layer is heated by compression,

Box 4-4   Orographic Effects: Windward Precipitation and Leeward Rain Shadows

Orographic lifting is a significant factor in the development of windward precipitation and leeward rain shadows. A simplified hypothetical situation, illustrated in **Figure 4-C**, shows prevailing winds forcing warm moist air over a 3000-meter-high mountain range. As the unsaturated air ascends the windward side of the range, it cools at a rate of 10°C per 1000 meters (dry adiabatic rate) until it reaches the dew-point temperature of 20°C. Because the dew-point temperature is reached at 1000 meters, we can say that this height represents the lifting condensation level and the height of the cloud base. Notice in that above the lifting condensation level, latent heat is released, which results in a slower rate of cooling, called the *wet adiabatic rate.*

From the cloud base to the top of the mountain, water vapor within the rising air is condensing to form more and more cloud droplets. As a result, the windward side of the mountain range experiences abundant precipitation.

For simplicity, we will assume that the air that was forced to the top of the mountain is cooler than the surrounding air and hence begins to flow down the leeward slope of the mountain. As the air descends, it is compressed and *heated* at the dry adiabatic rate. Upon reaching the base of the mountain range, the temperature of the descending air has risen to 40°C, or 10°C warmer than the temperature at the base of the mountain on the windward side. The higher temperature on the leeward side is a result of the latent heat that was released during condensation as the air ascended the windward slope of the mountain range.

Two factors account for the rain shadow commonly observed on leeward mountain slopes. First, water is extracted from air in the form of precipitation on the windward side. Second, the air on the leeward side is warmer than the air on the windward side. (Recall that an increase in temperature results in a drop in relative humidity.)

A classic example of windward precipitation and leeward rain shadows is found in western Washington State. As moist Pacific air flows inland over the Olympic and Cascade Mountains, orographic precipitation is abundant (**Figure 4-D**). By contrast, precipitation data for Sequim and Yakima indicate the presence of rain shadows on the leeward sides of these highlands.

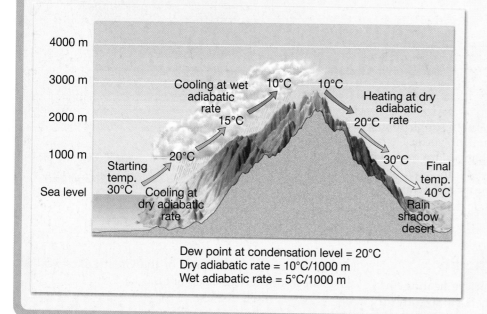

Dew point at condensation level = 20°C
Dry adiabatic rate = 10°C/1000 m
Wet adiabatic rate = 5°C/1000 m

FIGURE 4-C Orographic lifting and the formation of rain shadow deserts.

more so than the lower portion. Usually the air near the surface is not involved in the subsidence, and so its temperature remains unchanged. The net effect is to stabilize the air, because the air aloft is warmed more than the surface air. The warming effect of a few hundred meters of subsidence is enough to evaporate the clouds. Thus, one sign of subsiding air is a deep blue, cloudless sky.

Upward movement of air generally enhances instability, particularly when the lower portion of the rising layer has a higher moisture content than the upper portion, which is usually the situation. As the air moves upward, the lower portion becomes saturated first and cools at the lesser wet adiabatic rate. The net effect is to increase the environmental lapse rate within the rising layer. This process is especially important in producing the instability associated with thunderstorms. In addition, recall that conditionally unstable air can become unstable if lifted sufficiently.

In summary, the role of stability in determining our daily weather cannot be overemphasized. The air's stability, or lack of it, determines to a large degree whether clouds develop and pro-

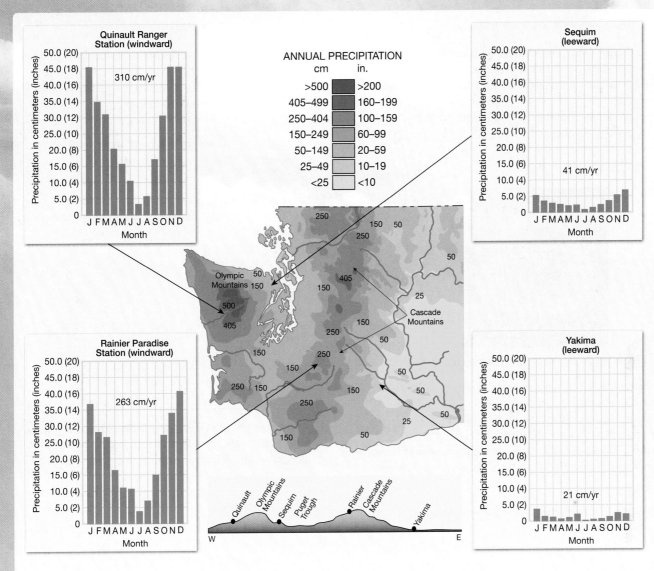

FIGURE 4-D Distribution of precipitation in western Washington State. Data from four stations provide examples of wetter windward locations and drier leeward rain shadows. (After Robert W. Christopherson)

duce precipitation and whether that precipitation will come as a gentle shower or a violent downpour. In general, when stable air is forced aloft, the associated clouds have little vertical thickness, and precipitation, if any, is light. In contrast, clouds associated with unstable air are towering and are frequently accompanied by heavy precipitation.

Concept Check 4.12

1 What weather conditions would lead you to believe that air is unstable?

2 What weather conditions would lead you to believe that air is stable?

3 List four ways instability can be enhanced.

4 List three ways stability can be enhanced.

Give It Some Thought

1. Refer to Figure 4-5 to complete the following.
   a. In which state of matter is water the most dense?
   b. In which state of matter are water molecules most energetic?
   c. In which state of matter is water compressible?

2. The accompanying photo shows a cup of hot coffee. What state of matter is the "steam" rising from the liquid? Explain your answer.

(Photo by Dmitry Kolmakov/Shutterstock)

3. The primary mechanism by which the human body cools itself is perspiration.
   a. Explain how perspiring cools the skin.
   b. Referring to the data for Phoenix, Arizona, and Tampa, Florida (Table A), in which city would it be easier to stay cool by perspiring? Explain your choice.

Table A

City	Temperature	Dew point temperature
Phoenix, AZ	101°F	47°F
Tampa, FL	101°F	77°F

4. As shown in the accompanying photo, during hot summer weather, many people put "koozies" around their beverages to keep their drinks cold. Describe at least two ways the koozies help keep beverages cold.

(Photo by JorgusPhotos/Fotolia.com)

5. Refer to Table 4-1 to answer this question. How much more water is contained in saturated air at a tropical location with a temperature of 40°C compared to a polar location with a temperature of –10°C?

6. Refer to the data for Phoenix, Arizona, and Bismarck, North Dakota (Table B), to complete the following:
   a. Which city has a higher relative humidity?
   b. Which city has the greatest quantity of water vapor in the air?

   c. In which city is the air closest to its saturation point with respect to water vapor?
   d. In which city does the air have the greatest holding capacity for water vapor?

Table B

City	Temperature	Dew point temperature
Phoenix, AZ	101°F	47°F
Bismark, ND	39°F	38°F

7. The accompanying graph shows how air temperature and relative humidity change on a typical summer day in the Midwest.

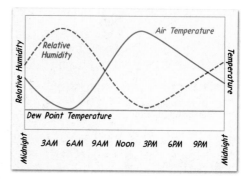

   a. Assuming that the dew-point temperature remained constant, what would be the best time of day to water a lawn to minimize evaporation of the water sprayed on the grass?
   b. Use this graph to explain why dew almost always forms early in the morning.

8. Are the atmospheric conditions illustrated in Figure 4-23 an example of absolute stability, absolute instability, or conditional instability.

9. This chapter examines four processes that cause air to rise. Describe how convective lifting is different from the other three processes.

10. Refer to the accompanying photograph and explain how the relative humidity inside the house compares to the relative humidity outside the house on this particular day.

(Photo by Clynt Garnham Housing/Alamy)

11. Imagine a parcel of air that has a temperature of 40°C at the surface and a dew-point temperature of 20°C at the lifting condensation level. Assume that the environmental lapse rate is 8°C/1000 meters, the dry adiabatic lapse rate is 10°C/1000 meters, and the wet adiabatic lapse rate is 6°C/1000 meters. In Table C, record the environmental temperature, the parcel temperature, and the temperature difference (parcel temperature minus the environmental temperature) and then determine whether the atmosphere is *stable* or *unstable* at each height.

Table C

Height (meters)	Parcel Temperature °C	Environmental Temperature °C	Temperature difference (Parcel–environment)	Stable or unstable
7000				
6000				
5000				
4000				
3000				
2000				
1000				
Surface	40 °C	40 °C	0 °C	Stable

a. What is the height of the lifting condensation level?
b. Does this example describe absolute stability, absolute instability, or conditional instability?
c. Would you forecast thunderstorms under these conditions?

12. Calculate the lifting condensation level (LCL) for the two following examples. (Assume that the dew point does not change until after the LCL is reached.)

	Surface Temperature	Surface Dew Point	LCL
Example A	35°C	20°C	_____
Example B	35°C	14°C	_____

What do these calculations tell you about the relationship between surface dew-point temperature and the height at which clouds develop?

MOISTURE AND ATMOSPHERIC STABILITY IN REVIEW

- The unending circulation of Earth's water supply is called the *hydrologic cycle*. The cycle illustrates the continuous movement of water from the oceans to the atmosphere, from the atmosphere to the land, and from the land back to the sea.
- *Water vapor,* an odorless, colorless gas, can change from one state of matter (solid, liquid, or gas) to another at the temperatures and pressures experienced on Earth.
- The processes involved in changes of state include *evaporation* (liquid to gas), *condensation* (gas to liquid), *melting* (solid to liquid), *freezing* (liquid to solid), *sublimation* (solid to gas), and *deposition* (gas to solid). During each change, *latent* (hidden, or stored) *heat* is either absorbed or released.
- *Humidity* is the general term used to describe the amount of water vapor in the air. The methods used to express humidity quantitatively include (1) *absolute humidity,* the mass of water vapor in a given volume of air; (2) *mixing ratio,* the mass of water vapor in a unit of air compared to the remaining mass of dry air; (3) *vapor pressure,* that part of the total atmospheric pressure attributable to its water-vapor content; (4) *relative humidity,* the ratio of the air's actual water-vapor content compared with the amount of water vapor required for saturation at that temperature; and (5) *dew point,* the temperature at which saturation occurs.
- When air is *saturated,* the pressure exerted by the water vapor, called the *saturation vapor pressure,* produces a balance between the number of water molecules leaving the surface of the

water and the number returning. Because the saturation vapor pressure is temperature dependent, at higher temperatures more water vapor is required for saturation to occur.
- Relative humidity can change in two ways: (1) when the amount of moisture in the air increases or decreases and (2) through temperature change. When the water vapor content of air remains at a constant level, a decrease in air temperature results in an increase in relative humidity, and an increase in temperature causes a decrease in relative humidity.
- The *dew-point temperature* (or simply *dew point*) is the temperature to which a parcel of air would need to be cooled to reach saturation. Unlike relative humidity, dew-point temperature is a measure of the air's actual moisture content. High dew-point temperatures equate to moist air, and low dew-point temperatures indicate dry air, regardless of the air's relative humidity.
- A variety of instruments called *hygrometers* are used to measure relative humidity.
- When air expands, it cools, and when air is compressed, it warms. Temperature changes produced in this manner, in which heat is neither added nor subtracted, are called *adiabatic temperature changes*. The rate of cooling or warming of vertically moving unsaturated ("dry") air is 10°C for every 1000 meters (5.5°F per 1000 feet), the *dry adiabatic rate*. At the *lifting condensation level* latent heat is released, and

the rate of cooling is reduced. The slower rate of cooling, called the *wet adiabatic rate* of cooling ("wet" because the air is saturated) varies from 5°C per 1000 meters to 9°C per 1000 meters.

- When air rises, it expands and cools adiabatically. If air is lifted sufficiently, it will cool to its dew-point temperature, and clouds will develop.
- Four mechanisms that cause air to rise are (1) *orographic lifting,* where air is forced to rise over a mountainous barrier; (2) *frontal wedging,* where warmer, less dense air is forced over cooler, denser air along a *front;* (3) *convergence,* a pile-up of horizontal airflow resulting in an upward flow; and (4) *localized convective lifting,* where unequal surface heating causes localized pockets of air to rise because of their buoyancy.
- *Stable air* resists vertical movement, whereas *unstable air* rises because of its buoyancy. The stability of air is determined

from the *environmental lapse rate,* the temperature of the atmosphere at various heights. The three fundamental conditions of the atmosphere are (1) *absolute stability,* when the environmental lapse rate is less than the wet adiabatic rate; (2) *absolute instability,* when the environmental lapse rate is greater than the dry adiabatic rate; and (3) *conditional instability,* when moist air has an environmental lapse rate between the dry and wet adiabatic rates.

- In general, when stable air is forced aloft, the associated clouds have little vertical thickness, and precipitation, if any, is light. In contrast, clouds associated with unstable air are towering and frequently accompanied by heavy rain.
- Any factor that causes air near the surface to become warmed in relation to the air aloft increases the air's instability. The opposite is also true; any factor that causes the surface air to be chilled results in the air becoming more stable.

VOCABULARY REVIEW

absolute humidity (p. 103)
absolute instability (p. 118)
absolute stability (p. 118)
adiabatic temperature change (p. 112)
calorie (p. 101)
condensation (p. 102)
conditional instability (p. 119)
convergence (p. 114)
deposition (p. 102)
dew point (p. 109)
dew-point temperature (p. 109)
dry adiabatic rate (p. 112)
entrainment (p. 112)

evaporation (p. 101)
front (p. 114)
frontal wedging (p. 114)
humidity (p. 103)
hydrogen bond (p. 100)
hydrologic cycle (p. 98)
hygrometer (p. 110)
latent heat (p. 101)
lifting condensation level (p. 112)
localized convective lifting (p. 115)
mixing ratio (p. 103)
orographic lifting (p. 113)
parcel (p. 112)

rain shadow desert (p. 113)
relative humidity (p. 106)
saturation (p. 104)
saturation vapor pressure (p. 104)
stable air (p. 117)
sublimation (p. 102)
subsidence (p. 122)
transpiration (p. 98)
unstable air (p. 117)
vapor pressure (p. 104)
wet adiabatic rate (p. 112)

PROBLEMS

1. Using Table 4-1, answer the following:
   a. If a parcel of air at 25°C contains 10 grams of water vapor per kilogram of air, what is its relative humidity?
   b. If a parcel of air at 35°C contains 5 grams of water vapor per kilogram of air, what is its relative humidity?
   c. If a parcel of air at 15°C contains 5 grams of water vapor per kilogram of air, what is its relative humidity?
   d. If the temperature of this same parcel of air dropped to 5°C, how would its relative humidity change?
   e. If 20°C air contains 7 grams of water vapor per kilogram of air, what is its dew point?

2. Using the standard tables (Appendix C Tables C-1 and C-2), determine the relative humidity and dew-point temperature if the dry-bulb thermometer reads 22°C and the wet-bulb thermometer reads 16°C. How would the relative humidity and dew point change if the wet-bulb reading were 19°C?

3. If unsaturated air at 20°C were to rise, what would its temperature be at a height of 500 meters? If the dew-point temperature at the lifting condensation level were 11°C, at what elevation would clouds begin to form?

4. Using Figure 4-31, answer the following. (*Hint:* Read Box 4-4.)
   a. What is the elevation of the cloud base?
   b. What is the temperature of the ascending air when it reaches the top of the mountain?
   c. What is the dew-point temperature of the rising air at the top of the mountain? (Assume 100 percent relative humidity.)

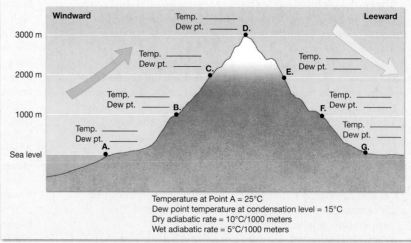

Figure 4-31 Orographic lifting problem.

d. Estimate the amount of water vapor that must have condensed (in grams per kilogram) as the air moved from the cloud base to the top of the mountain.

e. What will the temperature of the air be if it descends to point G? (Assume that the moisture that condensed fell as precipitation on the windward side of the mountain.)

f. What is the approximate capacity of the air to hold water vapor at point G?

g. Assuming that no moisture was added or subtracted from the air as it traveled downslope, estimate the relative humidity at point G.

h. What is the *approximate* relative humidity at point A? (Use the dew-point temperature at the lifting condensation level for the surface dew point.)

i. Give *two* reasons for the difference in relative humidity between points A and G.

j. Needles, California, is situated on the dry leeward side of a mountain range similar to the position of point G. What term describes this situation?

5. If you were to start with a 1-gallon pot of water having a temperature of 10°C and boil it away completely on a stove, it would take a considerable amount of time. Large amounts of energy from the burner would have to be conducted into the pot of water to bring it up to its boiling temperature (100°C), and even more energy would be required to convert it to a gas. If you could take all of the vaporized water that you boiled away (which now occupies part of the air in your kitchen) and instantly condensed it back into the pot, it would release enough energy to blow your house off its foundation. Do the calculation to prove that this statement is true by comparing the amount of energy that would be released to that contained in a stick of dynamite.

Important information:

1 gallon of water = 3785 grams

J = Joule, a unit of energy used in the SI system

4.186 J/g = amount of energy required to raise 1 gram of water 1°C

2260 J/g = energy required to vaporize 1 gram of liquid water when the water's temperature is 100°C

10°C = starting temperature of water

2.1×10^6 J = amount of energy contained in one stick of dynamite

How much energy (measured in sticks of dynamite) is needed to completely boil away 1 gallon of water? This is the same amount of energy that would be released if the water vapor condensed back into the pot

6. Figure 4-11 shows the nonlinear relationship between air temperature and the saturation mixing ratio. This relationship makes it possible for two unsaturated air parcels to mix and form a saturated parcel. For example, consider the following two parcels of air:

	A	**B**
Temperature	10°C	40°C
Relative humidity	75%	85%

a. Use Table 4-1 to find the *saturation mixing ratios* of parcels A and B.

b. What are the actual mixing ratios of Parcels A and B? Let's assume that the two air masses mix together and the resulting temperature is halfway between 10°C and 40°C, and the actual mixing ratio is halfway between the values you found for the two parcels in part(b).

c. What is the temperature of the combined parcel? _____ °C

d. What is the saturation missing ratio of the combined parcel? _____ g/kg

e. What is the actual mixing ratio of the combined parcel? _____ g/kg

f. How much does the actual mixing ratio differ from the saturation mixing ratio? _____ g/kg

g. Because relative humidity normally does not exceed 100 percent, what must happen to the excess water vapor in the combined parcel?

Can you describe an observable situation in which the mixing of two unsaturated parcels of air produces a parcel of saturated air?

MyMeteorologyLab™

Forms of Condensation and Precipitation

Clouds, fog, and the various forms of precipitation are among the most observable weather phenomena. The primary focus of this chapter is to provide a basic understanding of each. In addition to learning how clouds are classified and named, you will learn that the formation of an average raindrop involves complex processes requiring water from roughly a million tiny water droplets.

Cumulonimbus clouds are often associated with thunderstorms and severe weather. (Photo by Cusp/SuperStock)

Focus On Concepts

After completing this chapter, you should be able to:

- Explain the roles of adiabatic cooling and cloud condensation nuclei in cloud formation.

- Classify the 10 basic cloud types based on form and height.

- Contrast nimbostratus and cumulonimbus clouds and their associated weather.

- Name the basic types of fogs and describe how they form.

- Describe the collision–coalescence process and explain how it is different from the Bergeron process.

- Describe the atmospheric conditions that produce sleet, freezing rain (glaze), and hail.

- Summarize the various methods of precipitation measurement.

- Discuss several ways that people attempt to modify the weather.

Cloud Formation

A **cloud** is a visible aggregate of minute water droplets and/or ice crystals that are suspended in the atmosphere above Earth's surface. In addition to being prominent and sometimes spectacular features in the sky, clouds are of continual interest to meteorologists because they provide a visual indication of atmospheric conditions (Figure 5-1).

Condensation Aloft

Clouds form when water vapor condenses in the atmosphere due to *adiabatic cooling*. Recall that when a parcel of air ascends, it passes through regions of successively lower pressure. As a result, rising air expands and cools adiabatically. At a height called the *lifting condensation level*, the ascending parcel has cooled to its dew-point temperature and triggers condensation. In order for condensation to occur, two conditions must be met: The air must be *saturated* and there must be a *surface* on which the water vapor can condense.

During the formation of dew, objects at or near the ground, such as blades of grass, serve as the surface for water vapor condensation. However, when condensation occurs aloft, tiny particles known as **cloud condensation nuclei** serve this purpose. Without these nuclei, a relative humidity well in excess of 100 percent is necessary to produce cloud droplets. (At very low temperatures—low kinetic energies—water molecules will "stick together" in tiny clusters without the presence of condensation nuclei.) Cloud condensation nuclei include microscopic dust, smoke, and salt particles, all of which are profuse in the lower atmosphere. Consequently, in the troposphere the relative humidity seldom exceeds 100 percent.

Growth of Cloud Droplets

The most effective sites for condensation are particles called **hygroscopic (*water-seeking*) nuclei.** Common food items such as crackers and cereals are hygroscopic: when exposed to humid air, they absorb moisture and quickly become stale.

Over the ocean, salt particles are released into the atmosphere when sea spray evaporates. Because salt is hygroscopic, water droplets begin to form around sea salt particles at relative humidities less than 100 percent. As a result, the cloud droplets that form on salt are generally much larger than those that grow on **hydrophobic (*water-repelling*) nuclei.** Although hydrophobic particles are not efficient condensation nuclei, cloud droplets will form on them when the relative humidity reaches 100 percent.

Dust storms, volcanic eruptions, and pollen are major sources of cloud condensation nuclei. In addition, hygroscopic nuclei are introduced into the atmosphere as a by-product of combustion (burning) from such sources as forest fires, automobiles, and coal-burning furnaces. Because cloud condensation nuclei have a wide range of affinities for water, cloud droplets of various sizes often coexist in the same cloud—an important factor for the formation of precipitation.

Figure 5-1 Caught in a downpour. (AP Photo/Keystone, Marcel Bier)

Initially, the growth of cloud droplets occurs rapidly. However, the rate of growth slows as the available water vapor is consumed by the large number of competing droplets. The result is the formation of a cloud consisting of billions of tiny water droplets so minute that they remain suspended in air. Even in very moist air, the growth of cloud droplets by additional condensation is quite slow. Furthermore, the vast size difference between cloud droplets and raindrops (it takes about a million cloud droplets to form a single raindrop) suggests that condensation is not singularly responsible for the formation of drops (or ice crystals) large enough to fall to the ground without evaporating. We will investigate this idea later, when we examine the processes that generate precipitation.

Concept Check 5.1

1 Describe the process of cloud formation.

2 What role do cloud condensation nuclei play in the formation of clouds?

3 Define *hygroscopic nuclei*.

Cloud Classification

 Forms of Condensation and Precipitation
▸ Classifying Clouds

In 1803, Luke Howard, an English naturalist, published a cloud classification that serves as the basis of our present-day system. According to Howard's system, clouds are classified on the basis of two criteria: *form* and *height* (Figure 5-2). Three basic cloud forms or shapes are recognized:

- **Cirrus** clouds are high, white, and thin. They form delicate veil-like patches or wisplike strands and often have a feathery appearance. (*Cirrus* is Latin for "curl" or "filament.")
- **Cumulus** clouds consist of globular cloud masses that are often described as cottonlike in appearance. Normally cumulus clouds exhibit a flat base and appear as rising domes or towers. (*Cumulus* means "heap" or "pile" in Latin.)
- **Stratus** clouds are best described as sheets or layers (*strata*) that cover much or all of the sky. Although there

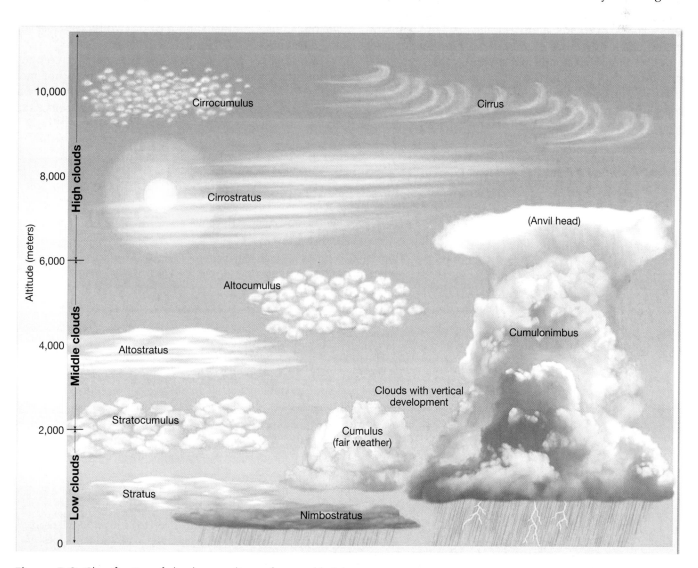

Figure 5-2 Classification of clouds according to form and height.

may be minor breaks, there are no distinct individual cloud units.

All clouds have at least one of these three basic forms, and some are a combination of two of them (for example, cirrocumulus clouds).

The second aspect of cloud classification—height—recognizes three levels: high, middle, and low. **High clouds** normally have bases above 6000 meters (20,000 feet); **middle clouds** generally occupy heights from 2000 to 6000 meters; and **low clouds** form below 2000 meters (6500 feet). These altitudes may vary somewhat according to season of the year and latitude. For example, at high (poleward) latitudes and during cold winter months, high clouds generally occur at lower altitudes. Further, some clouds extend upward to span more than one height range and are called **clouds of vertical development.**

The internationally recognized 10 cloud types are described in this section and summarized in Table 5-1.

High Clouds

The family of high clouds (above 6000 meters [20,000 feet]) include *cirrus, cirrostratus,* and *cirrocumulus*. Low temperatures and small quantities of water vapor present at high altitudes result in high clouds that are thin, white, and made up primarily of ice crystals (Box 5-1).

Cirrus clouds are composed of delicate, icy filaments. Winds aloft often cause these fibrous ice trails to bend or curl. Cirrus clouds with hooked filaments are called "mares' tails" (Figure 5-3a).

Cirrostratus are transparent, whitish cloud veils with fibrous or sometimes smooth appearance that may cover much or all of the sky. These clouds are easily recognized when they produce a halo around the Sun or Moon (Figure 5-3b). Occasionally, cirrostratus clouds are so thin and transparent that they are barely discernible.

Cirrocumulus clouds appear as white patches composed of small cells or ripples (Figure 5-3c). These small globules, which may be merged or separate, are often arranged in a pattern that resembles fish scales. When this occurs, it is commonly called "mackerel sky."

Although high clouds are generally not precipitation makers, when cirrus clouds give way to cirrocumulus clouds, they may warn of impending stormy weather. The following mariner's phrase is based on this observation: *Mackerel scales and mares' tails make lofty ships carry low sails.*

Middle Clouds

Clouds that form in the middle altitude range (2000–6000 meters [6500–20,000 feet]) are described with the prefix *alto* (meaning "middle") and include two types: *altocumulus* and *altostratus.*

Altocumulus tend to form in large patches composed of rounded masses or rolls that may or may not merge (Figure 5-4a). Because they are generally composed of water droplets rather than ice crystals, the individual cells usually

TABLE 5-1 Basic Cloud Types

Cloud Family and Height	Cloud Type	Characteristics
High clouds—above 6000 m (20,000 ft)	Cirrus (Ci)	Thin, delicate, fibrous, ice-crystal clouds. Sometimes appear as hooked filaments called "mares' tails" or cirrus uncinus (Figure 5-3a).
	Cirrostratus (Cs)	Thin sheet of white, ice-crystal clouds that may give the sky a milky look. Sometimes produce halos around the Sun and Moon (Figure 5-3b).
	Cirrocumulus (Cc)	Thin, white, ice-crystal clouds. In the form of ripples or waves, or globular masses all in a row. May produce a "mackerel sky." Least common of high clouds (Figure 5-3c).
Middle clouds—2000–6000 m (6500–20,000 ft)	Altocumulus (Ac)	White to gray clouds, often made up of separate globules; "sheepback" clouds (Figure 5-4a).
	Altostratus (As)	Stratified veil of clouds that is generally thin and may produce very light precipitation. When thin, the Sun or Moon may be visible as a "bright spot," but no halos are produced (Figure 5-4b).
Low clouds—below 2000 m (6500 ft)	Stratus (St)	Low uniform layer resembling fog but not resting on the ground. May produce drizzle.
	Stratocumulus (Sc)	Soft, gray clouds in globular patches or rolls. Rolls may join together to make a continuous cloud.
	Nimbostratus (Ns)	Amorphous layer of dark gray clouds. One of the primary precipitation-producing clouds (Figure 5-5).
Clouds of vertical development	Cumulus (Cu)	Dense, billowy clouds often characterized by flat bases. May occur as isolated clouds or closely packed (Figure 5-6).
	Cumulonimbus (Cb)	Towering cloud, sometimes spreading out on top to form an "anvil head." Associated with heavy rainfall, thunder, lightning, hail, and tornadoes (Figure 5-7).

(a) Cirrus

(b) Cirrostratus

Figure 5-3 Three basic cloud types make up the family of high clouds: (a) cirrus, (b) cirrostratus, and (c) cirrocumulus. (Photos (a) and (c) by E. J. Tarbuck, (b) by Bulus/DreamsTime)

(c) Cirrocumulus

have a more distinct outline. Altocumulus are sometimes confused with cirrocumulus (which are smaller and less dense) and stratocumulus (which are thicker).

Altostratus is the name given to a formless layer of grayish clouds that cover all or large portions of the sky. Generally, the Sun is visible through altostratus clouds as a bright spot but with the edge of its disc not discernible (Figure 5-4b). However, unlike cirrostratus clouds, altostratus do not produce halos. Infrequent precipitation in the form of light snow or drizzle may accompany these clouds. Altostratus clouds, commonly associated with approaching warm

fronts, thicken into a dark gray layer of nimbostratus clouds capable of producing copious rainfall.

Low Clouds

There are three members of the family of low clouds (below 2000 meters [6500 feet]): *stratus, stratocumulus,* and *nimbostratus.*

Stratus is a uniform layer that frequently covers much of the sky and, on occasion, may produce light precipitation. Stratus clouds that develop a scalloped bottom that appears as long parallel rolls or broken globular patches are called *stratocumulus* clouds.

Nimbostratus clouds derive their name from the Latin *nimbus,* "rain cloud," and *stratus,* "to cover with a layer" (Figure 5-5). As the name implies, nimbostratus clouds are one of the chief precipitation producers. Nimbostratus clouds form under stable conditions when air is forced to rise, as along a front. Such forced ascent of stable air leads to the formation of a stratified cloud layer that is large horizontally compared to its thickness. Precipitation associated with nimbostratus clouds is generally light to moderate but of long duration and widespread.

Clouds of Vertical Development

Clouds that do not fit into any of the three height categories but instead have their bases in the low height range and extend upward into the middle or high altitudes are referred to as *clouds of vertical development.*

The most familiar type, *cumulus* clouds, are individual masses that develop into vertical domes or towers having

(a)

(b)

Figure 5-4 Two forms of clouds are generated in the middle-altitude range. (a) Altocumulus tend to form in patches composed of rolls or rounded masses. (b) Altostratus occur as grayish sheets covering a large portion of the sky. When visible, the Sun appears as a bright spot through these clouds. (Photos by E. J. Tarbuck)

tops that resemble a head of cauliflower. Cumulus clouds most often form on clear days when unequal surface heating causes parcels of air to rise convectively above the lifting condensation level (Figure 5-6).

When cumulus clouds are present early in the day, we can expect an increase in cloudiness in the afternoon as solar heating intensifies. Furthermore, because small cumulus clouds (*cumulus humilis*) form on "sunny" days and rarely produce appreciable precipitation, they are often called "fair-weather clouds." However, when the air is unstable, cumulus clouds grow dramatically in height. As such a cloud grows, its top enters the middle height range, and it is called a *cumulus congestus*. Finally, if the cloud continues to grow and rain begins to fall, it is called a *cumulonimbus*.

Cumulonimbus are large, dense, billowy clouds of considerable vertical extent in the form of huge towers (Figure 5-7). In late stages of development, the upper part of a cumulonimbus turns to ice, appears fibrous, and frequently spreads out in the shape of an anvil. Cumulonimbus towers extend from a few hundred meters above the surface upward to 12 kilometers (7 miles) or, on rare oc-

Figure 5-5 Nimbostratus clouds are one of the chief precipitation producers. These dark gray layers often exhibit a ragged-appearing base. (Photo by E. J. Tarbuck)

Figure 5-6 Cumulus clouds. These small, white, billowy clouds generally form on sunny days and, therefore, are often called "fair-weather clouds." (Photo by E. J. Tarbuck)

Figure 5-7 Cumulonimbus clouds. These dense, billowy clouds have great vertical extent and can produce heavy precipitation. (Photo by Doug Millar/Science Source/Photo Researchers, Inc.)

casions, 20 kilometers (12 miles). These huge towers produce heavy precipitation with accompanying lightning, thunder, and occasionally hail. We consider the development of these important weather producers in Chapter 10.

Cloud Varieties

Variations exist in the 10 basic cloud types. These are named using adjectives that describe particular characteristics. For example, the term *uncinus*, meaning "hook shaped," is applied to streaks of cirrus clouds that are shaped like a comma resting on its side. Cirrus uncinus clouds are often precursors of bad weather.

When stratus or cumulus clouds appear broken (or fractured), the adjective *fractus* is used in their description. In addition, some clouds have rounded protuberances on their bottom surface, similar to a cow udder. When these structures are present, the term *mammatus* is applied. This configuration is usually associated with stormy weather and cumulonimbus clouds.

Lens-shaped clouds, referred to as *lenticular*, are common in rugged or mountainous topographies, where they are called *lenticular altocumulus* (Figure 5-8a). Al-

(a)

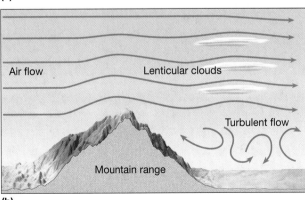

Air flow
Lenticular clouds
Turbulent flow
Mountain range

(b)

Figure 5-8 Lenticular clouds. (a) These lens-shaped clouds are relatively common in mountainous areas. (Photo by Dennis Tasa) (b) This diagram depicts the formation of lenticular clouds in the turbulent flow that develops in the lee of a mountain range.

Box 5-1   Aircraft Contrails and Cloudiness

You have undoubtedly seen a *contrail* (from *condensation trail*) in the wake of an aircraft flying on a clear day (Figure 5-A). Contrails are produced by jet aircraft engines that expel large quantities of hot, moist air. As this air mixes with the frigid air aloft, a streamlined cloud is produced.

Why do contrails occur on some occasions and not on others? Contrails form under the same conditions as any other cloud—that is, when the air reaches saturation and condensation nuclei exist in sufficient numbers. Most contrails form when the exhaust gases add sufficient water vapor to the air to cause saturation.

Contrails typically form above 9 kilometers (6 miles) where air temperatures are a frigid −50°C (−58°F) or colder. Thus, it is not surprising that contrails are composed of minute ice crystals. Most contrails have a very short life span. Once formed, these streamlined clouds mix with surrounding cold, dry air and ultimately sublimate. However, if the air aloft is near saturation, contrails may survive for long periods. Under these conditions, the upper airflow usually spreads the streamlike clouds into broad bands of clouds called *contrail cirrus*.

With the increase in air traffic during the past few decades, an overall increase in cloudiness has been recorded, particularly near major transportation hubs (Figure 5-B). This is most evident in the American Southwest, where aircraft contrails persist in otherwise cloudless skies.

In addition to the added cloud cover associated with increasing levels of jet aircraft traffic, other less noticeable effects are of concern. Research is currently under way to assess the impact of contrail-produced cirrus clouds on the planet's heat budget. Recall from Chapter 2 that high, thin clouds are effective transmitters of solar radiation (that is, most radiation reaches the surface) but are good absorbers of outgoing infrared radiation emitted from Earth's surface. As a consequence, cirrus clouds tend to have an overall warming effect. However, it appears that these human-induced cirrus clouds may actually lead to surface cooling rather than warming. Additional research is still needed to accurately determine the impact of contrails on climate change.

One effect of contrails that is known with some certainty is how these "artificial" clouds impact daily temperature ranges (that is, the difference between daily maximum and minimum temperatures). During the three-day commercial flight hiatus following the September 11 terrorist attacks, contrails all but disappeared. As a result, during these days of "clearer skies," the differences between the high and low temperatures increased by 2°F (1.1°C). Consequently, it appears that cities located near major air-traffic centers may experience lower daily highs and higher daily lows than would be the case if jet aircraft did not produce contrails.

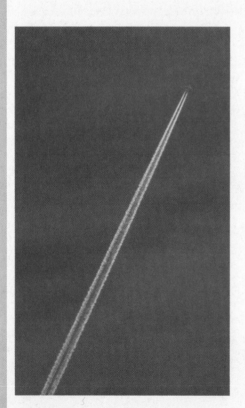

FIGURE 5-A Aircraft contrails. Condensation trails produced by jet aircraft often spread out to form broad bands of cirrus clouds. (Photo by Dennis Tasa)

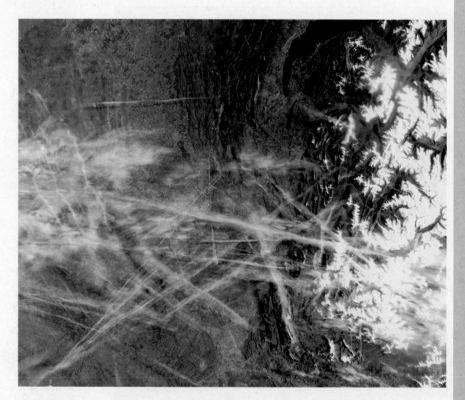

FIGURE 5-B This photograph, taken through the window of the International Space Station, shows contrails over the Rhone Valley in eastern France. It is estimated that these "artificial clouds" cover 0.1 percent of the planet's surface, and the percentages are far higher in some places, such as southern California and some parts of Europe, as illustrated here. (Photo courtesy of NASA)

EYE ON THE ATMOSPHERE

(amana images inc./Alamy)

The cap cloud perched on the top of this mountain may remain in place for hours. Cap clouds belong to a group referred to as *orographic clouds.*

Question 1 Describe how these clouds form, based on the fact that they are referred to as orographic clouds.

Question 2 Why does this cloud have a relatively flat bottom?

Question 3 Cap clouds are related to another cloud type that has a very similar shape. Can you name the cloud type?

though lenticular clouds can form whenever the airflow undulates sharply in the vertical direction, they most frequently form on the leeward side of mountains. As air passes over mountainous terrain, a wave pattern develops, as shown in Figure 5-8b. Clouds form where air is ascending, whereas areas with descending air are cloud free.

Concept Check 5.2

1 What are the two criteria by which clouds are classified?

2 Why are high clouds always thin in comparison to low and middle clouds?

3 Which cloud types are associated with the following characteristics: thunder, halos, precipitation, hail, mackerel sky, lightning, and mares' tails?

4 What do layered clouds indicate about the stability of the air? What do clouds of vertical development indicate about the stability of the air?

Types of Fog

GE**O**De ATMOSPHERE   Forms of Condensation and Precipitation
▶ Types of Fog

Fog is defined as *a cloud with its base at or very near the ground.* Physically, there are no differences between fog and a cloud; their appearances and structures are the same. The essential difference is the method and place of formation. While clouds result when air rises and cools adiabatically, fog results from cooling or when air becomes saturated through the addition of water vapor (evaporation fog).

Although fog is not inherently dangerous, it is generally considered an atmospheric hazard (Figure 5-9). During daylight hours, fog reduces visibility to 2 or 3 kilometers (1 or 2 miles). When the fog is particularly dense, visibility may be cut to a few dozen meters or less, making travel by any mode difficult and dangerous. Official weather stations report fog only when it is thick enough to reduce visibility to 1 kilometer or less.

Fogs Formed by Cooling

When the temperature of a layer of air in contact with the ground falls below its dew point, condensation produces fog. Depending on the prevailing conditions, fog formed by cooling are called either *radiation fog, advection fog,* or *upslope fog.*

Radiation Fog   As the name implies, **radiation fog** results from radiation cooling of the ground and adjacent air. It is a nighttime phenomenon requiring clear skies and a high relative humidity. Under clear skies, the ground and the air immediately above cools rapidly. Because of the high relative humidity, a small amount of cooling will lower the temperature to the dew point. If the air is calm, the fog is usually patchy and less than 1 meter deep. For radiation fog to be more extensive vertically, a light breeze of 3 to 5 kilometers (2 to 3 miles) per hour is necessary, to create enough turbulence to carry the fog upward 10 to 30 meters (30 to 100 feet) without dispersing it. High winds, on the other hand, mix the air with drier air above and disperse the fog.

Because the air containing the fog is relatively cold and dense, it flows downslope in hilly terrain. As a result,

Figure 5-9 Radiation fog. (a) Satellite image of dense fog in California's San Joaquin Valley on November 20, 2002. This early-morning radiation fog was responsible for several car accidents in the region, including a 14-car pileup. The white areas to the east of the fog are the snow-capped Sierra Nevadas. (NASA image) (b) Radiation fog can make a morning commute quite hazardous. (Photo by Tim Gainey/Alamy)

radiation fog is thickest in valleys, whereas the surrounding hills may remain clear (Figure 5-9a). Normally, radiation fog dissipates within one to three hours after sunrise—and is often said to "lift." However, the fog does not actually "lift." Instead, as the Sun warms the ground, the lowest layer of air is heated first, and the fog evaporates from the bottom up. The last vestiges of radiation fog may appear as a low layer of stratus clouds.

Advection Fog   When warm, moist air blows over a cold surface, it becomes chilled by contact with the cold surface below. If cooling is sufficient, the result will be a blanket of fog called **advection fog.** (The term *advection* refers to air moving horizontally.) A classic example is the frequent advection fog around San Francisco's Golden Gate Bridge (Figure 5-10). The fog experienced in San Francisco, California, as well as many other West Coast locations is produced when warm, moist air from the Pacific Ocean moves over the cold California Current.

A certain amount of turbulence is needed for proper development of advection fog; typically winds between 10 and 30 kilometers (6 and 18 miles) per hour are required. Not only does the turbulence facilitate cooling through a thicker layer of air, but it also carries the fog to greater heights. Thus, advection fogs often extend 300 to 600 meters (1000 to 2000 feet) above the surface and persist longer than radiation fogs. An example of such fog can be found at Cape Disap-

pointment, Washington—the foggiest location in the United States. The name is indeed appropriate because the station averages about 2552 hours of fog each year—equivalent to 106 days.

Advection fog is also a common wintertime phenomenon in the Southeast and Midwest when relatively warm, moist air from the Gulf of Mexico and Atlantic moves over cold and occasionally snow-covered surfaces to produce widespread foggy conditions. This type of advection fog tends to be thick and produce hazardous driving conditions.

Upslope Fog   As its name implies, **upslope fog** is created when relatively humid air moves up a gradually sloping landform or, in some cases, up the steep slopes of a mountain. Because of the upward movement, air expands and cools adiabatically. If the dew point is reached, an extensive layer of fog will form.

It is easy to visualize how upslope fog might form in mountainous terrain. However, in the United States upslope fog also occurs in the Great Plains, when humid Gulf air moves from the Gulf of Mexico toward the Rocky Mountains. (Recall that Denver, Colorado, is called the "mile-high city," and the Gulf of Mexico is at sea level.) Air flowing "up" the Great Plains expands and cools adiabatically by as much as 12°C (22°F), which can result in extensive upslope fog in the western plains.

Evaporation Fogs

When saturation occurs primarily because of the addition of water vapor, the resulting fogs are called *evaporation fogs.* Two types of evaporation fogs are recognized: *steam fog* and *frontal (precipitation) fog.*

Steam Fog   When cool air moves over warm water, enough moisture may evaporate from the water surface to saturate the air immediately above. As the rising water vapor meets the cold air, it condenses and rises with the air that is being warmed from below. Because the rising foggy air looks like the "steam" that forms above a hot cup of coffee, the phenomenon is called **steam fog** (Figure 5-11). Steam fog is a fairly common occurrence over lakes and rivers on clear, crisp mornings in the autumn when the water is still relatively warm but the air is rather cold. Steam fog is usually shallow because as it rises, the water droplets mix with the unsaturated air above and evaporate.

Figure 5-10 Advection fog rolling into San Francisco Bay. (Photo by Ed Pritchard/Getty Images)

In a few settings, steam fogs can be dense, especially during the winter, as cold arctic air pours off the continents and ice shelves toward the comparatively warm open ocean. The temperature contrast between the warm ocean and cold air has been known to exceed 30°C (54°F). The result is an intense steam fog produced as the rising water vapor saturates a large volume of air. Because of its source and appearance, this type of steam fog is given the name *arctic sea smoke.*

Students Sometimes Ask…

Why do I see my breath on cold mornings?

On cold days when you "see your breath," you are actually creating steam fog. The moist air that you exhale saturates a small volume of cold air, causing tiny droplets to form. Like most steam fogs, the droplets quickly evaporate as the "fog" mixes with the unsaturated air around it.

Figure 5-11 Steam fog rising from Sierra Blanca Lake, Arizona. (Photo by Michael Collier)

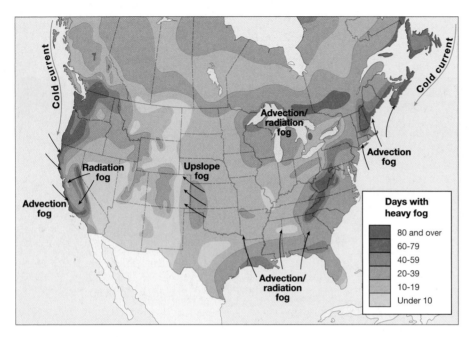

Figure 5-12 Map showing average numbers of days per year with heavy fog. Coastal areas, particularly the Pacific Northwest and New England, where cold currents prevail, have high occurrences of dense fog.

Frontal (Precipitation) Fog   Frontal, or precipitation, fog occurs when raindrops falling from relatively warm air above a frontal surface evaporate into the cooler air below and cause it to become saturated. Frontal fog is most common on cool days during extended periods of light rainfall.

The frequency of dense fog varies considerably from place to place (Figure 5-12). As might be expected, fog incidence is highest in coastal areas, especially where cold currents prevail, as along the Pacific and New England coasts. Relatively high frequencies are also found in the Great Lakes region and in the humid Appalachian Mountains of the East. In contrast, fogs are rare in the interior of the continent, especially in the arid and semiarid areas of the West (the yellow areas in Figure 5-12).

Concept Check 5.3

1 Distinguish between clouds and fog.

2 List five types of fog and discuss how they form.

3 What actually happens when a radiation fog "lifts"?

4 Why is there a relatively high frequency of dense fog along the Pacific Coast?

How Precipitation Forms

 Forms of Condensation and Precipitation
ATMOSPHERE   ▶ How Precipitation Forms

If all clouds contain water, why do some produce precipitation while others drift placidly overhead? This seemingly simple question perplexed meteorologists for many years.

Cloud droplets are miniscule—20 micrometers (0.02 millimeter) in diameter (Figure 5-13). (1 micrometer equals 0.001 millimeter.) In comparison, a human hair is about 75 micrometers in diameter. Because of their small size, cloud droplets fall in still air incredibly slowly. An average cloud droplet falling from a cloud base at 1000 meters would require several hours to reach the ground. However, it would never complete its journey. Instead, the cloud droplet would evaporate before it fell a few meters from the cloud base into the unsaturated air below.

How large must a cloud droplet grow in order to fall as precipitation? A typical raindrop has a diameter of about 2 millimeters, or 100 times that of the average cloud droplet (Figure 5-13). However, the *volume* of a typical raindrop is 1 million times that of a cloud droplet. Thus, for precipitation to form, cloud droplets must grow in volume by roughly 1 million times. You might suspect that additional condensation creates drops large enough to survive the descent to the surface. However, clouds consist of many billions of tiny cloud droplets that all compete for the available water. Thus, condensation provides an inefficient means of raindrop formation.

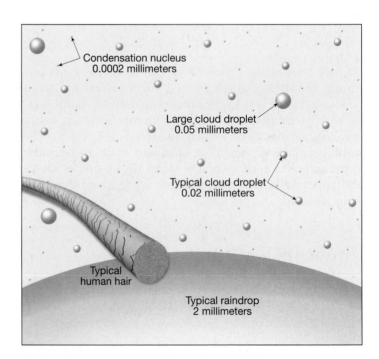

Figure 5-13 Comparative diameters of particles involved in condensation and precipitation processes.

Box 5-2   Science and Serendipity*

Nobel Laureate Irving Langmuir defined *serendipity* as "the art of profiting from unexpected occurrences." In other words, if you are observing something and the entirely unexpected happens, and if you see in this accident a new and meaningful discovery, then you have experienced serendipity. Several great discoveries in science have been serendipitous.

An excellent example of serendipity in science occurred when Tor Bergeron, the great Swedish meteorologist, discovered the importance of ice crystals in the initiation of precipitation in supercooled clouds. Bergeron's discovery occurred when he spent several weeks at a health resort at an altitude of 430 meters (1400 feet) on a hill near Oslo. During his stay, Bergeron noted that this hill was often "fogged in" by a layer of supercooled clouds (Figure 5-C). As he walked along a narrow road in the fir forest along the hillside, he noticed that the fog did not occupy the road at temperatures below −5°C but did enter it when the temperature was warmer than 0°C. (Drawings of the hill, trees, and fog for the two temperature regimes are shown in Figure 5-C.)

Bergeron immediately concluded that at temperatures below about −5°C the branches of the firs acted as freezing nuclei upon which some of the supercooled droplets crystallized. Once the ice crystals developed, they grew rapidly, at the expense of the remaining water droplets (Figure 5-C). The result was the growth of ice crystals (rime) on the branches of the firs, accompanied by a "clearing-off" between the trees and along the road.

From this experience, Bergeron realized that if ice crystals somehow were to appear in the midst of a cloud of supercooled droplets, they would grow rapidly as water molecules diffused toward them from the evaporating cloud droplets. This rapid growth

forms snow crystals that, depending on the air temperature beneath the cloud, fall to the ground as snow or rain. Bergeron had thus discovered one way that minuscule cloud droplets can grow large enough to fall as precipitation. (See the section "Precipitation from Cold Clouds: The Bergeron Process" on page 140.)

Serendipity influences the entire realm of science. Can we conclude that anyone who makes observations will necessarily make a major discovery? Not at all. A perceptive and inquiring mind is required, a mind that has been searching for order in a labyrinth of facts. As Langmuir said, the unexpected occurrence is not enough; you must know how to profit from it. Louis Pasteur observed that "in the field of observation, chance favors only the prepared mind." The discoverer of vitamin C, Nobel Laureate Albert Szent-Gyorgyi, remarked that discoveries are made by those who "see what everybody else has seen, and think what nobody else has thought." Serendipity is at the heart of science itself.

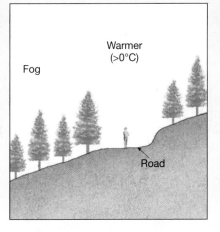 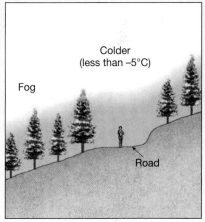

FIGURE 5-C Distribution of fog when the temperature is above freezing and when the temperature falls below −5°C.

*Based on material prepared by Duncan C. Blanchard.

Two processes are responsible for the formation of precipitation: the Bergeron process and the collision–coalescence process.

Precipitation from Cold Clouds: The Bergeron Process

You have seen TV coverage during which mountain climbers brave an intensely cold and ferocious snowstorm as they scale an ice-covered peak. Very similar conditions exist in the upper portions of towering cumulonimbus clouds, even on sweltering summer days. (In fact, in the upper troposphere where commercial aircraft cruise, the temperature typically approaches −50°C [−58°F] or lower.) Frigid conditions high in the troposphere provide an ideal environment to initiate precipitation. In fact, in the middle latitudes much of the rain that falls begins with the birth of snowflakes high in the cloud tops, where temperatures are considerably be-

low freezing. In the winter, even low clouds are cold enough to trigger precipitation.

Students Sometimes Ask...

Why does it often seem like the roads are slippery when it rains after a long dry period?

It appears that a buildup of debris on roads during dry weather can cause slippery conditions after a rainfall. One traffic study indicates that if it rains today, there will be no increase in risk of fatal crashes if it also rained yesterday. However, if it has been two days since the last rain, then the risk for a deadly accident increases by 3.7 percent. If it has been 21 days since the last rain, the risk increases by 9.2 percent.

The process that generates much of the precipitation in the middle latitudes is named the **Bergeron process** for its discoverer, the highly respected Swedish meteorologist Tor Bergeron (Box 5-2). The Bergeron process depends on

EYE ON THE ATMOSPHERE

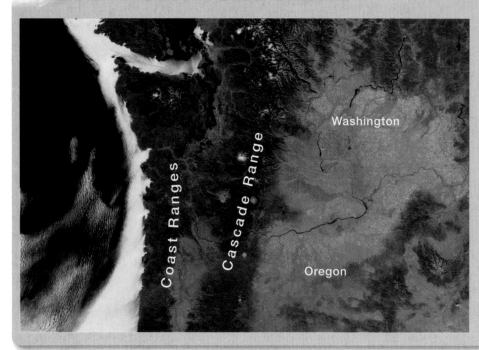

Coast Ranges

Cascade Range

Washington

Oregon

(NASA)

This satellite image shows a layer of low-lying clouds moving eastward into the bays and waterways of the northwestern United States. Lining the coast are the forested Coast Ranges and further inland is the Cascade Range, composed of several large volcanoes.

Question 1 Why do you think the clouds formed over the ocean, while the land areas are cloud free?

Question 2 Why is western Washington covered with lush vegetation while much of the eastern half of the state appears semiarid?

---

the coexistence of water vapor, liquid cloud droplets, and ice crystals. To understand how this mechanism operates, we must first examine two important properties of water. First, contrary to what you might expect, *cloud droplets do not freeze at 0°C (32°F)*. In fact, pure water suspended in air does not freeze until it reaches a temperature of nearly –40°C (–40°F). Water in the liquid state below 0°C (32°F) is referred to as **supercooled.** Supercooled water readily freezes if it impacts an object, which explains why airplanes collect ice when they pass through a liquid cloud made up of supercooled droplets. This also explains why *freezing rain,* or *glaze,* falls as a liquid but then turns to a sheet of ice when it strikes the pavement, tree branches, or car windshields.

In the atmosphere, supercooled droplets freeze on contact with solid particles that have a shape that closely resembles that of ice (silver iodide, for example). These materials are called **freezing nuclei.** The need for freezing nuclei to initiate freezing is similar to the role of condensation nuclei in the process of condensation.

In contrast to condensation nuclei, however, freezing nuclei are sparse in the atmosphere and do not generally become active until the temperature reaches –10°C (14°F) or below. Thus, at temperatures between 0 and –10°C, clouds consist mainly of supercooled water droplets. Between –10 and –20°C, liquid droplets coexist with ice crystals, and below –20°C (–4°F), clouds are generally composed entirely of ice crystals—for example, high-altitude cirrus clouds.

This brings us to a second important property of water: *The saturation vapor pressure above ice crystals is slightly lower than above supercooled liquid droplets.* This occurs because

ice crystals are solid, so the individual water molecules are held together more tightly than those forming a liquid droplet. As a result, it is easier for water molecules to escape (evaporate) from the supercooled liquid droplets. Consequently, when air is saturated (100 percent relative humidity) with respect to liquid droplets, it is supersaturated with respect to ice crystals. Table 5-2, for example, shows that at –10°C (14°F), when the relative humidity is 100 percent with respect to water, the relative humidity with respect to ice is about 110 percent.

With these facts in mind, we can now explain how precipitation is produced via the Bergeron process. Visualize a cloud at a temperature of –10°C (14°F), where each ice crystal (snow crystal) is surrounded by many thousands of liquid droplets (Figure 5-14). Because the air is saturated (100 percent) with respect to liquid water, it will be supersaturated (over 100 percent) with respect to the newly formed ice crystals. As a result of this supersaturated condition, the ice crys-

TABLE 5-2 Relative Humidity with Respect to Ice When Relative Humidity with Respect to Water Is 100 Percent

| Temperature (°C) | Relative Humidity with Respect To: | |
	Water	Ice
0	100%	100%
–5	100%	105%
–10	100%	110%
–15	100%	115%
–20	100%	121%

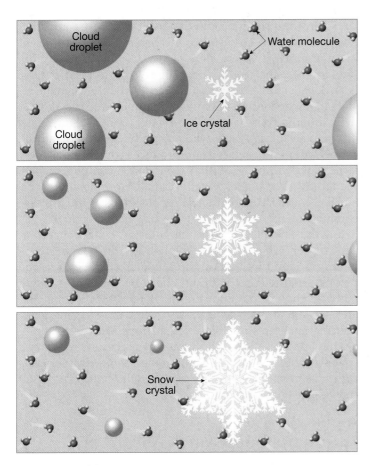

Figure 5-14 The Bergeron process. Ice crystals grow at the expense of cloud droplets until they are large enough to fall. The size of these particles has been greatly exaggerated.

tals collect water molecules, lowering the relative humidity of the air. In response, water droplets shrink (evaporate) to replenish the lost water vapor. Thus the growth of ice crystals is fed by the continued evaporation of liquid droplets.

When ice crystals become sufficiently large, they begin to fall. During their descent, the ice crystals grow as they intercept cloud droplets that freeze on them. Air movement will sometimes break up these delicate crystals, and the fragments will serve as freezing nuclei for other liquid droplets. A chain reaction ensues and produces many snow crystals, which, by accretion, form into larger masses called *snowflakes*. Large snowflakes may consist of many individual ice crystals.

The Bergeron process can produce precipitation throughout the year in the middle latitudes, provided that at least the upper portions of clouds are cold enough to generate ice crystals. The type of precipitation (snow, sleet, rain, or freezing rain) that reaches the ground depends on the temperature profile in the lower few kilometers of the atmosphere. When the surface temperature is above 4°C (39°F), snowflakes usually melt before they reach the ground and continue their descent as rain. Even on a hot summer day, a heavy rainfall may have begun as a snowstorm high in the clouds overhead.

Precipitation from Warm Clouds: The Collision–Coalescence Process

A few decades ago meteorologists believed that the Bergeron process was responsible for the formation of most precipitation except for light drizzle. Later it was discovered that copious rainfall is often associated with clouds located well below the freezing level (called *warm clouds*), especially in the tropics. This second mechanism for triggering precipitation is called the **collision–coalescence process.**

Research has shown that clouds made entirely of liquid droplets often contain some droplets larger than 20 micrometers (0.02 millimeter). These large droplets form when "giant" condensation nuclei are present or when hygroscopic particles exist (such as sea salt). Hygroscopic particles begin to remove water vapor from the air at relative humidities under 100 percent. Because the rate at which drops fall is dependent on their size, these "giant" droplets fall most rapidly. Table 5-3 summarizes drop size and falling velocities.

As the larger droplets fall through a cloud, they collide with smaller, slower droplets and coalesce. They become larger in the process, and fall even more rapidly (or, in an updraft, they rise more slowly) increasing their chances of collision and rate of growth (Figure 5-15a). After a million or so cloud droplets coalesce, a rain drop is are large enough to fall to the surface without evaporating.

Because of the huge number of collisions required for growth to raindrop size, clouds that have great vertical thickness and contain large cloud droplets have the best chance of producing precipitation. Updrafts also aid this process because the droplets can traverse the cloud repeatedly, which results in more collisions.

As raindrops grow in size, their fall velocity increases. This in turn increases the frictional resistance of the air, which causes the drop's "bottom" to flatten out (Figure 5-15b). As a drop approaches 4 millimeters in diameter, it develops a depression, as shown in Figure 5-15c. Raindrops can grow to a maximum of 5 millimeters when they fall at the rate of 33 kilometers (20 miles) per hour. At this size, the water's surface tension, which holds the drop together, is surpassed by the frictional drag of the air. The depression grows almost explosively, forming a donutlike ring that immediately breaks apart. The resulting breakup of a large rain-

TABLE 5-3 Fall Velocity of Water Drops

Types	Diameter (millimeters)	Fall Velocity (km/hr)	Fall Velocity (mi/hr)
Small cloud droplets	0.01	0.01	0.006
Typical cloud droplets	0.02	0.04	0.03
Large cloud droplets	0.05	0.3	0.2
Drizzle droplets	0.5	7	4
Typical rain drops	2.0	23	14
Large rain drops	5.0	33	20

Data from Smithsonian Meteorological Tables.

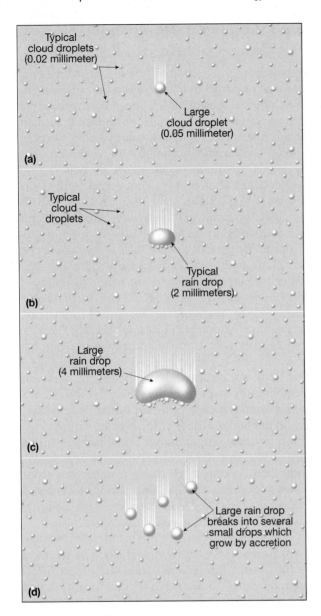

Figure 5-15 The collision–coalescence process.

From the preceding discussion, it should be apparent that the collision–coalescence mechanism is most efficient in environments where large cloud droplets are plentiful. The air over the tropics, particularly the tropical oceans, is an ideal setting: Very humid and relatively clean air results in fewer condensation nuclei compared to the air over more populated regions. With fewer condensation nuclei to compete for available water vapor (which is plentiful), condensation is fast paced and produces comparatively few large cloud droplets. Within developing cumulus clouds, the largest drops quickly gather smaller droplets to generate the warm afternoon showers associated with tropical climates.

In the middle latitudes the collision–coalescence process may contribute to the precipitation from a large cumulonimbus cloud by working in tandem with the Bergeron process—particularly during the hot, humid summer months. High in these towers the Bergeron process generates snow that melts as it passes below the freezing level. Melting generates relatively large drops with fast velocities. As these large drops descend, they overtake and coalesce with the slower and smaller cloud droplets that comprise much of the lower regions of the cloud. The result can be a heavy downpour.

Students Sometimes Ask...

What is the largest annual rainfall ever recorded?

The greatest recorded rainfall for a single 12-month period occurred at Cherrapunji, India, where an astounding 2647 centimeters (1042 inches)—over 86 feet—fell. Cherrapunji is located at an elevation of 1293 meters (4309 feet), where orographic lifting of the onshore monsoon winds greatly contributed to the total.

In summary, *two mechanisms are known to generate precipitation: the Bergeron process and the collision–coalescence process. The Bergeron process is dominant in the middle latitudes, where cold clouds (or cold cloud tops) are the rule. In the tropics, abundant water vapor and comparatively few condensation nuclei are more typical. This leads to the formation of fewer, larger drops with fast fall velocities that grow by collision and coalescence. Regardless of which process initiates precipitation, further growth in drop size occurs through collision–coalescence.*

drop produces numerous smaller drops that begin anew the task of sweeping up cloud droplets (Figure 5-15d).

The collision–coalescence process is not that simple, however. First, as the larger droplets descend, they produce an airstream around them similar to that produced by an automobile when driven rapidly down the highway. The airstream repels objects, especially small ones. Imagine driving on a summer night along a country road. The bugs in the air are like cloud droplets—most are pushed aside through the air. However, larger cloud droplets (bugs) have an increased chance of colliding with the giant droplet (car).

Further, collision does not guarantee coalescence. Experimentation has indicated that the presence of atmospheric electricity may be the key to what holds these droplets together once they collide. If a droplet with a negative charge collides with a positively charged droplet, their electrical attraction may bind them together.

Concept Check 5.4

1 Describe the steps in the formation of precipitation according to the Bergeron process. Be sure to include (a) the importance of supercooled cloud droplets, (b) the role of freezing nuclei, and (c) the difference in saturation vapor pressure between liquid water and ice.

2 How does the collision–coalescence process differ from the Bergeron process?

3 If snow is falling from a cloud, which process produced it? Explain.

Forms of Precipitation

GE**⬤**De Forms of Condensation and Precipitation
ATMOSPHERE ▶ Forms of Precipitation

Atmospheric conditions vary greatly both geographically and seasonally, resulting in several different types of precipitation (Figure 5-16). Rain and snow are the most common and familiar forms, but others, as listed in Table 5-4, are important as well. Sleet, freezing rain (glaze), and hail often produce hazardous weather. Although limited in occurrence and sporadic in both time and space, these forms, especially freezing rain and hail, may on occasion cause considerable damage.

Rain

In meteorology the term **rain** is restricted to drops of water that fall from a cloud and have a diameter of at least 0.5 millimeter. (Drizzle and mist have smaller droplets, and are therefore not considered rain.) Most rain originates in either nimbostratus clouds or in towering cumulonimbus clouds that are capable of producing unusually heavy rainfalls known as *cloudbursts.*

Fine, uniform droplets of water having a diameter less than 0.5 millimeter are called *drizzle.* Drizzle and small raindrops generally are produced in stratus or nimbostratus clouds, where precipitation may be continuous for several hours or for days on rare occasions.

As rain enters the unsaturated air below the cloud, it begins to evaporate. Depending on the humidity of the air and the size of the drops, the rain may completely evaporate before reaching the ground. This phenomenon produces *virga,* which appear as streaks of precipitation falling from a cloud that extend toward Earth's surface without reaching it (Figure 5-17). Similar to virga, ice crystals may sublimate when they enter the dry air below. These wisps of ice particles are called *fallstreaks.*

Precipitation containing the very smallest droplets able to reach the ground is called *mist.* Mist can be so fine that the tiny droplets appear to float, and their impact is almost imperceptible.

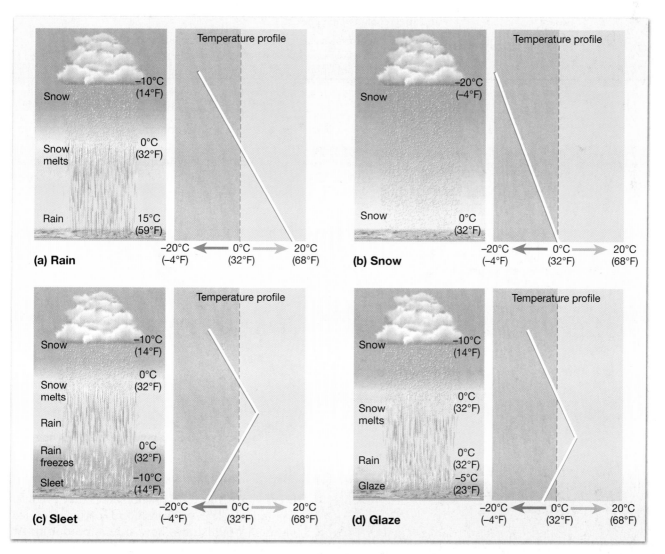

Figure 5-16 Four precipitation types and their temperature profiles.

TABLE 5-4 Types of Precipitation

Type	Approximate Size	State of Water	Description
Mist	0.005–0.05 mm	Liquid	Droplets large enough to be felt on the face when air is moving 1 meter/second. Associated with stratus clouds.
Drizzle	<0.5 mm	Liquid	Small, uniform droplets that fall from stratus clouds, generally for several hours.
Rain	0.5–5 mm	Liquid	Generally produced by nimbostratus or cumulonimbus clouds. When heavy, size can be highly variable from one place to another.
Sleet	0.5–5 mm	Solid	Small, spherical to lumpy ice particles that form when raindrops freeze while falling through a layer of subfreezing air. Because the ice particles are small, any damage is generally minor. Sleet can make travel hazardous.
Freezing Rain (Glaze)	Layers 1 mm–2 cm thick	Solid	Produced when supercooled raindrops freeze on contact with solid objects. Glaze can form a thick coating of ice that has sufficient weight to seriously damage trees and power lines.
Rime	Variable accumulations	Solid	Deposits usually consisting of ice feathers that point into the wind. These delicate frostlike accumulations form as supercooled cloud or fog droplets encounter objects and freeze on contact.
Snow	1 mm–2 cm	Solid	The crystalline nature of snow allows it to assume many shapes, including six-sided crystals, plates, and needles. Produced in supercooled clouds where water vapor is deposited as ice crystals that remain frozen during their descent.
Hail	5–10 cm or larger	Solid	Precipitation in the form of hard, rounded pellets or irregular lumps of ice. Produced in large convective, cumulonimbus clouds, where frozen ice particles and supercooled water coexist.
Graupel	2–5 mm	Solid	"Soft hail" that forms as rime collects on snow crystals to produce irregular masses of "soft" ice. Because these particles are softer than hailstones, they normally flatten out upon impact.

Snow

Snow is precipitation in the form of ice crystals or, more often, aggregates of ice crystals. The size, shape, and concentration of snowflakes depend to a great extent on the temperature at which they form.

Figure 5-17 Virga, Latin for "streak." In the arid west, rain frequently evaporates before reaching the ground. (Photo by James Steinberg/Photo Researchers, Inc.)

Recall that at very low temperatures, the moisture content of air is low. The result is the generation of very light and fluffy snow made up of individual six-sided ice crystals (Figure 5-18). This is the "powder" that downhill skiers covet. By contrast, at temperatures warmer than about −5°C (23°F), the ice crystals join together into larger clumps consisting of tangled aggregates of crystals. Snowfalls consisting of these composite snowflakes are generally heavy and have a high moisture content, which makes them ideal for making snowballs.

Students Sometimes Ask...

What is the snowiest city in the United States?
According to National Weather Service records, Rochester, New York, which averages nearly 239 centimeters (94 inches) of snow annually, is the snowiest city in the United States. However, Buffalo, New York, is a close runner-up.

Sleet and Freezing Rain or Glaze

Sleet is a wintertime phenomenon that involves the fall of clear to translucent particles of ice. Figure 5-19 illustrates how sleet is produced: An above-freezing air layer must overlie a subfreezing layer near the ground. When the raindrops, which

Figure 5-18 Snow crystals. All snow crystals are six sided, but they come in an infinite variety of forms. (Photo by Ted Kinsman/ Photo Researchers, Inc.)

are often melted snow, leave the warmer air and encounter the colder air below, they freeze and reach the ground as small pellets of ice roughly the size of the raindrops from which they formed.

Occasionally, the distribution of temperatures in a column of air is such that **freezing rain,** or **glaze,** results (Figure 5-16d). In these situations, the raindrops become supercooled because the subfreezing air near the ground is not thick enough to cause them to freeze. Upon striking objects on Earth's surface, these supercooled raindrops instantly turn to ice. The result can be a thick coating of glaze that has sufficient weight to break tree limbs, down power lines, and make walking and driving extremely hazardous.

In January 1998 an ice storm of historic proportions caused enormous damage in New England and southeastern Canada. Five days of freezing rain deposited a heavy layer of ice on exposed surfaces from eastern Ontario to the Atlantic coast. The 8 centimeters (3 inches) of precipitation caused trees, power lines, and high-voltage towers to collapse, leaving over 1 million households without power—many for nearly a month following the storm (Figure 5-20). At least 40 deaths were blamed on the storm, which caused damages in excess of $3 billion. Much of the damage was to the electrical grid, which one Canadian climatologist summed up this way: "What it took human beings a half-century to construct, took nature a matter of hours to knock down."

Hail

Hail is precipitation in the form of hard, rounded pellets or irregular lumps of ice. Hail is produced only in large cumulonimbus clouds where updrafts can sometimes reach speeds approaching 160 kilometers (100 miles) per hour and where there is an abundant supply of supercooled water. Figure 5-21a illustrates this process. Hailstones begin as small embryonic ice pellets (graupel) that grow by collecting supercooled droplets as they fall through the cloud. If they encounter a strong updraft, they may be carried upward again and begin the return downward journey. Each trip through the supercooled portion of the cloud results in an additional layer of ice. Hailstones can also form from a single descent through an updraft. Either way, the process continues until the hailstone grows too heavy to remain suspended by the thunderstorm's updraft or encounters a downdraft.

Hailstones may contain several layers that alternate between clear and milky ice (Figure 5-21b). High in the clouds, rapid freezing of small supercooled water droplets traps air bubbles, which cause the milky appearance. By contrast, clear ice is produced in the lower and warmer regions of the clouds, where colliding droplets wet the surface of the hailstones. As these droplets slowly freeze, they produce relatively bubble-free clear ice.

Most hailstones have diameters between 1 centimeter (pea size) and 5 centimeters (golf ball size), although some can be as big as an orange or larger. Occasionally, hailstones weighing a pound or more have been reported; most of these are composites of several stones frozen together.

The record for the largest hailstone ever found in the United States was set on July 23, 2010, in Vivian, South Dakota. The stone was over 20 centimeters (8 inches)

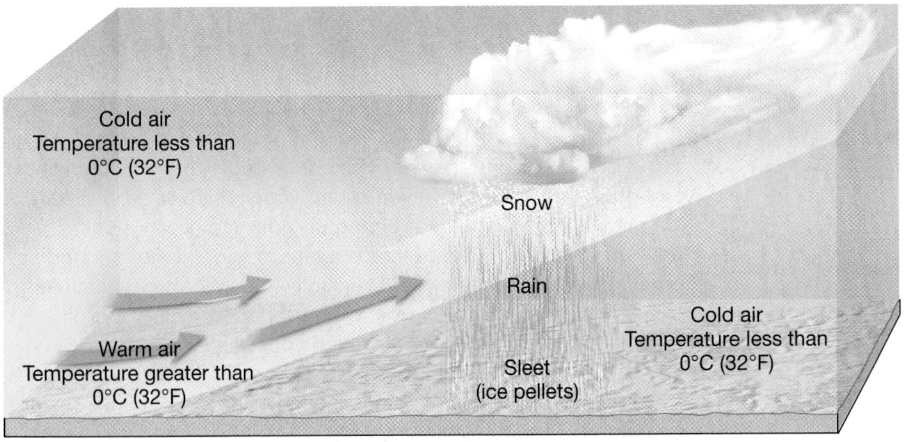

Cold air
Temperature less than
0°C (32°F)

Snow

Rain

Cold air
Temperature less than
0°C (32°F)

Warm air
Temperature greater than
0°C (32°F)

Sleet
(ice pellets)

Figure 5-19 Sleet forms when rain passes through a cold layer of air and freezes into ice pellets. This occurs most often in the winter, when warm air is forced over a layer of subfreezing air.

Figure 5-20 Glaze forms when supercooled raindrops freeze on contact with objects. In January 1998 an ice storm of historic proportions caused enormous damage in New England and southeastern Canada. Nearly five days of freezing rain (glaze) caused 40 deaths and more than $3 billion in damages, and it left millions of people without electricity—some for as long as a month. (Photo by Syracuse Newspapers/The Image Works)

Students Sometimes Ask...

What is the difference between a winter storm warning and a blizzard warning?

A winter storm warning is usually issued when heavy snow exceeding 6 inches in 12 hours or possible icing conditions are likely. It is interesting to note that in Upper Michigan and mountainous areas where snowfall is abundant, winter storm warnings are issued only if 8 or more inches of snow is expected in 12 hours. By contrast, blizzard warnings are issued for periods in which considerable falling and/or blowing snow will be accompanied by winds of 35 or more miles per hour. Thus, a blizzard is a type of winter storm in which winds are the determining factor, not the amount of snowfall.

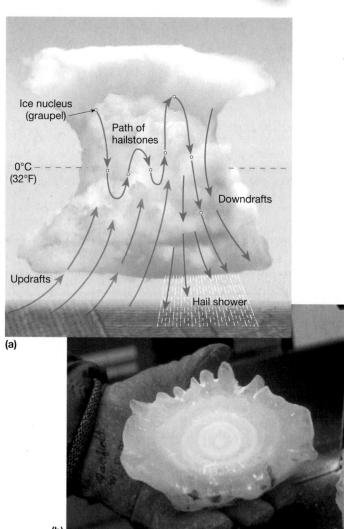

Figure 5-21 Hailstones. (a) Hailstones begin as small ice pellets that grow through the addition of supercooled water droplets as they move through a cloud. Strong updrafts may carry stones upward in several cycles, increasing the size of the hail by adding a new layer with each cycle. Eventually, the hailstones encounter a downdraft or grow too large to be supported by the updraft. (b) This cut hailstone, which fell over Coffeyville, Kansas, in 1970, weighed 0.75 kilogram (1.67 pounds). (Photo courtesy of University Corporation for Atmospheric Research/National Science Foundation/Visual Communications NCAR)

in diameter and weighed nearly 900 grams (2 pounds). The stone that held the previous record of 766 grams (1.69 pounds) fell in Coffeyville, Kansas, in 1970 (Figure 5-21b). The diameter of the stone found in South Dakota also surpassed the previous record of a 17.8-centimeter (7-inch) stone that fell in Aurora, Nebraska, in 2003. Even larger hailstones have reportedly been recorded in Bangladesh, where a 1987 hailstorm killed more than 90 people. It is estimated that large hailstones hit the ground at speeds exceeding 160 kilometers (100 miles) per hour.

SEVERE AND HAZARDOUS WEATHER — Worst Winter Weather

Extremes, whether they be the tallest building or the record low temperature for a location, fascinate us. When it comes to weather, some places take pride in claiming to have the worst winters on record. In fact, Fraser, Colorado, and International Falls, Minnesota, have both proclaimed themselves the "ice box of the nation." Although Fraser recorded the lowest temperature for the 48 contiguous states 23 times in 1989, its neighbor, Gunnison, Colorado, recorded the lowest temperature 62 times, far more than any other location.

Such facts do not impress the residents of Hibbing, Minnesota, where the temperature dropped to −38°C (−37°F) during the first week of March 1989. But this is mild stuff, say the old-timers in Parshall, North Dakota, where the temperature fell to −51°C (−60°F) on February 15, 1936. Not to be left out, Browning, Montana, holds the record for the most dramatic 24-hour temperature drop. Here the temperature plummeted 56°C (100°F), from a cool 7°C (44°F) to a frosty −49°C (−56°F) during a January evening in 1916.

Although impressive, the temperature extremes cited here represent only one aspect of winter weather. What about snowfall (Figure 5-D)? Cooke City holds the seasonal snowfall record for Montana, with 1062 centimeters (418.1 inches) during the winter of 1977–1978. But what about cities like Sault Ste. Marie, Michigan, and Buffalo, New York? The winter snowfalls associated with the Great Lakes are legendary. Even larger snowfalls occur in many sparsely inhabited mountainous areas.

Try telling residents of the eastern United States that heavy snowfall alone makes for the worst weather. A blizzard in March 1993 produced heavy snowfall along with hurricane-force winds and record low temperatures that immobilized much of the region from Alabama to the Maritime Provinces of eastern Canada. This event quickly earned the well-deserved title Storm of the Century.

FIGURE 5-D A winter blizzard of historic proportions struck Chicago Illinois, on February 2, 2011. (AP Photo/Kiichiro Sato).

So, determining which location has the worst winter weather depends on how you measure it. Most snowfall in a season? Longest cold spell? Coldest temperature? Most disruptive storm?

Winter Weather Events

Here are the meanings of some common terms that the National Weather Service uses for winter weather events.

Snow flurries Snow falling for short durations at intermittent periods and resulting in generally little or no accumulation.

Blowing snow Snow lifted from the surface by the wind and blown about to such a degree that horizontal visibility is reduced.

Drifting snow Significant accumulations of falling or loose snow caused by strong wind.

Blizzard A winter storm characterized by winds of at least 56 kilometers (35 miles) per hour for at least three hours. The storm must also be accompanied by low temperatures and considerable falling and/or blowing snow that reduces visibility to one-quarter mile or less.

Severe blizzard A storm with winds of at least 72 kilometers (45 miles) per hour, a great amount of falling or drifting snow, and temperatures −12°C (10°F) or lower.

Heavy snow warning A snowfall in which at least 4 inches (10 centimeters) in 12 hours or 6 inches (15 centimeters) in 24 hours is expected.

Freezing rain Rain falling in a liquid form through a shallow subfreezing layer of air near the ground. The rain (or drizzle) freezes on impact with the ground or other objects, resulting in a clear coating of ice known as *glaze*.

Sleet Also called *ice pellets*. Sleet is formed when raindrops or melted snowflakes freeze as they pass through a subfreezing layer of air near Earth's surface. Sleet does not stick to trees and wires, and it usually bounces when it hits the ground. An accumulation of sleet sometimes has the consistency of dry sand.

Travelers' advisory An alert issued to inform the public of hazardous driving conditions caused by snow, sleet, freezing precipitation, fog, wind, or dust.

Cold wave A rapid fall of temperature in a 24-hour period, usually signifying the beginning of a spell of very cold weather.

Wind chill A measure of apparent temperature that uses the effects of wind and temperature on the human body by translating the cooling power of wind to a temperature under calm conditions. It is an approximation only for the human body and has no meaning for cars, buildings, or other objects.

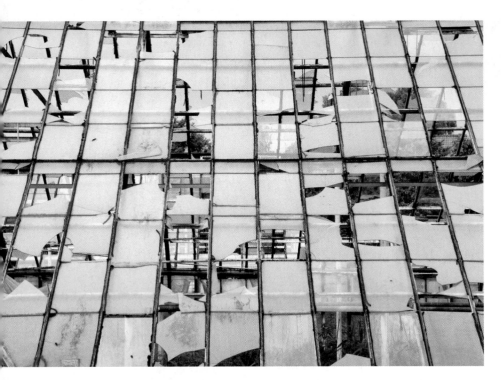

Figure 5-22 Hailstorm damage to a greenhouse. (Photo by McPHOTO/KPA/agefotostock)

Figure 5-23 Rime consists of delicate ice crystals that form when supercooled fog or cloud droplets freeze on contact with objects. (Photo by Siepman/Photolibrary)

The destructive effects of large hailstones are well known, especially to farmers whose crops have been devastated in a few minutes and to people whose windows, roofs, and cars have been damaged (Figure 5-22). In the United States, hail damage each year can run into the hundreds of millions of dollars. One of the costliest hailstorms to occur in North America took place June 11, 1990, in Denver, Colorado, with total damage estimated to exceed $625 million.

Rime

Rime is a deposit of ice crystals formed by the freezing of supercooled fog or cloud droplets on objects whose surface temperature is below freezing. When rime forms on trees, it adorns them with its characteristic ice feathers, which can be spectacular to observe (Figure 5-23). In these situations, objects such as pine needles act as freezing nuclei, causing the supercooled droplets to freeze on contact. On occasions when the wind is blowing, only the windward surfaces of objects will accumulate the layer of rime.

Concept Check 5.5

1 Compare and contrast rain, drizzle, and mist.

2 Describe sleet and freezing rain and the circumstances under which they form. Why does freezing rain result on some occasions and sleet on others?

3 How does hail form? What factors govern the ultimate size of hailstones?

Precipitation Measurement

The most common form of precipitation, rain, is probably the easiest to measure. Any open container that has a consistent cross section throughout can be a rain gauge (Figure 5-24a). In general practice, however, more sophisticated devices are used to measure small amounts of rainfall more accurately and to reduce loss from evaporation.

Standard Instruments

The **standard rain gauge** (Figure 5-24b) has a diameter of about 20 centimeters (8 inches) at the top. Once the water is caught, a funnel conducts the rain through a narrow opening into a cylindrical measuring tube that has a cross-sectional area only one-tenth as large as the receiver. Consequently, rainfall depth is magnified 10 times, which allows for accurate measurements to the nearest 0.025 centimeter

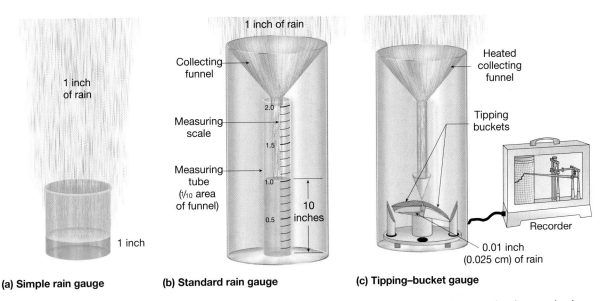

(a) Simple rain gauge   **(b) Standard rain gauge**   **(c) Tipping–bucket gauge**

Figure 5-24 Precipitation measurement. (a) The simplest gauge is any container left in the rain. (b) The standard rain gauge increases the height of water collected by a factor of 10, allowing for accurate rainfall measurement to the nearest 0.025 centimeter (0.01 inch). Because the cross-sectional area of the measuring tube is only one-tenth as large as the collector, rainfall is magnified 10 times. (c) The tipping-bucket rain gauge contains two "buckets" that each hold the equivalent of 0.025 centimeter (0.01 inch) of liquid precipitation. When one bucket fills, it tips and the other bucket takes its place. Each event is recorded as 0.01 inch of rainfall.

(0.01 inch). When the amount of rain is less than 0.025 centimeter (0.01 inch), it is generally reported as being a **trace of precipitation.**

In addition to the standard rain gauge, several types of recording gauges are routinely used. These instruments not only record the amount of rain but also its time of occurrence and intensity (amount per unit of time). Two of the most common gauges are the tipping-bucket gauge and the weighing gauge.

As Figure 5-24c illustrates, the **tipping-bucket gauge** consists of two compartments, each one capable of holding 0.025 centimeter (0.01 inch) of rain, situated at the base of a funnel. When one "bucket" fills, it tips and empties its water. Meanwhile, the other "bucket" takes its place at the mouth of the funnel. Each time a compartment tips, an electrical circuit is closed, and 0.025 centimeter (0.01 inch) of precipitation is automatically recorded on a graph.

A **weighing gauge** collects precipitation in a cylinder that rests on a spring balance. As the cylinder fills, the movement is transmitted to a pen that records the data.

Measuring Snowfall

When snow records are kept, two measurements are normally taken—depth and water equivalent (Figure 5-25). Usually, the depth of snow is measured with a calibrated stick. The actual measurement is not difficult, but choosing a representative spot can be. Even when winds are light or moderate, snow drifts freely. As a rule, it is best to take several measurements in an open place, away from trees and obstructions, and then average them. To obtain the water equivalent, samples may be melted and then weighed or measured as rain.

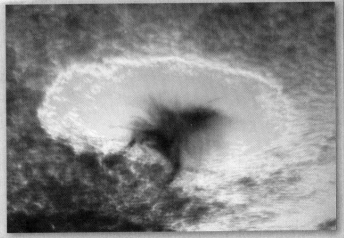

EYE ON THE ATMOSPHERE

This image shows a phenomenon called a *punch hole cloud* that was produced when a jet aircraft ascended through a cloud deck composed of supercooled water droplets. As the aircraft passed though the clouds, tiny particles in the jet- engine exhaust interacted with some of the supercooled water droplets—which froze instantly. The dark area in the center of the image is actually white and consists of large ice crystals that formed within the area of the cloud occupied by the punch hole and began to fall as precipitation. (Photo by H. Raab)

Question 1 What is the name of the processes whereby some cloud droplets freeze and grow in size at the expense of the remaining liquid cloud droplets?

Question 2 Explain why only clouds that are composed of supercooled droplets can develop "punch holes."

Question 3 Describe the role that a jet aircraft plays in the formation of punch hole clouds?

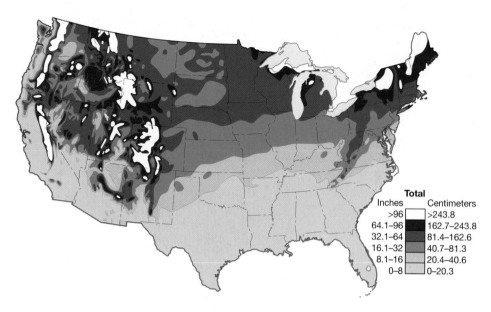

Figure 5-25 Average annual snowfall in the contiguous United States.

Precipitation Measurement by Weather Radar

The National Weather Service produces weather maps, like the one in Figure 5-26, based on images produced by *weather radar*. The development of weather radar has given meteorologists an important tool to track storm systems and the precipitation patterns they produce, even when the storms are as far as a few hundred kilometers away. A radar unit has a transmitter that sends out short pulses of radio waves. The specific wavelengths that are used depend on the objects the user wants to detect. When radar is used to monitor precipitation, wavelengths between 3 and 10 centimeters are employed.

These radio waves are able to penetrate small cloud droplets but are reflected by larger raindrops, ice crystals, and hailstones. The reflected signal, called an *echo*, is received and displayed on a TV monitor. Because the echo is "brighter" when the precipitation is more intense, modern radar is able to depict both the regional extent and rate of precipitation. Figure 5-26 is a typical radar display in which colors show precipitation intensity. As you will see in Chapters 10 and 12, weather radar can also measure the rate and direction of storm movement.

The quantity of water in a given volume of snow is not constant. A general ratio of 10 units of snow to 1 unit of water is often used when exact information is not available. You may have heard TV weathercasters use this ratio, saying, "Every 10 inches of snow equals 1 inch of rain." But the actual water content of snow may deviate widely from this figure. It may take as much as 30 centimeters of light and fluffy dry snow (30:1) or as little as 4 centimeters of wet snow (4:1) to produce 1 centimeter of water.

Concept Check 5.6

1 Although an open container can serve as a rain gauge, what advantages does a standard rain gauge provide?

2 Describe how recording rain gauges work.

3 What advantage does weather radar have over a standard rain gauge?

Figure 5-26 Doppler radar display produced by the National Weather Service. Colors indicate different intensities of precipitation. Note the band of heavy precipitation along the eastern seaboard. (Courtesy of NOAA)

Intentional Weather Modification

Intentional weather modification is deliberate human intervention to influence atmospheric processes that constitute the weather—that is, to alter the weather for human purposes. The desire to change or enhance certain weather phenomena dates back to ancient history, when people used prayer, wizardry, dances, and even black magic in attempts to alter the weather.

Weather-modification strategies fall into three broad categories. The first employs energy to forcefully alter the weather. Examples are the use of intense heat sources or the mechanical mixing of air (such as by helicopters) to disperse fog at some airports or keep fruit trees from freezing.

The second category involves modifying land and water surfaces to change their natural interaction with the lower atmosphere. One often discussed but untried example is the blanketing of a land area with a dark substance. The additional solar energy absorbed by this dark surface would warm the layer of air near the surface and encourage the development of updrafts that might aid cloud formation and ultimately produce precipitation.

The third category involves triggering, intensifying, or redirecting atmospheric processes. The seeding of clouds with agents such as dry ice (frozen carbon dioxide) and silver iodide to stimulate precipitation is the primary example. Because **cloud seeding** has shown some promising results and is a relatively inexpensive technique, it has been a primary focus of modern weather-modification technology.

Snow and Rain Making

The first breakthrough in weather modification came in 1946, when Vincent J. Schaefer discovered that dry ice, dropped into a supercooled cloud, spurred the growth of ice crystals. Recall that once ice crystals form in a supercooled cloud, they grow larger, at the expense of the remaining liquid cloud droplets and, upon reaching a sufficient size, fall as precipitation.

Later it was learned that silver iodide crystals could also be used for *cloud seeding*. Unlike dry ice, which simply chills the air, silver iodide crystals act as freezing nuclei. Because silver iodide can be easily delivered to clouds from burners on the ground or from aircraft, it is a more cost-effective alternative than dry ice (Figure 5-27).

For cloud seeding to trigger precipitation, certain atmospheric conditions must exist. Most importantly, a portion of the cloud must be supercooled—that is, consist of liquid droplets with temperatures at or below 0°C (32°F).

Seeding of winter clouds that form along mountain barriers (orographic clouds) has been tried on numerous occasions. These clouds are thought to be good candidates for seeding because, under normal conditions, only a small percentage of the water that condenses in cold orographic clouds actually falls as precipitation. Since 1977 Colorado's Vail and Beaver Creek ski areas have used this method to increase winter snows. An additional benefit to cloud seeding is that the increased precipitation, which melts and runs off during spring and summer months, can be collected in reservoirs for irrigation and hydroelectric power generation.

In recent years the seeding of warm convective clouds with hygroscopic (water-seeking) particles has received renewed attention. The interest in this technique arose when it was discovered that a pollution-belching paper mill near Nelspruit, South Africa, seemed to be triggering precipitation. Research aircraft flying through clouds near the paper mill collected samples of the particulate matter emitted from the mill. It turned out that the mill was emitting tiny salt crystals (potassium chloride and sodium chloride), which rose into the clouds. Because these salts attract moisture, they quickly form large cloud droplets, which grow into raindrops through the collision–coalescence process. Ongoing experiments being conducted over the arid landscape of northern Mexico are attempting to duplicate this process by seeding clouds using flares mounted on airplanes that spread hygroscopic salts. Although the results of these studies are promising, conclusive evidence that such seeding can increase rainfall over economically significant areas has not yet been established.

In the United States, more than 50 weather-modification projects are currently under way in 10 states. Some of the most promising results come from western Texas, where researchers estimate a 10 percent increase in rainfall from clouds seeded with silver iodide compared to those left unseeded.

Fog and Cloud Dispersal

One of the most successful applications of cloud seeding involves spreading dry ice (solid carbon dioxide) into layers of supercooled fog or stratus clouds to disperse them and

Figure 5-27 Cloud seeding using silver iodide flares is one way that freezing nuclei are supplied to clouds. (Courtesy of University Corporation for Atmospheric Research/National Science Foundation/National Center for Atmospheric Research)

Figure 5-28 Effects produced by seeding a cloud deck with dry ice. Within one hour a hole developed over the seeded area. (Photo courtesy of E. J. Tarbuck)

thereby improve visibility. Airports, harbors, and foggy stretches of interstate highway are obvious candidates. Such applications trigger a transformation in cloud composition from supercooled water droplets to ice crystals. The ice crystals then settle out, leaving an opening in the cloud or fog (Figure 5-28). The U.S. Air Force has practiced this technology for many years at airbases, and commercial airlines have used this method at selected foggy airports in the western United States.

In northern Utah, where supercooled fog can persist in mountain valleys for weeks, the state transportation department uses liquid carbon dioxide to disperse fog and improve visibility. When released, the liquid carbon dioxide evaporates quickly, thereby cooling the air and causing the supercooled droplets to freeze and precipitate as fine ice crystals.

Unfortunately, most fog does not consist of supercooled water droplets. The more common "warm fogs" are more expensive to combat because seeding will not diminish them. Successful attempts at dispersing warm fogs have involved mixing drier air from above into the fog. When the layer of fog is very shallow, helicopters have been used. By flying just above the fog, the helicopter creates a strong downdraft that forces drier air toward the surface, where it mixes with the saturated foggy air.

At some airports where warm fogs are common, it is becoming more common to heat the air and thus evaporate the fog. A sophisticated thermal fog-dissipation system, called Turboclair, was installed in 1970 at Orly Airport in Paris. It uses eight jet engines in underground chambers along the upwind edge of the runway. Although expensive to install, the system is capable of improving visibility for about 900 meters (3000 feet) along the approach and landing zones.

Hail Suppression

Each year hailstorms inflict an average of $500 million in property damage and crop loss in the United States (Figure 5-29). Occasionally, a single severe hailstorm can produce damages that exceed that amount. Examples include large hailstorms that occurred in Germany in 1984, Canada in 1991, and Australia in 1999; in addition, a 2001 Kansas City hailstorm inflicted an estimated $2 billion in damage. As a result, some of history's most interesting efforts at weather modification have focused on hail suppression.

Farmers who are desperate to find ways to save their crops have long believed that strong noises—explosions, cannon shots, or ringing church bells—can help reduce the amount of hail that is produced during a thunderstorm. In Europe it was common practice for village priests to ring church bells to shield nearby farms from hail. Although this practice was banned in 1780 due to bell ringers being killed by lightning, it is still employed in a few locations.

In 1880, a new hypothesis of hail formation sparked renewed interest in hail suppression. This untested proposal suggested that hailstones form in cold moist clouds during calm conditions. (We now know that large hailstones form in clouds with strong vertical air currents.) In this environment the pointed ends of tiny ice needles were thought to collect around a central embryo to form a spherical ice pel-

Figure 5-29 Hail damage to a soybean field north of Sioux Falls, South Dakota. (AP/Wide World Photos)

let. Once a layer was complete, another layer of ice crystals was added, which explains the layered structure of hailstones. It was also believed that the migration of the tiny crystals toward the embryo would not take place if there was chaos in the environment.

This incorrect hypothesis of hail formation led to the development of the first "anti-hail cannons" in the 1890s. These hail cannons were vertical muzzle-loading mortars that resembled huge megaphones. The one shown in Figure 5-30 weighted 9000 kilograms (10 tons), was 9 meters (30 feet) tall, and pivoted in all directions. When fired, the cannons produced a loud whistling noise and a large smoke ring that rose to a height of about 300 meters (1000 feet). The noise from the cannons was thought to disrupt the development of large hailstones.

During a two-year test of hail cannons near the town of Windisch-Feistritz, Austria, no hail was observed. At the same time, nearby provinces suffered severe hail damage. From this unscientific study, it was concluded that hail cannons were effective. Use of these devices spread to other areas in Europe where crops of great value frequently experienced hail losses. Soon much of Europe acquired "cannon fever," and by 1899 more than 2000 cannons were in use in Italy alone. Over the years, however, the anti-hail cannons proved to be ineffective, and the practice was largely abandoned.

In 1972 a French company began producing a modern version of the anti-hail cannon, and today several small companies still manufacture these devices. In the United States they can be found in scattered sites in Colorado, Texas, and California. A Japanese auto manufacturer has installed anti-hail cannons to protect new cars at a plant located in Mississippi.

Modern attempts at hail suppression have employed various methods of cloud seeding using silver iodide crystals to disrupt the growth of hailstones. In an effort to verify the effectiveness of cloud-seeding, the U.S. government established the National Hail Research Experiment in northeastern Colorado. This experiment included several randomized cloud-seeding experiments. An analysis of the data collected after three years revealed no statistically significant difference in the occurrence of hail between the seeded and non-seeded clouds, so the planned five-year experiment was abandoned. Nevertheless, cloud seeding is still employed today in many locations, including Colorado.

Frost Prevention

Frost, a strictly temperature-dependent phenomena, occurs when the air temperature falls to 0°C (32°F) or below. Deposits of ice crystals, called *white frost,* form only when the air becomes saturated.

A frost or freeze hazard can be generated in two ways: when a cold air mass moves into a region or when sufficient radiation cooling occurs on a clear night. Frost associated with an invasion of cold air is characterized by low daytime temperatures and long periods of freezing conditions that inflict widespread crop damage. By contrast, frost induced by radiation cooling is strictly a nighttime phenomenon that tends to be confined to low-lying areas. Obviously, the latter phenomenon is much easier to combat.

Several methods of frost prevention are being used with varying success. They either conserve heat (reduce the heat lost at night) or add heat to warm the lowermost layer of air.

Heat-conservation methods include covering plants with insulating material, such as paper or cloth, and generating particles that, when suspended in air, reduce the rate of radiation cooling. Warming methods employ *water sprinklers, air-mixing techniques,* and/or *orchard heaters.* Sprinklers (Figure 5-31a) add heat in two ways: first, from the warmth of the water and, more importantly, from the latent heat of fusion released when the water freezes. As long as an ice–water mixture remains on the plants, the latent heat released will keep the temperature from dropping below 0°C (32°F).

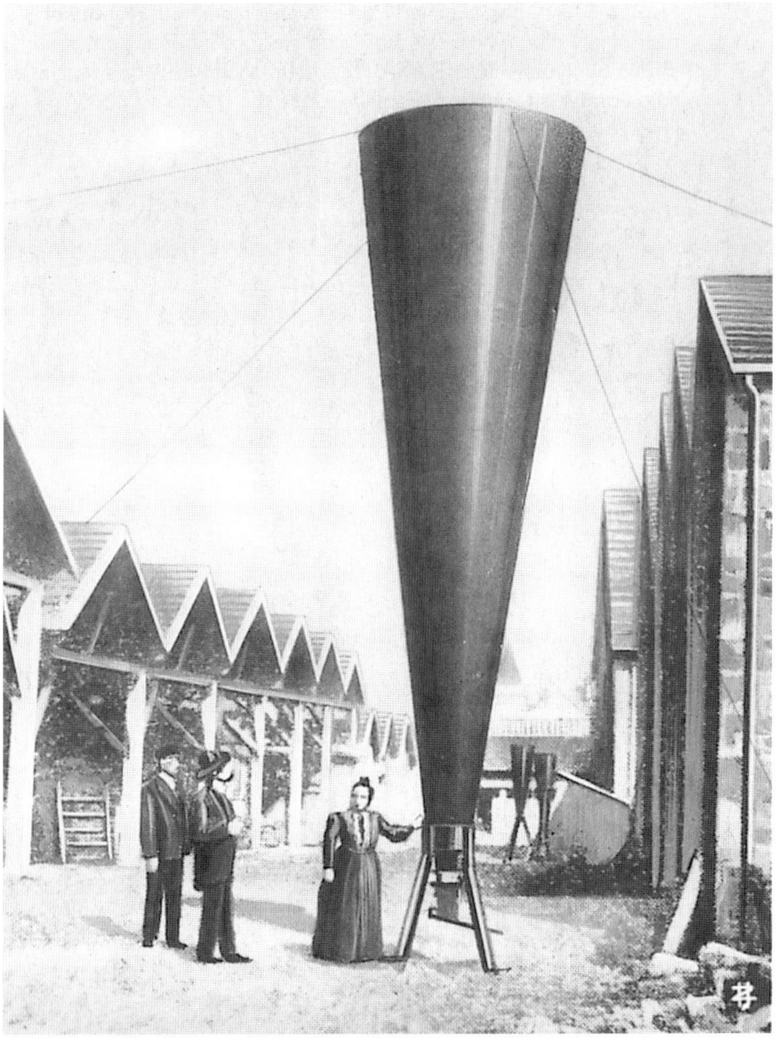

Figure 5-30 Around 1900, cannons like this fired smoke particles into thunderstorms to prevent the formation of hail. However, hail shooting proved to be ineffective, and the practice was largely abandoned after a decade. (Photo courtesy of J. Loreena Ivens, Illinois State Water Survey)

(a)

(b)

Figure 5-31 Two common frost-prevention methods. (a) Sprinklers distribute water, which releases latent heat as it freezes on citrus. (Photo by Ryan Anson/ZUMA Press/Newscom) (b) Wind machines mix warmer air aloft with cooler surface air. (Photo by ZUMA Press/Newscom)

Air mixing works best when the air temperature 15 meters (50 feet) above the ground is at least 5°C (9°F) warmer than the surface temperature. By using a wind machine, the warmer air aloft is mixed with the colder surface air (Figure 5-31b).

Orchard heaters probably produce the most successful results (Figure 5-32). However, because as many as 30 to 40 heaters are required per acre, fuel cost can be significant.

Concept Check 5.7

1 What is meant by the phrase "intentional weather modification."

2 In order for cloud seeding to work, what atmospheric condition must exist?

3 Describe one technique used to disperse "warm fog" to improve visibility.

4 How do frost and white frost differ from one another?

5 Describe how orchard heaters, sprinkling, and air mixing are used in frost prevention.

Figure 5-32 Orchard heaters used to prevent frost damage to pear trees, Hood River, Oregon. (Photo by Exactostock/Superstock)

Give It Some Thought

1. Clouds are classified into three major height categories: low, middle, and high. Explain why low clouds (or clouds of vertical development) are much more likely to produce precipitation than middle or high clouds.

2. Which of the three basic cloud forms (cirrus, cumulus, or stratus) is illustrated by each of the accompanying images labeled a–f?

(a)   (b)   (c)

(d)   (e)   (f)

3. Fog can be defined as a cloud with its base at or very near the ground, yet fogs and clouds form by different processes. Describe the similarities and differences in the formation of fogs and clouds.

4. Why does radiation fog form mainly on clear nights as opposed to cloudy nights?

5. The accompanying three images (a–c) depict three types of fog.
   a. Identify each fog type shown.
   b. Describe the mechanism that generated each of the fogs.

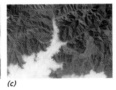

(a)   (b)   (c)

6. Assume that it is a midwinter day in central Illinois. Mild conditions prevail because of a steady wind from the south. As the day progresses, fog forms across a broad area. Identify the likely type of fog.

7. Imagine that you are driving in a hilly area early in the morning and encounter fog as you descend into the valleys, but as you drive out of the valleys, the conditions clear. Identify the likely type of fog.

8. Cloud droplets form and grow as water vapor condenses onto hydroscopic condensation nuclei. Research has shown that the maximum radius for cloud droplets is about 0.05 millimeter. However, typical raindrops are thousands of times larger. Describe in your own words one way in which cloud droplets become raindrops.

9. Describe or sketch the vertical temperature profile that would cause precipitation that began as snow to reach the ground as each of the following:
   a. Snow
   b. Rain
   c. Freezing rain

10. Why is it unlikely to have virga and fog occur simultaneously?

11. The accompanying images show two different types of precipitation after they have been deposited.
    a. Name the types of precipitation shown.
    b. Describe the nature (form) of each type of precipitation before it was deposited.
    c. In what way are these types of precipitation similar to one another?
    d. In what way are they different?

(a)   (b)

12. What are the advantages and disadvantages of using rain gauges compared to weather radar in measuring rainfall?

13. Weather radar provides information on the intensity as well as the total amount of precipitation. Table A shows the relationship between radar reflectivity values and rainfall rates. If radar measured a reflectivity value of 47 dBZ for 2½ hours at a particular location, how much rain has fallen there?

Table A Conversion of radar reflectivity to rainfall rate	
Radar Reflectivity (dBZ)	Rainfall Rate (inches/hr)
65	16+
60	8.0
55	4.0
52	2.5
47	1.3
41	0.5
36	0.3
30	0.1
20	trace

14. Cloud seeding efforts have met with varying degrees of success. If humans develop effective cloud seeding techniques that reliably increase rainfall in designated areas, what, if any, drawbacks might there be?

FORMS OF CONDENSATION AND PRECIPITATION IN REVIEW

- *Condensation* occurs when water vapor changes to a liquid. For condensation to take place, the air must be saturated and there must be a surface on which the vapor can condense.
- *Clouds,* visible aggregates of minute droplets of water and/or tiny crystals of ice, are one form of condensation. Clouds are classified on the basis of two criteria: form and height. The three basic cloud forms are *cirrus* (high, white, and thin), *cumulus* (globular, individual cloud masses), and *stratus* (sheets or layers). Cloud heights can be either *high,* with bases above 6000 meters (20,000 feet); *middle,* from 2000 to 6000 meters; or *low,* below 2000 meters (6500 feet). *Clouds of vertical development* have bases in the low height range and extend upward into the middle or high range. Based on the two criteria, 10 basic cloud types exist.
- *Fog,* generally considered an atmospheric hazard, is a cloud with its base at or very near the ground. Fogs formed by cooling include *radiation fog, advection fog,* and *upslope fog.* Fogs formed by the addition of water vapor are *steam fog* and *frontal fog.*
- *Dew* is the condensation of water vapor on objects that have radiated sufficient heat to lower their temperature below the dew point of the surrounding air. *White frost* forms when the dew point of the air is below freezing.
- For precipitation to form, millions of cloud droplets must coalesce into drops large enough to sustain themselves during their descent. The two mechanisms that have been proposed to explain this phenomenon are the *Bergeron process,* which produces precipitation from cold clouds primarily in the middle latitudes, and the warm-cloud process most associated with the tropics, called the *collision–coalescence process.*
- The two most common and familiar forms of precipitation are *rain* and *snow.* Other forms include *sleet, freezing rain (glaze), hail,* and *rime.*
- The most common instruments used to measure rain are the *standard rain gauge,* which is read directly, and the *tipping-bucket gauge* and *weighing gauge,* both of which record the amount and intensity of rain. The two most common measurements of snow are depth and water equivalent. Although the quantity of water in a given volume of snow is not constant, a general ratio of 10 units of snow to 1 unit of water is often used when exact information is not available.
- *Intentional weather modification* is deliberate human intervention to influence atmospheric processes that constitute the weather. Weather modification falls into three categories: (1) using energy to forcefully alter the weather; (2) modifying land and water surfaces to change their natural interaction with the lower atmosphere; and (3) triggering, intensifying, or redirecting atmospheric processes.

VOCABULARY REVIEW

advection fog (p. 138)
Bergeron process (p. 141)
cirrus (p. 131)
cloud (p. 130)
cloud condensation nucleus (p. 130)
cloud seeding (p. 153)
cloud of vertical development (p. 132)
collision–coalescence process (p. 143)
cumulus (p. 131)
fog (p. 137)
freezing nucleus (p. 142)
freezing rain, or glaze (p. 147)

frontal, or precipitation, fog (p. 140)
frost (p. 155)
hail (p. 147)
high cloud (p. 132)
hydrophobic nucleus (p. 130)
hygroscopic nucleus (p. 130)
intentional weather modification (p. 152)
low cloud (p. 132)
middle cloud (p. 132)
radiation fog (p. 137)
rain (p. 145)
rime (p. 150)

sleet (p. 146)
snow (p. 146)
standard rain gauge (p. 150)
steam fog (p. 138)
stratus (p. 131)
supercooled (p. 142)
tipping-bucket gauge (p. 151)
trace of precipitation (p. 151)
upslope fog (p. 138)
weighing gauge (p. 151)

PROBLEMS

1. Suppose that the air temperature is 20°C (68°F) and the relative humidity is 50 percent at 6:00 PM and that during the evening, the air temperature drops but its water vapor content does not change. If the air temperature drops 1°C (1.8°F) every two hours, will fog occur by sunrise (6:00 AM) the next morning? Explain your answer. (*Hint:* The data you need are found in Table 4–1, p. 105.)

2. By using the same conditions as in Problem 1, will fog occur if the air temperature drops 1°C (1.8°F) every hour? If so, when will it first appear? What name is given to fog of this type that forms because of surface cooling during the night?

3. Assuming that the air is still, how long would it take a large raindrop (5 mm) to reach the ground if it fell from a cloud base at 3000 meters? (See Table 5-3, p. 143.) How long would a typical raindrop (2 mm) take to fall to the ground from the same cloud? How long if it were a drizzle drop (0.5 mm)?

4. Assuming that the air is still, how long would it take for a typical cloud droplet (0.02 mm) to reach the ground if it fell from a cloud base at 1000 meters? (See Table 5-3.) It is very unlikely that a cloud droplet would ever reach the ground, even if the air were perfectly still. Explain why.

Figure 5-33 Cumulonimbus clouds are often associated with thunderstorms and severe weather. (Photo by Rolf Nussbaumer Photography/Alamy)

5. The dimensions of the large cumulonimbus cloud pictured in Figure 5-33 are roughly 12 kilometers high, 8 kilometers wide, and 8 kilometers long. Assume that the droplets in every cubic meter of the cloud total 0.5 cubic centimeter of water. How much liquid water does the cloud contain? How many gallons does it contain ()?

6. The record for the largest hailstone in the United States was set on July 23, 2010, in Vivian, South Dakota. This stone was 8 inches in diameter and 18.62 inches in circumference, and it weighed nearly 2 pounds. The maximum fall speed of a spherical hailstone of diameter d can be calculated using , where if d is in centimeters and V is in meters per second. Estimate the strength (speed) of the updrafts in this storm that were necessary to support the stone just before it began to fall. Convert your answer from meters per second to feet per second and then to miles per hour.

MyMeteorologyLab™

Log in to www.mymeteorologylab.com for animations, videos, MapMaster interactive maps, GEODe media, *In the News* RSS feeds, web links, glossary flashcards, self-study quizzes and a Pearson eText version of this book to enhance your study of *Forms of Condensation* and *Precipitation*.

Air Pressure and Winds

Of the various elements of weather and climate, changes in air pressure are the least perceptible to humans; however, they are very important in producing changes in our weather. For example, variations in air pressure generate winds that trigger changes in temperature and humidity. In addition, air pressure is a significant factor in weather forecasting and is closely tied to the other elements of weather (temperature, moisture, and wind) in cause-and-effect relationships.

Horizontal differences in air pressure created the winds that propelled these wind surfers. (Photo by Sam Pellissier / SuperStock)

Focus On Concepts

After completing this chapter, you should be able to:

- Define air pressure and explain why it changes with altitude.

- Explain the principle of the mercury barometer and how it differs from the aneroid barometer.

- Describe how horizontal pressure differences are generated and how differences in solar heating can cause global variations in air pressure.

- List and describe the three forces that act on the atmosphere to either create or alter winds.

- Explain why the winds aloft flow roughly parallel to the isobars, while surface winds travel at an angle across the isobars.

- Describe how horizontal airflow (wind) can generate vertical air motion.

- Contrast the weather associated with low-pressure centers (cyclones) and high-pressure centers (anticyclones).

- Describe how wind direction is expressed using compass directions.

Wind and Air Pressure

We know that air moves *vertically* if it is forced over a barrier or if it is warmer and thus more buoyant than surrounding air. But what causes air to move *horizontally*—the phenomenon we call **wind** (Figure 6-1)? Simply stated, *wind is the result of horizontal differences in atmospheric pressure.* Air flows from areas of higher pressure to areas of lower pressure. You may have experienced this condition when opening a can of something vacuum packed. The noise you hear is caused by air rushing from the area of higher pressure outside the can to the lower pressure inside. Wind is nature's attempt to balance inequalities in air pressure.

We live at the bottom of the atmosphere, and just as the creatures living at the bottom of the ocean are subjected to pressure exerted by water, humans are subjected to the pressure exerted by the weight of the atmosphere above. Although we do not generally notice the pressure exerted by the ocean of air around us (except when rapidly ascending or descending in an elevator or airplane), it is nonetheless substantial. The pressurized suits used by astronauts on space walks were designed to duplicate the atmospheric pressure experienced at Earth's surface. Without these protective suits to keep body fluids from boiling away, astronauts would perish in minutes.

Air pressure can be visualized by examining the behavior of gases. Gas molecules, unlike those of the liquid and solid phases, are not "bound" to one another but are freely moving about, filling all space available to them. When two gas molecules collide, which happens frequently under normal atmospheric conditions, they bounce off each other like elastic balls. If a gas is confined to a container, this motion is restricted by its sides, much as the walls of a hand-

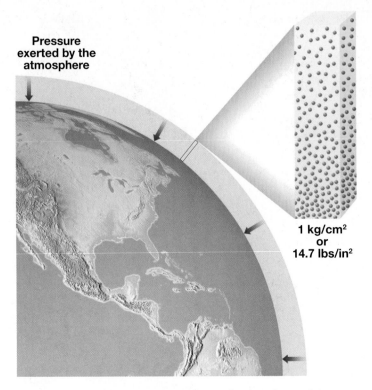

Pressure exerted by the atmosphere

1 kg/cm²
or
14.7 lbs/in²

Figure 6-2 Average air pressure at sea level is about 1 kilogram per square centimeter, or 14.7 pounds per square inch.

ball court redirect the motion of a handball. The continuous bombardment of gas molecules against the sides of the container exerts an outward force called *air pressure.* Although the atmosphere is without walls, it is confined from below by Earth's surface and effectively from above because the *force of gravity* prevents its escape. Thus, we define **atmospheric pressure**, or simply **air pressure**, as the force exerted against a surface by the continuous collision of gas molecules.

Average air pressure at sea level is about 1 kilogram per square centimeter, or 14.7 pounds per square inch. Specifically, a column of air 1 square inch in cross section, measured from sea level to the top of the atmosphere, would weigh about 14.7 pounds (Figure 6-2). (This is roughly the same pressure produced by a 1-square-inch column of water 10 meters [33 feet] in height.) The pressure that air exerts at Earth's surface is much greater than most people realize. For example, the air pressure exerted on the top of a small (50-centimeter-by-100-centimeter) school desk exceeds 5000 kilograms (11,000 pounds), or about the weight of a 50-passenger school bus. Why doesn't the desk collapse under the weight of the air above? Simply, air pressure is exerted in all directions—down, up, and sideways. Thus, the air pressure around all sides of the desk is exactly balanced.

Imagine the base of a tall aquarium with the same dimensions as the desktop. When this aquarium is filled to a height of 10 meters (33 feet), the water pressure at the bottom equals 1 atmosphere (14.7 pounds per square inch).

Figure 6-1 Strong winds created these waves that battered the coast of Great Britain near Tynemouth. Ocean waves are energy traveling along the interface between the ocean and atmosphere, often transferring energy from a storm far out at sea. (Photo by Geoff Love/Alamy)

Now imagine what would happen if this aquarium were placed on top of the desk so that all the force was directed downward? By contrast, if the desk is placed inside the aquarium and allowed to sink to the bottom, the desk survives because the water pressure is exerted in all directions, not just downward, as in our earlier example. The desk, like a human body, is built to withstand the pressure of 1 atmosphere.

Concept Check 6.1

1 What is wind and what is its basic cause?

2 What is standard sea-level pressure, in pounds per square inch?

3 Describe atmospheric pressure in your own words.

Measuring Air Pressure

 GE●De
ATMOSPHERE   Air Pressure and Winds
▶ Measuring Air Pressure

When describing atmospheric pressure, meteorologists use a unit of force from physics called the **newton.*** Under average conditions (at sea level) the atmosphere exerts a force of 101,325 newtons per square meter. To simplify this large number, the National Weather Service adopted the **millibar (mb),** which equals 100 newtons per square meter. Thus, standard sea-level pressure is stated as 1013.25 millibars (Figure 6-3).****

You may be acquainted with the expression "inches of mercury," which weather reporters use to describe atmospheric pressure. This expression dates from 1643, when Torricelli, a student of the famous Italian scientist Galileo, invented the **mercury barometer.** Torricelli correctly described the atmosphere as a vast ocean of air that exerts pressure on us and all things about us. To measure this force, he closed one end of a glass tube and filled it with mercury. He then inverted the tube into a dish of mercury (Figure 6-4). Torricelli found that the mercury flowed out of the tube until the weight of the mercury column was balanced by the pressure exerted on the surface of the mercury by the air above. In other words, the weight of the mercury in the column equaled the weight of a similar-diameter column of air that extended from the ground to the top of the atmosphere.

Torricelli noted that when air pressure increased, the mercury in the tube rose; conversely, when air pressure decreased, so did the height of the column of mercury. The length of the column of mercury, therefore, became the measure of the air pressure–"inches of mercury." With some refinements, the mercury barometer invented by Torricelli remains the standard pressure-measuring instrument.

---

*A newton is the force needed to accelerate a 1 kilogram mass 1 meter per second squared.

**The standard unit of pressure in the SI system is the pascal, which is the name given to a newton per square meter (N/m^2). In this notation a standard atmosphere has a value of 101,325 pascals, or 101.325 kilopascals. If the National Weather Service officially converts to the metric system, it will probably adopt this unit.

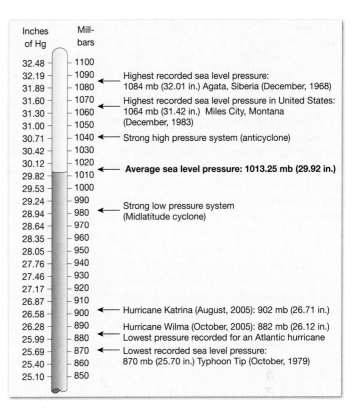

Figure 6-3 A comparison of atmospheric pressure in inches of mercury and millibars.

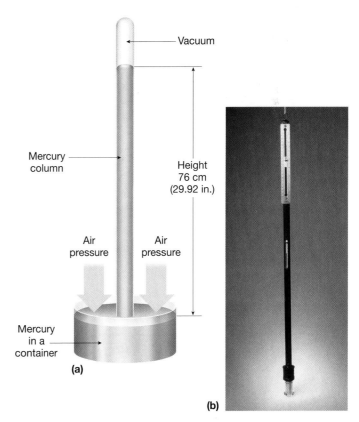

Figure 6-4 Mercury barometer. (a) The weight of the column of mercury is balanced by the pressure exerted on the dish of mercury by the air above. If the pressure decreases, the column of mercury falls; if the pressure increases, the column rises. (b) Image of a mercury barometer. (Photo by D. Winters/Photo Researchers, Inc.)

Standard atmospheric pressure at sea level equals 29.92 inches (760 millimeters) of mercury. In the United States the National Weather Service uses millibars on weather maps and charts, but it reports surface pressure to the public in inches of mercury.

The need for a smaller and more portable instrument for measuring atmospheric pressure led to the development of the **aneroid barometer** (*aneroid* means "without liquid"). Rather than a mercury column held up by air pressure, the aneroid barometer uses a partially evacuated metal chamber (Figure 6-5). The chamber, being very sensitive to pressure variations, changes shape, compressing as pressure increases and expanding as pressure decreases.

As shown in Figure 6-5, the face of an aneroid barometer intended for home use includes the words *fair, change, rain,* and *stormy*. Notice that "fair weather" corresponds with high-pressure readings, whereas "rain" is associated with low pressures. To "predict" weather in a local area, the change in air pressure over the past few hours is more important than the current pressure reading. Falling pressure is often associated with increasing cloudiness and the possibility of precipitation, whereas rising air pressure generally indicates clearing conditions.

Another advantage of the aneroid barometer is that it can easily be connected to a recording mechanism. This recording instrument, called a **barograph,** provides a con-

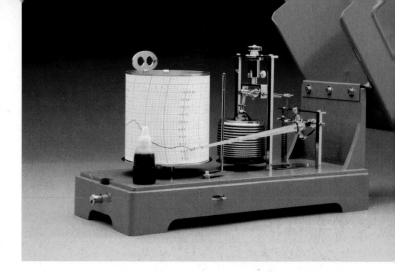

Figure 6-6 An aneroid barograph makes a continuous record of pressure changes. (Photo courtesy of Qualimetrics, Inc., Sacramento, California.)

tinuous record of pressure changes over time (Figure 6-6). Another important adaptation of the aneroid barometer is its use to indicate altitude for aircraft, mountain climbers, and mapmakers (see Box 6-1).

Concept Check 6.2

1 What is average sea-level pressure when measured in millibars? In inches of mercury?

2 Describe the operating principles of the mercury barometer and the aneroid barometer.

3 List two advantages of the aneroid barometer.

4 What can an aneroid barometer be used for other than measuring barometric pressure?

Students Sometimes Ask...

Why is mercury used in a barometer? I thought it was poisonous.

You are correct: Mercury poisoning can be quite serious. However, the mercury in a barometer is held in a reservoir where the chance of spillage is minimal. A barometer could be constructed using any number of different liquids, including water. The problem with a water-filled barometer is size. Because water is 13.6 times less dense than mercury, the height of a water column at standard sea-level pressure would be 13.6 times taller than that of a mercury barometer. Consequently, a water-filled barometer would need to be nearly 34 feet tall.

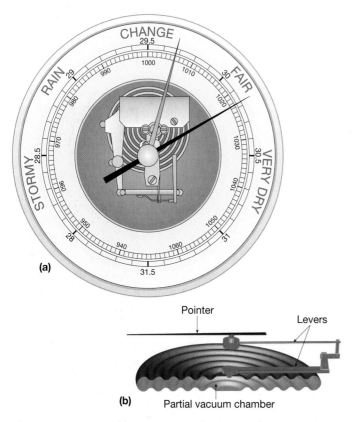

Figure 6-5 Aneroid barometer. (a) Illustration of an aneroid barometer. (b) The aneroid barometer has a partially evacuated chamber that changes shape, compressing as atmospheric pressure increases and expanding as pressure decreases.

Pressure Changes with Altitude

Just as scuba divers experience a decrease in water pressure as they rise toward the surface, we experience a decrease in air pressure as we ascend into the atmosphere. The relationship between air pressure and the air's density largely explains the drop in air pressure that occurs with altitude. Recall that at sea level a column of air weighs 14.7 pounds per square inch and therefore exerts that amount of pressure. As we ascend through the atmosphere, we find that the air becomes less dense because of the continual decrease in the amount (weight) of air above. Therefore, as would be expected, there is a corresponding *decrease in pressure with an increase in altitude.*

The fact that density decreases with altitude is why the term "thin air" is normally associated with mountainous regions. Except for the Sherpas (indigenous peoples of Nepal), most of the climbers who reach the summit of Mount Everest use supplementary oxygen for the final leg of the journey. Even with the aid of supplementary oxygen, many of these climbers experience periods of disorientation because of an inadequate supply of oxygen to their brains.

The decrease in pressure with altitude also affects the boiling temperature of water, which at sea level is 100°C (212°F). For example, in Denver, Colorado—the Mile High City—water boils at about 95°C (203°F). Although water comes to a boil faster in Denver than in San Diego, it takes longer to cook spaghetti in Denver because of its lower boiling temperature.

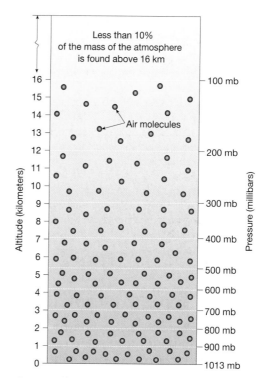

Figure 6-7 Illustration of the U.S. standard atmosphere. Each layer contains about 10 percent of the atmosphere. Notice that at roughly 5 kilometers, the pressure is about one-half of its sea-level value.

Recall from Chapter 1 that the rate at which pressure decreases with altitude is not constant. The rate of decrease is much greater near Earth's surface, where pressure is high, than aloft, where air pressure is low. A model of the **U.S. standard atmosphere,** shown in Figure 6-7, depicts the idealized vertical distribution of atmospheric pressure at various altitudes.

Near Earth's surface, air pressure decreases by about 10 millibars for every 100-meter increase in elevation, or about a 1-inch drop of mercury for every 1000-foot rise in elevation. Further, atmospheric pressure is reduced by approximately one-half for each 5-kilometer increase in altitude. Therefore, at 5 kilometers the pressure is 500 millibars, about one-half its sea-level value; at 10 kilometers it is one-fourth, at 15 kilometers it is one-eighth, and so forth. Thus, at the altitude at which commercial jets fly (10 kilometers), the air exerts a pressure equal to only one-fourth that at sea level.

(AP Photo/ Jae C. Hong)

In May and June 2011, the Wallow Fire burned more than 538 thousand acres (840 square miles) in southeastern Arizona. It was the largest wildfire in Arizona history.

Question 1 Can you suggest two ways in which wind helps sustain wildfires such as this?

Concept Check 6.3

1 Explain why air pressure decreases with an increase in altitude.

2 Why do climbers use supplemental oxygen when summiting Mount Everest?

3 What is the U.S. standard atmosphere?

Why Does Air Pressure Vary?

Why does atmospheric pressure vary daily, and why is that important? Recall that variations in air pressure cause the wind to blow, which in turn causes changes in temperature and humidity. In short, difference in air pressure create global winds that become organized into the systems that bring us our weather. Therefore, the National Weather Service closely monitors daily changes in air pressure because this information is critical to forecasting.

Although vertical pressure changes are important, meteorologists are much more interested in the horizontal pressure differences that occur daily from place to place around the globe.

Pressure differences from place to place are relatively small. Extreme readings are rarely more than 30 millibars (1 inch of mercury) above average sea-level pressure or 60 millibars (2 inches) below average sea-level pressure. Occasionally, the barometric pressure measured in severe storms, such as hurricanes, is lower (see Figure 6-3). As you will see, these small differences in air pressure can be sufficient to generate violent winds.

Influence of Temperature on Air Pressure

How do pressure differences arise? One of the easiest ways to envision this is to picture northern Canada in midwinter. Here the snow-covered surface is continually radiating heat to space, while receiving minimal incoming solar radiation. The frigid ground cools the air above so that daily lows of –34°C (–30°F) are common.

Recall that temperature is a measure of the average molecular motion (kinetic energy) of a substance. As a result, cold Canadian air is composed of comparatively slow-moving gas molecules that are packed closely together. These cold, dense air masses are associated with high surface pressures and are labeled *highs* (H) on weather maps. By contrast, during the summer, large areas of the American Southwest experience extremely high temperatures that are accompanied by low surface pressures labeled *lows* (L) on weather maps. Although a cold air column is generally

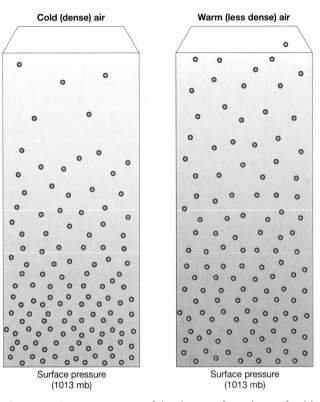

Figure 6-8 A comparison of the density of a column of cold air and a column of warm air. With an increase in altitude, air pressure drops more rapidly in the column of cold air.

associated with high surface pressure, and a warm air column is often associated with low surface pressure, other factors also influence surface pressures. These factors are examined in the next two sections.

Figure 6-8 shows another important difference between cold and warm air: *Air pressure drops more rapidly with altitude in a column of cold (dense) air than in a column of warm (less dense) air.* For purposes of illustration only, we assume that both columns of air exert the same surface pressure, and (although greatly exaggerated) differences in the spacing of air molecules represent differences in density. Looking at the line drawing halfway up the two columns, notice that there are more air molecules above this altitude in the warm column than in the cold column (Figure 6-8). As a result, *warm air aloft tends to exhibit a higher pressure than cold air at the same altitude.* This important concept has implications in aviation (see Box 6-1) and will be considered in more detail when we discuss the forces that generate airflow.

Influence of Water Vapor on Air Pressure

Although of less importance, factors other than temperature affect the amount of pressure a column of air will exert. For example, the amount of water vapor contained in a volume of air influences its density. Contrary to popular perception, water vapor *reduces* the density of air. The air may feel "heavy" on hot, humid days, but it is not. You can easily verify this fact for yourself by examining a periodic table of the elements and noting that the molecular weights of nitrogen (N_2) and oxygen (O_2) are greater than that of water vapor

Box 6-1 Air Pressure and Aviation

The cockpit of nearly every aircraft contains a *pressure altimeter*, an instrument that allows a pilot to determine the altitude of a plane. A pressure altimeter is essentially an aneroid barometer marked in meters instead of millibars and, as such, responds to changes in air pressure. For example, Figure 6-7 shows that the standard atmospheric pressure of about 800 millibars "normally" occurs at a height of 2000 meters. Therefore, when the pressure reaches 800 millibars, the altimeter will indicate an altitude of 2000 meters.

Because of temperature variations and moving pressure systems, actual conditions are usually different from that shown by an aircraft's altimeter. When the air is warmer than predicted by the standard atmosphere, the plane will fly higher than

the height indicated by the altimeter. By contrast, in cold air the plane will fly lower than indicated. This could be especially dangerous if the pilot is flying a small plane through mountainous terrain with poor visibility (Figure 6-A). To avoid dangerous situations, pilots make altimeter corrections before takeoffs and landings, and in some cases they make corrections en route as well.

Above about 5.5 kilometers (18,000 feet), pressure changes are more gradual, so corrections cannot be made as precisely as at lower levels. Consequently, commercial aircraft have their altimeters set at the standard atmosphere and fly paths of constant pressure, called *flight levels*, instead of constant altitude (Figure 6-B). In other words, when an aircraft flies at a constant altimeter setting, a pressure

variation will result in a change in the plane's altitude. When pressure increases (warm air column) along a flight path, the plane will climb, and when pressure decreases (cold air column), the plane will descend. There is little risk of midair collisions because nearby aircraft are assigned different flight levels in order to maintain sufficient separation.

Large commercial aircraft also use radar altimeters to measure heights above the terrain. The time required for a radio signal to reach the surface and return is used to accurately determine the height of the plane above the ground. This system is not without drawbacks. Because a radar altimeter provides the elevation above the ground rather than above sea level, a knowledge of the underlying terrain is required. However, radar altimeters are useful when measuring the height above ground level during landing.

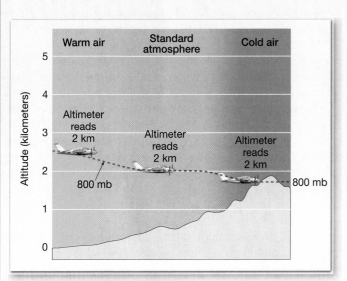

FIGURE 6-A Aircraft altimeters are calibrated based on the relationship between air pressure and altitude, using data provided by the standard atmosphere. However, when a plane enters a warm air column, its altitude is higher than indicated by the altimeter. By contrast, when a plane flies into a cold air mass, its altitude is lower than shown on the altimeter, which can be a problem in mountainous terrains.

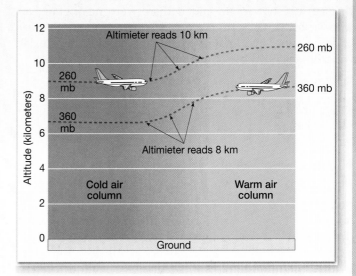

FIGURE 6-B Commercial aircraft above 5.5 kilometers (18,000 feet) generally fly paths of constant pressure instead of constant altitude.

(H_2O). In a mass of air the molecules of these gases are intermixed, and each takes up roughly the same amount of space. As the water content of an air mass increases, lighter water vapor molecules displace heavier nitrogen and oxygen molecules. Therefore, humid air is lighter (less dense) than dry air. Nevertheless, even very humid air is only about 2 percent less dense than dry air at the same temperature.

In summary, *cold, dry air produces higher surface pressures than warm, humid air. Further, a warm, dry air mass exhibits higher pressure than an equally warm, but humid, air mass. Conse-*

quently, differences in temperature and to a lesser extent moisture content are largely responsible for the pressure variations observed at Earth's surface.

Airflow and Pressure

The movement of air can also cause variations in air pressure. For example, in situations where there is a net flow of air into a region, a phenomenon called **convergence**, air accumulates. As it converges horizontally, the air is squeezed into a smaller space, which results in a more massive air column that exerts more pressure at the surface. By contrast, in regions where there is a net outflow of air, a situation referred to as **divergence,** the surface pressure drops. We will return to the important mechanisms of convergence and divergence, which help generate areas of high and low pressure, later in this chapter.

In summary, *the pressure at the surface will increase when there is a net convergence in a region and decrease when there is a net divergence.*

Concept Check 6.4

1 Explain why a cold, dry air mass produces a higher surface pressure than a warm, humid air mass.

2 If all other factors are equal, does a *dry* or *moist* air mass exert more air pressure? Explain.

3 Explain how horizontal *convergence* affects surface pressure.

Factors Affecting Wind

 Air Pressure and Winds
▶Factors Affecting Wind

If Earth did not rotate and if there were no friction, air would flow directly from areas of higher pressure to areas of lower pressure. However, because both factors exist, wind is controlled by a combination of forces, including:

1. Pressure gradient force
2. Coriolis force
3. Friction

Pressure Gradient Force

If an object experiences an unbalanced force in one direction, it will accelerate (experience a change in velocity). The force that generates winds results from horizontal pressure differences. When air is subjected to greater pressure on one side than on another, the imbalance produces a force that is directed from the region of higher pressure toward the area of lower pressure. Thus, pressure differences cause the wind to blow, and the greater these differences, the greater the wind speed.

Variations in air pressure over Earth's surface are determined from barometric readings taken at hundreds of weather stations. These pressure measurements are shown on surface weather maps using **isobars** (*iso* = "equal," *bar* = "pressure") lines connecting places of equal air pressure (Figure 6-9). The *spacing* of the isobars indicates the amount of pressure change occurring over a given distance and is called the **pressure gradient force.** Pressure gradient is analogous to gravity acting on a ball rolling down a hill. A steep pressure gradient, like a steep hill, causes greater acceleration of a parcel of air than does a weak pressure gradient (a gentle hill). Thus, the relationship between wind speed

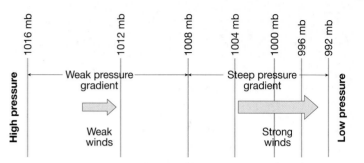

Figure 6-9 Isobars are lines connecting places of equal atmospheric pressure. The spacing of isobars indicates the amount of pressure change occurring over a given distance—called the *pressure gradient force.* Closely spaced isobars indicate a strong pressure gradient and high wind speeds, whereas widely spaced isobars indicate a weak pressure gradient and low wind speeds.

and the pressure gradient is straightforward: *Closely spaced isobars indicate a steep pressure gradient and strong winds; widely spaced isobars indicate a weak pressure gradient and light winds.* Figure 6-9 illustrates the relationship between the spacing of isobars and wind speed. Note also that the pressure gradient force is always directed at *right angles* to the isobars.

In order to draw isobars on a weather map to show air pressure patterns, meteorologists must compensate for the *elevation* of each station. Otherwise, all high-elevation locations, such as Denver, Colorado, would always be mapped as having low pressure. This compensation is accomplished by converting all pressure measurements to sea-level equivalents (Figure 6-10). Doing so requires meteorologists to determine the pressure that would be exerted by an imaginary column of air equal in height to the elevation of the recording station and adding it to the station's pressure reading. Because temperature greatly affects the density, and hence the weight of this imaginary column, temperature is also considered in the calculations. Thus, the corrected reading would give the pressure, as if it were taken at sea level under the same conditions.

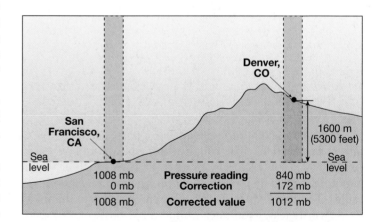

Figure 6-10 To compare atmospheric pressures, meteorologists first convert all pressure measurements to sea-level values. This is done by adding the pressure that would be exerted by an imaginary column of air (shown in red) to the station's pressure reading.

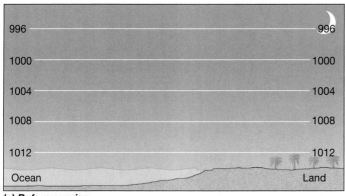

(a) Before sunrise

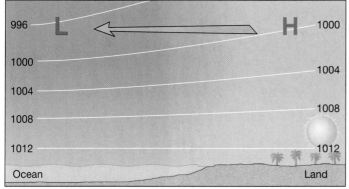

(b) Shortly after sunrise

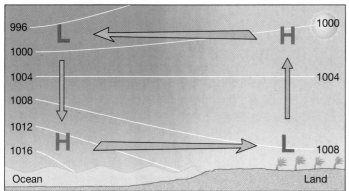

(c) Mid-afternoon sea breeze established

Figure 6-11 Cross sectional view illustrating the formation of a sea breeze. (a) Just before sunrise; (b) after sunrise; (c) sea breeze established.

How Temperature Differences Generate Wind
To illustrate how temperature differences can generate a horizontal pressure gradient and thereby create winds, consider a common example, the *sea breeze*. Figure 6-11a shows a vertical cross section of a coastal location before sunrise. At this time of day, we will assume that temperatures and pressures do not vary horizontally. Because there is no horizontal variation in pressure (zero horizontal pressure gradient) either aloft or at the surface, there is no wind.

After sunrise, the temperature of Earth's land surface begins to increase, while the temperature over the ocean remains nearly constant. As air over the land warms, it expands and forms a less dense air column. Although the air pressure at the surface remains essentially the same, this is

not true at higher altitudes. As we saw earlier, in Figure 6-8, warm air aloft tends to have higher air pressure compared to a column of cooler air. Consequently, above the land a high pressure develops, and the air aloft begins to flow away from the land (Figure 6-11b).

The mass transfer of air aloft creates a surface high-pressure area over the ocean, where the air is collecting, and a surface low over the land. The *surface* circulation that develops from this redistribution of mass is from the sea toward the land (sea breeze), as shown in Figure 6-11c. Thus, a simple thermal circulation develops, with a seaward flow aloft and a landward flow at the surface. Note that *vertical* movement is also required to make the circulation complete.

An important relationship exists between air pressure and temperature, as you saw in the preceding discussion. Temperature variations create pressure differences and ultimately wind. Daily temperature differences caused by unequal heating as in the sea breeze example tend to be confined to a zone only a few kilometers thick. On a global scale, however, variations in the amount of solar radiation received in the polar versus the equatorial latitudes generate the much larger pressure systems that in turn produce the planetary atmospheric circulation. Therefore, the underlying cause of global pressure differences and, by extension, wind is mainly the *unequal heating of Earth's land–sea surface.*

Isobars on a Surface Chart Figure 6-12 is a surface weather map that shows isobars and winds. Wind *direction* is shown as wind arrow shafts and *speed* as wind bars (see the accompanying key). Isobars, used to depict pressure patterns, are rarely straight or evenly spaced on surface maps. Consequently, wind generated by the pressure gradient force typically changes speed and direction as it flows.

The area of somewhat circular closed isobars represented by the red letter L is a *low-pressure system.* Low-pressure systems that occur in the middle latitudes are called **cyclones,** or **midlatitude cyclones,** to differentiate them from *tropical cyclones.* Midlatitude cyclones tend to produce stormy weather. (Tropical cyclones are also called *hurricanes* or *typhoons,* depending on their locations.)

In western Canada, a *high-pressure system,* denoted by the blue letter *H,* can also be seen. High-pressure areas such as this are called **anticyclones.** In contrast to cyclones, anticyclones tend to be associated with clearing conditions.

In summary, the *horizontal pressure gradient is the driving force of wind.* The magnitude of the pressure gradient force is shown by the spacing of isobars. The direction of force is always from areas of higher pressure toward areas of lower pressure and at right angles to the isobars.

Coriolis Force

The weather map in Figure 6-12 shows the typical airflow associated with surface high- and low-pressure systems. As expected, the air moves out of the regions of higher pressure and into the regions of lower pressure. However, the wind rarely crosses the isobars at right angles, as the pres-

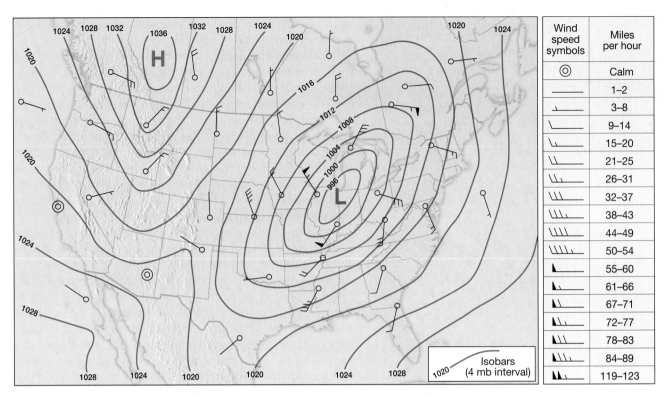

Wind speed symbols	Miles per hour
◎	Calm
———	1–2
⌐	3–8
⌐	9–14
⌐	15–20
⌐	21–25
⌐	26–31
⌐	32–37
⌐	38–43
⌐	44–49
⌐	50–54
◣	55–60
◣	61–66
◣	67–71
◣	72–77
◣	78–83
◣	84–89
◣◣	119–123

Figure 6-12 Isobars are used to show the distribution of pressure on daily weather maps. Isobars are seldom straight but usually form broad curves. Concentric isobars indicate cells of high and low pressure. The "wind flags" indicate the expected airflow surrounding pressure cells and are plotted as "flying" with the wind (that is, the wind blows toward the station circle). Notice on this map that the isobars are more closely spaced and the wind speed is faster around the low-pressure center than around the high.

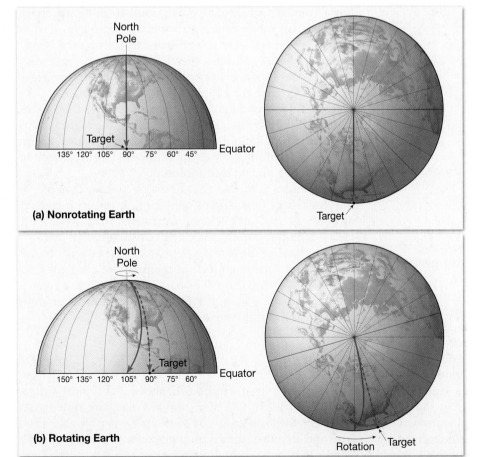

(a) Nonrotating Earth

(b) Rotating Earth

sure gradient force directs. This deviation is the result of Earth's rotation and has been named the **Coriolis force,** after the French scientist Gaspard-Gustave Coriolis, who first expressed its magnitude quantitatively. It is important to note that the Coriolis force cannot generate wind; rather, *it modifies airflow.*

The Coriolis force causes all free-moving objects, including wind, to be deflected to the *right* of their path of motion in the Northern Hemisphere and to the *left* in the Southern Hemisphere. The reason for this deflection can be illustrated by imagining the path of a rocket launched from the North Pole toward a target on the equator (Figure 6-13). If

Figure 6-13 The Coriolis force, illustrated using the one-hour flight of a rocket traveling from the North Pole to a location on the equator. (a) On a nonrotating Earth, the rocket would travel straight to its target. (b) However, Earth rotates 15° each hour. Thus, although the rocket travels in a straight line, when we plot the path of the rocket on Earth's surface, it follows a curved path that veers to the right of the target.

the rocket travels one hour toward its target, Earth would have rotated 15° to the east during its flight. To someone watching the rocket's path from the location of the intended target, it would look as if the rocket veered off its path and hit Earth 15° west of its target. The true path of the rocket was straight and would appear as such to someone in space looking toward Earth. Earth's rotation under the rocket produced the *apparent* deflection.

Note that the rocket did not hit its target and as such was deflected to the right of its path of motion because of the *counterclockwise* rotation of the Northern Hemisphere. *Clockwise* rotation produces a similar deflection in the Southern Hemisphere, but to the left of the path of motion.

Although it is usually easy for people to visualize the Coriolis deflection when the motion is from north to south, as in our rocket example, it is not as easy to see how a west-to-east flow would be deflected. Figure 6-14 illustrates this situation, using winds blowing eastward at four different latitudes (0°, 20°, 40°, and 60°). Notice that after a few hours the winds along the 20th, 40th, and 60th parallels appear to be veering off course. However, when viewed from space, it is apparent that these winds have maintained their original direction. It is the "change" of orientation of North America as Earth rotates on its axis that produces the deflection we observe in Figure 6-14.

We can also see in Figure 6-14 that the amount of deflection is greater at 60° latitude than at 40° latitude, and likewise greater at 40° than at 20°. Furthermore, there is no deflection observed for the airflow along the equator. We conclude, therefore, that the magnitude of the Coriolis force is dependent on latitude—it is strongest at the poles and weakens equatorward, where it eventually becomes nonex-

istent. We can also see that the amount of Coriolis deflection increases with wind speed because faster winds travel farther than slower winds in the same time period.

All "free-moving" objects are affected by the Coriolis force. This fact was dramatically discovered by the U.S. Navy at the beginning of World War II. During target practice, long-range guns on battleships continually missed their targets by as much as several hundred yards until ballistic corrections were made for the changing position of seemingly stationary targets. Across short distances, however, the Coriolis force is relatively small. Nevertheless, in the middle latitudes this deflecting force is great enough to potentially affect the outcome of a baseball game. A ball hit a horizontal distance of 100 meters (330 feet) in 4 seconds down the right field line will be deflected 1.5 centimeters (more than 1/2 inch) to the right by the Coriolis force. This could be enough to turn a potential home run into a foul ball!

In summary, *the Coriolis force acts to change the direction of a moving body to the right in the Northern Hemisphere and to the left in the Southern Hemisphere. This deflecting force (1) is always directed at right angles to the direction of airflow; (2) affects only wind direction, not wind speed; (3) is affected by wind speed (the stronger the wind, the greater the deflecting force); and (4) is strongest at the poles and weakens equatorward, becoming nonexistent at the equator.*

Friction

Earlier we stated that the pressure gradient force is the primary driving force of the wind. As an unbalanced force, it causes air to accelerate from regions of higher pressure

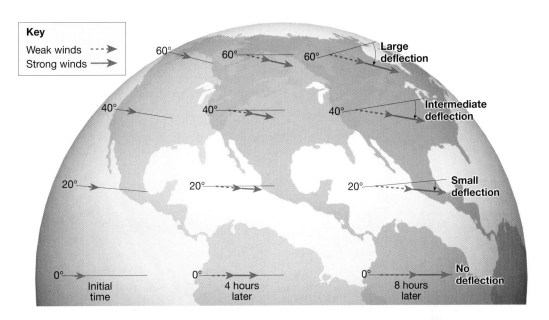

Figure 6-14 Coriolis deflection of winds blowing eastward at different latitudes. After a few hours the winds along the 20th, 40th, and 60th parallels appear to veer off course. This deflection (which does not occur at the equator) is caused by Earth's rotation, which changes the orientation of the surface over which the winds are moving.

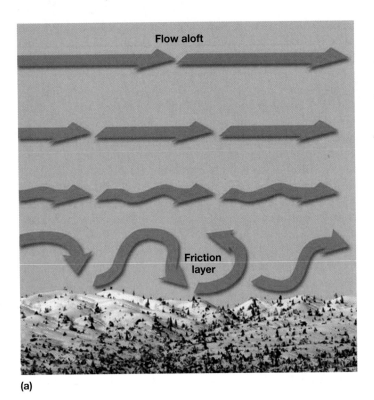

(a)

(b)

Figure 6-15 Friction as a factor in wind speed. (a) Wind increases in strength with an increase in altitude because it is less affected by friction from objects near Earth's surface. (b) This snow-covered tree shows the effects of strong winds in a high mountain setting. (Photo by E. J. Tarbuck)

toward regions of lower pressure. Thus, we might expect wind speeds to continually increase (accelerate) as long as this imbalance existed. But we know from personal experience that winds do not become faster indefinitely. Rather, friction, which acts to slow moving objects, comes into play.

Although friction significantly influences airflow near Earth's surface, its effect is negligible above a height of a few kilometers (Figure 6-15). Therefore, we will examine the flow aloft first (where the effect of friction is small) and then analyze surface winds (where friction significantly influences airflow).

Concept Check 6.5

1 List three forces that combine to direct horizontal airflow (wind).

2 What force is responsible for *generating* wind?

3 Write a generalization relating the spacing of isobars to wind speed.

4 Briefly describe how the Coriolis force modifies air movement.

5 Which two factors influence the magnitude of the Coriolis force?

Students Sometimes Ask...

I've been told that water goes down a sink in one direction in the Northern Hemisphere and in the opposite direction in the Southern Hemisphere. Is that true?

Not necessarily! The origin of this myth comes from applying a scientific principle to a situation where it does not fit. Recall that the Coriolis deflection causes cyclonic systems to rotate counterclockwise in the Northern Hemisphere and clockwise in the Southern Hemisphere. It was inevitable that someone would suggest (without checking) that a sink should drain in a similar manner. However, a cyclone is more than 1000 kilometers in diameter and may exist for several days. By contrast, a typical sink is less than 1 meter in diameter and drains in a matter of seconds. On this scale, the Coriolis force is minuscule. Therefore, the shape of the sink and how level it is has more to do with the direction of waterflow than the Coriolis force.

Winds Aloft

In this section we address airflow that occurs at least a few kilometers above Earth's surface, where the effects of friction are small enough to disregard.

Geostrophic Flow

Aloft, the Coriolis force is responsible for balancing the pressure gradient force and thereby directing airflow. Figure 6-

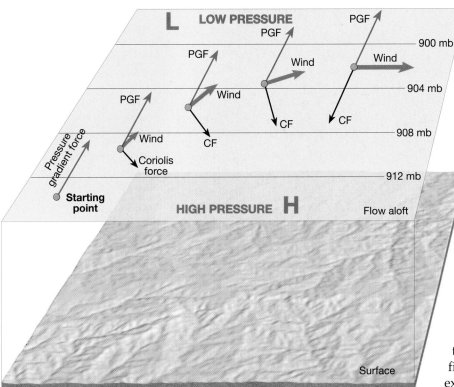

Figure 6-16 The geostrophic wind. The only force acting on a stationary parcel of air is the pressure gradient force. Once the air begins to accelerate, the Coriolis force deflects it to the right in the Northern Hemisphere. Greater wind speeds result in a stronger Coriolis force (deflection) until the flow is parallel to the isobars. At this point the pressure gradient force and Coriolis force are in balance, and the flow is called a *geostrophic wind*. It is important to note that in the "real" atmosphere, airflow is continually adjusting for variations in the pressure field. As a result, the adjustment to geostrophic equilibrium is much more irregular than shown.

of airflow aloft. Under natural conditions in the atmosphere, winds are never purely geostrophic. Nonetheless, the geostrophic model offers a useful estimation of the actual winds aloft. By measuring the pressure field (orientation and spacing of isobars) that exists aloft, meteorologists can determine both wind direction and speed (Figure 6-17).

As we have seen, wind direction is directly linked to the prevailing pressure pattern. Therefore, if we know the wind direction, we can also establish a rough approximation of the pressure distribution. This rather straightforward relationship between wind direction and pressure distribution was first formulated by the Dutch meteorologist Buys Ballot in 1857. Essentially, **Buys Ballot's law** states that in the Northern Hemisphere, *if you stand with your back to the wind, low pressure will be found to your left and high pressure to your right.* In the Southern Hemisphere, the situation is reversed.

Although Buys Ballot's law holds for airflow aloft, it must be used with caution when applied to surface winds. At the surface, friction and topography interfere with the idealized circulation. At the surface, if you stand with your back to the wind and then turn clockwise about 30°, low pressure will be to your left and high pressure to your right.

In summary, winds above a few kilometers can be considered geostrophic—that is, they flow in a straight path parallel to the isobars at speeds that can be calculated from the pressure gradient. The major discrepancy from true geostrophic winds involves the flow along highly curved paths, a topic considered next.

16 shows how a balance is achieved between these opposing forces. For illustration only, we assume a nonmoving parcel of air at the starting point in Figure 6-16. (Remember that air is rarely stationary in the upper atmosphere.) Because our parcel of air has no motion, the Coriolis force exerts no influence. Under the influence of the pressure gradient force, the parcel begins to accelerate directly toward the area of low pressure. As soon as the flow begins, the Coriolis force commences and causes a deflection to the right for winds in the Northern Hemisphere. As the parcel accelerates, the Coriolis force intensifies. (Recall that the magnitude of the Coriolis force is proportional to wind speed.) Thus, the increased speed results in further deflection.

Eventually the wind turns so that it is flowing parallel to the isobars. When this occurs, the pressure gradient force is balanced by the opposing Coriolis force, as shown in Figure 6-16. As long as these forces remain balanced, the resulting wind will continue to flow parallel to the isobars at a constant speed. Stated another way, the wind can be considered to be coasting (not accelerating or decelerating) along a pathway defined by the isobars.

Under these idealized conditions, when the Coriolis force is exactly equal in strength but acting in the opposite direction of the pressure gradient force, the airflow is said to be in *geostrophic balance*. The winds generated by this balance are called **geostrophic** ("turned by Earth") **winds.** Geostrophic winds flow in relatively straight paths, parallel to the isobars, with velocities proportional to the pressure gradient force. A steep pressure gradient creates strong winds, and a weak pressure gradient produces light winds.

It is important to note that the geostrophic wind is an idealized model that only approximates the actual behavior

Curved Flow and the Gradient Wind

A casual glance at an upper-level weather map shows that the isobars are not generally straight; instead, they make broad, sweeping curves (see Figure 6-17). Occasionally the isobars connect to form roughly circular cells of either high or low pressure. Thus, unlike geostrophic winds that flow along relatively straight paths, winds around cells of high or low pressure follow curved paths in order to parallel

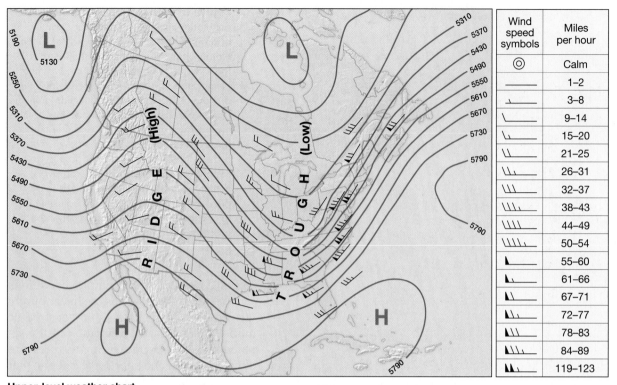

Upper-level weather chart

Representation of upper-level chart

Wind speed symbols	Miles per hour
◎	Calm
—	1–2
⌐	3–8
\	9–14
_	15–20
\\	21–25
\\\	26–31
\\\	32–37
\\\\	38–43
\\\\	44–49
\\\\\	50–54
▌	55–60
▌\	61–66
▌\\	67–71
▌\\\	72–77
▌\\\	78–83
▌\\\\	84–89
▌▌\	119–123

Figure 6-17 Upper-air weather chart. This simplified weather chart shows the direction and speed of the upper-air winds. Note from the flags that the airflow is almost parallel to the contours. Like most other upper-air charts, this one shows variations in the height (in meters) at which a selected pressure (500 millibars) is found instead of showing variations in pressure at a fixed height, like surface maps. Do not let this confuse you because there is a simple relationship between height contours and pressure. Places experiencing 500-millibar pressure at higher altitudes (toward the south on this map) are experiencing higher pressures than places where the height contours indicate lower altitudes. Thus, *higher-elevation* contours indicate *higher* pressures, and *lower-elevation* contours indicate *lower* pressures.

the isobars. Winds of this nature, which blow at a constant speed parallel to curved isobars, are called **gradient winds.**

Let us examine how the pressure gradient force and Coriolis force combine to produce gradient winds. Figure 6-18a shows the gradient flow around a center of low pressure. As soon as the flow begins, the Coriolis force causes the air to be deflected. In the Northern Hemisphere, where the Coriolis force deflects the flow to the right, the resulting wind blows counterclockwise about a low. Conversely, around a high-pressure cell, the outward-directed pressure gradient force is opposed by the inward-directed Coriolis force, and a clockwise flow results. Figure 6-18b illustrates this idea.

Because the Coriolis force deflects the winds to the left in the Southern Hemisphere, the flow is reversed—clockwise around low-pressure centers and counterclockwise around high-pressure centers.

Recall that meteorologists call centers of low pressure *cyclones* and the flow around them *cyclonic*. There are several types and scales of cyclones. Large low pressure systems that are major weather makers in the United States are called *midlatitude cyclones*. Other examples include *tropical cyclones* (hurricanes), which are generally smaller than midlatitude cyclones, and *tornadoes*, which are tiny and extremely in-

tense cyclonic storms. **Cyclonic flow** has the same direction of rotation as Earth: counterclockwise in the Northern Hemisphere and clockwise in the Southern Hemisphere. Centers of high pressure are called *anticyclones* and exhibit **anticyclonic flow** (opposite that of Earth's rotation). Whenever isobars curve to form elongated regions of low and high pressure, these areas are called **troughs** and **ridges**, respectively (see Figure 6-17). The flow around a trough is cyclonic; conversely, the flow around a ridge is anticyclonic.

Now let us consider the forces that produce the gradient flow associated with cyclonic and anticyclonic circulations. Wherever the flow is curved, a force has deflected the air (changed its direction), even when no change in speed results. This is a consequence of Newton's first law of motion, which states that a moving object will continue to move in a straight line unless acted upon by an unbalanced force. You have undoubtedly experienced the effect of Newton's law when the automobile in which you were riding made a sharp turn and your body tried to continue moving straight ahead (see Appendix E).

Referring to Figure 6-18a, we see that in a low-pressure center, the inward-directed pressure gradient force is opposed by the outward-directed Coriolis force. But to keep

Box 6-2   Do Baseballs Really Fly Farther at Denver's Coors Field?

Since Denver's Coors Field was built in 1995, it has become known as the "home-run hitter's ballpark" (Figure 6-C). This notoriety is warranted because Coors Field led all Major League ballparks in both total home runs and home runs per at-bat during seven of its first eight seasons.

In theory, a well-struck baseball should travel roughly 10 percent farther in Denver (elevation 5280 feet) than it would in a ballpark at sea level. This so-called elevation enhancement results from lower air density at mile-high Coors Field. According to Robert

Adair, Sterling Professor Emeritus of Physics at Yale University, a 400-foot blast in Atlanta could carry perhaps 425 feet in Denver, although Adair admits that calculating the actual difference is tricky for reasons related to fluid dynamics.

Recently a group of researchers at the University of Colorado at Denver tested the assumption that batted balls travel farther in the "thin air" at Coors Field than in ballparks near sea level. They concluded that the assumed elevation enhancement of fly ball distance has been greatly overestimated.

Instead, they suggest that the hitter-friendly conditions should be attributed to the prevailing weather conditions of the nearby Front Range of the Rocky Mountains and the effects of low air density on the act of pitching a baseball.

For example, wind can make or break a home run. According to Professor Adair, if there's a 10-mile-an-hour breeze behind a batter, it will add an extra 30 feet to a 400-foot home run. Conversely, if the wind is blowing in toward home plate at 10 miles per hour, the flight of the ball will be reduced by about 30 feet. During the summer months winds most frequently blow from the south and southwest in the Denver area. Because of the orientation of Coors Field, these winds blow toward the outfield, thus aiding the hitters rather than the pitchers.

The act of pitching a baseball is also greatly affected by air density. In particular, one factor that determines how much a curveball will curve is air density. At higher elevations, thinner air causes a ball to break less, which makes it easier for a batter to hit a pitch.

FIGURE 6-C Denver's Coors Field, home of the Colorado Rockies, is nearly 1 mile above sea-level. It is known as the "homerun hitter's ballpark." (Photo by Russell Lansford/Icon SMI/Newscom)

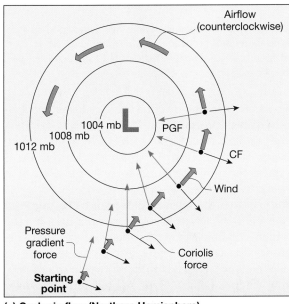

(a) Cyclonic flow (Northern Hemisphere)

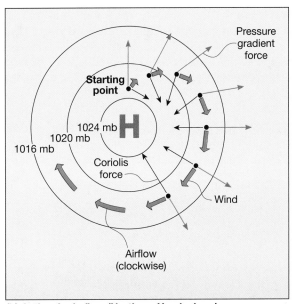

(b) Anticyclonic flow (Northern Hemisphere)

Figure 6-18 Idealized illustrations showing expected airflow aloft around a circular low-pressure center (a) and a high-pressure center (b). It is important to note that in the "real" atmosphere, airflow is continually adjusting for variations in the pressure field by slightly changing speed and direction.

the path curved (parallel to the isobars), the inward pull of the pressure gradient force must be strong enough to balance the Coriolis force as well as to turn (accelerate) the air inward. The inward turning of the air is called *centripetal acceleration*. Stated another way, the pressure gradient force must exceed the Coriolis force to overcome the air's tendency to continue moving in a straight line.*

The opposite situation exists in anticyclonic flows, where the inward-directed Coriolis force must balance the pressure gradient force as well as provide the inward acceleration needed to turn the air. Notice in Figure 6-18 that the pressure gradient force and Coriolis force are not balanced (the arrow lengths are different), as they are in geostrophic flow. This imbalance provides a change in direction (centripetal acceleration) that generates curved flow.

Concept Check 6.6

1 Explain the formation of geostrophic wind.

2 Describe how geostrophic winds blow relative to isobars.

3 Describe the direction of cyclonic and anticyclonic flow in both the Northern and Southern Hemispheres.

Despite the importance of centripetal acceleration in establishing curved flow aloft, near the surface friction comes into play and greatly overshadows this much weaker force. Consequently, except with rapidly rotating storms such as tornadoes and hurricanes, the effect of centripetal acceleration is negligible and therefore is not considered in the discussion of surface circulation.

Surface Winds

Friction as a factor affecting wind is important only within the first few kilometers of Earth's surface. We know that friction acts to slow the movement of air (Figure 6-19).

_____
*The tendency of a particle to move in a straight line when rotated creates an imaginary outward force called *centrifugal force*.

Figure 6-19 A snow fence slows the wind, thereby decreasing the wind's ability to transport snow. As a result, snow accumulates on the downwind side of the fence. (Photo by Scott T. Slattery/Shutterstock))

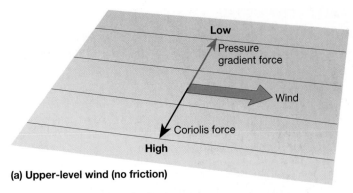

(a) Upper-level wind (no friction)

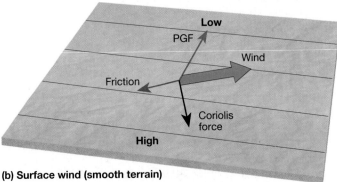

(b) Surface wind (smooth terrain)

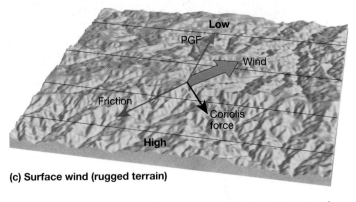

(c) Surface wind (rugged terrain)

Figure 6-20 Comparison between upper-level winds and surface winds, showing the effects of friction on airflow. Friction slows surface wind speed, which weakens the Coriolis force, causing the winds to cross the isobars.

By slowing air movement, friction also reduces the Coriolis force, which is proportional to wind speed. Because the pressure gradient force is not affected by wind speed, it wins the tug of war against the Coriolis force (Figure 6-20). The result is the movement of air at an angle across the isobars, toward the area of lower pressure.

The roughness of the surface determines the angle at which the air will flow across the isobars and influences the speed at which it will move. Over relatively smooth surfaces, where friction is low, air moves at an angle of 10° to 20° to the isobars and at speeds roughly two-thirds of geostrophic flow (Figure 6-20b). Over rugged terrain, where friction is high, the angle can be as great as 45° from the isobars, with wind speeds reduced by as much as 50 percent (Figure 6-20c).

We have learned that above the friction layer in the Northern Hemisphere, winds blow counterclockwise around

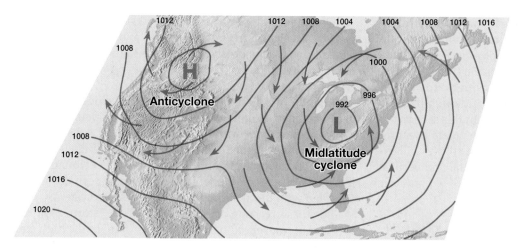

Figure 6-21 Cyclonic and anticyclonic winds in the Northern Hemisphere. Arrows show the winds blowing inward and counterclockwise around a low and outward and clockwise around a high.

a cyclone and clockwise around an anticyclone, with winds nearly parallel to the isobars. Combined with the effect of friction, we find that the airflow crosses the isobars at varying angles, depending on the terrain, but always from higher to lower pressure. In a cyclone, in which pressure decreases inward, friction causes a net flow *toward* its center (Figure 6-21). In an anticyclone, the opposite is true: The pressure decreases outward, and friction causes a net flow *away* from the center. Therefore, the resultant winds blow into and counterclockwise about a cyclone and outward and clockwise about an anticyclone (Figure 6-21). Of course, in the Southern Hemisphere the Coriolis force deflects the winds to the left and reverses the direction of flow. Regardless of hemisphere, however, friction causes a net inflow (*convergence*) around a cyclone and a net outflow (*divergence*) around an anticyclone.

Concept Check 6.7

1 Unlike winds aloft, which blow nearly parallel to the isobars, surface winds generally cross the isobars. Explain what causes this difference.

2 Prepare a diagram with isobars and wind arrows that shows the winds associated with surface cyclones and anticyclones in both the Northern and Southern Hemispheres.

How Winds Generate Vertical Air Motion

 Air Pressure and Winds
  ▶ Highs and Lows

So far we have discussed wind without regard to how airflow in one region might affect airflow elsewhere. As one researcher put it, a butterfly flapping its wings in South America can generate a tornado in the United States. Although this is an exaggeration, it suggests that airflow in one region might cause a change in weather at some later time at a different location.

Cold winds blowing from Greenland encountered moist air over the Greenland Sea, and their convergence generated parallel rows of clouds called *cloud streets* in the skies around Jan Mayen Island. The island acts as an obstacle that causes the wind to form spiraling eddies (called *van Karman vortices*) on the south (leeward) side of the island. (NASA)

Question 1 Based on the orientation of the cloud streets, what direction is the wind blowing over the Greenland Sea?

Question 2 Can you think of an analogy for the spiraling eddies that formed downwind of Jan Mayen Island?

Question 3 Eddies are also generated by bicycles, cars, and airplanes. Participants in bicycle and car races use these eddies, which reduce wind resistance, by traveling directly behind another competitor. What term is used to describe this competitive advantage?

Of particular importance is the question of how horizontal airflow (wind) relates to vertical flow. Although the rate at which air ascends or descends (except in violent storms) is slow compared to horizontal flow, it is very important as a weather maker. Rising air is associated with cloudy conditions and often precipitation, whereas subsidence produces adiabatic heating and clearing conditions. In this section we will examine how the movement of air can create pressure change and hence generate winds. In doing so, we will examine the interrelationship between horizontal and vertical flow and its effect on the weather.

Vertical Airflow Associated with Cyclones and Anticyclones

Let us first consider the situation of a surface low-pressure system (cyclone) in which the air is spiraling inward (see Figure 6-21). The net inward transport of air causes a shrinking of the area it occupies, a process called *horizontal convergence* (Figure 6-22). Whenever air converges horizontally, it piles up—that is, it increases in mass to allow for the decreased area it occupies. This process generates a "denser" column. We seem to have encountered a paradox: Low-pressure centers cause a net accumulation of air, which increases their pressure. Consequently, a surface cyclone should quickly eradicate itself in a manner not unlike what happens to the vacuum in a coffee can when it is opened.

You can see that for a surface low to exist for a substantial duration, compensation must occur aloft. For example, surface convergence could be maintained if *divergence* (spreading out) aloft occurred at a rate equal to the inflow below. Figure 6-22 illustrates the relationship between surface convergence (inflow) and the divergence aloft (outflow) needed to maintain a low-pressure center.

Divergence aloft may even exceed surface convergence, thereby accelerating vertical motion and intensifying surface inflow. Because rising air often results in cloud formation and precipitation, the passage of a low-pressure center is generally accompanied by "bad weather."

Like their cyclonic counterparts, anticyclones must also be maintained from above. Outflow near the surface is accompanied by convergence aloft and general subsidence of the air column (Figure 6-22). Because descending air is compressed and warmed, cloud formation and precipitation are less likely in an anticyclone. Thus, fair weather can usually be expected with the approach of a high-pressure system.

For these reasons, it is common to see "stormy" at the low end of household barometers and "fair" at the high end (Figure 6-23). By noting the pressure trend—rising, falling, or steady—we have a good indication of forthcoming weather. Such a determination, called the **pressure tendency,** or **barometric tendency,** is useful in short-range weather prediction. The generalizations relating cyclones and anticyclones to weather conditions are stated nicely in this verse (where "glass" refers to the barometer):

> When the glass falls low,
> Prepare for a blow;
> When it rises high,
> Let all your kites fly.

In conclusion, it should be obvious why local television weather broadcasters emphasize the positions and projected paths of cyclones and anticyclones. The "villain" on these weather programs is always the low-pressure system, which produces "foul" weather in any season. Lows move in roughly a west-to-east direction across the United States and require from a few days to more than a week for the journey. Because their paths can be erratic, accurate prediction of their migration is difficult, yet it is es-

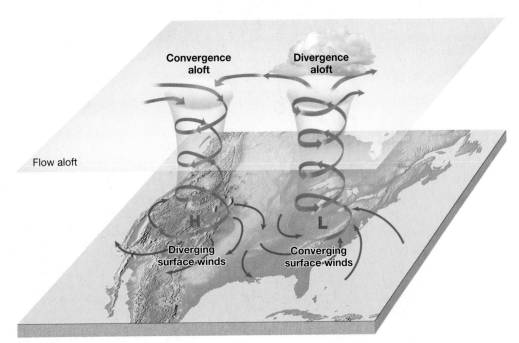

Figure 6-22 Airflow associated with surface cyclones (L) and anticyclones (H). A low, or cyclone, has converging surface winds and rising air, resulting in cloudy conditions. A high, or anticyclone, has diverging surface winds and descending air, which leads to clear skies and fair weather.

(a)

(b)

Figure 6-23 These two photographs illustrate the basic weather generalizations associated with pressure centers. (a) A rainy day in London. Low pressure systems are frequently associated with cloudy conditions and precipitation. (Photo by Lourens Smak/Alamy) (b) By contrast, clear skies and "fair" weather may be expected when an area is under the influence of high pressure. (Photo by Prisma Bildagentur/Alamy)

sential for short-range forecasting. Meteorologists must also determine whether the flow aloft will intensify an embryo storm or act to suppress its development.

Factors That Promote Vertical Airflow

Because of the close tie between vertical motion in the atmosphere and our daily weather, we will consider some other factors that contribute to surface convergence and surface divergence.

Friction can cause both convergence and divergence. When air moves from the relatively smooth ocean surface

to land, for instance, the increased friction causes an abrupt drop in wind speed. This reduction of wind speed downstream results in a pileup of air upstream. Thus, converging winds and ascending air accompany flow from the ocean to the land. This effect contributes to the cloudy conditions over the humid coastal regions of Florida. Conversely, when air moves from the land to the ocean, general divergence and subsidence accompany the seaward flow of air because of lower friction and increasing wind speed over the water. The result is often subsidence and clearing conditions.

Mountains also hinder the flow of air and cause divergence and convergence. As air passes over a mountain range, it is compressed vertically, which produces horizontal spreading (divergence) aloft. On reaching the leeward side of the mountain, the air experiences vertical expansion, which causes horizontal convergence. This effect greatly influences the weather in the United States east of the Rocky Mountains, as we shall examine later.

As a result of the connections between surface conditions and those aloft, significant research has been aimed at understanding atmospheric circulation, especially in the midlatitudes. Once we examine global atmospheric circulation in the next chapter, we will again consider the relationships between horizontal airflow (wind) and vertical motions (rising and descending air currents).

Concept Check 6.8

1 For surface low pressure to exist for an extended period, what condition must exist aloft?

2 What general weather conditions are to be expected when the pressure tendency is rising? When the pressure tendency is falling?

3 Converging winds and ascending air are often associated with the flow of air from the oceans onto land. Conversely, divergence and subsidence often accompany the flow of air from land to sea. What causes this convergence over land and divergence over the ocean?

Students Sometimes Ask ...

What causes "mountain sickness"?

When visitors drive up to a mountain pass above 3000 meters (10,000 feet) and take a walk, they typically notice shortness of breath and possibly fatigue. These symptoms are caused by breathing air with roughly 30 percent less oxygen than at sea level. At these altitudes our bodies try to compensate for the oxygen deficiency of the air by breathing more deeply and increasing the heart rate, thereby pumping more blood to the body's tissues. The additional blood is thought to cause brain tissues to swell, resulting in headaches, insomnia, and nausea—the main symptoms of *acute mountain sickness*. Mountain sickness is generally not life threatening and usually can be alleviated with a night's rest at a lower altitude. Occasionally people become victims of *high-altitude pulmonary edema*. This life-threatening condition involves a buildup of fluid in the lungs and requires prompt medical attention.

EYE ON THE ATMOSPHERE

These satellite images show four different tropical cyclones (Hurricanes) that occurred on different dates in different parts of the world. (NASA)

Question 1 For each storm, examine the cloud pattern and determine whether the flow is clockwise or counterclockwise.

Question 2 In which hemisphere is each storm located, Northern or Southern?

Wind Measurement

Two basic wind measurements—direction and speed—are important to weather observers. Winds are always labeled by the direction *from* which they blow. A north wind blows from the north toward the south; an east wind blows from the east toward the west. One instrument commonly used to determine wind direction, the **wind vane,** is often seen on the tops of buildings (Figure 6-24a). Sometimes the wind direction is shown on a dial that is connected to the wind vane. The dial indicates the direction of the wind either by points of the compass—that is, N, NE, E, SE, and so on—or

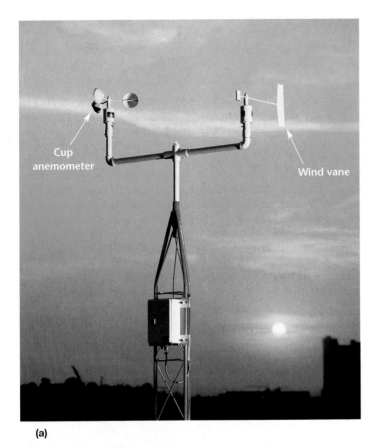

Figure 6-24 Wind measurement. (a) Wind vane (right) and cup anemometer (left). The wind vane shows wind direction, and the anemometer measures wind speed. (Photo by Belfort Instrument Company) (b) A *wind sock* is a device for determining wind direction and estimating wind speed. They are common sights at small airports and landing strips. (Photo by Lourens Smak/Almay Images)

by a scale of 0° to 360°. On the latter scale, 0° (or 360°) is north, 90° is east, 180° is south, and 270° is west.

When the wind consistently blows more often from one direction than from any other, it is called a **prevailing wind.** You may be familiar with the prevailing westerlies that dominate the circulation in the midlatitudes. In the United States, for example, these winds consistently move the "weather" from west to east across the continent. Embed-ded within this general eastward flow are cells of high and low pressure, with their characteristic clockwise and counterclockwise flow. As a result, the winds associated with the westerlies, as measured at the surface, often vary considerably from day to day and from place to place. A *wind rose* provides a method of representing prevailing winds by indicating the percentage of time the wind blows from various directions (Figure 6-25). The length of the lines on

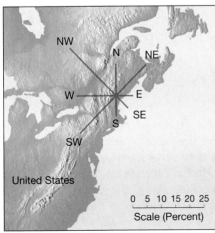

(a) Westerlies (winter)

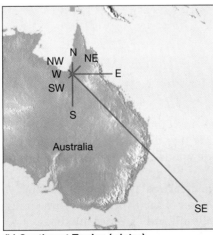

(b) Southeast Trades (winter)

Figure 6-25 The percentage of directional windflow is charted on a *wind rose* and may represent daily, weekly, monthly, seasonal, or annual totals. (a) Wind frequency for the winter in the eastern United States. (b) Wind frequency for the winter in northern Australia. Note the reliability of the southeast trades in Australia as compared to the westerlies in the eastern United States. (Data from G. T. Trewartha)

Box 6-3 Wind Energy: An Alternative with Potential

Air has mass, and when it moves (that is, when the wind blows), it contains the energy of that motion—kinetic energy. A portion of that energy can be converted into other forms—mechanical force or electricity—that we can use to perform work (Figure 6-D).

Mechanical energy from wind is commonly used for pumping water in rural or remote places. The "farm windmill," still a familiar sight in many rural areas, is an example. Mechanical energy converted from wind can also be used for other purposes, such as sawing logs, grinding grain, and propelling sailboats. By contrast, wind-powered electric turbines generate electricity for homes, businesses, and for sale to utilities.

Today, modern wind turbines are being installed at breakneck speed. In fact, worldwide, in 2010 the installed wind power capacity exceeded 203,000 megawatts, an increase of 28 percent over 2009.* This is equivalent to the total electrical demand of Italy—2 percent of global electricity production. Worldwide, wind energy installations have doubled every three years since 2000. The United States is the world's leading producer (22.3 percent), followed by China (16.3 percent), Germany (16.2 percent), and Spain (11.5 percent). Within the next decade, China

*One megawatt is enough electricity to supply 250–300 average American households.

is expected to produce the most wind-generated electricity.

Wind speed is a crucial element in determining whether a place is a suitable site for installing a wind-energy facility. Generally, a minimum average wind speed of 21 kilometers (13 miles) per hour is necessary for a large-scale wind-power plant to be profitable. A small difference in wind speed results in a large difference in energy production, and therefore a large difference in the cost of the electricity generated. For example, a turbine operating on a site with an average wind speed of 12 miles per hour would generate about 33 percent more electricity than one operating at 11 miles per hour. Also, there

FIGURE 6-D Farm windmills such as the one on the left remain familiar sights in some areas. Mechanical energy from wind is commonly used to pump water. (Photo by Mehmet Dilsiz/Shutterstock) The wind turbines on the right are operating near Palm Springs, California. Although California is the state where significant wind-power development began, it has been surpassed by Texas and Iowa. (Photo by John Mead/Science Photo Library/Photo Researchers, Inc.)

the wind rose indicates the percentage of time the wind blew from that direction. As seen in Figure 6-25b, the direction of airflow associated with the belt of trade winds is much more consistent than the westerlies.

Knowledge of the wind patterns for a particular area can be useful. For example, during the construction of an airport, the runways are aligned with the prevailing wind to assist in takeoffs and landings. Furthermore, prevailing winds greatly affect the weather and climate of a region. Mountain ranges that trend north–south, such as the Cascade Range of the Pacific Northwest, cause the ascent of the prevailing westerlies. Thus, the windward (west) slopes of these ranges are rainy, whereas the leeward (east) sides are dry.

is little energy to be harvested at low wind speeds: 6-mile-per-hour winds contain less than one-eighth the energy of 12-mile-per-hour winds.

The United States has tremendous wind energy resources (**Figure 6-E**). In 2010, 36 states had commercial facilities that produced electricity from wind power. Texas was number one in installed capacity, followed by Iowa, California, and Minnesota. Despite the fact that the modern U.S. wind industry began in California, 16 states have greater wind potential. As shown in **Table 6-A**, the top five states for wind energy potential are North Dakota, Texas, Kansas, South Dakota, and Montana. Although only a small fraction of U.S. electrical generation currently comes from wind energy, it has been estimated that wind energy potential equals more than twice the total electricity currently consumed.

TABLE 6-A The Leading States for Wind Energy Potential

Rank	State	Potential*	Rank	State	Potential*
1	North Dakota	1210	11	Colorado	481
2	Texas	1190	12	New Mexico	435
3	Kansas	1070	13	Idaho	73
4	South Dakota	1030	14	Michigan	65
5	Montana	1020	15	New York	62
6	Nebraska	868	16	Illinois	61
7	Wyoming	747	17	California	59
8	Oklahoma	725	18	Wisconsin	58
9	Minnesota	657	19	Maine	56
10	Iowa	551	20	Missouri	52

* The total amount of electricity that could potentially be generated each year, measured in *billions* of kilowatt hours. A typical American home would use several hundred kilowatt hours per month.

The U.S. Department of Energy has announced a goal of obtaining 5 percent of U.S. electricity from wind by the year 2020—a goal that seems consistent with the current growth rate of wind energy nationwide. Thus wind-generated electricity appears to be shifting from being an "alternative" to a "mainstream" energy source.

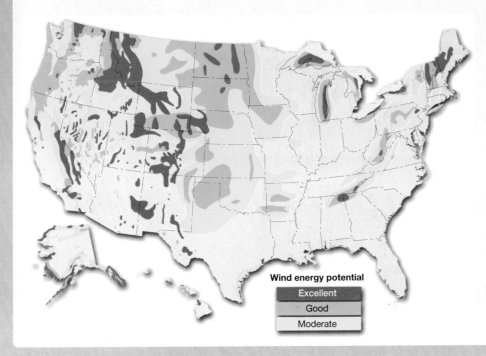

Wind energy potential

Excellent
Good
Moderate

FIGURE 6-E The wind energy potential for the United States. Large wind systems require average wind speeds of about 6 meters per second (13 miles per hour). In the key, "moderate" refers to regions that experience wind speeds of 6.4 to 6.9 meters per second (m/s), "good" means 7 to 7.4 m/s, and "excellent" means 7.5 m/s and higher.

Wind speed is often measured with a **cup anemometer,** which has a dial much like the speedometer of an automobile (Figure 6-24a). Sometimes an **aerovane** is used instead of a wind vane and cup anemometer. As illustrated in Figure 6-26, this instrument resembles a wind vane with a propeller at one end. The fin keeps the propeller facing into the wind, allowing the blades to rotate at a rate proportional to the wind speed. This instrument is commonly attached to a recorder that produces a continuous record of wind speed and direction. This information is valuable for determining locations where winds are steady and speeds are relatively high—potential sites for tapping wind energy (Box 6-3).

At small airstrips *wind socks* are frequently used (Figure 6-24b). A wind sock consists of a cone-shaped bag

Figure 6-26 An aerovane. (Photo by imagebroker/Alamy)

that is open at both ends and free to change position with shifts in wind direction. The degree to which the sock is inflated indicates the strength of the wind.

Recall that 70 percent of Earth's surface is covered by water, making conventional methods of measuring wind speed difficult. Weather buoys and ships at sea provide limited coverage, but the availability of satellite-derived wind data has dramatically improved weather forecasts. For example, wind speed and direction can be established using satellite images to track cloud movements. This is currently being accomplished by comparing a sequence of satellite images separated by intervals of 5 to 30 minutes—an innovation especially useful in predicting the timing and location of a hurricane making landfall (Figure 6-27).

Concept Check 6.9

1 A southwest wind blows from the _____ (direction) toward the _____ (direction).

2 When the wind direction is 315°, from what compass direction is it blowing?

3 What is the name of the prevailing winds that affect the contiguous United States?

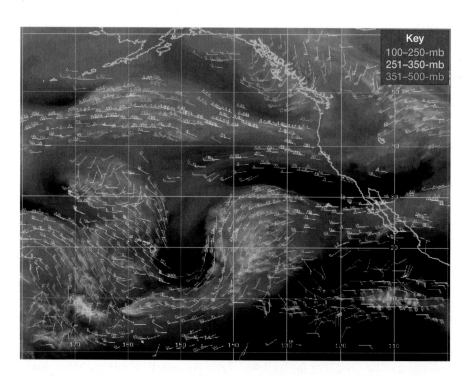

Figure 6-27 Upper-level winds obtained from the GOES meteorological satellite. (Image by NOAA)

Give It Some Thought

1. The accompanying satellite images show the cloud patterns associated with two low pressure systems.
   a. Determine the airflow around each of these pressure cells (clockwise or counterclockwise).
   b. Which one of these two areas of low pressure is located in the Northern Hemisphere?
   c. Can you identify which one of these storm systems is a tropical cyclone (hurricane)? (*Hint:* Compare these images to Figures 11-1 and 11-4.)

(a)　　　　(NASA)　　(b)　　　　(NASA)

2. Temperature variations create pressure differences, which in turn produce winds. On a small scale, the sea breeze is a good illustration of this principle. Prepare and label a sketch that illustrates the formation of a sea breeze.

3. Why is the Coriolis force considered an apparent force and not a "real" force?

4. Given the following descriptions, identify the direction (for example, east, west, northwest) in which the Coriolis force is acting on the moving object.
   a. A commercial jet flying from New York to Chicago
   b. A baseball thrown from south to north in South Dakota
   c. A blimp floating from St. Louis northeast toward Detroit
   d. A boomerang thrown from west to east in Australia
   e. A football thrown along the equator

5. Geostrophic winds are maintained by a balance between the pressure gradient force and the Coriolis force. Explain why gradient winds *cannot* achieve a similar balance between these two forces.

6. If a weather system with strong winds approached Lake Michigan from the west, how might the speed of the winds change as the system traversed the lake? Explain your answer.

7. The accompanying map is a simplified surface weather map for April 2, 2011, on which the centers of three pressure cells are numbered.
   a. Identify which of the pressure cells are anticyclones (highs) and which are midlatitude cyclones (lows).

b. Which pressure system has the steepest pressure gradient and hence exhibits the strongest winds?
c. Refer to Figure 6-3 to determine whether pressure system 3 should be considered strong or weak.

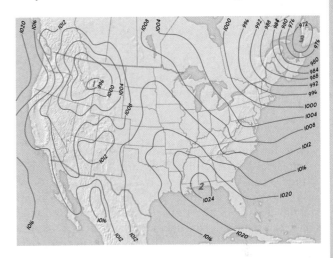

8. If you live in the Northern Hemisphere and are directly west of the center of a midlatitude cyclone, what is the probable wind direction?

9. Sketch a cross section of a low-pressure cell that shows surface flow (inward or outward), vertical flow, and flow aloft.

10. Given the following time line of barometric pressure readings, interpret the *likely* weather conditions, with regard to cloudiness and precipitation, for each of the following days:

    Day 1: Pressure steady at 1025 millibars

    Day 2: Pressure at 1010 millibars and falling

    Day 3: Pressure reaches four-day minimum of 992 millibars

    Day 4: Pressure at 1008 millibars and rising

11. If you wanted to erect wind turbines to generate electricity, would you search for a location that typically experiences a strong pressure gradient or a weak pressure gradient? Explain.

12. When designing an airport, it is important to have the runways positioned so that planes take off into the wind. Refer to the accompanying wind rose and discuss the orientation of the runway and the direction the planes would travel when they take off.

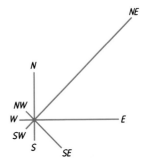

13. You and a friend are watching TV on a rainy day

when the weather reporter states, "The barometric pressure is 28.8 inches and rising." Hearing this, you say, "It looks like fair weather is on its way." How would you respond if your friend asked the following questions?

a. "I thought air pressure had something to do with the weight of air. How does 'inches' relate to weight?"

b. "Why do you think the weather is going to improve?"

AIR PRESSURE AND WINDS IN REVIEW

- *Wind*, the horizontal movement of air, is the result of horizontal differences in air pressure.

- *Air pressure* is the force exerted by the weight of air above. Average air pressure at sea level is about 1 kilogram per square centimeter, or 14.7 pounds per square inch, or 1013.25 millibars.

- Two instruments used to measure atmospheric pressure are the *mercury barometer*, where the height of a mercury column provides a measure of air pressure (standard atmospheric pressure at sea level equals 29.92 inches, or 760 millimeters), and the *aneroid barometer*, which uses a partially evacuated metal chamber that changes shape as air pressure changes.

- The pressure at any given altitude is equal to the weight of the air above that point. The rate at which pressure decreases with an increase in altitude is much greater near Earth's surface than aloft.

- Two factors that largely determine the amount of air pressure exerted by an air mass are temperature and humidity. A cold, dry air mass will produce higher surface pressures than a warm, humid air mass.

- Wind is controlled by a combination of (1) the *pressure gradient force*, (2) the *Coriolis force*, and (3) *friction*. The *pressure gradient force* is the primary driving force of wind that results from pressure differences that are depicted by the spacing of *isobars* on a map. Closely spaced isobars indicate a steep pressure gradient and strong winds; widely spaced isobars indicate a weak pressure gradient and light winds. The Coriolis force produces a deviation in the path of wind due to Earth's rotation (to the right in the Northern Hemisphere and to the left in the Southern Hemisphere). Friction, which significantly influences airflow near Earth's surface, is negligible above a height of a few kilometers.

- Above a height of a few kilometers, the Coriolis force is equal to and opposite the pressure gradient force, which results in *geostrophic winds*. Geostrophic winds flow in a nearly straight path, parallel to the isobars, with velocities proportional to the pressure gradient force.

- Winds that blow at a constant speed parallel to curved isobars are termed *gradient winds*. In low-pressure systems, such as *midlatitude cyclones*, the circulation of air, referred to as *cyclonic flow*, is counterclockwise in the Northern Hemisphere and clockwise in the Southern Hemisphere. Centers of high pressure, called *anticyclones*, exhibit *anticyclonic flow*, which is clockwise in the Northern Hemisphere and counterclockwise in the Southern Hemisphere.

- Near the surface, friction plays a major role in determining the direction of airflow. The result is a movement of air at an angle across the isobars, toward the area of lower pressure. Therefore, the resultant winds blow into and counterclockwise about a Northern Hemisphere surface cyclone. In a Northern Hemisphere surface anticyclone, winds blow outward and clockwise.

- A surface low-pressure system with its associated horizontal convergence is maintained or intensified by divergence (spreading out) aloft. Surface convergence in a cyclone accompanied by divergence aloft causes a net upward movement of air. Therefore, the passage of a low-pressure center is often associated with stormy weather. By contrast, fair weather can usually be expected with the approach of a high-pressure system.

- Two basic wind measurements—direction and speed—are important to a weather observer.

VOCABULARY REVIEW

aerovane (p. 183)
aneroid barometer (p. 164)
anticyclone (p. 169)
anticyclonic flow (p. 174)
atmospheric (air) pressure (p. 162)
barograph (p. 164)
barometric tendency (p. 178)
Buys Ballot's law (p. 173)
convergence (p. 167)
Coriolis force (p. 169)

cup anemometer (p. 183)
cyclone (p. 169)
cyclonic flow (p. 174)
divergence (p. 167)
geostrophic wind (p. 173)
gradient wind (p. 173)
isobar (p. 168)
mercury barometer (p. 163)
midlatitude cyclone (p. 169)
millibar (p. 163)

newton (p. 163)
pressure gradient force (p. 168)
pressure tendency (p. 178)
prevailing wind (p. 181)
ridge (p. 174)
trough (p. 174)
U.S. standard atmosphere (p. 165)
wind (p. 162)
wind vane (p. 180)

PROBLEMS

1. Figure 6-4, illustrates a simple mercury barometer. When a glass tube is completely evacuated of air and placed into the dish of mercury, the mercury rises to a height such that the force of the air pushing on the open dish matches the gravity pulling the mercury back down the tube. The density of mercury is 13,534 kilograms per cubic meter, which is very dense compared to water (1000 kilograms per cubic meter).

   In the mercury barometer, the mercury will rise to 29.92 inches under standard sea-level pressure. How tall would the barometer need to be if you used water instead of mercury?

2. Average air pressure at sea level is about 14.7 pounds per square inch. Using this information, how much does the entire atmosphere weigh? (*Hint:* The radius of Earth is 3963 miles.)

MyMeteorologyLab™

Circulation of the Atmosphere

Winds are generated by pressure differences that arise because of unequal heating of Earth's surface. Earth's winds blow in an unending attempt to balance these surface temperature differences. Because the zone of maximum solar heating migrates with the seasons—moving northward during the Northern Hemisphere summer and southward as winter approaches—the wind patterns that make up the general circulation also migrate latitudinally. This chapter focuses on global circulation and local wind systems, and it concludes with a discussion of global precipitation patterns.

Focus On Concepts

After completing this chapter, you should be able to:

- Distinguish between macroscale, mesoscale, and microscale winds and give an example of each.

- List four types of local winds and describe their formation.

- Describe or sketch the three-cell model of global circulation.

- Label and describe the basic characteristics of Earth's idealized zonal pressure belts on an appropriate map.

- Describe the seasonal change in Earth's global circulation that produces the Asian monsoon.

- Explain why the airflow aloft in the middle latitudes has a strong west-to-east component.

- Explain the origin of the polar jet stream and its relationship to midlatitude cyclonic storms.

- Sketch and label the major ocean currents on a world map.

- Describe the Southern Oscillation and its relationship to El Niño and La Niña.

- List the climate impacts to North America that are associated with El Niño and La Niña.

- Describe the relationships between global pressure systems and the global distribution of precipitation.

Scales of Atmospheric Motion

Earth's highly integrated wind system can be thought of as a series of deep rivers of air that encircle the planet. Embedded in the main currents are vortices of various sizes, including hurricanes, tornadoes, and midlatitude cyclones (Figure 7-1). Like eddies in a stream, these rotating wind systems develop and die out with somewhat predictable regularity. In general, the smallest eddies, such as dust devils, last only a few minutes, whereas larger and more complex systems, such as midlatitude cyclones and hurricanes, may survive for several days.

Residents of the United States and Canada are familiar with the term *westerlies*, which describes winds that predominantly blow across the midlatitudes from west to east. However, over short time periods, the winds may blow from any direction. You may recall being in a storm when shifts in wind direction and speed came in rapid succession. With such variations, how can we describe our winds as westerly? The answer lies in our attempt to simplify descriptions of the atmospheric circulation by sorting out events according to *size*. On the scale of a weather map, for instance, where observing stations are spaced about 150 kilometers (nearly 100 miles) apart, small whirl-winds that carry dust skyward are far too small to be identified. Instead, weather maps reveal larger-scale wind patterns, such as those associated with traveling cyclones and anticyclones.

In addition, equal consideration is given to the time frame in which the wind systems occur. In general, large weather patterns persist longer than their smaller counterparts. For example, dust devils usually last a few minutes. By contrast, midlatitude cyclones typically take a few days to cross the United States and sometimes dominate the weather for a week or longer.

Small- and Large-Scale Circulation

The wind systems shown in Figure 7-2 illustrate the three major categories of atmospheric circulation: microscale, mesoscale, and macroscale.

Microscale Winds The smallest scale of air motion is referred to as **microscale circulation.** These small, often chaotic winds normally last for seconds or at most minutes. Examples include simple gusts, which hurl debris into the air (Figure 7-2a), downdrafts, and small, well-developed vortices such as dust devils (see Box 7-1).

Figure 7-1 Gale force winds. (Photo by Emma Lee/Life Files/Photolibrary)

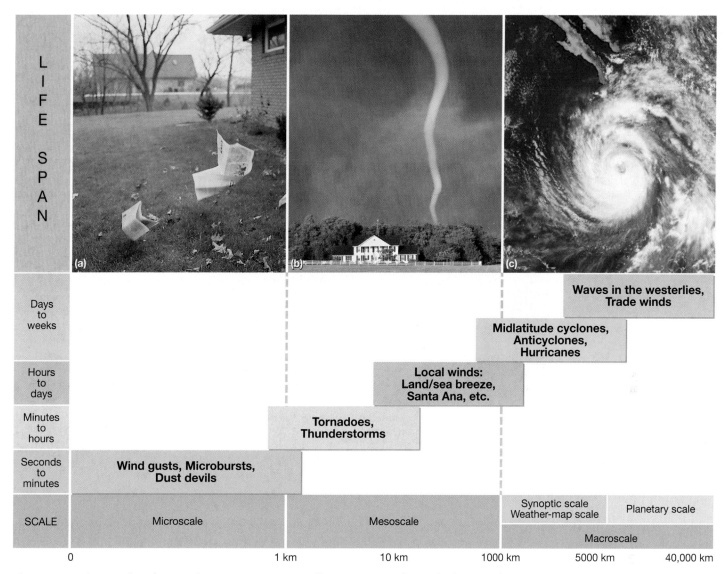

Figure 7–2 Three scales of atmospheric motion. (a) Gusts illustrate microscale winds. (b) Tornadoes exemplify mesoscale wind systems. (c) Satellite image of a hurricane, an example of macroscale circulation. (Photos by (a) E. J. Tarbuck, (b) Erin Nguyen/Photo Researchers Inc., (c) NASA/Science Photo Library/Photo Researchers, Inc.)

Mesoscale Winds **Mesoscale winds** generally last for several minutes and may exist for hours. These middle-size phenomena are usually less than 100 kilometers (62 miles) across. Further, some mesoscale winds—for example, thunderstorms and tornadoes—also have a strong vertical component (Figure 7-2b). It is important to remember that thunderstorms and tornadoes are often embedded within a larger cyclonic storm and thus move with it. Further, much of the vertical air movement within a midlatitude cyclone, or hurricane, is provided by thunderstorms, where updrafts in excess of 100 kilometers (62 miles) per hour have been measured. Land and sea breezes, as well as mountain and valley winds, also fall into this category and will be discussed in the next section, with other local winds.

Macroscale Winds The largest wind patterns, called **macroscale winds,** are exemplified by the westerlies and trade winds that carried sailing vessels back and forth across the Atlantic during the opening of the New World. These *planetary-scale* flow patterns extend around the entire globe and can remain essentially unchanged for weeks at a time.

A somewhat smaller macroscale circulation is called *synoptic scale*—also referred to as *weather-map scale*. Synoptic-scale wind systems are about 1000 kilometers in diameter and are easily identified on weather maps. Two well-known synoptic-scale systems are the individual traveling *midlatitude cyclones* and *anticyclones* that appear on weather maps as areas of low and high pressure, respectively. These weather producers are found in the middle latitudes and are sometimes hundreds to thousands of kilometers across.

Somewhat smaller macroscale systems are *tropical storms* and *hurricanes* that develop in late summer and early fall over warm tropical oceans (Figure 7-2c). Airflow in these systems is inward and upward, as in the larger midlatitude cyclones. However, the rate of horizontal flow associated

Box 7-1   Dust Devils

A common phenomenon in arid regions of the world are the whirling vortices called *dust devils* (Figure 7–A). Although they resemble tornadoes, dust devils are generally much smaller and less intense than their destructive cousins. Most dust devils are only a few meters in diameter and reach heights no greater than about 100 meters (300 feet). Further, these whirlwinds are usually short-lived phenomena that die out within minutes.

Unlike tornadoes, which are associated with convective clouds, dust devils form on days when clear skies dominate. Further, these whirlwinds develop from the ground upward, which is opposite of tornado formation. Because surface heating is critical to their formation, dust devils occur most frequently in the afternoon, when surface temperatures are highest.

Recall that when the air near the surface is considerably warmer than the air a few dozen meters overhead, the layer of air near Earth's surface becomes unstable. In this situation, warm surface air begins to rise, causing air near the ground to be drawn into the developing whirlwind. The rotating winds that are associated with dust devils are produced by the same phenomenon that causes ice skaters to spin faster as they pull their arms closer to their body. As the inwardly spiraling air rises, it carries sand, dust, and other loose debris dozens of meters into the air—making a dust devil visible.

Most dust devils are small and short-lived; consequently, they are not generally destructive. Occasionally, however, these whirlwinds grow to be 100 meters or more in diameter and over a kilometer high. With wind speeds that may reach 100 kilometers (60 miles) per hour, large dust devils can do considerable damage.

FIGURE 7–A   Dust devil. Although these whirling vortices resemble tornadoes, they have a different origin and are much smaller and less intense. (Courtesy of St. Meyers/Okapia/Photo Researchers, Inc.)

---

with hurricanes is usually more rapid than that of their more poleward cousins.

Structure of Wind Patterns

Although it is common practice to divide atmospheric motions according to size, remember that global winds are a composite of motion on all scales—much like a meandering river that contains large eddies composed of smaller eddies containing still smaller eddies. As an example, we will examine winds associated with hurricanes that form over the North Atlantic. When we view one of these tropical cyclones on a satellite image, the storm appears as a large whirling cloud migrating slowly across the ocean (Figure 7-2c). From this perspective, which is at the weather-map (synoptic) scale, the general counterclockwise rotation of the storm is easily seen.

In addition to their rotating motions, hurricanes often move from east to west or northwest. (Once hurricanes move into the belt of the westerlies, they tend to change course and move in a northeasterly direction.) This motion demonstrates that these large eddies are embedded in a still larger flow (planetary scale) that is moving westward across the tropical portion of the North Atlantic.

When we examine a hurricane more closely by flying an airplane through it, some of the small-scale aspects of the storm become noticeable. As the plane approaches the outer edge of the system, it is evident that the large rotating cloud that we see in the satellite image consists of many individual cumulonimbus towers (thunderstorms). Each of these mesoscale phenomena lasts for a few hours and must be continually replaced by new ones for the hurricane to persist. During the flight, we also realize that the individual thunderstorms are made up of even smaller-scale turbulences. The small thermals of rising and descending air in these clouds make for a rough flight.

In summary, *a typical hurricane exhibits several scales of motion, including many mesoscale thunderstorms, which consist of numerous microscale turbulences. Furthermore, the counterclockwise circulation of the hurricane (weather-map scale) is embedded in the global winds (planetary scale) that flow from east to west in the tropical North Atlantic.*

Concept Check 7.1

1 List the three major categories of atmospheric circulation and give at least one example of each.

2 Describe how the size of a wind system is related to its duration (life span).

3 What scale of atmospheric circulation includes midlatitude cyclones, anticyclones, and tropical cyclones (hurricanes)?

Local Winds

Local winds are examples of mesoscale winds (time frame of minutes to hours and size of 1 to 100 kilometers). Remember that most winds have the same cause: pressure differences that arise because of temperature differences caused by unequal heating of Earth's surface. Most local winds are linked to temperature and pressure differences that result from variations in topography or in local surface conditions.

Recall that winds are named for the direction *from which they blow*. This holds true for local winds. Thus, a sea breeze originates over water and blows toward the land, whereas a valley breeze blows upslope, away from its source.

Land and Sea Breezes

The daily temperature differences that develop between the sea and adjacent land areas, and the resulting pressure pattern that creates a sea breeze, were discussed in Chapter 6 (see Figure 6–11). Recall that land is heated more intensely during daylight hours than is an adjacent body of water. As a result, the air above the land heats and expands, creating an area of high pressure aloft. This, in turn, causes the air above the land to move seaward. This mass transfer of air aloft creates a surface low over the land. A **sea breeze** develops as cooler air over the the water moves landward toward the area of lower pressure (Figure 7-3a). At night, the reverse may take place; the land cools more rapidly than the sea, and a **land breeze** develops, with airflow off the land (Figure 7-3b).

A sea breeze has a significant moderating influence in coastal areas. Shortly after a sea breeze begins, air temperatures over the land may drop by as much as 5° to 10°C. However, the cooling effect of these breezes is generally noticeable for only 100 kilometers (60 miles) inland in the tropics and often less than half that distance in the middle latitudes. These cool sea breezes generally begin shortly before noon and reach their greatest intensity—about 10 to 20 kilometers per hour—in the midafternoon.

Smaller-scale sea breezes can also develop along the shores of large lakes. Cities near the Great Lakes, such as Chicago, benefit from the "lake effect" during the summer, when residents typically enjoy cooler temperatures near the lake compared to warmer inland areas. In many places sea breezes also affect the amount of cloud cover and rainfall. The peninsula of Florida, for example, experiences increased summer precipitation partly because of convergence associated with sea breezes from both the Atlantic and Gulf coasts.

The intensity and extent of land and sea breezes vary by location and season. Tropical areas where intense solar heating is continuous throughout the year experience more frequent and stronger sea breezes than midlatitude locations. The most intense sea breezes develop along tropical coastlines adjacent to cool ocean currents. In the middle latitudes, sea breezes are best developed during the warmest months, but land breezes are often missing because at night the land does not always cool below the temperature of the ocean surface.

Mountain and Valley Breezes

A daily wind similar to land and sea breezes occurs in mountainous regions. During the day, air along mountain slopes is heated more intensely than air at the same elevation over

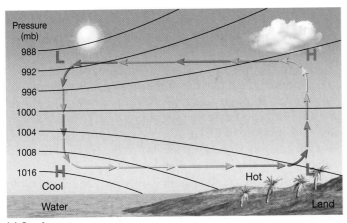

(a) Sea breeze

(b) Land breeze

Figure 7–3 Illustration of a sea breeze and a land breeze. (a) During the daylight hours, cooler and denser air over the water moves onto the land, generating a sea breeze. (b) At night the land cools more rapidly than the sea, generating an offshore flow called a land breeze.

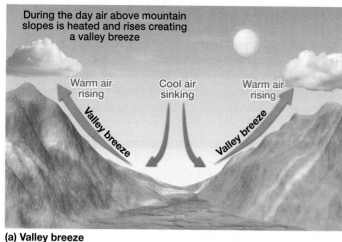

During the day air above mountain slopes is heated and rises creating a valley breeze

Warm air rising

Cool air sinking

Warm air rising

Valley breeze

Valley breeze

(a) Valley breeze

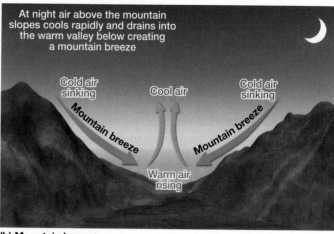

At night air above the mountain slopes cools rapidly and drains into the warm valley below creating a mountain breeze

Cold air sinking

Cool air

Cold air sinking

Mountain breeze

Mountain breeze

Warm air rising

(b) Mountain breeze

Figure 7–4 Valley and mountain breezes. (a) Heating during the daylight hours warms the air along the mountain slopes. This warm air rises, generating a valley breeze. (b) After sunset, cooling of the air near the mountain slopes can result in cool air draining into the valley, producing a mountain breeze.

the valley floor (Figure 7-4a). This warmer air glides up the mountain slope and generates a **valley breeze.** Valley breezes can often be identified by the cumulus clouds that develop over adjacent mountain peaks and may account for late afternoon thundershowers that occur on warm summer days (Figure 7-5).

After sunset the pattern is reversed. Rapid heat loss along the mountain slopes cools the air, which drains into the valley and causes a **mountain breeze** (Figure 7-4b). Similar cool air drainage can occur in hilly regions that have modest slopes. The result is that the coldest pockets of air are usually found in the lowest spots.

Like many other winds, mountain and valley breezes have seasonal preferences. Valley breezes are most common during warm seasons, when solar heating is most intense, whereas mountain breezes tend to be more frequent during cold seasons.

Chinook (Foehn) Winds

Warm, dry winds called **chinooks** sometimes move down the slopes of mountains in the United States. Similar winds in the Alps are called **foehns.** Such winds are usually created when a strong pressure gradient develops in a mountainous region. As the air descends the leeward slopes of the mountains, it is heated adiabatically (by compression). Because condensation may have occurred as the air ascended the windward side, releasing latent heat, the air descending the leeward side will be warmer and drier than at a similar elevation on the windward side.

Chinooks commonly flow down the east slopes of the Colorado Rockies in the winter and spring, when the affected area may be experiencing subfreezing temperatures. Thus, these dry, warm winds often bring drastic change. Within minutes of the arrival of a chinook, the temperature may climb 20°C (36°F). When the ground has a snow cover, these winds melt it rapidly, which explains why the Native American word *chinook,* meaning "snow-eater," has been applied to these winds. Chinook winds have been known to melt more than a foot of snow in a single day. A chinook that moved through Granville, North Dakota, on February 21, 1918, caused the temperature to rise from –33°F to 50°F—an increase of 83°F!

Chinooks are sometimes viewed as beneficial to ranchers east of the Rockies because they keep the grasslands clear of snow during much of the winter. However, this benefit is offset by the loss of moisture that the snow would bequeath to the land if it remained until the spring melt.

Another chinook-like wind that occurs in the United States is the **Santa Ana.** Found in southern California, the hot, dry Santa Ana winds greatly increase the threat of fire

Figure 7–5 The occurrence of a daytime upslope (valley) breeze is identified by cloud development on mountain peaks, sometimes becoming a midafternoon thunderstorm. (Photo by Herber Koeppel/Alamy)

in this already dry area (see Severe and Hazardous Weather: Santa Ana Winds and Wildfires on page 196).

Katabatic (Fall) Winds

In the winter, areas adjacent to highlands may experience a local wind called a **katabatic** or **fall wind.** These winds originate when cold dense air, situated over a highland area such as the ice sheets of Greenland and Antarctica, begins to move. Under the influence of gravity, the cold air cascades over the rim of a highland like a waterfall. Although the air is heated adiabatically, the initial temperatures are so low that the wind arrives in the lowlands still colder and more dense than the air it displaces. As this frigid air descends, it occasionally is channeled into narrow valleys, where it acquires velocities capable of great destruction.

A few of the better-known katabatic winds have local names. Most famous is the **mistral,** which blows from the French Alps toward the Mediterranean Sea. Another is the **bora,** which originates in the mountains of the Balkan Peninsula and blows to the Adriatic Sea.

Country Breezes

One mesoscale wind, called a **country breeze,** is associated with large urban areas. As the name implies, this circulation pattern is characterized by a light wind blowing into the city from the surrounding countryside. In cities, the massive buildings composed of rocklike materials tend to retain the heat accumulated during the day more than the open landscape of outlying areas (see Box 3–3 on the urban heat island). The result is that the warm, less dense air over cities rises, which in turn initiates the country-to-city flow. A country breeze is most likely to develop on a relatively clear, calm night. One of the unfortunate consequences of the country breeze is that pollutants emitted near the urban perimeter tend to drift in and concentrate near the city's center.

Concept Check 7.2

1 The most intense sea breezes develop along tropical coasts adjacent to cool ocean currents. Explain.

2 In what way are land and sea breezes similar to mountain and valley breezes?

3 What are chinook winds? Name two areas where they are common.

4 In what way are katabatic (fall) winds different from most other types of local winds?

5 Explain how cities create their own local winds.

Global Circulation

Our knowledge of global winds comes from two sources: the patterns of pressure and winds observed worldwide and theoretical studies of fluid motion. We first consider the classical model of global circulation that was developed largely from average worldwide pressure distribution. We will then modify this idealized model by adding more recently discovered aspects of the atmosphere's complex motions.

Single-Cell Circulation Model

One of the first contributions to the classical model of global circulation came from George Hadley in 1735. Well aware that solar energy drives the winds, Hadley proposed that the large temperature contrast between the poles and the equator creates a large *convection cell* in both the Northern and Southern Hemispheres (Figure 7-6).

In Hadley's model, warm equatorial air rises until it reaches the tropopause, where it spreads toward the poles. Eventually, this upper-level flow reaches the poles, where cooling causes it to sink and spread out at the surface as equatorward-moving winds. As this cold polar air approaches the equator, it is reheated and rises again. Thus, the

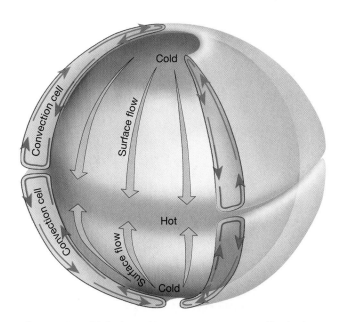

Figure 7-6 Global circulation on a nonrotating Earth. A simple convection system is produced by unequal heating of the atmosphere on a nonrotating Earth.

SEVERE AND HAZARDOUS WEATHER

Santa Ana Winds and Wildfires

Santa Ana winds are chinook-like winds that characteristically sweep through southern California and northwestern Mexico in the fall and winter. These hot, dry winds are infamous for fanning regional wildfires. Santa Ana winds are driven by strong high-pressure systems with subsiding air that tend to develop in the fall over the Great Basin. The clockwise flow from the anticyclone directs desert air from Arizona and Nevada westward toward the Pacific (Figure 7–B inset). The wind gains speed as it is funneled through the canyons of the Coast Ranges, in particular the Santa Ana Canyon, from which the winds derive their name. Adiabatic heating of this already warm, dry air as it descends mountain slopes further accentuates the already parched conditions. Vegetation, seared by the summer heat, is dried even further by these hot, dry winds.

> These hot, dry winds are infamous for fanning regional wildfires.

Although Santa Ana winds occur every year, they were particularly hazardous in the fall of 2003, and to a lesser extent in 2007, when hundreds of thousands of acres were scorched. In late October 2003, Santa Anas began blowing toward the coast of southern California, with speeds that sometimes exceeded 100 kilometers (60 miles) per hour. Much of this area is covered by brush known as chaparral and related shrubs. It didn't take much—a careless camper or motorist, a lightning strike, or an arsonist—to ignite fires. Soon a number of small fires occurred in portions of Los Angeles, San Bernardino, Riverside, and San Diego counties (Figure 7–B). Several developed quickly into wildfires that moved almost as fast as the ferocious Santa Ana winds sweeping through the canyons.

Within a few days more than 13,000 firefighters were on firelines that extended from north of Los Angeles to the Mexican border. Nearly two months later, when all the fires were officially extinguished, more than 742,000 acres had been scorched, more than 3000 homes destroyed, and 26 people killed (Figure 7–C). The Federal Emergency Management Agency put the dollar losses at over $2.5 billion. The 2003 southern California wildfires became the worst fire disaster in the state's history.

Strong Santa Ana winds, coupled with dry summers, have produced wildfires in southern California for millennia. These fires are nature's way of burning out chaparral thickets and sage scrub to prepare the land for new growth. When people began building homes and crowding into the fire-prone area between Santa Barbara and San Diego, the otherwise natural problems were compounded. Landscaped properties consisting of highly flammable eucalyptus and pine trees have further increased the hazard risk. In addition, fire prevention efforts have resulted in the accumulation of large quantities of flammable material over time, ultimately producing fewer but larger and more destructive fires. Clearly, wildfires will remain a major threat in these areas into the foreseeable future.

FIGURE 7–B Ten large wildfires rage across southern California in this image taken on October 27, 2003, by NASA's *Aqua* satellite. Inset shows an idealized high-pressure area composed of cool dry air that drives Santa Ana winds. Adiabatic heating causes the air temperature to increase and the relative humidity to decrease. (Photo courtesy of NASA)

FIGURE 7–C Flames from a wildfire fanned by Santa Ana winds move toward a home south of Valley Center, California, on October 27, 2003. (Photo by Denis Poroy/Associated Press)

circulation proposed by Hadley has upper-level air flowing poleward and surface air moving equatorward.

Although correct in principle, Hadley's model does not take into account Earth's rotation. (Hadley's model would better approximate the circulation of a *nonrotating* planet.)

Three-Cell Circulation Model

In the 1920s a three-cell circulation model was proposed. Although this model has been modified to fit upper-air observations, it remains a useful tool for examining global circulation. Figure 7-7 illustrates the idealized three-cell model and the surface winds that result.

In the zones between the equator and roughly 30° latitude north and south, the circulation closely resembles the model used by Hadley and is known as the **Hadley cell.** Near the equator, the warm rising air that releases latent heat during the formation of cumulus towers is believed to provide the energy that drives the Hadley cells. As the flow aloft moves poleward, the air begins to subside in a zone between 20° and 35° latitude. Two factors contribute to this general subsidence: (1) As upper-level flow moves away from the stormy equatorial region, radiation cooling becomes the dominant process. As a result, the air cools, becomes more dense, and sinks. (2) In addition, the Coriolis force becomes stronger with increasing distance from the equator, causing the poleward-moving upper air to be deflected into a nearly west-to-east flow by the time it reaches 30° latitude. This restricts the poleward flow of air. Stated

another way, the Coriolis force causes a general pileup of air (convergence) aloft. As a result, general subsidence occurs in the zones between 20° and 35° latitude.

The subsiding air between 20° and 35° latitude is relatively dry because it has released its moisture near the equator. In addition, the effect of adiabatic heating during descent further reduces the relative humidity of the air. Consequently, this zone of subsidence is the site of the world's subtropical deserts. The Sahara Desert of North Africa and the Great Australian Desert are examples of subtropical deserts located in these regions of subsiding air. Further, because surface winds are sometimes weak in the zones between 20° and 35° latitude, this belt was named the **horse latitudes** (Figure 7-7). The name stems from the fact that Spanish sailing ships crossing the Atlantic were sometimes becalmed and stalled for long periods of time in these waters. If food and water supplies for the horses on board became depleted, the Spanish sailors were forced to throw the horses overboard.

From the center of the horse latitudes, the surface flow splits into two branches—one flowing poleward and one flowing toward the equator. The equatorward flow is deflected by the Coriolis force to form the reliable **trade winds,** so called because they enabled early sailing ships to move goods between Europe and North America. In the Northern Hemisphere, the trades blow from the northeast, while in the Southern Hemisphere, the trades are from the southeast. The trade winds from both hemispheres meet near the equator, in a region that has a weak pressure gradient. This zone is called the **doldrums.** Here light winds and humid conditions provide the monotonous weather that is the basis for the expression "the doldrums."

In the three-cell model, the circulation between 30° and 60° latitude (north and south), called the **Ferrel cell,** was proposed by William Ferrel to account for the westerly surface winds in the middle latitudes (see Figure 7-7). These **prevailing westerlies** were known to Benjamin Franklin, perhaps the first American weather forecaster, who noted that storms migrated from west to east across the colonies. Franklin also observed that the westerlies were much more sporadic and, therefore less reliable than the trade winds for sail power. We now know that it is the migration of cyclones and anticyclones across the mid-latitudes that disrupts the general westerly flow at the surface. The Ferrel cell, however, is *not* a good model for the flow aloft because it predicts a flow from east to west, just opposite of what is observed. In fact, the westerlies become stronger with an increase in altitude and reach their maximum speeds 10-12 kilometers above Earth's surface in zones called *jet streams.* Because of the importance of the mid-latitude circulation in producing our daily weather, we will consider the westerlies in more detail later.

The circulation in a *polar cell* is driven by subsidence near the poles that produces a surface flow that moves equatorward; this is called the **polar easterlies** in both hemispheres. As these cold polar winds move equatorward,

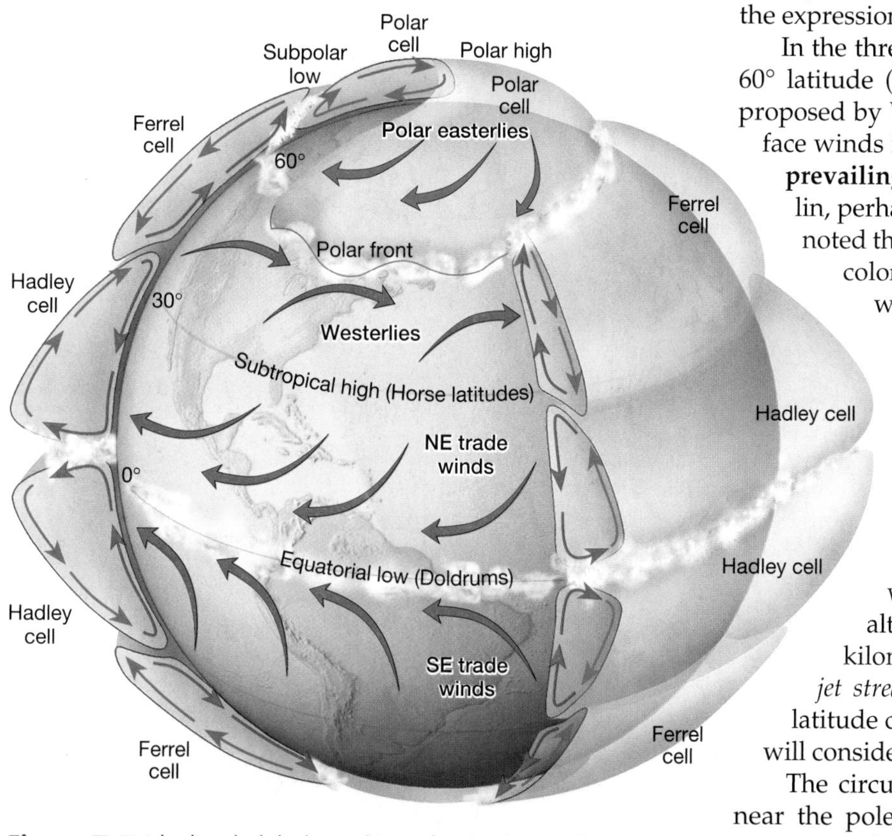

Figure 7-7 Idealized global circulation for the three-cell circulation model on a rotating Earth.

they eventually encounter the warmer westerly flow of the midlatitudes. The region where the flow of cold air clashes with warm air has been named the **polar front.** The significance of this region will be considered later.

Concept Check 7.3

1 Briefly describe the idealized global circulation proposed by George Hadley. What are the shortcomings of the Hadley model?

2 Which two factors cause air to subside between 20° and 35° latitude?

3 Referring to the idealized three-cell model of atmospheric circulation, most of the United States is situated in which belt of prevailing winds?

4 What wind belts are found between the equator and 30° latitude?

Pressure Zones Drive the Wind

Earth's global wind patterns are derived from a distinct distribution of surface air pressure. To simplify this discussion, we will first examine the idealized pressure distribution that would be expected if Earth's surface were uniform—that is, composed entirely of water or smooth land.

Idealized Zonal Pressure Belts

If Earth had a uniform surface, each hemisphere would have two east-west oriented belts of high pressure and two of low pressure (Figure 7-8a). Near the equator, the warm rising branch of the Hadley cells is associated with the low-pressure zone known as the **equatorial low.** This region of ascending moist, hot air is marked by abundant precipitation. Because this region of low pressure is where the trade winds

PROFESSIONAL PROFILE   Sally Benson: Climate and Energy Scientist

Sally Benson has spent her career researching solutions to some of the most pressing environmental problems of our time. In the mid-1970s, as a young scientist with Lawrence Berkeley National Laboratory, she tackled the first oil shortage by investigating ways to harness the power of geothermal energy.

> "I get to do important work . . . and have fun while I do it."

Ten years later, she was elbow-deep in the mud of Kesterson Reservoir in California's Central Valley. Irrigation runoff had caused selenium from local soil to accumulate in the water, causing local birds to hatch chicks with horrifying deformities. Benson's studies made her realize microbes could be environmentally friendly tools to clean up other sites with toxic metal contamination.

"The experience was so positive because I was doing this cutting-edge science at the same time regulators had to make a decision about how to clean up the site. I got an idea of the impact that science can make, and how research could get results," Benson says.

By the mid-1990s, Benson was directing all Earth science research at the laboratory. "I talked with many people, read a lot, and came to the conclusion that climate change was the most significant issue facing the world." Benson organized research programs that developed regional models of climate change to help residents plan for droughts and temperature shifts, and studied the carbon cycle of the oceans, among other projects.

In 2007, Benson was appointed director of the Global Climate and Energy Project at Stanford University, which seeks to develop energy sources that release fewer greenhouse gases. "Energy efficiency in lighting, heating and cooling systems, and autos makes all the sense in the world. But at the end of the day, we need to do a lot more than that. If our current understanding is correct, we need to cut overall emissions by 80 percent of today's levels," Benson states.

Benson sees many promising ways to achieve that goal. One is renewable energy. "Today's biofuels don't provide much advantage in terms of carbon dioxide emissions. But alternatives such as cellulosic ethanol (producing ethanol with plant fibers), where there are low emissions in the process of growing and making them, are incredibly important," Benson opines.

Another is to continue using some fossil fuels to run power plants and ships, but to capture the greenhouse gases they emit. "More than half of the electricity worldwide is produced by burning coal, a plentiful resource in many places. We need to find a way to make it carbon neutral," Benson says. One way to remove such emissions from the atmosphere altogether would be to inject them in aquifers deep within Earth. Benson herself has studied how to use technology developed by the oil and gas industry to select sites where the rock offers a good seal, inject the gas, and monitor the area for leaks.

"I get to do important work solving critical problems, get to be outdoors, do experiments at scale, and have fun while I do it. Interacting with the huge number of people impacted by these issues makes the Earth sciences very rewarding."

—Kathleen Wong

PROFESSOR SALLY BENSON is Executive Director of Stanford University's Global Climate and Energy Project, a $225 million effort to develop energy resources that are not harmful to the environment, especially energy sources that release fewer greenhouse gases. Dr. Benson has a multidisciplinary background with degrees in geology, materials science, and minerals engineering, and with applied research in hydrology and reservoir engineering. (Courtesy Dr. Sally Benson)

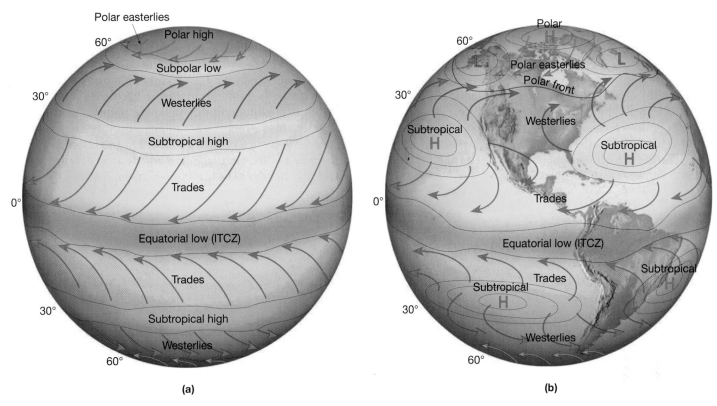

Figure 7–8 Idealized global circulation. (a) An imaginary uniform Earth with idealized zonal (continuous) pressure belts. (b) The actual Earth, with its large landmasses, disrupts the idealized zonal pattern. As a result, Earth's pressure patterns consist mainly of semipermanent high- and low-pressure cells.

converge, it is also referred to as the **intertropical convergence zone (ITCZ).** In Figure 7-9, the ITCZ is visible as a band of clouds near the equator.

In the belts about 20° to 35° on either side of the equator, where the westerlies and trade winds originate and go their separate ways, are the high-pressure zones known as the **subtropical highs.** In these zones a subsiding air column produces weather that is normally warm and dry.

Another low-pressure region is situated at about 50° to 60° latitude, in a position corresponding to the polar front. Here the polar easterlies and the westerlies clash in the low-pressure convergence zone known as the **subpolar low.** As you will see later, this zone is responsible for much of the stormy weather in the middle latitudes, particularly in the winter.

Finally, near Earth's poles are the **polar highs,** from which the polar easterlies originate (see Figure 7-8a). The polar highs exhibit high surface pressure mainly because of surface cooling. Because air near the poles is cold and dense, it exerts higher-than-average pressure.

Semipermanent Pressure Systems: The Real World

Up to this point, we have considered the global pressure systems as if they were continuous belts around Earth. How-

Figure 7–9 The intertropical convergence zone (ITCZ) is seen as the band of clouds that extends east–west just north of the equator. (Courtesy of NOAA)

ever, because Earth's surface is not uniform, the only true zonal distribution of pressure exists along the subpolar low in the Southern Hemisphere, where the ocean is continuous. To a lesser extent, the equatorial low is also continuous. At other latitudes, particularly in the Northern Hemisphere, where there is a higher proportion of land compared to ocean, the zonal pattern is replaced by semipermanent cells of high and low pressure.

The idealized pattern of pressure and winds for the "real" Earth is illustrated in Figure 7-8b. The pattern shown is always in a state of flux because of seasonal temperature changes, which serve to either strengthen or weaken these pressure cells. In addition, the position of these pressure systems moves either poleward or equatorward with the seasonal migration of the zone of maximum solar heating. As a consequence of these factors, Earth's pressure patterns vary in strength and location during the course of the year.

Views of average global pressure patterns and resulting winds for the months of January and July are shown in Figure 7-10. Notice on these maps that the observed pressure patterns are circular (or elongated) instead of zonal (east-west bands). The most prominent features on both

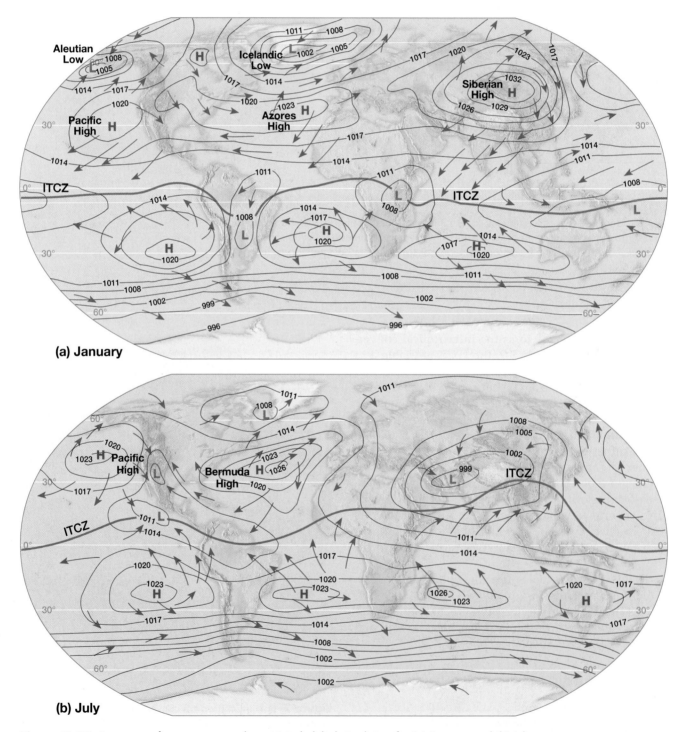

(a) January

(b) July

Figure 7–10 Average surface pressure and associated global circulation for (a) January and (b) July.

maps are the subtropical highs. These systems are centered between 20° and 35° latitude over the subtropical oceans.

When we compare Figures 7-10a (January) and 7-10b (July), we see that some pressure cells are year-round features—the subtropical highs, for example. Others, however, are seasonal. For example, the low-pressure cell over northern Mexico and the southwestern United States is a summer phenomenon and appears only on the July map. The main cause of these variations is the greater seasonal temperature fluctuations experienced over landmasses, especially in the middle and higher latitudes.

January Pressure and Wind Patterns The **Siberian high,** a very strong high-pressure center positioned over the frozen landscape of northern Asia, is the most prominent feature on the January pressure map (Figure 7-10a). A weaker polar high is located over the chilled North American continent. These cold anticyclones consist of very dense air that accounts for the significant weight of these air columns. In fact, the highest sea-level pressure ever measured, 1084 millibars (32.01 inches of mercury), was recorded in December 1968 at Agata, Siberia. Subsidence within these air columns results in clear skies and divergent surface flow.

As the Arctic highs strengthen over the continents, the subtropical anticyclones situated over the oceans become weaker. Further, the average position of the subtropical highs tends to be closer to the eastern shore of the oceans in January than in July. For example, notice in Figure 7-10a that the center of the subtropical high is located in the eastern part of the North Atlantic (sometimes called the **Azores high**).

Also shown on the January map but absent in July are two intense semipermanent low-pressure centers. Named the **Aleutian low** and the **Icelandic low,** these cyclonic cells are situated over the North Pacific and North Atlantic, respectively. They are not stationary cells but rather composites of numerous cyclonic storms that traverse these regions. In other words, so many midlatitude cyclones occur during the winter that these regions are almost always experiencing low pressure, hence the term *semipermanent*. As a result, the areas affected by the Aleutian and Icelandic lows are frequently cloudy and receive abundant winter precipitation.

Because a large number of cyclonic storms form over the North Pacific and travel eastward, the coastal areas of southern Alaska receive abundant precipitation. This fact is exemplified by the climate data for Sitka, Alaska, a coastal town that receives 215 centimeters (85 inches) of precipitation each year, more than five times that received in Churchill, Manitoba, Canada. Although both towns are situated at roughly the same latitude, Churchill is located in the continental interior, far removed from the influence of the cyclonic storms associated with the Aleutian low.

July Pressure and Wind Patterns The pressure pattern over the Northern Hemisphere changes dramatically in summer (see Figure 7-10b). High surface temperatures over the continents generate lows that replace wintertime highs. These thermal lows consist of warm ascending air that induces inward-directed surface flow. The strongest of these low-pressure centers develops over southern Asia, while a weaker thermal low is found in the southwestern United States.

Notice in Figure 7-10 that during the summer months, the subtropical highs in the Northern Hemisphere migrate westward and become stronger than during the winter months. These strong high-pressure centers dominate the summer circulation over the oceans and pump warm moist air onto the continents that lie to the west of these highs. This results in an increase in precipitation over parts of eastern North America and Southeast Asia.

During the peak of the summer season, the subtropical high found in the North Atlantic is positioned near the island of Bermuda, hence the name **Bermuda high.** In the Northern Hemisphere winter, the Bermuda high is located near Africa and is called the *Azores high* (see Figure 7-10).

Concept Check 7.4

1 What is the intertropical convergence zone (ITCZ)?

2 If Earth had a uniform surface, east–west belts of high and low pressures would exist. Name these zones and the approximate latitude in which each would be found.

3 During what season is the Siberian high strongest?

4 During what season is the Bermuda high strongest?

Monsoons

Large *seasonal* changes in Earth's global circulation are called monsoons. Contrary to popular belief, **monsoon** does not mean "rainy season"; rather, it refers to a particular wind system that reverses its direction twice each year. In general, winter is associated with winds that blow predominantly off the continents, called the *winter monsoon*. In contrast, in summer, warm moisture-laden air blows from the sea toward the land. Thus, the *summer monsoon* is usually associated with abundant precipitation over affected land areas and is the source of the misconception.

The Asian Monsoon

The best-known and best-developed monsoon circulation occurs in southern and southeastern Asia—affecting India and the surrounding areas as well as parts of China, Korea, and Japan. As with most other winds, the Asian monsoon is driven by pressure differences that are generated by unequal heating of Earth's surface.

As winter approaches, long nights and low sun angles result in the accumulation of frigid air over the vast landscape of northern Russia. This generates the cold Siberian high, which ultimately dominates Asia's winter circulation. The subsiding dry air of the Siberian high produces surface flow that moves across southern Asia, producing predominantly offshore winds (Figure 7-11a). By the time this flow reaches India, it has warmed considerably but remains extremely dry. For example, Kolkata, India, receives less than

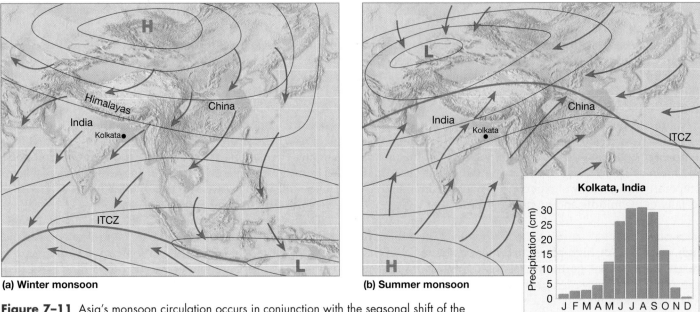

Figure 7–11 Asia's monsoon circulation occurs in conjunction with the seasonal shift of the intertropical convergence zone (ITCZ). (a) In January a strong high pressure develops over Asia. The resulting flow of cool air off the continent generates the dry winter monsoon. (b) With the onset of summer, the ITCZ migrates northward and draws warm, moist air onto the continent.

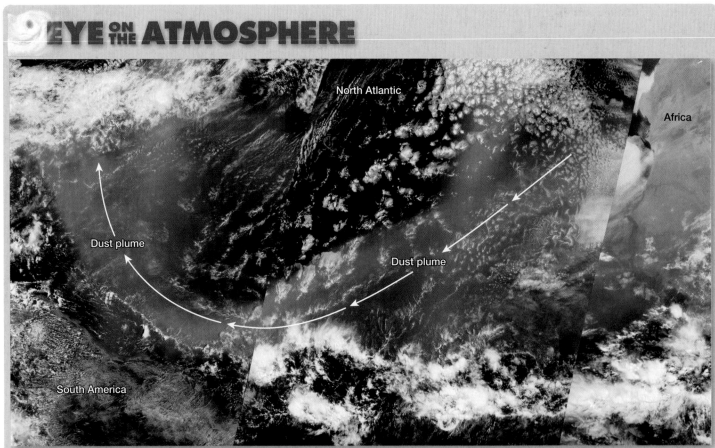

Composite image stitched together from a series of images collected by MODIS. (Courtesy of NASA)

This image shows a January dust storm in Africa that produced a dust plume that reached all the way to the northeast coast of South America. It has been estimated that plumes such as this transport about 40 million tons of dust from the Sahara Desert to the Amazon basin each year. The minerals carried by these dust plumes help to replenish nutrients in rainforest soils that are continually being washed out of the soils by heavy tropical rains.

Question 1 Notice that the dust plume is following a curved path. Does the atmospheric circulation carrying this dust plume exhibit a clockwise or counterclockwise rotation?

Question 2 What global pressure system was responsible for transporting this dust from Africa to South America?

2 percent of its annual precipitation in the cooler six months. The remainder comes in the warmer six months, with the vast majority falling from June through September.

In contrast, summertime temperatures in the interior of southern Asia often exceed 40°C (104°F). This intense solar heating generates a low-pressure area over the region similar to that associated with a sea breeze, but on a much larger scale. This in turn generates outflow aloft that encourages inward flow at the surface. With the development of the low-pressure center over southeastern Asia, moisture-laden air from the Indian and Pacific Oceans flows landward, thereby generating a pattern of precipitation typical of the summer monsoon.

One of the world's rainiest regions is found on the slopes of the Himalayas, where orographic lifting of moist air from the Indian Ocean produces copious precipitation. Cherrapunji, India, once recorded an annual rainfall of 25 meters (82.5 feet), most of which fell during the four months of the summer monsoon (Figure 7-11b).

The Asian monsoon is complex and strongly influenced by the seasonal change in solar heating received by the vast Asian continent. However, another factor, related to the annual migration of the Sun's vertical rays, also contributes to the monsoon circulation of southern Asia. As shown in Figure 7-11, the Asian monsoon is associated with a larger-than-average seasonal migration of the ITCZ. With the onset of summer, the ITCZ moves northward over the continent and is accompanied by peak rainfall. The opposite occurs in the Asian winter, as the ITCZ moves south of the equator.

Nearly half the world's population inhabits regions affected by the Asian monsoons. Many of these people depend on subsistence agriculture for their survival. The timely arrival of monsoon rains often means the difference between adequate nutrition and widespread malnutrition.

The North American Monsoon

Other regions experience seasonal wind shifts like those associated with the Asian monsoon. For example, a seasonal wind shift influences a portion of North America. Sometimes called the *North American monsoon*, this circulation pattern produces a dry spring followed by a relatively rainy summer that impacts large areas of the southwestern United States and northwestern Mexico.* This is illustrated by precipitation patterns observed in Tucson, Arizona,

Figure 7-12 Tucson, Arizona, has a summer precipitation maximum, the result of a monsoon circulation that draws moist air in from the Gulf of California and to a lesser extent from the Gulf of Mexico. (Photo by Clint Farlinger/Alamy)

which typically receives nearly 10 times more precipitation in August than in May. As shown in Figure 7-12, summer rains typically last into September before drier conditions are reestablished.

Summer daytime temperatures in the American Southwest, particularly in the low deserts, can be extremely high. This intense surface heating generates a low-pressure center over Arizona. The resulting circulation pattern brings warm, moist air from the Gulf of California and to a lesser extent the Gulf of Mexico (Figure 7-13). The supply of atmospheric moisture from nearby marine sources, coupled with the convergence and upward flow of the thermal low, is responsible for generating the precipitation this region experiences during the hottest months. Although often associated with the state of Arizona, this monsoon is actually strongest in northwestern Mexico and is also quite pronounced in New Mexico.

Concept Check 7.5

1 Define *monsoon*.

2 Explain the cause of the Asian monsoon. Which season (summer or winter) is the rainy season?

3 What areas of North America experience a pronounced monsoon circulation?

---

*This event is also called the *Arizona monsoon* and the *Southwest monsoon* because it has been extensively studied in this part of the United States.

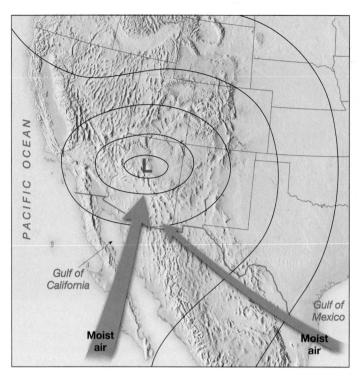

Figure 7-13 High summer temperatures over the southwestern United States create a thermal low that draws moist air from the Gulf of California and the Gulf of Mexico. This summer monsoon produces an increase in precipitation, which often comes in the form of thunderstorms, over the southwestern United States and northwestern Mexico.

The Westerlies

Prior to World War II, upper-air observations were scarce. Since then, aircraft and radiosondes have provided a great deal of data about the upper troposphere. Among the most important discoveries was that airflow aloft in the middle latitudes has a strong west-to-east component, thus the name *westerlies.*

Why Westerlies?

Let us consider the reason for the predominance of westerly flow aloft. In the case of the westerlies, the temperature contrast between the poles and equator drives these winds. Figure 7-14 illustrates the pressure distribution with height over the cold polar region as compared to the much warmer tropics. Recall that because cold air is more dense (compact) than warm air, air pressure decreases more rapidly in a column of cold air than in a column of warm air. The pressure surfaces (planes) in Figure 7-14 represent a simplified view of the pressure distribution we would expect to observe from pole to equator.

Over the equator, where temperatures are higher, air pressure decreases more gradually than over the cold polar regions. Consequently, at the same altitude above Earth's surface, higher pressure exists over the tropics and lower pressure is the norm above the poles. Thus, the pressure gradient aloft is directed from the equator (area of higher pressure) toward the poles (area of lower pressure).

Once the air from the tropics begins to advance poleward in response to this pressure-gradient force (red arrow in Figure 7-14), the Coriolis force causes a change in the direction of airflow. Recall that in the Northern Hemisphere, the Coriolis force causes winds to be deflected to the right. Eventually, a balance is reached between the poleward-directed pressure-gradient force and the equatorward-directed Coriolis force to generate a wind with a strong west-to-east component (Figure 7-14). Recall that such winds are called *geostrophic winds.* Because the equator-to-pole temperature gradient shown in Figure 7-14 is typical over the globe, a westerly flow aloft should be expected, and it does prevail on most occasions.

The pressure gradient also increases with altitude; as a result, so do wind speeds. This increase in wind speed occurs as we ascend into the tropopause but begins to decrease upward into the stratosphere. These zones of fast wind speeds are called *jet streams* and will be considered in the next section.

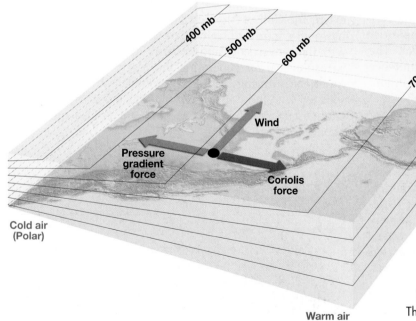

Figure 7-14 Idealized pressure gradient that develops aloft because of density differences between cold polar air and warm tropical air. Notice that the poleward-directed pressure-gradient force is balanced by an equatorward-directed Coriolis force. The result is a prevailing flow from west to east called the *westerlies.*

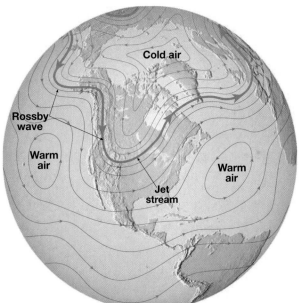

Figure 7–15 Idealized airflow of the westerlies at the 500-millibar level. The five long-wavelength undulations, called *Rossby waves*, compose this flow. The jet stream is the fast core of this wavy flow.

Waves in the Westerlies

Studies of upper-level wind charts show that the westerlies follow wavy paths that have long wavelengths. Much of our knowledge of these large-scale motions is attributed to C. G. Rossby, who first explained the nature of these waves. The longest wave patterns (called *Rossby waves*), shown in Figure 7-15, usually consist of four to six meanders that encircle the globe. Although the air flows eastward along this wavy path, these long waves tend to remain stationary or drift slowly from west to east.

Rossby waves can have a tremendous impact on our daily weather, especially when they meander widely from north to south. We will consider this role in the following section and again in Chapter 9.

Concept Check 7.6

1 Why is the flow aloft in the midlatitudes predominantly westerly?

2 What name is given to the long-wavelength flow that is apparent on upper air charts?

Jet Streams

Embedded within the westerly flow aloft are narrow ribbons of high-speed winds that typically meander for a few thousand kilometers

(Figure 7-16a). These fast streams of air, once considered analogous to jets of water, were named **jet streams.** Jet streams occur near the top of the troposphere and have widths that vary from less than 100 kilometers (60 miles) to over 500 kilometers (300 miles). Wind speeds often exceed 100 kilometers (60 miles) per hour and occasionally approach 400 kilometers (240 miles) per hour.

Although jet streams had been detected earlier, their existence was first dramatically illustrated during World War II. American bombers heading westward toward Japanese-occupied islands sometimes encountered unusually strong headwinds. On abandoning their missions, the planes experienced strong westerly tailwinds on their return flights.

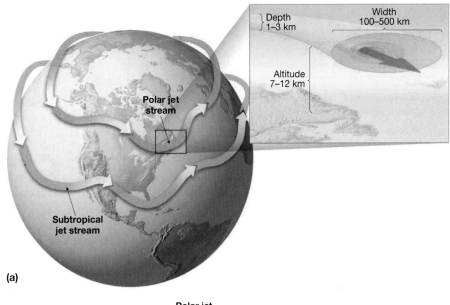

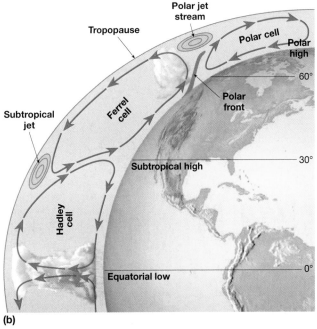

Figure 7–16 Jet streams. (a) Approximate positions of the polar and subtropical jet streams. (b) A cross-sectional view of the polar and subtropical jets in relation to the three-cell model of global circulation.

Modern commercial aircraft pilots use the strong flow within jet streams to increase their speed when making eastward flights around the globe. On westward flights, of course, they avoid these fast currents of air when possible.

The Polar Jet Stream

What is the origin of the distinctive energetic winds that exist within the somewhat slower general westerly flow? The key is that large temperature differences at the surface produce steep pressure gradients aloft and hence faster upper-air winds. In winter and early spring, it is not unusual to have a warm balmy day in southern Florida and near-freezing temperatures in Georgia, only a few hundred kilometers to the north. Such large wintertime temperature contrasts lead us to expect faster westerly flow at that time of year. In general, the fastest upper-air winds are located above regions of the globe having large temperature contrasts across very narrow zones.

These large temperature contrasts occur along linear zones called *fronts*. The most prevalent jet stream occurs along a major frontal zone called the *polar front* and is appropriately named the **polar jet stream,** or simply the *polar jet* (Figure 7-16b). Because this jet stream is often found in the middle latitudes, particularly in the winter, it is also known as the *midlatitude jet stream* (Figure 7-17). Recall that the polar front is situated between the cool winds of the polar easterlies and the relatively warm westerlies. Instead of flowing nearly straight west to east, the polar jet stream usually has a meandering path. Occasionally, it flows almost due north–south. Sometimes it splits into two jets that may, or may not, rejoin. Like the polar front, this zone of high-velocity airflow is not continuous around the globe.

On average, the polar jet travels at 125 kilometers (75 miles) per hour in the winter and roughly half that

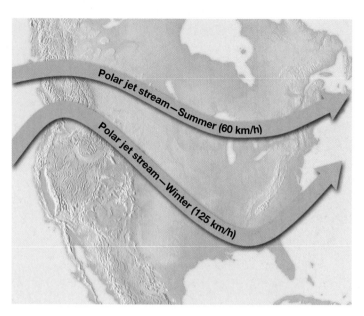

Figure 7-18 The position and speed of the polar jet stream changes with the seasons, migrating freely between about 30° and 70° latitude. Shown are flow patterns that are common for summer and winter.

speed in the summer (Figure 7-18). This seasonal difference is due to the much stronger temperature gradient that exists in the middle latitudes during the winter.

Because the location of the polar jet roughly coincides with that of the polar front, its latitudinal position migrates with the seasons. Thus, like the zone of maximum solar heating, the jet moves northward during summer and southward in winter. During the cold winter months, the polar jet stream may extend as far south as central Florida (Figure 7-18). With the coming of spring, the zone of maximum solar heating, and therefore the jet, begins a gradual northward migration. By midsummer, its *average* position is near the Canadian border, but it can be located much farther poleward.

The polar jet stream plays a very important role in the weather of the midlatitudes. In addition to supplying energy to help drive the rotational motion of surface storms, it also directs the paths of these storms. Consequently, determining changes in the location and flow pattern of the polar jet is an important part of modern weather forecasting. As the polar jet shifts northward, there is a corresponding change in the region where outbreaks of severe thunderstorms and tornadoes occur. For example, in February most thunderstorms and tornadoes occur in the states bordering the Gulf of Mexico. By midsummer the center of this activity shifts to the northern plains and Great Lakes states.

In addition, the location of the polar jet stream affects other surface conditions, particularly temperature and humidity. When it is situated substantially equatorward of your location, the weather will be colder and drier than normal. Conversely, when the polar jet moves poleward of your location, warmer and more humid conditions will prevail. Thus, depending on the position of the jet stream, the weather could be hotter, colder, drier, or wetter than normal.

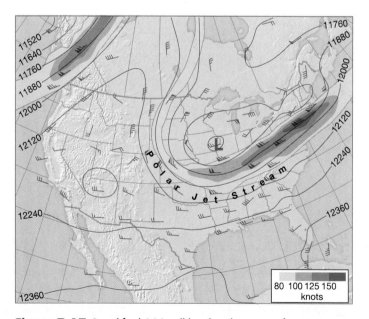

Figure 7-17 Simplified 200-millibar height-contour for January. The position of the jet stream core (or streak) is shown in dark pink.

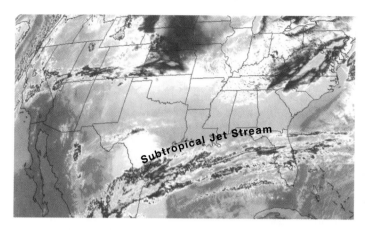

Figure 7–19 Infrared image of the subtropical jet stream shown as a band of cloudiness extending from Mexico to Florida. (Photo courtesy of NOAA)

Subtropical Jet Stream

A semipermanent jet over the subtropics is called the **subtropical jet stream** (Figure 7-16b). The subtropical jet is mainly a wintertime phenomenon. Somewhat slower than the polar jet, this west-to-east current is centered at 25° latitude (in both hemispheres) at an altitude of about 13 kilometers (8 miles). In the Northern Hemisphere, when the subtropical jet sweeps northward in the winter, it may bring warm humid conditions to the Gulf states, particularly southern Florida (Figure 7-19).

Jet Streams and Earth's Heat Budget

Let us return to the wind's function of maintaining Earth's heat budget by transporting heat from the equator toward the poles. Although the flow in the tropics (Hadley cell) is somewhat *meridional* (north to south), at most other latitudes the flow is *zonal* (west to east). The reason for the zonal flow, as we have seen, is the Coriolis force. The question we now consider is: *How can wind with a west-to-east flow transfer heat from south to north?*

The important function of heat transfer is accomplished by the wavy flow (Rossby waves) of the westerlies centered on the polar jet stream. There may be periods of a week or more when the flow in the polar jet is essentially west to east, as shown in Figure 7-20a. When this condition prevails, relatively mild temperatures occur south of the jet stream, and cooler

Figure 7–20 Cyclic changes that occur in the upper-level airflow of the westerlies. The flow, which has the jet stream as its axis, starts out nearly straight and then develops meanders and cyclonic activity that dominate the weather.

temperatures prevail to the north. Then, with minimal warning, the flow aloft may begin to meander and produce large-amplitude waves and a general north-to-south flow, as shown in Figure 7-20b and Figure 7-20c. Such a change causes cold air to advance equatorward and warm air to flow poleward. In addition, a cold air mass may become detached and produce an outbreak of cold air, as shown in Figure 7-20d.

This redistribution of energy eventually results in a weakened temperature gradient and a return to a flatter flow aloft (Figure 7-20d). Therefore, the wavy flow of the westerlies centered on the polar jet plays an important role in Earth's heat budget.

Concept Check 7.7

1 How are jet streams generated?

2 At what time of year should we expect the fastest polar jet streams? Explain.

3 Why is the polar jet sometimes referred to as the midlatitude jet stream?

4 Describe the expected winter temperatures in the north-central states when the polar jet stream is located over central Florida.

5 Explain how the wavy flow centered on the jet stream helps balance Earth's heat budget.

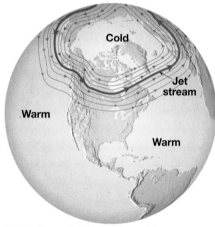

(a) Gently undulating upper airflow

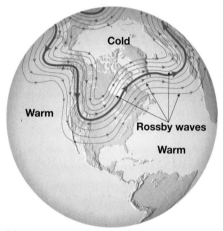

(b) Meanders form in jet stream

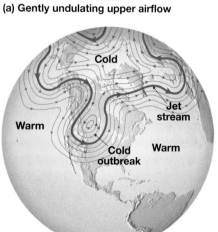

(c) Strong waves form in upper airflow

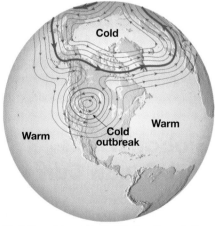

(d) Return to a period of flatter flow aloft

EYE ON THE ATMOSPHERE

Iceland

Volcanic ash plume

Norway

United Kingdom

(Image courtesy of NASA)

In mid-April 2010, Iceland's Eyjafjallajökull volcano produced an ash plume that rose nearly 50,000 feet into the atmosphere as it moved eastward across the North Atlantic. This event prompted authorities in the United Kingdom, Ireland, France, and Scandinavia to close airspace over their countries for fear that the volcanic ash, if sucked into an airplane's turbines, could cause engine failure.

Question 1 Based on the fact that Iceland is located in the zone of the polar easterlies (winds that blow from the northeast toward the southwest), explain why this ash plume moved primarily from west to east toward Northern Europe.

Students Sometimes Ask...

Why do pilots of commercial aircraft always remind passengers to keep their seat belts fastened, even in ideal flying conditions?

The reason for this request is a phenomenon known as *clear air turbulence*. Clear air turbulence occurs when airflow in two adjacent layers is moving at different velocities. This can happen when the air at one level is traveling in a different direction than the air above or below it. More often, however, it occurs when air at one level is traveling faster than air in an adjacent layer. Such movements create eddies (turbulence) that can cause the plane to move suddenly up or down.

Global Winds and Ocean Currents

Energy is passed from moving air to the surface of the ocean through friction. As a consequence, winds blowing steadily across the ocean drag the water along with them. Because winds are the primary driving force of surface ocean currents, a relationship exists between atmospheric circulation and oceanic circulation. A comparison of Figures 7-21 and 7-10 illustrates this.

As shown in Figure 7-21, north and south of the equator are two westward-moving currents, the North and South Equatorial Currents. These currents derive their energy principally from the trade winds that blow from the northeast and southeast, respectively, toward the equator. Because of the Coriolis force, surface currents are deflected poleward to form clockwise spirals in the Northern Hemisphere and counterclockwise spirals in the Southern Hemisphere. These nearly circular ocean currents, called *gyres*, are found in each of the major ocean basins centered around the subtropical high-pressure systems (Figure 7-21).

In the North Atlantic, the equatorial current is deflected northward through the Caribbean, where it becomes the *Gulf Stream*. As the Gulf Stream moves along the eastern coast of the United States, it is strengthened by the prevailing westerly winds. As it continues northeastward beyond the Grand Banks, it gradually widens and slows

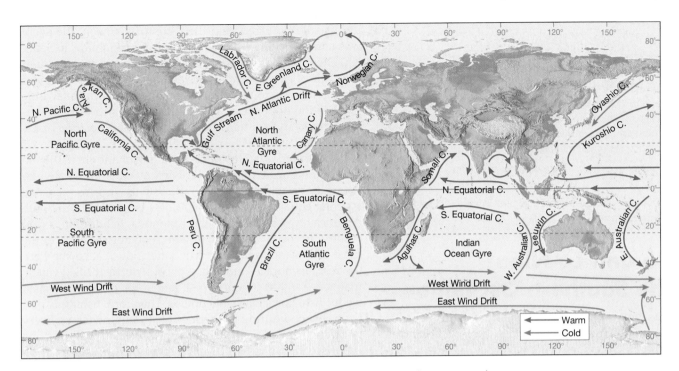

Figure 7–21 Major ocean currents. Poleward-moving currents are warm, and equatorward-moving currents are cold.

until it becomes a vast, slowly moving current known as the North Atlantic Drift. The North Atlantic Drift splits as it approaches Western Europe. Part moves northward past Great Britain toward Norway, while the other portion is deflected southward as the cool Canary Current. As the Canary Current moves south, it eventually merges with the North Equatorial Current.

The Importance of Ocean Currents

Currents have an important effect on climate. The moderating effect of poleward-moving warm ocean currents is well known. The North Atlantic Drift, an extension of the warm Gulf Stream, keeps Great Britain and much of northwestern Europe warmer than might be expected considering their latitudes.

In addition to influencing temperatures of adjacent land areas, cold currents have other climatic effects. For example, in tropical deserts that exist along the west coasts of continents, such as the Atacama in Peru and Chile, and the Namib in southern Africa, cold ocean currents have a dramatic impact. The aridity along these coasts is intensified because air flowing over the cold currents is chilled from below. This causes the air to become very stable and resist the upward movements necessary to create precipitation-producing clouds.

Cold currents may also cause the lowest layer of air to chill below its dew-point temperature. As a result, these desert areas are characterized by high relative humidities and abundant fog. Thus, not all tropical deserts are hot with low humidities and clear skies. Rather, the presence of cold currents transforms some tropical deserts into relatively cool, damp places that are often shrouded in fog.

Ocean currents also play a major role in maintaining Earth's heat balance. They accomplish this task by transferring heat from the tropics, where there is an excess of heat, to the polar regions, where a deficit exists. Ocean currents account for about one-quarter of this total heat transport, and winds make up the remainder.

Ocean Currents and Upwelling

Upwelling, the rising of cold water from deeper layers to replace warmer surface water, is a common wind-induced vertical movement. It is most characteristic along the eastern shores of the global oceans, most notably along the coasts of California, Peru, and West Africa.

Upwelling occurs in areas where winds blow parallel to the coast toward the equator. Because of the Coriolis force, the surface water is directed (deflected) away from the shore. As the surface layer moves away from the coast, it is replaced by water that "upwells" from below the surface. This slow upward flow from depths of 50 to 300 meters (165 to 1000 feet) brings water that is cooler than the water it replaces and creates a characteristic zone of relatively cool water adjacent to the shore.

Swimmers accustomed to the waters along the mid-Atlantic states of the United States might find a dip in the Pacific on central California's coast a chilling surprise. In August, when water temperatures along the Atlantic shore usually exceed 21°C (70°F), the surf along the coast of California is only about 15°C (60°F).

El Niño and La Niña and the Southern Oscillation

El Niño was first recognized by fishermen from Ecuador and Peru, who noted a gradual warming of waters in the eastern Pacific in December or January. Because the warm-ing usually occurred near the Christmas season, the event was named El Niño—"little boy," or "Christ child," in Span-ish. These periods of abnormal warming happen at irregular intervals of two to seven years and usually persist for spans of nine months to two years. **La Niña**, which means "little girl," is the opposite of El Niño and refers to colder-than-normal sea-surface temperatures along the coastline of Ec-uador and Peru.

As Figure 7-22a illustrates, the atmospheric circulation in the central Pacific during a La Niña event is dominated by strong trade winds. These wind systems, in turn, generate a strong equatorial current that flows westward from South America toward Australia and Indonesia. In addition, a cold ocean current is observed flowing equatorward along the coast of Ecuador and Peru. The latter flow, called the Peru Current, encourages upwelling of cold, nutrient-filled waters that serve as the primary food source for millions of small feeder fish, particularly anchovies. Therefore, fishing is par-ticularly good during the periods of strong upwelling. Every few years, however, the circulation associated with La Niña is replaced by an El Niño event (Figure 7-22b).

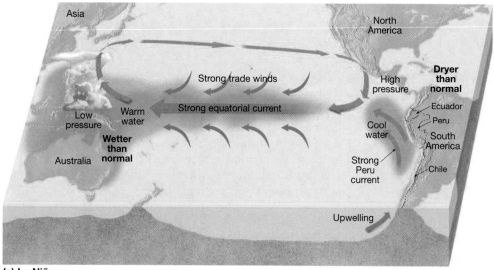

(a) La Niña

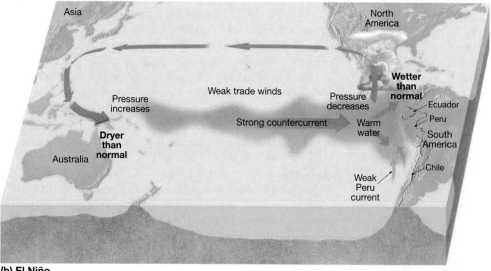

(b) El Niño

Figure 7-22 The relationship between El Niño, La Niña, and the Southern Oscillation is illustrated by these diagrams. (a) During a La Niña event, strong trade winds drive the equatorial currents toward the west. At the same time, the strong Peru Current causes upwelling of cold water along the west coast of South America. (b) When the Southern Oscillation occurs, the pressure over the eastern and western Pacific flip-flops. This causes the trade winds to diminish, leading to an eastward movement of warm water along the equator and the beginning of an El Niño. As a result, the surface waters of the central and eastern Pacific warm, with far-reaching consequences for weather patterns.

Impact of El Niño

El Niño is noted for its potentially catastrophic impact on the weather and economies of Peru, Chile, and Australia, among other countries. As shown in Figure 7-22b, during an El Niño, strong equatorial countercurrents amass large quantities of warm water that block the upwelling of colder, nutrient-filled water along the west coast of South America. As a result, the anchovies, which support the population of game fish, starve, devastating the fishing industry. At the same time, some inland areas of Peru and Chile that are normally arid receive above-average rainfall, which can cause major flooding. These climatic fluctuations have been known for years, but they were considered local phenomena.

Scientists now recognize that El Niño is part of the global atmospheric circulation pattern that affects the weather at great distances from Peru. One of the most severe El Niño events on record occurred in 1997–1998 and was responsible for a variety of weather extremes in many parts of the world. During the 1997–1998 El Niño episode, ferocious winter storms struck the California coast, causing unprecedented beach erosion, landslides, and floods. In the southern United States, heavy rains also brought floods to Texas and the Gulf states.

Although the effects of El Niño are somewhat variable, some locales appear to be affected more consistently. In particular, during the winter, warmer-than-normal conditions prevail in the north-central United States and parts of Canada (Figure 7-23a). In addition, significantly wetter winters are experienced in the southwestern United States and northwestern Mexico, while the southeastern United States experiences wetter and cooler conditions. In the western Pacific, drought conditions are observed in parts of Indonesia, Australia, and the Philippines (Figure 7-23a). One major benefit of El Niño is

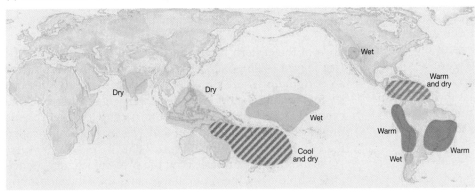

(a) El Niño: December to February

(b) El Niño: June to August

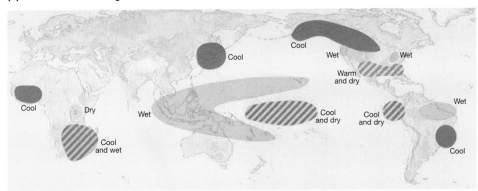

(c) La Niña: December to February

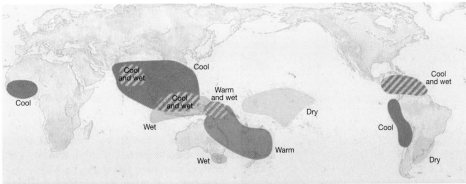

(d) La Niña: June to August

| | Cool | | Warm | | Dry | | Wet | | Cool and dry | | Cool and wet | | Warm and dry | | Warm and wet |

Figure 7–23 Climatic impacts of El Niño and La Niña in various locations during December through February and June through August. El Niño has the most significant impact on the climate of North America during the winter. In addition, El Niño affects the areas around the tropical Pacific in both winter and summer. Likewise, La Niña has its most significant impact on North America in the winter but affects other areas during all seasons.

a lower-than-average number of Atlantic hurricanes. El Niño is credited with suppressing hurricanes during the 2009 hurricane season, the least active in 12 years.

Impact of La Niña

La Niña was once thought to be the normal conditions that occur between two El Niño events, but meteorologists now consider La Niña an important atmospheric phenomenon in its own right. Researchers have come to recognize that when surface temperatures in the eastern Pacific are *colder than average*, a La Niña event is triggered and exhibits a distinctive set of weather patterns (Figure 7-22a).

Typical La Niña winter weather includes cooler and wetter conditions over the northwestern United States and especially cold winter temperatures in the Northern Plains states (Figure 7-23c). In addition, unusually warm conditions occur in the Southwest and Southeast. In the western Pacific, La Niña events are associated with wetter-than-normal conditions. The 2010–2011 La Niña contributed to a deluge in Australia, which resulted in one of the country's worst natural disasters: Large portions of the state of Queensland were extensively flooded (Figure 7-24). Another La Niña impact is more frequent hurricane activity in the Atlantic. A recent study concluded that the cost of hurricane damages in the United States is 20 times greater in La Niña years than in El Niño years.

Southern Oscillation

Major El Niño and La Niña events are intimately related to the large-scale atmospheric circulation. Each time an El Niño occurs, the barometric pressure drops over large por-

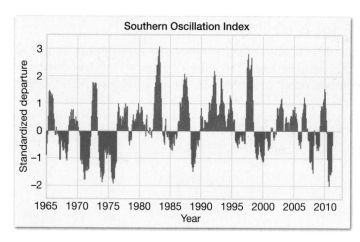

Figure 7–25 This graph illustrates the Southern Oscillation. Negative values (blue) represent the cold La Niña phase, whereas positive values (red) represent the warm El Niño phase. The graph was created by analyzing six variables, including sea-surface temperatures and sea-level pressures.

tions of the eastern Pacific and rises in the western Pacific (see Figure 7-22b). Then, as a major El Niño event comes to an end, the pressure difference between these two regions swings back in the opposite direction, triggering a La Niña event (Figure 7-22a). This seesaw pattern of atmospheric pressure between the eastern and western Pacific is called the **Southern Oscillation** (Figure 7-25).

Winds are the link between the pressure change associated with the Southern Oscillation and the ocean warming and cooling associated with El Niño and La Niña. The start of an El Niño event begins with a rise in surface pressure over Australia and Indonesia and a decrease in pressure

Figure 7–24 Flooding of large portions of Rockhampton, Queensland, Australia, January 2011. The Queensland floods of 2010–2011 have been attributed to one of the strongest La Niña events as far back as records have been kept. Unusually warm sea-surface temperatures around Australia contributed to the heavy rains. In other parts of Australia, this strong La Niña event brought relief from a decade-long drought. (Courtesy of NASA)

over the eastern Pacific (Figure 7-22b). During a strong El Niño, this pressure change causes the trade winds to weaken and countercurrents to develop and move warm water eastward. The resulting change in atmospheric circulation takes the rain with it, causing drought in the western Pacific and increased rainfall in the normally dry regions of Peru and Chile. The opposite circulation develops during La Niña events, as shown in Figure 7-22a. When a strong La Niña develops, the trade winds strengthen, causing dryer-than-normal conditions in the eastern Pacific, while extreme flooding may occur in Indonesia and northeastern Australia.

Concept Check 7.9

1 Describe how a major El Niño event tends to affect the weather in Peru and Chile as compared to Indonesia and Australia.

2 Describe the sea-surface temperatures on both sides of the tropical Pacific during a La Niña event.

3 How does a major La Niña event influence the hurricane season in the Atlantic Ocean?

4 Briefly describe the Southern Oscillation and how it is related to El Niño and La Niña.

5 Describe how an El Niño event might affect the climate in North America during the winter. Describe the same for a La Niña event.

Global Distribution of Precipitation

Figure 7-26 shows the distribution pattern for average annual precipitation over the globe. Although this map may appear complicated, the general features of the pattern can be explained using knowledge of global winds and pressure systems. In general, regions influenced by high pressure, with its associated subsidence and divergent winds, experience dry conditions. Conversely, regions under the influence of low pressure, with its converging winds and ascending air, receive ample precipitation. However, if the wind-pressure regimes were the only control of precipitation, the pattern shown in Figure 7-26 would be much simpler.

Air temperature is also important in determining precipitation potential. Because cold air has a low capacity for moisture compared with warm air, we would expect a latitudinal variation in precipitation, with low latitudes (warm regions) receiving the greatest amounts of precipitation and high latitudes (cold regions) receiving the least.

In addition to latitudinal variations in precipitation, the distribution of land and water complicates the precipitation pattern. Large landmasses in the middle latitudes commonly experience decreased precipitation toward their interiors. For example, North Platte, Nebraska, receives less than half the precipitation that falls on the coastal community of Bridgeport, Connecticut, despite being located at the same

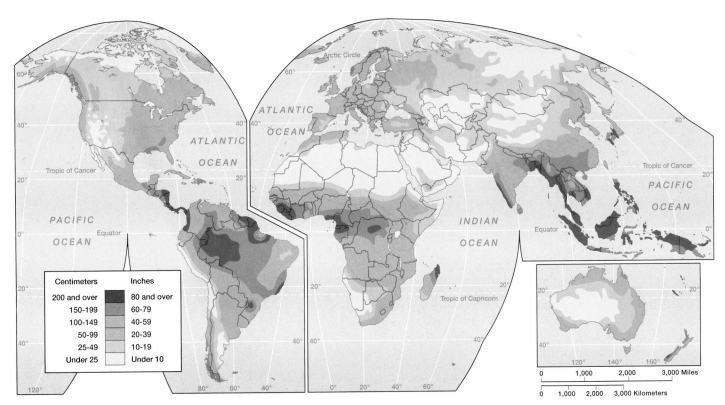

Figure 7-26 Global distribution of average annual precipitation.

EYE ON THE ATMOSPHERE

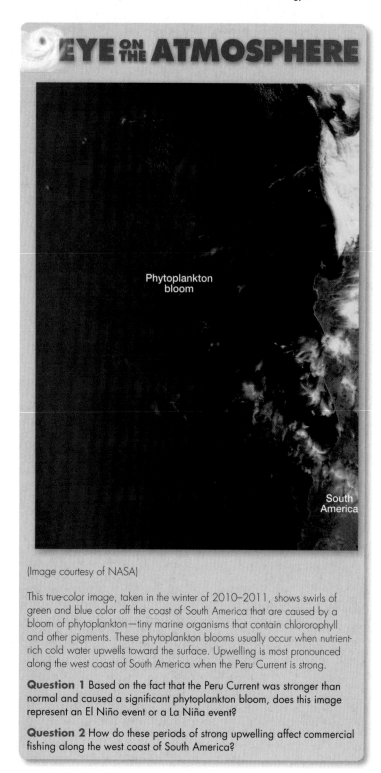

Phytoplankton bloom

South America

(Image courtesy of NASA)

This true-color image, taken in the winter of 2010–2011, shows swirls of green and blue color off the coast of South America that are caused by a bloom of phytoplankton—tiny marine organisms that contain chlororophyll and other pigments. These phytoplankton blooms usually occur when nutrient-rich cold water upwells toward the surface. Upwelling is most pronounced along the west coast of South America when the Peru Current is strong.

Question 1 Based on the fact that the Peru Current was stronger than normal and caused a significant phytoplankton bloom, does this image represent an El Niño event or a La Niña event?

Question 2 How do these periods of strong upwelling affect commercial fishing along the west coast of South America?

latitude. Furthermore, mountain barriers alter precipitation patterns. Windward mountain slopes receive abundant precipitation, whereas leeward slopes and adjacent lowlands are usually deficient in moisture.

Zonal Distribution of Precipitation

We will first examine the zonal distribution of precipitation that we would expect on a uniform Earth covered entirely in water and then add the variations caused by the distribution

of land and water. Recall from our earlier discussion that on a uniform Earth, four major pressure zones emerge in each hemisphere (see Figure 7-8a, page 199). These zones are the equatorial low (ITCZ), the subtropical high, the subpolar low, and the polar high. Also, remember that these pressure belts show a marked seasonal shift toward the summer hemisphere.

The idealized precipitation regimes expected from these pressure systems are shown in Figure 7-27. Near the equator the trade winds converge (ITCZ), resulting in heavy precipitation in all seasons. Poleward of the equatorial low in each hemisphere lie the belts of subtropical high pressure. In these regions, subsidence contributes to dry conditions throughout the year. Between the wet equatorial regime and the dry subtropical regime lies a zone that is influenced by both high- and low-pressure systems. Because the pressure systems migrate seasonally with the Sun, these transitional regions receive most of their precipitation in the summer, when they are under the influence of the ITCZ. They experience a dry season in the winter, when the subtropical high moves equatorward.

The midlatitudes receive most of their precipitation from traveling cyclonic storms (Figure 7-28). This region is the site of the polar front, the convergence zone between

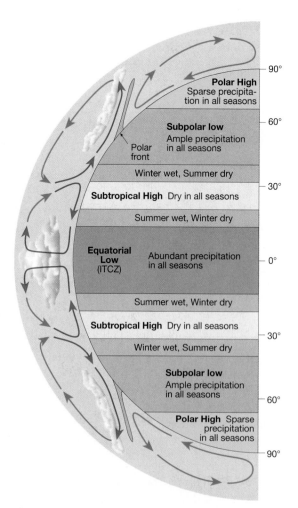

Figure 7–27 Zonal precipitation patterns.

Figure 7–28 Satellite image of a well-developed midlatitude cyclone over the British Isles. These traveling storms produce most of the precipitation in the middle latitudes. (Courtesy of European Space Agency/Science Photo Library/Photo Researchers, Inc.)

cold polar air and the warmer westerlies. Because the position of the polar front migrates freely between approximately 30° and 70° latitude, most midlatitude areas receive ample precipitation.

The polar regions are dominated by high pressure and cold air that holds little moisture. Throughout the year, these regions experience only meager precipitation.

Distribution of Precipitation over the Continents

The zonal pattern outlined in the previous section roughly approximates general global precipitation. Abundant precipitation occurs in the equatorial and midlatitude regions, whereas substantial portions of the subtropical and polar realms are relatively dry.

Numerous exceptions to this idealized zonal pattern are obvious in Figure 7-26. For example, several arid areas are found in the midlatitudes. The desert region of southern South America, known as Patagonia, is one example. Midlatitude deserts such as Patagonia exist mostly on the leeward (rain shadow) side of a mountain barrier or in the interior of a continent, cut off from a source of moisture.

The most notable anomaly in the zonal distribution of precipitation occurs in the subtropics. Here we find not only

many of the world's great deserts but also regions of abundant rainfall (Figure 7-26). This pattern results because the subtropical high-pressure centers that dominate the circulation in these latitudes have different characteristics on their eastern and western sides (Figure 7-29). Subsidence is most

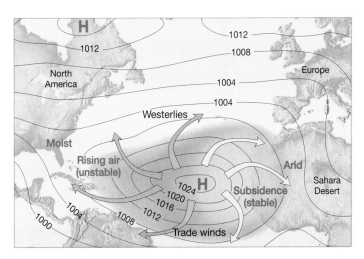

Figure 7-29 Characteristics of subtropical high-pressure systems. Subsidence on the east side of these systems produces stable conditions and aridity. Convergence and uplifting on the western flank promote uplifting and instability.

Box 7-2 Precipitation Regimes on a Hypothetical Continent

When we consider the influence of land and water on the distribution of precipitation, the pattern illustrated in **Figure 7–D** emerges. This figure is a highly idealized precipitation scheme for a hypothetical continent located in the Northern Hemisphere. The "continent" is shaped as it is to approximate the percentage of land found at various latitudes.

Notice that this hypothetical landmass has been divided into seven zones. Each zone represents a different precipitation regime. Stated another way, all locations within the same zone (precipitation regime) experience roughly the same precipitation pattern. As you will see, the odd shapes and sizes of these zones reflect the way precipitation is normally distributed across landmasses located in the Northern Hemisphere. The cities shown are representative of the precipitation pattern of each location. With this in mind, we will examine each precipitation regime (numbered 1 through 7) to look at the influences of land and water on the distribution of precipitation.

First, compare the precipitation patterns for the nine cities. Notice that the precipitation regimes for west coast locations correspond to the zonal pressure and precipitation pattern shown in Figure 7–27: zone 6 equates to the equatorial low (wet all year), zone 4 matches the subtropical high (dry all year), and so on.

For illustration, Aklavik, Canada, which is strongly influenced by the polar high, has scant moisture, whereas Singapore, near the equator, has plentiful rain every month. Also note that precipitation varies seasonally with both latitude and each city's location with respect to the ocean.

For example, precipitation graphs for San Francisco and Mazatlan, Mexico, illustrate the marked seasonal variations found in zones 3 and 5. Recall that it is the seasonal migration of the pressure systems that causes these fluctuations. Notice that zone 2 shows a precipitation maximum in the fall and early winter, as illustrated by data for Juneau, Alaska. This pattern occurs because cyclonic storms over the North Pacific are more prevalent in the coolest half of the year. Further, the average path of these storms moves equatorward during the winter, impacting locations as far south as southern California in midwinter.

The eastern half of the continent does not experience the zonal precipitation pattern that we observe on the western side. Only the dry polar regime (zone 1) is similar in size and position on both the eastern and western seaboards. The most noticeable departure from the zonal pattern is the absence of an arid subtropical region along the east coast. This is caused by the behavior of subtropical anticyclones located over the oceans. Recall that southern Florida,

which is located on the east coast of the North American continent, receives ample precipitation year-round, whereas the Baja Peninsula on the west coast is arid. Moving inland from the east coast, observe how precipitation decreases (by comparing total precipitation westward for New York, Chicago, and Omaha). This decrease, however, does not hold true for mountainous regions; even relatively small ranges, like the Appalachians, are able to extract a greater share of precipitation.

Another variation is seen with latitude. As we move poleward along the eastern half of our hypothetical continent, note the decrease in precipitation (by comparing Cancun, New York, and Aklavik). This is the decrease we would expect with cooler air temperatures and corresponding lower moisture capacities.

Consider next the continental interior, which shows a somewhat different precipitation pattern, especially in the middle latitudes. The precipitation regimes of the continental interior are affected to a degree by a type of monsoon circulation. Cold winter temperatures and flow off the land make winter the drier season (for example, Omaha and Chicago). In summer, the inflow of warm, moist air from the ocean is aided by a thermal low that develops over the land and causes a general increase in precipitation over the midcontinent.

---

pronounced on the eastern side, which results in stable atmospheric conditions. Because these anticyclones tend to crowd the eastern side of an ocean, particularly in the winter, we find that the western portions of the continents adjacent to subtropical highs are arid (Figure 7-29). Centered at approximately 25° north or south latitude on the west side of their respective continents, we find the Sahara Desert of North Africa, the Namib of southwest Africa, the Atacama of South America, the deserts of northwestern Mexico, and Australia.

On the western flanks of these highs, however, subsidence is less pronounced. In addition, the surface air that flows out of these highs often traverses large expanses of warm water. As a result, this air acquires moisture through evaporation that acts to enhance its instability. Consequently, landmasses located west of a subtropical high generally receive ample precipitation throughout the year. Southern Florida is a good example (Figure 7-29).

Concept Check 7.10

1 Are regions that are dry throughout the year dominated by high or low pressure?

2 Describe the precipitation pattern for locations near the equator and near the poles.

3 Earth's major subtropical deserts are located between about 20° and 35° latitude.
  a Name five subtropical deserts.
  b On what side (western or eastern) of the continent are they located?
  c What is the general name of the pressure systems responsible for subtropical deserts?

4 List two reasons that explain why polar regions experience meager precipitation.

5 What factors, in addition to global wind and pressure systems, influence the global distribution of precipitation?

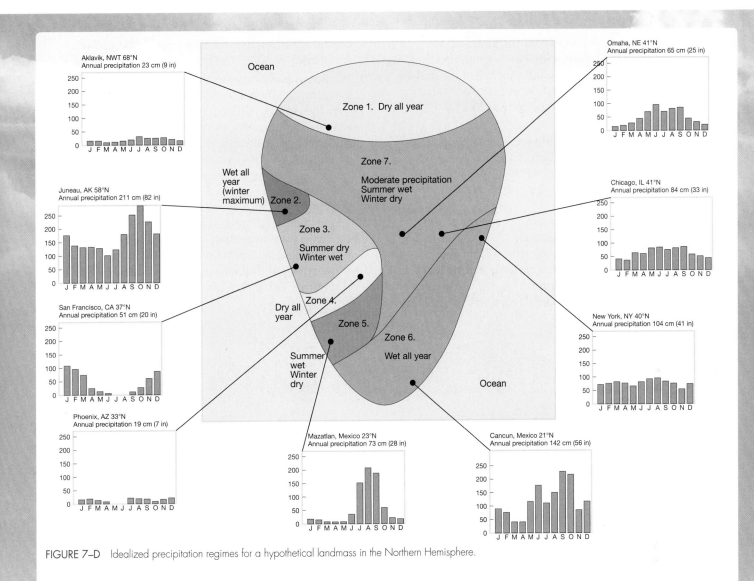

FIGURE 7–D   Idealized precipitation regimes for a hypothetical landmass in the Northern Hemisphere.

Give It Some Thought

1. It is a warm summer day, and you are shopping in downtown Chicago, just a few blocks from Lake Michigan. All morning the winds have been calm, suggesting that no major weather systems are nearby. By midafternoon, should you expect a cool breeze from Lake Michigan or a warm breeze originating from the rural areas outside the city?

2. Boulder, Colorado, in the eastern foothills of the Rocky Mountains, is experiencing a warm, dry January day with strong westerly winds. What type of local winds are likely responsible for these weather conditions?

3. Which of the local winds described in this chapter is not heavily dependent on differences in the rates at which various ground surfaces are heated?

4. The accompanying sketch shows the three-cell circulation model in the Northern Hemisphere. Match the appropriate number on the sketch to each of the following features:
   a. Hadley cell
   b. Equatorial low
   c. Polar front
   d. Ferrel cell
   e. Subtropical high
   f. Polar cell
   g. Polar high

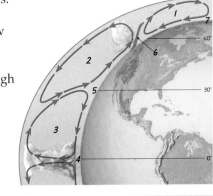

5. If Earth did not rotate on its axis and if its surface were completely covered with water, what direction would a boat drift if it started its journey in the middle latitudes of the Northern Hemisphere? (*Hint:* What would the global circulation pattern be like for a nonrotating Earth?)

6. Briefly explain each of the following statements that relate to the global distribution of surface pressure:
   a. The only true zonal distribution of pressure exists in the region of the subpolar low in the Southern Hemisphere.
   b. The subtropical highs are stronger in the North Atlantic in July than in January.
   c. The subpolar lows in the Northern Hemisphere are the result of individual cyclonic storms that are more common in the winter.
   d. A strong high-pressure cell develops in the winter over northern Asia.

7. On the accompanying image of Jupiter, notice the many zones of clouds that are produced by large convective cells similar to the Hadley cells on Earth. Given your understanding of the role of the Coriolis force in producing Earth's wind belts, do you think Jupiter rotates faster or slower on its axis than Earth? Explain.

Jupiter (NASA)

8. Refer to Figure 7–26, p. 213 and Figure 7-10, p. 200, to determine what aspect of global circulation (polar highs, equatorial low, etc.) is responsible for each of the following:
   a. Dry conditions over North Africa.
   b. The wet summer monsoon over southeast Asia.
   c. Dry conditions over west-central Australia.
   d. Wet conditions over northeastern South America.

9. Explain why the west coasts of continents generally experience cold ocean currents. How do these cold currents contribute to desert conditions in some coastal areas, such as the Atacama region of Peru and Chile?

10. The accompanying map shows sea-surface temperature anomalies (difference from normal) over the equatorial Pacific Ocean. Based on this map, answer the following questions:
    a. In what phase was the Southern Oscillation (El Niño or La Niña) when this image was made?
    b. Would the trade winds be strong or weak at this time?
    c. If you lived in Australia during this event, what weather conditions would you expect?
    d. If you were attending college in the southeastern United States during winter months, what type of weather conditions would you expect? (*Hint:* See Figure 7–23, p. 211.)

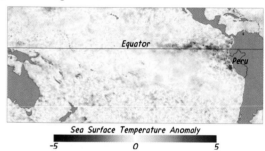

Sea Surface Temperature Anomaly
−5   0   5

11. Refer to Figure 7–D on page 217 to determine in which of the seven precipitation regimes you reside. What elements of the global atmospheric circulation interact to create the precipitation regime of your location?

12. The accompanying maps of Africa show the distribution of precipitation for July and January. Which map represents July and which represents January? How did you determine your answer?

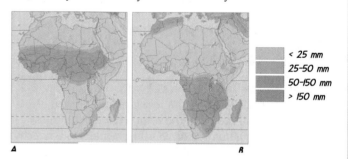

	< 25 mm
	25–50 mm
	50–150 mm
	> 150 mm

A   B

CIRCULATION OF THE ATMOSPHERE IN REVIEW

- The largest planetary-scale wind patterns, called *macroscale winds*, include the westerlies and trade winds. *Mesoscale winds*, such as thunderstorms, tornadoes, and land-sea breezes, influence smaller areas and often exhibit intense vertical flow. The smallest scale of air motion is the *microscale*, which includes gusts and dust devils.

- Most winds have the same cause: pressure differences that arise because of temperature differences that are caused by unequal heating of Earth's surface. In addition to *land* and *sea breezes* brought about by the daily temperature contrast between land and water, other mesoscale winds include *mountain breezes* and *valley breezes*, *chinook (foehn)* winds, *katabatic (fall)* winds, and *country breezes*.

- The three-cell circulation model for each hemisphere provides a simplified view of global circulation. According to this model, atmospheric circulation cells are located between the equator and 30° latitude, 30° and 60° latitude, and 60° latitude and the pole. The areas of general subsidence in the zone between 20° and 35° are called the *horse latitudes*. In each hemisphere, the equatorward flow from the horse latitudes forms the reliable *trade winds*. The circulation between 30° and 60° latitude (north and south) results in the *prevailing westerlies*. Air that moves equatorward from the poles produces the *polar easterlies* in both hemispheres.

- If Earth's surface were uniform, four latitudinally oriented belts of pressure would exist in each hemisphere—two high and two low. Beginning at the equator, the four belts would be the
  (1) *equatorial low*, also referred to as the *intertropical convergence zone (ITCZ)*, (2) *subtropical high*, at about 20° to 35° on either side of the equator, (3) *subpolar low*, situated at about 50° to 60° latitude, and (4) *polar high*, near Earth's poles.

- In reality, the only true zonal pattern of pressure exists along the subpolar low in the Southern Hemisphere. At other latitudes, particularly in the Northern Hemisphere, where there is a higher proportion of land compared to ocean, the zonal pattern is replaced by semipermanent cells of high and low pressure.

- The greatest seasonal change in Earth's global circulation is the development of *monsoons*, wind systems that exhibit a pronounced seasonal reversal in direction. The best-known and most pronounced monsoonal circulation is the *Asian monsoon*. The *North American monsoon*, a relatively small

seasonal wind shift, produces a dry spring followed by a comparatively rainy summer that impacts large areas of the southwestern United States and northwestern Mexico.

- The temperature contrast between the poles and the equator drives the *westerlies* in the middle latitudes. Embedded within the flow aloft are narrow ribbons of high-speed winds, called *jet streams*, that meander for thousands of kilometers. The key to the origin of polar jet streams is found in great temperature contrasts at the surface.

- Because winds are the primary driving force of ocean currents, a strong relationship exists between the oceanic circulation and the general atmospheric circulation. Because of the circulation associated with the subtropical highs, ocean currents form clockwise spirals in the Northern Hemisphere and counterclockwise spirals in the Southern Hemisphere.

- *El Niño* refers to episodes of ocean warming along the coasts of Ecuador and Peru. When surface temperatures in the eastern Pacific are colder than average, a *La Niña* event is triggered. These events are part of the global circulation and are related to a seesaw pattern of atmospheric pressure between the eastern and western Pacific called the *Southern Oscillation*. El Niño and La Niña events influence weather on both sides of the tropical Pacific Ocean as well as the United States.

- The general features of the global distribution of precipitation can be explained by global winds and pressure systems. In general, regions influenced by high pressure, with its associated subsidence and divergent winds, experience dry conditions. On the other hand, regions under the influence of low pressure and its converging winds and ascending air receive ample precipitation.

- On a uniform Earth throughout most of the year, heavy precipitation would occur in the equatorial region, the midlatitudes would receive most of their precipitation from traveling cyclonic storms, and polar regions would be dominated by cold air that holds little moisture.

VOCABULARY REVIEW

Aleutian low (p. 201)
Azores high (p. 201)
Bermuda high (p. 201)
bora (p. 195)
chinook (p. 194)
country breeze (p. 195)
doldrums (p. 197)
El Niño (p. 210)
equatorial low (p. 198)
Ferrel cell (p. 197)
foehn (p. 194)
Hadley cell (p. 197)
horse latitudes (p. 197)
Icelandic low (p. 201)

intertropical convergence zone (ITCZ) (p. 199)
jet stream (p. 205)
katabatic or fall wind (p. 195)
land breeze (p. 193)
La Niña (p. 210)
macroscale wind (p. 191)
mesoscale wind (p. 191)
microscale circulation (p. 190)
mistral (p. 195)
monsoon (p. 201)
mountain breeze (p. 194)
polar easterlies (p. 197)
polar front (p. 198)

polar high (p. 199)
polar jet stream (p. 206)
prevailing westerlies (p. 197)
Santa Ana (p. 194)
sea breeze (p. 193)
Siberian high (p. 201)
Southern Oscillation (p. 212)
subpolar low (p. 199)
subtropical high (p. 199)
subtropical jet stream (p. 207)
trade winds (p. 197)
upwelling (p. 209)
valley breeze (p. 194)

MyMeteorologyLab™

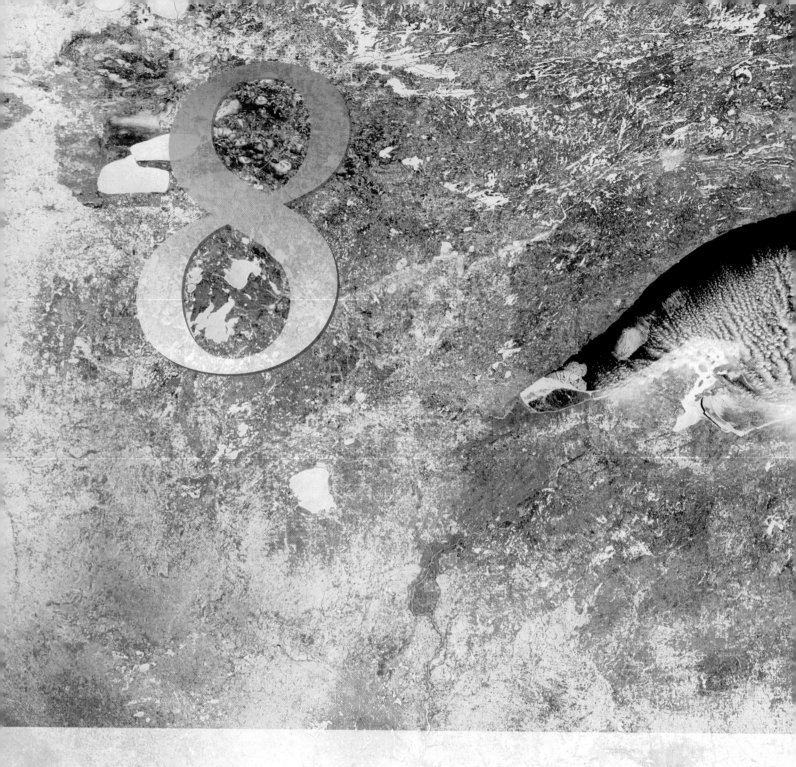

Air Masses

Most people living in the middle latitudes have experienced hot, "sticky" summer heat waves and frigid winter cold waves. In the case of a heat wave, after several days of sultry weather, the spell may come to an abrupt end that is marked by thundershowers, followed by a few days of relatively cool relief. In the case of a cold wave, thick stratus clouds and snow may replace the clear skies that had prevailed, and temperatures may climb to values that seem mild compared with what preceded them. In both examples, what was experienced was a period of generally uniform weather conditions followed by a relatively short period of change and the subsequent reestablishment of a new set of weather conditions that remained for perhaps several days before changing again.

Satellite image illustrating the process that leads to lake-effect snow storms, a phenomenon associated with cold, dry air masses. (NASA)

Focus On Concepts

After completing this chapter you should be able to:

- Define *air mass* and *air-mass weather*.

- List the basic criteria for an air-mass source region and explain why regions of high pressure are favored sites.

- On a map, locate and label the source regions that influence North America.

- Classify air masses.

- Describe the processes by which traveling air masses are modified and discuss two examples.

- Summarize the weather conditions associated with each of the air masses that influence North America in summer and winter.

221

What Is an Air Mass?

Basic Weather Patterns
▶ Air Masses

The weather patterns just described result from the movements of large bodies of air, called air masses. An **air mass,** as the term implies, is an immense body of air, usually 1600 kilometers (1000 miles) or more across and perhaps several kilometers thick, which is characterized by homogeneous physical properties (in particular, temperature and moisture content) at any given altitude. When this air moves out of its region of origin, it will carry these temperatures and moisture conditions elsewhere, eventually affecting a large portion of a continent (Figure 8-1).

An excellent example of the influence of an air mass is illustrated in Figure 8-2. Here a cold, dry mass from northern Canada moves southward. With a beginning temperature of −46°C (−51°F), the air mass warms 13°C (24°F), to −33°C (−27°F), by the time it reaches Winnipeg. It continues to warm as it moves southward through the Great Plains and into Mexico. Throughout its southward journey, the air mass becomes warmer, but it also brings some of the coldest weather of the winter to the places in its path. Thus, the air mass is modified, but it also modifies the weather in the areas over which it moves.

The horizontal uniformity of an air mass is not complete because it may extend through 20° or more of latitude and cover hundreds of thousands to millions of square kilometers. Consequently, small differences in temperature and humidity from one point to another at the same level are to be expected. Still, the differences observed within an air mass are small in comparison to the rapid rates of change experienced across air-mass boundaries.

Because it may take several days for an air mass to traverse an area, the region under its influence will probably experience generally constant weather conditions, a situation called **air-mass weather.** Certainly, some day-to-day variations may exist, but the events will be very unlike those in an adjacent air mass.

The air-mass concept is an important one because it is closely related to the study of atmospheric disturbances. Many significant middle-latitude disturbances originate along the boundary zones that separate different air masses.

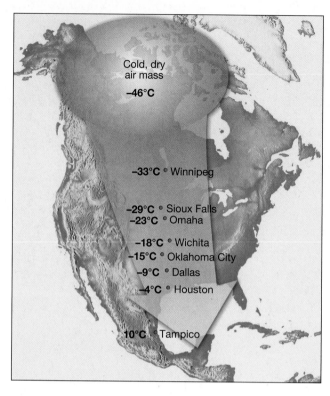

Figure 8-2 As this frigid Canadian air mass moved southward, it brought some of the coldest weather of the winter to the areas in its path. As it advanced out of Canada, the air mass slowly got warmer. Thus, the air mass was gradually modified at the same time that it modified the weather in the areas over which it moved. (From PHYSICAL GEOGRAPHY: A LANDSCAPE APPRECIATION, 9th Edition, by Tom L. McKnight and Darrell Hess, © 2008. Reprinted by permission of Pearson Education, Inc., Upper Saddle River, NJ.)

Concept Check 8.1

1 Define *air mass.*

2 What is *air-mass weather?*

Figure 8-1 Air masses that form over this high-latitude location in the South Atlantic will be cold and humid. (Photo by Michael Collier)

Source Regions

Basic Weather Patterns
▶ Air Masses

Where do air masses form? What factors determine the nature and degree of uniformity of an air mass? These two basic questions are closely related because the site where an air mass forms vitally affects the properties that characterize it.

Areas in which air masses originate are called **source regions.** Because the atmosphere is heated chiefly from below and gains its moisture by evaporation from Earth's surface, the nature of the source region largely determines the initial characteristics of an air mass. An ideal source region must meet two essential criteria. First, it must be an extensive and physically uniform area. A region having highly irregular topography or one that has a surface consisting of both water and land is not satisfactory.

The second criterion is that the area be characterized by a general stagnation of atmospheric circulation so that air will stay over the region long enough to come to some measure of equilibrium with the surface. In general, it means regions dominated by stationary or slow-moving anticyclones, with their extensive areas of calm or light winds.

Regions under the influence of cyclones are not likely to produce air masses because such systems are characterized by converging surface winds. The winds in lows are constantly bringing air with unlike temperature and humidity properties into the area. Because the time involved is not long enough to eliminate these differences, steep temperature gradients result, and air-mass formation cannot take place.

Figure 8-3 shows the source regions that produce the air masses that most often influence North America. The waters of the Gulf of Mexico and Caribbean Sea and similar regions in the Pacific west of Mexico yield warm air masses, as does the land area that encompasses the southwestern United States and northern Mexico. In contrast, the North Pacific, the North Atlantic, and the snow- and ice-covered areas comprising northern North America and the adjacent Arctic Ocean are major source regions for cold air masses. It is also clear that the size of the source regions and the intensity of temperatures change seasonally.

Notice in Figure 8-3 that major source regions are not found in the middle latitudes but instead are confined to subtropical and subpolar locations. The fact that the middle latitudes are the site where cold and warm air masses clash, often because the converging winds of a traveling cyclone draw them together, means that this zone lacks the conditions necessary to be a source region. Instead, this latitude belt is one of the stormiest on the planet.

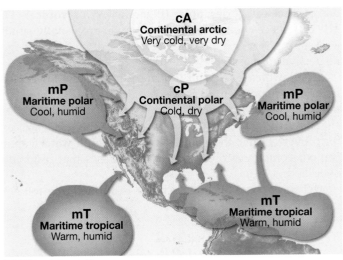

(a) Mid-winter pattern

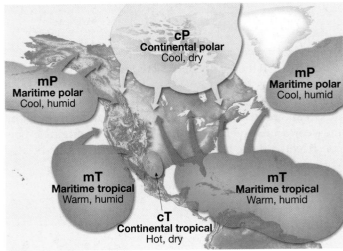

(b) Mid-summer pattern

Figure 8-3 Air-mass source regions for North America. Source regions are largely confined to subtropical and subpolar locations. The fact that the middle latitudes are the site where cold and warm air masses clash, often because the converging winds of a traveling cyclone draw them together, means that this zone lacks the conditions necessary to be a source region. The differences between polar and arctic are relatively small and serve to indicate the degree of coldness of the respective air masses. By comparing the summer and winter maps, it is clear that the extent and temperature characteristics of source regions fluctuate.

Classifying Air Masses

Basic Weather Patterns
▶ Air Masses

The classification of an air mass depends on the latitude of the source region and the nature of the surface in the area of origin—ocean or continent. The latitude of the source region indicates the temperature conditions within the air mass, and the nature of the surface below strongly influences the moisture content of the air.

Air masses are identified by two-letter codes. With reference to latitude (temperature), air masses are placed into

---

Concept Check 8.2

1 What two criteria must be met for an area to be an air-mass source region?

2 Why are regions that have a cyclonic circulation generally not conducive to air-mass formation?

one of three categories: **polar (P), arctic (A),** or **tropical (T).** The differences between polar and arctic are usually small and simply serve to indicate the degree of coldness of the respective air masses.

The lowercase letter **m** (for **maritime**) or the lowercase letter **c** (for **continental**) is used to designate the nature of the surface in the source region and hence the humidity characteristics of the air mass. Because maritime air masses form over oceans, they have a relatively high water-vapor content compared to continental air masses that originate over landmasses.

When this classification scheme is applied, the following air masses can be identified:

cA	continental	arctic
cP	continental	polar
cT	continental	tropical
mT	maritime	tropical
mP	maritime	polar

Notice that the list does not include mA (maritime arctic). These air masses are not listed because they seldom, if ever, form. Although arctic air masses form over the Arctic Ocean, this water body is largely ice covered throughout the year. Consequently, the air masses that originate here consistently have the moisture characteristics associated with a continental source region.

Concept Check 8.3

1 On what basis are air masses classified?

2 Compare the temperature and moisture characteristics of the following air masses: cA, cP, mP, mT, and cT.

3 Why is mA left out of the air-mass classification scheme?

Air-Mass Modification

After an air mass forms, it normally migrates from the area where it acquired its distinctive properties to a region with different surface characteristics. Once the air mass moves from its source region, it not only modifies the weather of the area it is traversing, but it is also gradually modified by the surface over which it is moving. This idea was clearly shown in Figure 8-2. Warming or cooling from below, the addition or loss of moisture, and vertical movements all act to bring about changes in an air mass. The amount of modification can be relatively small or, as the following example illustrates, the changes can be profound enough to alter completely the original identity of the air mass.

When cA or cP air moves over the ocean in winter, it undergoes considerable change (Figure 8-4). Evaporation from the water surface rapidly transfers large quantities of moisture to the once-dry continental air. Furthermore, because the underlying water is warmer than the air above, the air is also heated from below. This factor leads to instability and vertically ascending currents that rapidly transport heat and moisture to higher levels. In a relatively short span, cold, dry, and stable continental air is transformed into an unstable mP air mass.

When an air mass is colder than the surface over which it is passing, as in the preceding example, the lowercase letter *k* may be added after the air-mass symbol. If, however, an air mass is warmer than the underlying surface, the lowercase letter *w* is added. It should be remembered that the *k* or *w* suffix does not mean that the air mass itself is cold or warm. It means only that the air is *relatively* cold or warm in comparison with the underlying surface over which it is traveling. For example, an mT air mass from the Gulf of Mexico is usually classified as mTk as it moves over the southeastern states in summer. Although the air mass is warm, it is still cooler than the highly heated landmass over which it is passing.

The *k* or *w* designation gives an indication of the stability of an air mass and hence the weather that might be expected. An air mass that is colder than the surface is going to be warmed in its lower layers. This fact causes greater instability that favors the ascent of the heated lower air and creates the

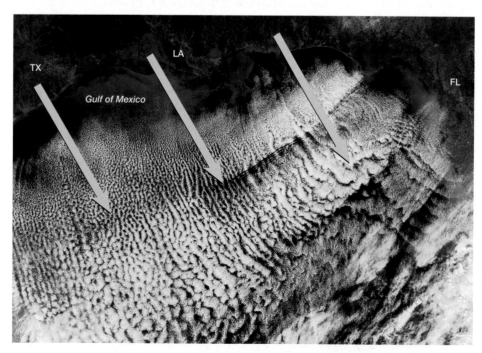

Figure 8-4 This satellite image from December 16, 2007, shows the modification of cold, dry, and cloud-free cP air as it moved over the Gulf of Mexico. The addition of heat and water vapor from the relatively warm water quickly modified the air mass, as evidenced by the development of clouds. (NASA)

possibility of cloud formation and precipitation. Indeed, a *k* air mass is often characterized by cumulus clouds, and if precipitation occurs, it will be of the shower or thunderstorm variety. Also, visibility is generally good (except in rain) because of the stirring and overturning of the air.

Conversely, when an air mass is warmer than the surface over which it is moving, its lower layers are chilled. A surface inversion that increases the stability of the air mass often develops. This condition does not favor the ascent of air, and so it opposes cloud formation and precipitation. Any clouds that do form will be stratus clouds, and precipitation, if any, will be light to moderate. Moreover, because of the lack of vertical movements, smoke and dust often become concentrated in the lower layers of the air mass and cause poor visibility. During certain times of the year, fogs, especially the advection type, may also be common in some regions.

In addition to modifications resulting from temperature differences between an air mass and the surface below, upward and downward movements induced by cyclones and anticyclones or topography can also affect the stability of an air mass. Such modifications are often called *mechanical* or *dynamic* and are usually independent of the changes caused by surface cooling or heating. For example, significant modification can result when an air mass is drawn into a low. Here convergence and lifting dominate and the air mass is rendered more unstable. Conversely, the subsidence associated with anticyclones acts to stabilize an air mass. Similar alterations in stability occur when an air mass is lifted over highlands or descends the leeward side of a mountain barrier. In the first case, the air's stability is reduced; in the second case, the air becomes more stable.

Concept Check 8.4

1 What do the lowercase letters *k* and *w* indicate about an air mass?

2 List the general weather conditions associated with *k* and *w* air masses.

3 How might vertical movements induced by a pressure system or topography act to modify an air mass?

Properties of North American Air Masses

Air masses frequently pass over us, which means that the day-to-day weather we experience often depends on the temperature, stability, and moisture content of these large bodies of air. In this section, we briefly examine the properties of the principal North American air masses. In addition, Table 8-1 serves as a summary.

Continental Polar (cP) and Continental Arctic (cA) Air Masses

Continental polar and continental arctic air masses are, as their classification implies, cold and dry. Continental polar air originates over the often snow-covered interior regions of Canada and Alaska, poleward of the 50th parallel. Continental arctic air forms farther north, over the Arctic basin and the Greenland ice cap (see Figure 8-3).

Continental arctic air is distinguished from cP air by its generally lower temperatures, although at times the differences may be slight. In fact, some meteorologists do not differentiate between cP and cA.

Winter During the winter, both cP and cA air masses are bitterly cold and very dry. Winter nights are long, and the daytime Sun is short-lived and low in the sky. Consequently, as winter advances, Earth's surface and atmosphere lose heat that, for the most part, is not replenished by incoming solar energy. Therefore, the surface reaches very low temperatures, and the air near the ground is gradually chilled to heights of 1 kilometer (0.6 mile) or more. The result is a strong and persistent temperature inversion, with the coldest temperatures near the ground. Marked stability is, therefore, the rule. Because the air is very cold and the surface below is frozen, the mixing ratio of these air masses is necessarily low, ranging from perhaps 0.1 gram per kilogram in cA up to 1.5 grams per kilogram in some cP air.

As wintertime cP or cA air moves outward from its source region, it carries its cold, dry weather to the United States, normally entering between the Great Lakes and the Rockies. Because there are no major barriers between the high-latitude source regions and the Gulf of Mexico, cP and cA air masses can sweep rapidly and with relative ease far southward into the United States. The winter cold waves experienced in much of the central and eastern United States are closely associated with such polar outbreaks. One such cold wave is described in "Severe and Hazardous Weather," page 227. Usually, the last freeze in spring and the first freeze in autumn can be correlated with outbreaks of polar or arctic air.

Students Sometimes Ask...

When a cold air mass moves south from Canada into the United States, how rapidly can temperatures change?

When a fast-moving frigid air mass advances into the northern Great Plains, temperatures have been known to plunge 20° to 30°C (40° to 50°F) in a matter of just a few hours. One notable example is a drop of 55.5°C (100°F), from 6.7°C to −48.8°C (44° to −56°F), in 24 hours at Browning, Montana, on January 23–24, 1916. Another remarkable example occurred on Christmas Eve, 1924, when the temperature at Fairfield, Montana, dropped from 17°C (63°F) at noon to −29°C (−21°F) at midnight—an amazing 46°C (83°F) change in just 12 hours.

TABLE 8-1 Weather Characteristics of North American Air Masses

Air Mass	Source Region	Temperature and Moisture Characteristics in Source Region	Stability in Source Region	Associated Weather
cA	Arctic basin and Greenland ice cap (winter only)	Bitterly cold and very dry in winter	Stable	Cold waves in winter
cP	Interior Canada and Alaska	Very cold and dry in winter	Stable entire year	a. Cold waves in winter b. Modified to cPk in winter over Great Lakes, bringing lake-effect snow to leeward shores
mP	North Pacific	Mild (cool) and humid entire year	Unstable in winter Stable in summer	a. Low clouds and showers in winter b. Heavy orographic precipitation on windward side of western mountains in winter c. Low stratus and fog along coast in summer; modified to cP inland
mP	Northwestern Atlantic	Cold and humid in winter Cool and humid in summer	Unstable in winter Stable in summer	a. Occasional nor'easter in winter b. Occasional periods of clear, cool weather in summer
cT	Northern interior Mexico and southwestern U.S. (summer only)	Hot and dry	Unstable	a. Hot, dry, and cloudless, rarely influencing areas outside source region b. Occasional drought to southern Great Plains
mT	Gulf of Mexico, Caribbean Sea, western Atlantic	Warm and humid entire year	Unstable entire year	a. In winter it usually becomes mTw, moving northward and bringing occasional widespread precipitation or advection fog b. In summer hot and humid conditions, frequent cumulus development, and showers or thunderstorms
mT	Eastern subtropical Pacific	Warm and humid entire year	Stable entire year	a. In winter it brings fog, drizzle, and occasional moderate precipitation to northwestern Mexico and the southwestern United States b. In summer occasionally reaches the western United States and is a source of moisture for infrequent convectional thunderstorms

Summer Because cA air is present principally in the winter, only cP air has any influence on our summer weather, and this effect is considerably reduced when compared with winter. During summer months the properties of the source region for cP air are very different from those during winter. Instead of being chilled by the ground, the air is warmed from below as the long days and higher Sun angle warm the snow-free land surface. Although summer cP air is warmer and has a higher moisture content than its wintertime counterpart, the air is still cool and relatively dry compared with air in areas farther south. Summer heat waves in the northern portions of the eastern and central

United States are often ended by the southward advance of cP air, which for a day or two brings cooling relief and bright, pleasant weather.

Lake-Effect Snow: Cold Air Over Warm Water

A glance at the chapter-opening image (pages 220–221) provides a perspective from space of the process discussed in this section. The skies over Lake Superior and Lake Michigan exhibit long rows of dense, white, snow-producing clouds. They formed as cold, dry cP air moved

SEVERE AND HAZARDOUS WEATHER The Siberian Express

The surface weather map for December 22, 1989, shows a very large high-pressure center covering the eastern two-thirds of the United States and a substantial portion of Canada (Figure 8-A). As is usually the case in winter, a large anticyclone such as this is associated with a huge mass of dense and bitterly cold arctic air. After such an air mass forms over the frozen expanses near the Arctic Circle, the winds aloft sometimes direct it toward the south and east. When an outbreak takes place, it is popularly called the "Siberian Express" by the news media, even though the air mass did not originate in Siberia.

November 1989 was unusually mild for late autumn. In fact, across the United States, more than 200 daily high-temperature records were set. December, however, was different. East of the Rockies, the month's weather was dominated by two arctic outbreaks. The second brought record-breaking cold.

More than 370 record low temperatures were reported.

Between December 21 and 25, as the frigid dome of high pressure advanced southward and eastward, more than 370 record low temperatures were reported. On December 21, Havre, Montana, had an overnight low of −42.2°C (−44°F), breaking a record set in 1884. Meanwhile, Topeka's −32.2°C (−26°F) was that city's lowest temperature for any date since record keeping had started 102 years earlier.

The three days that followed saw the arctic air migrate toward the south and east. By December 24, Tallahassee, Florida, had a low temperature of −10°C (14°F). In the center of the state the daily minimum at Orlando was −5.6°C (22°F). It was actually warmer in North Dakota on Christmas Eve than in central and northern Florida!

As we would expect, utility companies in many states reported record demand. When the arctic air advanced into Texas and Florida, agriculture was especially hard hit. Some Florida citrus growers lost 40 percent of their crop, and many vegetable crops were wiped out completely (Figure 8-B).

It was actually warmer in North Dakota on Christmas Eve than in central and northern Florida!

After Christmas the circulation pattern that brought this record-breaking Siberian Express from the arctic deep into the United States changed. As a result, temperatures for much of the country during January and February 1990 were well above normal. In fact, January 1990 was the second warmest January in 96 years. Thus, despite frigid December temperatures, the winter of 1989–1990 "averaged out" to be a relatively warm one.

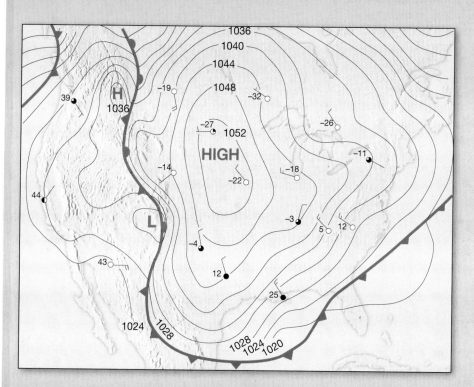

FIGURE 8-A A surface weather map for 7 AM EST, December 22, 1989. This simplified National Weather Service (NWS) map shows an intense winter cold wave caused by an outbreak of frigid continental arctic air. This event brought subfreezing temperatures as far south as the Gulf of Mexico. Temperatures on NWS maps are in degrees Fahrenheit.

FIGURE 8-B When an arctic air mass invades the citrus regions of Florida and Texas, even modern freeze controls may not be able to prevent significant losses. (Photo courtesy of Florida Department of Citrus)

from the land surface across the open water. Continental polar air masses are not, as a rule, associated with dense clouds and heavy precipitation. Yet during late autumn and winter a unique and interesting weather phenomenon takes place along the downwind shores of the Great Lakes.* Periodically, brief, heavy snow showers issue from dark clouds that move onshore from the lakes (see "Severe and Hazardous Weather," p. 230). Seldom do these storms move more than about 80 kilometers (50 miles) inland from the shore before the snows come to an end. These highly localized storms, occurring along the leeward shores of the Great Lakes, create what are known as **lake-effect snows.**

Lake-effect storms account for a high percentage of the snowfall in many areas adjacent to the lakes. The strips of land that are most frequently affected, called *snowbelts*, are shown in Figure 8-5. A comparison of average snowfall totals at Thunder Bay, Ontario, on the north shore of Lake Superior, and Marquette, Michigan, along the southern shore, provides another excellent example. Because Marquette is situated on the leeward shore of the lake, it receives substantial lake-effect snow and therefore has a much higher snowfall total than does Thunder Bay (Table 8-2).

What causes lake-effect snow? The answer is closely linked to the differential heating of water and land (Chapter 3) and to the concept of atmospheric instability (Chapter 4). During the summer months, bodies of water, including the Great Lakes, absorb huge quantities of energy from the Sun and from the warm air that passes over them. Although these water bodies do not reach particularly high temperatures, they nevertheless represent huge reservoirs of heat. The surrounding land, in contrast, cannot store heat nearly as effectively. Consequently, during autumn and winter, the temperature of the land drops quickly, whereas water bodies lose their heat more gradually and cool slowly.

---

*Actually, the Great Lakes are just the best-known example. Other large lakes can also experience this phenomenon.

TABLE 8-2 Monthly Snowfall at Thunder Bay, Ontario, and Marquette, Michigan

Thunder Bay, Ontario			
October	November	December	January
3.0 cm (1.2 in.)	14.9 cm (5.8 in.)	19.0 cm (7.4 in.)	22.6 cm (8.8 in.)

Marquette, Michigan			
October	November	December	January
5.3 cm (2.1 in.)	37.6 cm (14.7 in.)	56.4 cm (22.0 in.)	53.1 cm (20.7 in.)

From late November through late January the contrasts in average temperatures between water and land range from about 8°C in the southern Great Lakes to 17°C farther north. However, the temperature differences can be much greater (perhaps 25°C) when a very cold cP or cA air mass pushes southward across the lakes. When such a dramatic temperature contrast exists, the lakes interact with the air to produce major lake-effect storms. Figure 8-6 depicts the movement of a cP air mass across one of the Great Lakes. During its journey, the air acquires large quantities of heat and moisture from the relatively warm lake surface. By the time it reaches the opposite shore, this cPk air is humid and unstable, and heavy snow showers are likely.

Maritime Polar (mP) Air Masses

Maritime polar air masses form over oceans at high latitudes. As the classification indicates, mP air is cool to cold and humid, but compared with cP and cA air masses in winter, mP air is relatively mild because of the higher temperatures of the ocean surface as contrasted to the colder continents.

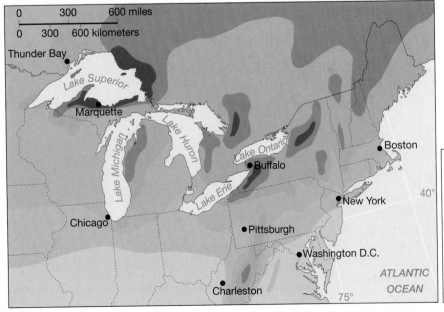

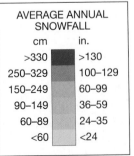

Figure 8-5 Average annual snowfall. The snowbelts of the Great Lakes region are easy to pick out on this snowfall map. (Data from NOAA)

AVERAGE ANNUAL SNOWFALL	
cm	in.
>330	>130
250–329	100–129
150–249	60–99
90–149	36–59
60–89	24–35
<60	<24

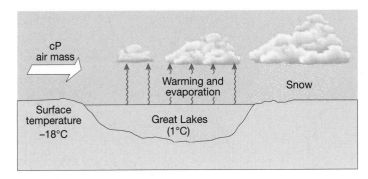

Figure 8-6 As continental polar air crosses the Great Lakes in winter, it acquires moisture and becomes unstable because of warming from below. "Lake-effect" snow showers on the downwind side of the lakes often result from this air-mass modification.

Students Sometimes Ask ...

I know that Buffalo, New York, is famous for its lake-effect snows. Just how bad can it get?

Because of Buffalo's location along the eastern shore of Lake Erie, it does indeed receive a great deal of lake-effect snow (see Figure 8-5). One of the most memorable events took place between December 24, 2001, and January 1, 2002. This storm, which set a record as the longest-lasting lake-effect event, buried Buffalo under 207.3 centimeters (81.6 inches) of snow. Prior to this storm, the record for the *entire month* of December had been 173.7 centimeters (68.4 inches)! The eastern shore of Lake Ontario was also hard hit, with one station recording more than 317 centimeters (125 inches) of snow.

Two regions are important sources for mP air that influences North America: the North Pacific and the northwestern Atlantic from Newfoundland to Cape Cod (see Figure 8-3). Because of the general west-to-east circulation in the middle latitudes, mP air masses from the North Pacific source region usually influence North American weather more than mP air masses generated in the northwest Atlantic. Whereas air masses that form in the Atlantic generally move eastward toward Europe, mP air from the North Pacific has a strong influence on the weather along the western coast of North America, especially in the winter.

Pacific mP Air Masses During the winter, mP air masses from the Pacific usually begin as cP air in Siberia (Figure 8-7). Although air rarely stagnates over this area, the source region is extensive enough to allow the air moving across it to acquire its characteristic properties. As the air advances eastward over the relatively warm water, active evaporation and heating occur in the lower levels. Consequently, what was once a very cold, dry, and stable air mass is changed into one that is mild and humid near the surface and relatively unstable. As this mP air arrives at the western coast of North America, it is often accompanied by low clouds and shower activity. When the mP air advances inland against the western mountains, orographic uplift can produce heavy rain or snow on the windward slopes of the mountains.

Summer brings a change in the characteristics of mP air masses from the North Pacific. During the warm season, the

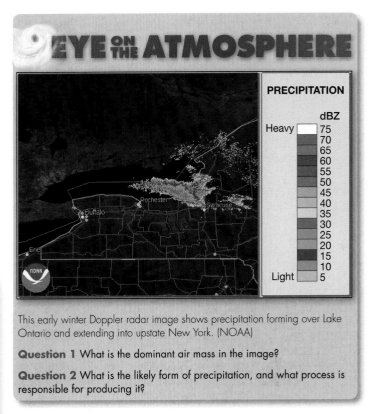

This early winter Doppler radar image shows precipitation forming over Lake Ontario and extending into upstate New York. (NOAA)

Question 1 What is the dominant air mass in the image?

Question 2 What is the likely form of precipitation, and what process is responsible for producing it?

ocean is cooler than the surrounding continents. In addition, the Pacific high lies off the western coast of the United States (see Figure 7-10). Consequently, there is almost continuous southward flow of moderate-temperature air. Although the air near the surface may often be conditionally unstable, the presence of the Pacific high means that there is subsidence and stability aloft. Thus, low stratus clouds and summer fogs characterize much of the western coast. Once summer mP air from the Pacific moves inland, it is heated at the surface over the hot and dry interior. The heating and resulting turbulence act to reduce the relative humidity in the lower layers, and the clouds dissipate.

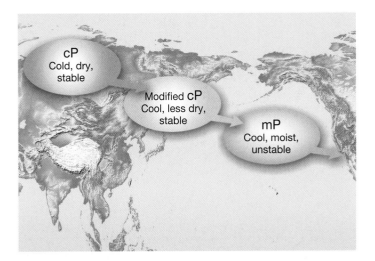

Figure 8-7 During winter, maritime polar (mP) air masses in the North Pacific usually begin as continental polar (cP) air masses in Siberia. The cP air is modified to mP as it slowly crosses the ocean.

SEVERE AND HAZARDOUS WEATHER

An Extraordinary Lake-Effect Snowstorm

Northeastern Ohio is part of the Lake Erie snowbelt, a zone that extends eastward into northwestern Pennsylvania and western New York (see Figure 8-5). Here snowfall is enhanced when cold winds blow from the west or northwest across the relatively warm unfrozen waters of Lake Erie. Average annual snowfall in northeastern Ohio is between 200 and 280 centimeters (80 to 110 inches) and increases to 450 centimeters (175 inches) in western New York.

Although residents here are accustomed to abundant snow storms, even they were surprised in November 1996 by an especially early and strong storm. During a six-day span, from November 9 to 14, northeastern Ohio experienced record-breaking lake-effect snows. Rather than raking fall leaves, people were shoveling sidewalks and clearing snow from overloaded roofs (Figure 8-C).

> ### Rather than raking fall leaves, people were shoveling sidewalks and clearing snow from overloaded roofs.

Persistent snow squalls (narrow bands of heavy snow) frequently produced accumulation rates of 5 centimeters (2 inches) per hour during the six-day siege. The snow was the result of especially deep and prolonged lower atmospheric instability created by the movement of cold air across a relatively warm Lake Erie. The surface temperature of the lake was 12°C (54°F), several degrees above normal. The air temperature at a height of 1.5 kilometers (nearly 5000 feet) was −5°C (23°F).

The northwest flow of cold air across Lake Erie and the 17°C lapse rate generated lake-effect rain and snow almost immediately.* During some periods loud thunder accompanied the snow squalls!

Data from the Cleveland National Weather Service Snow Spotters Network showed 100 to 125 centimeters (39 to 49 inches) of snow through the core of the Ohio snowbelt (Figure 8-D). The deepest accumulation was measured near Chardon, Ohio. Here the six-day total of 175 centimeters (68.9 inches) far exceeded the previous Ohio record of 107 centimeters (42 inches) set in 1901. In addition, the November 1996 snowfall total of 194.8 centimeters (76.7 inches) at the same site set a new monthly snowfall record for Ohio. The previous one-month record had been 176.5 centimeters (69.5 inches).

> ### During some periods loud thunder accompanied the snow squalls!

The impact of the storm was considerable. Ohio's governor declared a state of emergency on November 12, and National Guard troops were dispatched to assist in snow removal and aid in the rescue of snowbound residents. There was widespread damage to trees and shrubs across northeastern Ohio because the snow was particularly wet and dense and readily clung to objects. Press reports indicated that about 168,000 homes were without electricity, some for several days. Roofs of numerous buildings collapsed under the excessive snow loads. Although residents of the northeast Ohio snowbelt are winter storm veterans, the six-day storm in November 1996 will be long remembered as an extraordinary event.

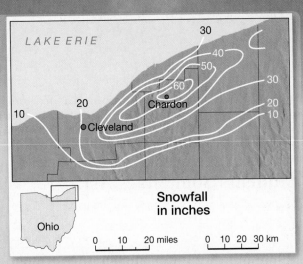

FIGURE 8-D Snowfall totals (in inches) for northeastern Ohio for the November 9–14, 1996, lake-effect storm. The deepest accumulation, 175 centimeters (nearly 69 inches, occurred near Chardon. (After National Weather Service)

FIGURE 8-C A six-day lake-effect snowstorm in November 1996 dropped 175 centimeters (nearly 69 inches) of snow on Chardon, Ohio, setting a new state record. (Photo by Tony Dejak/AP/Wide World Photos)

*Thomas W. Schmidlin and James Kasarik, "A Record Ohio Snowfall During 9–14 November 1996," *Bulletin of the American Meteorological Society,* Vol. 80, No. 6, June 1999, p. 1109.

Maritime Polar Air from the North Atlantic As is the case for mP air from the Pacific, air masses forming in the northwestern Atlantic source region were originally cP air masses that moved from the continent and were transformed over the ocean. However, unlike air masses from the North Pacific, mP air from the Atlantic only occasionally affects the weather of North America. Nevertheless, this air mass does have an effect when the northeastern United States is on the northern or northwestern edge of a passing low-pressure center. In winter, strong cyclonic winds can draw mP air into the region. Its influence is generally confined to the area east of the Appalachians and north of Cape Hatteras, North Carolina. The weather associated with a wintertime invasion of mP air from the Atlantic is known locally as a **nor'easter.** Strong northeast winds, freezing or near-freezing temperatures, high relative humidity, and the likelihood of precipitation make this weather phenomenon an unwelcome event. A classic case is presented in "Severe and Hazardous Weather," page 232.

Whereas mP air masses from the Atlantic may produce an occasional unwelcome nor'easter during the winter, summertime incursions of this air mass may bring pleasant weather. Like the Pacific source region, the northwestern Atlantic is dominated by high pressure during the summer (see Figure 7-10). Thus, the upper air is stable because of subsidence and the lower air is essentially stable because of the chilling effect of the relatively cool water. As the circulation on the southern side of the anticyclone carries this stable mP air into New England, and occasionally as far south as Virginia, the region enjoys clear, cool weather and good visibility.

Maritime Tropical (mT) Air Masses

Maritime tropical air masses affecting North America most often originate over the warm waters of the Gulf of Mexico, the Caribbean Sea, or the adjacent western Atlantic Ocean (see Figure 8-3). The tropical Pacific is also a source region for mT air. However, the land area affected by this latter source is small compared with the size of the region influenced by air masses produced in the Gulf of Mexico and adjacent waters.

As expected, mT air masses are warm to hot, and they are humid. In addition, they are often unstable. It is through invasions of mT air masses that the subtropics export much heat and moisture to the cooler and drier areas to the north. Consequently, these air masses are important to the weather whenever present because they are capable of contributing significant precipitation.

North Atlantic mT Air Maritime tropical air masses from the Gulf–Caribbean–Atlantic source region greatly affect the weather of the United States east of the Rocky Mountains. Although the source region is dominated by the North Atlantic subtropical high, the air masses **produced** are not stable but are neutral or unstable because the source

EYE ON THE ATMOSPHERE

This satellite image from December 27, 2010, shows a strong winter storm off the East Coast. (NASA)

Question 1 Can you identify the very center of the storm?

Question 2 What air mass is being drawn into the storm to produce the dense clouds in the upper right?

Question 3 What name is applied to a storm such as this?

Question 4 Farther south, the cold air mass over the southeastern states is cloud free. What is its likely classification? Explain how it is being modified as it moves over the Atlantic.

region is located on the weak western edge of the anticyclone, where pronounced subsidence is absent.

During winter, when cP air dominates the central and eastern United States, mT air only occasionally enters this part of the country. When an invasion does occur, the lower portions of the air mass are chilled and stabilized as it moves northward. Its classification is changed to mTw. As a result, the formation of convective showers is unlikely. Widespread precipitation does occur, however, when a northward-moving mT air mass is pulled into a traveling cyclone and forced to ascend. In fact, much of the wintertime precipitation over the eastern and central states results when mT air from the Gulf of Mexico is lifted along fronts in traveling cyclones.

Another weather phenomenon associated with a northward-moving wintertime mT air mass is advection fog. Dense fogs can develop as the warm, humid air is chilled as it moves over the cold land surface.

During the summer, mT air masses from the Gulf, Caribbean, and adjacent Atlantic affect a much wider area of North America and are present for a greater percentage of the time than during the winter. As a result, they exert

SEVERE AND HAZARDOUS WEATHER

A classic nor'easter moved up the east coast on January 12, 2011, dumping heavy snow on the New England states for the third time in three weeks (Figure 8-E). The storm began developing to the south a day earlier. As it moved northward along the coast, it merged with another system crossing from the Midwest.

The satellite image in Figure 8-F shows that the storm had a distinctive comma shape—which forms from the counterclockwise circulation around a low-pressure center. Cold, humid mP air from the North Atlantic was drawn toward the storm center, producing dense clouds, especially on the north and west sides of the storm. In parts of New England, snow fell as fast as 7.6 centimeters (3 inches) per hour. More than 61 centimeters (24 inches) fell in many areas by the evening of January 12.

The storm left more than 100,000 people without electricity.

Blizzard conditions—with visibility cut to less than 0.4 kilometer (0.25 mile) and gale-force winds for more than three hours—developed in parts of Connecticut and Massachusetts. The storm left more than 100,000 people without electricity and led to the shutdown of portions of Interstate 95 and the northeastern railroad service. Bradley International Airport near Hartford, Connecticut, set a one-day record with 57 centimeters (22 inches) of snow, and Wilmington, Vermont, received in excess of 91 centimeters (35 inches).

FIGURE 8-F Satellite image of a strong winter storm called a *nor'easter* along the coast of New England on January 12, 2011. In winter, a nor'easter exhibits a weather pattern in which strong northeast winds carry cold, humid mP air from the North Atlantic into New England and the middle Atlantic states. The combination of ample moisture and strong convergence can result in heavy snow. (NASA)

FIGURE 8-E Digging out in Boston following the January 12, 2011 blizzard. (Photo by Michael Dwyer/ Alamy)

a strong and often dominating influence over the summer weather of the United States east of the Rocky Mountains. This influence is due to the general sea-to-land airflow over the eastern portion of North America during the warm months, which brings more frequent incursions of mT air that penetrate much deeper into the continent than during the winter months. Consequently, these air masses are largely responsible for the hot and humid conditions that prevail over the eastern and central United States.

Initially, summertime mT air from the Gulf is unstable. As it moves inland over the warmer land, it becomes an mTk air mass as daytime heating of the surface layers further increases the air's instability. Because the relative humidity is high, only modest lifting is necessary to bring about active convection, cumulus development, and thunderstorm or shower activity (Figure 8-8). This is, indeed, a common warm-weather phenomenon associated with mT air.

It should also be noted here that air masses from the Gulf–Caribbean–Atlantic region are the primary source of much, if not most, of the precipitation received in the eastern two-thirds of the United States. Pacific air masses contribute little to the water supply east of the Rockies because the western mountains effectively "drain" the moisture from the air through numerous episodes of orographic uplift.

Figure 8-9, which shows the distribution of average annual precipitation for the eastern two-thirds of the United States by using **isohyets** (lines connecting places having equal rainfall), illustrates this situation nicely. The pattern of isohyets shows the greatest rainfall in the Gulf region and a decrease in precipitation with increasing distance from the mT source region.

North Pacific mT Air   Compared to mT air from the Gulf of Mexico, mT air masses from the Pacific source region have much less of an impact on North American weather. In winter, only northwestern Mexico and the extreme southwestern United States are influenced by air from the tropical Pacific. Because the source region lies along the eastern side of the Pacific anticyclone, subsidence aloft produces upper-level stability. When the air mass moves northward, cooling at the surface also causes the lower layers to become more stable, often resulting in fog or drizzle. If the air mass is lifted along a front or forced over mountains, moderate precipitation results.

There are times, however, when mT air from the subtropical North Pacific is involved in a weather phenomenon popularly known as the *Pineapple Express*. Unlike the *Siberian Express* described earlier in the chapter, which

Figure 8-8 As mT air from the Gulf of Mexico moves over the heated land in summer, cumulus development and afternoon showers frequently result. (Photo by Rod Planck/ Photo Researchers, Inc.)

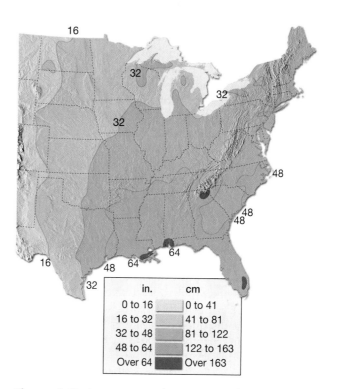

Figure 8-9 Average annual precipitation for the eastern two-thirds of the United States. Note the general decrease in yearly precipitation totals with increasing distance from the Gulf of Mexico, the source region for mT air masses. Isohyets are labeled in inches. (Courtesy of Environmental Data Service, NOAA)

Figure 8-10 (a) This satellite image of clouds over the Pacific Ocean on December 19, 2010, illustrates the "Pineapple Express," in which a strong jet stream carries mT air from the vicinity of Hawaii to California. (NASA) (b) The Pineapple Express battered much of California December 17–22, 2010, bringing as much as 50 centimeters (20 inches) of rain to the San Gabriel Mountains and more than 1.5 meters (5 feet) of snow to the Sierra Nevada. Southern California bore the brunt of the storms, as coastal and hillside areas experienced mudflows and floods. (Photo by Jebb Harris/*The Orange County Register*/Newscom)

(a)

delivers bitter-cold waves to the nation's midsection, the Pineapple Express can bring extraordinary rains to southern California and other West Coast locations.

Most precipitation along the West Coast results from wintertime storms that pass across the Gulf of Alaska. These storms are dominated by humid, cool mP air. However, in some years, a strong southern branch of the polar jet stream acts as a conduit that transports humid, warm mT air from the tropics near the Hawaiian islands northeastward to the west coast (Figure 8-10). This mT air feeds into storm systems that can bring torrential rains to lower elevations and heavy snows to the Sierra Nevada. Mudflows (popularly called mudslides) can result when the soaking rains saturate hillsides that have lost their anchoring vegetation to recent wildfires.

For many years the summertime influence of air masses from the tropical Pacific source region on the weather of the southwestern United States and northern Mexico was believed to be minimal. It was thought that the moisture for the infrequent summer thunderstorms that occur in the region came from occasional westward thrusts of mT air from the Gulf of Mexico. However, the Gulf of Mexico is no longer believed to be the primary supplier of moisture for the area west of the Continental Divide. Rather, it has been demonstrated that the tropical North Pacific west of central Mexico is a more important source of moisture for this area.

In summer the mT air moves northward from its Pacific source region up the Gulf of California and into the interior of the western United States (see Figure 7-13, page 204). This movement, which is confined largely to July and August, is essentially monsoonal in character. That is, the inflow of moist air is a response to thermally produced low pressure that develops over the heated landmass. The July–August rainfall maximum for Tucson is a response to this incursion of Pacific mT air (Figure 8-11).

(b)

Figure 8-11 (a) Cumulonimbus clouds developing over the Sonoran Desert in southern Arizona on a July afternoon. The source of moisture for these summer storms is maritime tropical air from the eastern North Pacific. (Photo by laurent baig/Alamy) (b) Monthly precipitation data for Tucson, Arizona. The July–August rainfall maximum in the desert Southwest results from the monsoonal flow of Pacific mT air into the region.

Continental Tropical (cT) Air Masses

North America narrows as it extends southward through Mexico; therefore, the continent has no extensive source region for continental tropical air masses. By checking the maps in Figure 8-3, you can see that only in summer do northern interior Mexico and adjacent parts of the arid southwestern United States produce hot, dry cT air. Because of the intense daytime heating at the surface, both a steep environmental lapse rate and turbulence extending to considerable heights are found. Nevertheless, although the air is unstable, it generally remains nearly cloudless because of extremely low humidity. Consequently, the prevailing weather is hot, with an almost complete lack of rainfall. Large daily temperature ranges are the rule. Although cT air masses are usually confined to the source region, occasionally they move into the southern Great Plains. If the cT air persists for long, drought may occur.

Concept Check 8.5

1 Which two air masses have the greatest influence on weather east of the Rocky Mountains? Explain your choice.

2 What air mass influences the weather of the Pacific Coast more than any other?

3 Why do cA and cP air masses often sweep far south into the United States?

4 Describe the modifications that occur as a cP air mass passes across a large ice-free lake in winter.

5 Why do mP air masses from the North Atlantic seldom affect the eastern United States?

6 What air mass and source region provide the greatest amount of moisture to the eastern and central United States?

Give It Some Thought

1. (a) We know that, during the winter, all polar (P) air masses are cold. Which should be colder: a wintertime mP air mass or a wintertime cP air mass? Briefly explain. (b) We expect tropical (T) air masses to be warm, but some are warmer than others. Which should be warmer: a summertime cT air mass or a summertime mT air mass? How did you figure this out?

2. Air mass source regions are large, relatively homogenous areas. As the accompanying map illustrates, the broad expanse between the Appalachians and the Rockies is such a zone, yet air masses do not form here. Why is this area not a source region?

3. What is the proper classification for an air mass that forms over the Arctic Ocean in winter: cA or mA? Explain your choice.

4. The Great Lakes are not the only water bodies associated with lake-effect snow. For example, large lakes in Canada also experience this phenomenon. Shown below are snowfall data for Fort Resolution, a settlement on the southeastern store of Great Slave Lake:

Winter Snowfall Data (centimeters) for Fort Resolution, Northwest Territories, Canada.

Sept.	Oct.	Nov.	Dec.	Jan.
2.5	13.7	36.6	19.6	15.4

During what month is snowfall greatest? Suggest an explanation as to why the maximum occurs when it does.

5. In each of the situations described here, indicate whether the air mass is becoming more stable or more unstable. Briefly explain each choice.
   a. An mT air mass moving northward from the Gulf of Mexico over the southeastern states in winter
   b. A cP air mass moving southward across Lake Superior in late November
   c. An mP air mass in the North Atlantic drawn into a low-pressure center off the coast of New England in January
   d. A wintertime cP air mass from Siberia moving eastward from Asia across the North Pacific

AIR MASSES IN REVIEW

- In the middle latitudes many weather events are associated with the movements of air masses. An *air mass* is a large body of air, usually 1600 kilometers (1000 miles) or more across and perhaps several kilometers thick, that is characterized by homogeneous physical properties (in particular temperature and moisture content) at any given altitude. A region under the influence of an air mass will probably experience generally constant weather conditions, a situation referred to as *air-mass weather.*

- Areas in which air masses originate, called *source regions,* must be extensive and physically uniform areas and be characterized by a general stagnation of atmospheric circulation.

- The classification of an air mass depends on the latitude of the source region and the nature of the surface in the area of origin—ocean or continent. Air masses are identified by two-letter codes. With reference to latitude (temperature), air masses are placed into one of three categories: *polar (P), arctic (A),* or *tropical (T).* A lowercase letter *(m* for *maritime* or *c* for *continental)* is placed in front of the uppercase letter

to designate the nature of the surface in the source region and therefore the humidity characteristics of the air mass. Using this classification scheme, the following air masses are identified: cA, cP, cT, mT, and mP.

- Changes to the stability of an air mass can result from temperature differences between an air mass and the surface and/or vertical movements induced by cyclones, anticyclones, or topography.

- The day-to-day weather we experience depends on the temperature, stability, and moisture content of the air mass we are experiencing. Among the numerous weather phenomena associated with various air masses are lake-effect snows, nor'easters, and the Pineapple Express.

VOCABULARY REVIEW

air mass (p. 222)
air-mass weather (p. 222)
arctic (A) air mass (p. 224)
continental (c) air mass (p. 224)

isohyet (p. 233)
lake-effect snow (p. 228)
maritime (m) air mass (p. 224)
nor'easter (p. 231)

polar (P) air mass (p. 224)
source region (p. 223)
tropical (T) air mass (p. 224)

PROBLEMS

1. Figure 8-12 shows the distribution of air temperatures (top number) and dew-point temperatures (lower number) for a December morning. Two well-developed air masses are influencing North America at this time. The air masses are separated by a broad zone that is not affected by either air mass. Draw lines on the map to show the boundaries of each air mass. Label each air mass with the proper classification.

2. Refer to Figure 8-5. Notice the narrow, north–south-oriented zone of relatively heavy snowfall east of Pittsburgh and Charleston. This region is too far from the Great Lakes to receive lake-effect snows. Speculate on a likely reason for the higher snowfalls here. Does your explanation explain the shape of this snowy zone?

3. Albuquerque, New Mexico, is situated in the desert Southwest. Its annual precipitation is just 21.2 centimeters (8.3 inches). Month-by-month data (in centimeters) are as follows:

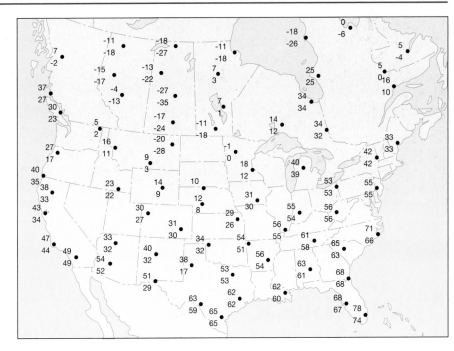

Figure 8-12 Map to accompany Problem 1.

Jan.	Feb.	Mar.	Apr.	May	Jun.	Jul.	Aug.	Sep.	Oct.	Nov.	Dec.
1.0	1.0	1.3	1.3	1.3	1.3	3.3	3.8	2.3	2.3	1.0	1.3

What are the two rainiest months? The pattern here is similar to the pattern in other southwestern cities, including Tuscon, Arizona. Briefly explain why the rainiest months occur when they do.

MyMeteorologyLab™

Midlatitude Cyclones

The winter of 1992–1993 came to a stormy conclusion in eastern North America during a weekend in mid-March. The daffodils were already in bloom across the South, and people were thinking about spring when the Blizzard of '93 struck on March 13 and 14. The huge storm brought record-low temperatures and barometric-pressure readings, accompanied by record-high snowfalls from Alabama to the Maritime Provinces of eastern Canada. The monster storm, with its driving winds and heavy snow, combined the attributes of both a hurricane and a blizzard as it moved up the spine of the Appalachians, lashing and burying a huge swath of territory. Although the atmospheric pressure at the storm's center was lower than the pressures at the centers of some hurricanes and the winds were frequently as strong as those in hurricanes, this was definitely not a tropical storm—but instead a classic winter cyclone.

Focus On Concepts

After completing this chapter, you should be able to:

- Compare and contrast typical weather associated with a warm front and a cold front.

- Explain the concept of occlusion and the weather associated with an occluded front.

- Describe the Norwegian cyclone model.

- Outline the stages in the life cycle of a typical midlatitude cyclone.

- Describe the weather associated with the passage of a mature midlatitude cyclone when the center of low pressure is 200 kilometers north of your location.

- Explain why divergence in the flow aloft is a necessary condition for the development and intensification of a midlatitude cyclone.

- List the primary sites for the development of midlatitude cyclones that affect North America.

- Describe blocking high-pressure systems and how they influence weather over the midlatitudes.

- Describe the weather associated with an occluded front over the north-central United States in winter.

- Explain the conveyor belt model of a midlatitude cyclone and sketch the three interacting air streams (conveyor belts) on which it is based.

In previous chapters we examined the basic elements of weather as well as the dynamics of atmospheric motions. Our knowledge of these diverse phenomena applies directly to an understanding of day-to-day weather patterns in the middle latitudes (Figure 9–1). For our purposes, *middle latitudes* refers to the area of North America roughly between southern Alaska and Florida—essentially the area of the westerlies where the primary weather producer is the **mid-latitude, or middle-latitude, cyclone.*** These are the same weather phenomena that weather reporters may call a *low-pressure system,* or simply a *low.* Because *weather fronts* are the major weather producers imbedded within midlatitude cyclones, we will begin our discussion with these basic structures (Figure 9–2).

---

*Midlatitude cyclones go by a number of different names, including wave cyclones, frontal cyclones, extratropical cyclones, low-pressure systems, and, simply, lows.

Frontal Weather

 GE●De ATMOSPHERE   Basic Weather Patterns ▶ Fronts

One of the most prominent features of middle-latitude weather is how suddenly and dramatically it can change (see Figure 9–1). Most of these sudden changes are associated with the passage of weather fronts. **Fronts** are boundary surfaces that separate air masses of different densities—one of which is usually warmer and contains more moisture than the other. However, fronts can form between any two contrasting air masses. When the vast sizes of air masses are considered, the zones (fronts) that separate them are relatively narrow and are shown as lines on weather maps.

Generally, the air mass located on one side of a front moves faster than the air mass on the other side. Thus, one air mass actively advances into the region occupied by another and collides with it. During World War I, Norwegian meteo-

Figure 9–1 This late spring storm was generated along a cold front. (Photo by Mike Hillingshead/ Photo Researchers, Inc.)

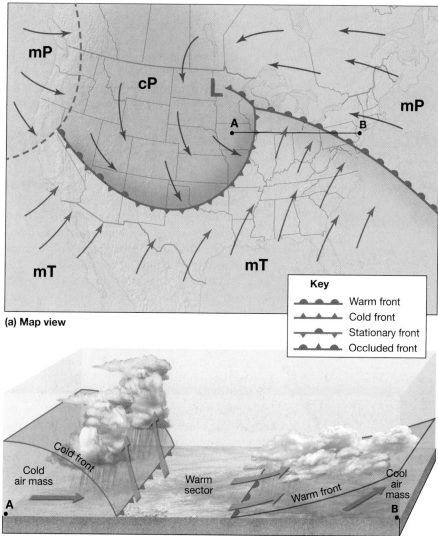

(a) Map view

Key
- ●●● Warm front
- ▲▲▲ Cold front
- ▲●▲● Stationary front
- ●▲●▲ Occluded front

Figure 9–2 Idealized structure of a midlatitude cyclone. (a) Map view showing fronts, air masses, and surface flow. (b) Three-dimensional view of the warm and cold fronts along a line from point A to point B.

(b) Three-dimentional view from points A to B

warm front is shown by a red line with red semicircles protruding into the area of cooler air. East of the Rockies, warm fronts are usually associated with maritime tropical (mT) air that enters the United States from the Gulf of Mexico and "glides" over cooler air positioned over land. The boundaries separating these air masses have very gradual slopes that average about 1:200 (height compared to horizontal distance). This means that if you traveled 200 kilometers (120 miles) ahead of the surface location of a warm front, the frontal surface would be 1 kilometer (0.6 mile) overhead.

As warm air ascends the retreating wedge of cold air, it expands and cools adiabatically. As a result, moisture in the ascending air condenses to generate clouds that may produce precipitation. The cloud sequence in Figure 9–3a typically precedes the approach of a warm front. The first sign of the approaching warm front is cirrus clouds that form 1000 kilometers (600 miles) or more ahead of the surface front. Another clue that a warm front is approaching is provided by aircraft contrails. On a clear day, when condensation trails persist for several hours, you can be fairly certain that comparatively warm, moist air is ascending overhead.

As the front nears, cirrus clouds grade into cirrostratus that gradually blend into denser sheets of altostratus. About 300 kilometers (180 miles) ahead of the front, thicker stratus and nimbostratus clouds appear and precipitation often commences.

Because warm fronts have relatively gentle slopes, the cloud deck that results from frontal lifting covers a large area and produces light-to-moderate precipitation for an extended duration (Figure 9–4). However, if the overriding air mass is relatively dry (low dew-point temperatures), there is minimal cloud development and no precipitation. During the hot summer months when moist conditionally unstable air is often forced aloft, towering cumulonimbus clouds and thunderstorms may occur (Figure 9–3b).

As you can see from Figure 9–3, the precipitation associated with a warm front occurs ahead of its surface position. Therefore, any precipitation that forms must fall through the cool layer below. During extended periods of light rainfall, enough of these raindrops may evaporate for saturation

rologists visualized these zones of air mass interactions as analogous to battle lines and tagged them "fronts," as in battlefronts. It is along these zones of "conflict" that midlatitude cyclones develop and produce much of the precipitation and severe weather in the belt of the westerlies.

As one air mass moves into the region occupied by another, minimal mixing occurs along the frontal surface. Instead, the air masses retain their identity as one is displaced upward over the other. No matter which air mass is advancing, it is always the warmer, less dense air that is forced aloft, whereas the cooler, denser air acts as a wedge on which lifting occurs. The term **overrunning** is applied to the process of warm air gliding up and over a cold air mass.

There are five basic types of fronts—*warm fronts, cold fronts, stationary fronts, occluded fronts,* and *drylines.*

Warm Fronts

When the surface position of a front moves so that warmer air invades territory formerly occupied by cooler air, it is called a **warm front** (Figure 9–3). On a weather map, a

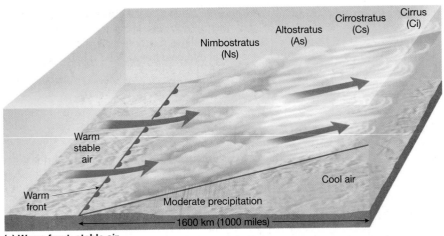

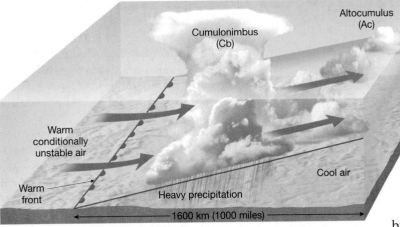

(a) Warm front, stable air

(b) Warm front, conditionally unstable air

Figure 9–3 Warm fronts. (a) Idealized clouds and weather associated with a warm front. During most of the year, warm fronts produce light-to-moderate precipitation over a wide area. (b) During the warm season, when conditionally unstable air is forced aloft, cumulonimbus clouds and thunderstorms often arise.

to occur, resulting in the development of a stratus cloud deck. These clouds occasionally grow rapidly downward, causing problems for pilots of small aircraft that require visual landings. One minute pilots may have adequate visibility and the next be in a cloud mass (frontal fog) that has the landing strip "socked in."

Occasionally during the winter, a relatively warm air mass is forced over a body of subfreezing air. This occurrence can create hazardous driving conditions. Raindrops become supercooled as they fall through the subfreezing air. Upon colliding with the road surface, these supercooled raindrops freeze to produce an icy layer called *freezing rain* or *glaze.*

With the passage of a warm front, temperatures gradually rise. As you would expect, the increase is most apparent when a large contrast exists between adjacent air masses. Moreover, a wind shift from the east or southeast to the south or southwest is generally noticeable. (The reason for this shift will be explained later.) The moisture content and stability of the encroaching warm air mass largely determine the time required for clear skies to return. During the summer, cumulus and occasionally cumulonimbus clouds are embedded in the warm unstable air mass that follows the front. These clouds may produce precipitation, which can be heavy but is usually scattered and short in duration. Table 9–1 shows the typical weather conditions that can be expected with the passage of a warm front in the Northern Hemisphere.

Figure 9–4 Rain associated with a warm front. (Photo by David Grossman/Alamy)

TABLE 9-1 Weather Typically Associated with a Warm Front (North America)

Weather Element	Before Passage	During Passage	After Passage
Temperature	Cool or cold	Rising	Warmer
Winds	East or southeast	Variable	South or southwest
Precipitation	Light-to-moderate rain, snow, or freezing rain in winter. Heavy rain possible in summer.	None or light rain	None, occasionally showers in summer
Clouds	Cirrus, cirrostratus, stratus, nimbostratus when air is stable. Cumulonimbus when air is conditionally unstable.	None, stratus, or fog	Clearing, cumulus, or cumulonimbus in summer
Pressure	Falling	Falling or steady	Falling then rising
Humidity	Moderate to high	Rising	High, particularly in summer

Cold Fronts

When cold air actively advances into a region occupied by warmer air, the zone of discontinuity is called a **cold front** (Figure 9–5). On a weather map, a cold front is shown by a blue line with blue triangles protruding into the area of warmer air. Air near the surface of a cold front advances more slowly than the air aloft because of friction. As a result, cold fronts steepen as they move. On the average, cold fronts are about twice as steep as warm fronts, having slopes of perhaps 1:100. In addition, cold fronts advance at speeds up to 80 kilometers (50 miles) per hour, about 50 percent faster than warm fronts. These two differences—steepness of slope and rate of movement—largely account for the more violent nature of cold-front weather compared to the weather generally accompanying a warm front.

As a cold front approaches, generally from the west or northwest, towering clouds can often be seen in the distance. Near the front, a dark band of ominous clouds foretells the ensuing weather. The forceful lifting of warm, moist air along a cold front is often rapid enough that the released latent heat increases the air's buoyancy sufficiently to render the air unstable. Heavy downpours and vigorous wind gusts associated with mature cumulonimbus clouds frequently result. Because a cold front produces roughly the same amount of lifting as a warm front, but over a shorter distance, the precipitation is generally more intense but of shorter duration (Figure 9–6). A marked temperature drop and wind shift from the southwest to the northwest usually accompany frontal passage. The reason for the wind shift will be explained later in this chapter.

The weather behind a cold front is dominated by subsiding air within a continental polar (cP) air mass. Thus, the drop in temperature is usually accompanied by clearing that begins soon after the front passes. Although subsidence causes adiabatic heating aloft, the effect on surface temperatures is minor. In winter the long, cloudless nights that follow the passage of a cold front allow for abundant radiation cooling that produces frigid surface temperatures. By contrast, the passage of a cold front during a summer heat wave produces a welcome change in conditions as hot, hazy, and sometimes polluted mT air is replaced by crisp, clear cP air.

When the air behind a cold front moves over a relatively warm surface, radiation emitted from Earth can heat the air enough to produce shallow convection. This in turn may generate low cumulus or stratocumulus clouds behind the front. However, subsidence aloft keeps these air masses relatively stable. Any clouds that form will not develop great vertical thickness and will seldom produce precipitation. One exception is the lake-effect snow discussed in Chapter 8, in which the cold air behind a front acquires heat and moisture as it traverses a comparatively warm body of water.

In North America, cold fronts form most commonly when a continental polar air mass clashes with maritime tropical air. However, wintertime cold fronts can form when even cold-

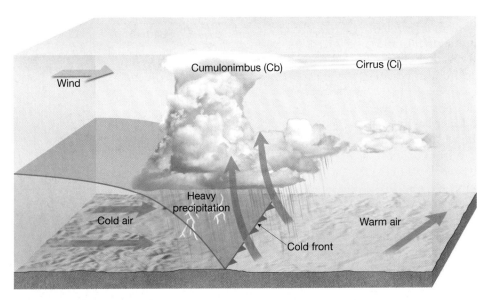

Figure 9–5 Fast-moving cold front and cumulonimbus clouds. Thunderstorms often occur if the warm air is unstable.

Figure 9–6 Cumulonimbus clouds along a cold front produced hail and heavy rain at this ballpark in Wichita, Kansas. Dozens of cars in the parking lot were damaged. (AP Photo/*The Wichita Eagle*, Fernando Salazar)

er, dryer continental arctic (cA) air invades a continental polar or maritime polar air mass. Over land, arctic cold fronts tend to produce very light snowfalls because the continental polar air masses they invade are quite dry. By contrast, the passage of an arctic cold front over a relatively warm water body may yield heavy snowfall and gusty winds. Table 9–2 shows the typical weather conditions associated with the passage of a cold front in North America.

The eastern seaboard of North America is sometimes affected by another type of cold front, called a **backdoor cold front.** Most cold fronts arrive from the west or northwest, whereas backdoor cold fronts come in from the east or northeast, hence their name. Driven by clockwise circulation from a strong high pressure center over northeastern Canada, colder, denser maritime polar (mP) air from the North Atlantic displaces warmer, lighter air over the continent, as shown in Figure 9–7. Backdoor fronts are primarily springtime events that tend to bring cold temperatures, low clouds, and drizzle, although thunder-

TABLE 9–2 Weather Typically Associated with a Cold Front (North America)

Weather Element	Before Passage	During Passage	After Passage
Temperature	Warm	Sharp drop	Colder
Winds	South or southwest	Variable and gusty	West or northwest
Precipitation	None or showers	Thunderstorms in summer, rain or snow in winter.	Clearing
Clouds	None, cumulus, or cumulonimbus	Cumulonimbus	None or cumulus in summer
Pressure	Falling then rising	Rising	Rising
Humidity	High, particularly in summer	Dropping	Low, particularly in winter

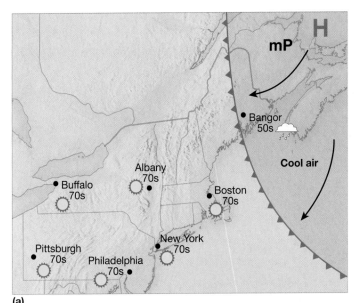

(a)

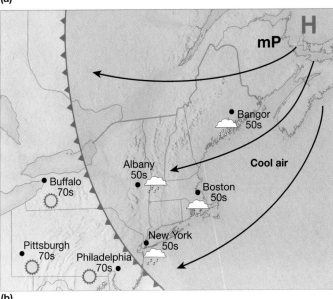

(b)

Figure 9–7 Weather associated with a backdoor cold front in the Northeast. In early spring a warm sunny day can become cool and damp as cool moist maritime polar air moves inland from the North Atlantic.

storms occur occasionally. Backdoor cold fronts are less frequent in summer, but when they occur, the cool air can provide welcome relief from midsummer heat waves in the northeastern United States.

Stationary Fronts

Occasionally, airflow on both sides of a front is neither toward the cold air mass nor toward the warm air mass. Rather, it is almost parallel to the line of the front. Consequently, the surface position of the front does not move, or it moves sluggishly. This condition is called a **stationary front.** On a weather map, a stationary front is shown with blue triangles pointing into the warm air and red semicircles pointing into

the cold air (see Figure 9–2a). Because some overrunning usually occurs along stationary fronts, gentle to moderate precipitation is likely. Stationary fronts may remain over an area for several days, in which case flooding is possible. When stationary fronts begin to move, they become cold or warm fronts, depending on which air mass advances.

Occluded Fronts

The fourth major type of front is the **occluded front,** in which a rapidly moving cold front overtakes a warm front, as shown in Figure 9–8a, b. As the cold air wedges the warm front upward, a new front forms between the advancing cold air and the air over which the warm front is gliding, a process known as **occlusion** (Figure 9–8b).

The weather of an occluded front is highly variable. Most precipitation is associated with the warm air that is being forced aloft (Figure 9–8c). When conditions are suitable, however, the newly formed front is capable of initiating precipitation of its own.

There are two types of occlusion—cold-type occluded fronts and warm-type occluded fronts. In the occluded front shown in Figure 9–9a, the air that had been behind the cold front is colder than the cool air it is overtaking. This is the most common type of occluded front east of the Rockies and is called a *cold-type occluded front.* Cold-type occluded fronts frequently produce thunderstorms and thus resemble cold fronts in the type of weather generated.

It is also possible for the air behind an advancing cold front to be warmer than the cold air it is overtaking. These *warm-type occluded fronts* (Figure 9–9b) frequently occur along the Pacific Coast, where milder maritime polar air invades more frigid polar air that had its origin over the continent.

Because of the complex nature of occluded fronts, they are often drawn on weather maps as either warm or cold fronts, depending on what kind of air is the aggressor. Sometimes, however, an occluded front is drawn as a purple line with alternating purple triangles and semicircles pointing in the direction of movement.

Drylines

Classifying fronts based solely on the temperature differences across the frontal boundary can be misleading. Humidity also influences the density of air. All other factors being equal, humid air is less dense than dry air. In the summer it is not unusual for a southeastward-moving air mass that originated over the northern Great Plains to displace warm, humid air over the lower Mississippi Valley. The front that develops is usually labeled a cold front, although the advancing air may not be any colder than the air it displaces. Simply, the drier air is denser and forcefully lifts the moist air in its path, much like a cold front. The passage of this type of frontal boundary is noticeable as a sharp drop in humidity, without an appreciable decrease in temperature.

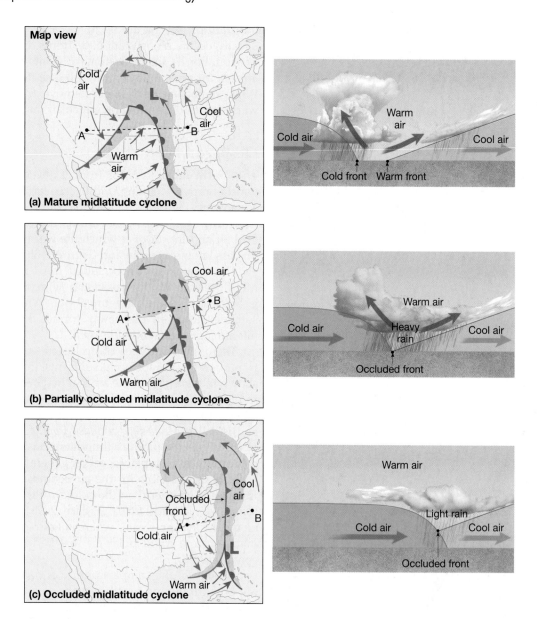

Figure 9–8 Stages in the formation of an occluded front and its relationship to an evolving midlatitude cyclone. After the warm air has been forced aloft, the system begins to dissipate. The shaded areas indicate regions where precipitation is most likely to occur.

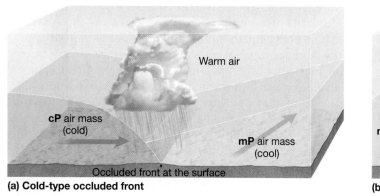

(a) Cold-type occluded front

(b) Warm-type occluded front

Figure 9–9 Occluded fronts of (a) the cold type and (b) the warm type.

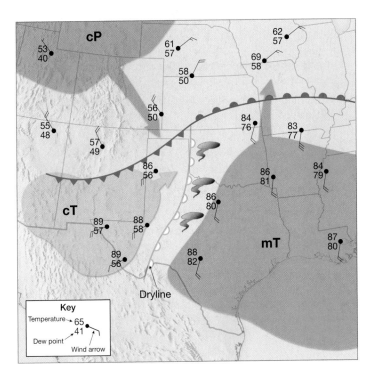

Figure 9–10 Dryline over Texas and Oklahoma generating thunderstorms and tornadic weather. Notice that the dry (low dew point) cT air is pushing eastward and displacing warm, moist mT air. The result is weather that resembles a rapidly moving cold front.

A related type of boundary, called a **dryline**, develops most often over the southern Great Plains. This occurs when dry, continental tropical (cT) air originating in the Southwest meets moist, maritime tropical (mT) air from the Gulf of Mexico. Drylines are spring and summer phenomena, that most often generate a band of severe thunderstorms along a line extending from Texas to Nebraska that moves eastward across the Great Plains. A dryline is easily identified by comparing the dew-point temperatures of the cT air west of the boundary with the dew points of the mT air mass to the east (Figure 9–10).

Concept Check 9.1

1 Compare the weather of a typical warm front with that of a typical cold front.

2 List two reasons why cold-front weather is usually more severe than warm-front weather.

3 Explain the basis for the following weather proverb:

Rain long foretold, long last;

Short notice, soon past.

4 What is a backdoor cold front?

5 How does a stationary front produce precipitation when its position does not change or when it changes very slowly?

6 In what way are drylines different from warm and cold fronts?

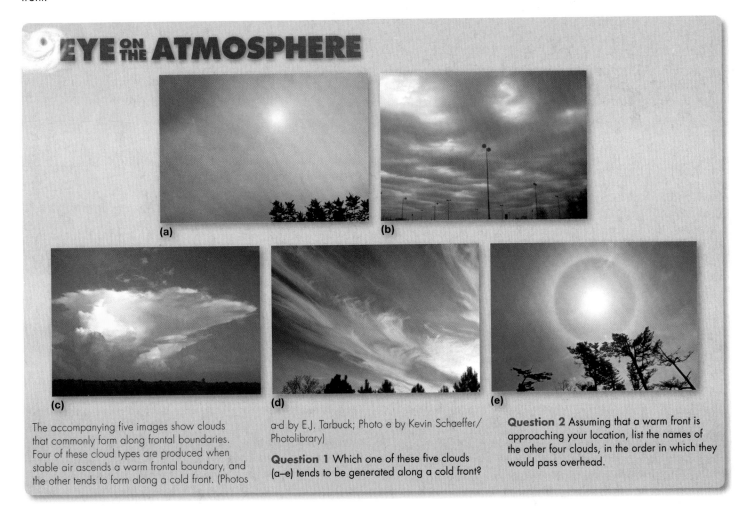

EYE ON THE ATMOSPHERE

(a)

(b)

(c)

(d)

(e)

The accompanying five images show clouds that commonly form along frontal boundaries. Four of these cloud types are produced when stable air ascends a warm frontal boundary, and the other tends to form along a cold front. (Photos

a–d by E.J. Tarbuck; Photo e by Kevin Schaeffer/ Photolibrary)

Question 1 Which one of these five clouds (a–e) tends to be generated along a cold front?

Question 2 Assuming that a warm front is approaching your location, list the names of the other four clouds, in the order in which they would pass overhead.

Midlatitude Cyclones and the Polar-Front Theory

Midlatitude cyclones are low-pressure systems with diameters that often exceed 1000 kilometers (600 miles) and travel from west to east across the middle latitudes in both hemispheres (see Figure 9–2). Lasting from a few days to more than a week, a midlatitude cyclone in the Northern Hemisphere has a counterclockwise circulation pattern with airflow directed inward toward its center. Most midlatitude cyclones have a cold front and a warm front extending from the central area of low pressure. Surface convergence and ascending air initiate cloud development that frequently produces precipitation.

As early as the 1800s, cyclones were known as the bearers of precipitation and severe weather. Thus, the barometer was established as the primary tool in "forecasting" day-to-day weather changes. However, this early method of weather prediction largely ignored the role of air mass interactions in the formation of these weather systems. Consequently, it was impossible to determine the conditions under which cyclone development was favorable.

The first comprehensive model of cyclone development and intensification was formulated by a group of Norwegian scientists during World War I. German influence restricted the Norwegians from receiving international communications—including critical weather reports pertaining to conditions in the Atlantic Ocean. To counter this deficiency, a closely spaced network of weather stations was established throughout Norway. Using this network, these Norwegian-trained meteorologists made great advances in broadening our understanding of the weather and, in particular, weather associated with a midlatitude cyclone. Included in this group were Vilhelm Bjerknes (pronounced *Bee-YURK-ness*), his son Jacob Bjerknes, Jacob's fellow student Halvor Solberg, and Swedish meteorologist Tor Bergeron (see Box 5–2). In 1921 the work of these scientists resulted in a publication outlining a compelling model of how midlatitude cyclones progress through stages of birth, growth, and decay. These insights, which marked a turning point in atmospheric science, became known as the **polar-front theory**—also referred to as the **Norwegian cyclone model.** Even without the benefit of upper-air charts, these skilled meteorologists presented a model that remains remarkably applicable in modern meteorology.

In the Norwegian cyclone model, midlatitude cyclones develop in conjunction with polar fronts. Recall that polar fronts separate cold polar air from warm subtropical air (see Chapter 7). During cool months polar fronts are generally well defined and form a nearly continuous band around Earth recognizable on upper-air charts. At the surface, this frontal zone is often broken into distinct segments separated by regions of gradual temperature changes. It is along these frontal zones that cold, equatorward-moving air collides with warm, poleward-moving air to produce most midlatitude cyclones.

Concept Check 9.2

1 Briefly describe the characteristics of a midlatitude cyclone.

2 According to the Norwegian cyclone model, where do midlatitude cyclones form?

Students Sometimes Ask...

What is an extratropical cyclone?

Meaning "outside the tropics," *extratropical* is simply another name for a midlatitude cyclone. The term *cyclone* refers to the circulation around any low-pressure center, regardless of its size or intensity. Hence, hurricanes and midlatitude cyclones are two types of cyclones. Whereas "extratropical cyclone" is another name for a midlatitude cyclone, the name "tropical cyclone" is often used to describe a hurricane.

Life Cycle of a Midlatitude Cyclone

GE●De
ATMOSPHERE
Basic Weather Patterns
▶ Introducing Middle-Latitude Cyclones

According to the Norwegian model, cyclones form along fronts and proceed through a generally predictable life cycle. This cycle can last from a few days to more than a week, depending on atmospheric conditions. One study found that more than 200 midlatitude cyclones form in the Northern Hemisphere annually.

Figure 9–11 shows six stages in the life of a typical midlatitude cyclone. The first stage in this process is called **cyclogenesis,** meaning *cyclone formation.*

Formation: The Clash of Two Air Masses

A midlatitude cyclone is born when two air masses of different densities (temperatures) are moving roughly parallel to a front but in opposite directions. In the classic polar-front model, this would be continental polar air associated with the *polar easterlies* on the north side of the front and maritime tropical air driven by the *westerlies* on the south side of the front.

Under suitable conditions the frontal surface that separates these two contrasting air masses becomes wave shaped and is usually several hundred kilometers long (Figure 9–11b). These waves are similar to, but much larger than, the waves that form on water. Some waves tend to dampen, or die out, whereas others grow in amplitude. As these storms intensify, or "deepen," the waves change shape, much like a gentle ocean swell does as it moves into shallow water and becomes a tall, breaking wave (Figure 9–11c).

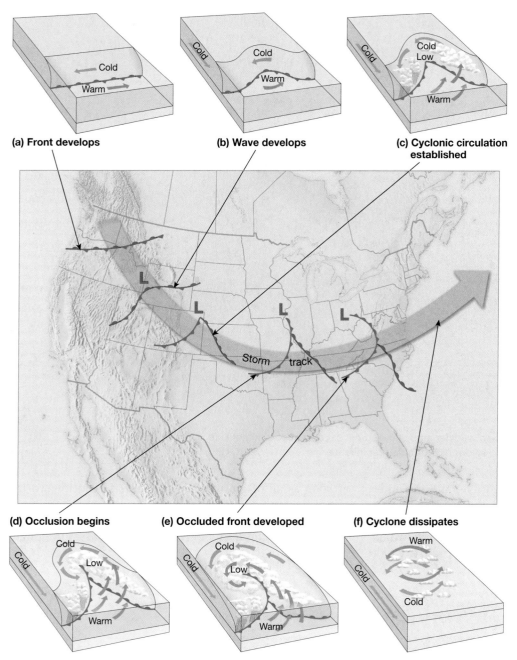

(a) Front develops

(b) Wave develops

(c) Cyclonic circulation established

(d) Occlusion begins

(e) Occluded front developed

(f) Cyclone dissipates

Figure 9–11 Stages in the life cycle of a midlatitude cyclone, as proposed by J. Bjerknes.

that the warm air is invading a region formerly occupied by cold air. Therefore, this must be a warm front. Similar reasoning indicates that to the left (west) of the wave front, cold air from the northwest is displacing the air of the warm sector and generating a cold front.

Mature Stage of a Midlatitude Cyclone

During the *mature stage* of a midlatitude cyclone the pressure surrounding the low continues to drop, causing winds to strengthen and frontal weather to develop. The weather can be quite varied, depending on the season and one's location in relation to the cyclonic storm. In winter, for example, a place directly north of a strong midlatitude cyclone might experience heavy snowfall and blizzard-like conditions, whereas the weather ahead of the warm front may be freezing rain.

Occlusion: The Beginning of the End

Usually, a cold front advances more rapidly than the warm front. As it moves, a cold front begins to overtake (lift) the warm front, as shown in Figure 9–11d, e. This process forms an *occluded front*, which grows in length as the warm sector is displaced aloft. As occlusion begins, the storm often intensifies. However, as more of the warm air is forced aloft, the surface pressure gradient weakens, as does the storm itself. Within one or two days the entire warm sector is forced aloft, and cold air surrounds the cyclone at the surface (Figure 9–11f). Thus, the horizontal temperature (density) difference that existed between the two contracting air masses is largely eliminated. At this point the cyclone has exhausted its source of energy. Friction slows the surface flow, and the once highly organized inward-directed, counterclockwise flow ceases to exist.

A simple analogy may help you visualize what happens to the cold and warm air masses in the preceding discussion. Imagine a large water tank that has a vertical divider separating it into two equal parts. Half of the tank

Development of Cyclonic Flow

As a wave evolves, warm air advances poleward to form a warm front, while cold air moves equatorward to form a cold front. This change in the direction of the surface flow is accompanied by a readjustment in the pressure pattern and results in somewhat circular isobars, with the lowest pressure located at the crest of the wave. The resulting flow is an inward-directed, counterclockwise circulation that can be seen clearly on the weather map shown in Figure 9–12. Once the cyclonic circulation develops, general convergence results in forceful lifting, especially where warm air is overrunning colder air. Figure 9–12 illustrates that the air in the warm sector (over the southern states) is flowing northeastward, toward cooler air that is moving toward the northwest. Because the warm air is moving perpendicular to the front, we can conclude

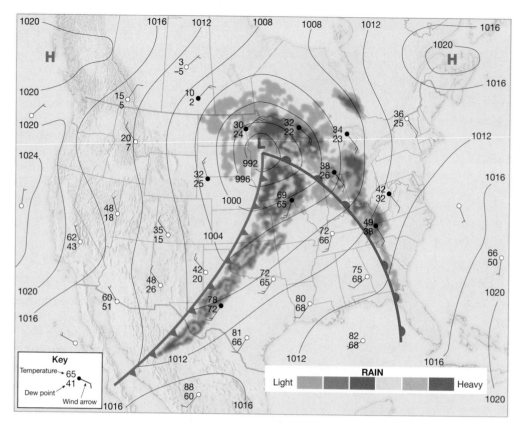

Figure 9–12 Simplified weather map showing the circulation of a midlatitude cyclone. The colored areas indicate regions of probable precipitation.

is filled with hot water containing red dye, and the other half is filled with blue-colored icy cold water. Now imagine what happens when the divider is removed. The cold, dense water will flow under the less dense warm water, displacing it upward. This rush of water will come to a halt as soon as all of the warm water is displaced toward the top of the tank. In a similar manner, a midlatitude cyclone dies when all of the warm air is displaced aloft and the horizontal discontinuity between the air masses no longer exists.

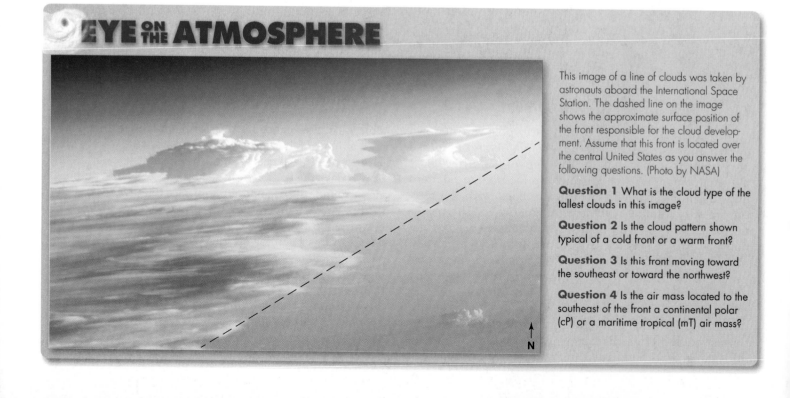

EYE ON THE ATMOSPHERE

This image of a line of clouds was taken by astronauts aboard the International Space Station. The dashed line on the image shows the approximate surface position of the front responsible for the cloud development. Assume that this front is located over the central United States as you answer the following questions. (Photo by NASA)

Question 1 What is the cloud type of the tallest clouds in this image?

Question 2 Is the cloud pattern shown typical of a cold front or a warm front?

Question 3 Is this front moving toward the southeast or toward the northwest?

Question 4 Is the air mass located to the southeast of the front a continental polar (cP) or a maritime tropical (mT) air mass?

Figure 9–13 illustrates a mature midlatitude cyclone; note the distribution of clouds and thus the regions of possible precipitation. Compare this map to the satellite image of a cyclone in Figure 9–26 (page 264). It is easy to see why we often refer to the cloud pattern of a cyclone as having a "comma" shape.

Guided by the westerlies aloft, cyclones generally move eastward across the United States. Therefore, we can expect the first signs of a cyclone's arrival to appear in the western sky. Upon reaching the Mississippi Valley, however, cyclones often begin a more northeasterly trajectory and occasionally move directly northward. Typically, a midlatitude cyclone requires two or more days to pass completely over a region. During that span, abrupt changes in atmospheric conditions may occur, particularly in late winter and spring, when the greatest temperature contrasts occur across the midlatitudes.

Using Figure 9–13 as a guide, let us examine these weather producers and the changes you might expect if a midlatitude cyclone passed over your area. To facilitate our discussion, profiles of the storm are provided above and below the map. They correspond to lines *A–E* and *F–G* on the map. (Remember that these storms move from west to east; therefore, the right side of the cyclone shown in Figure 9–13 will be the first to pass over the particular region.)

First, imagine the change in weather as you move from right to left along profile *A–E* (see bottom of Figure 9–13). At point *A* the sighting of cirrus clouds is the first sign of the approaching cyclone. These high clouds can precede the surface front by 1000 kilometers (600 miles) or more and are normally accompanied by falling pressure. As the warm front advances, lowering and thickening of the cloud deck occurs. Within 12 to 24 hours after the first sighting of cirrus clouds, light precipitation usually commences (point *B*). As the front nears, the rate of precipitation increases, the

---

Idealized Weather of a Midlatitude Cyclone

Basic Weather Patterns
▶ Introducing Middle-Latitude Cyclones

The Norwegian model is a valuable tool for interpreting midlatitude weather patterns, so keeping it in mind may help you understand, and possibly anticipate, changes in daily weather.

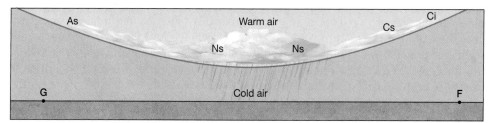

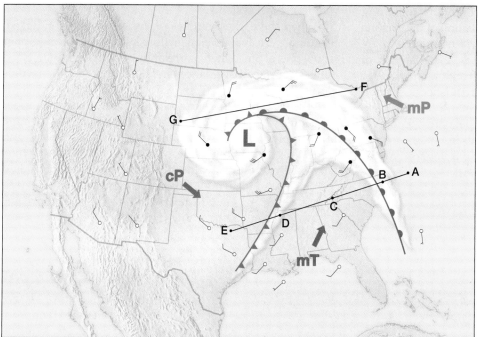

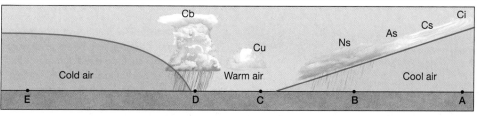

Figure 9–13 Cloud patterns typically associated with a mature midlatitude cyclone. The middle section is a map view. Note the cross-section lines (*F–G*, *A–E*). Above the map is a vertical cross section along line *F–G*. Below the map is a cross section along *A–E*. For cloud abbreviations, refer to Figures 9–3 and 9–5.

temperature rises, and winds begin to change from an easterly to a southwesterly flow.

With the passage of the warm front, the area behind (west of) the front, called the *warm sector,* is under the influence of a maritime tropical air mass (point C). Depending on the season, the region affected by this part of the cyclone experiences warm to hot temperatures, southwesterly winds, fairly high humidities, and clear to partly cloudy skies containing cumulus or cumulus congestus clouds.

The warm conditions associated with the warm sector pass quickly and are replaced by gusty winds and precipitation generated along the cold front. The approach of a rapidly advancing cold front is marked by a wall of rolling black clouds (point D). Severe weather accompanied by heavy precipitation, and occasionally hail or a tornado, can be expected.

The passage of the cold front is easily detected by a dramatic shift in wind direction. The warm flow from the south or southwest is replaced by cold winds from the west to northwest, resulting in a pronounced decrease in temperature. Also, rising pressure hints of the subsiding cool, dry air behind the cold front. Once the front passes, the skies clear quickly as cooler, drier air invades the region (point E). A day or two of almost cloudless blue skies is often experienced, unless another cyclone is edging into the region.

A very different set of weather conditions prevails in the portion of the cyclone located north of the center of low pressure—profile F–G at the top of Figure 9–13. In this part of the storm temperatures remain cool. The first hints of the approaching low-pressure center are a continual drop in air pressure and increasingly overcast conditions that bring varying amounts of precipitation. This section of the cyclone most often generates snow during the winter months.

Once the process of occlusion begins, the character of the storm changes. Because occluded fronts tend to move more slowly than other fronts, the entire wishbone-shaped frontal structure shown in Figure 9–11 rotates counterclockwise. As a result, the occluded front appears to "bend over backward." This effect adds to the misery of the region influenced by the occluded front because it lingers over the area longer than the other fronts (Box 9–1).

Concept Check 9.4

1 Briefly describe the weather associated with the passage of a mature midlatitude cyclone when the center of low pressure is located about 200 to 300 kilometers north of your location.

2 If the midlatitude cyclone described in Question 1 took three days to pass your location, on which day would the temperatures be the warmest? On which day would the temperatures be the coldest?

3 What winter weather is expected with the passage of a mature midlatitude cyclone when the center of low pressure is located about 100 to 200 kilometers south of a city located near the Great Lakes?

Flow Aloft and Cyclone Formation

The polar-front model shows that cyclogenesis (cyclone formation) occurs where a frontal surface is distorted such that it takes on the shape of an ocean wave. Several surface factors are thought to produce this wave in a frontal zone. Topographic irregularities (such as mountains), temperature contrasts (as between sea and land), or ocean-current influences can disrupt the general zonal (west to east) flow sufficiently to produce a wave along a front. In addition, a strong jet stream in the flow aloft frequently precedes the formation of a surface cyclone. This fact strongly suggests that upper-level flow contributes to the formation of these rotating storm systems.

When the earliest studies of cyclones were conducted, few data were available on airflow in the middle and upper troposphere. Since then, correlations have been established between surface disturbances and the flow aloft. When winds aloft exhibit a relatively straight zonal flow, little cyclonic activity occurs at the surface. However, when the upper air begins to meander widely from north to south, forming high-amplitude waves of alternating troughs (lows) and ridges (highs), cyclonic activity intensifies. Moreover, when surface cyclones form, almost invariably they are centered below the jet stream core and downwind from an upper-level low (Figure 9–14).

Cyclonic and Anticyclonic Circulation

Before discussing how surface cyclones are generated and supported by the flow aloft, let us re-

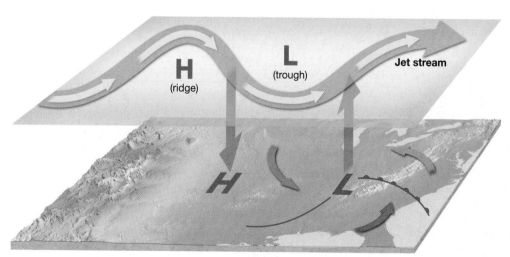

Figure 9–14 Relationship between the meandering flow in the jet stream aloft and cyclone development at the surface. Midlatitude cyclones tend to form downstream of an upper-level low (trough).

Box 9–1 Winds as a Forecasting Tool

> "Every wind has its weather."
> *Francis Bacon, English philosopher*
> *and scientist (1561–1626)*

People living in the middle latitudes know well that during the winter, north winds can chill a person to the bone (Figure 9–A). Conversely, a sudden change to a southerly flow can bring welcome relief from these frigid conditions. Especially observant people (or those who have completed a meteorology course!) might notice that when the wind direction switches from southeast to south, foul weather often follows. By contrast, a change in wind direction from the southwest to the northwest is usually accompanied by clearing conditions. Just how are the winds and the forthcoming weather related?

Modern weather forecasts require the processing capabilities of high-speed computers as well as the expertise of people with considerable professional training. Nevertheless, some reasonable insights into the impending weather can be gained through careful observation. The two most significant weather elements for this purpose are barometric pressure and wind direction. Recall that anticyclones (high-pressure cells) are associated with clear skies and that cyclones (low-pressure cells) frequently bring clouds and precipitation. Thus, by noting whether the barometer is rising, falling, or steady, we have some indication of the forthcoming weather. For example, rising pressure indicates the approach of a high-pressure system and generally fair weather.

The use of winds in weather forecasting is also straightforward. Because cyclones are the "villains" in the weather game, we are most concerned with the circulation around these storm centers. In particular, changes in wind direction that occur with the passage of warm and cold fronts are useful in predicting impending weather. Notice in Figure 9–13 that with the passage of both the warm and cold fronts, the wind arrows change positions in a clockwise manner. For example, with the passage of the cold front, the wind shifts from southwest to northwest. From nautical terminology, the word *veering* is applied to a clockwise wind shift such as this. Because clearing conditions normally occur with the passage of either front, veering winds are indicators that the weather will improve.

In contrast, the area in the northern portion of the cyclone will experience winds that shift in a counterclockwise direction, as can be seen in Figure 9–13. Winds that shift in this manner are said to be *backing*. With the approach of a midlatitude cyclone, backing winds indicate cool temperatures and continued foul weather.

A summary of the relationship among barometer readings, winds, and the impending weather is provided in Table 9–A. Although this information is applicable in a very general way to much of the United States, local influences must be taken into account. For example, a rising barometer and a change in wind direction from southwest to northwest is usually associated with the passage of a cold front and indicates that clearing conditions should follow. However, in the winter, residents of the southeast shore of one of the Great Lakes may not be so lucky. As cold, dry northwest winds cross large expanses of open water, they acquire heat and moisture from the relatively warm lake surface. By the time this air reaches the leeward shore, it is often humid and unstable enough to produce heavy lake-effect snow (see Chapter 8).

FIGURE 9–A Blizzard conditions made for hazardous driving along Highway 30 in western Iowa. (Photo by Mike Hollingshead/Photo Researchers, Inc.)

TABLE 9–A Wind, Barometric Pressure, and Impending Weather

Changes in Wind Direction	Barometric Pressure	Pressure Tendency	Impending Weather
Any direction	1023 mb and above (30.20 in.)	Steady or rising	Continued fair with no temperature change
SW to NW	1013 mb and below (29.92 in.)	Rising rapidly	Clearing within 12 to 24 hours and colder
S to SW	1013 mb and below (29.92 in.)	Rising slowly	Clearing within a few hours and fair for several days
SE to SW	1013 mb and below (29.92 in.)	Steady or slowly falling	Clearing and warmer, followed by possible precipitation
E to NE	1019 mb and above (30.10 in.)	Falling slowly	In summer, with light wind, rain may not fall for several days; in winter, rain within 24 hours
E to NE	1019 mb and above (30.10 in.)	Falling rapidly	In summer, rain probable within 12 to 24 hours; in winter, rain or snow with strong winds likely
SE to NE	1013 mb and below (29.92 in.)	Falling slowly	Rain will continue for 1 to 2 days
SE to NE	1013 mb and below (29.92 in.)	Falling rapidly	Stormy conditions followed within 36 hours by clearing and, in winter, colder temperatures

Source: Adapted from the National Weather Service.

view the nature of cyclonic and anticyclonic winds. Recall that airflow around a surface low is inward and leads to mass convergence (coming together). Because the accumulation of air is accompanied by a corresponding increase in surface pressure, we might expect a surface low-pressure center to "fill" rapidly and be eliminated, just as the vacuum in a coffee can is quickly equalized when opened. However, cyclones often exist for a week or longer. For this to happen, surface convergence must be offset by outflow aloft (Figure 9–15). As long as divergence (spreading out) aloft is equal to or greater than the surface inflow, the low pressure can be sustained.

Because cyclones are bearers of stormy weather, they have received far more attention than their counterparts, anticyclones. Yet the close relationship between them makes it difficult to totally separate a discussion of these two pressure systems. The surface air that feeds a cyclone, for example, generally originates as surface air flowing out of an anticyclone (Figure 9–15). Consequently, cyclones and anticyclones are typically adjacent to one another. Like a cyclone, an anticyclone depends on the flow aloft to maintain its circulation. In an anticyclone, divergence at the surface is balanced by convergence aloft and general subsidence of the air column (Figure 9–15).

Divergence and Convergence Aloft

Because divergence aloft is essential to cyclogenesis, a basic understanding of its role is important. Divergence aloft does not involve the outward flow in all directions, as in a surface anticyclone. Instead, the winds aloft generally flow from west to east along sweeping curves. How does zonal flow aloft cause upper-level divergence?

One mechanism responsible for divergence aloft is a phenomenon known as *speed divergence*. Wind speeds can change dramatically in the vicinity of the jet stream. On entering a zone of high wind speed, air accelerates and

stretches out (divergence). In contrast, when air enters a zone of slower wind speed, an air pileup (convergence) results. Analogous situations occur every day on a toll highway. When exiting a toll booth and entering the zone of maximum speed, we find automobiles diverging (increasing the number of car lengths between them). As automobiles slow to pay the toll, they experience convergence (coming together).

In addition to speed divergence, other factors contribute to divergence (or convergence) aloft. These include *directional divergence* and *directional convergence*, which result from changes in wind direction. For example, directional convergence (also called *confluence*) is the result of air being funneled into a restricted area. On an upper-air chart, convergence occurs in regions where height contours move progressively closer together. Returning to our interstate highway analogy, convergence occurs where a crowded three-lane highway is reduced to two lanes due to construction. Divergence, the opposite phenomenon, is analogous to the location where the two lanes change back to three.

The combined effect of these and other factors is that an area of upper-air divergence and surface cyclonic circulation generally develop downstream from an upper-level trough, as illustrated in Figure 9–15. Consequently, in the United States surface cyclones generally form downstream (east) of an upper-level trough. As long as divergence aloft exceeds convergence at ground level, surface pressures will fall, and the cyclonic storm will intensify.

Conversely, the zone in the jet stream that experiences convergence and anticyclonic rotation is located downstream from a ridge (Figure 9–15). The accumulation of air in this region of the jet stream leads to subsidence and increased surface pressure and, hence, is a favorable site for the development of a surface anticyclone.

Because of the significant role that the upper-level flow has on cyclogenesis, it should be evident that any attempt at weather prediction must account for the airflow aloft. This is why TV weather reporters frequently illustrate the flow within the jet stream.

In summary, the jet stream contributes to the formation and intensification of surface low- and high-pressure systems. Changes in wind speeds and/or directions cause air either to pile up (convergence) or spread out (divergence). Upper-level convergence is favored downstream (east) from a ridge, whereas divergence occurs downstream from an upper-level trough. At the surface,

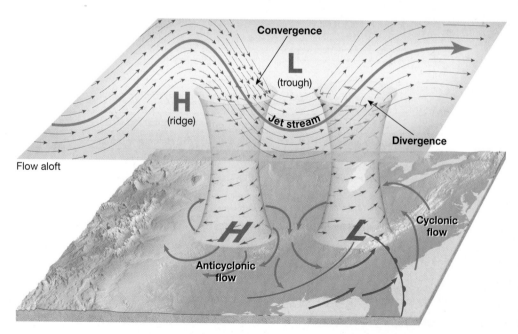

Figure 9–15 Idealized depiction of the support that divergence and convergence aloft provide to cyclonic and anticyclonic circulation at the surface.

below regions of upper-level convergence, are areas of high pressure (anticyclones), whereas upper-level divergence supports the formation and development of surface cyclonic systems (lows).

Concept Check 9.5

1 Briefly explain how the flow aloft maintains cyclones at the surface.

2 What is speed divergence? Speed convergence?

3 Given an upper-air chart, where do forecasters usually look to find favorable sites for cyclogenesis? Where do anticyclones usually form in relation to the wavy flow aloft?

Where Do Midlatitude Cyclones Form?

The development of midlatitude cyclones does not occur uniformly over Earth's surface but tends to favor certain locations, such as the leeward (eastern) sides of mountains and along coastal areas. In general, midlatitude cyclones form in areas where significant temperature contrasts occur in the lower troposphere. Figure 9–16 shows the areas of greatest cyclone development over North America and adjacent oceans. Notice that a prime site for cyclone formation is along the east side of the Rocky Mountains. Other important sites are in the North Pacific and the Gulf of Mexico.

Midlatitude cyclones that influence western North America originate over the North Pacific. Many of these systems move northeastward toward the Gulf of Alaska. However, during the winter months these storms travel farther southward and often reach the West Coast of the contiguous 48 states, occasionally traveling as far south as southern California. These low-pressure systems provide the winter rainy season that affects much of the West Coast.

Most Pacific storms diminish in strength as they cross the Rockies, but they often redevelop on the eastern side of these mountains. A common area for redevelopment is Colorado, but other sites exist as far south as Texas and as far north as Alberta, Canada. Cyclones that redevelop over the Great Plains generally migrate eastward until they reach the central United States, where they follow a northeastward or even northward trajectory. Many of these cyclones traverse the Great Lakes region, making it one of the most storm-ridden portions of the country. In addition, some storms develop off the coast of the Carolinas and tend to move northward with the warm Gulf Stream, bringing stormy conditions to the entire Northeast.

Patterns of Movement

Once formed, most midlatitude cyclones tend to travel in an easterly direction across North America and then follow a more northeasterly path into the North Atlantic (Figure 9–17). However, numerous exceptions occur.

Panhandle Hook One well-known storm pattern anomaly is known as a *Panhandle hook.* The "hook" describes the curved path these storms follow (Figure 9–17). Developing in southern Colorado near the Texas and Oklahoma panhandles, these cyclones first travel toward the southeast and then bend and travel sharply northward across Wisconsin and into Canada.

Alberta Clipper An *Alberta Clipper* is a cold, windy cyclonic storm that forms on the eastern side of the Canadian Rockies in the province of Alberta (Figure 9–17). Noted for their speed, they are called "clippers" because in Colonial times the fastest vehicles were small ships by the same name. Alberta Clippers dive southeastward into Montana or the Dakotas and then track across the Great Lakes, bringing dramatically lower temperatures. Winds associated with clippers frequently exceed 50 kilometers (30 miles) per hour. Be-

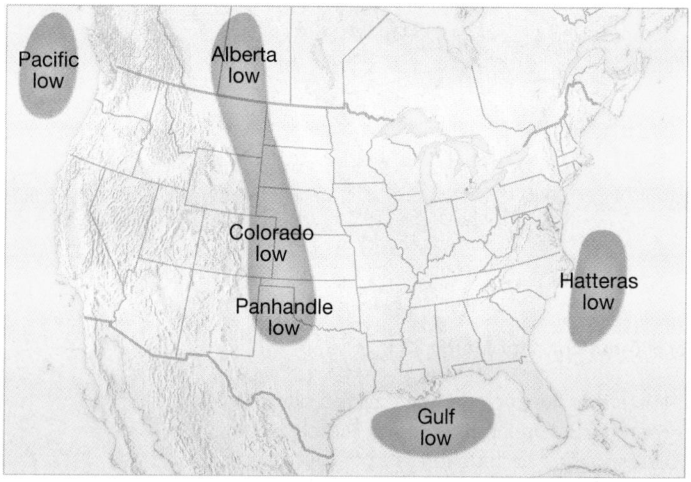

Figure 9–16 Major sites of cyclone formation.

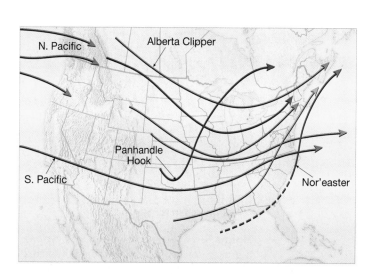

Figure 9–17 Typical paths of cyclonic storms that affect the lower 48 states.

cause clippers move rapidly and remain great distances from the mild waters of the Gulf of Mexico, they tend to be moisture deprived. As a result, they do not drop large amounts of snow. Instead, they may leave a few inches in a narrow band from the Dakotas to New York over a span as short as two days. However, because these winter storms are relatively frequent occurrences, they make a significant contribution to the total winter snowfall in the northern tier of states.

Nor'easter From the Mid-Atlantic coast to New England, the classic storm is called a *nor'easter* (see Figure 9–17). These storms are called nor'easters because the winds preceding them in coastal areas are from the northeast. They are most frequent and violent between September and April, when cold air pouring south from Canada meets relatively warm humid air from the Atlantic. Once formed, a nor'easter follows the coast, often bringing rain, sleet, and heavy snowfall to the Northeast. Because the circulation produces strong onshore winds, these storms can cause considerable coastal erosion, flooding, and property damage. The popular book and movie *The Perfect Storm* was based on a true story of a fishing boat that was caught in an intense nor'easter in October 1991. A brief case study of a classic nor'easter is found in Chapter 8, page 232.

Flow Aloft and Cyclone Migration

Recall that the wavy flow aloft is important to the development and evolution of a surface cyclonic storm. In addition, the flow in the middle and upper troposphere strongly influences the rate at which these pressure systems advance and the direction they follow. Generally, surface cyclones move in the same direction as winds aloft at the 500-millibar level, but at about one-quarter to one-half the speed. Normally these systems travel at 25 to 50 kilometers (15 to 30 miles) per hour, advancing roughly 600 to 1200 kilometers (400 to 800 miles) daily. The faster speeds occur during the coldest months, when temperature gradients are greatest.

One of the most difficult tasks in weather forecasting is predicting the paths of cyclonic storms. We have seen that the flow aloft tends to steer developing pressure systems. Let us examine an example of this steering effect by seeing how changes in the upper-level flow correspond to changes in the path taken by a cyclone.

Figure 9–18a illustrates the changing position of a midlatitude cyclone over a four-day period. Notice in Figure 9–18b that on March 21, the 500-millibar contours are relatively flat. Also notice that for the following two days, the cyclone moves in a southeasterly direction. By March 23 the 500-millibar contours make a sharp bend northward on

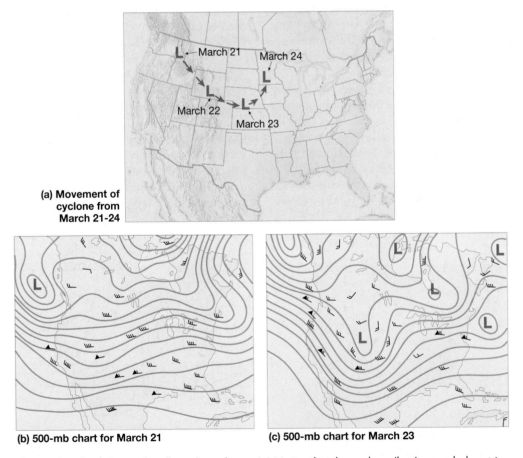

(a) Movement of cyclone from March 21-24

(b) 500-mb chart for March 21

(c) 500-mb chart for March 23

Figure 9–18 Steering of midlatitude cyclones. (a) Notice that the cyclone (low) moved almost in a straight southeastward direction on March 21 and March 22. On the morning of March 23, it abruptly turned northward. This change in direction corresponded to the change from (b) rather straight contours on the upper-air chart for March 21 to (c) curved contours on the chart for March 23.

EYE ON THE ATMOSPHERE

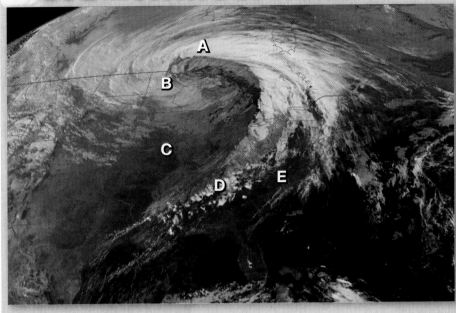

This midlatitude cyclone swept across the central United States and produced strong wind gusts (up to 78 miles per hour), rain, hail, and snow, and it spawned 61 tornadoes on October 26, 2010. The cyclone set a record for the lowest pressure over land (not associated with a hurricane) ever recorded in the continental United States: 28.21 inches of mercury. This pressure corresponds to that measured in a typical category 3 hurricane. Use the letters A–E, which represent various parts of the storm, to answer the following. (Courtesy of NASA)

Question 1 What part of the storm is dominated by a maritime tropical (mT) air mass?

Question 2 Where is the area dominated by a continental polar (cP) air mass located?

Question 3 Where is the location of lowest pressure?

Question 4 What area of the storm experienced the heaviest snowfall?

Question 5 Where did the strongest thunderstorms occur when this image was produced?

---

the eastern side of a trough situated over Wyoming (Figure 9–18c). Likewise, the next day the path of the cyclone makes a similar northward migration.

Although this is an oversimplified example, it illustrates the "steering" effect of upper-level flow. Our example examines the influence of upper airflow on cyclonic movement after the fact. To make useful predictions of cyclone movements, accurate appraisals of changes in the westerly flow aloft are required. For this reason, predicting the behavior of the wavy flow in the middle and upper troposphere is an important part of modern weather forecasting.

Concept Check 9.6

1 List four locations where midlatitude cyclones that affect North America tend to form.

2 Where do the midlatitude cyclones that affect the Pacific Coast of the United States originate?

3 How did the *Alberta Clipper* get its name?

4 What portion of the United States is most affected by *nor'easters*?

5 Describe the motion of a midlatitude cyclone in relation to the flow at the 500-millibar level.

Anticyclonic Weather and Atmospheric Blocking

Due to the gradual subsidence within them, anticyclones generally produce clear skies and calm conditions. Because

these high-pressure systems are not associated with stormy weather, both their development and movement have not been studied as extensively as those of midlatitude cyclones. However, anticyclones do not always bring desirable weather. Large anticyclones often develop over the Arctic during the winter. These cold high-pressure centers are known to migrate as far south as the Gulf Coast, where they can impact the weather over as much as two-thirds of the United States (Figure 9–19). This dense frigid air often brings record-breaking cold temperatures.

Occasionally, large anticyclones persist over a region for several days or even weeks. Once in place, these stagnant anticyclones block or redirect the migration of midlatitude cyclones. Thus, they are sometimes called *blocking highs*. During these events, one section of the nation is kept dry for a week or more while another region remains continually under the influence of cyclonic storms. Such a situation prevailed during the summer of 1993, when a strong high-pressure system became anchored over the southeastern United States and caused migrating storms to stall over the Midwest. The result was severe drought in the southeast and the most devastating flooding on record for the central and upper Mississippi Valley (see *Severe and Hazardous Weather*, page 260).

Low-pressure systems can also generate a blocking pattern. These lows, called *cut-off low pressure systems*, are literally cut off from the west-to-east flow in the jet stream (Figure 9–20). Without a connection to the prevailing flow aloft, these lows remain over the same area for days, often producing dreary weather and large quantities of precipitation.

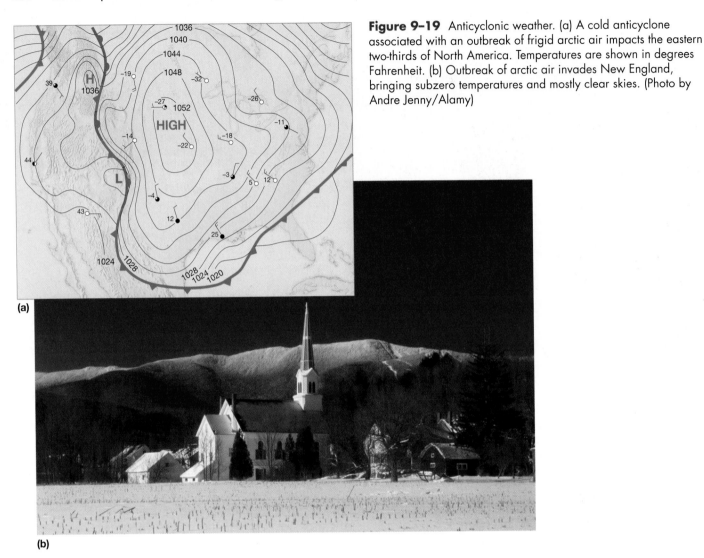

(a)

(b)

Figure 9–19 Anticyclonic weather. (a) A cold anticyclone associated with an outbreak of frigid arctic air impacts the eastern two-thirds of North America. Temperatures are shown in degrees Fahrenheit. (b) Outbreak of arctic air invades New England, bringing subzero temperatures and mostly clear skies. (Photo by Andre Jenny/Alamy)

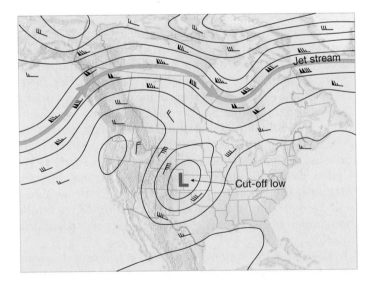

Figure 9–20 Cut-off low pressure systems are literally cut off from the west-to-east flow in the jet stream. As a result, these systems can spin for days over the same area and are capable of producing very large quantities of precipitation.

Concept Check 9.7

1 Describe the weather associated with a strong anticyclone that penetrates the southern tier of the United States in winter.

2 What is a blocking high-pressure system?

3 What type of weather does a cut-off low-pressure system typically bring to an area?

Case Study of a Midlatitude Cyclone

 GE◯De ATMOSPHERE Basic Weather Patterns
▶ In the Lab: Examining a Mature Middle-Latitude Cyclone

To visualize the weather one might expect from a strong late-winter midlatitude cyclone, we are going to look at the evolution of a storm as it migrated across the United States

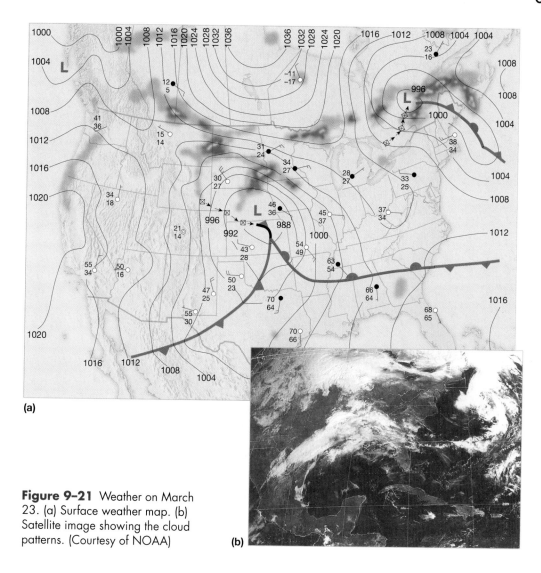

Figure 9–21 Weather on March 23. (a) Surface weather map. (b) Satellite image showing the cloud patterns. (Courtesy of NOAA)

during the latter half of March. This cyclone reached the U.S. West Coast on March 21, a few hundred kilometers northwest of Seattle, Washington. Like many other Pacific storms, this one rejuvenated over the western United States and moved eastward into the Plains States. By the morning of March 23, it was centered over the Kansas–Nebraska border (Figure 9–21). At this time, the central pressure had reached 985 millibars, and its well-developed cyclonic circulation exhibited a warm front and a cold front.

During the next 24 hours, the forward motion of the storm's center became sluggish. The storm curved slowly northward through Iowa, and the pressure deepened to 982 millibars (Figure 9–22). Although the storm center advanced slowly, the associated fronts moved vigorously toward the east and somewhat northward. The northern sector of the cold front overtook the warm front and generated an occluded front, which by the morning of March 24 was oriented nearly east–west (Figure 9–22).

At that point the storm produced one of the worst blizzards ever to hit the north-central states. While the winter storm was brewing in the north, the cold front marched from western Texas (on March 23) to the Atlantic Ocean (March 25). During its two-day trek to the ocean, this violent cold front generated numerous severe thunderstorms and 19 tornadoes.

By March 25 the low pressure had diminished in intensity (1000 millibars) and had split into two centers (Figure 9–23). Although the remnant low from this system, which was situated over the Great Lakes, generated some snow for the remainder of March 25, it had completely dissipated by the following day.

Now that you have read an overview of this storm, let us revisit this cyclone's passage in detail, using the weather charts for March 23 through March 25 (see Figures 9–21 to 9–23). The weather map for March 23 depicts a classic developing cyclone. The warm sector of this system, as exemplified at Fort Worth, Texas, is under the influence of a warm, humid air mass having a temperature of 70°F and a dew-point temperature of 64°F. Notice that winds in the warm sector are from the south and are overrunning cool-

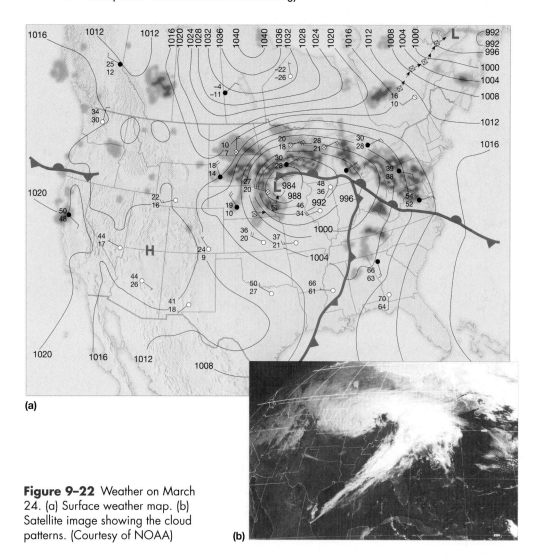

Figure 9–22 Weather on March 24. (a) Surface weather map. (b) Satellite image showing the cloud patterns. (Courtesy of NOAA)

er air situated north of the warm front. In contrast, the air behind the cold front is 20° to 40°F cooler than the air of the warm sector and is flowing from the northwest, as depicted by the data for Roswell, New Mexico.

On the morning of March 23, little activity was occurring along the fronts. As the day progressed, however, the storm intensified and changed dramatically. The map for March 24 illustrates the highly developed midlatitude cyclone. The extent and spacing of the isobars indicate a strong system that affected the circulation of the entire eastern two-thirds of the United States. A glance at the winds reveals a robust counterclockwise flow converging on the low.

The activity in the cold sector of the storm just north of the occluded front produced one of the worst March blizzards ever to affect the north-central United States. In the Duluth–Superior area of Minnesota and Wisconsin, winds up to 81 miles per hour were measured. Unofficial estimates of wind speeds in excess of 100 miles per hour were made on the aerial bridge connecting these cities. Winds blew 12 inches of snow into 10- to 15-foot drifts, and some roads were closed for three days (Figure 9–24). One large restaurant in Superior was destroyed by fire because drifts prevented firefighting equipment from reaching the blaze.

By late afternoon on March 23, the cold front had generated a hailstorm in parts of eastern Texas. As the cold front moved eastward, it affected the entire southeastern United States except southern Florida. Throughout this region, numerous thunderstorms were spawned. Although high winds, hail, and lightning caused extensive damage, the 19 tornadoes generated by the storm caused even greater death and destruction.

The path of the front can be easily traced from the reports of storm damage. By the evening of March 23, hail and wind damage were reported as far east as Mississippi and Tennessee. Early on the morning of March 24, golf ball–size hail was reported in downtown Selma, Alabama. About 6:30 AM that day, the "Governor's Tornado" struck Atlanta, Georgia, where the storm produced its worst destruction. Damage was estimated at over $50 million, 3 lives were lost, and 152 people were injured. The 12-mile path of the Governor's Tornado cut through an affluent residential area of Atlanta that included the governor's mansion (hence the name of the storm). The final damage (hail and a small tornado) along the cold front was reported at 4:00 AM on March 25 in northeastern Florida. Thus, a day and a half and some 1200 kilometers after the cold front became active in

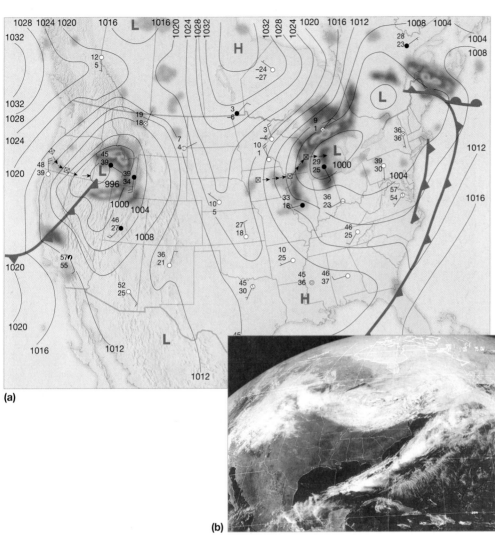

(a)

(b)

Figure 9–23 Weather on March 25. (a) Surface weather map. (b) Satellite image showing the cloud patterns. (Courtesy of NOAA)

Figure 9–24 A paralyzing blizzard strikes the north-central United States. (Photo by Mike McCleary/Bismarck Tribune)

SEVERE AND HAZARDOUS **WEATHER**   The Midwest Floods of 2008 and 1993

Floods are part of the natural behavior of streams. They are also among the most deadly and destructive of natural hazards. Flash floods in small valleys are frequently triggered by torrential rains that last just a few hours (see *Severe and Hazardous Weather* in Chapter 10, page 279). By contrast, major regional floods in large river valleys often result from an extraordinary series of precipitation events over a broad region for an extended time span. The Midwest floods of 2008 and 1993 are examples of the latter situation.

Midwest Floods—June 2008

June 2008 saw record-breaking floods in many parts of the Midwest. June was a very wet month across a significant part of this region, with numerous heavy rain events resulting in record flooding in Iowa, Wisconsin, Indiana, and Illinois. Rainfall totals were more than twice the average in many places. Martinsville, Indiana, for example, reported a total of 20.11 inches of rain for the month—an amount equal to about half its annual rainfall and more than double its previous record high for a single month.

FIGURE 9–B The flooding Cedar River covers a large portion of Cedar Rapids, Iowa, on June 14, 2008. (AP Photo/Jeff Roberson)

Rainfall totals were more than twice the average in many places. Martinsville, Indiana, for example, reported a total of 20.11 inches of rain for the month—an amount equal to about half its annual rainfall and more than double its previous record high for a single month.

During the two months prior to the floods, the jet stream had regularly been dipping southward into the central United States, placing the favored storm track

FIGURE 9–C Water rushes through a break in an artificial levee in Monroe County, Illinois, in 1993. During the record-breaking floods of 2008 and 1993, many levees could not withstand the force of the floodwaters. Sections of many weakened structures were overtopped or simply collapsed. (Photo by James A. Finley/AP/Wide World Photos)

for rainy low pressure centers over the Midwest. As a result, at least 65 locations in the Midwest set new June rainfall records, while more than 100 other stations had rainfall totals that ranked second to fifth on record. Prior to the heavy June rains, many of these stations had an exceptionally wet winter and spring. Soils that were already waterlogged provided no place for additional rain to go, forcing the flow across the surface and causing streams to rise.

> The June 2008 floods represented the costliest weather disaster in Indiana history, while Iowa suffered widespread losses, with 83 of its 99 counties declared disaster areas.

The June 2008 floods represented the costliest weather disaster in Indiana history, while Iowa suffered widespread losses, with 83 of its 99 counties declared disaster areas. Nine of Iowa's rivers were at or above previous record flood levels, and millions of acres of productive farmland (an estimated 16 percent of the state's total) were submerged. Residential and commercial areas were evacuated, mostly in Cedar Rapids, where more than 400 city blocks were under water (Figure 9–B). The list of affected towns and rural areas throughout the Midwest was astonishing. Many riverside communities were flooded when swollen streams breached levees that had been built to protect the towns (Figure 9–C).

Overall agricultural losses exceeded $7 billion, and property damages topped $1 billion. The Midwest regional floods of 2008 were the worst in 15 years.

The Great Flood of 1993

Although flood levels in some parts of the Midwest in June 2008 were the highest ever recorded, the overall impact of this event did not approach that of the Great Flood of 1993. Perhaps the country's worst nonhurricane flooding disaster, the 1993 event caused an estimated $27 billion in damages (adjusted to 2007 dollars).

Unprecedented rainfall produced the wettest spring and early summer of the twentieth century for the Upper Mississippi River basin (Figure 9–D). Soils throughout the Midwest were already saturated from ample rainfall in summer and fall 1992, and soils remained moist as winter began. This pattern continued into spring and summer 1993. Rainfall in the Upper Mississippi River basin during April and May was 40 percent higher than the long-term average. As the deluge continued through July 1993, much of the basin received precipitation between two and three times the norm.

A stationary weather pattern over the United States was responsible for the Midwest's persistent drenching rains. Most of the

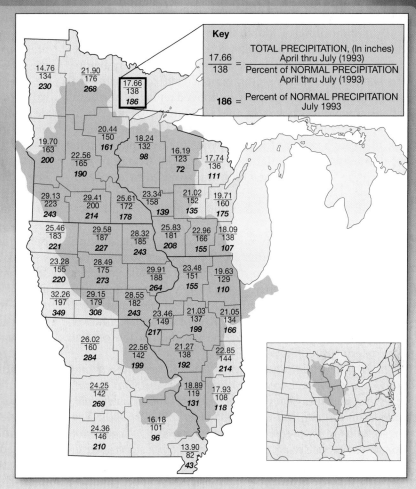

FIGURE 9–D Precipitation data for the Upper Mississippi River basin between April 1 and July 31, 1993. The magnitude of rainfall over such a large area resulted in flooding of extraordinary and catastrophic proportions on the Mississippi and many of its tributaries. (Data from the Illinois State Water Survey)

showers and thunderstorms developed in the boundary area between cooler air over the Northern Plains and warm, very humid air over the South. This front oscillated north and south over the Midwest during much of June and July. Meanwhile, a strong high-pressure system (the "Bermuda High") became anchored over the southeastern United States, blocking the progression of weather systems through the eastern half of the nation.

Some of the individual station reports were nothing short of astounding. In northwestern Missouri, Skidmore reported 25.35 inches of rain in July, where normal July rainfall is about 4 inches. Alton, in northwestern Iowa, had July rains that totaled 20.41 inches, and Leon in south-central Iowa reported 20.68 inches. Rainfall totals such as these leave no doubt that the record-breaking flooding on the Mississippi and other rivers in the Midwest was directly related to the exceptional rainfall during the spring and early summer.

Texas, the front left the United States to vent its energy over the Atlantic Ocean.

By the morning of March 24, cold polar air had penetrated deep into the United States behind the cold front (see Figure 9–22). Fort Worth, Texas, which just the day before had been in the warm sector, experienced cool northwest winds. Subfreezing temperatures moved as far as northern Oklahoma. Notice, however, that by March 25, Fort Worth was again experiencing a southerly flow. We can conclude that this was a result of the decaying cyclone that no longer dominated the circulation in the region. Also notice on the map for March 25 that a high was situated over eastern Louisiana. The clear skies and weak surface winds associated with the center of a subsiding air mass are well illustrated here.

You may have already noticed another cyclone moving in from the Pacific on March 25, as the earlier storm exited to the east. This new storm developed in a similar manner but was centered farther north. As you might guess, another blizzard hammered the Northern Plains, and a few tornadoes spun through Texas, Arkansas, and Kentucky, while precipitation dominated the weather in the central and eastern United States.

In summary, this example demonstrates the effect of a spring cyclone on midlatitude weather. Within three days, Fort Worth, Texas, temperatures changed from warm to unseasonably cold. Thunderstorms with hail were followed by cold clear skies. You can see how the north–south temperature gradient, which is most pronounced in the spring, generates these intense storms. Recall that it is the role of these storms to transfer heat from the tropics poleward. Because of Earth's rotation, however, this latitudinal heat exchange is complex. If Earth rotated more slowly, a more leisurely north–south flow would exist that might reduce the temperature gradient, in which case the midlatitudes would not experience stormy weather.

Concept Check 9.8

1 Briefly describe the weather at Fort Worth, Texas, from March 23 through March 25, during the passage of a well-developed spring storm.

2 How did the weather in the Duluth–Superior area of Minnesota and Wisconsin compare to that in the Fort Worth area during March 23–25?

A Modern View: The Conveyor Belt Model

The Norwegian cyclone model has proven to be a valuable tool for describing the formation and development of midlatitude cyclones. Although modern ideas have not completely replaced this model, upper-air and satellite data provide meteorologists with a more thorough understanding of the structure and evolution of these storm systems. Armed with this additional information, meteorologists have developed another model to describe the circulation of a midlatitude cyclone.

The new way of describing the flow within a cyclone calls for a new analogy. Recall that the Norwegian model describes cyclone development in terms of the interactions of air masses along frontal boundaries, similar to armies clashing along battlefronts. The new model employs a modern example from industry—conveyor belts. Just as conveyor belts transport goods or people from one location to another, atmospheric conveyor belts transport air with distinct characteristics from one location to another.

The modern view of cyclogenesis, called the **conveyor belt model,** provides a good picture of the airflow within a cyclonic system. It consists of three interacting air streams: two that originate near the surface and ascend and a third that originates in the uppermost troposphere. A schematic representation of these air streams is shown in Figure 9–25.

The *warm conveyor belt* (shown in red) carries warm, moist air from the Gulf of Mexico into the warm sector of the midlatitude cyclone (Figure 9–25). As this air stream flows northward, convergence causes it to slowly ascend. When it reaches the sloping boundary of the warm front, it rises even more rapidly over the cold air that lies beyond (north of) the front. During its ascent, the warm, humid air cools adiabatically and produces a wide band of clouds and precipitation. Depending on atmospheric conditions, drizzle, rain, freezing rain (glaze), and snow are possible. When this air stream reaches the middle troposphere, it begins to turn right (eastward) and eventually joins the general west-to-east flow aloft. The warm conveyor belt is the main precipitation-producing air stream in a midlatitude cyclone.

The *cold conveyor belt* (blue arrow) is airflow that starts at the surface ahead (north) of the warm front and flows westward toward the center of the cyclone (Figure 9–25). Flowing beneath the warm conveyor belt, this air is moistened by the evaporation of raindrops falling through it. (Near the Atlantic Ocean this conveyor belt has a marine origin and feeds significant moisture into the storm.) Convergence causes this air stream to rise as it nears the center of the cyclone. During its ascent, this air cools adiabatically, becomes saturated, and contributes to the cyclone's precipitation. As the cold conveyor belt reaches the middle troposphere, some of the flow rotates cyclonically around the low to produce the distinctive "comma head" of the mature storm system (Figure 9–26). The remaining flow turns right (clockwise) and becomes incorporated into the general westerly flow. Here it parallels the flow of the warm conveyor belt and may generate precipitation.

The third air stream, called the *dry conveyor belt*, is shown as a yellow arrow in Figure 9–25. Whereas both the warm and cold conveyor belts begin at the surface, the dry air stream originates in the uppermost troposphere.

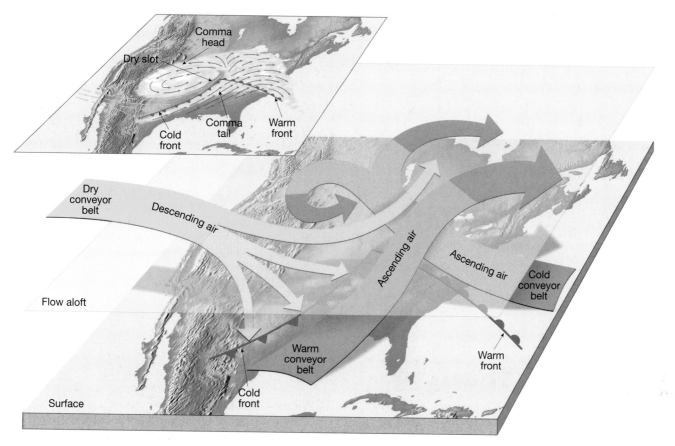

Figure 9–25 Schematic drawing of the circulation of a mature midlatitude cyclone, showing the warm conveyor belt (red), cold conveyor belt (blue), and dry conveyor belt (yellow). The inset shows the cloud cover produced by the warm and cold conveyor belts and the dry slot produced by the dry conveyor belt.

As part of the upper-level westerly flow, the dry conveyor belt is relatively cold and dry. As this air stream enters the cyclone, it splits. One branch descends behind the cold front, resulting in the clear, cool conditions normally associated with the passage of a cold front. In addition, this flow maintains the strong temperature contrast observed across the cold front. The other branch of the dry conveyor belt maintains its westerly flow and forms the *dry slot* (cloudless area) that separates the head and tail of a comma cloud pattern (Figure 9–26).

Figure 9–26 Satellite view of a mature midlatitude cyclone over the eastern half of the United States. It is easy to see why we often refer to the cloud pattern of a cyclone as having a "comma" shape. (Courtesy of NASA)

In summary, the conveyor belt model of a midlatitude cyclone provides a three-dimensional picture of the major circulation of these storm systems. It also accounts for the distribution of precipitation and the comma-shaped cloud pattern characteristic of mature cyclonic storms.

Concept Check 9.9

1 Briefly describe the conveyor belt model of midlatitude cyclones.

2 What is the source of air that is carried by the warm conveyor belt?

3 In what way is the dry conveyor belt different from both the warm and cold conveyor belts?

Students Sometimes Ask...

Sometimes when there is a major flood, it is described as a 100-year flood. What does that mean?

The phrase "100-year flood" is misleading because it leads people to believe that such an event happens only once every 100 years. The truth is that an uncommonly big flood can happen any year. The phrase "100-year flood" is really a statistical designation, indicating that there is a 1-in-100 chance that a flood of a certain size will happen during any year. Perhaps a better term would be the "1-in-100-chances flood."

Many flood designations are reevaluated and changed over time as more data are collected or when a river basin is altered in a way that affects the flow of water. Dams and urban development are examples of some human influences in a basin that affect floods.

Give It Some Thought

1. Refer to the accompanying weather map and answer the following questions.
   a. What is the probable wind direction at each of the cities?
   b. What air mass is likely to be influencing each city?
   c. Identify the cold front, warm front, and occluded front.
   d. What is the barometric tendency at city A and at city C?
   e. Which of the three cities is coldest? Which is warmest?

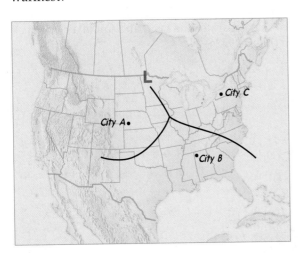

2. The following questions refer to midlatitude cyclones and the frontal weather associated with them.
   a. Describe the weather behind a cold front. With what type of pressure system are these weather conditions associated?
   b. What is the source region for the mT air mass that produces most of the clouds and precipitation in the eastern two-thirds of the United States?
   c. With what type of severe weather would a meteorologist be most concerned when a stationary front is present over a region? (*Hint:* This severe weather type claims the most lives, on average, each year in the United States.)

3. The accompanying illustration shows the fronts associated with a midlatitude cyclone. Match the numbered regions with the type of precipitation that is most typically associated with each region during winter.
   a. Thunderstorms
   b. Light-to-moderate rain
   c. Sleet or freezing rain (glaze)
   d. Heavy snow

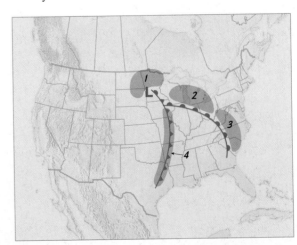

4. Which half of the United States (western or eastern) would you expect to have the greatest number of occluded fronts? Explain.

5. The accompanying diagrams show surface temperatures (in degrees Fahrenheit) for noon and 6 PM on January 29, 2008. On this day, an incredibly powerful weather front moved through Missouri and Illinois.

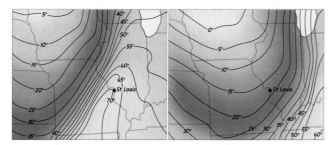

   a. What type of front passed through the Midwest?
   b. Describe how the temperature changed in St. Louis, Missouri, over the six-hour period that began at noon and ended at 6 PM.
   c. Describe the likely shift in wind direction in St. Louis during this time period.

6. Sketch a vertical cross section of a cold front and a warm front and include the following elements.
   a. Shape and slope of each type of front.
   b. Air mass on both sides of each front.
   c. Type of clouds typically associated with each front.
   d. Type of rainfall associated with each front.
   e. Characterization of temperature and humidity associated with each front.

7. The accompanying satellite image shows the comma-shaped cloud pattern of a well-developed midlatitude cyclone over the United States in late winter.

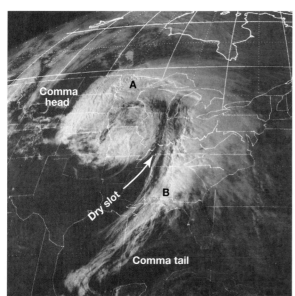

(Courtesy of NOAA)

   a. Describe the weather in northern Minnesota (A), which is located in the comma head of the winter storm.
   b. What is the probable weather in eastern Alabama (B), which is located in the comma tail?

8. Why are occurrences of midlatitude cyclones less common over the United States in late summer than in winter and spring?

9. Write a statement explaining the relationship between the circulation around a surface low-pressure system and the flow aloft, using terms such as *ascending air*, *descending air*, *divergence*, and *convergence*.

10. The accompanying map shows the path of the jet stream over the United States.

   a. In what region of the country or states would you expect the center of a low pressure system to form?
   b. In what region of the country or states would you expect a high-pressure system to be located?
   c. Where is the upper-air low (trough) located?
   d. Where is the upper-air high (ridge) located?

11. Refer to the *Severe and Hazardous Weather* material on page 262. What is the major difference between weather systems that cause flash floods in small watersheds and weather systems that cause regional floods in large watersheds?

12. Distinguish between veering and backing winds (see Box 9–1).

MIDLATITUDE CYCLONES IN REVIEW

- *Fronts* are boundary surfaces that separate air masses of different densities, one usually warmer and more humid than the other. As one air mass moves into another, the warmer, less dense air mass is forced aloft in a process called *overrunning*. The five types of fronts are *warm front, cold front, stationary front, occluded front,* and *dryline.*

- The primary weather producer in the middle latitudes is the *midlatitude,* or *middle-latitude, cyclone.* Midlatitude cyclones are large low-pressure systems with diameters often exceeding 1000 kilometers (600 miles) that generally travel from west to east. They last a few days to more than a week, have a counterclockwise circulation pattern with a flow inward toward their centers, and have a cold front and a warm front extending from the central area of low pressure.

- According to the *polar front theory,* midlatitude cyclones form along fronts and proceed through a generally predictable life cycle. Along the polar front, where two air masses of different densities are moving parallel to the front and in opposite directions, *cyclogenesis* (cyclone formation) occurs, and the frontal surface takes on a wave shape that is usually several hundred kilometers long. Once a wave forms, warm air advances poleward, invading the area formerly occupied by colder air. This change in the direction of the surface flow causes a readjustment in the pressure pattern that results in somewhat circular isobars, with the low pressure centered at the apex of the wave. Usually, the cold front advances faster than the warm front and gradually closes the warm sector and lifts the warm front. This process is known as *occlusion.* Eventually, all of the warm sector is forced aloft, and cold air surrounds the cyclone at low levels. At this point, the cyclone has exhausted its source of energy and the once highly organized counterclockwise flow ceases to exist.

- Guided by the westerlies aloft, cyclones generally move eastward across the United States. Airflow aloft (divergence and convergence) plays an important role in maintaining cyclonic and anticyclonic circulation. In cyclones, divergence aloft supports the inward flow at the surface

- During the colder months, when temperature gradients are steepest, cyclonic storms move at their fastest rate. Furthermore, the westerly airflow aloft tends to steer these developing pressure systems in a general west-to-east direction.

- Due to the gradual subsidence within them, anticyclones generally produce clear skies and generally calm conditions.

- The modern view of cyclogenesis, called the *conveyor belt model,* consists of three interacting air streams: two that originate at the surface and ascend and a third that originates aloft.

VOCABULARY REVIEW

backdoor cold front (p. 244)
cold front (p. 243)
conveyor belt model (p. 264)
cyclogenesis (p. 248)
dryline (p. 247)

front (p. 240)
midlatitude (midlatitude) cyclone
(p. 240)
Norwegian cyclone model (p. 248)
occluded front (p. 245)

occlusion (p. 245)
overrunning (p. 241)
polar-front theory (p. 248)
stationary front (p. 245)
warm front (p. 241)

PROBLEMS

1. Refer to Table A, which provides weather observations for Champaign, Illinois, over a three-day period as a midlatitude cyclone passed through the area. Complete the following questions, keeping in mind the general wind and temperature changes expected with the passage of fronts.

   a. During which day and at approximately what time did the warm front pass through Champaign?

   b. List two lines of evidence indicating that a warm front passed through Champaign.

   c. Explain the slight drop in temperature experienced between midnight and 6:00 AM on Day 2.

   d. On what day and time did the cold front pass through Champaign?

   e. List two changes that indicate the passage of the cold front.

   f. Did the thunderstorms in Champaign occur with the passage of the warm front, the cold front, or the occluded front?

2. In the spring and summer, a warm moist (mT) air mass from the Gulf of Mexico occasionally collides with a warm dry (cT) air mass from the desert Southwest. These air masses meet over Texas, Oklahoma, and Kansas, producing what meteorologists call a dryline. Thunderstorms that erupt along drylines can produce some of the most severe storms on the planet. When these two air masses meet, stormy weather is triggered when the less dense air mass overruns the denser air mass. Which of these two air masses is denser? Assume that the air temperature of the two air masses is the same. (*Hint:* Find the molecular weight of dry air [N_2 and O_2] with no water vapor [H_2O] and then compare that to what the molecular weight would be if about 4 percent of the N_2 and O_2 molecules were replaced by the H_2O molecules.)

3. If you were located 400 kilometers ahead of the surface position of a typical warm front (with slope 1:200), how high would the frontal surface be above you?

4. Calculate the distance you are in front of a typical warm front when the first cirrus clouds begin to form. (*Hint:* Refer to Figure 5–2 to find the minimum height range for high clouds.)

TABLE A Weather Data for Champaign, Illinois

	Temperature (°F)	Wind Direction	Weather and Precipitation
Day 1			
00:00	46	E	Partly Cloudy
3:00 am	46	ENE	Partly Cloudy
6:00 am	48	E	Overcast
9:00 am	49	ESE	Drizzle
12:00 pm	52	ESE	Light rain
3:00 pm	53	SE	Rain
6:00 pm	68	SSW	Partly cloudy
9:00 pm	67	SW	Partly cloudy
Day 2			
00:00	66	SW	Partly cloudy
3:00 am	64	SW	Mostly sunny
6:00 am	63	SSW	Mostly sunny
9:00 am	69	SW	Mostly sunny
12:00 pm	72	SSW	Mostly sunny
3:00 pm	76	SW	Mostly sunny
6:00 pm	74	SW	Cloudy
9:00 pm	64	W	Thunderstorm, gusty winds
Day 3			
00:00	52	WNW	Isolated thunderstorms
3:00 am	48	WNW	Cloudy
6:00 am	42	NW	Partly cloudy
9:00 am	39	NW	Mostly sunny
12:00 pm	38	NW	Sunny
3:00 pm	40	NW	Sunny
6:00 pm	42	NW	Sunny
9:00 pm	40	NW	Sunny

MyMeteorologyLab™

Log in to www.mymeteorologylab.com for animations, videos, MapMaster interactive maps, GEODe media, *In the News* RSS feeds, web links, glossary flashcards, self-study quizzes and a Pearson eText version of this book to enhance your study of *Midlatitude Cyclones*.

Thunderstorms and Tornadoes

The subject of this chapter and the following chapter is severe and hazardous weather. In this chapter we examine the severe local weather produced in association with cumulonimbus clouds—namely, thunderstorms and tornadoes. In Chapter 11 the focus will turn to the large tropical cyclones we call hurricanes.

Occurrences of severe weather are more fascinating than ordinary weather phenomena. The lightning display generated by a thunderstorm can be a spectacular event that elicits both awe and fear. Of course, hurricanes and tornadoes also attract a great deal of much-deserved attention. A single tornado outbreak or hurricane can cause billions of dollars in property damage as well as many deaths.

Focus On Concepts

After completing this chapter you should be able to:

- Distinguish among three types of storm-producing cyclones.

- List the basic requirements for thunderstorm formation and locate places on a map that exhibit frequent thunderstorm activity.

- Sketch diagrams that illustrate the stages of an air-mass thunderstorm.

- Outline the characteristics of a severe thunderstorm.

- Discuss the formation of a supercell thunderstorm and distinguish between squall lines and mesoscale convective complexes.

- Explain what causes lightning and thunder.

- Describe the structure and basic characteristics of tornadoes.

- Summarize the atmospheric conditions and locations that are favorable to the formation of tornadoes.

- Distinguish between a tornado watch and warning and describe the role of Doppler radar in the warning process.

What's in a Name?

In Chapter 9 we examined the midlatitude cyclones that play an important role in causing day-to-day weather changes, but the use of the term *cyclone* is often confusing. For many people the term implies only an intense storm, such as a tornado or a hurricane. When a hurricane unleashes its fury on India, Bangladesh, or Myanmar, for example, it is usually reported in the media as a cyclone (the term denoting a hurricane in that part of the world). For example, in February 2011, the floods and destruction caused when Cyclone Yasi struck Australia were major news stories (Figure 10–1a).

(a)

Figure 10–1 (a) Sometimes the use of the term *cyclone* can be confusing. In South Asia and Australia, the term *cyclone* is applied to storms that are called *hurricanes* in the United States. This image shows Cyclone Yasi, which struck eastern Australia in Feburary 2011. (NASA) (b) In parts of the Great Plains, *cyclone* is a synonym for *tornado*. The nickname for the athletic teams at Iowa State University is the *Cyclones*.* (Image courtesy of Iowa State University)

———
*Iowa State University is the only Division I school to use *Cyclones* as its team name. The pictured logo was created to better communicate the school's image by combining the mascot, a cardinal bird named Cy, and the Cyclone team name, a swirling twister in the bird's tail.

Similarly, tornadoes are referred to as cyclones in some places. This custom is particularly common in portions of the Great Plains of the United States. Recall that in the *Wizard of Oz*, Dorothy's house was carried from her Kansas farm to the land of Oz by a cyclone. Indeed, the nickname for the athletic teams at Iowa State University is the *Cyclones* (Figure 10–1b). Although hurricanes and tornadoes are, in fact, cyclones, the vast majority of cyclones are not hurricanes or tornadoes. The term *cyclone* simply refers to the circulation around any low-pressure center, no matter how large or intense it is.

Tornadoes and hurricanes are both smaller and more violent than midlatitude cyclones. A midlatitude cyclone may have a diameter of 1600 kilometers (1000 miles) or more. By contrast, hurricanes average only 600 kilometers (375 miles) across, and tornadoes, with a diameter of just 0.25 kilometer (0.16 mile), are much too small to show up on a weather map.

(b)

The thunderstorm, a much more familiar weather event, hardly needs to be distinguished from tornadoes, hurricanes, and midlatitude cyclones. Unlike the flow of air about these storms, the circulation associated with thunderstorms is characterized by strong up-and-down movements. Winds in the vicinity of a thunderstorm do not follow the inward spiral of a cyclone, but they are typically variable and gusty.

Although thunderstorms form "on their own," away from cyclonic storms, they also form in conjunction with cyclones. For instance, thunderstorms are frequently spawned along the cold front of a midlatitude cyclone, and on rare occasions, a tornado may descend from the thunderstorm's cumulonimbus tower. Hurricanes also generate widespread thunderstorm activity. Thus, thunderstorms are related in some manner to all three types of cyclones mentioned here.

Concept Check 10.1

1 List and briefly describe three different ways in which the term *cyclone* is used.

2 Compare the wind speeds and sizes of middle-latitude cyclones, tornadoes, and hurricanes.

Thunderstorms

Almost everyone has observed various small-scale phenomena that result from the vertical movements of relatively warm, unstable air. Perhaps you have seen a dust devil over an open field on a hot day, whirling its dusty load to great heights, or maybe you have seen a bird glide effortlessly skyward on an invisible thermal. These examples illustrate the dynamic thermal instability that occurs during the development of a *thunderstorm.* A **thunderstorm** is a storm that generates lightning and thunder. It frequently produces gusty winds, heavy rain, and hail. A thunderstorm may be produced by a single cumulonimbus cloud and influence only a small area, or it may be associated with clusters of cumulonimbus clouds covering a large area.

Thunderstorms form when warm, humid air rises in an unstable environment. Various mechanisms can trigger the upward air movement needed to create thunderstorm-producing cumulonimbus clouds. One mechanism, the unequal heating of Earth's surface, significantly contributes to the formation of *air-mass thunderstorms.* These storms are associated with the scattered puffy cumulonimbus clouds that commonly form *within* maritime tropical air masses and produce scattered thunderstorms on summer days. Such storms are usually short lived and seldom produce strong winds or hail.

In contrast, thunderstorms in a second category not only benefit from uneven surface heating but are associated with the lifting of warm air, as occurs along a front or a mountain slope. Moreover, diverging winds aloft frequently contribute to the formation of these storms because they tend to draw air from lower levels upward beneath them. Some of the thunderstorms in this second category may produce high winds, damaging hail, flash floods, and tornadoes. Such storms are described as *severe.*

At any given time there are an estimated 2000 thunderstorms in progress. As we would expect, the greatest proportion of these storms occur in the tropics, where warmth, plentiful moisture, and instability are always present. About 45,000 thunderstorms take place each day, and more than 16 million occur annually around the world. The lightning from these storms strikes Earth 100 times each second (Figure 10–2).

Annually the United States experiences about 100,000 thunderstorms and millions of lightning strikes. A glance at Figure 10–3 shows that thunderstorms are most frequent in Florida and the eastern Gulf Coast region, where activity is recorded between 70 and 100 days each year. The region on the east side of the Rockies in Colorado and New Mexico is next, with thunderstorms occurring 60 to 70 days annually. Most of the rest of the nation experiences thunderstorms 30 to 50 days a year. Clearly, the western margin of the United States has little thunderstorm activity. The same is true for the northern tier of states and for Canada, where warm, moist, unstable maritime tropical (mT) air seldom penetrates.

Concept Check 10.2

1 What are the primary requirements for the formation of thunderstorms?

2 Where would you expect thunderstorms to be most common on Earth? In the United States?

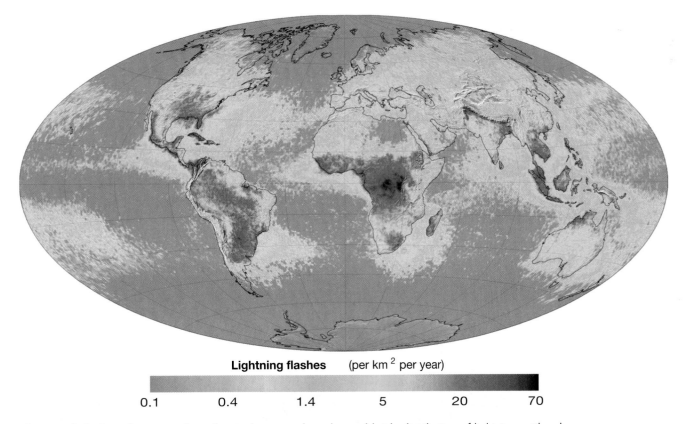

Lightning flashes (per km² per year)

0.1   0.4   1.4   5   20   70

Figure 10–2 Data from space-based optical sensors show the worldwide distribution of lightning, with color variations indicating the average annual number of lightning flashes per square kilometer. (NASA image)

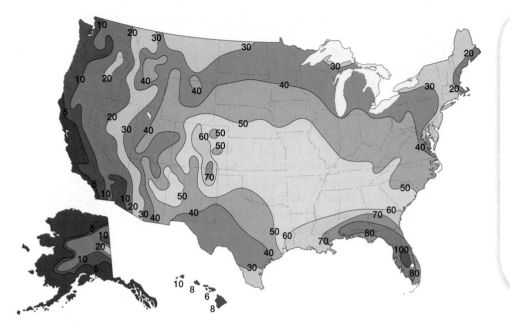

Figure 10–3 Average number of days per year with thunderstorms. The humid subtropical climate that dominates the southeastern United States receives much of its precipitation in the form of thunderstorms. Most of the Southeast averages 50 or more days each year with thunderstorms. (Environmental Data Service, NOAA)

Students Sometimes Ask...

What's the difference between a thundershower and a thunderstorm?

Technically, there is no difference. The term *thundershower* implies a relatively weak storm with light to moderate rainfall and low levels of lightning activity. However, there are no defined parameters that distinguish between the two terms. To avoid confusion, the National Weather Service does not use the term *thundershower*. If a shower is strong enough to produce lightning, even just one bolt, it's called a *thunderstorm*.

Air-Mass Thunderstorms

In the United States **air-mass thunderstorms** frequently occur in mT air that moves northward from the Gulf of Mexico. These warm, humid air masses contain abundant moisture in their lower levels and can be rendered unstable when heated from below or lifted along a front. Because mT air most often becomes unstable in spring and summer, when it is warmed from below by the heated land surface, it is during these seasons that air-mass thunderstorms occur most frequently. They also have a strong preference for midafternoon, when surface temperatures are highest. Because local differences in surface heating aid the growth of air-mass thunderstorms, they generally occur as scattered, isolated cells instead of being organized in relatively narrow bands or other configurations.

Stages of Development

Important field experiments that were conducted in Florida and Ohio in the late 1940s probed the dynamics of air-mass thunderstorms. This pioneering work, known as the *Thunderstorm Project*, was prompted by a number of thunderstorm-related airplane crashes. It involved the use of radar, aircraft, radiosondes, and an extensive network of surface instruments. The research produced a three-stage model of the life cycle of an air-mass thunderstorm that remains basically unchanged after more than 70 years. The three stages are depicted in Figure 10–4.

Cumulus Stage Recall that an air-mass thunderstorm is largely a product of the uneven heating of the surface, which leads to rising currents of air that ultimately produce a cumulonimbus cloud. At first the buoyant thermals produce fair-weather cumulus clouds that may exist for just minutes before evaporating into the drier surround-

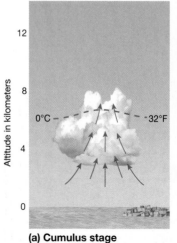
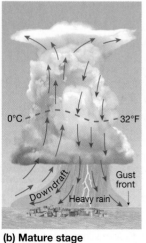
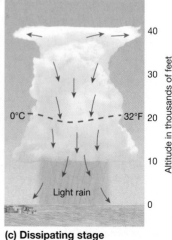

(a) Cumulus stage   **(b) Mature stage**   **(c) Dissipating stage**

Figure 10–4 Stages in the development of an air-mass thunderstorm. (a) During the cumulus stage, strong updrafts build the storm. (b) The mature stage is marked by heavy precipitation and cool downdrafts in part of the storm. (c) When the warm updrafts disappear completely, precipitation becomes light, and the cloud begins to evaporate.

ing air (Figure 10–5a). This initial cumulus development is important because it moves water vapor from the surface to greater heights. Ultimately, the air becomes sufficiently humid that newly forming clouds do not evaporate but instead continue to grow vertically.

The development of a cumulonimbus tower requires a continuous supply of moist air. The release of latent heat allows each new surge of warm air to rise higher than the last, adding to the height of the cloud. This phase in the development of a thunderstorm, called the **cumulus stage,** is dominated by updrafts (see Figure 10–4a).

Once the cloud passes beyond the freezing level, the Bergeron process begins producing precipitation. (The Bergeron process is discussed in Chapter 5.) Eventually, the accumulation of precipitation in the cloud is too great for the updrafts to support. The falling precipitation causes drag on the air and initiates a downdraft.

The creation of the downdraft is further aided by the influx of cool, dry air surrounding the cloud, a process termed **entrainment.** This process intensifies the downdraft because the air added during entrainment is cool and therefore heavy; possibly of greater importance, it is dry. It thus causes some of the falling precipitation to evaporate (a cooling process), thereby cooling the air within the downdraft.

Mature Stage As the downdraft leaves the base of the cloud, precipitation is released, marking the beginning of the cloud's **mature stage** (see Figure 10–4b). At the surface the cool downdrafts spread laterally and can be felt before the actual precipitation reaches the ground. The sharp, cool gusts at the surface are indicative of the downdrafts aloft. During the mature stage, updrafts exist side by side with downdrafts and continue to enlarge the cloud. When the cloud grows to the top of the unstable region, often located at the base of the stratosphere, the updrafts spread laterally and produce the characteristic anvil top. Generally, ice-laden cirrus clouds make up the top and are spread downwind by strong winds aloft. The mature stage is the most active period of a thunderstorm. Gusty winds, lightning, heavy precipitation, and sometimes small hail are common.

Dissipating Stage Once downdrafts begin, the vacating air and precipitation encourage more entrainment of the cool, dry air surrounding the cell. Eventually, the downdrafts dominate throughout the cloud and initiate the **dissipating stage** (see Figure 10–4c). The cooling effect of falling precipitation and the influx of colder air aloft mark the end of the thunderstorm activity. Without a supply of moisture from updrafts, the cloud will soon evaporate. An interesting fact is that only a modest portion—on the order of 20 percent—of the moisture that condenses in an air-mass thunderstorm actually leaves the cloud as precipitation. The remaining 80 percent evaporates back into the atmosphere.

It should be noted that within a single air-mass thunderstorm there may be several individual *cells*—that is, zones of adjacent updrafts and downdrafts. When you view a thunderstorm, you may notice that the cumulonimbus cloud consists of several towers (Figure 10–5b). Each tower may represent an individual cell that is in a somewhat different part of its life cycle.

(a)

(b)

Figure 10–5 (a) At first, buoyant thermals produce fair-weather cumulus clouds that soon evaporate into the surrounding air, making it more humid. As this process of cumulus development and evaporation continues, the air eventually becomes sufficiently humid so that newly forming clouds do not evaporate but continue to grow. (Photo by Henry Lansford/Photo Researchers, Inc.) (b) This developing cumulonimbus cloud became a towering August thunderstorm over central Illinois. (Photo by E. J. Tarbuck)

To summarize, the stages in the development of an air-mass thunderstorm are as follows:

1. The *cumulus stage,* in which updrafts dominate throughout the cloud and growth from a cumulus to a cumulonimbus cloud occurs.

2. The *mature stage,* the most intense phase, with heavy rain and possibly small hail, in which downdrafts are found side by side with updrafts.

3. The *dissipating stage,* dominated by downdrafts and entrainment, causing evaporation of the structure.

Occurrence

Mountainous regions, such as the Rockies in the West and the Appalachians in the East, experience a greater number of air-mass thunderstorms than do the Plains states. The air near the mountain slope is heated more intensely than air at the same elevation over the adjacent lowlands. A general upslope movement then develops during the daytime that can sometimes generate thunderstorm cells. These cells may remain almost stationary above the slopes below.

Although the growth of thunderstorms is aided by high surface temperatures, many thunderstorms are not generated solely by surface heating. For example, many of Florida's thunderstorms are triggered by the convergence associated with sea-to-land airflow (see Figure 4–22, page 114). Many thunderstorms that form over the eastern two-thirds of the United States occur as part of the general convergence and frontal wedging that accompany passing midlatitude cyclones. Near the equator, thunderstorms commonly form in association with the convergence along the equatorial low (also called the *intertropical convergence zone*). Most of these thunderstorms are not severe, and their life cycles are similar to the three-stage model described for air-mass thunderstorms.

Concept Check 10.3

1 During what seasons and at what time of day is air-mass thunderstorm activity greatest? Why?

2 Why does entrainment intensify thunderstorm downdrafts?

3 Summarize the three stages of an air-mass thunderstorm.

Severe Thunderstorms

Severe thunderstorms are capable of producing heavy downpours and flash flooding as well as strong, gusty, straight-line winds; large hail; frequent lightning; and perhaps tornadoes. For the National Weather Service to officially classify a thunderstorm as *severe,* the storm must have winds in excess of 93 kilometers (58 miles) per hour (50 knots) or produce hailstones larger than 1.9 centimeters (0.75 inch) in diameter or generate a tornado. Of the estimated 100,000 thunderstorms that occur annually in the United States, about 10 percent (10,000 storms) reach severe status.

As you learned in the preceding section, air-mass thunderstorms are localized, relatively short-lived phenomena that dissipate after a brief, well-defined life cycle. They actually extinguish themselves because downdrafts cut off the supply of moisture necessary to maintain the storm. For this reason, air-mass thunderstorms seldom, if ever, produce severe weather. By contrast, other thunderstorms do not quickly dissipate but instead may remain active for hours. Some of these larger, longer-lived thunderstorms attain severe status.

Why do some thunderstorms persist for many hours? A key factor is the existence of strong vertical wind shear—that is, changes in wind direction and/or speed at different heights. When such conditions prevail, the updrafts that provide the storm with moisture do not remain vertical but become tilted. Because of this, the precipitation that forms high in the cloud falls into the downdraft rather than into the updraft, as occurs in air-mass thunderstorms. This allows the updraft to maintain its strength and continue to build upward. Sometimes the updrafts are sufficiently strong that the cloud top is able to push its way into the stable lower stratosphere, a situation called *overshooting* (Figure 10–6).

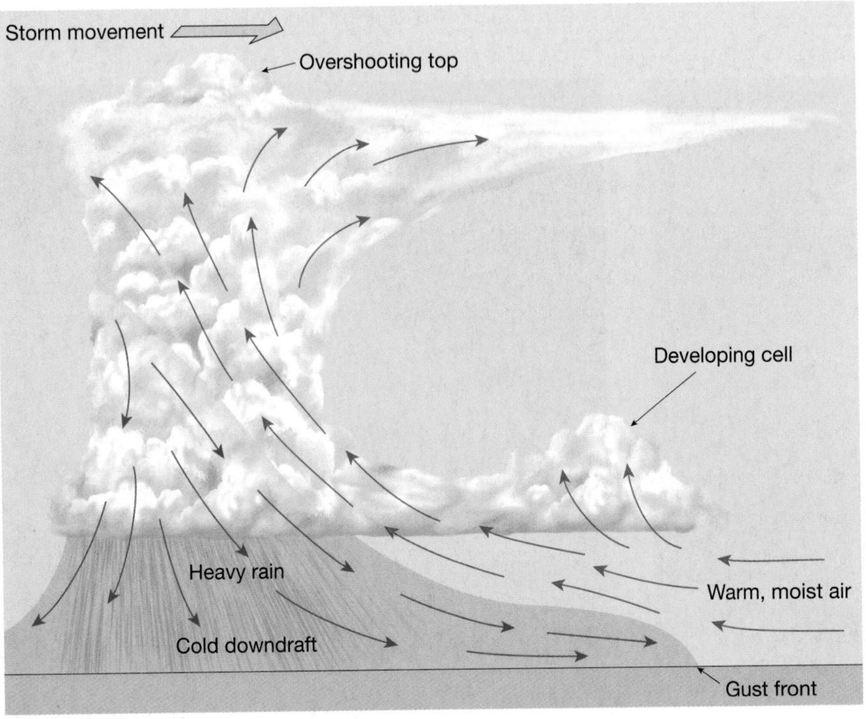

Figure 10–6 Diagram of a well-developed cumulonimbus tower, showing updrafts, downdrafts, and an overshooting top. Precipitation forming in the tilted updraft falls into the downdraft. Beneath the cloud, the denser cool air of the downdraft spreads out along the ground. The leading edge of the outflowing downdraft acts to wedge moist surface air into the cloud. Eventually the outflow boundary may become a gust front that initiates new cumulonimbus development.

Labels in figure: Storm movement · Overshooting top · Developing cell · Heavy rain · Warm, moist air · Cold downdraft · Gust front

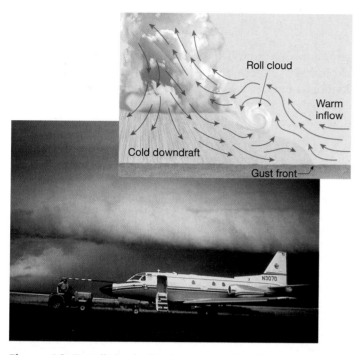

Figure 10–7 Roll clouds, like this one at Miles City, Montana, are sometimes produced along a gust front in an eddy between the inflow and the downdraft. (Photo by National Science Foundation/ National Center for Atmospheric Research)

Beneath the cumulonimbus tower, where downdrafts reach the surface, the denser cool air spreads out along the ground. The leading edge of this outflowing downdraft acts like a wedge, forcing warm, moist surface air into the thunderstorm. In this way, the downdrafts act to maintain the updrafts, which in turn sustain the thunderstorm.

By examining Figure 10–6, you can see that the outflowing cool air of the downdraft acts as a "mini cold front" as it advances into the warmer surrounding air. This outflow boundary is called a **gust front**. As the gust front moves across the ground, the very turbulent air sometimes picks up loose dust and soil, making the advancing boundary visible. Frequently a *roll cloud* may form as warm air is lifted along the leading edge of the gust front (Figure 10–7). The advance of the gust front can provide the lifting needed for the formation of new thunderstorms many kilometers away from the initial cumulonimbus clouds.

Concept Check 10.4

1 How does a severe thunderstorm differ from an air-mass thunderstorm?

2 Describe how downdrafts in a severe thunderstorm act to maintain updrafts.

3 What is a gust front?

Supercell Thunderstorms

Some of our most dangerous weather is caused by a type of thunderstorm called a **supercell**. Few other weather phenomena are as awesome (Figure 10–8). An estimated

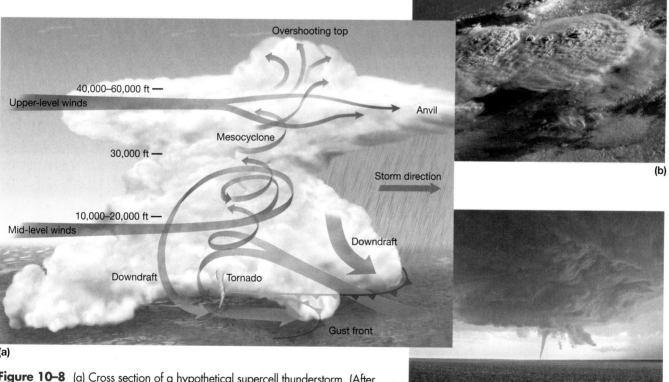

Figure 10–8 (a) Cross section of a hypothetical supercell thunderstorm. (After National Weather Service) (b) This photo of a cluster of supercell thunderstorms along the Manitoba–Minnesota border in September 1994 was taken from space by an astronaut. (NASA) (c) View from ground level of a towering supercell thunderstorm that has produced a tornado. (Photo by Mike Theiss/National Geographic Stock)

2000 to 3000 supercell thunderstorms occur annually in the United States. They represent just a small fraction of all thunderstorms, but they are responsible for a disproportionate share of the deaths, injuries, and property damage associated with severe weather. Fewer than half of all supercells produce tornadoes, yet virtually all of the strongest and most violent tornadoes are spawned by supercells.

A supercell consists of a single, very powerful cell that at times can extend to heights of 20 kilometers (65,000 feet) and persist for many hours. These massive clouds have diameters ranging between about 20 and 50 kilometers (12 and 30 miles).

Despite the single-cell structure of supercells, these storms are remarkably complex. The vertical wind profile may cause the updraft to rotate. For example, this could occur if the surface flow is from the south or southeast and the winds aloft increase in speed and become more westerly with height. If a thunderstorm develops in such a wind environment, the updraft is made to rotate. It is within this column of cyclonically rotating air, called the **mesocyclone,** that tornadoes often form (see Figure 10–8c).*

The huge quantities of latent heat needed to sustain a supercell require special conditions that keep the lower troposphere warm and moisture rich. Studies suggest that the existence of an inversion layer a few kilometers above the surface helps to provide this basic requirement. Recall that temperature inversions represent very stable atmospheric conditions that restrict vertical air motions. The presence of an inversion seems to aid the production of a few very large thunderstorms by inhibiting the formation of many smaller ones (Figure 10–9). The inversion prevents the mixing of warm, humid air in the lower troposphere with cold, dry air above. Consequently, surface heating continues to increase the temperature and moisture content of the layer of air trapped below the inversion. Eventually, the inversion is locally eroded by

---

*More on mesocyclones can be found in the section on "Tornado Development."

strong mixing from below. The unstable air below "erupts" explosively at these sites, producing unusually large cumulonimbus towers. It is from such clouds, with their concentrated, persistent updrafts, that supercells form.

> ## Students Sometimes Ask...
>
> What would it be like inside a towering cumulonimbus cloud?
> The words *violent* and *dangerous* come to mind! Here is an example that involved a German paragliding champion who got pulled into an erupting thunderstorm near Tamworth, New South Wales, Australia, and blacked out.* Updrafts lofted her to 32,600 feet, covered her with ice, and pelted her with hailstones. She finally regained consciousness around 22,600 feet. Her GPS equipment and a computer tracked her movements as the storm carried her away. Shaking vigorously and with lightning all around, she managed to slowly descend and eventually land 40 miles from where she started.
>
> *Reported in the *Bulletin of the American Meteorological Society,* Vol. 88, No. 4 (April 2007): 490.

Because the atmospheric conditions favoring the formation of severe thunderstorms often exist over a broad area, they frequently develop in groups that consist of many individual storms clustered together. Sometimes these clusters occur as elongate bands called *squall lines.* At other times the storms are organized into roughly circular clusters known as *mesoscale convective complexes.* No matter how the cells are arranged, they are not simply clusters of unrelated individual storms. Rather, they are related by a common origin, or they occur in a situation in which some cells lead to the formation of others.

Squall Lines

A **squall line** is a relatively narrow band of thunderstorms, some of which may be severe, that develops in the warm sector of a midlatitude cyclone, usually 100 to 300 kilometers (60 to 180 miles) in advance of the cold front. The

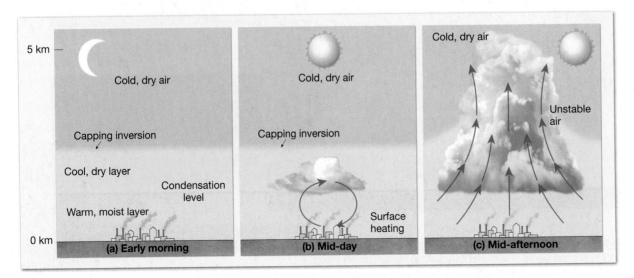

Figure 10–9 The formation of severe thunderstorms can be enhanced by the existence of a temperature inversion located a few kilometers above the surface.

SEVERE AND HAZARDOUS WEATHER

Flash Floods—Thunderstorms' Number-One Killer

Tornadoes and hurricanes are nature's most awesome storms. Because of this status, they are logically the focus of much well-deserved attention. Yet, in most years these dreaded events are not responsible for the greatest number of storm-related deaths. That distinction is reserved for floods (Figure 10–A). For the 10-year period 2001–2010, the number of storm-related deaths in the United States from flooding averaged 71 per year. By contrast, tornado fatalities averaged 56 annually. The average number of fatalities for hurricanes was dramatically affected by Hurricane Katrina in 2005 (more than 1000). For all years other than 2005 represented in Figure 10–A, the average number of hurricane fatalities was fewer than 20.

Floods are part of the *natural* behavior of streams. Most have a meteorologic origin related to atmospheric processes that can vary greatly in both time and space. Major regional floods in large river valleys often result from an extraordinary series of precipitation events over a broad region for an extended time span. Examples of such floods are discussed in the preceding chapter. By contrast, just an hour or two of intense thunderstorm activity can trigger flash floods in small valleys. Such situations are described here.

Flash floods are local floods of great volume and short duration. The rapidly rising surge of water usually occurs with little ad-

FIGURE 10–B A raging flash flood deposited this pile of debris and washed out a road near Hagerman, New Mexico, in July 2009. (AP Photo/*Roswell Daily Record*, Mark Wilson)

Floods are part of the natural behavior of streams.

vance warning and can destroy roads, bridges, homes, and other substantial structures (Figure 10–B). The amount of water flowing in the channel quickly reaches a maximum and then diminishes rapidly. Flood flows often contain large quantities of sediment and debris as they sweep channels clean.

Several factors influence flash flooding. Among them are rainfall intensity and duration, surface conditions, and topography. Urban areas are susceptible to flash floods because a high percentage of the surface area is composed of impervious roofs, streets, and parking lots, where runoff is very rapid (Figure 10–C). In fact, a recent study indicated that the area of impervious surfaces in the United States (excluding Alaska and Hawaii) amounts to more than 112,600 square kilometers (nearly 44,000 square miles), which is slightly smaller than the area of the state of Ohio.*

Frequently, flash floods result from the torrential rains associated with a slow-moving severe thunderstorm or take place when a series of thunderstorms repeatedly pass over the same location. Sometimes they are triggered by heavy rains from hurricanes and tropical storms. Occasionally, floating debris or ice can accumulate at a natural or artificial obstruction and restrict the flow of water. When such temporary dams fail, torrents of water can be released as flash floods.

Flash floods can take place in almost any area of the country. They are particularly common in mountainous terrain, where steep

Flash floods are local floods of great volume and short duration.

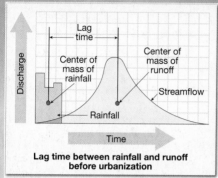

Lag time between rainfall and runoff before urbanization

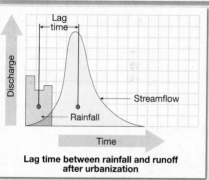

Lag time between rainfall and runoff after urbanization

FIGURE 10–C Urban areas are susceptible to flash floods because runoff following heavy rains is rapid due to the high percentage of the surface that is impervious. The graphs demonstrate the effect of urbanization on streamflow in a hypothetical setting. Notice that the water level in the stream does not rise at the onset of precipitation because time is required for water to move from the place where it fell to the stream. This span is called the *lag time*. Notice that when an area changes from rural to urban, the lag time is shortened and the flood peak gets higher. (After L. B. Leopold, U.S. Geological Survey)

slopes can quickly channel runoff into narrow valleys. The hazard is most acute when the soil is already nearly saturated from earlier rains or consists of impermeable materials.

Why do so many people perish in flash floods? Aside from the factor of surprise (many are caught sleeping), people do not appreciate the power of moving water. Just 15 centimeters (6 inches) of fast-moving flood water can knock a person down. Most automobiles will float and be swept away in only 0.6 meter (2 feet) of water. *More than half of all U.S. flash-flood fatalities are auto related!* Clearly, people should never attempt to drive over a flooded road.

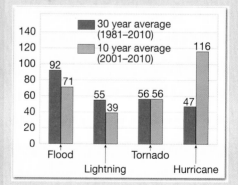

FIGURE 10–A Average annual storm-related deaths in the United States for two time spans. The average number of hurricane fatalities was dramatically affected by a single event—Hurricane Katrina in 2005 (more than 1000 deaths). For all of the other years represented on this graph, the average number of hurricane-related fatalities was fewer than 20 per year. (Data from National Weather Service)

*C. C. Elvidge, et al., "U.S. Constructed Area Approaches the Size of Ohio," in *EOS, Transactions, American Geophysical Union*, Vol. 85, No. 24 (15 June 2004): 233.

Figure 10–10 The dark overcast of a mammatus sky, with its characteristic downward-bulging pouches, sometimes precedes a squall line. When a mammatus formation develops, it is usually after a cumulonimbus cloud reaches its maximum size and intensity. Its presence is generally a sign of an especially vigorous thunderstorm.

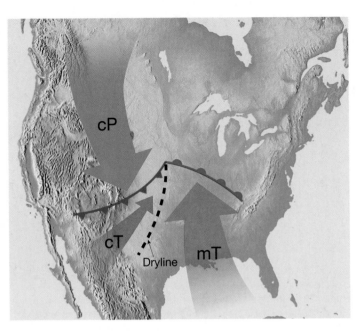

Figure 10–11 Squall-line thunderstorms frequently develop along a dryline, the boundary separating warm, dry continental tropical air and warm, moist maritime tropical air.

linear band of cumulonimbus development might stretch for 500 kilometers (300 miles) or more and consist of many individual cells in various stages of development. An average squall line can last for 10 hours or more, and some have been known to remain active for more than a day. Sometimes the approach of a squall line is preceded by a *mammatus sky*, consisting of dark cloud rolls that have downward pouches (Figure 10–10).

Most squall lines are not the product of forceful lifting along a cold front. Some develop from a combination of warm, moist air near the surface and an active jet stream aloft. The squall line forms when the divergence and resulting lift created by the jet stream is aligned with a strong, persistent low-level flow of warm, humid air from the south.

A squall line with severe thunderstorms can also form along a boundary called a **dryline,** a narrow zone along which there is an abrupt change in moisture. It forms when continental tropical (cT) air from the southwestern United States is pulled into the warm sector of a midlatitude cyclone, as shown in Figure 10–11. The denser cT air acts to lift the less dense mT air with which it is converging.* In this case, both cloud formation and storm development along the cold front are minimal because the front is advancing into dry cT air.

Drylines most frequently develop in the western portions of Texas, Oklahoma, and Kansas. Such a situation is illustrated by Figure 10–12. The dryline is easily identified

*Warm, dry air is more dense than warm, humid air because the molecular weight of water vapor (H_2O) is only about 62 percent as great as the molecular weight of the mixture of gases that make up dry air.

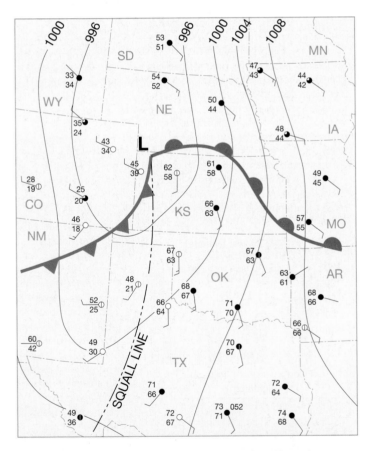

Figure 10–12 The squall line of this middle-latitude cyclone was responsible for a major outbreak of tornadoes. The squall line separates dry (cT) and very humid (mT) air. The dryline is easily identified by comparing dew-point temperatures on either side of the squall line. The dew point (°F) is the lower number at each station.

by comparing the dew-point temperatures on either side of the squall line. The dew points in the mT air to the east are 30° to 45°F higher than those in the cT air to the west. Much severe weather was generated as this extraordinary squall line moved eastward, including 55 tornadoes over a six-state region.

Mesoscale Convective Complexes

A **mesoscale convective complex (MCC)** consists of many individual thunderstorms organized into a large oval to circular cluster. A typical MCC is large, covering an area of at least 100,000 square kilometers (39,000 square miles). The usually slow-moving complex may persist for 12 hours or more (Figure 10–13).

MCCs tend to form most frequently in the Great Plains. When conditions are favorable, an MCC develops from a group of afternoon air-mass thunderstorms. In the evening, as the local storms decay, the MCC starts developing. The transformation of afternoon air-mass thunderstorms into an MCC requires a strong low-level flow of very warm and moist air. This flow enhances instability, which in turn spurs convection and cloud development. As long as favorable conditions prevail, MCCs remain self-propagating as gust fronts from existing cells lead to the formation of new powerful cells nearby. New thunderstorms tend to develop near the side of the complex that faces the incoming low-level flow of warm, moist air.

Although mesoscale convective complexes sometimes produce severe weather, they are also beneficial because they provide a significant portion of the growing-season rainfall to the agricultural regions of the central United States.

Figure 10–13 This satellite image shows a mesoscale convective complex (MCC) over the eastern Dakotas. (NOAA)

ning. Why? The answer relates to the wide geographic occurrence and frequency of lightning. Lightning strikes the ground tens of millions of times each year. Because lightning is so widespread and strikes the ground with such great frequency, it is not possible to warn each person of every flash. For this reason, lightning is the most dangerous weather hazard many people encounter during the year (Table 10–1).

A storm is classified as a thunderstorm only after thunder is heard. Because thunder is produced by lightning,

---

Concept Check 10.5

1 What is a dryline?

2 Briefly describe the formation of a squall line along a dryline.

3 What circumstances favor the development of mesoscale convective complexes?

Lightning and Thunder

In most years, **lightning** ranks second only to floods in the number of storm-related deaths in the United States. Although the number of reported lightning deaths in the United States averages about 60 per year, some go unreported. It is estimated that as many as 100 people are killed and more than 1000 injured by lightning every year in the United States.

Watches, warnings, and forecasts are routinely issued for floods, tornadoes, and hurricanes, but not for light-

TABLE 10–1 Lightning Casualties in the United States, by Location or Activity

Rank	Location/Activity	Relative Frequency
1	Open areas (including sports fields)	45%
2	Going under trees to keep dry	23%
3	Water-related activities (swimming, boating, and fishing)	14%
4	Golfing (while in the open)	6%
5	Farm and construction vehicles (with open exposed cockpits)	5%
6	Corded telephone (number-one indoor source of lightning casualties)	4%
7	Golfing (while mistakenly seeking "shelter" under trees)	2%
8	Using radios and radio equipment	1%

SOURCE: National Weather Service.

SEVERE AND HAZARDOUS WEATHER   Downbursts

Downbursts are strong localized zones of sinking air that originate in the lower part of some cumulus and cumulonimbus clouds. Downbursts are different from typical thunderstorm downdrafts in that they are more intense and concentrated over smaller areas (Figure 10–D). Their small horizontal dimension, usually less than 4 kilometers (2.5 miles), is the reason the alternate term *microburst* is often used. When a downburst reaches the ground, air spreads out in all directions, similar to a jet of water from a faucet splashing in a sink. Within minutes the downburst dissipates, while the outflow of air at the ground continues to expand. The straight-line winds from a downburst can exceed 160 kilometers (100 miles) per hour and can cause damages equivalent to that of a weak tornado.

> **The straight-line winds from a downburst can exceed 160 kilometers (100 miles) per hour and can cause damages equivalent to that of a weak tornado.**

The acceleration of air in a downburst occurs when evaporating raindrops cool the air. Remember, the colder the air, the denser it is, and the denser the air, the faster it will "fall." A second mechanism that contributes to downbursts is the drag of falling precipitation. Although a single raindrop does not exert much drag, the force of millions of falling drops can be substantial.

Downbursts are extremely hazardous to aircraft

The violent winds of a downburst can be dangerous and destructive. For example, in July 1993 millions of trees were uprooted by a downburst near Pak Wash, Ontario. In July 1984, 11 people drowned when a downburst caused a 28-meter-long (90-foot-long) sternwheeler boat to capsize in the Tennessee River. Downbursts are extremely hazardous to aircraft, especially during takeoffs and landings, when planes are nearest the ground. In just a matter of minutes, a cloud can change from being dominated by updrafts to having a downburst. Imagine an aircraft attempting to land and being confronted by a downburst, as in Figure 10–E. As the airplane flies into the

FIGURE 10–D The rain shaft extending downward from a cumulonimbus cloud identifies a powerful downburst near Denver's Stapleton Airport. (Photo by Wendy Schreiber/University Corporation for Atmospheric Research)

downburst, it initially encounters a strong headwind, which tends to carry it upward. To reduce the lift, the pilot points the nose of the aircraft downward. Then, seconds later, a tailwind is encountered. Now, because the wind is moving with the airplane, the amount of air flowing over the wings and providing lift is dramatically reduced,

causing the craft to suddenly lose altitude and possibly crash. This serious aviation hazard has been reduced significantly because systems to detect the wind shifts associated with downbursts were developed and deployed at major airports. Moreover, pilots receive training on how to handle downbursts during takeoffs and landings.

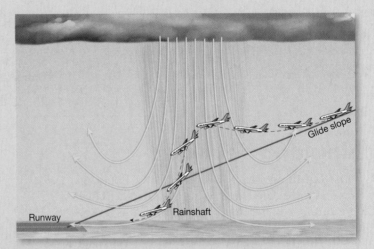

FIGURE 10–E The arrows in this sketch represent the downward and outward movement of air in a downburst. An airplane passing through a downburst as it attempts to land initially experiences a strong headwind and lift. That is followed by an abrupt descent, caused by the downward motion of air, and a rapid loss of air speed.

Figure 10–14 Summertime lightning display. (Photo by agefotostock/SuperStock)

lightning must also be present (Figure 10–14). Lightning is similar to the electrical shock you may have experienced when touching a metal object on a very dry day. However, the intensity is dramatically different.

During the formation of a large cumulonimbus cloud, a separation of charge occurs; this means that part of the cloud develops an excess negative charge, whereas another part acquires an excess positive charge. The objective of lightning is to equalize these electrical differences by producing a negative flow of current from the region of excess negative charge to the region with excess positive charge or vice versa. Because air is a poor conductor of electricity (good insulator), the electrical potential (charge difference) must be very high before lightning will occur.

The most common type of lightning occurs between oppositely charged zones *within* a cloud or between clouds. About 80 percent of all lightning is of this type. It is often called *sheet lightning* because it produces a bright but diffuse illumination of the parts of the cloud in which the flash occurred. Sheet lightning is not a unique form; rather, it is ordinary lightning in which the flash is obscured by the clouds. The second type of lightning, in which the electrical discharge occurs between the cloud and Earth's surface, is often more dramatic. This *cloud-to-ground lightning* represents about 20 percent of lightning strokes and is the most damaging and dangerous form.

What Causes Lightning?

The origin of charge separation in clouds, although not fully understood, must hinge on rapid vertical movements within clouds because lightning occurs primarily in the violent mature stage of a cumulonimbus cloud. In the mid-latitudes the formation of these towering clouds is chiefly a summertime phenomenon, which explains why lightning is seldom observed there in the winter. Furthermore, lightning rarely occurs before the growing cloud penetrates the 5-kilometer level, where sufficient cooling begins to generate ice crystals.

Some cloud physicists believe that charge separation occurs during the formation of ice pellets. Experimentation shows that as droplets begin to freeze, positively charged ions are concentrated in the colder regions of the droplets, whereas negatively charged ions are concentrated in the warmer regions. Thus, as a droplet freezes from the outside in, it develops a positively charged ice shell and a negatively charged interior. As the interior begins to freeze, it expands and shatters the outside shell. The small positively charged ice fragments are carried upward by turbulence, and the relatively heavy droplets eventually carry their negative charge toward the cloud base. As a result, the upper part of the cloud is left with a positive charge, and the lower portion of the cloud maintains an overall negative charge, with small positively charged pockets (Figure 10–15).

As the cloud moves, the negatively charged cloud base alters the charge at the surface directly below by repelling negatively charged particles. Thus, the surface beneath the cloud acquires a net positive charge. These charge differences build to millions and even hundreds of millions of volts before a lightning stroke acts to discharge the negative region of the cloud by striking the positive area of the ground below, or, more frequently, the positively charged portion of that cloud or a nearby cloud.

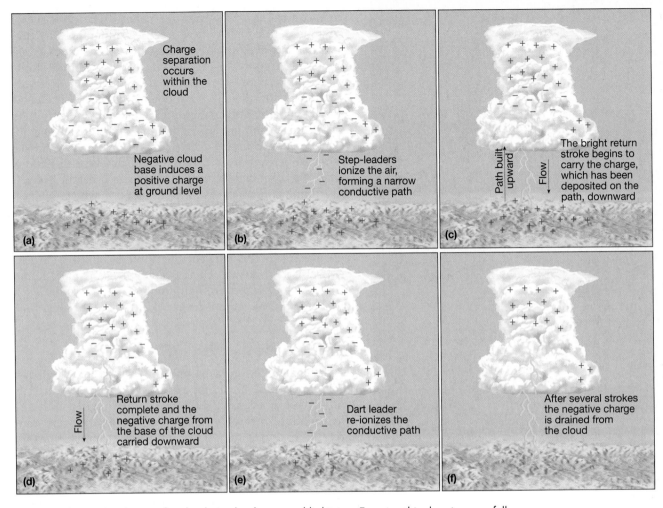

Figure 10–15 Discharge of a cloud via cloud-to-ground lightning. Examine this drawing carefully while reading the text.

Lightning Strokes

Cloud-to-ground strokes are of most interest and have been studied in detail. Moving-film cameras have greatly aided in these studies (Figure 10–16). They show that the lightning we see as a single flash is really several very rapid strokes between the cloud and the ground. We call the total discharge—which lasts only a few tenths of a second and appears as a bright streak—the **flash.** Individual components that make up each flash are termed **strokes.** Each stroke is separated by roughly 50 milliseconds, and there are usually three to four strokes per flash.* When a lightning flash appears to flicker, it is because your eyes discern the individual strokes that make up this discharge. Moreover, each stroke consists of a downward propagating leader that is immediately followed by a luminous return stroke.

Each stroke is believed to begin when the electrical field near the cloud base frees electrons in the air immediately below, thereby ionizing the air (Figure 10–15). Once ionized, the air becomes a conductive path with a radius of roughly 10 centimeters and a length of 50 meters. This path is called a **leader.** During this electrical breakdown,

Figure 10–16 Multiple lightning stroke of a single flash as recorded by a moving-film camera. (Courtesy of E. J. Tarbuck)

---

*One millisecond equals one one-thousandth ($^1/_{1000}$) of a second.

EYE ON THE ATMOSPHERE

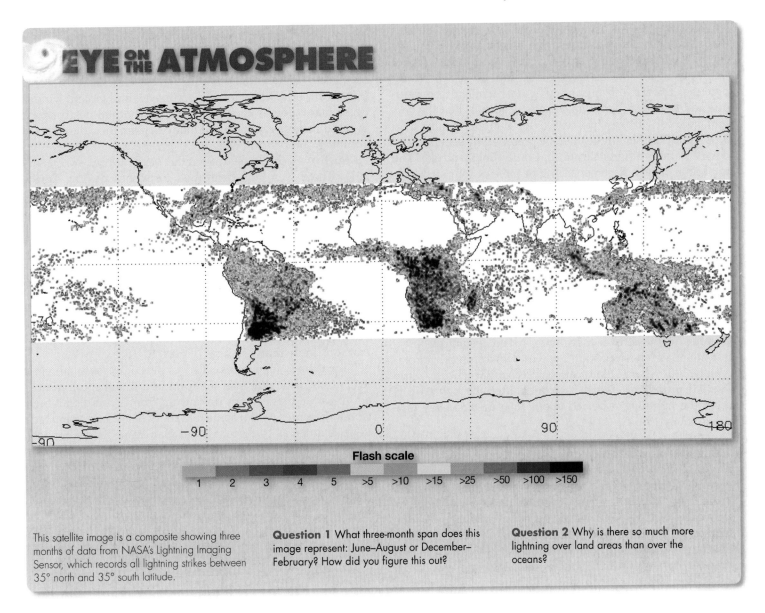

Flash scale

1　2　3　4　5　>5　>10　>15　>25　>50　>100　>150

This satellite image is a composite showing three months of data from NASA's Lightning Imaging Sensor, which records all lightning strikes between 35° north and 35° south latitude.

Question 1 What three-month span does this image represent: June–August or December–February? How did you figure this out?

Question 2 Why is there so much more lightning over land areas than over the oceans?

the mobile electrons in the cloud base begin to flow down this channel. This flow increases the electrical potential at the head of the leader, which causes a further extension of the conductive path through further ionization. Because this initial path extends itself earthward in short, nearly invisible bursts, it is called a **step leader.** Once this channel nears the ground, the electrical field at the surface ionizes the remaining section of the path. With the path completed, the electrons that were deposited along the channel begin to flow downward. The initial flow begins near the ground.

As the electrons at the lower end of the conductive path move earthward, electrons positioned successively higher up the channel begin to migrate downward. Because the path of electron flow is continually being extended upward, the accompanying electrical discharge has been appropriately named a **return stroke.** As the wave front of the return stroke moves upward, the negative charge that was deposited on the channel is effectively lowered to the ground. This intense return stroke illuminates the conductive path

and discharges the lowest kilometer or so of the cloud. During this phase, tens of coulombs of negative charge are lowered to the ground.*

Students Sometimes Ask...

What are the chances of being struck by lightning?

Most people greatly underestimate the probability of being involved in a lightning strike. According to the National Weather Service, the chance of an individual in the United States being killed or injured during a given year is 1 in 240,000. Assuming a life span of 80 years, a person's odds over his or her lifetime becomes 1 in 3000. Assuming that the average person has 10 family members and others with whom he or she is close, the chances are 1 in 300 that a lightning strike will closely affect a person during his or her lifetime.

---

*A coulomb is a unit of electrical charge equal to the quantity of charge transferred in 1 second by a steady current of 1 ampere.

The first stroke is usually followed by additional strokes that apparently drain charges from higher areas within the cloud. Each subsequent stroke begins with a **dart leader** that once again ionizes the channel and carries the cloud potential toward the ground. The dart leader is continuous and less branched than the step leader. When the current between strokes has ceased for periods greater than 0.1 second, further strokes will be preceded by a step leader whose path is different from that of the initial stroke. The total time of each flash consisting of three or four strokes is about 0.2 second.

Thunder

The electrical discharge of lightning superheats the air immediately around the lightning channel. In less than a second the temperature rises by as much as 33,000°C. When air is heated this quickly, it expands explosively and produces the sound waves we hear as **thunder.** Because lightning and thunder occur simultaneously, it is possible to estimate the distance to the stroke. Lightning is seen instantaneously, but the relatively slow sound waves, which travel approximately 330 meters (1000 feet) per second, reach us a little later. If thunder is heard 5 seconds after the lightning is seen, the lightning occurred about 1650 meters away (approximately 1 mile).

The thunder that we hear as a rumble is produced along a long lightning path located at some distance from the observer. The sound that originates along the path nearest the observer arrives before the sound that originated farthest away. This factor lengthens the duration of the thunder. Reflection of the sound waves by mountains or buildings further delays their arrival and adds to this effect. When lightning occurs more than 20 kilometers (12 miles) away, thunder is rarely heard. This type of lightning, popularly called *heat lightning,* is no different from the lightning that we associate with thunder.

Concept Check 10.6

1 How is thunder produced?

2 Which is more common: sheet lightning or cloud-to-ground lightning?

3 What is heat lightning?

Students Sometimes Ask...

About how many people who are struck by lightning are actually killed?

According to the National Weather Service, about 10 percent of lightning strike victims are killed; 90 percent survive. However, survivors are not unaffected. Many suffer severe, life-long injury and disability.

Tornadoes

Tornadoes are local storms of short duration that must be ranked high among nature's most destructive forces (Figure 10–17). Their sporadic occurrence and violent winds cause many deaths each year. The nearly total destruction in some stricken areas has led many to liken their passage to bombing raids during war.

Such was the case during the very stormy spring of 2011. There were 753 confirmed tornadoes during April, setting a record for the number of tornadoes in a single month. The deadliest and most destructive outbreak occurred between April 25 and 28, when 326 confirmed tornadoes struck, mostly in the South. The loss of life and property were extraordinary. Estimated fatalities numbered between 350 and 400, and damages were in the billions of dollars. The tornado that struck Tuscaloosa, Alabama, was especially notable for the death and destruction it caused. A few weeks later, on May 22, another outbreak struck the Midwest. Joplin, Missouri, was in the direct path of a storm that took more than 150 lives (Figure 10–18).

Tornadoes, sometimes called *twisters* or *cyclones,* are violent windstorms that take the form of a rotating column of air, or *vortex,* that extends downward from a cumulonimbus cloud. Pressures within some tornadoes have been estimated to be as much as 10 percent lower than immediately outside the storm. Drawn by the much lower pressure in the center of the vortex, air near the ground rushes into the tornado from all directions. As the air streams inward, it is spiraled upward around the core until it eventually merges with the airflow of the parent thunderstorm deep in the cumulonimbus tower.

Figure 10–17 An EF4 tornado near Manchester, South Dakota, on June 24, 2003. A tornado is a violently rotating column of air in contact with the ground. The air column is visible when it contains condensation or when it contains dust and debris. Often the appearance is the result of both. When the column of air is aloft and does not produce damage, the visible portion is properly called a *funnel cloud.* (Photo by Carsten Peter/National Geographic Stock)

(a)

(b)

Figure 10–18 A devastating multivortex tornado rated EF5 struck Joplin, Missouri, on Sunday, May 22, 2011. There were more than 150 tornado deaths and nearly 1000 injured. (a) An aerial view on Tuesday morning, May 24th. (Photo by J. B. Forbes/MCT/Newscom) (b) A ground-level view showing total destruction. (Photo by c51/ZUMA Press/Newscom)

suction vortices that orbit the center of the larger tornado circulation (Figure 10–19). The tornadoes in this latter category are called **multiple-vortex tornadoes.** Suction vortices have diameters of only about 10 meters (30 feet) and usually form and die out in less than a minute. They can occur in all sorts of tornado sizes, from huge "wedges" to narrow "ropes." Suction vortices are responsible for most of the narrow, short swaths of extreme damage that sometimes occur through tornado tracks. It is now believed that most reports of several tornadoes at once—from news accounts and early-twentieth-century tornado tales—actually were multiple-vortex tornadoes.

Because of the tremendous pressure gradient associated with a strong tornado, maximum winds can sometimes exceed 480 kilometers (300 miles) per hour. For example, using Doppler radar observations, scientists measured wind speeds of 486 kilometers (302 miles) per hour in a devastating tornado that struck the Oklahoma City area in May 1999. Reliable wind-speed measurements using traditional anemometers are lacking.

Most records of changes in atmospheric pressure associated with the passage of a tornado are estimates based on a few storms that happened to pass a nearby weather station or were studied by storm-chasing meteorologists with mobile equipment. Many attempts have been made to deploy instruments in the path of a tornado, but only a few have met with success. One of the most successful

Because of the rapid drop in pressure, air sucked into the storm expands and cools adiabatically. If the air cools below its dew point, the resulting condensation creates a pale and ominous-appearing cloud that may darken as it moves across the ground, picking up dust and debris. Occasionally, when the inward spiraling air is relatively dry, no condensation funnel forms because the drop in pressure is not sufficient to cause the necessary adiabatic cooling. In such cases, the vortex is made visible only by the material that it vacuums from the surface and carries aloft.

A tornado may consist of a single vortex, but within many stronger tornadoes are smaller intense whirls called

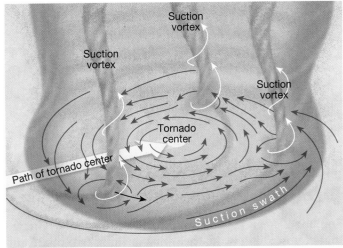

Figure 10–19 Some tornadoes have multiple suction vortices. These small and very intense vortices are roughly 10 meters (30 feet) across and move in a counterclockwise path around the tornado center. Because of this multiple-vortex structure, one building might be heavily damaged and another one, just 10 meters away, might suffer little damage.

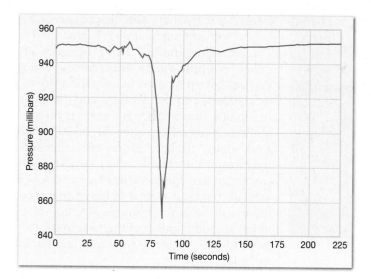

Figure 10–20 This graph shows the dramatic pressure change of 100 millibars in just 40 seconds that occurred as a violent tornado passed directly over a specially designed probe that was placed in the storm's path just moments earlier. Manchester, South Dakota, June 24, 2003. Figure 10-17 is a photo of this tornado.

attempts occurred in the small village of Manchester, South Dakota, on June 24, 2003. Meteorologists chasing a supercell thunderstorm deployed a specially designed probe directly in the path of a violent tornado. The tornado passed directly over the instrument package, which measured a sharp drop of 100 millibars over a span of about 40 seconds (Figure 10–20).* Gathering such data is challenging because tornadoes are short lived, highly localized, and dangerous. The development of Doppler radar, however, has improved our ability to study tornado-producing thunderstorms. As you will see, this technology is allowing meteorologists to expand our understanding from a safe distance.

Concept Check 10.7

1 Why do tornadoes have such high wind speeds?

2 Describe the formation of the visible funnel cloud that most tornadoes exhibit.

3 What are multiple-vortex tornadoes? Are all tornadoes of this type?

The Development and Occurrence of Tornadoes

Tornadoes form in association with severe thunderstorms that produce high winds, heavy (sometimes torrential) rainfall, and often damaging hail. Although hail may or may not precede a tornado, the portion of the thunderstorm ad-

jacent to large hail is often the area where strong tornadoes are most likely to occur.

Fortunately, fewer than 1 percent of all thunderstorms produce tornadoes. Nevertheless, a much higher number must be monitored as potential tornado producers. Although meteorologists are still not sure what triggers tornado formation, it is apparent that they are the product of the interaction between strong updrafts in a thunderstorm and the winds in the troposphere.

Tornado Development

Tornadoes can form in any situation that produces severe weather, including cold fronts, squall lines, and tropical cyclones (hurricanes). Usually the most intense tornadoes are those that form in association with supercells. An important precondition linked to tornado formation in severe thunderstorms is the development of a mesocyclone. A *mesocyclone* is a vertical cylinder of rotating air, typically about 3 to 10 kilometers (2 to 6 miles) across, that develops in the updraft of a severe thunderstorm. The formation of this large vortex often precedes tornado formation by 30 minutes or so.

Mesocyclone formation depends on the presence of vertical wind shear. Moving upward from the surface, winds change direction from southerly to westerly, and the wind speed increases. The speed wind shear (that is, stronger winds aloft and weaker winds near the surface) produces a rolling motion about a horizontal axis, as shown in Figure 10–21a. If conditions are right, strong updrafts in the storm tilt the horizontally rotating air to a nearly vertical alignment (see Figure 10–21b). This produces the initial rotation within the cloud interior.

At first the mesocyclone is wider, shorter, and rotating more slowly than will be the case in later stages. Subsequently, the mesocyclone is stretched vertically and narrowed horizontally, causing wind speeds to accelerate in an inward vortex (just as spinning ice skaters accelerate by pulling arms in, or as a sink full of water accelerates as it spirals down a drain).* Next, the narrowing column of rotating air stretches downward until a portion of the cloud protrudes below the cloud base to produce a very dark, slowly rotating *wall cloud*. Finally, a slender and rapidly spinning vortex emerges from the base of the wall cloud to form a *funnel cloud*. If the funnel cloud makes contact with the surface, it is then classified as a *tornado*.

The formation of a mesocyclone does not necessarily mean that tornado formation will follow. Only about half of all mesocyclones produce tornadoes. The reason for this is not understood. Because this is the case, forecasters cannot determine in advance which mesocyclones will spawn tornadoes.

Tornado Climatology

Severe thunderstorms—and hence tornadoes—are most often spawned along the cold front or squall line of a mid-

---

*J. J. Lee, T. M. Samaras, and C. R. Young, "Pressure Measurements at the Ground in an F-4 Tornado," in *Preprints of the 22nd Conference on Severe Local Storms*, Hyannis, MA, October 4, 2004.

*For more on this concept, see Box 11–1, on the conservation of angular momentum.

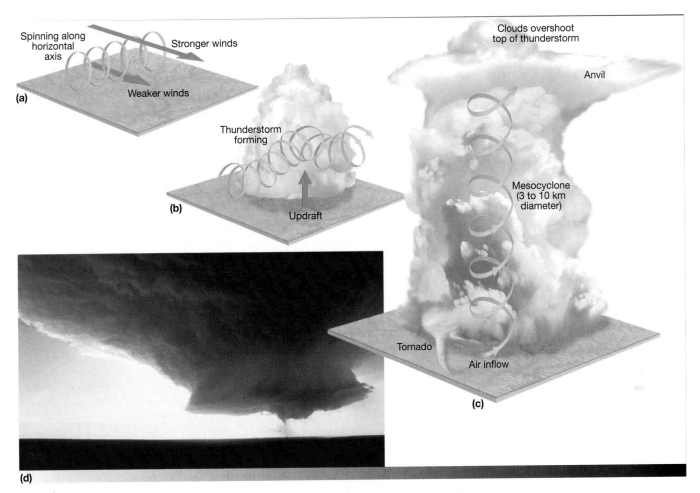

Figure 10–21 The formation of a mesocyclone often precedes tornado formation. (a) Winds are stronger aloft than at the surface (called *speed wind shear*), producing a rolling motion about a horizontal axis. (b) Strong thunderstorm updrafts tilt the horizontally rotating air to a nearly vertical alignment. (c) The mesocyclone, a vertical cylinder of rotating air, is established. (d) If a tornado develops, it will descend from a slowly rotating wall cloud in the lower portion of the mesocyclone. This supercell tornado hit the Texas Panhandle in May 1996. (Photo by Warren Faidley/Photolibrary)

latitude cyclone or in association with supercell thunderstorms. Throughout the spring, air masses associated with mid-latitude cyclones are most likely to have greatly contrasting conditions. Continental polar air from Canada may still be very cold and dry, whereas maritime tropical air from the Gulf of Mexico is warm, humid, and unstable. The greater the contrast, the more intense the storm tends to be.

These two contrasting air masses are most likely to meet in the central United States because there is no significant natural barrier separating the center of the country from the arctic or the Gulf of Mexico. Consequently, this region generates more tornadoes than any other part of the country or, in fact, the world. Figure 10–22, which depicts the average annual tornado incidence in the United States over a 27-year period, readily substantiates this fact.

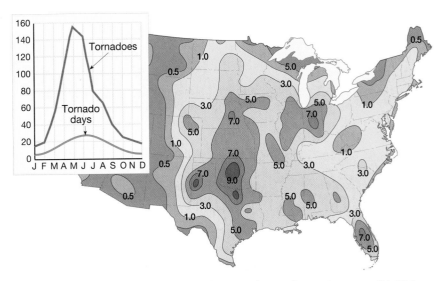

Figure 10–22 The map shows average annual tornado incidence per 26,000 square kilometers (10,000 square miles) for a 27-year period. The graph shows average number of tornadoes and tornado days each month in the United States for the same period. (After NOAA)

SEVERE AND HAZARDOUS WEATHER

Surviving a Violent Tornado

About 11 AM on Tuesday, July 13, 2004, much of northern and central Illinois was put on a tornado watch.* A large supercell had developed in the northwestern part of the state and was moving southeast into a very unstable environment (Figure 10–F). A few hours later, as the supercell entered Woodford County, rain began to fall, and the storm showed signs of becoming severe. The National Weather Service issued a *severe thunderstorm warning* at 2:29 PM CDT. Minutes afterward, a tornado developed. Twenty-three minutes later, the quarter-mile-wide twister had carved a 9.6-mile-long path across the rural Illinois countryside.

Fewer than 1 percent of tornadoes attain this level of severity.

What, if anything, made this storm special or unique? After all, it was just one of a record-high 1819 tornadoes reported in the United States in 2004. For one, this tornado attained EF-4 status for a portion of its life.** The National Weather Service estimated that maximum winds reached 240 miles per hour. Fewer than 1 percent of tornadoes attain this level of severity. How-

*Watches and warnings are discussed later in this chapter on page 296.
**EF-4 is a reference to the Enhanced Fujita Scale of tornado intensity. See Table 10–3, page 293.

FIGURE 10–F On July 13, 2004, an EF-4 tornado cut a 23-mile path through the rural countryside near the Woodford County town of Roanoke, Illinois. The Parsons Manufacturing plant was just west of town.

ever, what was most remarkable is that no one was killed or injured when the Parsons Manufacturing facility west of the small town of Roanoke took a direct hit while the storm was most intense. At the time, 150 people were in three buildings comprising the plant. The 250,000-square-foot facility was flattened, cars were twisted into gnarled masses, and debris was strewn for miles (Figure 10–G).

How did 150 people escape death or injury?

How did 150 people escape death or injury? The answer is foresight and plan-

ning. More than 30 years earlier, company owner Bob Parsons was inside his first factory when a small tornado passed close enough to blow out windows. Later, when he built a new plant, he made sure that the restrooms were constructed to double as tornado shelters, with steel-reinforced concrete walls and 8-inch-thick concrete ceilings. In addition, the company developed a severe weather plan. When the severe thunderstorm warning was issued at 2:29 PM on July 13, the emergency response team leader at the Parsons plant was immediately notified. A few moments later he went outside and observed a rotating wall cloud with a developing funnel cloud. He radioed back to the office to institute the company's severe weather plan. Employees were told to immediately go to their designated storm shelters. Everyone knew where to go and what to do because the plant conducted semi-annual tornado drills. All 150 people reached shelters in less than four minutes. The emergency response team leader was the last person to reach a shelter, less than 2 minutes before the tornado destroyed the plant at 2:41 PM.

The total number of tornado deaths in 2004 for the entire United States was just 36. The toll could have been much higher. The building of tornado shelters and the development of an effective severe storm plan made the difference between life and death for 150 people at Parsons Manufacturing.

FIGURE 10–G The quarter-mile-wide tornado had wind speeds reaching 240 miles per hour. The destruction at Parsons Manufacturing was devastating. (Photos courtesy of NOAA)

An average of nearly 1300 tornadoes were reported annually in the United States between 2000 and 2010. Still, the actual numbers that occur from one year to the next vary greatly. During the 11-year span just mentioned, for example, yearly totals ranged from a low of 935 in the year 2002 to a high of 1819 in 2004. Tornadoes occur during every month of the year. April through June is the period of greatest tornado frequency in the United States, and December and January are the months of lowest activity. Of the nearly 40,522 confirmed tornadoes reported over the contiguous 48 states during the 50-year period 1950–1999, an average of almost 6 per day occurred during May. At the other extreme, a tornado was reported only about every other day in December and January.

More than 40 percent of all tornadoes take place during the spring. Fall and winter, by contrast, together account for only 19 percent (Figure 10–22). In late January and February, when the incidence of tornadoes begins to increase, the center of maximum frequency lies over the central Gulf states. During March this center moves eastward, to the southeastern Atlantic states, with tornado frequency reaching its peak in April.

During May and June the center of maximum frequency moves through the southern Great Plains and then to the Northern Plains and Great Lakes area. This drift is due to the increasing penetration of warm, moist air while contrasting cool, dry air still surges in from the north and northwest. Thus, when the Gulf states are substantially under the influence of warm air after May, there is no cold-air intrusion to speak of, and tornado frequency drops. Such is the case across the country after June. Winter cooling permits fewer and fewer encounters between warm and cold air masses, and tornado frequency returns to its lowest level by December.

Profile of a Tornado

The average tornado has a diameter between 150 and 600 meters (500 to 2000 feet), travels across the landscape at approximately 45 kilometers (30 miles) per hour, and cuts a path about 26 kilometers (16 miles) long. Because many tornadoes occur slightly ahead of a cold front, in the zone of southwest winds, most move toward the northeast. The Illinois example shown in Figure 10–23 demonstrates this fact well. The figure also shows that many tornadoes do not fit the description of the "average" tornado.

Students Sometimes Ask...

What is "Tornado Alley"?

Tornado Alley is a nickname used by the popular media and others that refers to the broad swath of high tornado occurrence in the central United States (see Figure 10–22). The heart of Tornado Alley stretches from the Texas Panhandle through Oklahoma and Kansas to Nebraska. It's important to remember that violent (killer) tornadoes occur outside of Tornado Alley every year. Tornadoes can occur almost anywhere in the United States.

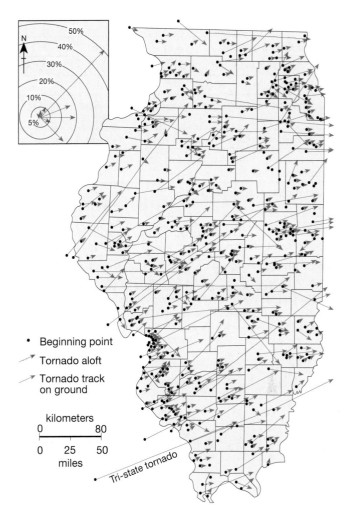

- • Beginning point
- ⟋ Tornado aloft
- ⟋ Tornado track on ground

kilometers
0 — 80

0   25   50
miles

Tri-state tornado

Figure 10–23 Paths of Illinois tornadoes (1916–1969). Because most tornadoes occur slightly ahead of a cold front, in the zone of southwest winds, they tend to move toward the northeast. Tornadoes in Illinois verify this. More then 80 percent of the tornadoes in this diagram exhibited directions of movement toward the northeast through east. (Paths of Illinois Tornadoes (1916-1969). John W. Wilson and Stanley Changnon, 1971. Reproduced by permission of the Illinois State Water Survey.)

Of the hundreds of tornadoes reported in the United States each year, more than half are comparatively weak and short lived. Most of these small tornadoes have lifetimes of three minutes or less and paths that seldom exceed 1 kilometer (0.6 mile) in length and 100 meters (330 feet) in width. Typical wind speeds are on the order of 150 kilometers (90 miles) per hour or less. On the other end of the tornado spectrum are the infrequent and often long-lived violent tornadoes (Table 10–2). Although large tornadoes constitute only a small percentage of the total reported, their effects are often devastating. Such tornadoes may exist for periods in excess of three hours and produce an essentially continuous damage path more than 150 kilometers (90 miles) long and perhaps 1 kilometer (0.6 mile) or more wide. Maximum winds range beyond 480 kilometers (300 miles) per hour (Figure 10–24).

TABLE 10–2 Tornado Extremes

Tornado Characteristic	Value	Date	Location
World's deadliest single tornado	1,300 dead, 12,000 injured	April 26, 1989	Salturia and Manikganj, Bangladesh
U.S. deadliest single tornado	695 dead	March 18, 1925	MO–IL–IN
U.S. deadliest tornado outbreak	747 dead	March 18, 1925	MO–IL–IN (includes the Tri-State Tornado deaths)
Biggest 24-hour total tornado outbreak	147 tornadoes	April 3–4, 1974	13 central U.S. states
Calendar month with greatest number of tornadoes	753 tornadoes	April 2011	United States
Widest tornado (diameter*)	Nearly 4000 m (2.5 miles) in width	May 22, 2004	Hallam, NE; EF-4 tornado
Highest recorded tornadic wind speed†	135 m/s (302 mph)	May 8, 1999	Bridge Creek, OK
Highest-elevation tornado	3650 m (12,000 ft)	Jul 7, 2004	Sequoia National Park, CA
Longest tornado transport	A personal check carried 359 km (223 miles)	April 11, 1991	Stockton, KS, to Winnetoon, NE

SOURCE: National Weather Service, Storm Prediction Center, as reported in *Bulletin of the American Meteorological Society*, Vol. 88, No. 6 (June 2007): 857.
*"Widest" is now defined by the National Weather Service as the maximum width of tornado damage.
†By necessity, this value is restricted to the small number of tornadoes sampled by mobile Doppler radars.

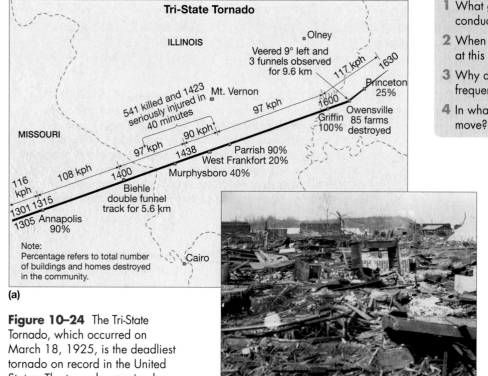

(a)

Figure 10–24 The Tri-State Tornado, which occurred on March 18, 1925, is the deadliest tornado on record in the United States. The tornado remained on the ground for more than 350 kilometers (219 miles), killing 695 and injuring 2027 people. The numbers immediately above the track line are a reference to time of day (24-hour clock). (After John W. Wilson and Stanley A. Changnon, Jr., *Illinois Tornadoes*, Illinois State Water Survey Circular 103, 1971, p. 32) (b) This historic image shows Murphysboro, Illinois, on March 21, 1925, three days after the Tri-State Tornado devastated the community. (Photo by Edeans)

Tornado Destruction

Because tornadoes generate the strongest winds in nature, they have accomplished many seemingly impossible tasks, such as driving a piece of straw through a thick wooden plank and uprooting huge trees (Figure 10–25). Although it may seem impossible for winds to cause some of the fantastic damage attributed to tornadoes, tests in engineering facilities have repeatedly demonstrated that winds in excess of 320 kilometers (200 miles) per hour are capable of incredible feats.

There is a long list of documented examples. In 1931 a tornado actually carried an 83-ton railroad coach and its 117 passengers 24 meters (80 feet) through the air and dropped them in a ditch. A year later, near Sioux Falls,

(a)

(b)

Figure 10–25 (a) The force of the wind during a tornado near Wichita, Kansas, in April 1991 was enough to drive this piece of metal into a utility pole. (Photo by John Sokich/NOAA) (b) The remains of a truck wrapped around a tree in Bridge Creek, Oklahoma, on May 4, 1999, following a major tornado outbreak. (AP Photo/L. M. Otero)

South Dakota, a steel beam 15 centimeters (6 inches) thick and 4 meters (13 feet) long was ripped from a bridge, flew more than 300 meters (nearly 1000 feet), and perforated a 35-centimeter-thick (14-inch-thick) hardwood tree. In 1970 an 18-ton steel tank was carried nearly 1 kilometer (0.6 mile) at Lubbock, Texas. Fortunately, the winds associated with most tornadoes are not this strong.

Tornado Intensity

Most tornado losses are associated with a few storms that strike urban areas or devastate entire small communities. The amount of destruction wrought by such storms depends to a significant degree (but not completely) on the strength of the winds. A wide spectrum of tornado strengths, sizes, and lifetimes are observed. The commonly used guide to tornado intensity is the **Enhanced Fujita intensity scale,** or the **EF-scale** for short (Table 10–3). Because tornado winds

cannot be measured directly, a rating on the EF-scale is determined by assessing the damage produced by a storm. Although widely used, the EF-scale is not perfect. Estimating tornado intensity based on damage alone does not take into account the structural integrity of the objects hit by a tornado. A well-constructed building can withstand very high winds, whereas a poorly built structure can suffer devastating damage from the same or even weaker winds.

The drop in atmospheric pressure associated with the passage of a tornado plays a minor role in the damage process. Most structures have sufficient venting to allow for the sudden drop in pressure. Opening a window, once thought to be a way to minimize damage by allowing inside and outside atmospheric pressure to equalize, is no longer recommended. In fact, if a tornado gets close enough to a structure for the pressure drop to be experienced, the strong winds probably will have already caused significant damage.

Although the greatest part of tornado damage is caused by violent winds, most tornado injuries and deaths result from flying debris. On average, tornadoes cause more deaths each year than any other weather events except lightning and flash floods. For the United States, the average annual death toll from tornadoes is about 60 people. However, the actual number of deaths each year can depart significantly from the average. On April 3–4, 1974, for ex-

TABLE 10–3 Enhanced Fujita Intensity Scale*

| Scale | Wind Speed | | Damage |
	Km/Hr	Mi/Hr	
EF-0	105–137	65–85	*Light.* Some damage to siding and shingles.
EF-1	138–177	86–110	*Moderate.* Considerable roof damage. Winds can uproot trees and overturn single-wide mobile homes. Flagpoles bend.
EF-2	178–217	111–135	*Considerable.* Most single-wide mobile homes destroyed. Permanent homes can shift off foundations. Flagpoles collapse. Softwood trees debarked.
EF-3	218–265	136–165	*Severe.* Hardwood trees debarked. All but small portions of houses destroyed.
EF-4	266–322	166–200	*Devastating.* Complete destruction of well-built residences, large sections of school buildings.
EF-5	>322	>200	*Incredible.* Significant structural deformation of mid- and high-rise buildings.

*The original Fujita scale was developed by T. Theodore Fujita in 1971 and put into use in 1973. The Enhanced Fujita Scale is a revision that was put into use in February 2007. Winds speeds are estimates (not measurements) based on damage and represent three-second gusts at the point of damage. More information about the criteria used to evaluate tornado intensity can be found at http://www.spc.noaa.gov/efscale/.

ample, an outbreak of 147 tornadoes brought death and destruction to a 13-state region east of the Mississippi River. More than 300 people died, and nearly 5500 people were injured (Figure 10–26).

Loss of Life

The proportion of tornadoes that result in the loss of life is small. In 2010, there were 1543 tornadoes reported in the United States. Of this total, just 22 were "killer tornadoes," a number that is about average. Put another way, in most years slightly less than 2 percent of all reported tornadoes in the United States are "killers." Although the percentage of tornadoes that result in death is small, every tornado is potentially lethal. If you examine Figure 10–27, which compares tornado fatalities with storm intensities, the results are quite interesting. It is clear from this graph that the majority (63 percent) of all tornadoes are weak and that the number of storms decreases as tornado intensity increases. The distribution of tornado fatalities, however, is just the opposite. Although only 2 percent of tornadoes are classified as violent, they account for nearly 70 percent of the deaths.

If there is some question about the causes of tornadoes, there certainly is none about the destructive effects of these

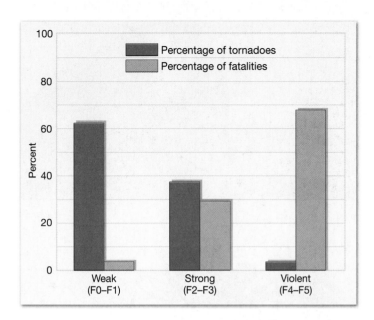

Figure 10–27 Percentage of tornadoes in each intensity category and percentage of fatalities associated with each category. Because this study was completed prior to the adoption of the EF-scale, storm intensities are expressed using the F-scale. Wind speeds (mph) for this scale are: F0 (<72), F1 (72–112), F2 (113–157), F3 (158–206), F4 (207–260), F5 (>260) (From Joseph T. Schaefer et al., "Tornadoes—When, Where and How Often," *Weatherwise*, 33, no. 2 (1980): 57)

violent storms. A severe tornado leaves the affected area stunned and disorganized and may require a response of the magnitude demanded in war.

Concept Check 10.9

1 Name the scale commonly used to rate tornado intensity. How is a rating on this scale determined?

2 What is the approximate number of tornado fatalities in an average year in the United States?

3 In an average year, about what percentage of tornadoes in the United States are "killer tornadoes"?

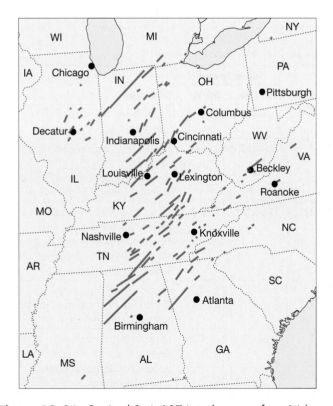

Figure 10–26 On April 3–4, 1974, in the span of just 16 hours, 147 tornadoes hit 13 states. Many of the tornado paths are shown here. It was the largest and costliest outbreak on record. The storms took 315 lives and injured 5500 more. Forty-eight tornadoes were killers, with 7 rated F5 on the Fujita scale and 23 rated F4. (This event took place when tornado intensities were described using the Fujita scale (F-scale), the predecessor of the EF-scale. Wind speeds (mph) for the F-scale are: F0 (<72), F1 (72–112), F2 (113–157), F3 (158–206), F4 (207–260), F5 (>260).) (After NOAA)

Tornado Forecasting

Because severe thunderstorms and tornadoes are small and short-lived phenomena, they are among the most difficult weather features to forecast precisely. Nevertheless, the prediction, detection, and monitoring of such storms are among the most important services provided by professional meteorologists. The timely issuance and dissemination of watches and warnings are both critical to the protection of life and property.

The Storm Prediction Center (SPC) located in Norman, Oklahoma, is part of the National Weather Service (NWS) and the National Centers for Environmental Prediction (NCEP). Its mission is to provide timely and accu-

PROFESSIONAL PROFILE

Warren Faidley: Storm Chaser

Warren Faidley is a storm chaser. As an extreme weather photojournalist, he has survived more blizzards, tornadoes, hailstorms, and flying debris than he would like to remember. His images of town-dwarfing tornadoes and hurricane destruction have been used in movies and magazines, news programs, and textbooks. As a frequent witness to many violent weather events, he is often interviewed for news programs and storm documentaries.

Faidley's calendar revolves around storms: Spring is tornado season, summer is lightning season, and late summer to fall is hurricane season. All the while, he analyzes weather charts, second-guesses forecasts, and consults Doppler radar data the way most people consult city maps. He spends his days zigzagging across the farm roads and lonely highways of Oklahoma, Kansas, and other states to approach storms in progress.

> It's one part science and meteorology, and another part artistry.

Getting near a storm is only half the job. "At the same time, I need to make images that convey what it looks like when a 2×4 goes through the side of a car." One trick is finding spots of color like a red barn and green fields against gray storm clouds and a gray sky. "It's one part science and meteorology, and another part artistry," Faidley says.

Faidley's life has been entwined with extreme weather since childhood. He nearly drowned after being swept away in a flash flood at age 12, and he steered his bicycle into dust devils as a teen. By the mid-1980s, after earning a degree in journalism, he decided to become the country's first full-time weather photojournalist.

His first break was a bolt from the blue. Faidley snapped a photo of a white-hot arc of lightning striking a light pole, suffusing the Arizona night in an eerie purple glow. Another fork hit perilously near Faidley, almost killing him. But the episode ended on a happy note. *Life* magazine published the lightning photo in 1989, launching Faidley's freelance career.

In 1992, Faidley followed up the lightning photo by obtaining some of the few existing shots of Hurricane Andrew in progress. Faidley hid under a shed in south Florida while the category 5 storm howled past.

> There are moments of terror but also moments of absolute beauty.

Storm chasing offers Faidley a heady mix of adrenaline and grace. "It's very awe-inspiring. There are moments of terror but also moments of absolute beauty. Capturing a picture of an orange sky cut in half by the emerald green of a coming storm is fantastic."

Faidley is no mere thrill-seeker. He does a great deal of advance planning and takes the precautions required to come back alive. In the late 1990s, he designed and built the first tornado-resistant chase vehicle, installing impact-resistant glass, a NASCAR-type roll cage, and other safety features on his SUV. "The purpose isn't to enable us to do something stupid, like penetrate a tornado.

Rather, it's to offer us safety in case something unexpected happens such as a sign blowing off a motel and careening down the road." He is always aware of escape roads whenever he's in storm country, and he speaks regularly to the public about the importance of staying informed and knowing how to respond during a violent weather episode.

When Faidley first began storm chasing in the 1980s, moment-by-moment weather information was hard to come by. Live weather radar on the Internet did not exist back then. Instead, he got to know National Weather Service forecasters and learned storm meteorology from them on the fly. Today, Faidley watches forecasts weeks ahead to ensure that he's within driving range when the looming clouds appear.

"I live a barnstorming, gypsy life driven by visual instinct," Faidley says. "The canvas keeps changing but the canvas wants to kill you. It's a juggling act."

Kathleen M. Wong

WARREN FAIDLEY is a well-known storm chaser and weather photographer. He is the author of *The Ultimate Storm Survival Handbook* and the autobiographical book *Storm Chaser*. (Photo courtesy of Warren Faidley)

rate forecasts and watches for severe thunderstorms and tornadoes.

Severe thunderstorm outlooks are issued several times daily. *Day 1* outlooks identify those areas likely to be affected by severe thunderstorms during the next 6 to 30 hours, and *day 2* outlooks extend the forecast through the following day. Both outlooks describe the type, coverage, and intensity of the severe weather expected. Many local NWS field of-

EYE ON THE ATMOSPHERE

This is a satellite image showing a portion of the diagonal path left by a tornado as it moved across northern Wisconsin in 2007. (NASA)

Question 1 Toward what direction did the storm advance, the northeast or the southwest?

Question 2 Did the tornado more likely occur ahead of or behind a cold front? Explain.

Question 3 Is it more probable that the storm took place in March or June? Why is the date you selected more likely?

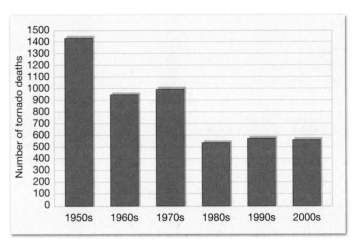

Figure 10–28 Number of tornado deaths in the United States by decade, 1950–2009. Even though the population has risen sharply since 1950, there has been a general downward trend in tornado deaths. (Data from NOAA)

fices also issue severe weather outlooks that provide a more local description of the severe weather potential for the next 12 to 24 hours.

Tornado Watches and Warnings

Tornado watches alert the public to the possibility of tornadoes over a specified area for a particular time interval. Watches serve to fine-tune forecast areas already identified in severe weather outlooks. A typical watch covers an area of about 65,000 square kilometers (25,000 square miles) for a four- to six-hour period. A tornado watch is an important part of the tornado alert system because it sets in motion the procedures necessary to deal adequately with detection, tracking, warning, and response. Watches are generally reserved for organized severe weather events where the tornado threat will affect at least 26,000 square kilometers (10,000 square miles) and/or persist for at least three hours. Watches typically are not issued when the threat is thought to be isolated and/or short lived.

Whereas a tornado watch is designed to alert people to the possibility of tornadoes, a **tornado warning** is issued by local offices of the National Weather Service when a tornado has actually been sighted in an area or is indicated by weather radar. It warns of a high probability of imminent danger. Warnings are issued for much smaller areas than watches, usually covering portions of a county or counties. In addition, they are in effect for much shorter periods, typi-

cally 30 to 60 minutes. Because a tornado warning may be based on an actual sighting, warnings are occasionally issued after a tornado has already developed. However, most warnings are issued prior to tornado formation, sometimes by several tens of minutes, based on Doppler radar data and/or spotter reports of funnel clouds or cloud-base rotation.

If the direction and the approximate speed of the storm are known, an estimate of its most probable path can be made. Because tornadoes often move erratically, the warning area is fan shaped downwind from the point where the tornado has been spotted. Improved forecasts and advances in technology have contributed to a significant decline in tornado deaths over the past 50 years. Figure 10–28 illustrates this trend. During a span when the U.S. population grew rapidly, tornado deaths trended downward.

As noted earlier, the probability of one place being struck by a tornado, even in the area of greatest frequency, is slight. Nevertheless, although the probabilities may be small, tornadoes have provided many mathematical exceptions. For example, the small town of Codell, Kansas, was hit three years in a row—1916, 1917, and 1918—and each time on the same date, May 20! Needless to say, tornado watches and warnings should never be taken lightly.

Doppler Radar

The installation of **Doppler radar** across the United States significantly improved our ability to track thunderstorms and issue warnings based on their potential to produce tornadoes (Figure 10–29). Conventional weather radar works by transmitting short pulses of electromagnetic energy. A small fraction of the waves that are sent out is scattered by a storm and returned to the radar. The strength of the returning signal indicates rainfall intensity, and the time difference between the transmission and return of the signal indicates the distance to the storm.

However, to identify tornadoes and severe thunderstorms, we must be able to detect the characteristic circulation

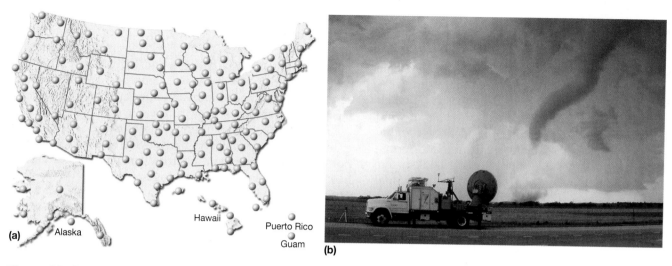

(a) Alaska Hawaii Puerto Rico Guam (b)

Figure 10–29 (a) Doppler radar sites in the United States. If you go to http://radar.weather.gov, you will see a similar map. You can click on any site to see the current National Weather Service Doppler radar display. (After NOAA) (b) Doppler on Wheels is a portable unit researchers use in field studies of severe weather events. (University Corporation for Atmospheric Research)

patterns associated with them. Conventional radar cannot do so except occasionally, when spiral rain bands occur in association with a tornado and give rise to a hook-shaped echo.

Doppler radar not only performs the same tasks as conventional radar but also has the ability to detect motion directly (Figure 10–30). The principle involved is known as the *Doppler effect* (Figure 10–31). Air movement in clouds is determined by comparing the frequency of the reflected signal to that of the original pulse. The movement of precipitation toward the radar increases the frequency of reflected pulses, whereas motion away from the radar decreases the frequency. These frequency changes are then interpreted in terms of speed toward or away from the Doppler radar unit. This same principle allows police radar to determine the speed of moving cars. Unfortunately, a single Doppler radar unit cannot detect air movements that occur parallel to it. Therefore, when a more complete picture of the winds within a cloud mass is desired, it is necessary to use two or more Doppler units.

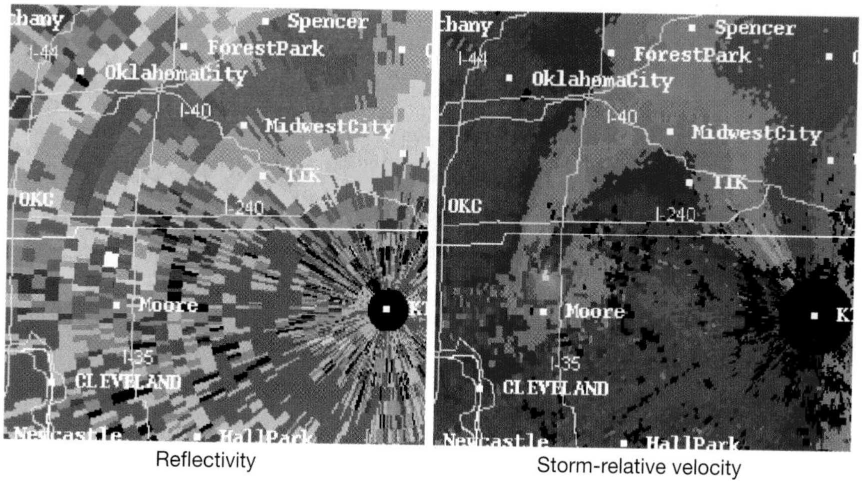

Reflectivity   Storm-relative velocity

Figure 10–30 This is a dual Doppler radar image of an EF-5 tornado near Moore, Oklahoma, on May 3, 1999. The left image (reflectivity) shows precipitation in the supercell thunderstorm. The right image shows motion of the precipitation along the radar beam—that is, how fast rain or hail is moving toward or away from the radar. In this example, the radar was unusually close to the tornado—close enough to make out the signature of the tornado itself. (Most of the time only the weaker and larger mesocyclone is detected.) (After NOAA)

Students Sometimes Ask...

How dangerous is it to be in a mobile home during a tornado?

Mobile homes represent a relatively small fraction of all residences in the United States. Yet, according to the National Weather Service, during the span 2000–2010, 52 percent of all tornado fatalities (314 of 604) occurred in mobile homes.

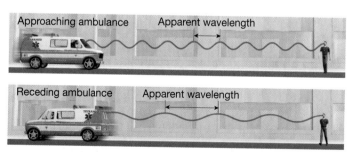

Approaching ambulance   Apparent wavelength

Receding ambulance   Apparent wavelength

Figure 10–31 This everyday example of the Doppler effect illustrates the apparent lengthening thand shortening of wavelengths caused by the relative movement between a source and an observer.

Doppler radar can detect the initial formation and subsequent development of the mesocyclone within a severe thunderstorm that frequently precedes tornado development. Almost all (96 percent) mesocyclones produce damaging hail, severe winds, or tornadoes. Those that produce tornadoes (about 50 percent) can sometimes be distinguished by their stronger wind speeds and their sharper gradients of wind speeds. Mesocyclones can sometimes be identified within parent storms 30 minutes or more before tornado formation, and if a storm is large, at distances up to 230 kilometers (140 miles). In addition, when close to the radar, individual tornado circulations may sometimes be detected. Ever since the implementation of the national Doppler network, the average lead time for tornado warnings has increased from less than 5 minutes in the late 1980s to about 13 minutes today.

Doppler radar is not without problems. One concern relates to the weak tornadoes that rank at or near the bottom of the Enhanced Fujita Intensity Scale. Because Doppler radar makes forecasting and detection of these tornadoes possible, the potential exists for numerous warnings being issued for tornadoes that do little or no damage. This could desensitize the public to the dangers of more rare, life-threatening tornadoes.

It should also be pointed out that not all tornado-bearing storms have clear-cut radar signatures and that other storms can give false signatures. Detection, therefore, is sometimes a subjective process, and a given display could be interpreted in several ways. Consequently, trained observers will continue to form an important part of the warning system in the foreseeable future.

Although some operational problems exist, the benefits of Doppler radar are many. As a research tool, it is not only providing data on the formation of tornadoes but is also helping meteorologists gain new insights into thunderstorm development, the structure and dynamics of hurricanes, and air-turbulence hazards that plague aircraft. As a practical tool for tornado detection, it has significant advantages over a system that uses conventional radar.

Concept Check 10.10

1 Distinguish between a tornado watch and a tornado warning.

2 What advantages does Doppler radar have over conventional radar?

Give It Some Thought

1. If you are a resident of central Ohio and hear that a cyclone is approaching, should you immediately seek shelter? What if you live in western Iowa?

2. Which one of the locations shown on the accompanying map is more likely to have dryline thunderstorms? Why is this the case?

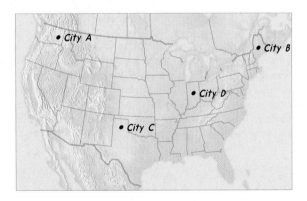

3. Sinking air warms by compression (adiabatically), yet thunderstorm downdrafts are usually cold. Explain this apparent contradiction.

4. Studies have linked the formation of supercell thunderstorms to the existence of temperature inversions. However, cumulonimbus clouds form in an unstable environment, whereas temperature inversions are associated with very stable atmospheric conditions. Explain the connection between these two phenomena.

5. The accompanying table lists the number of tornadoes reported in the United States by decade. Propose a reason that might explain why the total for the 2000s was so much higher than for the 1950s.

Number of U.S. Tornadoes Reported, by Decade

1950–1959	4896
1960–1969	6813
1970–1979	8579
1980–1989	8196
1990–1999	12,138
2000–2009	12,729

6. Figure 10–28 shows that the number of tornado deaths in the United States in the 2000s was less than 40 percent the number that occurred in the 1950s, even though there was a significant rise in the population during that span. To what can you attribute this decline in the death toll?

7. As you will learn in Chapter 11, the intensity of a hurricane is monitored and reported as the storm approaches. However, the intensity of a tornado is not determined and reported until *after* the storm passes. Why is this the case?

THUNDERSTORMS AND TORNADOES IN REVIEW

- Although tornadoes and hurricanes are, in fact, cyclones, the vast majority of cyclones are not hurricanes or tornadoes. The term *cyclone* refers to the circulation around *any* low-pressure center, no matter how large or intense it is.

- Dynamic thermal instability occurs during the development of thunderstorms, which form when warm, humid air rises in an unstable environment. A number of mechanisms, such as unequal heating of Earth's surface or lifting of warm air along a front or mountain slope, can trigger the upward air movement needed to create thunderstorm-producing cumulonimbus clouds.

- *Air-mass thunderstorms* frequently occur in maritime tropical (mT) air during the spring and summer. Generally, three stages are involved in the development of these storms—the *cumulus stage, mature stage,* and *dissipating stage.*

- Mountainous regions, such as the Rockies and the Appalachians, experience a greater number of air-mass thunderstorms than do the Plains states. Many thunderstorms that form over the eastern two-thirds of the United States occur as part of the general convergence and frontal wedging that accompany passing midlatitude cyclones.

- *Severe thunderstorms* are capable of producing heavy downpours and flash flooding as well as strong, gusty, straight-line winds. They are influenced by strong vertical wind shear—that is, changes in wind direction and/or speed between different heights and updrafts that become tilted and continue to build upward. Downdrafts from the thunderstorm cells reach the surface and spread out to produce an advancing wedge of cold air, called a *gust front.*

- Some of the most dangerous weather is produced by a type of thunderstorm called a *supercell,* a single, very powerful thunderstorm cell that at times may extend to heights of 20 kilometers (65,000 feet) and persist for many hours. These cells may produce a *mesocyclone,* a column of cyclonically rotating air, within which tornadoes sometimes form.

- *Squall lines* are relatively narrow, elongated bands of thunderstorms that develop in the warm sector of a midlatitude cyclone, usually in advance of a cold front. A *mesoscale convective complex* (MCC) consists of many individual thunderstorms that are organized into a large oval to circular cluster. They form most frequently in the Great Plains from groups of afternoon air-mass thunderstorms.

- *Thunder* is produced by *lightning.* Lightning equalizes the electrical difference associated with the formation of a large cumulonimbus cloud by producing a negative flow of current from the region of excess negative charge to the region with excess positive charge or vice versa. The most common type of lightning, often called *sheet lightning,* occurs within and between clouds. The less common, but more dangerous, type of lightning is *cloud-to-ground lightning.*

- The origin of charge separation in clouds, although not fully understood, hinges on rapid vertical movements within a cloud. The lightning we see as single flashes is really several very rapid strokes between the cloud and the ground. When air is heated by the electrical discharge of lightning, it expands explosively and produces the sound waves we hear as *thunder.*

- *Tornadoes* are violent windstorms that take the form of a rotating column of air, or *vortex,* that extends downward from a cumulonimbus cloud. Some tornadoes consist of a single vortex. Within many stronger tornadoes, called *multiple-vortex tornadoes,* are smaller intense whirls called *suction vortices* that rotate within the main vortex. Pressures within some tornadoes have been estimated to be as much as 10 percent lower than immediately outside the storm. Because of the tremendous pressure gradient associated with a strong tornado, maximum winds approach 480 kilometers (300 miles) per hour.

- Severe thunderstorms, and hence tornadoes, are most often spawned along the cold front or squall line of a midlatitude cyclone or in association with supercell thunderstorms. Tornadoes can also form in association with tropical cyclones (hurricanes). April through June is the period of greatest tornado activity, but tornadoes occur during every month of the year. The average tornado has a diameter between 150 and 600 meters, travels across the landscape toward the northeast at approximately 45 kilometers per hour, and cuts a path about 26 kilometers long.

- Most tornado damage is caused by tremendously strong winds. One commonly used guide to tornado intensity is the *Enhanced Fujita Intensity Scale (EF-scale)*. A rating on the EF-scale is determined by assessing damages produced by a storm.

- Because severe thunderstorms and tornadoes are small and short-lived phenomena, they are among the most difficult weather features to forecast precisely. When weather conditions favor the formation of tornadoes, a *tornado watch* is issued to alert the public to the possibility of tornadoes over a specified area for a particular time interval. A *tornado warning* is issued by local offices of the National Weather Service when a tornado has been sighted in an area or is indicated by weather radar. With its ability to detect the movement of precipitation within a cloud, *Doppler radar* technology has greatly advanced the accuracy of tornado warnings.

VOCABULARY REVIEW

air-mass thunderstorm (p. 274)
cumulus stage (p. 275)
dart leader (p. 286)
dissipating stage (p. 275)
Doppler radar (p. 296)
dryline (p. 280)
Enhanced Fujita intensity scale (EF-scale)
   (p. 293)
entrainment (p. 275)
flash (p. 284)

gust front (p. 277)
leader (p. 284)
lightning (p. 281)
mature stage (p. 275)
mesocyclone (p. 278)
mesoscale convective complex (MCC)
   (p. 281)
multiple-vortex tornado (p. 287)
return stroke (p. 284)
severe thunderstorm (p. 276)

squall line (p. 278)
step leader (p. 284)
stroke (p. 284)
supercell (p. 277)
thunder (p. 286)
thunderstorm (p. 273)
tornado (p. 286)
tornado warning (p. 296)
tornado watch (p. 296)

PROBLEMS

1. If thunder is heard 15 seconds after lightning is seen, about how far away was the lightning stroke?

2. Examine the upper-left portion of Figure 10–23 and determine the percentage of tornadoes that exhibited directions of movement toward the E through NNE.

3. Figures 10–32 and 10–33 represent two common ways that U. S. tornado statistics are graphically presented to the public. Which four states experience the greatest number of tornadoes? Are these the states with the greatest tornado threat? Which map is most useful for depicting the tornado hazard in the United States? Does the map in Figure 10–22 have an advantage over either or both of these maps?

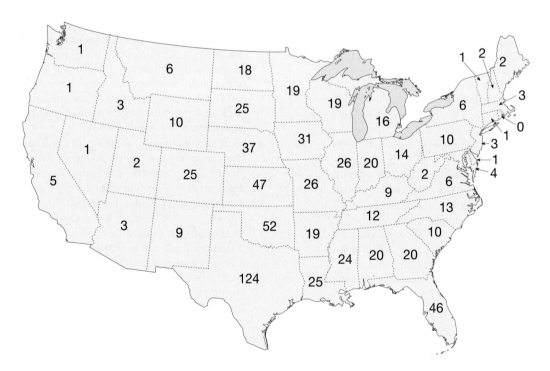

Figure 10–32 Annual average number of tornadoes by state for a 45-year period.

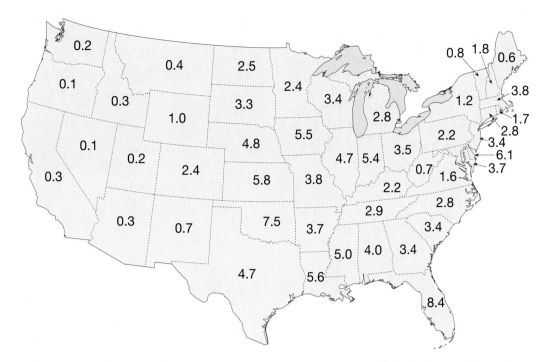

Figure 10–33 Annual average number of tornadoes per 10,000 square miles by state for a 45-year period.

MyMeteorologyLab™

Log in to www.mymeteorologylab.com for animations, videos, MapMaster interactive maps, GEODe media, *In the News* RSS feeds, web links, glossary flashcards, self-study quizzes, and a Pearson eText version of this book to enhance your study of *Thunderstorms and Tornadoes*.

Hurricanes

The whirling tropical cyclones that occasionally have wind speeds exceeding 300 kilometers (185 miles) per hour are known in the United States as *hurricanes*—the greatest storms on Earth. Hurricanes are among the most destructive of natural disasters. When a hurricane reaches land, it is capable of annihilating low-lying coastal areas and killing thousands of people. On the positive side, hurricanes provide essential rainfall over many areas they cross. Consequently, a resort owner along the Florida coast may dread the coming of hurricane season, whereas a farmer in Japan may welcome its arrival.

Aftermath of Hurricane Ike, September 14, 2008. The eye of the storm passed directly over Galveston, Texas. (Photo by Smiley N. Pool/Rapport Press/Newcom)

Focus On Concepts

After completing this chapter, you should be able to:

- Define *hurricane* and describe the basic structure and characteristics of this storm.

- Identify areas of hurricane formation on a world map and discuss the conditions that promote hurricane formation.

- Distinguish among *tropical depression*, *tropical storm*, and *hurricane*.

- List and discuss the factors that cause hurricanes to diminish in intensity.

- Use the Saffir–Simpson scale and explain how hurricane intensity is determined.

- Discuss the three broad categories of hurricane destruction, including an example of each.

- List four tools that provide data used to track hurricanes and develop forecasts.

- Contrast *hurricane watch* and *hurricane warning* and relate these concepts to hurricane forecasts.

Profile of a Hurricane

Many view the weather in the tropics with favor—and rightfully so. Places such as islands in the South Pacific and the Caribbean are known for their lack of significant day-to-day variations. Warm breezes, steady temperatures, and rains that come as heavy but brief tropical showers are expected. It is ironic that these relatively tranquil regions occasionally produce some of the most violent storms on Earth (Figure 11–1).

Hurricanes are intense centers of low pressure that form over tropical or subtropical oceans and are characterized by intense convective (thunderstorm) activity and strong cyclonic circulation. Sustained winds must equal or exceed 119 kilometers (74 miles) per hour. Unlike midlatitude cyclones, hurricanes lack contrasting air masses and fronts. Rather, the source of energy that produces and maintains hurricane-force winds is the huge quantity of latent heat liberated during the formation of the storm's cumulonimbus towers.

These intense tropical storms are known in various parts of the world by different names. In the northwestern Pacific, they are called *typhoons*, and in the southwestern Pacific and

Hurricane Igor
8–21 September 2010

— Hurricane
— Tropical storm
— Tropical depression
---- Extratropical

Figure 11–1 Satellite image of Hurricane Igor on September 16, 2010. Maximum sustained winds were 213 kilometers (132 miles) per hour. The inset image in the upper right is a digital photograph of Igor's well-developed eye, taken from the International Space Station. The storm had its greatest impact in Newfoundland, where most of the damage resulted from flooding triggered by Igor's heavy rains. (NASA)

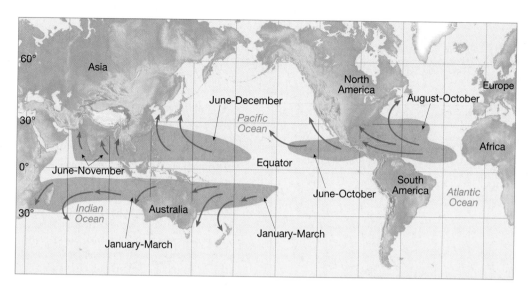

Figure 11-2 This world map shows the regions where most hurricanes form as well as their principal months of occurrence and the most common tracks they follow. Hurricanes do not develop within about 5° of the equator because the Coriolis force is too weak in that region. Because warm surface ocean temperatures are necessary for hurricane formation, hurricanes seldom form poleward of 20° latitude nor over the cool waters of the South Atlantic and the eastern South Pacific.

the Indian Ocean, they are called *cyclones*. In the following discussion, these storms will be referred to as hurricanes. The term *hurricane* is derived from Huracan, a Carib god of evil.

Most hurricanes form between the latitudes of 5° and 20° over all the tropical oceans except only rarely in the South Atlantic and the eastern South Pacific (Figure 11–2). The western North Pacific has the greatest number of storms, averaging 20 per year. Fortunately for those living in the coastal regions of the southern and eastern United States, only about five hurricanes, on average, develop each year in the warm sector of the North Atlantic.

Although many tropical disturbances develop each year, only a few reach hurricane status. By international

agreement, a hurricane has sustained wind speeds of at least 119 kilometers (74 miles) per hour and a rotary circulation.* Mature hurricanes average about 600 kilometers (375 miles) across, although they can range in diameter from 100 kilometers (60 miles) up to about 1500 kilometers (930 miles). From the outer edge of a hurricane to the center, the barometric pressure has sometimes dropped 60 millibars, from 1010 to 950 millibars.

A steep pressure gradient like that shown in Figure 11–3 generates the rapid, inward spiraling winds of a hurricane. As the air moves closer to the center of the storm, its ve-

---

Sustained winds are defined as the wind speed averaged over a one-minute interval.

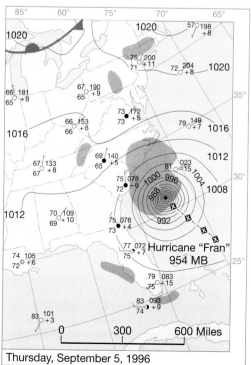

Thursday, September 5, 1996

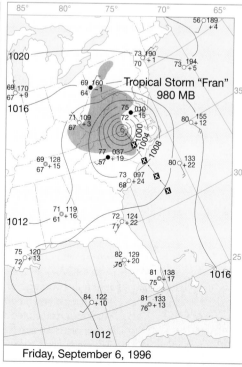

Friday, September 6, 1996

Figure 11-3 Weather maps showing Hurricane Fran at 7 AM, on two successive days, September 5 and 6, 1996. On September 5, winds exceeded 190 kilometers per hour. As the storm moved inland, heavy rains caused flash floods, killed 30 people, and caused more than $3 billion in damages. The station information plotted off the Gulf and Atlantic coasts is from data buoys, which are remote floating instrument packages.

Box 11-1   The Conservation of Angular Momentum

Why do winds blowing around a storm move faster near the center and more slowly near the edge? To understand this phenomenon, we must examine the *law of conservation of angular momentum*. This law states that the product of the velocity of an object around a center of rotation (axis) and the distance of the object from the axis is constant.

Picture an object on the end of a string being swung in a circle. If the string is pulled inward, the distance of the object from the axis of rotation decreases and the speed of the spinning object increases. The change in radius of the rotating mass is balanced by a change in its rotational speed.

Another common example of the conservation of angular momentum occurs when a figure skater starts whirling on the ice with both arms extended (**Figure 11–A**). Her arms are traveling in a circular path about an axis (her body). When the skater pulls her arms inward, she decreases the radius of the circular path of her arms. As a result, her arms go faster and the rest of her body must follow, thereby increasing her rate of spinning.

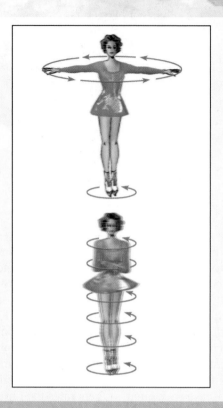

In a similar manner, when a parcel of air moves toward the center of a storm, the product of its distance and velocity must remain unchanged. Therefore, as air moves inward from the outer edge, its rotational velocity must increase.

Let us apply the law of conservation of angular momentum to the horizontal movement of air in a hypothetical hurricane. Assume that air with a velocity of 5 kilometers per hour begins 500 kilometers from the center of the storm. By the time it reaches a point 100 kilometers from the center, it will have a velocity of 25 kilometers per hour (assuming that there is no friction). If this same parcel of air were to continue to advance toward the storm's center until its radius was just 10 kilometers, it would be traveling at 250 kilometers per hour. Friction reduces these values somewhat.

FIGURE 11–A When the skater's arms are extended, she spins slowly. When her arms are pulled in, she spins much faster.

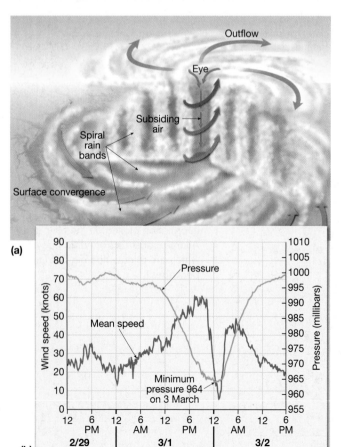

locity increases. This acceleration is explained by the law of conservation of angular momentum (Box 11–1).

As the inward rush of warm, moist surface air approaches the core of the storm, it turns upward and ascends in a ring of cumulonimbus towers (Figure 11–4a). This doughnut-shaped wall of intense convective activity surrounding the center of the storm is called the **eye wall.** It is here that the greatest wind speeds and heaviest rainfall occur. Surrounding the eye wall are curved bands of clouds that trail away in a spiral fashion. Near the top of the hurricane the airflow is outward, carrying the rising air away

Figure 11–4 (a) Cross section of a hurricane. Note that the vertical dimension is greatly exaggerated. The eye, the zone of relative calm at the center of the storm, is a distinctive hurricane feature. Sinking air in the eye warms by compression. Surrounding the eye is the eye wall, the zone where winds and rain are most intense. Tropical moisture spiraling inward creates rain bands that pinwheel around the storm center. Outflow of air at the top of the hurricane is important because it prevents the convergent flow at lower levels from "filling in" the storm. (After NOAA) (b) Measurements of surface pressure and wind speed during the passage of Cyclone Monty at Mardie Station in Western Australia between February 29 and March 2, 2004. The strongest winds are associated with the eye wall, and the weakest winds and lowest pressure are found in the eye. In this part of the world the term *cyclone* is used instead of *hurricane.* (Data from World Meteorological Organization)

from the storm center, thereby providing room for more inward flow at the surface.

At the very center of the storm is the **eye** of the hurricane. This well-known feature is a zone where precipitation ceases and winds subside. The graph in Figure 11–4b shows changes in wind speed and air pressure as Cyclone Monty came ashore at Mardie Station in Western Australia between February 29 and March 2, 2004. The very steep pressure gradient and strong winds associated with the eye wall are evident, as is the relative calm of the eye.

The eye offers a brief but deceptive break from the extreme weather in the enormous curving wall clouds that surround it. The air within the eye gradually descends and heats by compression, making it the warmest part of the storm. Although many people believe that the eye is characterized by clear blue skies, this is usually not the case because the subsidence in the eye is seldom strong enough to produce cloudless conditions. Although the sky appears much brighter in this region, scattered clouds at various levels are common.

Concept Check 11.1

1 Define *hurricane*. What other names are used for this storm?

2 In what latitude zone do hurricanes develop?

3 Distinguish between the eye and the eye wall of a hurricane. How do conditions differ in these zones?

Hurricane Formation and Decay

A hurricane is a heat engine that is fueled by the latent heat liberated when huge quantities of water vapor condense. The amount of energy produced by a typical hurricane in just a single day is truly immense. The release of latent heat warms the air and provides buoyancy for its upward flight. The result is to reduce the pressure near the surface, which in turn encourages a more rapid inflow of air. To get this engine started, a large quantity of warm, moist air is required, and a continuous supply is needed to keep it going.

As the graph in Figure 11–5 illustrates, hurricanes most often form in late summer and early fall. It is during this span that sea-surface temperatures reach 27°C (80°F) or higher and are thus able to provide the necessary heat and moisture to the air (Figure 11–6). This ocean-water temperature requirement accounts for the fact that hurricane formation over the relatively cool waters of the South Atlantic and the eastern South Pacific is extremely rare. For the same reason, few hurricanes form poleward of 20° latitude (see Figure 11–2). Although water temperatures are sufficiently high, hurricanes do not form within 5° of the equator because the Coriolis force is too weak in that region to initiate the necessary rotary motion.

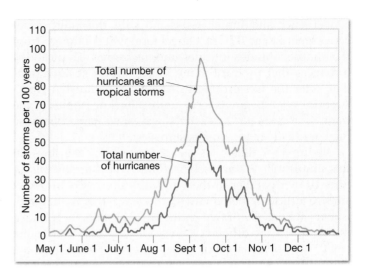

Figure 11–5 Frequency of tropical storms and hurricanes from May 1 through December 31 in the Atlantic basin. The graph shows the number of storms to be expected over a span of 100 years. The period from late August through October is clearly the most active. (After National Hurricane Center/NOAA)

Hurricane Formation

Many tropical storms achieve hurricane status in the western parts of oceans, but their origins often lie far to the east. In such locations, disorganized arrays of clouds and thunderstorms, called **tropical disturbances,** sometimes develop and exhibit weak pressure gradients and little or no rotation. Most of the time these zones of convective activity die out. However, tropical disturbances occasionally grow larger and develop strong cyclonic rotation.

Several different situations can trigger tropical disturbances. They are sometimes initiated by the powerful

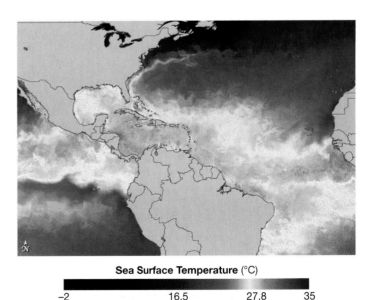

Figure 11–6 Among the necessary ingredients for a hurricane is warm ocean temperatures above 27°C (80°F). This color-coded satellite image from June 1, 2010, shows sea-surface temperatures at the beginning of hurricane season. (NASA)

convergence and lifting associated with the intertropical convergence zone (ITCZ). Others form when a trough from the middle latitudes intrudes into the tropics. Tropical disturbances that produce many of the strongest hurricanes that enter the western North Atlantic and threaten North America often begin as large undulations or ripples in the trade winds known as **easterly waves,** so named because they gradually move from east to west.

Figure 11–7 illustrates an easterly wave. The lines on this simple map are not isobars. Rather, they are *streamlines,* lines drawn parallel to the wind direction used to depict surface airflow. When middle-latitude weather is analyzed, isobars are usually drawn on the weather map. By contrast, in the tropics the differences in sea-level pressure are quite small, so isobars are not always useful. Streamlines are helpful because they show where surface winds converge and diverge.

To the east of the wave axis the streamlines move poleward and get progressively closer together, indicating that the surface flow is convergent. Convergence encourages air to rise and form clouds. Therefore, the tropical disturbance is located on the east side of the wave. To the west of the wave axis, surface flow diverges as it turns toward the equator. Consequently, clear skies are the rule here.

Easterly waves frequently originate as disturbances in Africa. As these storms head westward with the prevailing trade winds, they encounter the cold Canary Current (see Figure 3–9, p. 72). If the disturbance survives the trip across the cold stabilizing waters of the current, it is rejuvenated by the heat and moisture of the warmer water of the mid-Atlantic. From this point on, a disturbance may develop

into a more intense and organized system; some of these may reach hurricane status.

Even when conditions seem to be right for hurricane formation, many tropical disturbances do not strengthen. One circumstance that may inhibit further development is a temperature inversion called a *trade wind inversion.* It forms in association with the subsidence that occurs in the region influenced by the subtropical high.* A strong inversion diminishes the ability of air to rise and thus inhibits the development of strong thunderstorms. Another factor that works against the strengthening of tropical disturbances is strong upper-level winds. When such winds are present, a strong flow aloft disperses the latent heat released from cloud tops—heat that is essential for continued storm growth and development.

What happens on those occasions when conditions favor hurricane development? As latent heat is released from the clusters of thunderstorms that make up the tropical disturbance, areas within the disturbance get warmer. As a consequence, air density lowers and surface pressure drops, creating a region of weak low pressure and cyclonic circulation. As pressure drops at the storm center, the pressure gradient steepens. In response, surface wind speeds increase and bring additional supplies of moisture to nurture storm growth. The water vapor condenses, releasing latent heat, and the heated air rises. Adiabatic cooling of rising air triggers more condensation and the release of more latent heat, which causes a further increase in buoyancy. And so it goes.

Meanwhile, higher pressure develops at the top of the developing tropical depression.** This causes air to flow outward (diverge) from the top of the storm. Without this outward flow up top, the inflow at lower levels would soon raise surface pressures and thwart further storm development.

Although many tropical disturbances occur each year, only a few develop into full-fledged hurricanes. Recall that tropical cyclones are called hurricanes only when their winds reach 119 kilometers (74 miles) per hour. By international agreement, lesser tropical cyclones are given different names, based on the strength of their winds. When a cyclone's strongest winds do not exceed 63 kilometers (39 miles) per hour, it is called a **tropical depression.** When sustained winds are between 63 and 119 kilometers (39 and 74 miles) per hour, the cyclone is termed a **tropical storm.** It is during this phase that a name is given (Andrew, Fran, Opal, and so on). If the tropical storm becomes a hurricane, the name remains the same (Box 11–2). Each year between 80 and 100 tropical storms develop around the world. Of them, usually half or more eventually reach hurricane status.

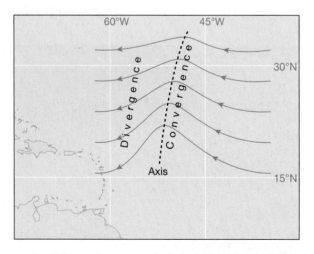

Figure 11–7 Easterly wave in the subtropical North Atlantic. Streamlines show low-level airflow. To the east of the wave axis, winds converge as they move slightly poleward. To the west of the axis, flow diverges as it turns toward the equator. Tropical disturbances are associated with the convergent flow of the easterly wave. Easterly waves extend for 2000 to 3000 kilometers (1200 to 1800 miles) and move from east to west with the trade winds at rates between 15 and 35 kilometers (10 and 20 miles) per hour. At this rate, it takes an imbedded tropical disturbance a week or 10 days to move across the North Atlantic.

---

*In the troposphere, a temperature inversion exists when temperatures in a layer of air increase with an increase in altitude rather than decrease with height, which is usually the case. For more on how subsidence can produce an inversion, see the section "Inversions Aloft" in Chapter 13, p. 370.

**See Figure 6–11 and the discussion of the sea–land breeze in the section "Pressure Gradient Force" in Chapter 6.

Box 11–2 Naming Tropical Storms and Hurricanes

Tropical storms are named to provide ease of communication between forecasters and the general public regarding forecasts, watches, and warnings. Tropical storms and hurricanes can last a week or longer, and two or more storms can occur in the same region at the same time. Thus, names can reduce the confusion about what storm is being described.

During World War II, tropical storms were informally assigned women's names (perhaps after wives and girlfriends) by U.S. Army Corps and Navy meteorologists who were monitoring storms over the Pacific. From 1950 to 1952, tropical storms in the North Atlantic were identified by the phonetic alphabet—Able, Baker, Charlie, and so forth. In 1953 the U.S. Weather Bureau (now the National Weather Service) switched to women's names.

The practice of using feminine names continued until 1978, when a list containing both male and female names was adopted for tropical cyclones in the eastern Pacific. In the same year the World Meteorological Organization (WMO) accepted a proposal that both male and female names be adopted for Atlantic hurricanes, beginning with the 1979 season.

The WMO has created six lists of names for tropical storms over ocean areas. The names used for Atlantic, Gulf of Mexico, and Caribbean hurricanes are shown in Table 11–A. The names are ordered alphabetically and do not contain names that begin with the letters Q, U, X, Y, and Z because of the scarcity of names beginning with those letters. When a tropical depression reaches tropical storm status, it is assigned the next unused name on the list. At the beginning of the next hurricane season, names from the next list are selected, even though many names may not have been used the previous season.

The names for Atlantic storms are used over again at the end of each six-year cycle unless a hurricane was particularly noteworthy. This is to avoid confusion when storms are discussed in future years. For example, five names from the list for the extraordinary 2005 Atlantic hurricane season were retired—Dennis, Katrina, Rita, Stan, and Wilma. They were replaced by Don, Katia, Rina, Sean, and Whitney on the list for the year 2011.

TABLE 11–A Tropical Storm and Hurricane Names for the Atlantic, Gulf of Mexico, and Caribbean Sea*

2011	2012	2013	2014	2015	2016
Arlene	Alberto	Andrea	Arthur	Ana	Alex
Bret	Beryl	Barry	Bertha	Bill	Bonnie
Cindy	Chris	Chantal	Cristobal	Claudette	Colin
Don	Debby	Dorian	Dolly	Danny	Danielle
Emily	Ernesto	Erin	Edouard	Erika	Earl
Franklin	Florence	Fernand	Fay	Fred	Fiona
Gert	Gordon	Gabrielle	Gonzalo	Grace	Gaston
Harvey	Helene	Humberto	Hanna	Henri	Hermine
Irene	Isaac	Ingrid	Isaias	Ida	Ian
Jose	Joyce	Jerry	Josephine	Joaquin	Julia
Katia	Kirk	Karen	Kyle	Kate	Karl
Lee	Leslie	Lorenzo	Laura	Larry	Lisa
Maria	Michael	Melissa	Marco	Mindy	Matthew
Nate	Nadine	Nestor	Nana	Nicholas	Nicole
Ophelia	Oscar	Olga	Omar	Odette	Otto
Philippe	Patty	Pablo	Paulette	Peter	Paula
Rina	Rafael	Rebekah	Rene	Rose	Richard
Sean	Sandy	Sebastien	Sally	Sam	Shary
Tammy	Tony	Tanya	Teddy	Teresa	Tobias
Vince	Valerie	Van	Vicky	Victor	Virginie
Whitney	William	Wendy	Wilfred	Wanda	Walter

*If the entire alphabetical list of names for a given year is exhausted, the naming system moves on to using letters of the Greek alphabet (alpha, beta, gamma, and so on). This issue never arose until the record-breaking 2005 hurricane season, when Tropical Storm Alpha, Hurricane Beta, tropical storms Gamma and Delta, Hurricane Epsilon, and Tropical Storm Zeta occurred after Hurricane Wilma.

Students Sometimes Ask...

When is hurricane season?

Hurricane season is different in different parts of the world. People in the United States are usually most interested in Atlantic storms. The Atlantic hurricane season officially extends from June through November. More than 97 percent of tropical activity in that region occurs during this six-month span. The "heart" of the season is August through October (see Figure 11–5, p. 307). During these three months, 87 percent of the minor hurricane (category 1 and 2) days and 96 percent of the major hurricane (category 3, 4, and 5) days occur. Peak activity is in early to mid-September.

Hurricane Decay

Hurricanes diminish in intensity whenever they (1) move over ocean waters that cannot supply warm, moist tropical air; (2) move onto land; or (3) reach a location where the large-scale flow aloft is unfavorable. Richard Anthes describes the possible fate of hurricanes in the first category as follows:

Many hurricanes approaching the North American or Asian continents from the southeast are turned toward the northeast, away from the continents, by the steering effect of an upper-level trough. This recurvature carries the storms toward higher latitudes where the ocean temperatures are cooler and an encounter with cool, dry polar air masses is more likely. Often the tropical cyclone and a polar front interact, with cold air entering the tropical cyclone from the west. As the release of latent heat is diminished, the upper-level divergence weakens, mean temperatures in the core fall and the surface pressure rises.*

Whenever a hurricane moves onto land, it loses its punch rapidly. For example, in Figure 11–3 notice how the isobars show a much weaker pressure gradient on September 6, after Hurricane Fran moved ashore, than on September 5, when it was over the ocean. The most important reason for this rapid demise is the fact that the storm's source of warm, moist air is cut off. When an adequate supply of water vapor does not exist, condensation and the release of latent heat must diminish.

In addition, the increased surface roughness over land results in a rapid reduction in surface wind speeds. This factor causes the winds to move more directly into the center of the low, thus helping to eliminate the large pressure differences.

Concept Check 11.2

1 What is the source of energy that drives a hurricane?

2 Why do hurricanes *not* form near the equator? Explain the lack of hurricanes in the South Atlantic and eastern South Pacific.

3 During what months do most tropical storms and hurricanes in the Atlantic basin occur? Why are these months favored?

4 List two factors that inhibit the strengthening of tropical disturbances.

5 Distinguish between *tropical depression, tropical storm,* and *hurricane.*

6 Why does the intensity of a hurricane diminish rapidly when it moves onto land?

Hurricane Destruction

The vast majority of hurricane-related deaths and damage are caused by relatively infrequent yet powerful storms. Table 11–1 lists the deadliest hurricanes to strike the United States between 1900 and 2011. The storm that pounded an unsuspecting Galveston, Texas, in 1900 was not just the deadliest U.S. hurricane ever but the deadliest natural disaster *of any kind* to affect the United States. Of course, the deadliest and most costly storm in recent memory occurred in August 2005, when Hurricane Katrina devastated the Gulf Coast of Louisiana, Mississippi, and Alabama.** Although hundreds of thousands fled before the storm made landfall, thousands of others were caught by the storm. In addition to the human suffering and

EYE ON THE ATMOSPHERE

This hurricane occurred in the Atlantic in 2004. (NASA)

Question 1 Did the storm occur in the North Atlantic or the South Atlantic? How did you figure this out?

Question 2 Did the storm more likely occur in March or September?

Question 3 Are hurricanes in this region common or rare? Explain.

TABLE 11–1 The 10 Deadliest Hurricanes to Strike the U.S. Mainland 1900–2011

Rank	Hurricane	Year	Category	Deaths
1.	Texas (Galveston)	1900	4	8000*
2.	Southeastern Florida (Lake Okeechobee)	1928	4	2500–3000
3.	Katrina	2005	4	1833
4.	Audrey	1957	4	At least 416
5.	Florida Keys	1935	5	408
6.	Florida (Miami)/ Mississippi/ Alabama/Florida (Pensacola)	1926	4	372
7.	Louisiana (Grande Isle)	1909	4	350
8.	Florida Keys/ South Texas	1919	4	287
9. (tie)	Louisiana (New Orleans)	1915	4	275
9. (tie)	Texas (Galveston)	1915	4	275

Source: National Weather Service/National Hurricane Center, *NOAA Technical Memorandum NWS TPC-5.*

*This number may actually have been as high as 10,000–12,000.

Tropical Cyclones: Their Evolution, Structure, and Effects, Meteorological Monographs, Vol. 19, no. 41 (1982), p. 61. Boston: American Meteorological Society.
**Many images of this storm can be seen in "Severe and Hazardous Weather: Hurricane Katrina from Space," which begins on p. 318.

tragic loss of life that were left in the wake of Hurricane Katrina, the financial losses caused by the storm are practically incalculable, Up until August 2005, the $25 billion in damages associated with Hurricane Andrew in 1992 represented the costliest natural disaster in U.S. history. This figure was exceeded many times over when Katrina's economic impact was calculated. The final accounting likely exceeded $100 billion.

Although the amount of damage caused by a hurricane depends on several factors, including the size and population density of the area affected and the nearshore bottom configuration, certainly the most significant factor is the strength of the storm itself.

Saffir–Simpson Scale

Based on the study of past storms, the **Saffir–Simpson scale** was established to rank the relative intensities of hurricanes (Table 11–2). Predictions of hurricane severity and damage are usually expressed in terms of this scale. When a tropical storm becomes a hurricane, the National Weather Service assigns it a scale (category) number. Category assignments are based on observed conditions at a particular stage in the life of a hurricane and are viewed as estimates of the amount of damage a storm would cause if it were to make landfall without changing size or strength. As conditions change, the category of a storm is reevaluated so that public safety officials can be kept informed. By using the Saffir–Simpson scale, the disaster potential of a hurricane can be monitored and appropriate precautions can be planned and implemented.

A rating of 5 on the scale represents the worst storm possible, and a 1 is least severe. Storms that fall into category 5 are rare. Only three storms this powerful are known to have hit the continental United States: Andrew struck Florida in 1992, Camille pounded Mississippi in 1969, and a Labor Day hurricane struck the Florida Keys in 1935. Damage caused by hurricanes can be divided into three classes: (1) storm surge, (2) wind damage, and (3) inland freshwater flooding.

Storm Surge

Without question, the most devastating damage in the coastal zone is caused by storm surge. It not only accounts for a large share of coastal property losses but is also responsible for 90 percent of all hurricane-caused deaths. A **storm surge** is a dome of water 65 to 80 kilometers (40 to 50 miles) wide that sweeps across the coast near the point where the eye makes landfall. If all wave activity were smoothed out, the storm surge would be the height of the water above normal tide level (Figure 11–8). In addition, tremendous wave activity is superimposed on the surge. We can easily imagine the damage that this surge of water could inflict on low-lying coastal areas (Figure 11–9). The worst surges occur in places like the Gulf of Mexico, where the continental shelf is very shallow and gently sloping. In addition, local features such as bays and rivers can cause the surge height to double and increase in speed.

In the delta region of Bangladesh, for example, most of the land is less than 2 meters (6.5 feet) above sea level. When a storm surge superimposed on normal high tide

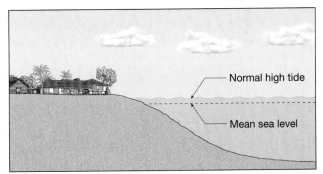
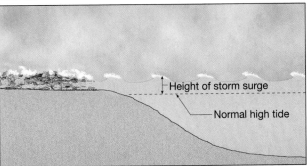

Figure 11–8 Superimposed upon high tide, a storm surge can devastate a coastal area. The worst storm surges occur in coastal areas where there is a very shallow and gently sloping continental shelf extending from the beach. The Gulf Coast is such a place.

Students Sometimes Ask...

Are larger hurricanes stronger than smaller hurricanes?

Not necessarily. Actually, there is very little correlation between intensity (either measured by maximum sustained winds or by central pressure) and size (either measured by the radius of gale-force winds or the radius of the outer closed isobar). Hurricane Andrew is a good example of a very intense storm (category 5) that was also relatively small (gale-force winds extended only 150 kilometers [90 miles] from the eye). Research has shown that changes in intensity and size are essentially independent of one another.

TABLE 11–2 Saffir–Simpson Hurricane Scale*

Scale Number (category)	Central Pressure (millibars)	Winds (km/hr)	Storm Surge (meters)	Damage
1	≥980	119–153	1.2–1.5	Minimal
2	965–979	154–177	1.6–2.4	Moderate
3	945–964	178–209	2.5–3.6	Extensive
4	920–944	210–250	3.7–5.4	Extreme
5	<920	>250	>5.4	Catastrophic

*A more complete version of the Saffir–Simpson hurricane scale can be found in Appendix F, p. 481.

Figure 11–9 Crystal Beach, Texas, on September 16, 2008, three days after Hurricane Ike came ashore. At landfall the storm had sustained winds of 165 kilometers (105 miles) per hour. The extraordinary storm surge caused much of the damage pictured here. (Photo by Earl Nottingham/Associated Press)

ter of the storm acts as a partial vacuum that allows the ocean to rise up in response. However, this effect is relatively insignificant. The most important factor responsible for the development of a storm surge is the piling up of ocean water by strong onshore winds. Gradually the hurricane's winds push water toward the shore, causing sea level to elevate while also churning up violent wave activity.

As a hurricane advances toward the coast in the Northern Hemisphere, storm surge is always most intense on the right side of the eye, where winds are blowing *toward* the shore. In addition, on this side of the storm the forward movement of the hurricane also contributes to the storm surge. In Figure 11–10, assume that a hurricane with peak winds of 175 kilometers (109 miles) per hour is moving toward the shore at 50 kilometers (31 miles) per hour. In this case, the net wind speed on the right side of the advancing storm is 225 kilometers (140 miles) per hour. On the left side, the hurricane's winds are blowing opposite the direction of storm movement, so the net winds are *away* from the coast, at 125 kilometers (78 miles) per hour. Along the shore facing the left side of the oncoming hurricane, the water level may actually decrease as the storm makes landfall.

inundated that area on November 13, 1970, the official death toll was 200,000; unofficial estimates ran to 500,000. It was one of the worst natural disasters of modern times. In May 1991 a similar event again struck Bangladesh. This time the storm took the lives of at least 143,000 people and devastated coastal towns in its path.

A common misconception about the cause of hurricane storm surges is that the very low pressure at the cen-

Figure 11–10 Winds associated with a Northern Hemisphere hurricane that is advancing toward the coast. This hypothetical storm, with peak winds of 175 kilometers (109 miles) per hour, is moving toward the coast at 50 kilometers (31 miles) per hour. On the right side of the advancing storm, the 175-kilometer-per-hour winds are in the same direction as the movement of the storm (50 kilometers per hour). Therefore, the *net* wind speed on the right side of the storm is 225 kilometers (140 miles) per hour. On the left side, the hurricane's winds are blowing opposite the direction of storm movement, so the *net* winds of 125 kilometers (78 miles) per hour are away from the coast. Storm surge will be greatest along the part of the coast hit by the right side of the advancing hurricane.

Wind Damage

Destruction caused by wind is perhaps the most obvious of the classes of hurricane damage. Debris such as signs, roofing materials, and small items left outside become dangerous flying missiles in hurricanes. For some structures, the force of the wind is sufficient to cause total ruin. Just read the descriptions of category 3, 4, and 5 storms in Appendix F, p. 481. Mobile homes are particularly vulnerable. High-rise buildings are also susceptible to hurricane-force winds. Upper floors are most vulnerable because wind speeds usually increase with height. Recent research suggests that people should stay below the tenth floor but remain above any floors at risk for flooding. In regions with good building codes, wind damage is usually not as catastrophic as storm-surge damage. However, hurricane-force winds affect a much larger area than storm surge and can cause huge economic losses. For example, in 1992 it was largely the winds associated with Hurricane Andrew that produced more than $25 billion of damage in southern Florida and Louisiana.

A hurricane may produce tornadoes that contribute to the storm's destructive power. Studies have shown that more than half of the hurricanes that make landfall produce at least one tornado. In 2004 the number of tornadoes associated with tropical storms and hurricanes was extraordinary. Tropical Storm Bonnie and five landfalling hurricanes—Charley, Frances, Gaston, Ivan, and Jeanne—produced nearly 300 tornadoes that affected the southeast and mid-Atlantic states (Table 11–3). Hurricane Frances produced the most tornadoes ever reported from one hurricane. The large number of hurricane-generated tornadoes in 2004 helped make this a record-breaking year for tornadoes—surpassing the previous record by more than 300.*

Heavy Rains and Inland Flooding

The torrential rains that accompany most hurricanes represent a third significant threat—flooding. The 2004 hurricane season was very deadly, with a loss of life that exceeded 3000 people. Nearly all of the deaths occurred in Haiti, as a result of flash floods and mudflows caused by the heavy rains associated with then Tropical Storm Jeanne.

TABLE 11–3 Number of Tornadoes Spawned by Hurricanes and Tropical Storms in the United States, 2004

Tropical Storm Bonnie	30
Hurricane Charley	25
Hurricane Frances	117
Hurricane Gaston	1
Hurricane Ivan	104
Hurricane Jeanne	16

*K. L. Gleason, et al., "U.S. Tornado Records" in *Bulletin of the American Meteorological Society,* Vol. 86, No. 6 (June 2005), p. 551.

Hurricane Agnes (1972) illustrates that even modest storms can have devasting results. Although it was only a category 1 storm on the Saffir–Simpson scale, it was one of the costliest hurricanes of the twentieth century, creating more than $2 billion in damage and taking 122 lives. The greatest destruction was attributed to flooding in the northeastern portion of the United States, especially in Pennsylvania, where record rainfalls occurred. Harrisburg received nearly 32 centimeters (12.5 inches) in 24 hours, and western Schuykill County measured more than 48 centimeters (19 inches) during the same span. Agnes's rains were not as devastating elsewhere. Prior to reaching Pennsylvania, the storm caused some flooding in Georgia, but most farmers welcomed the rain because dry conditions had been plaguing them earlier. In fact, the value of the rains to crops in the region far exceeded the losses caused by flooding.

Another well-known example is Hurricane Camille (1969). Although this storm is best known for its exceptional storm surge and the devastation it brought to coastal areas, the greatest number of deaths associated with this storm occurred in the Blue Ridge Mountains of Virginia two days after Camille's landfall. Many areas received more than 25 centimeters (10 inches) of rain, and severe flooding took more than 150 lives.

To summarize, extensive damage and loss of life in the coastal zone can result from storm surge, strong winds, and torrential rains. When loss of life occurs, it is commonly caused by storm surge, which can devastate entire barrier islands or zones within a few blocks of the coast. Although wind damage is usually not as catastrophic as storm surge, it affects a much larger area. Where building codes are inadequate, economic losses can be especially severe. Because hurricanes weaken as they move inland, most wind damage occurs within 200 kilometers (125 miles) of the coast. Far from the coast, a weakening storm can produce extensive flooding long after the winds have diminished below hurricane levels. Sometimes the damage from inland flooding exceeds storm-surge destruction.

Concept Check 11.3

1 What is the purpose of the Saffir–Simpson scale?

2 What are the three broad categories of hurricane damage? Provide brief examples of each.

3 Which side of an advancing hurricane in the Northern Hemisphere has the strongest winds and highest storm surge—right or left? Why does one side of the storm have stronger winds than the other side?

Estimating the Intensity of a Hurricane

The Saffir–Simpson hurricane scale appears to be a straightforward tool. However, the accurate observations needed to correctly portray hurricane intensity at the surface are

SEVERE AND HAZARDOUS WEATHER

Cyclone Nargis

All three classes of hurricane damage came into play in May 2008, when the first cyclone of the season in the northern Indian Ocean struck Myanmar (also called Burma). The satellite image in Figure 11–B shows the storm over the Bay of Bengal and the map shows the path of the storm and the rainfall it produced. The storm devastated much of the fertile Irrawaddy Delta (Figure 11–C). According to a United Nations estimate, more than 90 percent of the dwellings were destroyed in the hardest-hit areas. In addition, death and destruction also occurred in portions of Yangon, the nation's largest city. In all, about 30,000 square kilometers (nearly 11,600 square miles) were significantly affected—an area that was home to nearly one-quarter of Myanmar's 57 million people.

> More than 90 percent of the dwellings were destroyed in the hardest-hit areas.

A storm surge of 3.6 meters (12 feet) and torrential rains (up to 600 mm [2 feet]

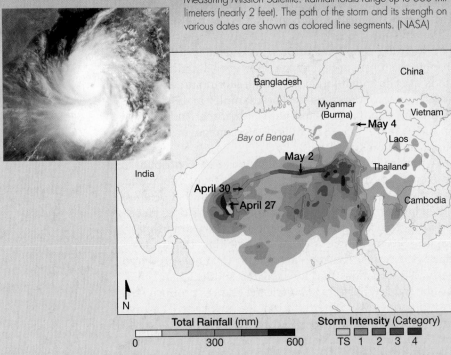

FIGURE 11–B The small satellite image shows Cyclone Nargis hovering over the Bay of Bengal on May 1, 2008. At one point Nargis was a category 4 storm on the Saffir–Simpson scale, with winds of 210 kilometers (131 miles) per hour. Although it lost strength before coming ashore on May 2 as a category 2 storm, Nargis still had powerful winds and torrential rains. (NASA) The map shows rainfall accumulations along the path of Cyclone Nargis between April 27 and May 4, 2008, using data from the Tropical Rainfall Measuring Mission Satellite. Rainfall totals range up to 600 millimeters (nearly 2 feet). The path of the storm and its strength on various dates are shown as colored line segments. (NASA)

Total Rainfall (mm)
0 300 600

Storm Intensity (Category)
TS 1 2 3 4

in some places) brought widespread flooding. Powerful winds contributed to the destruction. The fact that this low-lying region was densely populated and had a high proportion of poorly constructed dwellings exacerbated the situation. The government reported the official death toll to be nearly 85,000, with an additional 54,000 people unaccounted for. It is likely that more than 100,000 perished. It was the deadliest cyclone to hit Asia since 1991, when 143,000 people died in Bangladesh. Many who survived the storm lost their homes; an estimated 2 million people were displaced.

FIGURE 11–C The storm surge and heavy rains produced by Cylone Nargis combined to cause extensive flooding on the low-lying Irrawaddy Delta, killing thousands and displacing as many as 2 million people. (Photo by Mandalay Gazette, HO/AP Photo)

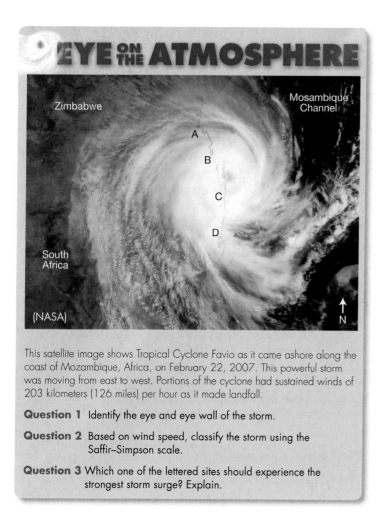

EYE ON THE ATMOSPHERE

This satellite image shows Tropical Cyclone Favio as it came ashore along the coast of Mozambique, Africa, on February 22, 2007. This powerful storm was moving from east to west. Portions of the cyclone had sustained winds of 203 kilometers (126 miles) per hour as it made landfall.

Question 1 Identify the eye and eye wall of the storm.

Question 2 Based on wind speed, classify the storm using the Saffir–Simpson scale.

Question 3 Which one of the lettered sites should experience the strongest storm surge? Explain.

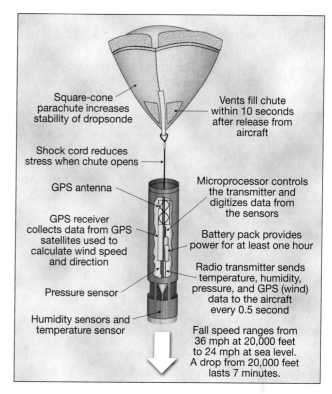

Figure 11–11 The Global Positioning System (GPS) dropwindsonde is frequently just called a dropsonde. This cylindrical instrument package is roughly 7 centimeters (2.75 inches) in diameter, 40 centimeters (16 inches) long, and weighs about 0.4 kilogram (0.86 pound). The instrument package is released from an aircraft and falls through the storm via a parachute, making and transmitting measurements of temperature, pressure, winds, and humidity every half-second. (After NASA)

sometimes difficult to obtain. Estimating hurricane intensity is difficult because direct surface observations in the eye wall are rarely available. Therefore, winds in this most intense part of the storm have to be estimated. One of the best ways to estimate surface intensity is to adjust the wind speeds measured by reconnaissance aircraft.

Winds aloft are stronger than winds at the surface. Therefore, the adjustment of values determined for winds aloft to values expected at the surface involves *reducing* the measurements made aloft. However, until the late 1990s, determining the proper adjustment factor was problematic because surface observations in the eye wall were too limited to establish a broadly accepted relationship between flight-level and surface winds. In the early 1990s the reduction factors commonly used ranged from 75 to 80 percent (that is, surface wind speeds were assumed to be between 75 and 80 percent of the speed at 3000 meters [10,000 feet]). Some scientists and engineers even maintained that surface winds were as low as 65 percent of flight-level winds.

Beginning in 1997 a new instrument, called a *Global Positioning System (GPS) dropwindsonde*, came into use (sometimes just called a *dropsonde*). After being released from the aircraft, this package of instruments, slowed by a small parachute, drifts downward through the storm (Figure 11–11).

During the descent, it continuously transmits data on temperature, humidity, air pressure, wind speed, and wind direction. The development of this technology provided, for the first time, a way to accurately measure the strongest winds in a hurricane, from flight level all the way to the surface.

Over a span of several years, hundreds of GPS dropwindsondes were released in hurricanes. The data accumulated from these trials showed that the speed of surface winds in the eye wall averaged about 90 percent of the flight-level winds, not 75 to 80 percent. Based on this new understanding, the National Hurricane Center now uses the 90 percent figure to estimate a hurricane's maximum surface winds from flight-level observations. This means that the winds in some storms in the historical record were underestimated.

For example, in 1992 the surface winds in Hurricane Andrew were estimated to be 233 kilometers (145 miles) per hour. This was 75 to 80 percent of the value measured at 3000 meters by the reconnaissance aircraft. When scientists at the National Hurricane Center reevaluated the storm using the 90 percent value, they concluded that the maximum-sustained surface winds had been 266 kilometers (165 miles) per hour—33 kilometers (20 miles) per hour faster than the original 1992 estimate. Consequently, in August 2002 the intensity of Hurricane Andrew was officially changed from category 4 to category 5. The upgrade makes Hurricane Andrew only the third category 5 storm on record to strike the continental United States. (The other two were the Florida Keys 1935 Hurricane and Hurricane Camille in 1969.)

Detecting, Tracking, and Monitoring Hurricanes

Figure 11–12 shows the paths followed by some notable Atlantic hurricanes. What determines these tracks? The storms can be thought of as being steered by the surrounding environmental flow throughout the depth of the troposphere. The movement of hurricanes has been likened to a leaf being carried along by the currents in a stream, except that for a hurricane the "stream" has no set boundaries.

In the latitude zone equatorward of about 25° north, tropical storms and hurricanes commonly move to the west, with a slight poleward component. This occurs because of the semipermanent cell of high pressure (called the *Bermuda High*) that is positioned poleward of the storm (see Figure 7–10b, p. 200). On the equatorward side of this high-pressure center, easterly winds prevail, guiding the storms westward. If this high-pressure center is weak in the western Atlantic, the storm often turns northward. On the poleward side of the Bermuda High, westerly winds prevail, steering the storm back toward the east. Often it is difficult to determine whether the storm will curve back out to sea or whether it will continue straight ahead and make landfall.

A location only a few hundred kilometers from a hurricane—just a day's striking distance away—may experience clear skies and virtually no wind. Before the age of weather satellites, such a situation made it difficult to warn people of impending storms. The worst natural disaster in U.S. history came as a result of a hurricane that struck an unprepared Galveston, Texas, on September 8, 1900. The strength of the storm, together with the lack of adequate warning, caught the population by surprise and cost the lives of 6000 people in the city and at least 2000 more elsewhere (Figure 11–13).*

In the United States, early warning systems have greatly reduced the number of deaths caused by hurricanes. At the same time, however, an astronomical rise has occurred in the amount of property damage. The primary reason for this latter trend is the rapid population growth and accompanying development in coastal areas.

The Role of Satellites

Today many different tools provide data that are used to detect and track hurricanes. This information is used to develop forecasts and to issue watches and warnings. The greatest single advancement in tools used for observing tropical cyclones has been the development of meteorological satellites.

Because the tropical and subtropical regions that spawn hurricanes consist of enormous areas of open ocean, conventional observations are limited. The need for meteorological data from these vast regions is now met primarily by satellites. Even before a storm begins to develop cyclonic flow and the spiraling cloud bands so typical of a hurricane, the storm can be detected and monitored by satellites.

*For a fascinating account of the Galveston storm, read *Isaac's Storm* by Erik Larson (New York: Crown Publishers, 1999).

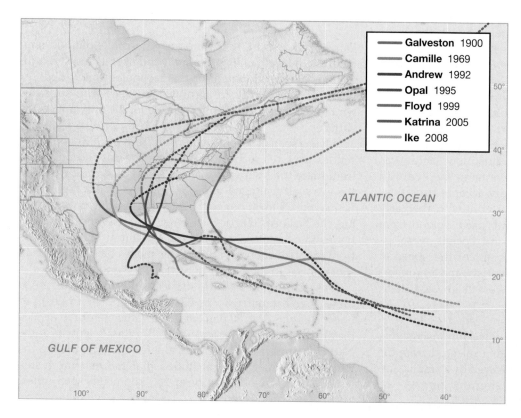

—— Galveston	1900
—— Camille	1969
—— Andrew	1992
—— Opal	1995
—— Floyd	1999
—— Katrina	2005
—— Ike	2008

ATLANTIC OCEAN

GULF OF MEXICO

Figure 11–12 This map shows a variety of tracks for some memorable hurricanes. To examine the history of many important storms, visit this interesting interactive site: http://www.csc.noaa.gov/hurricanes/.

Figure 11–13 Aftermath of the Galveston hurricane of 1900. Entire blocks were swept clean, and mountains of debris accumulated around the few remaining buildings. (AP Photos)

The advent of weather satellites has largely solved the problem of detecting tropical storms and has significantly improved monitoring. The box "Severe and Hazardous Weather: Hurricane Katrina from Space" (p. 318) provides excellent examples. However, satellites are remote sensors, and it is not unusual for wind-speed estimates to be off by tens of kilometers per hour and for storm-position estimates to have errors. It is still not possible to precisely determine detailed structural characteristics. A combination of observing systems is necessary to provide the data needed for accurate forecasts and warnings.

Aircraft Reconnaissance

Aircraft reconnaissance represents a second important source of information about hurricanes. Ever since the first experimental flights into hurricanes were made in the 1940s, the aircraft and the instruments employed have become quite sophisticated (Figure 11–14). When a hurricane is within range, specially instrumented aircraft can fly directly into a threatening storm and accurately measure details of its position and current state of development. Data transmission can be made directly from an aircraft in the midst of a storm to the forecast center, where input from many sources is collected and analyzed.

Measurements from reconnaissance aircraft are limited because they cannot be taken until a hurricane is relatively close to shore. Moreover, measurements are not taken continuously or throughout the storm. Rather, the aircraft provides sample "snapshots" of small parts of the hurricane. Nevertheless, the data collected are critical in analyzing the current characteristics needed to forecast the future behavior of a storm.

A major contribution to hurricane forecasting and warning programs has been an improved understanding of the structure and characteristics of these storms. Although advancements in remote sensing from satellites have been made, measurements from reconnaissance aircraft will be

Figure 11–14 In the Atlantic basin, most operational hurricane reconnaissance is carried out by the U.S. Air Force Reconnaissance Squadron based at Keesler AFB, Mississippi. Pilots fly through the hurricane to its center, measuring all basic weather elements as well as providing an accurate location of the eye. They use planes like the one in the background of this image. The National Oceanic and Atmospheric Administration uses small, specially-equipped jets (foreground) mostly on research missions to aid scientists in better understanding these storms. (Photo by NOAA)

SEVERE AND HAZARDOUS WEATHER

Hurricane Katrina from Space

Satellites allow us to track the formation, movement, and growth of hurricanes. In addition, their specialized instruments provide data that can be transformed into images that allow scientists to analyze the internal structure and workings of these huge storms. The images and captions in this box provide a unique perspective of Hurricane Katrina, the most devastating storm to strike the United States in more than a century.

Figure 11–D is from NASA's Terra satellite and is a relatively "traditional" image that shows Katrina approaching the Gulf coast.

Figure 11–E is a color-enhanced infrared (IR) image from the GOES-East satellite. Recall from Chapter 2 that the wavelengths of radiation emitted by an object are temperature dependent. Longer IR wavelengths indicate cooler temperatures, and shorter wavelengths are associated with warmer temperatures. The high tops of towering cumulonimbus clouds are colder than the tops

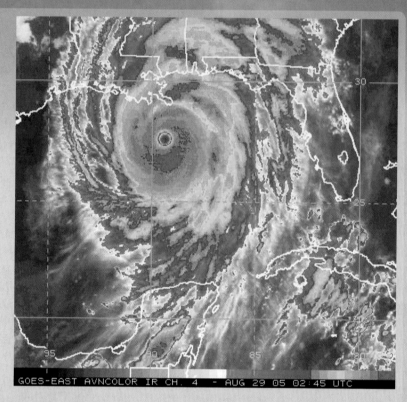

GOES-EAST AVNCOLOR IR CH. 4 - AUG 29 05 02:45 UTC

FIGURE 11–E Color-enhanced infrared image from the GOES-East satellite of Hurricane Katrina several hours before it made landfall on August 29, 2005. The most intense activity is associated with red and orange. (NOAA)

FIGURE 11–D This image shows Katrina at 1 PM on Sunday, August 28, 2005, as a massive storm covering much of the Gulf of Mexico. After passing over Florida as a category 1 hurricane, Katrina entered the Gulf and intensified into a category 5 storm with winds of 257 kilometers (160 miles) per hour and even stronger gusts. Air pressure at the center of the storm measured 902 millibars. When Katrina came ashore the next day, it was a slightly less vigorous category 4 storm. (NASA image)

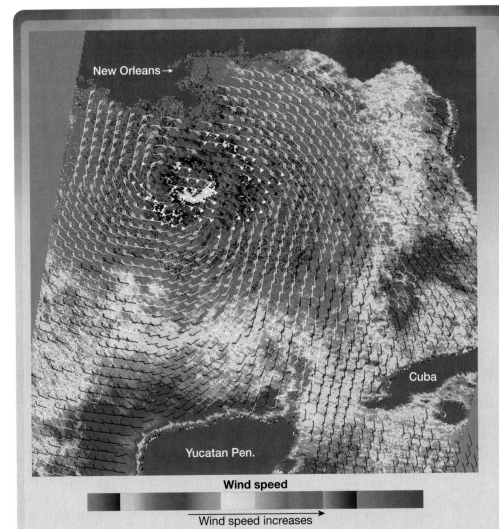

New Orleans→

Cuba

Yucatan Pen.

Wind speed

Wind speed increases

FIGURE 11–F NASA's QuikSCAT satellite was the source of data for this image of Hurricane Katrina on August 28, 2005. The image depicts relative wind speeds. The strongest winds, shown in shades of purple, circle a well-defined eye. The barbs show wind direction. (NASA image)

of clouds at lower altitudes (less vertically developed clouds). In this image of Katrina taken a few hours before landfall, the highest (coldest) cloud tops and thus the most intense storms are easily seen. Meteorologists use color-enhanced imagery to aid with satellite interpretation. The colors enable them to easily and quickly see features that are of special interest.

Figure 11–F from NASA's QuikSCAT satellite is very different in appearance. It provides a detailed look at Katrina's surface winds shortly before the storm made landfall. This image depicts relative wind speeds rather than actual values. The satellite sends out high-frequency radio waves, some of which bounce off the ocean and return to the satellite. Rough, storm-tossed seas create a strong signal, whereas a smooth surface returns a weaker signal. In order to match wind speeds with the type of signal that returns to the satellite, scientists compare wind measurements taken by data buoys in the ocean to the strength of the signal received by the satellite. When there are too few data buoy measurements to compare to the satellite data, exact wind speeds cannot be determined. Instead, the image provides a clear picture of relative wind speeds.

Figure 11–G shows the Multi-satellite Precipitation Analysis (MPA) of the storm. This image, which also depicts the track of the storm, shows the overall pattern of rainfall. It was constructed from data collected over several days by the Tropical Rainfall Measuring Mission (TRMM) satellite and other satellites.

Figure 11–H provides yet another satellite perspective. Seeing the pattern of rainfall in different parts of a hurricane is very useful to forecasters because it helps determine the strength of the storm. Scientists have developed a way to process data from the Precipitation Radar (PR) aboard the TRMM satellite within three hours and display it in 3-D. Every time the satellite passes over a named tropical cyclone anywhere in the world, the PR instrument sends data to create a 3-D snapshot of the storm. Such an image provides information on how heavily the rain is falling in different parts of the storm, such

Key to symbols

+ Location of storm center at 8 A.M. EDT
O Tropical depression
⑥ Tropical storm
⑥ Hurricane

Total rainfall

| 8 | 16 | 24 | 32 cm |
| 3.2 | 6.4 | 9.6 | 12.8 in. |

FIGURE 11–G Storm track and rainfall values for Hurricane Katrina for the period August 23 to 31, 2005. Rainfall amounts are derived from satellite data. The highest totals (dark red) exceeded 30 centimeters (12 inches) over northwestern Cuba and the Florida Keys. Amounts over southern Florida (green to yellow) were 12–20 centimeters (5–8 inches). Rainfall along the Mississippi coast (yellow to orange) was between 15–23 centimeters (6–9 inches). After coming ashore, Katrina moved through Mississippi, western Tennessee, and Kentucky and into Ohio. Because the storm moved rapidly, rainfall totals (green to blue) in these areas were generally less than 13 centimeters (5 inches). (NASA image)

as the eye wall versus the outer rain bands. It also gives a 3-D look at the cloud heights and "hot towers" inside the storm. Higher hot towers in the eye wall usually indicate a strengthening storm.

FIGURE 11–H Tropical Rainfall Measuring Mission (TRMM) Satellite image of Hurricane Katrina early on August 28, 2005. This cutaway view of the inner portion of the storm shows cloud height on one side and rainfall rates on the other. The tall rain columns provide a clue that the storm is strengthening. Two isolated towers (in red) are visible: one in an outer rain band and the other in the eye wall. The eye wall tower rises 16 kilometers (10 miles) above the ocean surface and is associated with an area of intense rainfall. Towers this tall near the core are often an indication that a storm is intensifying. Katrina grew from a category 3 to a category 4 storm soon after this image was received. (NASA image)

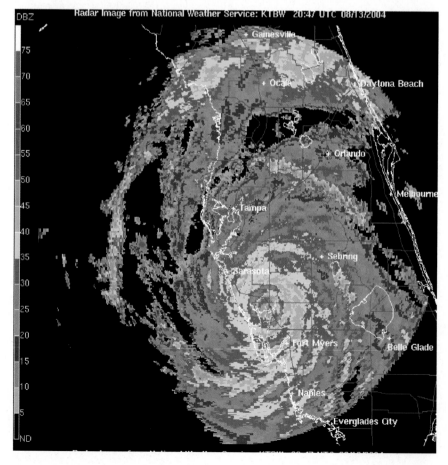

required for the foreseeable future to maintain the present level of accuracy for forecasts of potentially dangerous tropical storms.

Radar and Data Buoys

Radar is a third basic tool in the observation and study of hurricanes (Figure 11–15). When a hurricane nears the coast, it is monitored by land-based Doppler weather radar.* Doppler radar provides detailed information on hurricane wind fields, rainfall intensity, and storm movement. As a result, local National Weather Service offices are able to provide short-term warnings for floods, tornadoes, and high winds for specific areas. Sophisticated mathematical calculations provide forecasters with important information derived from the radar data, such as estimates of rainfall amounts. A limitation of radar is

*For a more complete discussion of this important tool, see the section "Doppler Radar" in Chapter 10.

Figure 11–15 Doppler radar image of Hurricane Charley over Charlotte Harbor, Florida, just after landfall on August 13, 2004. The range of coastal radar is about 320 kilometers (200 miles). (NOAA/National Weather Service)

that it cannot "see" farther than about 320 kilometers (200 miles) from the coast, and hurricane watches and warnings must be issued long before a storm comes into range.

VORTRAC—A New Hurricane Tracking Technique

Rapidly intensifying storms can catch vulnerable coastal areas by surprise. In 2007 Hurricane Humberto struck near Port Arthur, Texas, after unexpectedly strengthening from a tropical depression to a hurricane in less than 19 hours. In 2004, parts of the southwest coast of Florida were caught off guard when Hurricane Charley's top winds increased from 175 to more than 230 kilometers (110 to 145 miles) per hour in just 6 hours as the storm approached land.

A new technique known as VORTRAC (*Vortex Objective Radar Tracking and Circulation*) was successfully tested in 2007 and put into use in 2008. It is intended to help improve short-term hurricane warnings by capturing sudden intensity changes in potentially dangerous hurricanes during the critical time when storms are nearing land.

Prior to the development of VORTRAC, the only way to monitor the intensity of a landfalling hurricane was to rely on aircraft that dropped instrument packages (dropwindsondes) into the storm. This can be done only every hour or two, which means that a sudden drop in barometric pressure and the accompanying increase in winds could be difficult to detect in a timely manner.

VORTRAC uses that portion of the Doppler radar network that is along the Gulf and Atlantic coastline from Maine to Texas. Each unit can measure winds blowing toward or away from it, but until VORTRAC was established, no single radar could estimate a hurricane's rotational winds and central pressure. The technique uses a series of mathematical formulas that combine data from a single radar with knowledge of Atlantic hurricane structure in order to map the storm's rotational winds. VORTRAC also infers the barometric pressure in the eye of the hurricane, which is a reliable indicator of its strength (Figure 11–16). Each radar can sample conditions out to a distance of 190 kilometers (120 miles), and forecasters using VORTRAC can update the status of a storm about every six minutes.

Data Buoys Data buoys represent a fourth method of gathering data for the study of hurricanes (see Figure 12–3, p. 330). These remote, floating instrument packages are positioned in fixed locations all along the Gulf Coast and Atlantic Coast of the United States. When you examine the weather maps in Figure 11–3, you can see data buoy information plotted at several offshore stations. Ever since the early 1970s, data provided by these units have become a dependable and routine part of daily weather analysis as well as an important element of the hurricane warning system. The buoys represent the only means of making nearly continuous direct measurements of surface conditions over ocean areas.

Hurricane Watches and Warnings

Using input from the observational tools that were just described in conjunction with sophisticated computer models,

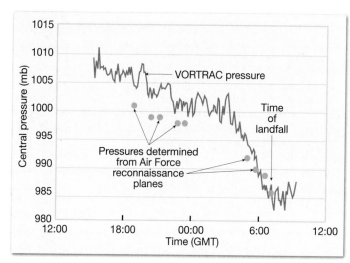

Figure 11–16 This graph displays the results of a test of the VORTRAC system on Hurricane Humberto, a storm that intensified rapidly as it approached the Texas/Louisiana coast on September 13, 2007. The VORTRAC estimates agreed closely with Humberto's actual central pressure, as measured by reconnaissance planes just before landfall. VORTRAC also captured the storm's rapid intensification, indicating that the pressure began falling quickly near the end of a four-hour span when no aircraft data were available. (After M. Bell/National Center for Atmospheric Research)

meteorologists attempt to forecast the movements and intensity of a hurricane. The goal is to issue timely watches and warnings.

A **hurricane watch** is an announcement that hurricane conditions are *possible* within a specified coastal area. Because preparedness activities become difficult once winds reach tropical storm force, a hurricane watch is issued 48 hours in advance of the anticipated onset of tropical storm–force winds. By contrast, a **hurricane warning** is issued 36 hours in advance and indicates that hurricane conditions are *expected* somewhere within a specified coastal area. A hurricane warning can remain in effect when dangerously high water or a combination of dangerously high water and exceptionally high waves continue, even though winds may be less than hurricane force.

Two factors are especially important in the watch-and-warning decision process. First, adequate lead time must be provided to protect life and, to a lesser degree, property. Second, forecasters must attempt to keep overwarning at a minimum. This, however, can be a difficult task. Clearly, the decision to issue a warning represents a delicate balance between the need to protect the public on the one hand and the desire to minimize the degree of overwarning on the other.

Hurricane Forecasting*

Hurricane forecasts are a basic part of any warning program. Several aspects can be part of such a forecast. We certainly want to know where a storm is headed. The predicted path

---

*Based on "Hurricane Forecasting in the United States: An Information Statement of the AMS," *Bulletin of the American Meteorological Society,* Vol. 88, No. 6 (June 2007), pp. 950–953.

PROFESSIONAL PROFILE

Daniel Brown: Senior Hurricane Specialist, National Hurricane Center

They say "to everything there is a season." Well, to Daniel Brown there are two seasons: hurricane and non-hurricane.

During hurricane season, Brown—a senior hurricane specialist and the warning coordination meteorologist at the National Hurricane Center in Miami—monitors weather conditions in the Atlantic and east Pacific basins. If a storm is brewing, Brown is one of the operational forecasters charged with determining what its track, strength, and size will be for the upcoming five days. This, he says, is the most exciting aspect of his job. "I love walking into the office every day and looking at satellite imagery to see what's going on around the globe."

The National Hurricane Center uses data from satellites, airplane reconnaissance, and sophisticated computer models when issuing storm watches and warnings. Advances in technology have helped this effort. "We are much better than we were 20 years ago at predicting the path of a storm," Brown says.

> We are much better than we were 20 years ago at predicting the path of a storm.

Even so, the National Hurricane Center can't rely on technology alone to make good predictions. Sometimes, for example, instruments that measure winds at the surface of the ocean might conflict with satellite data. "When data conflict, that's where the hurricane specialist's experience comes in. Is the storm a category 1, or is it a category 2? The specialist can determine why one measurement might not agree with another and make the best determination," says Brown.

The off-season is also busy. "A lot happens outside of hurricane season," says Brown, referring to the National Hurricane

Center's outreach, training, and education effort. "Run from water. Hide from wind" is a mantra Brown gives to the public and emergency management personnel on hurricane preparedness. "Communicating risk to people is a difficult challenge. With our outreach and education effort, we are always looking for new ways to reach people, especially if they live on the coast or in an evacuation zone." Social media now play a role. "Twitter and Facebook have been positive avenues for reaching thousands of people."

Communicating risk to people is a difficult challenge.

The National Hurricane Center in Miami is also a Regional Specialized Meteorological Center (RSMC), part of the United Nations's World Meteorological Organization. Brown and his colleagues provide forecasts to all countries in the Atlantic basin and also teach courses to international forecasters and emergency management personnel.

Brown credits his father for sparking a serious interest in meteorology. When Brown was growing up, his father was an avid weather watcher and never lost interest in learning more about the weather. By high school, Brown knew that he wanted to pursue a career in meteorology.

Brown majored in meteorology at the University of North Carolina at Asheville. While there, he worked as an intern at NOAA's National Climatic Data Center in Asheville for a year and a half. In 1993, one year after Hurricane Andrew struck Florida, Brown got an internship at the Tropical Analysis and Forecast Branch (TAFB) of the National Hurricane Center. He also worked for three years in the Miami National Weather Service Forecast Office (WFO), where he issued the

warning for a tornado that struck downtown Miami.

After working for the WFO, he went back to TAFB and became a hurricane specialist in 2006. At TAFB, Brown has been involved in improving the National Hurricane Center's storm tracking time frames. The center used to issue watches 36 hours in advance and warnings 24 hours in advance. Thanks to more sophisticated computer models, the National Hurricane Center now issues watches and warnings 12 hours earlier, helping emergency management personnel to prepare sooner.

After a storm is over, Brown and his colleagues put together a Tropical Cyclone Report. "It's basically the 'what happened' of the storm," he says. They reanalyze all the data collected during the storm and compile size and intensity estimates for every six hours of the storm's life. The report goes into the long-term climate record and helps the National Hurricane Center determine forecasting errors and areas for improvement.

DANIEL BROWN, Meteorologist Dan Brown is a Senior Hurricane Specialist at the National Hurricane Center in Miami, Florida. (NOAA)

of a storm is called the *track forecast*. Of course, there is also interest in knowing the intensity (strength of the winds), probable rainfall amounts, and likely size of the storm surge.

Track Forecasts The track forecast is probably the most basic information because accurate prediction of other storm characteristics is of little value if there is significant uncertainty about where the storm is going. Accurate track forecasts are important because they can lead to timely evacuations from the surge zone, where the greatest number of deaths usually occur. Fortunately, track forecasts have been steadily improving. During the span 2001–2005, forecast er-

rors were roughly half of what they were in 1990. During the very active 2004 and 2005 Atlantic hurricane seasons, 12- to 72-hour track forecast accuracy was at or near record levels. Consequently, the length of official track forecasts issued by the National Hurricane Center was extended from three days to five days (Figure 11–17). Current five-day track forecasts today are as accurate as three-day forecasts were 15 years ago. Much of the progress is due to improved computer models and a dramatic increase in the quantity of satellite data from over the oceans.

Despite improvements in accuracy, forecast uncertainty still requires that hurricane warnings be issued for relatively

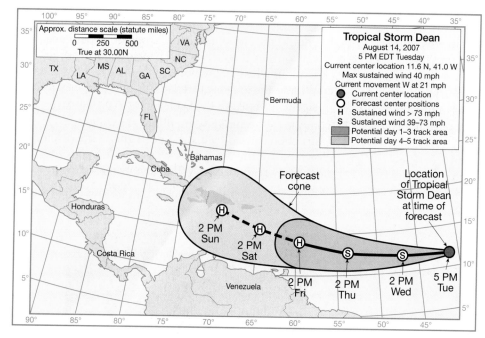

Figure 11-17 Five-day track forecast for Tropical Storm Dean issued at 5 PM Tuesday, August 14, 2007. When a hurricane track forecast is issued by the National Hurricane Center, it is termed a *forecast cone*. The cone represents the probable track of the center of the storm and is formed by enclosing the area swept out by a set of circles along the forecast track (at 12 hours, 24 hours, 36 hours, and so on). The size of each circle gets larger with time. Based on statistics from 2003–2007, the entire track of an Atlantic tropical cyclone can be expected to remain entirely within the cone roughly 60 to 70 percent of the time. (National Weather Service/National Hurricane Center)

large coastal areas. During the span 2000–2005, the average length of coastline under a hurricane warning in the United States was 510 kilometers (316 miles). This represents a significant improvement over the preceding decade, when the average was 730 kilometers (452 miles). Nevertheless, only about one-quarter of an average warning area experiences hurricane conditions.

Forecasting Other Characteristics In contrast to the improvements in track forecasts, errors in forecasts of hurricane intensity (wind speeds) have not changed significantly in 30 years. Accurate predictions of rainfall as hurricanes make landfall also remains elusive. However, accurate predictions of impending storm surge are possible when good information regarding the storm's track and surface wind structure are known and reliable data regarding coastal and offshore (underwater) topography are available.

Concept Check 11.5

1 What was the worst natural disaster in United States history? Why is such an event unlikely to occur in the United States again?

2 List four tools that provide data used to track hurricanes and develop forecasts.

3 Distinguish between a *hurricane watch* and a *hurricane warning*.

4 What is a track forecast? Why are such forecasts important?

Give It Some Thought

1. Why might people in some parts of the world welcome the arrival of hurricane season?

2. Hurricanes are sometimes referred to as "heat engines." What is the "fuel" that provides the energy for these high-powered engines?

3. The accompanying world map shows the tracks and intensities of nearly 150 years of tropical cyclones. It is based on all storm tracks available from the National Hurricane Center and the Joint Typhoon Warning Center.

   a. What area has experienced the greatest number of category 4 and 5 storms?

   b. Why do hurricanes *not form* in the very heart of the tropics, astride the equator?

   c. Explain the absence of storms in the South Atlantic and eastern South Pacific.

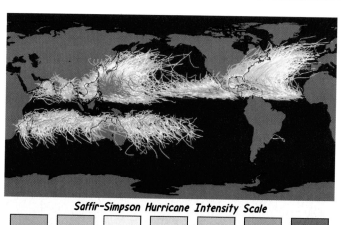

4. Refer to the graph in Figure 11–4b. Explain why wind speeds are greatest when the slope of the pressure curve is steepest.

5. Although observational tools and hurricane forecasts continue to improve, the potential for loss of life due to hurricanes is likely growing. Suggest a reason for this apparent contradiction.

6. Assume that it is late September 2016, and Hurricane Gaston, a category 5 storm, is projected to follow the path shown on the accompanying map. Answer the following questions.

a. Name the stages of development that Gaston must have gone through to become a hurricane. At what point did it receive its name?

b. Should the city of Houston expect to experience Gaston's fastest winds and greatest storm surge? Explain why or why not.

c. What is the greatest threat to life and property if this storm approaches the Dallas–Fort Worth area?

7. Examine the accompanying photo showing destruction caused by a strong category 4 hurricane. Which one of the three basic classes of damage was most likely responsible? What is your reasoning?

(Barry Williams. AFP/Getty Images/Newscom)

8. A television meteorologist is able to inform viewers about the intensity of an approaching hurricane. However, the meteorologist can report the intensity of a tornado only *after* it has occurred. Why is this true?

HURRICANES IN REVIEW

- Most hurricanes form between latitudes 5° and 20° over all tropical oceans except the South Atlantic and eastern South Pacific. The North Pacific has the greatest number of storms, averaging 20 per year. In the western Pacific, hurricanes are called *typhoons*, and in the Indian Ocean, they are referred to as *cyclones*.

- A steep pressure gradient generates the rapid, inward spiraling winds of a hurricane. As the warm, moist air approaches the core of the storm, it turns upward and ascends in a ring of cumulonimbus towers and forms a doughnut-shaped wall called the *eye wall*. At the very center of the storm, called the *eye*, the air gradually descends, precipitation ceases, and winds subside.

- A hurricane is a heat engine fueled by the latent heat liberated when huge quantities of water vapor condense. Hurricanes develop most often in late summer, when ocean waters have reached temperatures of 27°C (80°F) or higher and are thus able to provide the necessary heat and moisture to the air. The initial stage of a tropical storm's life cycle, called a *tropical disturbance*, is a disorganized array of clouds that exhibits a weak pressure gradient and little or no rotation. Tropical disturbances that produce many of the strongest hurricanes that enter the western North Atlantic and threaten North America often begin as large undulations or ripples in the trade winds known as *easterly waves*.

- Each year, only a few tropical disturbances develop into full-fledged hurricanes (storms that require minimum wind speeds of 119 kilometers per hour to be so termed). When a cyclone's strongest winds do not exceed 63 kilometers per hour, it is called a *tropical depression*. When winds are between 63 and 119 kilometers per hour, the cyclone is termed a *tropical storm*. Hurricanes diminish in intensity whenever they (1) move over cool ocean waters that cannot supply warm, moist tropical air; (2) move onto land; or (3) reach a location where large-scale flow aloft is unfavorable.

- Although damages caused by a hurricane depend on several factors, including the size and population density of the area affected and the nearshore ocean bottom configuration, the most significant factor is the strength of the storm itself. The *Saffir–Simpson scale* ranks the relative intensities of hurricanes. A 5 on the scale represents the strongest storm possible, and a 1 is the least severe. Damage caused by hurricanes can be divided into three categories: (1) *storm surge*, (2) *wind damage*, and (3) *inland freshwater flooding*.

- North Atlantic hurricanes develop in the trade winds, which generally move these storms from east to west. Because of early warning systems that help detect and track hurricanes, the number of deaths associated with these violent storms have been greatly reduced. Because the tropical and subtropical regions that spawn hurricanes consist of enormous areas of open oceans, meteorological data from these vast regions are provided primarily by

satellites. Other important sources of hurricane information are *aircraft reconnaissance, radar*, and remote, floating instrument platforms called *data buoys.*

- Hurricane watches and warnings alert coastal residents to possible or expected hurricane conditions. Two important factors in the watch-and-warning decision process are (1) providing adequate lead time and (2) attempting to keep overwarning to a minimum.

VOCABULARY REVIEW

easterly wave (p. 308)
eye (p. 307)
eye wall (p. 306)
hurricane (p. 304)

hurricane warning (p. 321)
hurricane watch (p. 321)
Saffir–Simpson scale (p. 311)
storm surge (p. 311)

tropical depression (p. 308)
tropical disturbance (p. 307)
tropical storm (p. 308)

PROBLEMS

Questions 1–5 refer to the weather maps of Hurricane Fran in Figure 11–3.

1. On which of the two days were Fran's wind speeds probably highest? How were you able to determine this?

2. **a.** How far did the center of the hurricane move during the 24-hour period represented by these maps?

   b. At what rate did the storm move during this 24-hour span? Express your answers in miles per hour.

3. The midlatitude cyclone shown in Figure 9–21 has an east–west diameter of approximately 1200 miles (when the 1008-millibar isobar is used to define the outer boundary of the low). Measure the diameter (north–south) of Hurricane Fran on September 5. Use the 1008-millibar isobar to represent the outer edge of the storm. How does this figure compare to the midlatitude cyclone?

4. Determine the pressure gradient for Hurricane Fran on September 5. Measure from the 1008-millibar isobar at Charleston to the center of the storm. Express your answer in millibars per 100 miles.

5. The weather map in Figure 9–21 shows a well-developed midlatitude cyclone. Calculate the pressure gradient of this storm from the 1008-millibar isobar at the Wyoming–Idaho border to the center of the low. Assume that the pressure at the center of the storm is 986 millibars and the distance is 625 miles. Express your answer in millibars per 100 miles. How does this answer compare to your answer to Problem 4?

6. Hurricane Rita was a major storm that struck the Gulf Coast in late September 2005, less than a month after Hurricane Katrina. Figure 11–18 is a graph showing changes in air pressure and wind speed from the storm's beginning as an unnamed tropical disturbance north of the Dominican Republic on September 18, until its last remnants faded away

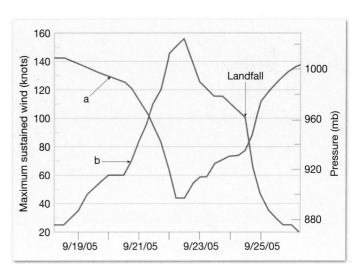

Figure 11–18 Graph to accompany Problem 6.

in Illinois on September 26. Use the graph to answer these questions.

   a. Which line represents air pressure, and which line represents wind speed? How did you figure this out?

   b. What was the storm's maximum wind speed, in knots? Convert this answer to kilometers per hour by multiplying by 1.85.

   c. What was the lowest pressure attained by Hurricane Rita?

   d. Using wind speed as your guide, what was the highest category reached on the Saffir–Simpson scale? On what day was this status reached?

   e. When landfall occurred, what was the category of Hurricane Rita?

MyMeteorologyLab™

12

Weather Analysis and Forecasting

People expect thorough and accurate weather forecasts. The desire for reliable weather predictions ranges from NASA's need to evaluate conditions leading up to the launch of a satellite to families wanting to know if the upcoming weekend weather will be suitable for a beach outing. Such diverse industries as airlines and fruit growers depend heavily on accurate weather forecasts. In addition, the designs of buildings, oil platforms, and industrial facilities rely on a sound knowledge of the atmosphere in its most extreme forms, including thunderstorms, tornadoes, and hurricanes. We are no longer satisfied with receiving only short-range predictions but expect accurate long-range forecasts as well.

Focus On Concepts

After completing this chapter, you should be able to:

- Explain the mission of the National Weather Service.
- Distinguish between weather analysis and weather forecasting.
- Describe the basis of numerical weather prediction.
- List and distinguish among the various methods of weather forecasting.
- Interpret what a particular pattern in the flow aloft indicates about surface weather.

- Differentiate between weather forecasts and 30- and 90-day outlooks.
- Explain why the percentage of correct forecasts is not always a good measure of forecast skill.
- Discuss the advantages and disadvantages of infrared imagery and visible imagery generated by weather satellites.

What is a weather forecast? Simply, a **weather forecast** is a scientific estimate of the weather conditions at some future time (Figure 12–1). Forecasts are usually expressed in terms of the most significant weather variables, which include temperature, cloudiness, humidity, precipitation, wind speed, and wind direction. As the following quote illustrates, weather forecasting is a formidable task:

> Imagine a system on a rotating sphere that is 8000 miles wide, consists of different materials, different gases that have different properties (one of the most important of which, water, exists in different concentrations), heated by a nuclear reactor 93 million miles away. Then, just to make life interesting, this sphere is oriented such that, as it revolves around the nuclear reactor, it is heated differently at different times of the year. Then, someone is asked to watch the mixture of gases, a fluid only 20 miles deep, that covers an area of 250 million square miles, and to predict the state of the fluid at one point on the sphere 2 days from now. This is the problem weather forecasters face.*

---

*Robert T. Ryan, "The Weather Is Changing . . . or Meteorologists and Broadcasters, the Twain Meet," *Bulletin of the American Meteorological Society* 63, no. 3 (March 1982), 308.

The Weather Business: A Brief Overview

The U.S. governmental agency responsible for gathering and disseminating weather-related information is the **National Weather Service (NWS),** a branch of the *National Oceanic and Atmospheric Administration* (*NOAA*). The mission of the NWS is as follows:

> The National Weather Service (NWS) provides weather, hydrologic and climate forecasts and warnings for the United States, its territories, adjacent waters and ocean areas, for the protection of life and property and the enhancement of the national economy. NWS data and products form a national information database and infrastructure that can be used by other governmental agencies, the private sector, the public, and the global community.

Perhaps the most important services provided by the NWS are forecasts and warnings of hazardous weather, including thunderstorms, floods, hurricanes, tornadoes, winter weather, and extreme heat. According to the Federal Emergency Management Agency, 80 percent of all declared emergencies are weather related. In a similar vein, the

Figure 12–1 Accurate weather forecasts are important to many human activities, such as the launch of the space shuttle Atlantis on July 8, 2011. (NASA)

Department of Transportation reports that more than 6000 vehicular fatalities per year can be attributed to weather. As global population increases, the economic impact of weather-related phenomena also escalates. As a result, the NWS is under greater pressure to provide more accurate and longer-range forecasts.

To produce even a short-range forecast is an enormous task. It involves complicated and detailed procedures, including collecting weather data, transmitting those data to a central location, and compiling them on a global scale. After processing, these data are analyzed so that an accurate assessment of the current conditions can be made. In the United States, weather information from around the world is collected by the **National Centers for Environmental Prediction,** located in Camp Springs, Maryland, near Washington, D.C. This branch of the NWS prepares weather maps, charts, and forecasts on national as well as global scales. These forecasts are disseminated to regional **Weather Forecast Offices,** where they are used to produce local and regional weather forecasts.

The final phase in the weather business is the dissemination of a wide variety of forecasts. Each of the 125 Weather Forecast Offices regularly issues regional and local forecasts, aviation forecasts, and warnings covering its forecast area. The *local forecast* seen on The Weather Channel or your local TV station is derived from a forecast issued by one of the offices operated by the NWS (Figure 12–2). Further, all the data and products (maps, charts, and forecasts) provided by the NWS are available at no cost to the general public and to private forecasting services, such as AccuWeather and WeatherData.

The demand for highly visual forecasts containing computer-generated graphics has increased proportionately with the use of personal computers and the Internet. The animated depictions of the weather that appear on most local newscasts are produced by the private sector because this task is outside the mission of the NWS. The private sector has also assumed responsibility for customizing NWS products to create a variety of specialized weather reports tailored for specific audiences. In a farming community, for example, the weather reports might include frost warnings, while winter forecasts in Denver, Colorado, include the snow conditions at area ski resorts.

It is important to note that despite the valuable role that the private sector plays in disseminating weather-related information to the public, the NWS is the *official* voice in the United States for issuing warnings during hazardous and life-threatening weather situations. Two major weather centers operated by the NWS serve critical functions in this regard. The Storm Prediction Center in Norman, Oklahoma, maintains a constant vigil for severe thunderstorms and tornadoes (see Chapter 10). Hurricane watches and warnings for the Atlantic, Caribbean, Gulf of Mexico, and eastern Pacific are issued by the National Hurricane Center/Tropical Prediction Center in Miami, Florida (see Chapter 11).

In summary, the process of providing weather forecasts and warnings throughout the United States occurs in three stages. First, to create an image of the current state of the atmosphere, data are collected and analyzed on a global scale. Second, the NWS employs a variety of techniques to establish the future state of the atmosphere; a process called weather forecasting. Third, forecasts are disseminated to the public, mainly through the private sector. The National Weather Service serves as the sole entity responsible for issuing watches and warnings of extreme weather events.

Figure 12–2 The forecast seen on your local TV station is derived from one issued by a NWS Weather Forecast Office. (Photo by Exactostock/SuperStock)

Concept Check 12.1

1 What is the mission of the National Weather Service?

2 Approximately what percentage of all declared emergencies are weather related?

3 What branch of the National Weather Service produces local and regional weather forecasts?

4 List the three steps involved in providing weather forecasts.

Weather Analysis

Before weather can be predicted, the forecaster must have an accurate picture of current atmospheric conditions. This formidable task, called **weather analysis**, involves collecting, transmitting, and compiling billions of pieces of observational data. Because the atmosphere is ever changing, this must be accomplished quickly. High-speed supercomputers have greatly aided weather analysts.

Figure 12–3 Data buoy used to record atmospheric conditions over a section of the global ocean. The data this buoy collects are transmitted via satellite to a land-based station. (Photo by Michael Dwyer/Alamy)

database. The Federal Aviation Administration (FAA), in cooperation with the NWS, also operates observation stations at most metropolitan airports. Together, the NWS and the FAA operate nearly 900 **Automated Surface Observing Systems (ASOS).** These automated systems provide weather observations that include temperature, dew point, wind, visibility, and sky conditions and can also detect conditions such as rain or snow (Figure 12–4). Automation often assists, or, in some cases, replaces human observers because it can provide information from remote areas. However, some research has shown that human observers are more reliable for determining certain weather elements, such as cloud cover and sky conditions.

Observations Aloft Upper-air observations are essential for producing reliable forecasts (see Figure 1–24, p. 28). Worldwide, nearly 900 balloon-borne instrument packages, called *radiosondes*, are launched twice daily at 0000 and 1200 Greenwich Mean Time (6:00 PM and 6:00 AM Central Standard Time). Most of the upper-air observation stations are located in the Northern Hemisphere, with 92 stations operated by the National Weather Service.

Gathering Data

A vast network of weather stations is required to collect the data needed for generating even the shortest-range forecasts. On a global scale the **World Meteorological Organization (WMO),** an agency of the United Nations, is responsible for the international exchange of weather data. This task requires that data from the more than 185 participating nations and territories be standardized.

Surface Observations Worldwide, more than 10,000 observation stations on land, 7000 ships at sea, and hundreds of data buoys and oil platforms report atmospheric conditions four times daily, at 0000, 0600, 1200, and 1800 Greenwich Mean Time (Figure 12–3). These data are rapidly sent around the globe, using a communications system dedicated to weather information.

In the United States, 125 Weather Forecast Offices are responsible for gathering and transmitting weather information to a central

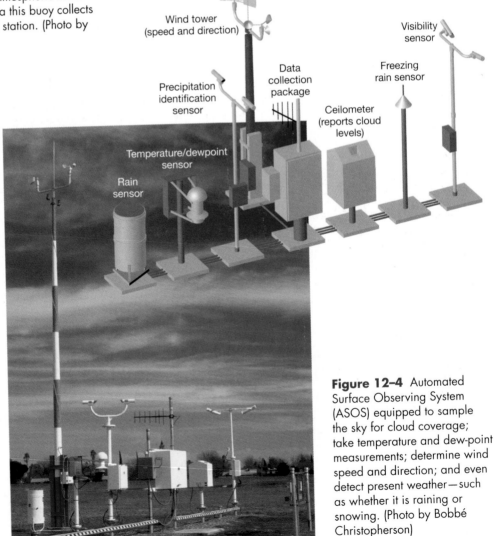

Figure 12–4 Automated Surface Observing System (ASOS) equipped to sample the sky for cloud coverage; take temperature and dew-point measurements; determine wind speed and direction; and even detect present weather—such as whether it is raining or snowing. (Photo by Bobbé Christopherson)

A **radiosonde** is a lightweight instrument package containing sensors that measure temperature, humidity, and pressure as a weather balloon rises through the atmosphere. By tracking the radiosonde, wind speed and direction at various altitudes can be determined. A radiosonde flight typically lasts about 90 minutes, during which it may ascend to heights of more than 35 kilometers (about 115,000 feet). Because pressure decreases with an increase in altitude, the balloon eventually stretches to its breaking point and bursts. When this occurs, a small parachute opens, and the instrument package slowly descends to Earth. If you find a radiosonde, follow the mailing instructions as the instruments can be reused.

Acquiring upper-air data over the ocean is problematic. Only a few ships launch radiosondes. Some commercial aircraft contribute upper-air information over the ocean by regularly reporting wind, temperature, and, occasionally, turbulence along their flight routes.

A number of technical advances have been made that improved our ability to make observations aloft. Special radar units called **wind profilers** are used to measure wind speed and direction up to 10 kilometers (6 miles) above the surface. These measurements can be taken every 6 minutes, in contrast to the 12-hour interval between balloon launches. In addition, satellites and weather radar have become invaluable tools for making weather observations. The importance of these modern technologies is considered later in this chapter.

Despite the advances being made in the collection of weather data, two difficulties remain. First, observations may be inaccurate due to instrument malfunctions or data transmission errors. Second, observations are difficult, or in some cases impossible, to obtain in some regions—particularly over oceans and in mountainous areas.

Weather Maps: Pictures of the Atmosphere

Once this large body of data has been collected, analysts display it in a format that can be comprehended easily by forecasters. This step is accomplished by placing the information on a number of **synoptic weather maps** (Box 12–1). These maps are called synoptic, which means "coincident in time," because they display a synopsis of the weather conditions at a given moment. These weather charts provide a symbolic representation of the state of the atmosphere. Thus, to the trained eye, a weather map is a snapshot that shows the status of the atmosphere, including data on temperature, humidity, pressure, and wind.

More than 200 surface maps and charts covering several levels of the atmosphere are produced each day by the NWS and its forecast centers. This task was once completed by hand, but computers now analyze and plot the data systematically. Typically, lines and symbols are used to depict weather patterns (Figure 12–5a). Once a map is generated, an analyst fine-tunes it and corrects any errors or omissions.

In addition to the surface map, twice-daily upper-air charts are drawn at 850-, 700-, 500-, 300-, and 200-millibar levels. Recall that on these charts, height contours (in meters or tens of meters) instead

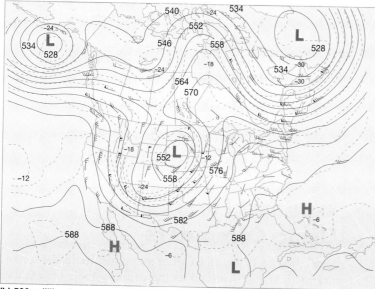

(a) Surface weather map.

(b) 500-millibar level chart.

Figure 12–5 Simplified synoptic weather maps. (a) Surface weather map for 7:00 AM Eastern Standard Time depicting a well-developed middle-latitude cyclone. (b) A 500-millibar-level map, with height contours in tens of meters, for the same time.

Box 12–1 Constructing a Synoptic Weather Chart

Production of surface weather charts starts with plotting the data from selected observing stations. By international agreement, data are plotted using the symbols illustrated in Figure 12–A. Data that are usually plotted include temperature, dew point, pressure and its tendency, cloud cover (height, type, and amount), wind speed and direction, and weather, both current and past. These data are always plotted in the same position around the station symbol for consistent reading. (The only exception to this arrangement is the wind arrow because it is oriented with the direction of airflow.) Using Figure 12–A, for example, you can see that the temperature is plotted in the upper-left corner of the sample model, and it will always appear in that location. A more complete weather station model and a key for decoding weather symbols are found in Appendix B.

The data are plotted as shown in Figure 12–B (left). Once data have been plotted, isobars and fronts are added to the weather chart (see Figure 12–B right). Isobars are usually plotted on surface maps at intervals of 4 millibars (1004, 1008, 1012, etc.). The positions of the isobars are estimated as accurately as possible from the pressure readings available. Note in

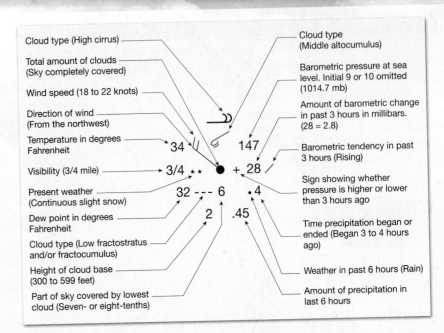

FIGURE 12–A A specimen station model showing the placement of commonly plotted data. (Abridged from the International Code)

Figure 12–B (right) that the 1012-millibar isobar is about halfway between two stations that report 1010 millibars and 1014 millibars. Frequently, observational errors and other complications require an analyst to

smooth the isobars so that they conform to the overall picture. Many irregularities in the pressure field are caused by local influences that have little bearing on the larger circulation depicted on the charts. Once the isobars

---

of isobars are used to depict the pressure field. These charts also contain isotherms (equal temperature lines) shown as dashed lines labeled in degrees Celsius. This series of upper-air charts provides a three-dimensional view of the atmosphere. Figure 12–5 shows a simplified version of a surface map, as well as a 500-millibar-height contour chart covering the same time period.

In summary, the analysis phase involves collecting and compiling billions of pieces of observational data describing the current state of the atmosphere. These data are then displayed on a number of different weather maps that show current weather patterns at predetermined levels throughout the atmosphere.

Concept Check 12.2

1 What does the process of weather analysis involve?

2 Weather Forecast Offices are responsible for collecting and transmitting weather data. In what other ways are these tasks accomplished?

3 What role do radiosondes play in weather forecasting?

4 Briefly explain the role of synoptic weather maps in weather forecasting.

Students Sometimes Ask...

Who was the first weather forecaster?

In the United States, Benjamin Franklin is often credited with making the first long-term weather predictions in his *Poor Richard's Almanac*, beginning in 1732. However, these forecasts were based primarily on *folklore* rather than weather data. Nevertheless, Franklin may have been the first to document that storm systems move. In 1743, while living in Philadelphia, rainy weather prevented Franklin from viewing an eclipse. Through later correspondence with his brother, he learned that the eclipse had been visible in Boston, but within a few hours that city also experienced rainy weather conditions. These observations led Franklin to the conclusion that the storm, which prevented his viewing of the eclipse in Philadelphia, moved up the east coast to Boston.

Weather Forecasting Using Computers

Until the late 1950s, all weather maps and charts were plotted manually and served as the primary tools for making weather forecasts. Forecasters used various techniques to

are drawn, centers of high and low pressure are designated.

Because fronts are boundaries that separate contrasting air masses, they can often be identified on weather charts by locating zones exhibiting abrupt changes in conditions. Because several elements change across a front, all are examined so that the frontal position is located most accurately. Some of the most easily recognized changes that can aid in identifying a front on a surface chart are as follows:

1. Marked temperature contrast over a short distance
2. Wind shift (in a clockwise direction) over a short distance by as much as 90°
3. Humidity variations commonly occurring across a front that can be detected by examining dew-point temperatures
4. Clouds and precipitation patterns that give clues to the position of fronts

Notice in Figure 12–B that all the conditions listed are easily detected across the frontal zone. However, not all fronts are as easily defined as the one on our sample map. In some cases, surface contrasts on opposite sides of the front are subdued. When this happens, charts of the upper air, where flow is less complex, become an important tool for detecting fronts.

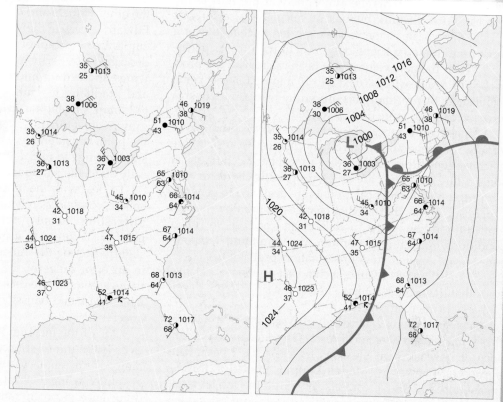

FIGURE 12–B Simplified weather charts. (left) Stations, with data for temperature, dew point, wind direction, wind speed, sky cover, and barometric pressure. (right) Same chart showing isobars and fronts.

extrapolate future conditions from the patterns depicted on the most recent weather charts. One method involved matching current conditions with similar, well-established patterns from the past. From such comparisons, meteorologists predicted how current systems might change in the hours and days to come. As this practice evolved, "rules of thumb" were established to aid forecasters. Applying these rules to current weather charts became the foundation of weather forecasting and still plays an important role in making short-range (24 hours or less) predictions.

Later, computers were used to plot data and produce surface and upper-air charts. As technologies improved, computers eventually replaced manual creation of weather forecasts. Computers have played a key role in improving the accuracy and detail of weather forecasts and in lengthening the period for which useful guidance can be given.

Numerical Weather Prediction

Numerical weather prediction is the technique used to forecast weather using mathematical models designed to represent atmospheric processes. (The word *numerical* is somewhat misleading here because all types of weather forecasting are based on some quantitative data and therefore could fit under this heading.) Numerical weather prediction relies on the fact that the behavior of atmospheric gases is governed by a set of physical principles or laws that can be expressed as mathematical equations (Box 12–2, page 337). If we can solve these equations, we will have a description of the future state (a forecast) of the atmosphere, derived from the current state, which we can interpret in terms of weather—temperature, moisture, cloud cover, and wind. This method is analogous to using a computer to predict the future positions of Mars, using Newton's laws of motion and knowing the planet's current position.

Highly refined computer models that attempt to mimic the behavior of the "real" atmosphere are used in numerical weather prediction. All numerical simulations are based on the same governing equations but differ in the way the equations are applied and in the parameters that are used. For example, some models use a very closely spaced set of data points and cover a specific concentrated area, whereas others describe the atmosphere more broadly on a global scale. In the United States, several different numerical models are used.

Numerical weather forecasting begins by entering current atmospheric variables (temperature, wind speed, humidity, and pressure) into a computer simulation. This set of values represents the atmosphere at the start of the forecast. After literally billions of calculations, a forecast of how these basic elements are expected to change over a short time frame (perhaps only 5 to 10 minutes) is generated. When the new values have been calculated, the process begins again with the next 5 to 10 minute forecast being established. By repeating this procedure many times over, a forecast for six or more days is created.

Using mathematical models, the NWS produces a variety of generalized forecast charts. Because these machine-generated maps predict atmospheric conditions at some future time, they are called **prognostic charts**, or simply **progs.** Most numerical models are designed to generate prognostic charts that predict changes in the flow pattern aloft. In addition, some models create forecasts for other conditions, including maximum and minimum temperatures, wind speeds, and precipitation probabilities. Even the simplest models require such a vast number of calculations that they could not have been used prior to the development of supercomputers.

Once generated, a statistical analysis is used to modify these machine-generated forecasts by making comparisons of how accurate previous forecasts have been. This approach, known as **Model Output Statistics (MOS),** (pronounced "moss"), corrects for errors the model tends to make consistently. For example, certain forecast models may predict too much rain, overly strong winds, or temperatures that are too high or too low. MOS forecasts form the baseline on which forecasters from the NWS, as well as private forecast companies, try to make improvements. This final step is performed by humans, using their knowledge of meteorology and making allowances for known model shortcomings (Figure 12–6).

In summary, meteorologists use equations to create mathematical models of the atmosphere. By utilizing data on initial atmospheric conditions, they solve these equations to predict a future state. Of course, this is a complicated process because Earth's atmosphere is a very complex, dynamic system that can only be roughly approximated using mathematical models. Further, because of the nature of the equations, tiny differences in the data can yield huge differences in outcomes. Nevertheless, these models produce surprisingly good results—much better than those made without them.

Ensemble Forecasting

One of the most significant challenges for weather forecasters is the apparently chaotic behavior of the atmosphere. Specifically, two very similar atmospheric disturbances may, over time, develop into two very different weather patterns. One may intensify, becoming a major disturbance, while the other withers and dissipates. To demonstrate this point, Edward Lorenz at the Massachusetts Institute of Technology employed a metaphor known as the *butterfly effect.* Lorenz described a butterfly in the Amazon rain forest fluttering its wings and setting into motion a subtle breeze that travels and gradually magnifies over time and space. According to Lorenz's metaphor, two weeks later this faint breeze has grown into a tornado over Kansas. Obviously, by stretching the point considerably, Lorenz tried to illustrate that a very small change in initial atmospheric conditions can dramatically affect the resulting weather pattern elsewhere.

To deal with the inherent chaotic behavior of the atmosphere, forecasters rely on a technique known as **ensemble forecasting.** Simply, this method involves producing a number of forecasts using the same computer model but slightly altering the initial conditions, while remaining within an error range of the observational instruments. Essentially, ensemble forecasting attempts to assess how the inevitable errors and omissions in weather measurements might affect the result.

One of the most important outcomes of ensemble forecasts is the information they provide about forecast uncertainty. For example, assume that a prognostic chart that was generated using the best available weather data predicts the occurrence of precipitation over a wide area of the southeastern United States within 24 hours. Now let's say that the same calculations are performed several times in succession, each time making minor adjustments to the initial conditions. If most of these prognostic charts also predict a pattern of precipitation in the Southeast, the forecaster will place a high degree of confidence in the forecast. On the

Figure 12–6 Forecasters at the National Weather Service provide nearly 2 million predictions annually to the public and commercial interests. (Photo by Ryan McGinnis/Alamy)

other hand, if the prognostic charts generated by the ensemble method differ significantly, there will be far less confidence in the forecast.

Role of the Forecaster

Despite faster computers, constant improvements in numerical models, and significant technological advances, prognostic charts provide only a generalized picture of atmospheric behavior. As a result, human forecasters, using their knowledge of meteorology as well as judgments based on experience, continue to serve a vital role in creating weather forecasts, particularly short-range forecasts.

When formulated, a variety of prognostic charts are sent to the regional weather forecast offices of the NWS. The responsibility of the forecasters is to blend numerical predictions with local conditions and regional weather quirks to produce site-specific forecasts. This task is further complicated by the availability of multiple prognostic charts. For example, generally two different numerical models are employed to predict the minimum temperature for a given day. One method works better on some days than on others and performs better in some locales than in others. It is up to the forecaster to select the "best" model each day, or perhaps to blend the data from both models.

Often, forecasters add extra detail to the model forecast. Isolated summer thunderstorms, for example, are on a scale too small for the computer models to adequately resolve. In addition, weather phenomena such as tornadoes and thunderstorm downbursts cannot be predicted using available forecasting techniques (Figure 12–7). Therefore, emphasis is placed on using satellites and weather radar to detect and track these phenomena.

Concept Check 12.3

1 Briefly describe the basis of numerical weather prediction.

2 How are prognostic charts different from synoptic weather maps?

3 What do computer-generated numerical models try to predict?

4 What is *ensemble forecasting*?

5 What additional information does an ensemble forecast provide over a traditional numerical weather prediction?

Other Forecasting Methods

Although machine-generated prognostic charts form the basis of modern forecasting, other methods are available to meteorologists. These methods, which have "stood the test of time," include persistence forecasting, climatological forecasting, analog methods, and trend forecasting.

Persistence Forecasting

Perhaps the simplest forecasting technique, called **persistence forecasting,** is based on the tendency of weather to remain unchanged for several hours or even days. If it is raining at a particular location, for example, it might be reasonable to assume that it will still be raining in a few hours. Persis-

Figure 12–7 Mesoscale phenomena such as this tornado are too small to appear on computer-generated prognostic charts. Detection of such events relies heavily on weather radar and geostationary satellites. (Photo by A. T. Wilett/Alamy)

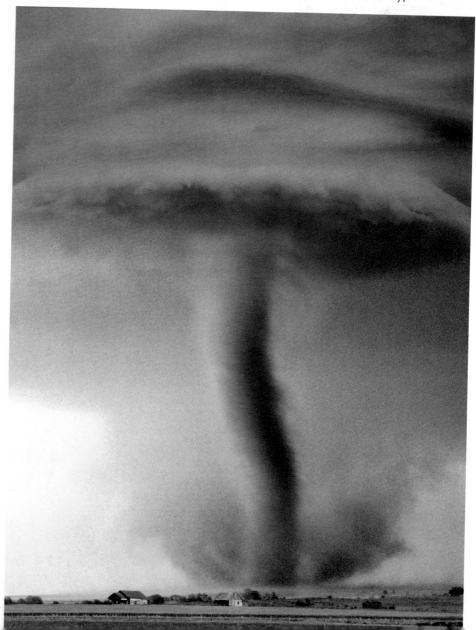

EYE ON THE ATMOSPHERE

This wintertime satellite image shows a large comma-shaped cloud pattern moving from the Pacific toward California. The green and turquoise water color off the southern coast of California indicates the presence of a bloom of phytoplankton (microscopic floating aquatic plants). Also notice the greenish-colored area south of the Salton Sea, a region classified as having a desert climate. (Image courtesy of NASA)

Question 1 What name is given to the storm system that resulted in the comma-shaped cloud pattern in the image?

Question 2 Describe the likely weather in the Monterey, CA area as this system passed by.

Question 3 Why do the Sierra Nevadas appear white?

Question 4 The phytoplankton bloom that is visible in this image is located in the area of the cold California current. How may this current, which is moving southward along the California coast, have contributed to this biological activity near the ocean surface?

Question 5 What is a possible explanation for the greenish-colored area south of the Salton Sea?

tence forecasts do not account for changes that might occur in the intensity or direction of a weather system, nor can they predict the formation or dissipation of storms. Because of these limitations and the rapidity with which weather systems change, persistence forecasts usually diminish in accuracy within 6 to 12 hours, or one day at the most.

Climatological Forecasting

Another relatively simple way of generating forecasts uses climatological data—average weather statistics accumulated over many years. This method is known as **climatological forecasting.** Consider, for example, that Yuma, Arizona, experiences sunshine approximately 90 percent of its daylight hours; thus, forecasters predicting sunshine every day of the year would be correct about 90 percent of the time. Likewise, forecasters in Portland, Oregon, would be correct about 90 percent of the time by predicting overcast skies in December.

Climatological forecasting is particularly useful when making agricultural business decisions. For example, in the relatively dry north-central portion of Nebraska known as the Sand Hills, the implementation of center-pivot irrigation made growing corn more feasible. However, farmers were faced with the question of which corn hybrid to plant. A high-yield variety widely used in southeastern Nebraska (its warmest region) seemed to be the logical choice. However, review of local climatological data showed that because of the cooler temperatures in the Sand Hills, corn planted in late April would not mature until late September, when there is a

50 percent probability of an autumn frost. Farmers used this important climate information to select a hybrid that was better suited for the shorter growing season of the Sand Hills.

One interesting use of climatological data is the prediction of a "white Christmas"—that is, a Christmas with 1 inch or more of snow on the ground. As Figure 12–8 illustrates, northern Minnesota and northern Maine have more than a 90 percent chance of experiencing a "white Christmas." By contrast, those who are in southern Florida for the holidays have a minuscule chance of experiencing snow.

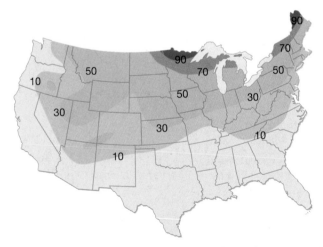

Figure 12–8 Probability (in percent) of a "white Christmas"—that is, a Christmas in which at least 1 inch of snow is on the ground. (Data from the U.S. Department of Commerce)

Box 12–2 Numerical Weather Prediction

Gregory J. Carbone*

During the past several centuries, the physical laws governing the atmosphere have been refined and expressed through mathematical equations. In the early 1950s, meteorologists began using computers, which provided an efficient means of solving these mathematical equations, to forecast the weather. The goal of such numerical weather prediction is to predict changes in large-scale atmospheric flow patterns. The equations relate to many of the processes already discussed in this book. Two *equations of motion* describe how horizontal air motion changes over time, taking into account pressure gradient, Coriolis, and frictional forces. The *hydrostatic equation* describes vertical motion in the atmosphere. The *first law of thermodynamics* is used to predict changes in temperature that result from the addition or subtraction of heat or from expansion and compression of air. Two equations account for the *conservation of mass* and the *conservation of water*. Finally, the *ideal gas law* or *equation of state* shows the relationship among three fundamental variables—temperature, density, and pressure.

Weather-prediction models begin with observations describing the current state of the atmosphere. They use the equations to compute new values for each variable of interest, usually at 5- to 10-minute intervals, called the *time step*. Predicted values serve as the initial conditions for the next series of computations and are made for specific locations and levels of the atmosphere. Each model has a spatial resolution that describes the distance between prediction points. Based on solving fundamental equations repeatedly, a model predicts the future state of the atmosphere. The model output is provided to weather forecasters for fixed intervals, such as 12, 24, 36, 48, and 72 hours in the future.

Despite the sophistication of numerical weather-prediction models, most still produce forecast errors. Three factors in particular restrict their accuracy: inadequate representation of physical processes, errors in initial observations, and inadequate model resolution. Whereas the models are grounded on sound physical laws and capture the major characteristics of the atmosphere, they necessarily simplify the workings of a very complex system. The representation of land–surface processes and topography are just two examples of features that are incompletely treated in

current numerical models. Errors in the initial observations fed into the computer will be amplified over time because numerical weather prediction models include many nonlinear relationships. Finally, physical conditions at all spatial scales can influence atmospheric changes. Yet the spatial resolution of numerical models is too coarse to capture many important processes. In fact, the atmospheric system moves at scales too small to ever be observed and incorporated explicitly into models.

A simple example illustrates how misrepresentation of physical processes or observation error might lead to inaccurate predictions. Figure 12–C is based on an equation used to predict future values of a given variable Y. The equation is written as

$$Y_{t+1} = (a \times Y_t) - Y_t^2$$

where Y represents the value of some variable at time t. Y_{t+1} represents the same variable at the next time step, and a represents a constant coefficient.

Notice that each predicted value serves as the initial value for the next calculation, the same way in which output from a

numerical weather-prediction model provides input for subsequent computations. The purple line in Figure 12–C shows the equation solution over a number of time steps, given an initial value of (for example, a meteorological observation) $Y_{t=0} = 1.5$ and a coefficient value of $a = 3.75$. The graph illustrates how the precision of our equations describing the evolving state of the atmosphere may affect predictions. The blue line represents values of Y that result from the adjustment of a from 3.75 to 3.749. Similarly, we can demonstrate how a very small observation error could amplify over time by adjusting $Y_{t=0}$ from 1.5 to 1.499. The red line in the graph shows how an incremental change in the initial value affects predictions. Small errors may make very little difference in the early stages of our prediction, but such errors amplify dramatically over time. Because we cannot observe many small-scale features of the atmosphere, nor incorporate all of its processes into computer models, weather forecasts have a theoretical limit.

*Professor Carbone is a faculty member in the Department of Geography at the University of South Carolina.

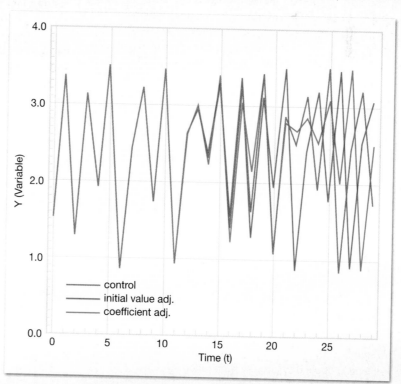

FIGURE 12–C Tiny errors may not significantly influence the early stages of a prediction, but with time, such errors amplify dramatically.

Analog Method

A somewhat more complex way to predict the weather is the **analog method,** which is based on the assumption that weather repeats itself, at least in a general way. Thus, forecasters attempt to find well-established weather patterns from the past that match (are analogous to) a current event. From such a comparison, forecasters predict how the current weather might evolve.

Prior to the advent of computer modeling, the analog method was the backbone of weather forecasting, and an analog method called *pattern recognition* remains an important tool with which short-range machine-generated forecasts are improved.

Trend Forecasting

Another related method, **trend forecasting,** involves determining the speed and direction of features such as fronts, cyclones, and areas of clouds and precipitation. Using this information, forecasters attempt to extrapolate the future position of these weather phenomena. For example, if a line of thunderstorms is moving toward the northeast at 35 miles per hour, a trend forecast would predict that the storm will reach a community located 70 miles to the northeast in about two hours.

Because weather events tend to increase or decrease in speed and intensity or change direction, trend forecasting is most effective over periods of just a few hours. Thus, it works particularly well for forecasting severe weather events that are expected to be short-lived, such as hailstorms, tornadoes, and downbursts, because warnings for such events must be issued quickly and must be site specific. The techniques used for this work, often called **nowcasting,** are heavily dependent on weather radar and geostationary satellites. These tools are important in detecting areas of heavy precipitation or clouds that are capable of triggering severe conditions. Nowcasting techniques use highly interactive computers capable of integrating data from a variety of sources. Prompt forecasting of tornadic winds is one example of the critical nature of nowcasting techniques.

Concept Check 12.4

1 If it is snowing today, what weather might be predicted for tomorrow if the technique called persistence forecasting is employed?

2 What type of weather phenomena are typically forecast using nowcasting techniques?

3 Briefly describe the basis of the analog method of weather forecasting.

4 What term is applied to the very short-range forecasting technique that relies heavily on weather radar and satellites?

5 What do forecasters call the technique that predicts the weather based on average weather statistics accumulated over many years?

Upper Airflow and Weather Forecasting

In Chapter 9 the strong connection between cyclonic disturbances at the surface and the wavy flow of the westerlies aloft was established. The importance of this connection cannot be overstated, particularly as it applies to weather forecasting. In order to understand the development of thunderstorms or the formation and movement of midlatitude cyclones, meteorologists must know what is occurring aloft as well as at the surface.

Upper-Level Maps

Upper-air maps are generated twice daily, at 0000 and 1200 GMT (midnight and noon GMT). Recall that these charts are drawn at 850-, 700-, 500-, 300-, and 200-millibar (mb) levels as height contours (in meters or tens of meters), which are analogous to isobars used on surface maps. These maps also contain isotherms, depicted as dashed lines, and some show humidity as well as wind speed and direction.

850- and 700-Millibar Maps   Figure 12–9a illustrates a standard 850-mb map, which is similar in general layout to a 700-mb map. Both maps show height contours using solid black lines at 30-meter intervals and isotherms labeled in degrees Celsius. If relative humidity is included, levels above 70 percent are shown in green. Wind data are plotted using black arrows.

The 850-mb map depicts the atmosphere at an average height of about 1500 meters (1 mile) above sea level. In nonmountainous areas, this level is above the layer where daily temperature fluctuations are strongly influenced by the warming and cooling of Earth's surface. (In areas where elevations are high, such as Denver, Colorado, the 850-mb level represents surface conditions.)

Forecasters regularly examine the 850-mb map to find areas of *cold-air* and *warm-air advection.* Cold-air advection occurs where winds blow across isotherms from colder areas toward warmer areas. The 850-mb map in Figure 12–9b shows a blast of cold air moving south from Canada into the Mississippi Valley. Notice that the isotherms (red dashed lines) in this region of the country are packed tightly together, which means the affected area will experience a rapid drop in temperature.

By contrast, warm-air advection is occurring along the eastern seaboard of the United States. In addition to moving warm air into a cooler region, warm-air advection is generally associated with widespread lifting in the lower troposphere. If the humidity in the region of warm-air advection is relatively high, then lifting could result in cloud formation and possible precipitation (see Figure 12–9c).

Temperatures at the 850-mb level also provide useful information. During the winter, for example, forecasters use the 0°C isotherm as the boundary between areas of rain and areas of snow or sleet. Further, because air

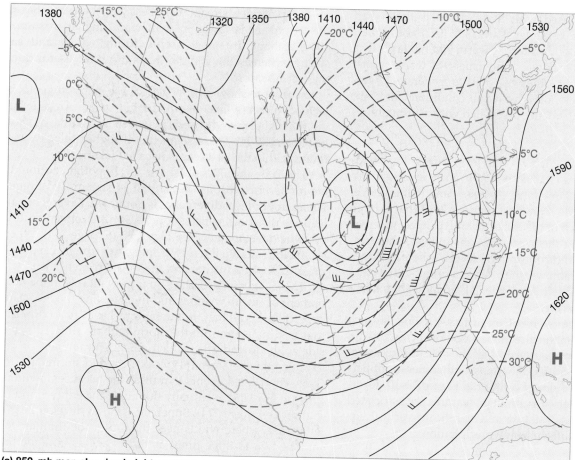

(a) 850–mb map showing height contours and isotherms in °C

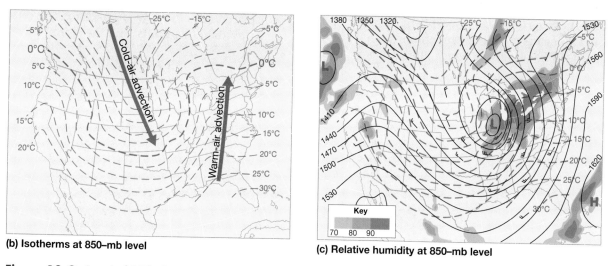

(b) Isotherms at 850–mb level

(c) Relative humidity at 850–mb level

Figure 12–9 A typical 850-mb map. (a) The solid lines are height contours spaced at 30-meter intervals, and the dashed lines are isotherms in degrees Celsius. (b) Areas of warm- and cold-air advection are shown with colored arrows. (c) Regions where the relative humidity is greater than 70 percent are depicted in shades of green.

at the 850-mb level does not experience the daily cycle of temperature changes that occurs at the surface, these maps provide a way to estimate the daily maximum surface temperature. In summer, the maximum surface temperature is usually 15°C (27°F) higher than the temperature at the 850-mb level. In winter, maximum surface temperatures tend to be about 9°C (16°F) warmer than at the 850-mb level.

The 700-mb flow, which occurs about 3 kilometers (2 miles) above sea level, serves as the steering mechanism for air mass thunderstorms. Thus, winds at this level are used to predict the movement of these weather producers.

One rule of thumb used with these charts is that when temperatures at the 700-mb level are 14°C (57°F) or higher, thunderstorms will not develop. A warm 700-mb layer acts as a lid inhibiting the upward movement of warm, moist surface air that might otherwise rise to generate towering cumulonimbus clouds. The warm conditions aloft are caused by general subsidence associated with a strong high-pressure center.

500-Millibar Maps The 500-mb level is found approximately 5.5 kilometers (18,000 feet) above sea level—where about half of Earth's atmosphere is below and half is above. Notice in the example shown in Figure 12–10 that a large trough occupies much of the eastern United States, whereas a ridge is influencing the West. Troughs indicate the presence of a storm at midtropospheric levels, whereas ridges are associated with calm weather. When a prog predicts that a trough will increase in magnitude, it is an indication that a storm is likely to intensify.

To estimate the movement of surface cyclonic storms, which tend to travel in the direction of the flow at the 500-mb level but at roughly a quarter to half the speed, forecasters find it useful to rely on 500-mb maps. Sometimes an upper-level low (shown as a closed contour in Figure 12–10) forms within a trough. When present, these features are associated with counterclockwise rotation and significant vertical lifting that typically results in heavy precipitation.

300- and 200-Millibar Maps Two of the regularly generated upper-level maps, the 300-mb and the 200-mb charts, represent zones near the top of the troposphere. Here, at altitudes above 10 kilometers (6 miles), temperatures can reach a frigid –60°C (–75°F)—levels at which the details

of the polar jet stream can best be observed. Because the jet stream is lower in winter and higher in summer, 300-mb maps are most useful during winter and early spring, whereas the 200-mb maps are most useful during the warm season.

In order to show airflow aloft, these maps often plot **isotachs,** which are lines of equal wind speed. Areas exhibiting the highest wind speeds may also be colored, as shown in Figure 12–11. These segments of higher-velocity winds found within the jet stream are called **jet streaks.**

The 200- and 300-mb charts are important to forecasters for several reasons. Recall that jet streams have regions where upper-level divergence is dominant. Divergence aloft leads to rising air, which supports surface convergence and cyclonic development (see Figure 9–15, page 254). When a jet streak is strong, it tends to energize a trough, causing the pressure to drop further and the storm to strengthen.

The jet stream also plays an important role in the development of severe weather and in extending the life span of supercells. Recall that thunderstorms have areas of updrafts that feed moisture into the storm and are situated side by side with downdrafts, which cause entrainment of cool, dry air into the storm. Typically, thunderstorms are relatively short-lived events that dissipate because downdrafts grow in size and eventually dominate the entire cloud (see Figure 10–4, page 274). Supercells tend to develop near the jet stream, where winds near the top of the thunderstorm may be two to three times as fast as winds near the base. This tilts the thunderstorm as it grows vertically, separating the area of updrafts from the area of downdrafts. When the updrafts are displaced from the downdrafts, the rising cells are not canceled by the downdrafts, and the storm grows in intensity.

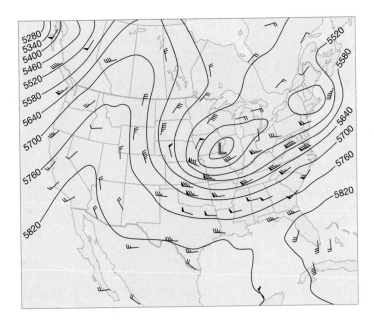

Figure 12–10 A 500-mb map. Notice the well-developed trough of low pressure in central and eastern North America and the ridge of high pressure over the western portion of the continent. Height contours are spaced at 60-meter intervals.

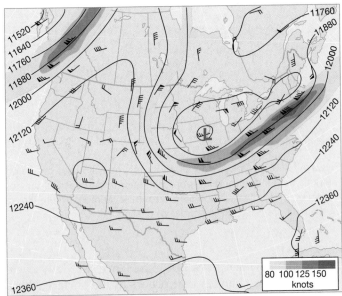

Figure 12–11 A 200-mb map showing the location of the jet stream (pink) and a jet streak (red). Height contours are spaced at 120-meter intervals.

The Connection Between Upper-Level Flow and Surface Weather

Thus far, we have considered the relationship between upper-level flow and short-lived weather phenomena—for example, how winds aloft steer and support the growth of thunderstorms. Next, we will examine long-period changes in the wavy flow of the westerlies and see how these changes impact our weather. Recall that midlatitude winds are generated by temperature contrasts that are most extreme where warm tropical air clashes with cold polar air. The boundary between these two contrasting air masses is the site of the polar jet stream, which is embedded in the meandering westerly flow near the top of the troposphere.

Occasionally, the flow within the westerlies is directed along a relatively straight path from west to east, a pattern referred to as *zonal*. Storms embedded in this zonal flow move quickly across the country, particularly in winter. This situation results in rapidly changing weather conditions in which periods of light to moderate precipitation are followed by brief periods of fair weather.

More often, however, the flow aloft consists of long-wave troughs and ridges that have large components of north–south flow, a pattern that meteorologists refer to as *meridional*. Typically, these airflow patterns slowly drift from west to east, but occasionally they stall and sometimes may even reverse direction. As this wavy pattern gradually migrates eastward, cyclonic storms embedded in the flow are carried across the country.

Figure 12–12 shows a trough that on February 1, 2003, was centered off the Pacific Coast of the United States. Over the next four days the trough grew in strength as it moved eastward into the Ohio Valley. This change in intensity is shown by the height contours, which are more closely packed around the trough on February 4 than on February 1. Embedded in this upper-level trough was a cyclonic system, which grew into the major storm that is shown on the surface map in Figure 12–12. By February 5, this disturbance generated considerable precipitation over much of the eastern United States. (Notice that the center of the surface low is located somewhat to the east of the 500-mb trough—as is typical of these weather patterns.)

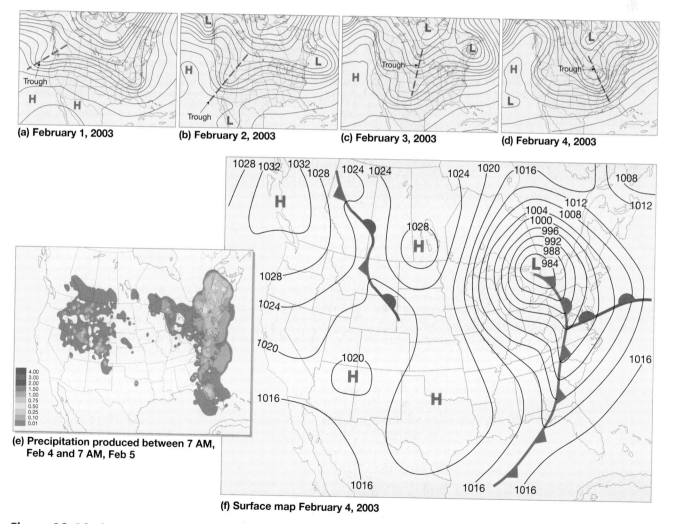

(a) February 1, 2003 (b) February 2, 2003 (c) February 3, 2003 (d) February 4, 2003

(e) Precipitation produced between 7 AM, Feb 4 and 7 AM, Feb 5

(f) Surface map February 4, 2003

Figure 12–12 The movement and intensification of an upper-level trough over a four-day period from February 1 to February 4, 2003. The surface map shows a strong cyclonic storm that formed and moved in conjunction with this trough.

EYE ON THE ATMOSPHERE

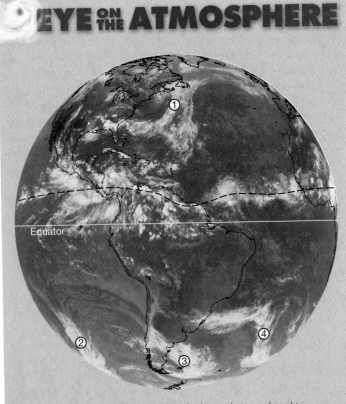

This is a whole-disk satellite image of Earth obtained using infrared imagery. On this type of image, high cold clouds that tend to produce precipitation appear to be white, whereas lower clouds appear in gray tones. The warmest cloud-free areas appear the darkest. (Courtesy of NOAA)

Question 1 What phenomenon produced the band of clouds that is centered on the dashed line that encircles Earth?

Question 2 Based on the position of this band of clouds, is it winter or summer in the Northern Hemisphere?

Question 3 What weather phenomena are associated with the four groups of clouds that are numbered?

Question 4 What region of Earth was warm and cloud free when this image was obtained?

In general, high-amplitude patterns have the potential for generating extreme conditions. During winter, strong troughs tend to spawn large snowstorms, whereas in summer they are associated with severe thunderstorms and tornado outbreaks. By contrast, high-amplitude ridges are associated with record heat in summer and tend to bring mild conditions in winter.

Sometimes these "looped" patterns stall over an area, causing surface patterns to change minimally from one day to the next or, in extreme cases, from one week to the next. Locations in and just east of a stationary or slow-moving trough experience extended periods of rainy or stormy conditions. By contrast, areas in and just east of a stagnant ridge experience prolonged periods of unseasonably warm, dry weather.

The strength and position of the polar jet stream fluctuate seasonally. The jet is strongest (faster wind speeds)

in winter and early spring, when the contrast between the warm tropics and the cold polar realm is greatest. As summer approaches, the temperature gradient diminishes, and the westerlies weaken. The position of the jet stream also changes with the migration of the Sun's vertical rays. With the approach of winter, the mean position of the jet moves toward the equator. By midwinter, it may penetrate as far south as central Florida. Because the paths of midlatitude cyclones shift with the flow aloft, the southern states typically encounter most of their severe weather in winter and spring. (Hurricane season is the exception.) During summer, because of a poleward shift of the jet, the northern states and Canada experience an increase in the number of severe storms and tornadoes. This more northerly storm track also carries most Pacific storms toward Alaska during the warm months, producing a rather long, dry summer season for much of the Pacific Coast to the south.

The position of the jet stream relative to where you live is important for other reasons. In winter, if the jet stream moves considerably south of your location, your area will experience a "cold snap," as frigid air moves in from Canada. During the summer, by contrast, the core of the jet tends to be positioned over Canada, while warm, moist tropical air dominates much of the United States east of the Rockies. If, during the winter or early spring, strong midlatitude cyclones carried by the flow aloft pass immediately south of your location, you can expect heavy snowfall and cooler conditions. In summer, if the jet is overhead, periods of heavy rain, and perhaps hail and tornadic winds, may occur.

An Extreme Winter Consider an example of the influence of the flow aloft for an extended period during an atypical winter. During a normal January, an upper-air ridge tends to be situated over the Rocky Mountains, while a trough extends across the eastern two-thirds of the United States. In January 1977, however, the normal flow pattern was greatly accentuated, as illustrated in Figure 12–13. The greater amplitude of the upper-level flow caused an almost continuous influx of cold air into the Deep South, producing record-low temperatures throughout much of the eastern and central United States (Figure 12–14). Therefore, people consumed above-average amounts of natural gas to heat their homes. As a result, many industries implemented employee layoffs in order to conserve their dwindling natural-gas supplies. Ohio was severely affected—such that four-day workweeks and massive shutdowns were ordered.

While most of the East and South were in the deep freeze, the westernmost states were being influenced by a strong ridge of high pressure that delivered an extended period of mild temperatures and clear skies. Unfortunately, the ridge of high pressure blocked the movement of Pacific storms that usually provide much-needed winter precipitation. The shortage of moisture was especially serious in California, where January is the middle of its three-month rainy season.

Throughout most of the western states, the winter rain and snow that supply water for summer irrigation were far

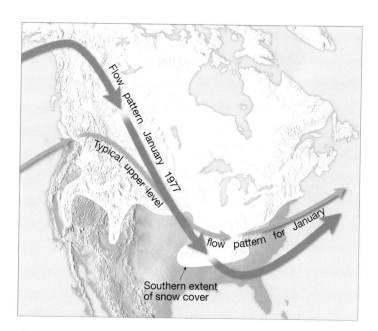

Figure 12–13 The unusually high amplitude experienced in the flow pattern of the prevailing westerlies during the winter of 1977 brought warmth to Alaska, drought to the West, and frigid temperatures to the central and eastern United States.

below normal. This dilemma was compounded by the fact that many reservoirs were almost empty due to the fact that the previous year was exceptionally dry. Although much of the country was concerned about economic disaster caused either by a lack of moisture or frigid temperatures, the highly accentuated flow pattern channeled unseasonably warm air into Alaska. Even Fairbanks, which generally experiences temperatures as low as –40°C (–40°F), had a mild January, with numerous days above freezing.*

In summary, the wavy flow aloft governs, to a large extent, the overall magnitude and distribution of weather disturbances observed in the midlatitudes. Thus, accurate forecasts depend on our ability to predict long- and short-term changes in the upper-level flow. Although the task of forecasting long-term variations remains beyond the capabilities of meteorologists, it is hoped that future research will allow forecasters to answer questions such as: Will next winter be colder or warmer than recent winters? and Will the Southwest experience a drought next year?

*For an excellent review of the winter weather of 1976–1977, see Thomas Y. Canby, "The Year the Weather Went Wild," *National Geographic* 152, no. 6 (1977), 798–892.

Figure 12–14 Arctic air invades the eastern United States. (Photo by Michael Stubblefield/Alamy)

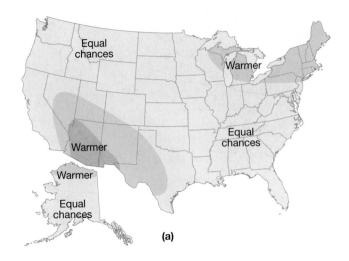

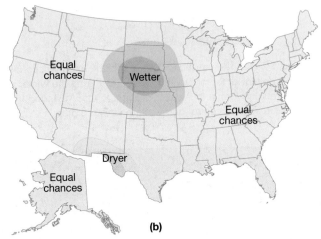

Figure 12–15 Extended forecast (90-day outlook) for (a) temperature and (b) precipitation for September, October, and November 2011. "Equal chances" means that there are not climatic signals for either above- or below-normal conditions.

Long-Range Forecasts

The National Weather Service issues a number of computer-generated forecasts for time spans ranging from a few hours to more than two weeks. Beyond 7 days, however, the accuracy of these forecasts diminishes considerably. In addition, the Climate Prediction Center, a branch of the NWS, produces *30- and 90-day outlooks*. These are not weather forecasts in the usual sense. Instead, they offer insights into whether it will be drier or wetter and colder or warmer than normal in a particular region of the country.

A series of 13 90-day outlooks are produced in one-month increments. The series begins with the outlook for the next three months—for example, September, October, and November. Next, a separate 90-day outlook for the period beginning one month later (October, November, and December) is constructed. Figure 12–15 shows the temperature and precipitation 90-day outlooks issued for September, October, and November 2011. The temperature outlook for this period calls for warmer-than-usual conditions across much of the southwestern, north-central, and northeastern United States (Figure 12–15a). The remainder of the country is labeled "equal chance," which indicates that there are not climatic signals for either above- or below-normal conditions during the forecast period. The precipitation outlook shown in Figure 12–15b calls for wetter-than-normal conditions across most of Kansas, Nebraska, and South Dakota, whereas drier-than-normal conditions are expected in a portion of the Southwest from Texas to southeastern Arizona.

Monthly and seasonal outlooks of this type are generated using a variety of criteria. Meteorologists consider the climatology of each region—the 30-year average of variables such as temperature and precipitation. Factors such as snow and ice cover in the winter and persistently dry or wet soils in the summer are taken into account. Forecasters also consider current patterns of temperature and precipitation. For example, during 2011 and the few preceding years, the southwestern United States experienced below-normal levels of precipitation. Based on climatological data, the weather patterns that produce these conditions tend to gradually shift toward the norm rather than changing abruptly. Thus, forecasters predicted that this below-

normal trend would continue for at least the first several months of 2012.

Recently, sea-surface temperatures in the equatorial Pacific Ocean have been shown to have predictive value in forecasting temperature and precipitation patterns in various locations around the globe—particularly in the winter season. For a discussion of this relationship, see the section "El Niño and La Niña and the Southern Oscillation" in Chapter 7.

Despite improved seasonal outlooks, the reliability of extended forecasts has been disappointing. Although some forecast skill is observed during the late winter and late summer, outlooks prepared for the "transition months," when the weather can fluctuate wildly, have shown little reliability.

Forecast Accuracy

Never, no matter what the progress of science, will honest scientific men who have a regard for their professional reputations venture to predict the weather.

Dominique François Arago
French physicist (1786–1853)

Some might argue that Arago's statement is still true! Nevertheless, a great deal of scientific and technological progress has been made in the two centuries since he made this observation. Accurate weather forecasts and warnings of extreme weather conditions are provided by the National Weather Service (Figure 12–16). Their forecasts are used by government agencies to protect life and property, by electric utilities and farmers, by the construction industry, travelers, and airlines—by nearly everyone.

How does the NWS measure forecast accuracy, a concept referred to as **skill**? When determining the occurrence of precipitation, for example, NWS forecasts are correct more than 80 percent of the time. Does this mean the NWS is doing a great job? Not necessarily. When establishing the skill of a forecaster, we need to examine more than the percentage of accurate forecasts. For example, measurable precipitation in Los Angeles is recorded only 11 days each year, on average. Therefore, the chance of rain in Los Angeles is 11 out of 365, or about 3 percent. Knowing this, a forecaster could predict no rain for every day of the year and be correct 97 percent of the time! Although the accuracy would be high, it would not indicate skill.

Any measure of forecasting skill must consider climatic data. Thus, for forecasters to exhibit forecasting skill, they must do better than forecasts that are based simply on climatic averages. In the Los Angeles example, the forecaster must be able to predict rain on at least a few of the rainy days to demonstrate forecasting skill.

The only aspect of a weather forecast that is expressed as a *percentage probability* is rainfall. Statistical data are studied to determine how often precipitation occurred under similar conditions. Although the occurrence of precipitation can be forecast with more than 80 percent accuracy, predictions of the amount, time of occurrence, and duration are not as reliable.

Figure 12–16 Meteorologist at the National Hurricane Center, in Miami, Florida, examining satellite imagery of Hurricane Ivan as the eye crosses the Alabama coastline on September 16, 2004. (Photo by Andy Newman/Associated Press)

How skillful are the weather forecasts provided by the National Weather Service? In general, very short-range forecasts (0 to 12 hours) have demonstrated considerable skill, especially for predicting the formation and movement of large weather systems such as midlatitude cyclones. Over the past 20 years, the accuracy of short-range forecasts (12 to 72 hours) at predicting precipitation has improved significantly. However, specific distribution of precipitation is tied to mesoscale structures, such as individual thunderstorms, that cannot be predicted by the current numerical models. Thus, a forecast could predict 2 inches of rain for an area, but one town might receive a trace, whereas a neighboring community may be deluged with 4 inches.

By contrast, predictions of maximum and minimum temperatures and wind tend to be quite accurate. Medium-range forecasts (3 to 7 days) have also shown significant improvement in the past few decades. Large cyclonic storms are often predicted a few days in advance. Yet, beyond day 7, the predictability of day-to-day weather using these modern methods proves no more accurate than projections made from climatic data.

Several factors account for the limited range of modern forecasting techniques. As noted earlier, the network of observing stations is limited in its coverage of our vast planet. Not only are large areas of Earth's land–sea surface monitored inadequately, but globally, data from the middle and upper troposphere are meager. Moreover, the physical laws of nature are difficult to apply to chaotic systems such as the atmosphere, and the current models of the atmosphere remain incomplete.

Nevertheless, numerical weather prediction has greatly improved the forecaster's ability to project changes in the upper-level flow. When the flow aloft can be tied more closely to surface conditions, weather forecasts are likely to demonstrate greater skill.

In summary, the accuracy of weather forecasts covering short- and medium-range periods has improved steadily over the past few decades, particularly in forecasting the evolution and movement of midlatitude cyclones. Technological developments and improved computer models, as well as an increased understanding of how the atmosphere behaves, have added greatly to this success. The ability to accurately predict daily weather beyond day 7, however, remains relatively low.

Concept Check 12.7

1 What is meant by the term forecast *skill*?

2 Give an example of why the percentage of correct forecasts is not always a good measure of forecasting skill.

3 The forecast for which weather element is expressed as a percentage probability?

4 List two reasons modern forecasting techniques do not accurately predict the weather beyond a few days for most regions of the United States.

Satellites in Weather Forecasting

Meteorology entered the space age on April 1, 1960, when the first weather satellite, *TIROS 1*, was launched. (*TIROS* stands for Television and Infra-Red Observation Satellite.) In its short life span of only 79 days, *TIROS 1* obtained thousands of images and established that satellites are useful in the study of our planet. Subsequently, more than 30 versions of *TIROS* have been launched by the United States—the most recent in 2009. In addition to specialized television cameras, each satellite is equipped with infrared sensors capable of "seeing" cloud coverage at night.

TIROS 1, like all other weather satellites in this series, was placed into a polar orbit—one that circles Earth from north-to-south (Figure 12–17a). **Polar-orbiting satellites** orbit Earth at low altitudes (about 850 kilometers) and require only 100 minutes per orbit. By properly orienting the orbits, these satellites drift about 15° westward over Earth's surface during each orbit. Thus, they are able to

Figure 12–17 Weather satellites. (a) Polar-orbiting satellites about 850 kilometers (530 miles) above Earth travel over the North and South Poles. (b) A geostationary satellite moves west to east above the equator, at a distance of 35,000 kilometers (22,300 miles) and at the same rate that Earth rotates.

(a) Polar-orbiting satellites

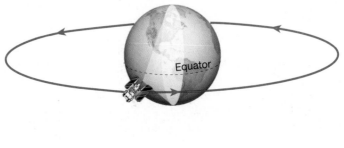

(b) Geostationary satellite

PROFESSIONAL PROFILE

Harold Brooks: Research Meteorologist

Modern weather forecasting relies on far-seeing satellites and complex computer algorithms. Yet anyone who has wished for an umbrella after a forecast called for "sunny, hot" weather knows that predicting the weather is an imperfect science. Harold Brooks, a research meteorologist at NOAA's National Severe Storms Laboratory in Norman, Oklahoma, is helping change this scenario. His primary areas of expertise—forecast evaluation and severe thunderstorm conditions—seek to make weather forecasting both more useful and accurate.

> Forecasts are not homework problems where someone knows the answers beforehand.

"Forecasts are not homework problems where someone knows the answers beforehand. Sometimes we get surprised and discover *we don't know why* the weather turned out as it did. The way we understood how the atmosphere works doesn't cover this particular case. That becomes a research problem," Brooks says.

One question Brooks studies is What constitutes a *good* weather forecast? "There are two primary things we look at: how closely the observations look like our forecast and how much value a user gets from the forecast." When the National Weather Service puts out a tornado warning for an area, Brooks wants to know whether the storm occurred or not. He also wants to know how people interpreted the warning. Brooks says, "The National Weather Service warning has a precise meaning. But it isn't clear whether users of those products have the same meaning in mind. The question becomes How do we relate the forecast to what my next-door neighbor would think? That's a hard problem—one that deals with social science."

Getting forecasts right, however, is a high-stakes business. Accurate forecasts can save both lives and dollars, but misses can cause frustration and lead people to disregard future warnings.

Brooks's other major area of interest is where and when severe thunderstorms occur. One recent project involved estimating the probability that severe tornado conditions will occur on any day of the year anywhere in the United States. The analysis relies on storm data that go back many decades. Storm records, however, are inherently biased: More reports exist for places that are more densely populated, and population changes from decade to decade. Brooks and colleagues have developed ways to smooth the data in both space and time, using statistics.

The result: a map that insurers, engineers, emergency agencies, and the public can use to estimate tornado risk. The map also broadens the reach of tornado forecasting, Brooks says. "We can draw better conclusions about the environmental conditions in which those storms form. We can use this information to estimate where severe thunderstorms have actually occurred around the world in the past and where they will happen in the future."

Brooks's interest in tornadoes may stem from growing up in St. Louis, Missouri—tornado country. He majored in physics and math as an undergraduate, and he landed a summer internship at NASA's Goddard Space Flight Center. While there, Brooks helped prepare models of global climate during the last ice age. The work earned him an invitation to graduate school. He earned a master's degree and then a doctorate modeling severe thunderstorms. That's when Brooks began working as a postdoctoral researcher at the National

Severe Storm Laboratory, where he remains today.

After more than 20 years in weather research, the atmosphere still surprises Brooks. "Extremely rare events of violent weather on the planet almost all come out of what are fairly normal and undramatic processes. Every once in a while, these come together to challenge your notions of how the planet functions."

> Extremely rare events of violent weather on the planet almost all come out of what are fairly normal and undramatic processes.

HAROLD BROOKS is a researcher at the National Severe Storms Laboratory in Norman, Oklahoma. (Photo courtesy of Harold Brooks)

obtain images of the entire planet twice each day and provide coverage of a large region in a few hours.

Another class of weather satellites, **geostationary satellites,** were first placed in orbit over the equator in 1966 (Figure 12–17b). These satellites, as their name implies, remain fixed over a point on Earth because their rate of travel keeps pace with Earth's rate of rotation. In order to remain positioned over a given site, however, the satellite must orbit at a greater distance from Earth's surface (about 35,000 kilometers, [22,000 miles]) than polar-orbiting satellites. At this altitude, the speed required to keep a satellite

in orbit will also keep it moving in time with the rotating Earth. At this distance, the images obtained are less detailed than those captured by satellites that orbit at lower altitudes.

What Type of Images Do Weather Satellites Provide?

The most recent generation of weather satellites, known as *Geostationary Operational Environmental Satellites* (GOES), provide visible, infrared, and water-vapor images for North America and adjacent ocean areas. These are the

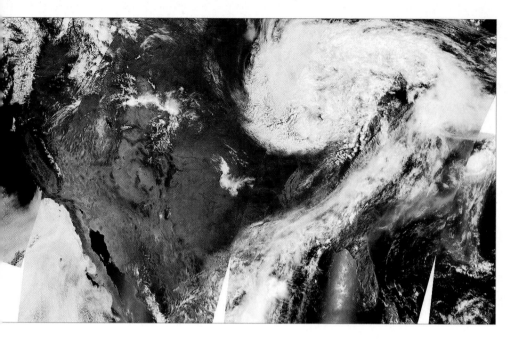

Figure 12–18 Weather satellites are invaluable tools for tracking storms and gathering atmospheric data. This is a true-color composite image of a mature midlatitude cyclone produced on June 11, 2011, by an instrument called MODIS (Moderate Resolution Imaging Spectroradiometer). (Courtesy of NASA)

satellite images often used on The Weather Channel. Other countries, including the European Union, India, China, Japan, and Russia, have launched similar weather satellites. Collectively, these satellites allow meteorologists to track the movement of large weather systems that cannot be adequately followed by weather radar or polar-orbiting satellites. Swirling cloud masses are easily tracked by satellites as they migrate across even the most remote portions of our planet (Figure 12–18). These satellites also play a very important role in monitoring the development and movement of tropical storms and hurricanes.

Visible Light Images *GOES* satellites are equipped with instruments that detect sunlight reflected from Earth, producing *visible light images* such as the one shown in Figure 12–19. (Visible light images, also called visible images, are similar to pictures taken with an ordinary camera.) Because visible light imagery records the intensity of light reflected from cloud tops and other surfaces, these images are similar to black-and-white photos of Earth. The bright white areas are mainly cloud tops or snow- and ice-covered areas that are strong reflectors of light. Land surfaces appear dark gray because they reflect less light than clouds, and the oceans, which reflect very little light, appear nearly black.

Visible images are useful in identifying cloud shapes, organizational patterns, and thickness. In general, the thicker a cloud is, the higher its albedo and the brighter it will appear in the image. In addition, cumulonimbus clouds have a bumpy appearance, whereas stratus clouds form a continuous flat-looking blanket. Figure 12–19b is a visible light image that shows a close-up of towering cumulonimbus clouds with overshooting tops that have penetrated the temperature inversion that lies above the troposphere.

Infrared Images In contrast to visible images, *infrared images* are obtained from radiation emitted (rather than reflected) from objects and, when compared to visible

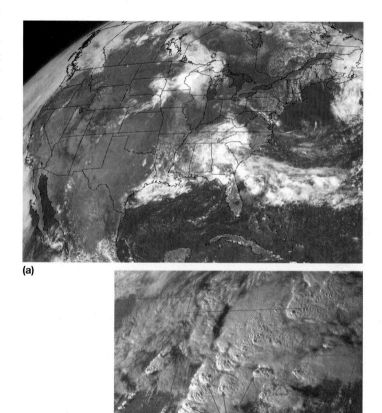

Figure 12–19 Visible light images. (a) Visible light image provided by *GOES* satellite of cloud distribution about midday on July 15, 2011. This image records sunlight reflected off various Earth surfaces and appears much like a black-and-white image. (b) Close-up visible image centered on the state of Iowa showing towering cumulonimbus clouds with overshooting tops. (Courtesy of NOAA)

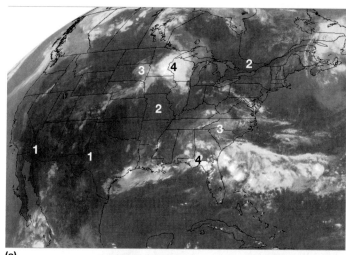

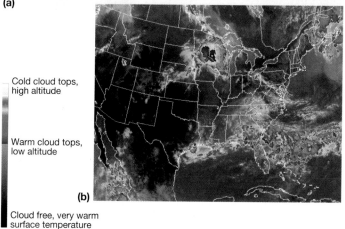

Cold cloud tops,
high altitude

Warm cloud tops,
low altitude

(b)

Cloud free, very warm
surface temperature

Figure 12–20 Infrared satellite images. (a) Infrared image provided by *GOES* satellite about midday on July 15, 2011. Numbers 1–4 in this infrared image represent areas that: (1) are cloud free; (2) contain low clouds (cumulus) with cloud tops below 2000 meters; (3) are blanketed by cloud tops between 2000–6000 meters, and; (4) the infrared sensor is measuring high cold cloud tops that are likely associated with towering cumulonimbus clouds and thunderstorms. (b) Color-enhanced infrared image for the same time period as part (a). The dark-reddish color indicates the location of towering cumulonimbus clouds. (Courtesy of NOAA)

southeastern United States are also easily identifiable on the infrared image in Figure 12–20.

To help meteorologists interpret infrared images, false colorization is sometimes used. The coldest, and thus highest, cloud tops, which typically identify areas of heavy precipitation, are shown in strong colors. Notice the reddish areas over the Minnesota–Wisconsin border in Figure 12–20b. This is the same area that appears as bright white in the infrared image in Figure 12–20a.

Visible Versus Infrared Imagery A major advantage of infrared imagery is its ability to detect energy radiated skyward both day and night. As a result, the movement of storms can be monitored 24 hours a day. In addition, infrared images are useful for distinguishing colder high cloud tops from warmer, lower clouds.

Visible images, which require reflected sunlight, are not available at night. However, an advantage of visible images is their high resolution—which makes detection of smaller features possible. Figure 12–21 is a close-up of a portion

images, are especially useful in determining which clouds are most likely to produce precipitation and/or stormy conditions. Infrared images are produced by computers that are programmed to display warm objects such as Earth's surface in black and colder objects such as high, cold cloud tops in white. Because high cloud tops are colder than low cloud tops, towering cumulonimbus clouds that usually generate heavy precipitation appear very bright, whereas midlevel clouds appear somewhat light gray. Low fair-weather cumulus, stratus clouds, and fog are all relatively warm and therefore appear dark gray or may not be discernable from Earth's surface on infrared images.

Compare Figure 12–19a, a visible image, to Figure 12–20a, which is an infrared image of the same region. Both images were acquired on July 15, 2011, a very hot summer day that produced thunderstorms over parts of Minnesota and Wisconsin, as well as in the southeast. As you can see, the clouds over eastern South Dakota that appear white on the visible image appear gray on the infrared image. These areas are blanketed by middle clouds between 2000–6000 meters. By contrast, the clouds over eastern Minnesota and western Wisconsin appear bright in both the visible and infrared images. This region has towering cumulonimbus clouds and was experiencing air mass thunderstorms that brought heavy rain to the area. The areas of isolated storm cells in the

Figure 12–21 Close-up *GOES* visible image of a portion of north-central United States about midday on July 15, 2011.

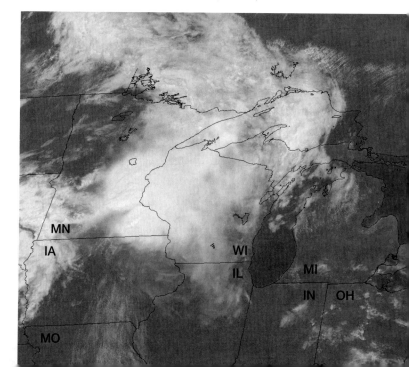

EYE ON THE ATMOSPHERE

(a)

(b)

These accompanying satellite images of the eastern Pacific and a portion of North America were obtained at the same time on July 14, 2001. One of these is an infrared image and the other is a visible light image. (Courtesy of NOAA)

Question 1 On which image (a or b) are the cloud shapes and patterns more easily recognized?

Question 2 Which one of these images was obtained using infrared imagery? How can you tell?

Question 3 What type of clouds are found in the western Great Plains (just east of the Rockies)?

Question 4 Are the clouds over the eastern Pacific mainly high clouds or middle- to low-level clouds?

of the visible image in Figure 12–19. Notice the bands of fair-weather cumulus clouds over eastern Missouri and Iowa and the bumpy tops of the developing cumulonimbus clouds over southeastern Minnesota. Visible images are ideal for showing cloud patterns and storm systems such as midlatitude cyclones and hurricanes.

Water-Vapor Images *Water-vapor images* provide yet another way to view Earth. Most of Earth's radiation with a wavelength of 6.7 micrometers is emitted by water vapor. Satellites equipped with detectors for this narrow band of radiation are, in effect, mapping the concentration of water vapor in the atmosphere. Bright-white regions in Figure 12–22 represent regions of high water-vapor concentration, whereas dark areas indicate drier air. Because most fronts occur between air masses having contrasting moisture conditions, water-vapor images are valuable tools for locating frontal boundaries.

Other Satellite Measurements

Ingenious developments have made weather satellites more than high-tech TV cameras pointed at Earth from space. For example, some satellites are equipped with instruments designed to measure temperatures and moisture at various altitudes. Other satellites allow us to measure rainfall intensity and amounts, both at the surface and

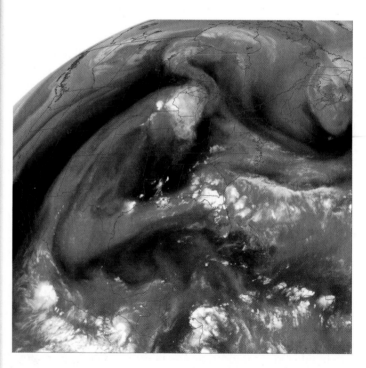

Figure 12–22 Water-vapor image from the *GOES* for July 15, 2011. The greater the intensity of white, the greater the atmosphere's water-vapor content. Black areas are driest. (Courtesy of NOAA)

above the surface, and to monitor winds in regions where few, if any, conventional instruments are available. Images from the *Tropical Rainfall Measuring Mission (TRMM)* illustrate satellite-based precipitation measurement (see Box 1.1 page 10).

Concept Check 12.8

1 How do satellites help identify clouds that are the most likely to produce precipitation?

2 What advantage do geostationary satellites have over polar-orbiting satellites? Name one disadvantage.

3 What advantages do infrared images have over visible light images? Name one disadvantage.

Students Sometimes Ask...

When were the first U.S. weather maps produced?

Meteorological observations were made by the U.S. Weather Bureau, the predecessor to the National Weather Service, for the first time on November 1, 1870. These observations led to the production of the first U.S. Daily Weather Map in 1871. The early maps included isobars and described the weather conditions at select locations but had little predictive value. It wasn't until the late 1930s that air mass and frontal-analysis techniques were used by the U.S. Weather Bureau. On August 1, 1941, a new Daily Weather Map was introduced. In addition to including various types of fronts, this map used the station model—a group of symbols indicating weather conditions at nearly 100 locations (see Box 12–1). Although the appearance has changed over the years, the basic structure of the Daily Weather Map remains the same today as it did more than 70 years ago.

Give It Some Thought

1. Discuss the difference between weather analysis and weather forecasting.

2. The accompanying radar image shows the precipitation pattern (reds and yellow indicate heavy precipitation, and greens and blues indicate light-to-moderate precipitation) associated with a strong midlatitude cyclone. Use the technique called trend forecasting and assume that the cyclone maintains its current strength for the next 24 hours to complete the following.

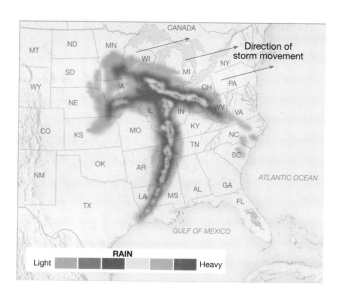

a. What state(s) will likely experience thunderstorms associated with a cold front in the next 6 to 12 hours?

b. Will the temperatures in Alabama more likely rise or fall during the next 24 hours? Explain.

c. Will Pennsylvania more likely be warmer or colder in 24 hours than it was when the image was produced? Explain.

d. In which state, New York or Georgia, were cirrus clouds more likely overhead when the image was produced?

3. Explain the difference between weather forecasts and 90-day outlooks.

4. The accompanying map is a simplified upper-air chart that shows the flow aloft and the position of the jet stream. Weather patterns such as this are relatively common over North America. When answering the following questions, assume that long-range weather forecasters had evidence that this flow pattern was going to persist for much of the spring and early summer.

a. Would the 90-day outlook for the southeastern United States predict wetter-than-normal, drier-than-normal, or equal chance?

b. Would the 90-day outlook for the southwestern United States predict wetter-than-normal, drier-than-normal, or equal chance?

c. Explain how you arrived at your answers.

5. Describe the basis of numerical weather prediction.

6. Use the accompanying weather station model to determine the following weather elements for this location (*Hint:* see Box 12–1 and Appendix B.):

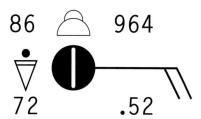

a. Wind direction and speed
b. Temperature
c. Dew-point temperature
d. Sky cover
e. Barometric pressure
f. Cloud type
g. Current weather
h. Amount of precipitation in the past six hours

7. What is the difference between numerical weather prediction and the analog methods of weather forecasting?

8. The accompanying infrared satellite image shows a portion of North America at midday. Match the numbers on the image to descriptions a–d.

a. These areas are cloud free, and the infrared sensor is measuring very warm surface temperatures.

b. These areas contain low clouds (cumulus), with cloud tops below 2000 meters.

c. These areas are blanketed by middle clouds between 2000 and 6000 meters.

d. These are areas where the infrared sensor is measuring high, cold cloud tops that are likely associated with towering cumulonimbus clouds and thunderstorms.

9. Sketch a hypothetical orbit for a polar-orbiting and a geostationary satellite on the same diagram (two orbits around Earth). Make sure the orbits are drawn to indicate relative height above Earth's surface.

a. List one advantage of a polar-orbiting satellite.

b. List one advantage of a geostationary satellite.

c. Explain why geostationary satellites orbit at about 36,000 kilometers (22,000 miles) above Earth's surface.

10. The accompanying maps show the positions of the jet stream core on two different midwinter days. On which of these days would the southeastern United States be warmer? Explain your choice.

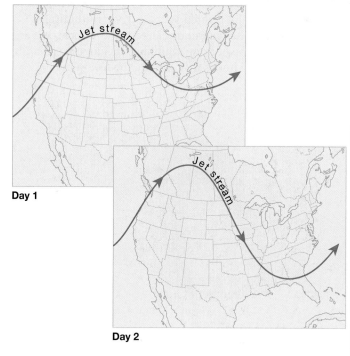

Day 1

Day 2

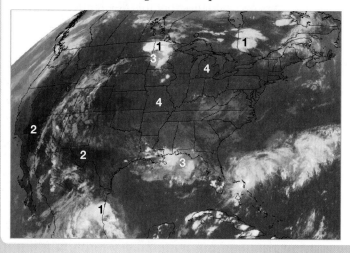

WEATHER ANALYSIS AND FORECASTING IN REVIEW

- In the United States, the government agency responsible for gathering and disseminating weather-related information is the *National Weather Service* (*NWS*). Perhaps the most important services provided by the NWS are forecasts and warnings of hazardous weather, including thunderstorms, floods, hurricanes, tornadoes, winter storms, and extreme heat.

- The process of providing weather forecasts and warnings throughout the United States occurs in three stages. First, data are collected and analyzed on a global scale. Second, a variety of techniques are used to establish the future state of the atmosphere—a process called *weather forecasting*. Finally, forecasts are disseminated to the public, mainly through the private sector.

- Assessing current atmospheric conditions, called *weather analysis*, involves collecting, transmitting, and compiling billions of pieces of observational data. On a global scale, the *World Meteorological Organization* is responsible for gathering, plotting, and distributing weather data.

- Weather data are displayed on a synoptic weather map, which shows the current status of the atmosphere and includes data on temperature, humidity, pressure, and wind.

- The methods used in weather forecasting include *numerical weather prediction, persistence forecasting, climatological forecasting*, the *analog method*, and *trend forecasting* (*nowcasting*). *Numerical weather prediction* (*NWP*) is the backbone of modern forecasting.

- Meteorologists are aware of the strong correlation between cyclonic disturbances at the surface and the daily and seasonal fluctuations in the wavy flow of the westerlies aloft. Upper-air maps are generated twice daily and provide forecasters with a picture of conditions at various levels in the troposphere. These maps provide information that can be used to predict daily maximum temperature, locate areas where precipitation is likely to occur, and determine whether a thunderstorm will develop into a supercell.

- *Long-range weather forecasting* relies heavily on statistical averages obtained from past weather events, also referred to as *climatic data*. Weekly, monthly, and seasonal weather outlooks prepared by the National Weather Service are not weather forecasts in the usual sense. Rather, they indicate whether or not a region will experience near-normal precipitation and temperatures.

- Over the past several decades, improvements in observing systems, computer models of physical processes, and assimilation of data into numerical weather prediction systems have produced steady improvement in the ability to predict the evolution of larger-scale weather systems as well as day-to-day variations in temperature, precipitation, cloudiness, and air quality.

- *Polar-orbiting satellites* and *geostationary satellites* are important tools that allow meteorologists to monitor even the most remote parts of the globe and to track the movement of large weather systems. *GOES* satellites provide visible, infrared, and water vapor images.

VOCABULARY REVIEW

analog method (p. 338)
Automated Surface Observing Systems (ASOS) (p. 330)
climatological forecasting (p. 336)
ensemble forecasting (p. 334)
geostationary satellite (p. 347)
isotach (p. 340)
jet streak (p. 340)
Model Output Statistics (MOS) (p. 334)

National Centers for Environmental Prediction (p. 329)
National Weather Service (NWS), (p. 328)
nowcasting (p. 338)
numerical weather prediction (p. 333)
persistence forecasting (p. 335)
polar-orbiting satellite (p. 346)
prognostic chart (prog) (p. 334)
radiosonde (p. 331)

skill (p. 345)
synoptic weather map (p. 331)
trend forecasting (p. 338)
weather analysis (p. 329)
weather forecast (p. 328)
Weather Forecast Office (p. 329)
wind profiler (p. 331)
World Meteorological Organization (WMO) (p. 330)

PROBLEMS

1. The map in Figure 12–23 has several weather stations plotted on it. Using the weather data for a typical day in March, which are given in Table 12–1, complete the following:

Figure 12–23 Map to accompany Problem 1.

a. On a copy of Figure 12–23, plot the temperature, wind direction, pressure, and sky coverage by using the international symbols given in Appendix B.

b. Using Figure 12–B as a guide, complete this weather map by adding isobars at 4-millibar intervals, the cold front and the warm front, and the symbol for low pressure.

c. Apply your knowledge of the weather associated with a midlatitude cyclone in the spring of the year and describe the likely clouds and precipitation (if any) at each of the following locations:

   1. Philadelphia, Pennsylvania
   2. Quebec, Canada
   3. Toronto, Canada
   4. Sioux City, Iowa

2. Many weather reports include a seven-day outlook. Check such a report and jot down the forecast for the last (seventh) day. Then, each day thereafter, write down the forecast for the day in question. Finally, record what actually occurred on that day. Contrast the forecast seven days ahead with what actually took place. How accurate (or inaccurate) was the seven-day forecast for the day you selected? How accurate was the five-day forecast for this day? The two-day forecast?

TABLE 12–1 Weather Data for a Typical March Day

Location	Temperature (°F)	Pressure (mb)	Wind Direction	Sky Cover (tenths)
Wilmington, NC	57	1009	SW	7
Philadelphia, PA	59	1001	S	10
Hartford, CT	47	1001	SE	Sky obscured
International Falls, MN	–12	1008	NE	0
Pittsburgh, PA	52	995	WSW	10
Duluth, MN	–1	1006	N	0
Sioux City, IA	11	1010	NW	0
Springfield, MO	35	1011	WNW	2
Chicago, IL	34	985	NW	10
Madison, WI	23	995	NW	10
Nashville, TN	40	1008	SW	10
Louisville, KY	40	1002	SW	5
Indianapolis, IN	35	994	W	10
Atlanta, GA	49	1010	SW	7
Huntington, WV	52	998	SW	6
Toronto, Canada	44	985	E	9
Albany, NY	50	998	SE	7
Savanna, GA	63	1012	SW	10
Jacksonville, FL	66	1013	WSW	10
Norfolk, VA	67	1005	S	10
Cleveland, OH	49	988	SW	4
Little Rock, AK	37	1014	WSW	0
Cincinnati, OH	41	997	WSW	10
Detroit, MI	44	984	SW	10
Montreal, Canada	42	993	E	10
Quebec, Canada	34	999	NE	Sky obscured

MyMeteorologyLab™

Log in to www.mymeteorologylab.com for animations, videos, MapMaster interactive maps, GEODe media, In the News RSS feeds, web links, glossary flashcards, self-study quizzes and a Pearson eText version of this book to enhance your study of *Weather Analysis and Forecasting*.

Air Pollution

Air pollution and meteorology are linked in two ways. One concerns the influence that weather conditions have on the dilution and dispersal of air pollutants. The second connection is the reverse and deals with the effect that air pollution has on weather and climate. The first of these associations is examined in this chapter. The second and equally important relationship is discussed in Chapter 14 and is the focus of several sections and special-interest boxes.[*]

[*]See Box 3–3, "How Cities Influence Temperature: The Urban Heat Island," and Box 13–1, "Air Pollution Changing the Climate of Cities." In addition, see the section "Country Breezes" in Chapter 7.

This crowded freeway reminds us that cars and trucks are a major source of air pollution. (Photo by Ashley Cooper/Alamy)

Focus On Concepts

After completing this chapter you should be able to:

- List several natural sources of air pollution and identify those that are human accentuated.

- Distinguish between primary and secondary pollutants.

- List the major primary pollutants and summarize their effects on people and the environment.

- Discuss different meanings for the term *smog* and describe the formation of photochemical smog.

- Summarize trends in air quality since 1970.

- Describe the influence of wind as a factor influencing air quality.

- Sketch a graph or diagram that shows a temperature inversion and relate it to mixing depth.

- Contrast surface temperature inversions and inversions aloft.

- Discuss the formation of acid precipitation and list some of its effects on the environment.

The Threat of Air Pollution

Air pollution is a continuing threat to our health and welfare. According to a National Research Council report, in the United States an estimated 1.8 to 3.1 years of life are lost by people living in the most polluted cities due to chronic exposure to particulates. In addition, more than 4000 premature deaths occur each year because of the elevated surface ozone concentrations commonly observed in the United States.* Air pollutants also have a negative impact on crop production, costing U.S. agriculture more than $1 billion annually. In other parts of the world, especially developing countries, the negative impact of air pollution on life and agriculture is even more serious.

An average adult requires about 13.5 kilograms (30 pounds) of air each day compared with about 1.2 kilograms (2.6 pounds) of food and 2 kilograms (4.4 pounds) of water. The cleanliness of air, therefore, should certainly be as important to us as the cleanliness of our food and water.

Air is never perfectly clean. Many natural sources of air pollution have always existed (Figure 13–1). Ash and gases from volcanic eruptions, salt particles from breaking waves, pollen and spores released by plants, smoke

_____

*As reported in *Bulletin of the American Meteorological Society,* Vol. 89, No. 6, June 2008. Also note that particulate matter is discussed in the section "Primary Pollutants" and that surface ozone is a significant component of urban smog and is discussed in the section "Secondary Pollutants."

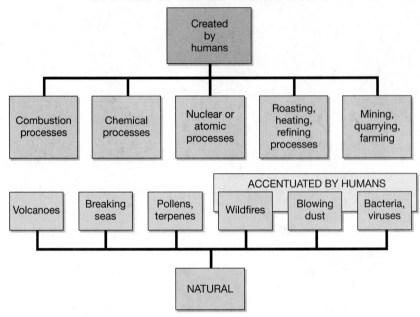

SOURCES OF PRIMARY POLLUTANTS

Figure 13–2 Sources of primary pollutants.

from forest fires and brush fires, and windblown dust are all examples of "natural air pollution" (Figure 13–2). Ever since people have been on Earth, however, they have added to the frequency and intensity of some of these natural pollutants, especially the last two. For example, the dust storm in Figure 13–3 occurred when strong winds raised dry soil from plowed farm fields.

With the discovery of fire came an increased number of accidental as well as intentional burnings. Even today, in many parts of the world, fire is used to clear land for agricultural purposes (the so-called slash-and-burn method), filling the air with smoke and reducing visibility. When people clear the land of its natural vegetative cover for any purpose, soil is exposed and blown into the air. Yet when we consider the air in a modern-day industrial city, these human-accentuated forms of pollution, although significant, may seem minor by comparison.

Although some types of air pollution are relatively recent creations, others have been around for centuries. Smoke pollution, for example, plagued London for centuries. In 1661, when John Evelyn wrote *Fumifugium,* or *The Inconvenience of Aer and Smoak of London Dissipated, Together with Some Remidies Humbly Proposed,* the problem of foul air obviously plagued Londoners. In his book, Evelyn notes that a traveler, although many miles from London, "sooner smells than sees the city to which he repairs." In fact, London continued to have severe air pollution problems well into the twentieth century. It was only after a devastating smog disaster in 1952 that truly decisive action was taken to clean the air.

London, however, has not monopolized the air pollution scene. With the coming of the Industrial Revolution, many cities began to experience major air pollution. Instead of just simply accelerating natural sources, people found

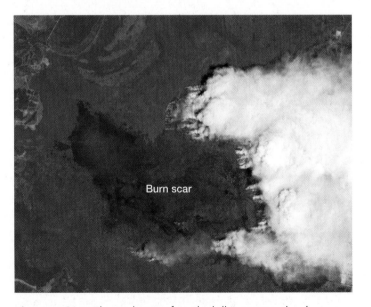

Burn scar

Figure 13–1 These plumes of smoke billowing into the sky from a wildfire in southern Georgia are an example of natural air pollution. Lightning started the fire on April 28, 2011, and on May 8, when this satellite image was acquired, nearly 62,000 acres had burned. (NASA)

Figure 13–3 An example of natural air pollution that has been accentuated by human activities. This dust storm near Elkhart, Kansas, in May 1937, occurred because the natural vegetative cover that anchored the soil had been removed from a marginal environment so that the land could be farmed. Severe drought made the plowed fields vulnerable to strong winds. It was because of storms like this that portions of the Great Plains were called the *Dust Bowl* in the 1930s. (Photo reproduced from the collection of the Library of Congress)

many new ways to pollute the air (see Figure 13–2) and many new things with which to pollute it. In the mid- to late nineteenth century, the populations of many American and European cities swelled as people sought work in the growing numbers of new foundries and steel mills. As a result, the urban environment became increasingly fouled by the fumes of industry. In *Hard Times*, Charles Dickens vividly describes the scene in a late-nineteenth-century factory town:

> It was a town of machinery and tall chimneys out of which interminable serpents of smoke trailed themselves forever and ever, and never got uncoiled. It had a black canal in it, and a river that ran purple with ill-smelling dye.

It is clear that poor air quality was not the only environmental pollution that plagued these places! However, it should be noted that the rapid rise in urban air pollution was not necessarily viewed with great alarm. Rather, chimneys belching forth smoke and soot were a symbol of growth and prosperity (Figure 13–4). The following quotation from an 1880 speech by the well-known lawyer and orator Robert Ingersoll, for example, is reported to have elicited great cheering and cries of "Good! Good!" from the audience: "I want the sky to be filled with the smoke of American industry and upon that cloud of smoke will rest forever the bow of perpetual promise. That is what I am for." With the rapid growth of the world's population and accelerated industrialization, the quantities of atmospheric pollutants increased drastically.

One of the most tragic air pollution episodes ever occurred in London in December 1952 and is the focus of *Severe and Hazardous Weather: The Great Smog of 1952* on

page 361. More than 4000 people died as a result of this five-day ordeal. The people who suffered most were those with respiratory and heart problems, primarily elderly individuals. Extreme air pollution darkened London again in 1953 and 1962 and affected New York City in 1953, 1963, and 1966. Since these events, the passage of legislation, the development of regulations and standards, and the advancement of control technology have reduced the frequency and severity of such episodes. Nevertheless, health authorities are equally concerned about the slow and subtle effects on our lungs and other organs of air-pollution levels that are much lower but that are present every day, year after year.

Concept Check 13.1

1 Describe the impact of air pollution on human health.

2 List several examples of natural air pollution. List three that are human accentuated.

Students Sometimes Ask...

What is haze?

Haze is a reduction in visibility caused when light encounters atmospheric particulate matter and gases. Some light is absorbed by the particles and gases, and some is scattered away before it reaches an observer. More pollutants mean more absorption and scattering of light, which limits the distance we can see and can also degrade the color, clarity, and contrast of what we can see. Visibility impairment is one of the most obvious effects of air pollution. It not only occurs in urban areas but is a serious issue in many of our best-known national parks and wilderness areas. The same fine particles that are linked to serious health problems can also significantly affect our ability to see.

Figure 13–4 Stacks belching smoke and soot such as these were once a sign of economic prosperity. (Photo by EverettCollection/Superstock)

Sources and Types of Air Pollution

Air pollutants are airborne particles and gases that occur in concentrations that endanger the health and well-being of organisms or disrupt the orderly functioning of the environment. Pollutants can be grouped into two categories: primary and secondary. **Primary pollutants** are emitted directly from identifiable sources. They pollute the air immediately upon being emitted. **Secondary pollutants,** in contrast, are produced in the atmosphere when certain chemical reactions take place among primary pollutants. The chemicals that make up smog are important examples. In some cases the effects of primary pollutants on human health and the environment are less severe than the effects of the secondary pollutants they form.

Primary Pollutants

Figure 13–5 depicts the major primary pollutants as percentages (by weight). Sources vary for each pollutant. For example, electricity generation is the most significant source of sulfur dioxide. By contrast, on-road vehicles are the number-one source of carbon monoxide, nitrogen oxides, and volatile organic compounds. Compared to other sources, the tens of millions of cars and trucks on our streets and highways are clearly the greatest contributors. What follows is a brief survey and description of the major primary pollutants.

Particulate Matter *Particulate matter (PM)* is the general term used for a mixture of solid particles and liquid droplets found in the air. Some particles are large or dark enough to be seen as soot or smoke. Others are so small that they can be detected only with an electron microscope. These particles come in a wide range of sizes: *Fine* particles are less than 2.5 micrometers in diameter, and coarser-size particles are larger than 2.5 micrometers. These particles originate from many different stationary and mobile sources as well as from natural sources (Figure 13–6). Fine particles ($PM_{2.5}$) result from fuel combustion from motor vehicles, power generation, and industrial facilities, as well as from residential fireplaces and woodstoves. Coarse particles (PM_{10}) are generally emitted from sources such as vehicles traveling on unpaved roads, materials handling, and crushing and grinding operations, as well as windblown dust. Some particles are emitted directly from their sources, such as smokestacks and cars. In other cases, gases such as sulfur dioxide interact with other compounds in the air to form fine particles.

Particulates are frequently the most obvious form of air pollution because they reduce visibility and leave deposits of dirt on the surfaces with which they come in contact. In addition, particulates may carry any or all of the other pollutants dissolved in or absorbed on their surfaces.

Originally, total suspended particulate (TSP) was the indicator used to represent this category. It included all particles up to 45 micrometers in diameter. In 1987 the U.S. Environmental Protection Agency (EPA) set new standards related only to particles smaller than 10 micrometers (identified as PM_{10}). Then, in 1997, the EPA revised its standards for particulate matter again so that they were based on $PM_{2.5}$. This change was in response to a large amount of research that analyzed the health effects of particulates.

Inhalable particulate matter includes both fine and coarse particles. These particles can accumulate in the respiratory system and are associated with numerous health effects. Exposure to coarse particles is primarily associated with the aggravation of respiratory conditions, such

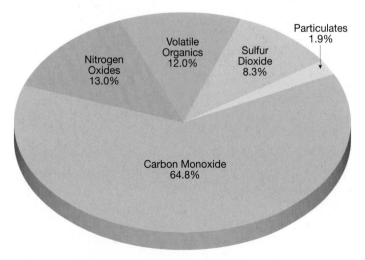

Figure 13–5 Emissions estimates of primary pollutants for the United States in 2009. Percentages are calculated on the basis of weight. The total weight in 2009 was 107 million tons. (Data from U.S. Environmental Protection Agency)

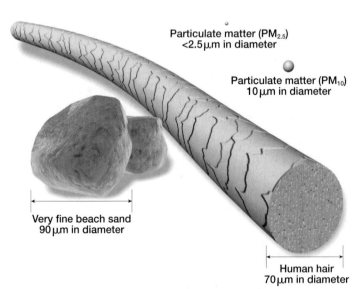

Particulate matter ($PM_{2.5}$)
<2.5 μm in diameter

Particulate matter (PM_{10})
10 μm in diameter

Very fine beach sand
90 μm in diameter

Human hair
70 μm in diameter

Figure 13–6 Particulate matter (PM) is a complex mixture of extremely small particles and liquid droplets. The size of particles is directly linked to their potential for causing health problems. Particles that are 10 micrometers in diameter or smaller frequently pass through the nose and throat and enter the lungs. Once inhaled, these particles can cause serious health effects. PM_{10} stands for "inhalable coarse particles" that are larger than 2.5 micrometers and smaller than 10 micrometers. $PM_{2.5}$ stands for "fine particles" that are 2.5 micrometers and smaller.

SEVERE AND HAZARDOUS WEATHER   The Great Smog of 1952

For centuries the fogs of Britain's major cities were polluted with smoke. London was especially well known for its poor air quality. In the early 1800s London's smoke-laden fog came to be known as a "London particular" (that is, a London characteristic). In 1853 Charles Dickens used the term in *Bleak House* and provided graphic descriptions of London's foul air in several of his novels.

One of London's most infamous air-pollution episodes occurred over a five-day span in December 1952. From December 5–9 acrid yellow smog shrouded the city, bringing premature death to thousands and inconvenience to millions (Figure 13–A). What conditions were responsible for this extraordinary event? As in practically all other major air-pollution episodes, it was a combination of emissions and meteorological circumstances.

The weather was unusually cold, and the people of London were burning large quantities of coal to warm their homes. The smoke that poured from these chimneys as well as from the chimneys of London's many factories was not dispersed but rather accumulated in a shallow zone of very stable air. The "lid" for this trap was a substantial temperature inversion associated with a high-pressure center that had become established over the southern British Isles.*

FIGURE 13–A London's infamous Great Smog of 1952 persisted for five days and was responsible for thousands of deaths. (Photo by Mirrorpix/Newscom)

The smog was so thick that people could not see their own feet.

Fog developed during the day on Friday, December 5. It was not especially dense at first, and it had a smoky character. Beneath the well-developed temperature inversion, a very light wind stirred the saturated air to create a fog layer that was 100 to 200 meters thick. With nightfall came sufficient radiation cooling to produce an even denser fog. Of course, more and more smoke continued to collect in the saturated air. Visibility dropped to just a few meters in many areas. On Saturday, December 6, the weak winter Sun could not "burn away" the fog. The foul yellow mixture of smoke and fog got to be so thick that night that pedestrians could not find their way, even in familiar surroundings. The smog was so thick that people could not see their own feet!

Winds did not sweep away the foul air until Tuesday, December 9. In central London the visibility remained below 500 meters continuously for 114 hours and below 50 meters for 48 hours straight. At Heathrow Airport the visibility was less than 10 meters for 48 hours beginning on the morning of December 6.

The huge quantities of pollutants released into the atmosphere during the event were estimated to be 1000 tons of smoke particles, 2000 tons of carbon dioxide, 140 tons of hydrochloric acid, and 14 tons of fluorine compounds. Moreover, 370 tons of sulfur dioxide were converted into 800 tons of sulfuric acid.

This infamous episode of December 1952 came to be known as "The Great Smog." Others called it "The Big Smoke." Even in a city where "pea soup" fogs were relatively common, this is a legendary event. No matter what name is used, this exceptional air pollution episode is commonly viewed as the watershed event that gave rise to modern air pollution control in Great Britain as well as elsewhere in Western Europe and North America.

Experts agree that The Great Smog killed about 4000 people in December 1952 alone (Figure 13–B). Furthermore, some researchers believe that an additional 8000 Londoners died in January and February 1953 due to the delayed effects of the smog or to lingering pollution. Other analyses disagree and blame the excess deaths on influenza. Yet others suggest that many of the

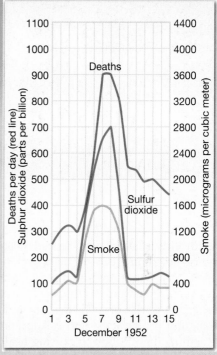

FIGURE 13–B Deaths and pollution during the Great Smog of 1952. Smoke and sulfur dioxide were monitored at various sites. The daily averages for 10 of these sites are shown here.

deaths may have resulted from an interaction between the smog and the flu. Referring to these differences of opinion, an article characterizes the situation this way:

> The debate reveals how much is unknown even today about the effects of smog, which continues to menace big cities, especially in developing countries with weak air-pollution laws.**

The debate about the effects of London's famous air-pollution episode of 1952 is more than just an academic exercise. Although London no longer experiences "great smogs," foul air continues to plague big cities. The World Health Organization estimates that outdoor air pollution is responsible for about 800,000 deaths worldwide each year. "Some lessons of the Big Smoke still haven't been learned."***

---

*The link between temperature inversions and air pollution episodes is explored in greater detail later in the chapter.

**Richard Stone, "Counting the Cost of London's Killer Smog," *Science*, December 13, 2002, vol. 298, pp. 2106–2107.

***Richard Stone, p. 2107.

EYE ON THE ATMOSPHERE

This satellite image from NASA's *Aqua* satellite shows Los Angeles County on September 2, 2009. Dingy gray smoke from several major wildfires pollutes the air. The well-developed cumulonimbus clouds that are also present owe their origin to the wildfires. (NASA)

Question 1 According to Figure 13–2, wildfires are identified as a natural source of primary pollutants. This fact implies that the smoke in this image is a "natural pollutant." However, these fires had a human origin. Is this air pollution "natural"? What term or phrase best fits this example?

Question 2 How could the wildfires have been responsible for the development of cumulonimbus clouds?

as asthma. Fine particles are most closely associated with health effects such as increased hospital admissions and emergency room visits for heart and lung disease, increased respiratory symptoms and disease, decreased lung function, and even premature death. Sensitive groups that appear to be at greatest risk of such effects include elderly people, children, and individuals with cardiopulmonary disease, such as asthma. In addition to causing health problems, particulate matter is the major cause of reduced visibility in many parts of the United States. Airborne particles can also cause damage to paints and building materials (see Box 13–1).

Sulfur Dioxide *Sulfur dioxide* (SO_2) is a colorless and corrosive gas that originates largely from the combustion of sulfur-containing fuels, primarily coal and oil. Important sources include power plants (Figure 13–7), smelters, petro-

leum refineries, and pulp and paper mills. Once SO_2 is in the air, it is frequently transformed into sulfur trioxide (SO_3), which reacts with water vapor or water droplets to form sulfuric acid (H_2SO_4). Very tiny particles act as a medium on which the acidic sulfate ion (SO_4^{-2}) is carried over long distances in the atmosphere. When it is "washed out" of the air or deposited on surfaces, it contributes to a serious environmental problem known as *acid precipitation.* This issue is the subject of a later section.

High concentrations of SO_2 can result in temporary breathing impairment for asthmatic children and adults who are active outdoors. Short-term exposures of asthmatic individuals to elevated SO_2 levels while at moderate exertion may result in reduced lung function that may be accompanied by such symptoms as wheezing, chest tightness, or shortness of breath. Other effects that have been associated with longer-term exposures to high concentrations of SO_2, in conjunction with high levels of particulate matter, include respiratory illness and aggravation of existing cardiovascular disease.

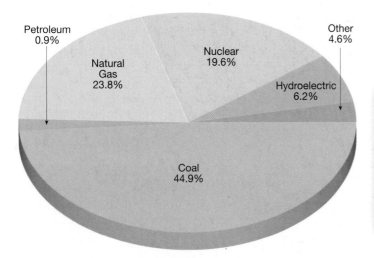

Figure 13–7 Sources of energy for generation of electricity in the United States (2010). A major source of sulfur dioxide pollution is coal-burning power plants that produce nearly 45 percent of the electricity generated in the United States. (Energy Information Administration)

Students Sometimes Ask...

You mentioned the burning of coal as a significant source of SO_2, but do we burn very much coal anymore?

Yes, indeed. Nearly 45 percent of the electricity produced in the United States comes from coal combustion (see Figure 13–7). In addition, coal is the fuel of choice for many heat-intensive processes, such as producing steel, aluminum, concrete, and wallboard. Many other countries are even more reliant on coal for their energy needs than the United States.

Nitrogen Oxides *Nitrogen oxides* are gases that form during the high-temperature combustion of fuel when nitrogen in the fuel or the air reacts with oxygen. Motor vehicles and power plants are the primary sources. These gases also form

Box 13–1 Air Pollution Changing the Climate of Cities

In Box 3–3 (page 84) you saw that air pollution in cities contributes to the heat island by inhibiting the loss of long-wave radiation at night. Studies of urban climate have also shown that pollutants may have a "cloud seeding" effect that increases precipitation in and downwind of cities. These influences, however, are not the only ways in which pollutants influence urban climate.

The blanket of particulates over most large cities significantly reduces the amount of solar radiation reaching the surface. In some cities, the overall reduction in the receipt of solar energy is 15 percent or more, whereas short-wavelength ultraviolet is decreased by up to 30 percent. This weakening of incoming solar energy is variable. During air pollution episodes, the decrease is much greater than for periods when air quality is good (Figure 13–C). Furthermore, particulates most effectively reduce solar radiation near the ground when the Sun angle is low. This occurs because the length of the path through the polluted air increases as the Sun angle drops. Thus, for a given quantity of particulate matter, solar energy will be reduced by the largest percentage in high-latitude cities and during the winter.

When compared to surrounding rural areas, the relative humidity in cities is generally from 2 to 8 percent lower. One reason is that cities are hotter. Remember from Chapter 4 that as air temperature increases, capacity also rises and relative humidity drops. A second reason is that less water

FIGURE 13–C An air pollution episode in Shanghai, China. It is not difficult to understand why the amount of solar radiation reaching the surface is reduced in cities. (Photo by Doable/Amana Images/Glow Images)

vapor is supplied to city air by evaporation from the surface. Evaporation is reduced in cities because rain water rapidly runs off, frequently into subsurface storm sewers.

Although relative humidity tends to be lower in cities, the occurrences of clouds and fogs are greater. What is the cause of this apparent paradox? A likely contributing factor is the large quantity of condensation nuclei produced by human activities in urban areas. When hygroscopic (water-seeking) nuclei are plentiful, water vapor readily condenses on them, even when the air is not quite saturated.

naturally when certain bacteria oxidize nitrogen-containing compounds.

The initial product formed is nitric oxide (NO). When NO oxidizes further in the atmosphere, nitrogen dioxide (NO_2) forms. Commonly, the general term NO_x is used to describe these gases. Although NO_x forms naturally, its concentration in cities is 10 to 100 times higher than in rural areas. Nitrogen dioxide has a distinctive reddish-brown color that frequently tints polluted city air and reduces visibility. When concentrations are high, NO_2 can also contribute to lung and heart problems. When air is humid, NO_2 reacts with water vapor to form nitric acid (HNO_3). Like sulfuric acid, this corrosive substance also contributes to the acid rain problem. Moreover, because nitrogen oxides are highly reactive gases, they play an important part in the formation of smog.

Volatile Organic Compounds *Volatile organic compounds (VOCs)*, also called *hydrocarbons*, encompass a wide array of solid, liquid, and gaseous substances that are composed

exclusively of hydrogen and carbon. Large quantities occur naturally, with methane (CH_4) being the most abundant. Methane, however, does not interact chemically with other substances and has no negative health effects. In cities, the incomplete combustion of gasoline in motor vehicles is the principal source of reactive VOCs. Although some hydrocarbons from other sources are cancer-causing agents, most of the VOCs in city air do not, by themselves, appear to pose significant environmental problems. However, as you will see in a later discussion, when VOCs react with certain other pollutants (especially nitrogen oxides), noxious secondary pollutants result.

Carbon Monoxide *Carbon monoxide (CO)* is a colorless, odorless, and poisonous gas produced by incomplete burning of carbon in fuels such as coal, oil, and wood. It is the most abundant primary pollutant, with more than three-quarters of U.S. emissions coming from highway vehicles and nonroad equipment.

Although CO is quickly removed from the atmosphere, it can nevertheless be dangerous. Carbon monoxide enters the bloodstream through the lungs and reduces oxygen delivery to the body's organs and tissues. Because it cannot be seen, smelled, or tasted, CO can have an effect on people without their realizing it. In small amounts, it causes drowsiness, slows reflexes, and impairs judgment. If concentrations are sufficiently high, CO can cause death. Carbon monoxide poses a serious health hazard where concentrations can reach high levels, as in poorly ventilated tunnels and underground parking facilities.

The world map in Figure 13–8 shows average CO concentrations for April 2010. In different parts of the world and in different seasons, the concentrations and sources change significantly. For example, in Africa, seasonal shifts in CO are tied to the widespread agricultural burning that shifts north and south of the equator with the seasons. Fires are also an important source of CO in other regions, such as the Amazon and Southeast Asia. In the United States, Europe, and China, on the other hand, the highest CO concentrations occur in and near urban areas where there are high concentrations of motor vehicles and factories. Wildfires burning over large areas of North America and Russia in some years can also be important sources.

Lead *Lead* (Pb) is very dangerous because it accumulates in the blood, bones, and soft tissues. It can impair the functioning of many organs. Even at low doses, lead exposure is associated with damage to the nervous systems of young children.

In the past, automotive sources were the major contributor of lead emissions to the atmosphere because lead was added to gasoline to prevent engine knock. Since the EPA-

TABLE 13-1 Air Quality and Emissions Trends

	Percentage Change in Concentrations	
	1980–2009	1990–2009
NO_2	–48	–40
O_3 8-hour	–30	–21
SO_2	–76	–65
PM_{10} 24-hour	—*	–38
$PM_{2.5}$ annual	—*	–27
CO	–80	–65
Pb	–93	–73

	Percentage Change in Emissions	
	1980–2009	1990–2009
NO_x	–48	–44
VOCs	–57	–43
SO_2	–65	–61
PM_{10}	–83	–67
$PM_{2.5}$	—*	–50
CO	–61	–51
Pb	–97	–60

*Data not available.

mandated phaseout of leaded gasoline, lead concentrations in the air of U.S. cities have shown a dramatic decline (Table 13–1). Occasional violations of the lead air quality standard still occur near large industrial sources such as lead smelters.

Students Sometimes Ask...

What are toxic air pollutants?

They are chemicals in the air that are known or suspected to cause cancer or other serious health effects, such as reproductive problems or birth defects. These substances are also commonly called "hazardous air pollutants" and "air toxics." The U.S. Environmental Protection Agency regulates 188 toxic air pollutants. Examples include benzene, which is found in gasoline; perchorethlyene, which is emitted by some dry-cleaning facilities; and methylene chloride, which is used as a solvent and paint stripper.

Secondary Pollutants

Recall that secondary pollutants are not emitted directly into the air but form in the atmosphere when reactions take place among primary pollutants. The sulfuric acid described earlier is one example of a secondary pollutant. After the primary pollutant, sulfur dioxide, is emitted into the atmosphere, it combines with oxygen to produce sulfur trioxide, which then combines with water to create the irritating and corrosive sulfuric acid.

Air pollution in urban and industrial areas is often called **smog.** The term, coined in 1905 by Harold A. Des Veaux, a London physician, was created by combining the words

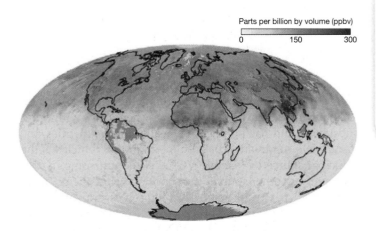

Parts per billion by volume (ppbv)

0   150   300

Figure 13–8 This map shows average concentrations of carbon monoxide (CO) in the troposphere in April 2010 at an altitude of about 3600 meters (12,000 feet). The data were collected by a sensor aboard NASA's *Terra* satellite. Concentrations of CO are expressed in parts per billion by volume (ppbv). Yellow areas have little or no CO. Progressively higher concentrations are shown in orange and red. Places where the sensor did not collect data, perhaps due to clouds, are gray. Satellite observations often show that pollution emitted in one locale can travel great distances and impact air quality far from the original source. (NASA)

smoke and *fog*. Des Veaux's term was indeed an apt description of London's principal air pollution threat, which was associated with the products of coal burning coupled with periods of high humidity.

Today, however, smog is used as a synonym for general air pollution and does not necessarily imply the smoke–fog combination. Therefore, when greater clarity is desired, we sometimes find the word "smog" preceded by such modifiers as "London-type," "classical," "Los Angeles–type," or "photochemical." The first two modifiers refer to the original meaning of the word, and the last two to air quality problems created by secondary pollutants.

Photochemical Reactions Many reactions that produce secondary pollutants are triggered by strong sunlight and so are called **photochemical reactions.** One common example occurs when nitrogen oxides absorb solar radiation, initiating a chain of complex reactions. When certain volatile organic compounds are present, the result is the formation of a number of undesirable secondary products that are very reactive, irritating, and toxic. Collectively, this noxious mixture of gases and particles is called *photochemical smog.* One of the substances it usually contains is called PAN (peroxyacetyl nitrate), which damages vegetation and irritates the eyes. The *major* component in photochemical smog is ozone. Recall from Chapter 1 that ozone is formed by natural processes in the stratosphere. However, when produced near Earth's surface, ozone is considered a pollutant.

Because the reactions that create ozone are stimulated by strong sunlight, the formation of this pollutant is limited to daylight hours. Peaks occur in the afternoon following a series of hot, sunny, calm days. As you might expect, ozone levels are highest during the warmer summer months. The "ozone season" varies from one part of the country to another. Although May through October is typical, areas in the Sunbelt of the American South and Southwest may experience problems throughout the year. By contrast, northern states have shorter ozone seasons, such as May through September for North Dakota.

The health and environmental effects of ozone are well documented. For example, according to the U.S. Environmental Protection Agency, health effects attributed to ozone exposure include significant decreases in lung function and increased respiratory symptoms, such as chest pain and cough. Exposure to ozone can make people more susceptible to respiratory infection, result in lung inflammation, and aggravate preexisting respiratory diseases such as asthma. These effects generally occur while individuals are actively exercising, working, or playing outdoors. Children, active outdoors during the summer when ozone levels are at their highest, are most at risk of experiencing such effects. In

addition, longer-term exposures to moderate levels of ozone present the possibility of irreversible changes in lung structure, which could lead to premature aging of the lungs.

Ozone also affects vegetation and ecosystems, leading to reductions in agricultural crop and commercial forest yields, reduced growth and survivability of tree seedlings, and increased plant susceptibility to disease, pests, and other environmental stresses such as harsh weather. In long-lived species, these effects may become evident only after many years or even decades, thus having the potential for long-term effects on forest ecosystems. Ground-level ozone damage to the foliage of trees and other plants can also decrease the aesthetic value of natural areas.

Volcanic Smog (Vog) Although smog is largely a human-induced atmospheric hazard, nature is capable of creating it as well. A prime example occurs in active volcanic areas such as Hawaii. The satellite image in Figure 13–9 shows a dense haze hanging over the Hawaiian Islands. It is a natural phenomenon that has come to be called *vog*, short for *volcanic smog*. In this region vog forms when sulfur dioxide from Kilauea Volcano combines with oxygen and water vapor in the presence of sunlight. The tiny sulfate particles that make up vog reflect light well, so that it shows up easily when viewed from space. On December 2, 2008, when this event occurred, SO_2 concentrations reached unhealthy levels in Hawaii Volcanoes National Park near the summit of Kilauea. In addition to reducing visibility, vog can aggravate respiratory problems such as asthma. Vog is not confined to Hawaii; it can and does occur in other volcanic areas.

Figure 13–9 This image from NASA's *Aqua* satellite was acquired on December 3, 2008. The dense haze over the Hawaiian Islands is called *vog,* short for *volcanic smog.* It forms when sulfur dioxide (SO_2) from Kilauea Volcano on the Big Island of Hawaii combines with oxygen and water vapor in the atmosphere to produce this natural form of smog. Vog is a relatively common occurrence in this region, although it usually is not this thick and widespread. (NASA)

Trends in Air Quality

Although Table 13–1 shows us that considerable progress has been made in controlling air pollution, the quality of the air we breathe still remains a serious public health problem. Economic activity, population growth, meteorological conditions, and regulatory efforts to control emissions all influence the trends in air pollutant emissions. Up until the 1950s the greatest influences on emissions were related to the economy and population growth. Emissions grew as the economy and population increased. Emissions fell in periods of economic recession. For example, dramatic declines in emissions in the 1930s were due to the Great Depression (Figure 13–10). Emissions also increase as a result of shifts in the demand for various products. For instance, the tremendous upsurge in demand for gasoline following World War II increased emissions associated with petroleum refining and on-road vehicles.

In the 1950s the states issued air-pollution statutes generally targeted toward smoke and particulate emissions. It was not until the passage of the federal Clean Air Act in 1970, that major strides were made in reducing air pollution. This legislation created the Environmental Protection Agency (EPA) and charged it with establishing air quality and emissions standards.

Establishing Standards

The Clean Air Act of 1970 mandated the setting of standards for four of the primary pollutants—particulates, sulfur dioxide, carbon monoxide, and nitrogen oxides—as well

as the secondary pollutant ozone. At the time, these five pollutants were recognized as being the most widespread and objectionable. Today, with the addition of lead, they are known as the *criteria pollutants* and are covered by the National Ambient Air Quality Standards (Table 13–2). The primary standard for each pollutant shown in Table 13–2 is based on the highest level that can be tolerated by humans without noticeable ill effects, minus a 10 to 50 percent margin for safety.

For some of the pollutants, both long-term and short-term levels are set. Short-term levels are designed to protect against acute effects, whereas the long-term standards were established to guard against chronic effects. *Acute* refers to pollutant levels that may be life-threatening within a period of hours or days. *Chronic* pollutant levels cause gradual deterioration of a variety of physiological functions over a span of years. It should be pointed out that standards are established using human health criteria and not according to their impact on other species or on atmospheric chemistry. Since the original Clean Air Act of 1970 was implemented, it has been amended, and its regulations and standards have been revised periodically.

By the year 2009 nearly 81 million people in the United States resided in counties that did not meet one or more air quality standards (Figure 13–11). It is clear from Figure 13–11 why the EPA describes ozone as our "most pervasive ambient air pollution problem." The number of people living in counties that exceeded the ozone standard is greater than the total number of those living in counties affected by the other five pollutants.

The fact that air quality standards have not yet been met in some places does not mean that progress has not been made. The United States has made significant strides in reducing air pollution. In 2009 emissions of the five major primary pollutants shown in Figure 13–5 totaled about 107 million tons. By contrast, in 1970, when the Clean Air Act first became law, the same five pollutants totaled about 301 million tons. The 2009 total is nearly 65 percent lower than the 1970 level. As Table 13–1 indicates, downward trends in all pollutants are substantial. This improvement in air quality has been achieved during a time when urban growth has been great. However, methods of control have not been as effective as expected in upgrading urban air quality.

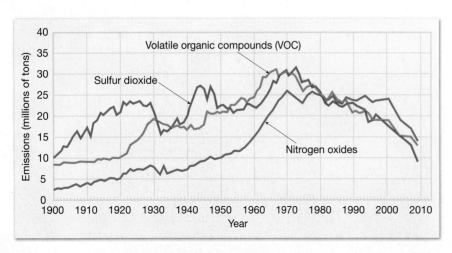

Figure 13–10 Trends in national emissions 1900–2009. Prior to the post-1970 period, economic activity and population growth were major factors influencing the emissions of air pollutants. Emissions grew as the economy and population increased, and emissions declined during economic downturns. For example, dramatic declines in emissions in the 1930s were due to the Great Depression. Since 1970, much of the downward trend in emissions has been due to the Clean Air Act. (After U.S. Environmental Protection Agency, Office of Air Quality Planning and Standards)

TABLE 13–2 National Ambient Air Quality Standards

Pollutant	Standard Value	
Carbon monoxide (CO)		
8-hour average	9 ppm*	$\left(10\ \text{mg/m}^3\right)$
1-hour average	35 ppm	$\left(40\ \text{mg/m}^3\right)$**
Nitrogen dioxide (NO₂)		
Annual arithmetic mean	0.053 ppm	$\left(100\ \mu\text{g/m}^3\right)$***
Ozone (O₃)		
1-hour average	0.12 ppm	$\left(235\ \mu\text{g/m}^3\right)$
8-hour average	0.08 ppm	$\left(157\ \mu\text{g/m}^3\right)$
Lead (Pb)		
Quarterly average		$0.15\ \mu\text{g/m}^3$
Particulate < 10 micrometers (PM₁₀)		
24-hour average		$150\ \mu\text{g/m}^3$
Particulate < 2.5 micrometers (PM₂.₅)		
Annual arithmetic mean		$15\ \mu\text{g/m}^3$
24-hour average		$35\ \mu\text{g/m}^3$
Sulfur dioxide (SO₂)		
Annual arithmetic mean	0.03 ppm	$\left(80\ \mu\text{g/m}^3\right)$
24-hour average	0.14 ppm	$\left(365\ \mu\text{g/m}^3\right)$
1-hour average	75 ppb†	

*ppm, parts per million.

**mg/m³, milligrams per cubic meter of air. A milligram is one-thousandth of a gram.

***μg/m³, micrograms per cubic meter. A microgram is one-millionth of a gram.

†ppb, parts per billion.

Source: U.S. Environmental Protection Agency, Office of Air Quality Planning and Standards.

An important reason for the slower-than-expected progress in improving air quality is related to growth. For example, *on a per-car basis*, emissions of primary pollutants have dramatically improved. However, at the same time this was occurring, the U.S. population increased by 50 percent and vehicle miles traveled went up by 168 percent (Figure 13–12). In other words, pollution controls have improved air quality, but the positive effects have been partly offset by an increase in the number of vehicles on the road.

Air Quality Index

The **Air Quality Index (AQI)** is a standardized indicator for reporting daily air quality to the general public. Simply, it is an attempt to answer the question: How clean or polluted is the air today? It provides information on what

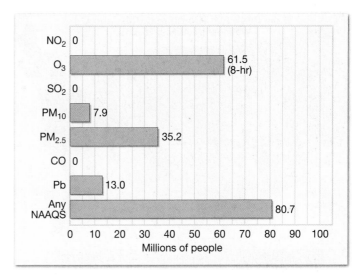

Figure 13–11 Number of people living in counties with air quality concentrations above the levels of the National Ambient Air Quality Standards (NAAQS) in 2009. For example, 7.9 million people live in counties where PM₁₀ concentrations exceed the national standard. Despite substantial progress in reducing emissions, there were still nearly 81 million people nationwide who lived in counties with monitored air quality levels above the primary national standards. (After U.S. Environmental Protection Agency)

health effects people might experience within a few hours or days of breathing polluted air. The EPA calculates the AQI for five major pollutants regulated by the Clean Air Act: ground-level ozone, particulate matter, carbon monoxide, sulfur dioxide, and nitrogen dioxide. Ground-level ozone and airborne particulates are the pollutants that pose the greatest risk to human health in the United States.

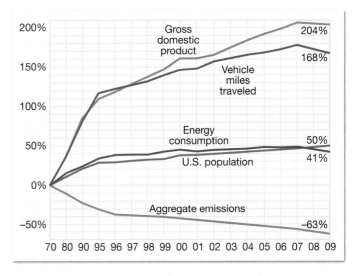

Figure 13–12 Comparison of growth areas and emissions. Between 1970 and 2009 gross domestic product increased 204 percent, vehicle miles traveled increased 168 percent, energy consumption increased 41 percent, and U.S. population increased 50 percent. At the same time, total emissions of the six principal air pollutants decreased about 63 percent. (After U.S. Environmental Protection Agency)

Daily Peak AQI (Combined PM$_{2.5}$ and O$_3$)
Wednesday, August 25, 2010

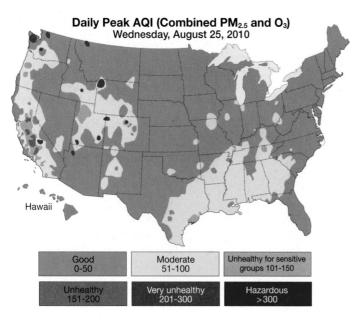

Hawaii

| Good 0–50 | Moderate 51–100 | Unhealthy for sensitive groups 101–150 |
| Unhealthy 151–200 | Very unhealthy 201–300 | Hazardous >300 |

Figure 13–13 The national Air Quality Index (AQI) forecast map for August 25, 2010. Specific colors for each AQI category make interpreting the map relatively easy. To check the current map, go to www.airnow.gov.

The AQI scale runs from 0 to 500 (Figure 13–13). The higher the value, the greater the level of air pollution and the greater the health concern. An AQI value of 100 generally corresponds to the national air quality standard for the pollutant. Values below 100 are usually considered satisfactory. When values exceed 100, air quality is considered to be unhealthy—at first for sensitive groups and then, as values increase, for everyone else. Specific colors are assigned to each AQI category to make it easier for people to quickly understand whether air pollution is reaching unhealthy levels in their communities. To access the current AQI, visit www.airnow.gov.

occur, they do not generally result from a drastic increase in the output of pollutants; instead, they occur because of changes in certain atmospheric conditions. *Severe and Hazardous Weather: Viewing an Air Pollution Episode from Space* on page 369 provides a good example.

Perhaps you have heard the phrase "The solution to pollution is dilution." To a significant degree, it is true. If the air into which the pollution is released is not dispersed, the air will become more toxic. Two of the most important atmospheric conditions affecting the dispersion of pollutants are (1) the strength of the wind and (2) the stability of the air. These factors are critical because they determine how rapidly pollutants are diluted by mixing with the surrounding air after leaving the source.

Wind As a Factor

The manner in which wind speed influences the concentration of pollutants is shown in Figure 13–14. Assume that a burst of pollution leaves a chimney stack every second. If the wind speed were 10 meters per second (23 miles per hour), the distance between each pollution "cloud" would be 10 meters. If the wind is reduced to 5 meters per second, the distance between "clouds" will be 5 meters. Consequently, because of the direct effect of wind speed, the concentration of pollutants is twice as great with the 5 meters per second wind as with the 10 meters per second wind. It is easy to understand why air pollution problems seldom occur when winds are strong but rather are associated with periods when winds are weak or calm.

A second aspect of wind speed influences air quality. The stronger the wind, the more turbulent the air. Thus, strong winds mix polluted air more rapidly with the surrounding air, thereby causing the pollution to be more dilute. Conversely, when winds are light, there is little turbulence, and the concentration of pollutants remains high.

Concept Check 13.3

1 When was the Clean Air Act established? What are its *criteria pollutants*?

2 Compare emissions of primary pollutants in 1970 with emissions in 2009.

3 What is the *Air Quality Index*?

Meteorological Factors Affecting Air Pollution

The most obvious factor influencing air pollution is the quantity of contaminants emitted into the atmosphere. Still, experience shows that even when emissions remain relatively steady for extended periods, wide variations in air quality often occur from one day to the next. Indeed, when air pollution episodes

Figure 13–14 Effect of wind speed on the dilution of pollutants. The concentration of pollutants increases as wind speed decreases.

SEVERE AND HAZARDOUS WEATHER

Viewing an Air Pollution Episode from Space

In early October 2010, a high-pressure system settled in over eastern China, and air quality began to deteriorate. By October 9 and 10, China's National Environmental Monitoring Center declared air quality to be *poor* to *hazardous* around Beijing and in 11 eastern provinces. Citizens were advised to take measures to protect themselves. Visibility was reduced to 100 meters (330 feet) in some areas, and news outlets reported that at least 32 people died in traffic accidents caused by the poor visibility. Thousands suffered with asthma and other respiratory difficulties.

Instruments on NASA's *Aqua* and *Terra* satellites captured the natural-color view of this air pollution episode shown in Figure 13–D. The milky white and gray covering the right portion of the image is smog, while the brighter white patches are clouds. Two other images from NASA's *Aura* satellite show levels of aerosols (Figure 13–E)

and sulfur dioxide (Figure 13–F). The primary source of sulfur dioxide is coal-burning power plants and smelters. Peak concentrations were 6 to 8 times the normal levels for China and 20 times the normal levels for the United States. The Aerosol Index indicates the presence of ultraviolet light–absorbing particles—largely smoke from agricultural burning and industrial processes. Figure 13–E shows that some areas had an index of 3.5. At an index value of 4, aerosols are so dense that you would have difficulty seeing the midday sun.

On October 11, the weather changed. The stagnant air associated with high pressure was replaced when a cold front brought cleansing rain and strong winds that cleared the sky. Clearly, this air pollution episode could not have occurred without emissions from human activities. However, it is also apparent that atmospheric conditions play a key role in causing variations in air quality.

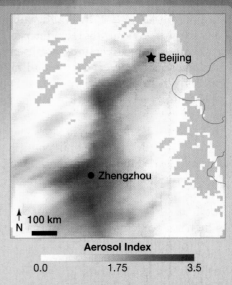

Aerosol Index

0.0   1.75   3.5

FIGURE 13–E This satellite image shows the extremely high levels of aerosols associated with the October 2010 air pollution episode in China. Gray areas lack data. (NASA)

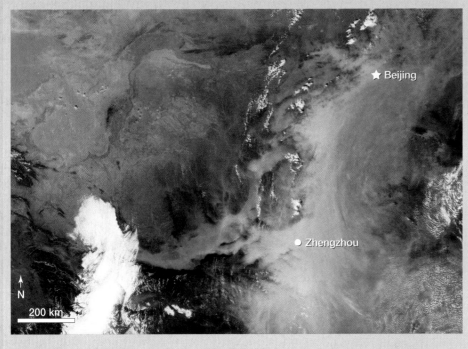

FIGURE 13–D Serious air pollution plagues a portion of China on October 8, 2010. (NASA)

SO₂ (dobson units)

0.0   4.0   8.0

FIGURE 13–F Sulfur dioxide (SO_2) levels were high during the October 2010 air pollution episode in China. Gray areas lack data. (NASA)

The Role of Atmospheric Stability

Whereas wind speed governs the amount of air into which pollutants are initially mixed, atmospheric stability determines the extent to which vertical motions will mix the pollution with cleaner air above. The vertical distance between Earth's surface and the height to which convectional movements extend is called the **mixing depth.** Generally, the greater the mixing depth, the better the air quality. When the mixing depth is several kilometers, pollutants are mixed through a large volume of cleaner air and dilute rapidly. When the mixing depth is shallow, pollutants are confined to a much smaller volume of air, and concentrations can reach unhealthy levels.

When air is stable, convectional motions are suppressed and mixing depths are small. Conversely, an unstable atmosphere promotes vertical air movements and greater mixing depths. Because heating of Earth's surface by the Sun enhances convectional movements, mixing depths are usually greater during the afternoon hours. For the same reason, mixing depths during the summer months are typically greater than during the winter months.

Temperature inversions represent situations in which the atmosphere is very stable and mixing depths are significantly restricted. Warm air overlying cooler air acts as a lid and prevents upward movement, leaving the pollutants trapped in a relatively narrow zone near the ground. This effect is dramatically illustrated by the photograph in Figure 13–15. Most of the air pollution episodes cited earlier were linked to the occurrence of temperature inversions that remained in place for many hours or days.

Surface Temperature Inversions Solar heating can result in high surface temperatures during the late morning and afternoon that increase the environmental lapse rate and render the lower air unstable. During nighttime hours, however, just the opposite situation may occur: Temperature inversions, which result in very stable atmospheric conditions, can develop close to the ground. These surface inversions form because the ground is a more effective radiator than the air

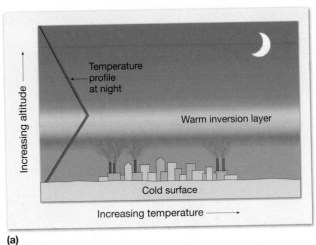

(a)

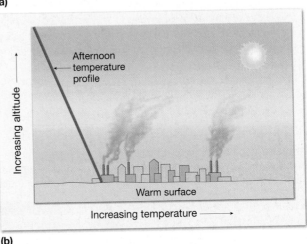

(b)

Figure 13–16 (a) A generalized temperature profile of a surface inversion. (b) The temperature profile changes after the Sun has heated the surface.

Figure 13–15 Air pollution in downtown Los Angeles. Temperature inversions act as lids that trap pollutants below. (Photo by Walter Bibikow/agefotostock)

above. Therefore, radiation from the ground to a clear night sky causes more rapid cooling at the surface than higher in the atmosphere. Consequently, the coldest air is found next to the ground, yielding a vertical temperature profile resembling the one shown in the upper portion of Figure 13–16. Once the Sun rises, the ground is heated, and the inversion disappears.

Although usually rather shallow, surface inversions may be deep in regions where the land surface is uneven. Because cold air is denser than warm air, the chilled air near the surface gradually drains from the uplands and slopes into adjacent lowlands and valleys. As might be expected, this deeper surface inversion will not dissipate as quickly after sunrise. Thus, although valleys are often preferred sites for manufacturing because they afford easy access to water transportation, they are also more likely to experience relatively thick surface inversions that in turn will have a negative effect on air quality.

Inversions Aloft Many extensive and long-lived air-pollution episodes are linked to temperature inversions that develop in association with the sinking air that characterizes centers of high air pressure (anticyclones). As the air sinks to lower altitudes, it is compressed and so its temperature rises.

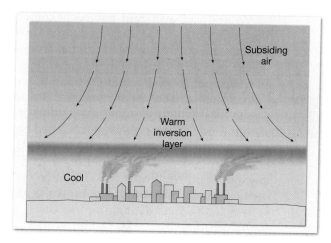

Figure 13–17 Inversions aloft frequently develop in association with slow-moving centers of high pressure, where the air aloft subsides and warms by compression. The turbulent surface zone does not subside as much. Thus, an inversion often forms between the lower turbulent zone and the subsiding layers above.

Because turbulence is almost always present near the ground, this lowermost portion of the atmosphere is generally prevented from participating in the general subsidence. Thus, an inversion develops aloft between the lower turbulent zone and the subsiding warmer layers above (Figure 13–17).

The air pollution that plagues Los Angeles is frequently related to inversions associated with the subsiding eastern portion of the subtropical high in the North Pacific. In addition, the adjacent cool waters of the Pacific Ocean and the mountains surrounding the city compound the problem. When winds move cool air from the Pacific into Los Angeles, the warmer air that is pushed aloft creates or strengthens an inversion aloft that acts as an effective lid. Because the surrounding mountains keep the smog from moving farther inland, air pollution is trapped in the basin until a change in weather brings relief. Clearly, the geographic setting of a place can significantly contribute to air quality problems. The Los Angeles area is an excellent example.

Students Sometimes Ask...

I've heard that in some places wood-burning fireplaces and stoves can be significant sources of air pollution. Is that actually the case?

Yes. Woodsmoke can build up in areas where it is generated and expose people to high levels of air pollution, especially on cold nights when there is a temperature inversion. Woodsmoke contains significant quantities of particulates and much higher levels of hazardous air pollutants, including some cancer-causing chemicals, than smoke from oil- and gas-fired furnaces. For example, in the Bay Area of California, woodsmoke is the largest source of particulate pollution. On an average winter night, 40 percent or more of the fine particulates are from woodsmoke. Some communities now ban the installation of conventional fireplaces or woodstoves that are not EPA certified. In addition, in many localities people are asked to refrain from burning wood on evenings when air pollution levels are expected to be high.

In summary, we have seen that when the wind is strong and an unstable environmental lapse rate prevails, the diffusion of pollutants is rapid, and high pollution concentrations will not occur except perhaps near a major source. In contrast, when an inversion exists and winds are light, diffusion is inhibited, and high pollution concentrations are to be expected in areas where there are sources. Air pollution is especially acute in urban areas experiencing frequent and prolonged temperature inversions.

Concept Check 13.4

1 Are most air pollution episodes triggered by a dramatic increase in the output of pollutants? Explain.

2 Describe two ways in which wind influences air quality.

3 What is *mixing depth*? How does it relate to air quality?

4 Contrast the formation of a surface temperature inversion with an inversion aloft.

5 How does the geographic setting of Los Angeles influence air pollution there?

Acid Precipitation

As a consequence of burning large quantities of fossil fuels, primarily coal and petroleum products, millions of tons of sulfur and nitrogen oxides are released into the atmosphere each year in the United States. In 2009 the total was 23 million tons. The major sources of these emissions include power-generating plants, industrial processes such as ore smelting and petroleum refining, and motor vehicles of all kinds. Through a series of complex chemical reactions, some of these pollutants are converted into acids that then fall to Earth's surface as rain or snow. This is referred to as *wet deposition*. By contrast, in areas where the weather is dry, the acid-producing chemicals may become incorporated into dust or smoke and fall to the ground as *dry deposition*. Dry-deposited particles and gases can be washed from the surface by rain, making the runoff more acidic. About half of the acidity in the atmosphere falls back to the surface as dry deposition.

In 1852 the English chemist Angus Smith coined the term *acid rain* to refer to the effect that industrial emissions had on precipitation in the English Midlands. A century and a half later this phenomenon is not only the focus of research for many environmental scientists but also a topic with substantial international political importance. Although Smith clearly realized that acid rain causes environmental damage, large-scale effects were not recognized until the middle part of the twentieth century. Eventually, widespread public concern in the late 1970s led to significant government-sponsored studies of the problem. Such research activities continue to examine this unresolved environmental problem.

Extent and Potency of Acid Precipitation

Rain is naturally somewhat acidic. When carbon dioxide from the atmosphere dissolves in water, it becomes weak carbonic acid. Small amounts of other naturally occurring

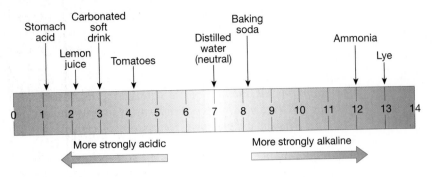

Figure 13–18 The *pH scale* is a common measure of the degree of acidity or alkalinity of a solution. The scale ranges from 0 to 14, with a value of 7 denoting a solution that is neutral. Values below 7 indicate greater acidity, and numbers above 7 indicate greater alkalinity. The pH values of some familiar substances are shown on the diagram. Although distilled water is neutral (pH 7), rainwater is naturally acidic. It is important to note that the pH scale is logarithmic; that is, each whole number increment indicates a tenfold difference. Thus, pH 4 is 10 times more acidic than pH 5 and 100 times (10 × 10) more acidic than pH 6.

acids also contribute to the acidity of precipitation. It was once thought that unpolluted rain has a pH of about 5.6 on the pH scale (Figure 13–18). However, studies in uncontaminated remote areas have shown that precipitation usually has a pH closer to 5. Unfortunately, in most areas within several hundred kilometers of large centers of human activity, precipitation has much lower pH values. This rain or snow is called **acid precipitation.**

Widespread acid rain has been known in Northern Europe and eastern North America for some time (Figure 13–19). Studies have also shown that acid rain occurs in many other regions, including western North America, Japan, China, Russia, and South America. In addition to local pollution sources, a portion of the acidity found in the northeastern United States and eastern Canada originates hundreds of kilometers away, in industrialized regions to the south and southwest. This situation occurs because many pollutants remain in the atmosphere for periods as long as five days, during which time they may be transported great distances.

One contributing factor is, of all things, a technology that is used to reduce pollution in the immediate vicinity of a source. Taller chimney stacks improve local air quality by releasing pollutants into the stronger and more persistent winds that exist at greater heights. Although such stacks enhance dilution and dispersion, they also promote the long-distance transport of these unwanted emissions. In this way, individual stack plumes with pollution concentrations considered too dilute to be a direct health or environmental threat locally contribute to inter-regional pollution problems. Unfortunately, because atmospheric processes in eastern North America lead to a thorough mixing of pollutants, it is not yet possible to distinguish clearly

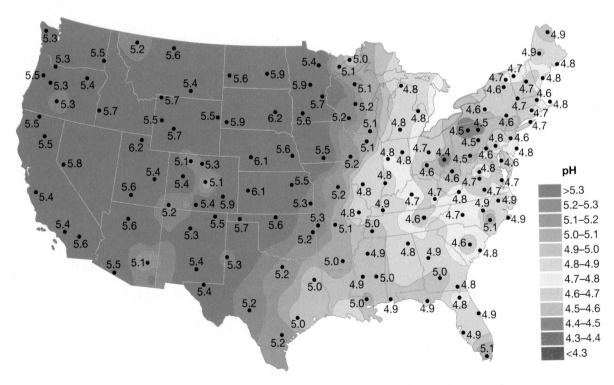

Figure 13–19 Precipitation pH values for the United States in the year 2008. Unpolluted rain has a pH of about 5. In the United States acid precipitation is most severe in the Northeast. (Data from National Atmospheric Deposition Program)

EYE ON THE ATMOSPHERE

Air pollution creates a translucent veil over China's Sichuan basin on May 6, 2011. The dull gray haze contrasts with bright white clouds and lingering snow on nearby mountains. The pollution results in part from urban and industrial sources in which coal burning is a significant energy source. A week earlier visibility here was good, and this region was *not* experiencing an air pollution episode. (NASA)

Question 1 What secondary pollutant might you expect to be present in the polluted air on May 6?

Question 2 If we assume that the output of pollutants in this region does not change significantly from week to week, why was the air relatively clear one week and polluted the next?

between the relative impact of distant sources compared with local sources.

Effects of Acid Precipitation

Acid rain looks, feels, and tastes just like clean rain. The harm to people from acid rain is not direct. Walking in acid rain, or even swimming in an acid lake, is no more dangerous than walking or swimming in water that has not been affected by acid rain. However, the sulfur dioxide and nitrogen oxide pollutants that cause acid rain do damage human health. These gases interact in the atmosphere to form fine sulfate and nitrate particles that can be transported long distances by winds and inhaled into people's lungs. These particles have been shown to aggravate heart problems and lung disorders such as asthma and bronchitis.

The damaging effects of acid rain on the environment are believed to be considerable in some areas and imminent in others. The best-known effect of acid precipitation is the lowering of pH in thousands of lakes and streams in Scandinavia and eastern North America. Accompanying this have been substantial increases in dissolved aluminum that is leached from the soil by the acidic water and that, in turn, is toxic to fish. Consequently, some lakes are virtually devoid of fish, whereas others are approaching this condition. Furthermore, ecosystems are characterized by many interactions at many levels of organization, which means that evaluating the effects of acid precipitation on these complex systems is difficult and expensive—and far from complete.

Even within small areas, the effects of acid precipitation can vary significantly from one lake to another. Much of this variation is related to the nature of the soil and rock materials in the area surrounding the lake. Because minerals such as calcite in some rocks and soils can neutralize acid solutions, lakes surrounded by such materials are less likely to become acidic. In contrast, lakes that lack this buffering material can be severely affected. Even so, over a period of time, the pH of lakes that have not yet been acidified may drop as the buffering material in the surrounding soil becomes depleted.

In addition to lakes no longer being able to support fish, research indicates that acid precipitation may also reduce agricultural crop yields and impair the productivity of forests. Acid rain not only harms the foliage but also damages roots and leaches nutrient minerals from the soil (Figure 13–20). Finally, acid precipitation promotes the corrosion of metals and contributes to the destruction of stone structures (Figure 13–21).

Figure 13–20 Damage to forests by acid precipitation is well documented in Europe and eastern North America. These trees in the Appalachian Mountains of North Carolina are one example. (Photo by Andre Jenny Stock Connection Worldwide/Newscom)

Figure 13–21 Acid rain accelerates the chemical weathering of stone monuments and structures. (Photo by Adam Hart-Davis/ Science Photo Library/Photo Researchers, Inc.)

In summary, acid precipitation involves the delivery of acidic substances through the atmosphere to Earth's surface. These compounds are introduced into the air as by-products of combustion and industrial activity. The atmosphere is both the avenue by which offending compounds travel from sources to the sites where they are deposited and the medium in which the combustion products are transformed into acidic substances. In addition to its detrimental impact on aquatic systems, acid precipitation has a number of other harmful effects.

The emission reductions in nitrogen oxides and sulfur dioxide noted earlier in the chapter have not only contributed to improved air quality but have also led to reductions in the acidity of precipitation in many areas (see Table 13–1 and Figure 13–10). Long-term monitoring of lakes and streams has shown that some acid-sensitive waters have experienced the beginnings of recovery. Nevertheless, although progress has been made, acid precipitation remains a complex and global environmental problem.

EYE ON THE ATMOSPHERE

This tall smokestack is part of a coal-fired electricity-generating plant located in a valley in the rolling hills of West Virginia. Tall stacks are commonly associated with such power plants as well as with many factories. An extensive radiation fog is hugging the ground. (Photo by Michael Collier)

Question 1 Is the time of day shown here more likely early morning or midafternoon? Explain.

Question 2 Sketch a simple graph illustrating the likely vertical temperature profile at the time the photo was taken.

Question 3 What are two reasons that air quality in the local area is better because the stack is present?

Concept Check 13.5

1 Which primary pollutants are associated with the formation of acid precipitation?

2 How much more acidic is a substance with a pH of 4 than a substance with a pH of 6?

3 Based on the map in Figure 13–19, where in the United States is precipitation most acidic?

4 What are some environmental effects of acid precipitation?

Give It Some Thought

1. In Chapter 1 you learned that we should be concerned about ozone depletion in the atmosphere. Based on what is presented in this chapter, it seems like getting rid of ozone would be a good idea. Can you clarify this apparent contradiction?

2. Table 13–1 shows trends in air quality and emissions. Explain why ozone (O_3) appears on the "Percentage change in concentrations" portion of the table but does not appear on the "Percentage change in emissions" portion.

3. The average motor vehicle today emits *much* less pollution than did the vehicles of 30 or 40 years ago. Why have the positive effects of this sharp reduction *not* been as great as we might have expected? Include information from one of the graphs in this chapter in your explanation.

4. Motor vehicles are a major source of air pollutants. Using electric cars, such as the one pictured here, is one way that emissions from this source can be reduced. Although these vehicles emit little or no pollution directly into the air, can they still be connected to the emission of primary pollutants? If so, explain.

(Photo by David Pearson/Alamy)

5. Assume that you are at an airport in a large urban area. The city is experiencing an air quality alert, and haze reduces visibility. As your plane climbs after takeoff, the air suddenly becomes much cleaner. Provide a likely explanation for the sudden change. You may wish to include a sketch.

6. As the accompanying photo illustrates, air pollution episodes reduce the amount of sunlight reaching Earth's surface. In which of these situations would the reduction likely be greatest (assuming that the level of pollutants is identical in each example)? Explain your choice.
   a. High-latitude city in summer
   b. High-latitude city in winter
   c. Low-latitude city in summer
   d. Low-latitude city in winter

(Photo by DC Premiumstock/Alamy)

7. A glance at a cross-section of a warm or cold front (see Figures 9–3 and 9–5) shows warm air *above* cool or cold air; that is, it shows a temperature inversion. Although temperature inversions are associated with fronts, they have little adverse effect on air quality. Why is this the case?

8. Ozone is sometimes called a "summer pollutant." Why is summertime also "ozone season"?

9. The use of tall smokestacks improves local air quality. However, tall stacks may contribute to pollution problems elsewhere. Explain.

AIR POLLUTION IN REVIEW

- Air is never perfectly clean. Volcanic ash, salt particles, pollen and spores, smoke, and windblown dust are all examples of "natural air pollution." Although some types of air pollution are recent creations, others, such as London's infamous smoke pollution, have been around for centuries. One of the most tragic air-pollution episodes occurred in London in December 1952, when more than 4000 people died.

- *Air pollutants* are airborne particles and gases that occur in concentrations that endanger the health and well-being of organisms or disrupt the orderly functioning of the environment. Pollutants can be grouped into two categories: (1) *primary pollutants,* which are emitted directly from identifiable sources, and (2) *secondary pollutants,* which are produced in the atmosphere when certain chemical reactions take place among primary pollutants. The major primary pollutants are particulate matter (PM), sulfur dioxide, nitrogen oxides, volatile organic compounds (VOCs), carbon monoxide, and lead. Atmospheric sulfuric acid is one example of a secondary pollutant. Air pollution in urban and industrial areas is often called *smog. Photochemical smog,* a noxious mixture of gases and particles, is produced when strong sunlight triggers *photochemical reactions* in the atmosphere. A major component of photochemical smog is *ozone.*

- Although considerable progress has been made in controlling air pollution, the quality of the air we breathe remains a serious public health problem. Economic activity, population growth, meteorological conditions, and regulatory efforts to control emissions all influence the trends in air pollution. The *Clean Air Act of 1970* mandated the setting of standards for four of the primary pollutants—particulates, sulfur dioxide, carbon monoxide, and nitrogen oxides—as well as the secondary pollutant ozone. In 2009 emissions of the major primary pollutants in the United States were about 65 percent lower than in 1970.

- When air pollution episodes take place, they do not generally result from a drastic increase in the output of pollutants; instead, they occur because of changes in certain atmospheric conditions. Two of the most important atmospheric conditions affecting the dispersion of pollutants are (1) the strength of the wind and (2) the stability of the air. The direct effect of wind speed is to influence the concentration of pollutants. Atmospheric stability determines the extent to which vertical motions will mix the pollution with cleaner air above the surface layer. The vertical distance between Earth's surface and the height to which convectional movements extend is called the *mixing depth.* Generally, the greater the mixing depth, the better the air quality. *Temperature inversions* represent a situation in which the atmosphere is very stable and the mixing depth is significantly restricted. When an inversion exists and winds are light, diffusion is inhibited and high pollution concentrations are to be expected in areas where pollution sources exist.

- In most areas within several hundred kilometers of large centers of human activity, the pH value of precipitation is lower than the usual value found in unpopulated areas. This acidic rain or snow, formed when sulfur and nitrogen oxides produced as by-products of combustion and industrial activity are converted into acids during complex atmospheric reactions, is called *acid precipitation.* The atmosphere is the avenue by which offending compounds travel from sources to the sites where they are deposited, and it is also the medium in which the combustion products are transformed into acidic substances. The damaging effects of acid precipitation on the environment include the lowering of pH in thousands of lakes in Scandinavia and eastern North America. Besides producing water that is toxic to fish, acid precipitation has also detrimentally altered complex ecosystems.

VOCABULARY REVIEW

acid precipitation (p. 372)
air pollutants (p. 360)
Air Quality Index (AQI) (p. 367)

mixing depth (p. 369)
photochemical reaction (p. 365)
primary pollutant (p. 360)

secondary pollutant (p. 360)
smog (p. 364)
temperature inversion (p. 370)

MyMeteorologyLab™

Log in to www.mymeteorologylab.com for animations, videos, MapMaster interactive maps, GEODe media, *In the News* RSS feeds, web links, glossary flashcards, self-study quizzes and a Pearson eText version of this book to enhance your study of *Air Pollution.*

14

The Changing Climate

The focus of this chapter and Chapter 15 is *climate*, the long-term aggregate of weather. Climate is more than just an expression of average atmospheric conditions. In order to accurately portray the character of a place or an area, variations and extremes must also be included. Anyone who has the opportunity to travel around the world will find such an incredible variety of climates that it is hard to believe they could all occur on the same planet.

Focus On Concepts

After completing this chapter you should be able to:

- List the five parts of the climate system and provide examples of each.

- Explain why unraveling past climate changes is important and discuss several ways in which such changes are detected.

- Discuss four hypotheses that relate to natural causes of climate change.

- Describe several ways in which humans are changing the composition of the atmosphere.

- Review the atmosphere's responses to human-caused changes in the composition of the atmosphere.

- Contrast positive- and negative-feedback mechanisms and provide examples of each.

- Discuss several possible consequences of global warming.

The Climate System

Climate has a significant impact on people, and we are learning that people also have a strong influence on climate. In fact, today global climate change caused by humans is making headlines. Why is climate change newsworthy? The reason is that research focused on human activities and their impact on the environment has demonstrated that people are inadvertently changing the climate. Unlike changes in the geologic past, which were natural variations, modern climate change is dominated by human influences that are sufficiently large that they exceed the bounds of natural variability. Moreover, these changes are likely to continue for many centuries. The effects of this venture into the unknown with climate could be very disruptive not only to humans but to many other life-forms as well. The latter portion of this chapter examines the ways in which humans may be changing global climate.

Throughout this book you have frequently been reminded that Earth is a complex system that consists of many interacting parts. A change in any one part can produce changes in any or all of the other parts—often in ways that are neither obvious nor immediately apparent. This fact is reinforced when it comes to the study of climate and climate change.

To understand and appreciate climate, it is important to realize that climate involves more than just the atmosphere (Figure 14–1):

> The atmosphere is the central component of the complex, connected, and interactive global environmental system upon which all life depends. Climate may be broadly

defined as the long-term behavior of this environmental system. To understand fully and to predict changes in the atmospheric component of the climate system, one must understand the sun, oceans, ice sheets, solid earth, and all forms of life.*

Indeed, we must recognize that there is a **climate system** that includes the atmosphere, hydrosphere, solid Earth, biosphere, and cryosphere. *Cryosphere* refers to the portion of Earth's surface where water is in solid form. This includes snow, glaciers, sea ice, freshwater ice, and frozen ground (termed *permafrost*). The climate system *involves the exchanges of energy and moisture that occur among the five spheres.* These exchanges link the atmosphere to the other spheres so that the whole functions as an extremely complex interactive unit. Changes in the climate system do not occur in isolation. Rather, when one part of the climate system changes, the other components react. The major components of the climate system are shown in Figure 14–2.

The climate system provides a framework for the study of climate. The interactions and exchanges among the parts of the climate system create a complex network that links the five spheres. This well-established relationship will be demonstrated often as we study climate change and world climates.

Concept Check 14.1

1 List the five parts of the climate system.

Figure 14–1 The five major parts of the climate system are represented in this image from Alaska's Denali National Park. (Photo by Arterra Picture Library/Alamy)

How Is Climate Change Detected?

Climate not only varies from place to place but is naturally variable over time. Over the great expanse of Earth history, and long before humans were roaming the planet, there were many shifts—from warm to cold and from wet to dry and back again. Practically every place on our planet has experienced wide swings in climate, such as from ice ages to conditions associated with subtropical coal swamps or desert dunes. How do we know about these changes? What are the causes? The first portion of this chapter takes a look at how scientists decipher Earth's climate history and then explores some significant natural causes of climate change.

---

*The American Meteorological Society and the University Corporation for Atmospheric Research, "Weather and the Nation's Well-Being," *Bulletin of the American Meteorological Society*, 73, no. 12 (December 1991), 2038.

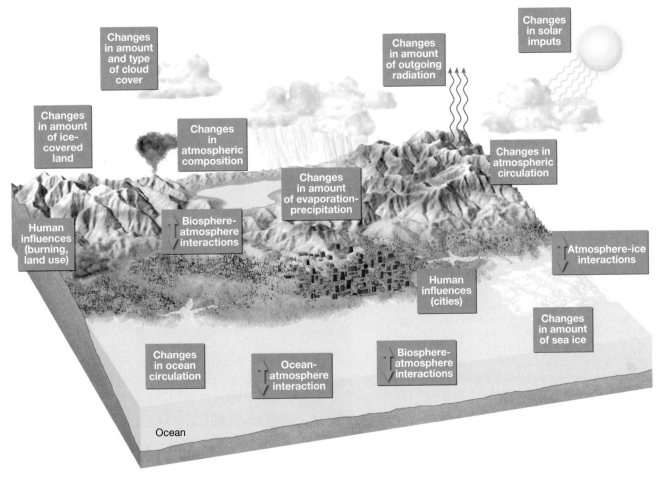

Figure 14–2 Schematic view showing several components of Earth's climate system. Many interactions occur among the various components on a wide range of space and time scales, making the system extremely complex.

High-technology and precision instrumentation are now available to study the composition and dynamics of the atmosphere. But such tools are recent inventions and therefore have been providing data for only a short time span. To understand fully the behavior of the atmosphere and to anticipate future climate change, we must somehow discover how climate has changed over broad expanses of time.

Instrumental records go back only a couple centuries, at best, and the further back we go, the less complete and more unreliable the data become. To overcome this lack of direct measurements, scientists must decipher and reconstruct past climates by using indirect evidence. **Proxy data** come from natural recorders of climate variability, such as seafloor sediments, glacial ice, fossil pollen, and tree-growth rings, as well as from historical documents (Figure 14–3). Scientists who analyze proxy data and reconstruct past climates are engaged in the study of **paleoclimatology.** The main goal of such work is to understand the climate of the past in order to assess the current and potential future climate in the context of natural climate variability. In the following discussion, we will briefly examine some of the important sources of proxy data.

Figure 14–3 Ancient bristlecone pines in California's White Mountains. The study of tree-growth rings is one way that scientists reconstruct past climates. Some of these trees are more than 4000 years old. (Photo by Bill Stevenson/Photolibrary)

Seafloor Sediment—A Storehouse of Climate Data

We know that the parts of the Earth system are linked so that a change in one part can produce changes in any or all of the other parts. In this section you will see how changes in atmospheric and oceanic temperatures are reflected in the nature of life in the sea.

Most seafloor sediments contain the remains of organisms that once lived near the sea surface (the ocean–atmosphere interface). When such near-surface organisms die, their shells slowly settle to the floor of the ocean, where they become part of the sedimentary record. These seafloor sediments are useful recorders of worldwide climate change because the numbers and types of organisms living near the sea surface change with the climate:

> We would expect that in any area of the ocean/atmosphere interface the average annual temperature of the surface water of the ocean would approximate that of the contiguous atmosphere. The temperature equilibrium established between surface seawater and the air above it should mean that... changes in climate should be reflected in changes in organisms living near the surface of the deep sea.... When we recall that the seafloor sediments in vast areas of the ocean consist mainly of shells of pelagic foraminifers, and that these animals are sensitive to variations in water temperature, the connection between such sediments and climatic change becomes obvious.*

Thus, in seeking to understand climate change, scientists have become increasingly interested in the huge reservoir of data concealed in seafloor sediments. Since the late 1960s the United States has been involved in major international projects. The pioneering program was the Deep Sea Drilling Project, with its research vessel the *Glomar Challenger,* that began in 1968. In 1983 the Deep Sea Drilling Project was replaced by the Ocean Drilling Program and a new drilling ship, the *JOIDES Resolution.* In October 2003 the Integrated Ocean Drilling Program began. This newest international effort does not rely on just one drilling ship but uses multiple vessels for exploration. One of the new additions is the massive 210-meter-long (nearly 700-foot-long) *Chikyu,* which began operations in 2007 (Figure 14–4). The primary objective of this program is to collect cores from sites where they were previously unobtainable. The new data will expand our understanding of many aspects of the Earth system, including climate-change patterns.

One notable example of the importance of seafloor sediments to our understanding of climate change relates to unraveling the fluctuating atmospheric conditions of the Ice Age. The records of temperature changes contained in cores of sediment from the ocean floor have proven critical to our present understanding of this recent span of Earth history.

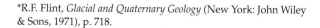

*R.F. Flint, *Glacial and Quaternary Geology* (New York: John Wiley & Sons, 1971), p. 718.

Figure 14–4 The *Chikyu* (meaning "Earth" in Japanese), the world's most advanced scientific drilling vessel. It can drill as deep as 7000 meters (nearly 23,000 feet) below the seabed in water as deep as 2500 meters (8200 feet). It is part of the Integrated Ocean Drilling Program. (AP Photo/Itsuo Inouye)

Oxygen-Isotope Analysis

Oxygen-isotope analysis is based on precise measurement of the ratio between two isotopes of oxygen: ^{16}O, which is the most common, and the heavier ^{18}O. A molecule of H_2O can form from either ^{16}O or ^{18}O. But the lighter isotope, ^{16}O, evaporates more readily from the oceans. Because of this, precipitation (and hence the glacial ice that it may form) is enriched in ^{16}O. This leaves a greater concentration of the heavier isotope, ^{18}O, in the ocean water. Thus, during periods when glaciers are extensive, more of the lighter ^{16}O is tied up in ice, so the concentration of ^{18}O in seawater increases. Conversely, during warmer interglacial periods when the amount of glacial ice decreases dramatically, more ^{16}O is returned to the sea, so the proportion of ^{18}O relative to ^{16}O in ocean water also drops. Now, if we had some ancient recording of the changes of the $^{18}O/^{16}O$ ratio, we could determine when there were glacial periods and therefore when the climate grew cooler.

Fortunately, we do have such a recording. As certain microorganisms secrete their shells of calcium carbonate ($CaCO_3$), the prevailing $^{18}O/^{16}O$ ratio is reflected in the composition of these hard parts. When the organisms die, their hard parts settle to the ocean floor, becoming part of the sediment layers there. Consequently, periods of glacial

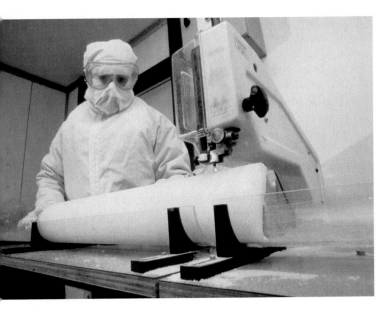

Figure 14–5 Scientist slicing an ice core sample from Antarctica for analysis. He is wearing protective clothing and a mask to minimize contamination of the sample. Chemical analysis of ice cores can provide important data about past climates. (Photo by British Antarctic Survey/Photo Researchers, Inc.)

activity can be determined from variations in the oxygen isotope ratio found in shells of certain microorganisms buried in deep-sea sediments.

The $^{18}O/^{16}O$ ratio also varies with temperature. Thus, more ^{18}O is evaporated from the oceans when temperatures are high, and less is evaporated when temperatures are low. Therefore, the heavy isotope is more abundant in the precipitation of warm eras and less abundant during colder periods. Using this principle, scientists studying the layers of ice and snow in glaciers have been able to produce a record of past temperature changes.

Climate Change Recorded in Glacial Ice

Ice cores are an indispensable source of data for reconstructing past climates (Figure 14–5). Research based on vertical cores taken from the Greenland and Antarctic ice sheets has changed our basic understanding of how the climate system works.

Scientists collect samples with a drilling rig, like a small version of an oil drill. A hollow shaft follows the drill head into the ice, and an ice core is extracted. In this way, cores that sometimes exceed 2000 meters (6500 feet) in length and may represent more than 200,000 years of climate history are acquired for study (Figure 14–6a).

The ice provides a detailed record of changing air temperatures and snowfall. Air bubbles trapped in the ice record variations in atmospheric composition. Changes in carbon dioxide and methane are linked to fluctuating temperatures. The cores also include atmospheric fallout such as wind-blown dust, volcanic ash, pollen, and modern-day pollution.

Past temperatures are determined by *oxygen-isotope analysis*. Using this technique, scientists are able to produce a record of past temperature changes. A portion of such a record is shown in Figure 14–6b.

Tree Rings—Archives of Environmental History

If you look at the end of a log, you will see that it is composed of a series of concentric rings. These *tree rings* become

(a)

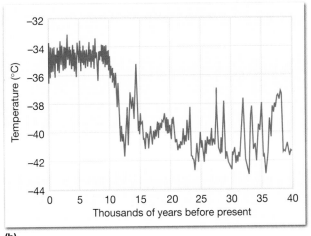

(b)

Figure 14–6 (a) The National Ice Core Laboratory is a physical plant for storing and studying cores of ice taken from glaciers around the world. These cores represent a long-term record of material deposited from the atmosphere. The lab provides scientists with the capability to conduct examinations of ice cores, and it preserves the integrity of these samples in a repository for the study of global climate change and past environmental conditions. (Photo by USGS/National Ice Core Laboratory) (b) This graph showing temperature variations over the past 40,000 years is derived from oxygen-isotope analysis of ice cores recovered from the Greenland ice sheet. (After U.S. Geological Survey)

(a)

(b)

Figure 14–7 (a) Each year a growing tree produces a layer of new cells beneath the bark. If the tree is felled and the trunk examined, each year's growth can be seen as a ring. Because the amount of growth (thickness of a ring) depends upon precipitation and temperature, tree rings are useful records of past climate. (Photo by Daniele Taurino/Dreamstime) (b) Scientists are not limited to working with trees that have been cut down. Small, nondestructive core samples can be taken from living trees. (Photo by Gregory K. Scott/Photo Researchers, Inc.)

larger in diameter outward from the center (Figure 14–7a). Every year in temperate regions, trees add a layer of new wood under the bark. Characteristics of each tree ring, such as size and density, reflect the environmental conditions (especially climate) that prevailed during the year when the ring formed. Favorable growth conditions produce a wide ring; unfavorable ones produce a narrow ring. Trees growing at the same time in the same region show similar tree-ring patterns.

Because a single growth ring is usually added each year, the age of the tree when it was cut can be determined by counting the rings. If the year of cutting is known, the age of the tree and the year in which each ring formed can be determined by counting back from the outside ring. The dating and study of annual rings in trees is called *dendrochronology*. Scientists are not limited to working with trees that have been cut down. Small, nondestructive core samples can be taken from living trees (Figure 14–7b).

To make the most effective use of tree rings, extended patterns known as *ring chronologies* are established. They are produced by comparing the patterns of rings among trees in an area. If the same pattern can be identified in two

samples, one of which has been dated, the second sample can be dated from the first by matching the ring patterns common to both. Tree-ring chronologies extending back for thousands of years have been established for some regions. To date a timber sample of unknown age, its ring pattern is matched against the reference chronology.

Tree-ring chronologies are unique archives of environmental history and have important applications in such disciplines as climate, geology, ecology, and archaeology. For example, tree rings are used to reconstruct climate variations within a region for spans of thousands of years prior to human historical records. Knowledge of such long-term variations is of great value in making judgments regarding the recent record of climate change.

Other Types of Proxy Data

Other sources of proxy data that are used to gain insight into past climates include fossil pollen, corals, and historical documents.

Fossil Pollen Climate is a major factor influencing the distribution of vegetation, so the nature of the plant community occupying an area is a reflection of the climate. Pollen and spores are parts of the life cycles of many plants, and because they have very resistant walls, they are often the most abundant, easily identifiable, and best-preserved plant remains in sediments (Figure 14–8). By analyzing pollen from accurately dated sediments, it is possible to obtain high-resolution records of vegetational changes in an area. From such information, past climates can be reconstructed.

Figure 14–8 This false-color image from an electron microscope shows an assortment of pollen grains. Note how the size, shape, and surface characteristics differ from one species to another. Analysis of the types and abundance of pollen in lake sediments and peat deposits provides information on how climate has changed over time. (Photo by David Scharf/Science Photo Library)

Corals Coral reefs consist of colonies of corals that live in warm, shallow waters and form atop the hard material left behind by past corals. Corals build their hard skeletons from calcium carbonate ($CaCO_3$) extracted from seawater. The carbonate contains isotopes of oxygen that can be used to determine the temperature of the water in which the coral grew. The portion of the skeleton that forms in winter has a density that is different from that formed in summer because of variations in growth rates related to temperature and other environmental factors. Thus, corals exhibit seasonal growth bands very much like those observed in trees. The accuracy and reliability of the climate data extracted from corals has been established by comparing recent instrumental records to coral records for the same period. Oxygen-isotope analysis of coral growth rings can also serve as a proxy measurement for precipitation, particularly in areas where large variations in annual rainfall occur.

Think of coral as a paleothermometer that enables us to answer important questions about climate variability in the world's oceans. The graph in Figure 14–9 is a 350-year sea-surface temperature record based on oxygen-isotope analysis of a core extracted from a reef in the Galapagos Islands.

Historical Data Historical documents sometimes contain helpful information. Although it might seem that such records should readily lend themselves to climate analysis, that is not the case. Most manuscripts were written for purposes other than climate description. Furthermore, writers understandably neglected periods of relatively stable atmospheric

Figure 14–10 Historical records can sometimes be helpful in the analysis of past climates. The date for the beginning of the grape harvest in the fall is an integrated measure of temperature and precipitation during the growing season. These dates have been recorded for centuries in Europe and provide a useful record of year-to-year climate variations. (Photo by SGM/agefotostock)

conditions and mention only droughts, severe storms, memorable blizzards, and other extremes. Nevertheless, records of crops, floods, and the migration of people have furnished useful evidence of the possible influences of changing climate (Figure 14–10).

Concept Check 14.2

1 What are proxy data, and why are they necessary in the study of climate change?

2 Why are seafloor sediments useful in the study of past climates?

3 Explain how past temperatures are determined using oxygen-isotope analysis.

4 Aside from seafloor sediments, list four sources of proxy climate data.

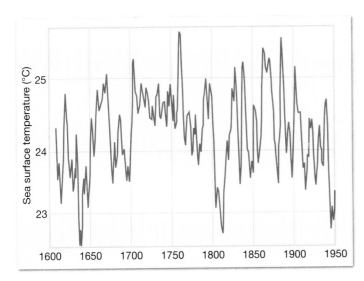

Figure 14–9 Coral colonies thrive in warm, shallow, tropical waters. The tiny invertebrates extract calcium carbonate from seawater to build hard parts. They live atop the solid foundation left by past coral. Chemical analysis of the changing composition of coral reefs with depth can provide useful data on past near-surface temperatures. This graph shows a 350-year record of sea-surface temperatures obtained by oxygen-isotope analysis of coral from the Galapagos Islands. (National Climatic Data Center)

Natural Causes of Climate Change

A great variety of hypotheses have been proposed to explain climate change. Several have gained wide support, only to lose it and then sometimes to regain it again. Some explanations are controversial. This is to be expected because planetary atmospheric processes are so large-scale and complex that they cannot be reproduced physically in laboratory experiments. Rather, climate and its changes must be simulated mathematically (modeled) using powerful computers.

In this section we examine several current hypotheses that have earned serious consideration from the scientific community. These describe "natural" mechanisms of climatic change, causes that are unrelated to human activities:

- Plate tectonics (rearranging Earth's continents, moving them closer or farther from the equator and the poles)
- Volcanic activity (changing the reflectivity and composition of the atmosphere)
- Variations in Earth's orbit (the natural, cyclic change in our planet's orbit, axial tilt, and wobble)
- Solar variability (whether the Sun varies in its radiation output and whether sunspots affect the output)

A later section examines human-made climate changes, including the effect of rising carbon dioxide levels caused primarily by our burning of fossil fuels.

As you read this section, you will find that more than one hypothesis may explain the same climate change. In fact, several mechanisms may interact to shift climate. Also, no single hypothesis can explain climate change on all time scales. A proposal that explains variations over millions of years generally cannot explain fluctuations over hundreds of years. If our atmosphere and its changes ever become fully understood, we will probably see that climate change is caused by many of the mechanisms discussed here, plus new ones yet to be proposed.

Plate Tectonics and Climate Change

Over the past several decades, a revolutionary idea has emerged from the science of geology: **plate-tectonics theory.** This theory has gained nearly universal acceptance in the scientific community. It states that the outer portion of Earth is made up of several vast rigid slabs, called *plates*, which move in relation to one another over a weak plastic rock layer below. They move incredibly slowly. The average rate at which plates move relative to each other is roughly the same rate at which human fingernails grow—perhaps a few centimeters a year.

Most of the largest plates include an entire continent plus a lot of seafloor. Thus, as plates ponderously shift, continents also change position. Not only does this theory allow geologists to understand and explain many processes and features of Earth's continents and oceans, it also provides climate scientists with a probable explanation for some hitherto unexplainable climate changes.

For example, evidence of extensive glacial activity in portions of present-day Africa, Australia, South America, and India indicate that these regions experienced an ice age about 250 million years ago. This finding puzzled scientists for many years. How could the climate in these presently warm latitudes once have been frigid, like Greenland and Antarctica?

Until the development of the plate-tectonics theory, no reasonable explanation existed. Today scientists realize that the areas containing these ancient glacial features were joined as a single "supercontinent" that was located toward the South Pole (Figure 14–11a). Later, as the plates spread apart, portions of the landmass, each moving on a different plate, slowly migrated toward their present locations. Thus, large fragments of glaciated terrain ended up in widely scattered subtropical locations (Figure 14–11b).

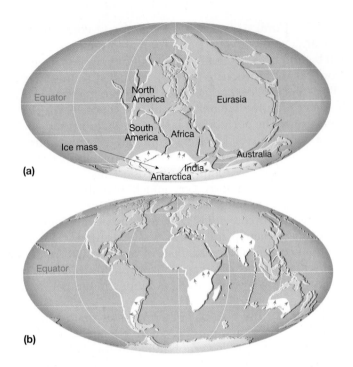

Figure 14–11 (a) The supercontinent Pangaea, showing the area covered by glacial ice 300 million years ago. (b) The continents as they are today. The white areas indicate places where evidence of former ice sheets exists.

It is now understood that during the geologic past, plate movements accounted for many other dramatic climate changes, as landmasses shifted in relation to one another and moved to different latitudes. Changes in oceanic circulation must also have occurred, altering the transport of heat and moisture and, hence, the climate as well.

Because the rate of plate movement is so slow, appreciable changes in the positions of the continents occur only over *great* spans of geologic time. Thus, climate changes brought about by plate movements are extremely gradual and happen on a scale of millions of years. As a result, the theory of plate tectonics is not useful for explaining climate variations that occur on shorter time scales, such as tens, hundreds, or thousands of years. Other explanations must be sought to explain these changes.

Volcanic Activity and Climate Change

The idea that explosive volcanic eruptions might alter Earth's climate was first proposed many years ago. It is still regarded as a plausible explanation for some aspects of climate variability. Explosive eruptions emit huge quantities of gases and fine-grained debris into the atmosphere (Figure 14–12). The greatest eruptions are sufficiently powerful to inject material high into the stratosphere, where it spreads around the globe and remains for many months or even years.

The basic premise is that the suspended volcanic material will filter out a portion of the incoming solar radiation, which in turn will lower temperatures in the troposphere.

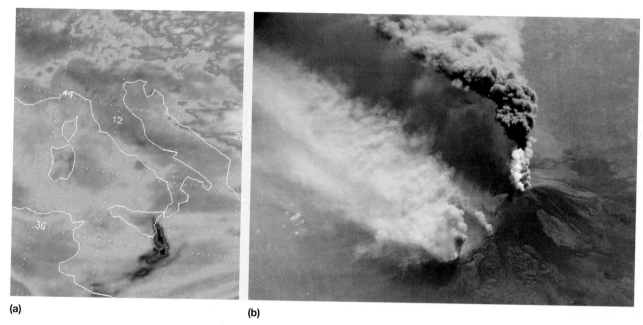

Figure 14–12 Mount Etna, a volcano on the island of Sicily, erupting in late October 2002. Mount Etna is Europe's largest and most active volcano. (a) This image from the Atmospheric Infrared Sounder on NASA's *Aqua* satellite shows the sulfur dioxide (SO_2) plume in shades of purple and black. Climate may be affected when large quantities of SO_2 are injected into the atmosphere. (b) This photo of Mount Etna looking southeast was taken by a member of the International Space Station. It shows a plume of volcanic ash streaming southeastward from the volcano. (Images courtesy of NASA)

More than 200 years ago Benjamin Franklin used this idea to argue that material from the eruption of a large Icelandic volcano could have reflected sunlight back to space and therefore might have been responsible for the unusually cold winter of 1783–1784.

Perhaps the most notable cool period linked to a volcanic event is the "year without a summer" that followed the 1815 eruption of Mount Tambora in Indonesia. The eruption of Tambora is the largest of modern times. During April 7–12, 1815, this nearly 4000-meter-high (13,000-foot-high) volcano violently expelled more than 100 cubic kilometers (24 cubic miles) of volcanic debris. The impact of the volcanic aerosols on climate is believed to have been widespread in the Northern Hemisphere. From May through September 1816 an unprecedented series of cold spells affected the northeastern United States and adjacent portions of Canada. There was heavy snow in June, and there was frost in July and August. Abnormal cold was also experienced in much of Western Europe.

Three major volcanic events have provided considerable data and insight regarding the impact of volcanoes on global temperatures. The eruptions of Washington State's Mount St. Helens in 1980, the Mexican volcano El Chichón in 1982, and Mount Pinatubo in the Philippines in 1991 gave scientists an opportunity to study the atmospheric effects of volcanic eruptions with the aid of more sophisticated technology than had been available in the past. Satellite images and remote-sensing instruments allowed scientists to closely monitor the effects of the clouds of gases and ash that these volcanoes emitted.

Mount St. Helens When Mount St. Helens erupted, there was immediate speculation about the possible effects on our climate. Could such an eruption cause our climate to change? There is no doubt that the large quantity of volcanic ash emitted by the explosive eruption had significant local and regional effects for a short period. Still, studies indicated that any longer-term lowering of hemispheric temperatures was negligible. The cooling was so slight—probably less than 0.1°C (0.2°F)—that it could not be distinguished from other natural temperature fluctuations.

El Chichón Two years of monitoring and studies following the 1982 El Chichón eruption indicated that its cooling effect on global mean temperature was greater than that of Mount St. Helens, on the order of 0.3° to 0.5°C (0.5° to 0.9°F). The eruption of El Chichón was *less explosive* than the Mount St. Helens blast, so why did it have a greater impact on global temperatures? The reason is that the material emitted by Mount St. Helens was largely fine ash that settled out in a relatively short time. El Chichón, on the other hand, emitted far greater quantities of sulfur dioxide gas (an estimated 40 times more) than Mount St. Helens. This gas combines with water vapor in the stratosphere to produce a dense cloud of tiny sulfuric-acid particles (Figure 14–13a). The particles, called *aerosols*, take several years to settle out completely. They lower the troposphere's mean temperature because they reflect solar radiation back into space (Figure 14–13b).

We now understand that volcanic clouds that remain in the stratosphere for a year or more are composed largely of

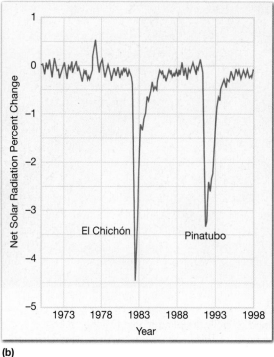

(a)

(b)

Figure 14–13 (a) This satellite image shows a plume of white haze from Anatahan Volcano, blanketing a portion of the Philippine Sea following a large eruption in April 2005. The haze is *not* volcanic ash. Rather, it consists of tiny droplets of sulfuric acid formed when sulfur dioxide from the volcano combined with water in the atmosphere. The haze is bright and reflects sunlight back to space. (NASA) (b) Net solar radiation at Hawaii's Mauna Loa Observatory relative to 1970 (zero on the graph). The eruptions of El Chichón and Mt. Pinatubo clearly caused temporary drops in solar radiation reaching the surface. (Earth System Research Laboratory/NOAA)

sulfuric-acid droplets and not of dust, as was once thought. Thus, the volume of fine debris emitted during an explosive event is not an accurate criterion for predicting the global atmospheric effects of an eruption.

Mount Pinatubo The Philippines volcano Mount Pinatubo erupted explosively in June 1991, injecting 25 million to 30 million tons of sulfur dioxide into the stratosphere. The event provided scientists with an opportunity to study the climatic impact of a major explosive volcanic eruption, using NASA's spaceborne Earth Radiation Budget Experiment. During the next year the haze of tiny aerosols increased albedo and lowered global temperatures by 0.5°C (0.9°F).

The Impact It may be true that the impact on global temperature of eruptions like El Chichón and Mount Pinatubo is relatively minor, but many scientists agree that the cooling produced could alter the general pattern of atmospheric circulation for a limited period. Such a change, in turn, could influence the weather in some regions. Predicting or even identifying specific regional effects still presents a considerable challenge to atmospheric scientists.

The preceding examples illustrate that the impact on climate of a single volcanic eruption, no matter how great, is relatively small and short-lived. The graph in Figure 14–13b reinforces this point. Therefore, if volcanism is to have a pronounced impact over an extended period, many great eruptions rich in sulfur dioxide and closely spaced in time, need to occur. If this happens, the stratosphere will be loaded with enough gases and volcanic dust to seriously diminish the amount of solar radiation reaching the surface. Because no such period of explosive volcanism is known to

have occurred in historic times, it is most often mentioned as a possible contributor to prehistoric climatic shifts. Another way in which volcanism may influence climate is described in Box 14–1.

Students Sometimes Ask...

Could a meteorite colliding with Earth cause the climate to change?

Yes, it is possible. For example, the most strongly supported hypothesis for the extinction of dinosaurs and many other organisms about 65 million years ago is related to such an event. When a large (about 10 kilometers in diameter) meteorite struck Earth, huge quantities of debris were blasted high into the atmosphere. For months the encircling dust cloud greatly restricted the amount of light reaching Earth's surface. Without sufficient sunlight for photosynthesis, delicate food chains collapsed. When the sunlight returned, more than half of the species on Earth, including the dinosaurs and many marine organisms, had become extinct.

Variations in Earth's Orbit

Geologic evidence indicates that the Ice Age that began about 3 million years ago was characterized by numerous glacial advances and retreats associated with periods of global cooling and warming. Today scientists understand that the climate oscillations that characterized the Ice Age are linked to variations in Earth's orbit. This hypothesis was first developed and strongly advocated by the Serbian scientist Milutin Milankovitch and is based on the premise that variations in incoming solar radiation are a principal factor controlling Earth's climate.

Milankovitch formulated a comprehensive mathematical model based on the following elements (Figure 14–14):

1. Variations in the shape (**eccentricity**) of Earth's orbit about the Sun

2. Changes in **obliquity**—that is, changes in the angle that the axis makes with the plane of Earth's orbit

3. The wobbling of Earth's axis, called **precession**

Using these factors, Milankovitch calculated variations in the receipt of solar energy and the corresponding surface

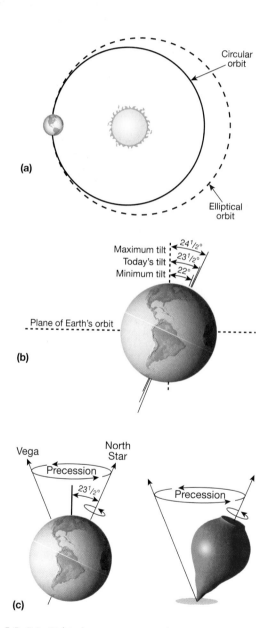

(a)

(b)

(c)

Figure 14–14 Orbital variations. (a) The shape of Earth's orbit changes during a cycle that spans about 100,000 years. It gradually changes from nearly circular to one that is more elliptical and then back again. This diagram greatly exaggerates the amount of change. (b) Today the axis of rotation is tilted about 23.5° to the plane of Earth's orbit. During a cycle of 41,000 years, this angle varies from 21.5° to 24.5°. (c) Precession. Earth's axis wobbles like that of a spinning top. Consequently, the axis points to different spots in the sky during a cycle of about 26,000 years.

temperature of Earth back into time, in an attempt to correlate these changes with the climate fluctuations of the Ice Age. It should be noted that these factors cause little or no variation in the *total* solar energy reaching the ground. Instead, their impact is felt because they change the degree of contrast between the seasons. Somewhat milder winters in the middle to high latitudes mean greater snowfall totals, whereas cooler summers bring a reduction in snowmelt.

Among the studies that added credibility to the foregoing hypothesis is one that examined deep-sea sediments.* Through oxygen-isotope analysis and statistical analyses of climatically sensitive microorganisms, the study established a chronology of temperature change going back nearly 500,000 years. This time scale of climate change was then compared to astronomical calculations of eccentricity, obliquity, and precession to determine whether a correlation did indeed exist.

Although the study was involved and mathematically complex, its conclusions were straightforward. The authors found that major variations in climate over the past several hundred thousand years were closely associated with changes in the geometry of Earth's orbit. Cycles of climate change were shown to correspond closely with the periods of obliquity, precession, and orbital eccentricity. More specifically, they stated: "It is concluded that changes in the earth's orbital geometry are the fundamental cause of the succession of Quaternary ice ages."**

Also, the study went on to predict the future trend of climate, toward a cooler climate and extensive glaciation in the Northern Hemisphere. But there are two qualifications: (1) The prediction applied only to the *natural* component of climate change and ignored any human influence, and (2) it was a forecast of *long-term trends* because it must be linked to factors that have periods of 20,000 years and longer. Thus, even if the prediction is correct, it contributes little to our understanding of climate changes over briefer periods of tens to hundreds of years because the cycles are too long for this purpose. Since the time of this study, subsequent research has supported its basic conclusions, namely:

> Orbital variations remain the most thoroughly examined mechanism of climatic change on time scales of tens of thousands of years and are by far the clearest case of a direct effect of changing insolation on the lower atmosphere of Earth.***

If orbital variations explain alternating glacial–interglacial periods, a question immediately arises: Why have glaciers been absent throughout most of Earth's history? Prior to the plate-tectonics theory, there was no widely accepted answer. Today we have a plausible answer. Because ice sheets can form only on the continents, landmasses must

---

*J. D. Hays, J. Imbrie, and N. J. Shackleton, "Variations in the Earth's Orbit: Pacemaker of the Ice Ages," *Science*, 194, no. 4270 (1976), 1121–1132.

**J. D. Hays, et al., p. 1131. The term *Quaternary* refers to the period on the geologic time scale that encompasses the last 2.6 million years.

***National Research Council, *Solar Variability, Weather, and Climate* (Washington, D.C.: National Academy Press, 1982), p. 7.

Box 14–1   Volcanism and Climate Change in the Geologic Past

The Cretaceous Period is the last period of the Mesozoic Era, the era of *middle life* that is often called the Age of Dinosaurs. It began about 145 million years ago and ended about 65 million years ago, with the extinction of the dinosaurs (and many other life-forms as well).

The Cretaceous climate was among the warmest in Earth's long history. Dinosaurs, which are associated with mild temperatures, ranged north of the Arctic Circle. Tropical forests existed in Greenland and Antarctica, and coral reefs grew as much as 15° latitude closer to the poles than at present. Deposits of peat that would eventually form widespread coal beds accumulated at high latitudes. Sea level was as much as 200 meters (650 feet) higher than today, indicating that there were no polar ice sheets.

What was the cause of the unusually warm climates of the Cretaceous Period? Among the significant factors that may have contributed was an enhancement of the greenhouse effect due to an increase in the amount of carbon dioxide in the atmosphere.

Where did the additional CO_2 that contributed to the Cretaceous warming come from? Many geologists suggest that the probable source was volcanic activity. Carbon dioxide is one of the gases emitted during volcanism, and there is now considerable geologic evidence that the Middle Cretaceous was a time when there was an unusually high rate of volcanic activity. Several huge oceanic lava plateaus were produced

on the floor of the western Pacific during this span. These vast features were associated with hot spots, zones where mobile plumes of material rise to the surface from deep in Earth's interior. Massive outpourings of lava over millions of years would have been accompanied by the release of huge quantities of CO_2, which in turn would have enhanced the atmospheric greenhouse effect. *Thus, the warmth that characterized the Cretaceous may have had is origins deep in Earth's interior.*

Other probable consequences of this extraordinarily warm period are linked to volcanic activity. For example, the high global temperatures and enriched atmospheric CO_2 in the Cretaceous led to increases in the quantity and types of phytoplankton (tiny, mostly microscopic plants, such as algae) and other life-forms in the ocean. The expansion in marine life is reflected in the widespread chalk deposits associated with the Cretaceous Period (Figure 14–A). Chalk consists of the calcium-carbonate ($CaCO_3$)–rich hard parts of microscopic marine organisms. Oil and gas originate from the alteration of biological remains (chiefly phytoplankton). Some of the world's most important oil and gas fields are in marine sediments of the Cretaceous, a consequence of the greater abundance of marine life during that warm time.

This list of possible consequences linked to the extraordinary period of volcanism during the Cretaceous Period is far from

FIGURE 14–A These famous chalk deposits, known as the White Cliffs of Dover, are associated with the expansion of marine life that occurred during the exceptional warmth of the Cretaceous Period. (Photo by Charles Bowman/ Photolibrary)

complete, yet it serves to illustrate the interrelationships among parts of the Earth system. Materials and processes that at first might seem to be completely unrelated turn out to be linked. Here you have seen how processes originating deep in Earth's interior are connected directly or indirectly to the atmosphere, the oceans, and the biosphere.

---

exist somewhere in the higher latitudes before an ice age can commence. Thus, it is likely that ice ages have occurred only when Earth's shifting crustal plates carried the continents from tropical latitudes to more poleward positions.

Solar Variability and Climate

Among the most persistent hypotheses of climate change have been those based on the idea that the Sun is a variable star and that its output of energy varies through time. The effect of such changes would seem direct and easily understood: Increases in solar output would cause the atmosphere to warm, and reductions would result in cooling. This notion is appealing because it can be used to explain climate change of any length or intensity. However, no major *long-term* variations in the total intensity of solar radiation have yet been measured outside the atmosphere. Such measurements were not even possible until satellite technology be-

came available. Now that it is possible, we will need many years of records before we begin to sense how variable (or invariable) energy from the Sun really is.

Several proposals for climate change, based on a variable Sun, relate to sunspot cycles. The most conspicuous and best-known features on the surface of the Sun are dark blemishes called **sunspots** (Figure 14–15). Sunspots are huge magnetic storms that extend from the Sun's surface deep into the interior. Moreover, these spots are associated with the Sun's ejection of huge masses of particles that, on reaching Earth's upper atmosphere, interact with gases there to produce auroral displays (see Figure 1–26, page 30).

Along with other solar activity, the numbers of sunspots seem to increase and decrease in a regular way, creating a cycle of about 11 years. The graph in Figure 14–16 shows the annual number of sunspots, beginning in the early 1700s. However, this pattern does not always occur. There have

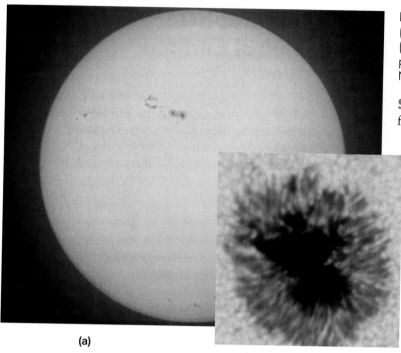

(a)

(b)

Figure 14–15 (a) Large sunspot group on the solar disk. (*Celestron 8* photo courtesy of Celestron International) (b) Sunspots having visible umbra (dark central area) and penumbra (lighter area surrounding umbra). (Courtesy of National Optical Astronomy Observatories)

Sunspots and Temperature Studies have identified prolonged periods when sunspots were absent or nearly so. Moreover, these spans correspond closely with cold periods in Europe and North America. Conversely, periods characterized by plentiful sunspots have correlated well with warmer times in these regions.

Referring to these matches, some scientists have suggested that such correlations make it appear that changes on the Sun are an important cause of climate change. But other scientists question this notion. Their hesitation stems in part from subsequent investigations using different climate records from around the world that failed to find a significant correlation between solar activity and temperature variations.

Sunspots and Drought A second possible Sun–climate connection, on a time scale different from the preceding example, relates to variations in precipitation rather than temperature. An extensive study of tree rings revealed a recurrent period of about 22 years in the pattern of droughts in the western United States. This periodicity coincides with the 22-year magnetic cycle of the Sun mentioned earlier.

Commenting on this possible connection, a panel of the National Research Council pointed out:

> No convincing mechanism that might connect so subtle a feature of the sun to drought patterns in limited regions has yet appeared. Moreover, the cyclic pattern of droughts found in tree rings is itself a subtle feature that shifts from place to place within the broad region of the study.*

been periods when the Sun was essentially free of sunspots. In addition to the well-known 11-year cycle, there is also a 22-year cycle. This longer cycle is based on the fact that the magnetic polarities of sunspot clusters reverse every successive 11 years.

Interest in possible Sun–climate effects has been sustained by an almost continuous effort to find correlations on time scales ranging from days to tens of thousands of years. Two widely debated examples are briefly described here.

Possible connections between solar variability and climate would be much easier to determine if researchers could identify physical linkages between the Sun and the lower atmosphere. But despite much research, no connection between solar variations and weather has yet been well established. Apparent correlations have almost always faltered when put to critical statistical examination or when tested with different data sets. As a result, the subject has been characterized by ongoing controversy and debate.

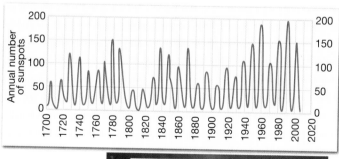

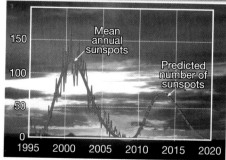

Figure 14–16 Mean annual sunspot numbers.

Concept Check 14.3

1 How does the theory of plate tectonics help us understand the cause of ice ages?

2 Describe and briefly explain the effect on global temperatures of the eruptions of El Chichón and Mount Pinatubo.

3 List and briefly describe the three variations in Earth's orbit that may cause global temperatures to vary.

4 List two examples of climate change linked to solar variability.

*National Research Council, p. 7.

EYE ON THE ATMOSPHERE

This volcanic eruption from Japan's Mount Kirishima began on January 26, 2011. In this view, the volcano is spewing ash 1500 meters (5000 feet) into the atmosphere. It was the largest eruption at this site since 1959. Scientists familiar with this volcanic area predicted that the eruption could last for a year or more. (Photo by Kyodo/AP Images)

Question 1 How might the volcanic ash from this eruption influence air temperatures?

Question 2 Would this effect likely be long lasting—perhaps for years? Explain.

Question 3 What "invisible" volcanic emission might have a greater effect than the volcanic ash?

Students Sometimes Ask...

Is there a connection between changes in the Sun's brightness and recent evidence of global warming?

Based on recent research using satellite data, it appears that the answer is no.* The scientists who carried out this detailed analysis state that the variations measured since 1978, which is as far back as these measurements go, are too small to have contributed appreciably to accelerated global warming over the past 30 years.

*P. Foukal, et al., "Variations in Solar Luminosity and Their Effect on the Earth's Climate," *Nature*, 443 (September 14, 2006), 161–166.

Human Impact on Global Climate

So far you have examined potential causes of climate change that are natural. In this section we discuss how humans contribute to global climate change. One impact largely results from the addition of carbon dioxide and other greenhouse gases to the atmosphere. A second impact is related to the addition of human-generated aerosols to the atmosphere.

Human influence on regional and global climate did not just begin with the onset of the modern industrial period. There is evidence that people have been modifying the environment over extensive areas for thousands of years. The use of fire and the overgrazing of marginal lands by domesticated animals have both reduced the abundance and distribution of vegetation. By altering ground cover, humans have modified such important climatological factors as surface albedo, evaporation rates, and surface winds. Commenting on this aspect of human-induced climate modification, the authors of one study observed: "In contrast to the prevailing view that only modern humans are able to alter climate, we believe it is more likely that the human species has made a substantial and continuing impact on climate since the invention of fire."**

Subsequently, these ideas were reinforced and expanded upon by a study that used data collected from Antarctic ice cores. This research suggested that humans may have started to have a significant impact on atmospheric composition and global temperatures thousands of years ago:

> Humans started slowly ratcheting up the thermostat as early as 8000 years ago, when they began clearing forests for agriculture, and 5000 years ago with the arrival of wet-rice cultivation. The greenhouse gases carbon dioxide and methane given off by these changes would have warmed the world.***

Concept Check 14.4

1 How might people have altered climate thousands of years ago?

Carbon Dioxide, Trace Gases, and Climate Change

In Chapter 1 you learned that carbon dioxide (CO_2) represents only about 0.039 percent of the gases that make up clean, dry air. Nevertheless, it is a very significant component meteorologically. Carbon dioxide is influential because it is

---

**C. Sagan, et al., "Anthropogenic Albedo Changes and the Earth's Climate," *Science*, 206, no. 4425 (1980), 1367.

***"An Early Start for Greenhouse Warming?" *Science*, 303 (January 16, 2004). This article is a report on a paper given by paleoclimatologist William Ruddiman at a meeting of the American Geophysical Union in December 2003.

transparent to incoming short-wavelength solar radiation, but it is not transparent to some of the longer-wavelength outgoing Earth radiation. A portion of the energy leaving the ground is absorbed by atmospheric CO_2. This energy is subsequently reemitted, part of it back toward the surface, thereby keeping the air near the ground warmer than it would be without CO_2.

Thus, along with water vapor, carbon dioxide is largely responsible for the *greenhouse effect* of the atmosphere. Carbon dioxide is an important heat absorber, and it follows logically that any change in the air's CO_2 content could alter temperatures in the lower atmosphere.

Rising CO_2 Levels

Earth's tremendous industrialization of the past two centuries has been fueled—and still is fueled—by burning fossil fuels: coal, natural gas, and petroleum (Figure 14–17). Combustion of these fuels has added great quantities of carbon dioxide to the atmosphere.

The use of coal and other fuels is the most prominent means by which humans add CO_2 to the atmosphere, but it is not the only way. The clearing of forests also contributes substantially because CO_2 is released as vegetation is burned or decays. Deforestation is particularly pronounced in the tropics, where vast tracts are cleared for ranching and agriculture or subjected to inefficient commercial logging operations. All major tropical forests—including those in South America, Africa, Southeast Asia, and Indonesia—are disappearing. According to United Nations estimates,

Figure 14–18 Clearing the tropical rain forest is a serious environmental issue. In addition to causing a loss of biodiversity, tropical deforestation is a significant source of carbon dioxide. Fires are frequently used to clear the land. This scene is in Brazil's Amazon basin. (Photo by Pete Oxford/Nature Picture Library)

nearly 10.2 million hectares (25.1 million acres) of tropical forest were permanently destroyed each year during the decade of the 1990s. Between the years 2000 and 2005, the average figure increased to 10.4 million hectares (25.7 million acres) per year. Figure 14–18 provides an example of deforestation in Brazil's Amazon basin.

Some of the excess CO_2 is taken up by plants or is dissolved in the ocean. It is estimated that 45 percent remains in the atmosphere. Figure 14–19 is a graphic record of changes in atmospheric CO_2 extending back 400,000 years. Over this long span, natural fluctuations varied from about 180 to 300 parts per million (ppm). As a result of human activities, the present CO_2 level is about 30 percent higher than its highest level over the past 600,000 years. The rapid increase in atmospheric CO_2 since the onset of industrialization is obvious. The annual rate at which atmospheric carbon dioxide concentrations are growing has been increasing over the past several decades.

The Atmosphere's Response

Given the increase in the atmosphere's carbon dioxide content, have global temperatures actually increased? The answer is yes. A 2007 report by the Intergovernmental Panel on Climate Change (IPCC) put it this way: "Warming of the climate system is unequivocal, as is now evident from observations of increases in global average air and ocean temperatures, widespread melting of snow and ice, and rising global sea level."[*] It is *very likely* that most of the observed increase in global average temperatures since the mid-twentieth century is due to the observed in-

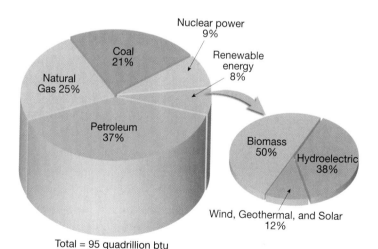

Total = 95 quadrillion btu

Figure 14–17 Paralleling the rapid growth of industrialization, which began in the nineteenth century, has been the combustion of fossil fuels, which has added great quantities of carbon dioxide to the atmosphere. The pie chart shows energy consumption in the United States for 2009. The total was 95 quadrillion Btu. A quadrillion is 10 raised to the 12th power, or a million million. A quadrillion Btu is a convenient unit for referring to U.S. energy use as a whole. Fossil fuels (petroleum, coal, and natural gas) represent about 83 percent of the total. (Data from U.S. Department of Energy)

[*]IPCC, "Summary for Policy Makers," in *Climate Change 2007: The Physical Science Basis* (New York: Cambridge University Press, 2007), p. 4.

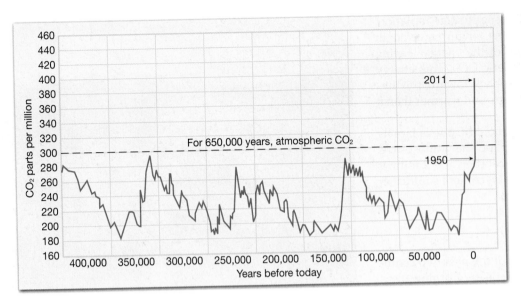

Figure 14-19 Carbon dioxide concentrations over the past 400,000 years. Most of these data come from the analysis of air bubbles trapped in ice cores. The record since 1958 comes from direct measurements of atmospheric CO_2 taken at Mauna Loa Observatory, Hawaii (see chapter opening photo). The rapid increase in CO_2 concentrations since the onset of the Industrial Revolution is obvious. (NOAA)

crease in human-generated greenhouse gas concentrations. As used by the IPCC, *very likely* indicates a probability of 90–99 percent. Global warming since the mid-1970s is now about 0.6°C (1°F), and total warming in the past century is about 0.8°C (1.4°F). The upward trend in surface temperatures is shown in Figure 14–20a. The world map in Figure 14–20b compares surface temperatures for 2010 to the base period (1951–1980). You can see that the greatest warming has been in the Arctic and neighboring high-latitude regions. Here are some related facts:

- When we consider the time span for which there are instrumental records (since 1850), 15 of the past 16 years (1995–2010) rank among the 16 warmest.
- Global mean temperature is now higher than at any other time in at least the past 500 to 1000 years.
- The average temperature of the global ocean has increased to depths of at least 3000 meters (10,000 feet).

Students Sometimes Ask...

Figure 14–17 shows biomass as a form of renewable energy. What exactly is biomass?

Biomass refers to organic matter that can be burned directly as fuel or converted into a different form and then burned. Biomass is a relatively new name for the oldest human fuels. Examples include firewood, charcoal, crop residues, and animal waste. Biomass burning is especially important in emerging economies.

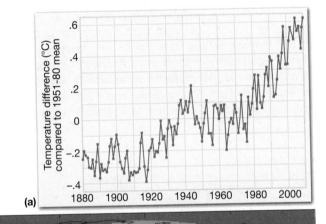

(a)

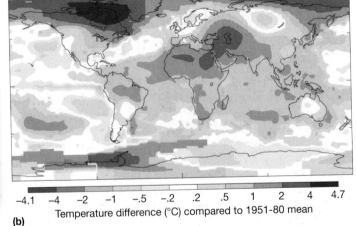

(b)

Figure 14-20 The year 2010 was tied with 2005 as the warmest year in the period of instrumental data. (a) The graph depicts global temperature change in °C since the year 1880. (b) The world map shows how temperatures in 2010 deviated from the mean for the 1951–1980 base period. The high latitudes in the Northern Hemisphere clearly stand out. (NASA/Goddard Institute for Space Studies)

Are these temperature trends caused by human activities, or would they have occurred anyway? The scientific consensus of the IPCC is that human activities "very likely" were responsible for most of the temperature increase since 1950.

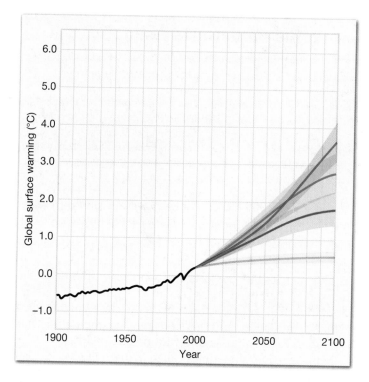

Figure 14–21 The left half of the graph (black line) shows global temperature changes for the twentieth century. The right half shows projected global warming in different emissions scenarios. The shaded zone adjacent to each colored line shows the uncertainty range for each scenario. The basis for comparison (0.0 on the vertical axis) is the global average for the period 1980–1999. The orange line represents the scenario in which carbon dioxide concentrations were held constant at the values for the year 2000. (From Intergovernmental Panel on Climate Change, *Climate Change 2007: Synthesis Report*, "Summary for Policymakers," Table SPM-5, p. 7.)

What about the future? Projections for the years ahead depend in part on the quantities of greenhouse gases that are emitted. Figure 14–21 shows the best estimates of global warming for several different scenarios. The 2007 IPCC report also states that if there is a doubling of the pre-industrial level of carbon dioxide (280 ppm) to 560 ppm, the "likely" temperature increase will be in the range of 2° to 4.5°C (3.5° to 8.1°F). The increase is "very unlikely" (1 to 10 percent probability) to be less than 1.5°C (2.7°F), and values higher than 4.5°C (8.1°F) cannot be excluded.

The Role of Trace Gases

Carbon dioxide is not the only gas contributing to a possible global increase in temperature. In recent years atmospheric scientists have come to realize that the industrial and agricultural activities of people are causing a buildup of several trace gases that may also play a significant role. The substances are called *trace gases* because their concentrations are so much smaller than that of carbon dioxide. The trace gases that are most important are methane (CH_4), nitrous oxide (N_2O), and chlorofluorocarbons (CFCs). These gases absorb wavelengths of outgoing radiation from Earth that would otherwise escape into space. Although individually

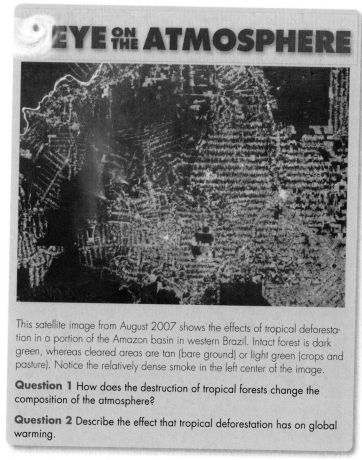

EYE ON THE ATMOSPHERE

This satellite image from August 2007 shows the effects of tropical deforestation in a portion of the Amazon basin in western Brazil. Intact forest is dark green, whereas cleared areas are tan (bare ground) or light green (crops and pasture). Notice the relatively dense smoke in the left center of the image.

Question 1 How does the destruction of tropical forests change the composition of the atmosphere?

Question 2 Describe the effect that tropical deforestation has on global warming.

their impact is modest, taken together these trace gases play a significant role in warming the troposphere.

Methane is present in much smaller amounts than CO_2, but its significance is greater than its relatively small concentration of about 1.7 ppm (parts per million) would indicate (Figure 14–22). The reason is that methane is about

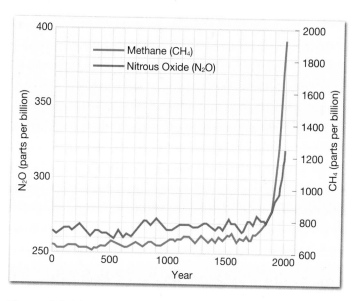

Figure 14–22 Increases in the concentrations of two trace gases, methane (CH_4) and nitrous oxide (N_2O), that contribute to global warming. The sharp rise during the industrial era is obvious. (U.S. Global Change Research Program)

20 times more effective than CO_2 at absorbing infrared radiation emitted by Earth.

Methane is produced by *anaerobic* bacteria in wet places where oxygen is scarce. (*Anaerobic* means "without air," specifically oxygen.) Such places include swamps, bogs, wetlands, and the guts of termites and grazing animals such as cattle and sheep. Methane is also generated in flooded paddy fields ("artificial swamps") used for growing rice (Figure 14–23). Mining of coal and drilling for oil and natural gas are other sources because methane is a product of their formation.

Figure 14–23 The mining of coal and drilling for oil and natural gas are sources of methane. Methane is also produced by anaerobic bacteria in wet places, where oxygen is scarce. (*Anaerobic* means "without air," specifically oxygen.) Such places include swamps, bogs, wetlands, and the guts of termites and grazing animals such as cattle and sheep. Methane is also generated in flooded paddy fields ("artificial swamps") used for growing rice. (Photo by Robert Harding Picture Library/SuperStock)

Students Sometimes Ask...

What is the Intergovernmental Panel on Climate Change?
Recognizing the problem of potential global climate change, the World Meteorological Organization and the United Nations' Environment Program established the *Intergovernmental Panel on Climate Change* (IPCC) in 1988. The IPCC assesses the scientific, technical, and socioeconomic information that is relevant to an understanding of human-induced climate change. This authoritative group provides advice to the world community through periodic reports that assess the state of knowledge of causes of climate change. More than 1250 authors and 2500 scientific reviewers from more than 130 countries contributed to the IPCC's most recent report, *Climate Change 2007: The Fourth Assessment Report.*

The increase in the concentration of methane in the atmosphere has been in step with the growth in human population. This relationship reflects the close link between methane formation and agriculture. As population has risen, so have the number of cattle and rice paddies.

Nitrous oxide, sometimes called "laughing gas," is also building in the atmosphere, although not as rapidly as methane. The increase is believed to result primarily from agricultural activity. When farmers use nitrogen fertilizers to boost crop yield, some of the nitrogen enters the air as nitrous oxide. This gas is also produced by high-temperature combustion of fossil fuels. Although the annual release into the atmosphere is small, the lifetime of a nitrous oxide molecule is about 150 years! If the use of nitrogen fertilizers and fossil fuels grows at projected rates, nitrous oxide may make a contribution to greenhouse warming that approaches half that of methane.

Unlike methane and nitrous oxide, chlorofluorocarbons (CFCs) are not naturally present in the atmosphere. As you learned in Chapter 1, CFCs are manufactured chemicals with many uses that have gained notoriety because they are responsible for ozone depletion in the stratosphere. The role of CFCs in global warming is less well known. CFCs are very effective greenhouse gases. They were not developed until the 1920s and were not used in great quantities until the 1950s, but they already contribute to the greenhouse effect at a level equal to methane. Although the Montreal Protocol represents strong corrective action, CFC levels will *not* drop rapidly. (See the section on the Montreal Protocol in Chapter 1.) CFCs remain in the atmosphere for decades, so even if all CFC emissions were to stop immediately, the atmosphere would not be free of them for many years.

Carbon dioxide is clearly the most important single cause for the projected global greenhouse warming. However, it is not the only contributor. When the effects of all human-generated greenhouse gases other than CO_2 are added together and projected into the future, their collective impact significantly increases the impact of CO_2 alone.

Sophisticated computer models show that the warming of the lower atmosphere caused by CO_2 and trace gases will not be the same everywhere. Rather, the temperature response in polar regions could be two to three times greater than the global average. One reason for this is that the polar troposphere is very stable, which suppresses vertical mixing and thus limits the amount of surface heat that is transferred upward. In addition, an expected reduction in sea ice would contribute to the greater temperature increase. This topic will be explored more fully in the next section.

Concept Check 14.5

1 Why has the CO_2 level of the atmosphere been increasing for nearly 200 years?

2 How are temperatures in the lower atmosphere likely to change as CO_2 levels continue to increase?

3 Aside from CO_2, what trace gases are contributing to global temperature change?

Students Sometimes Ask...

If Earth's atmosphere had no greenhouse gases, what would surface-air temperatures be like?

Cold! Earth's average surface temperature would be a chilly −18°C (−0.4°F) instead of the relatively comfortable 14.5°C (58°F) that it is today.

Climate-Feedback Mechanisms

Climate is a very complex interactive physical system. Thus, when any component of the climate system is altered, scientists must consider many possible outcomes. These possible outcomes are called **climate-feedback mechanisms.** They complicate climate-modeling efforts and add greater uncertainty to climate predictions.

Types of Feedback Mechanisms

What climate-feedback mechanisms are related to carbon dioxide and other greenhouse gases? One important mechanism is that warmer surface temperatures increase evaporation rates. This, in turn, increases the water vapor in the atmosphere. Remember that water vapor is an even more powerful absorber of radiation emitted by Earth than is carbon dioxide. Therefore, with more water vapor in the air, the temperature increase caused by carbon dioxide and trace gases is reinforced.

Figure 14–20b showed that high-latitude regions warmed more in 2010 than lower-latitude areas. This was not just the case in 2010; it has been true for many years. Scientists who model global climate change indicate that the temperature increase at high latitudes may be two to three times greater than the global average. This assumption is based in part on the likelihood that the area covered by sea ice will decrease as surface temperatures rise. Because ice reflects a much larger percentage of incoming solar radiation than does open water, the melting of sea ice replaces a highly reflecting surface with a relatively dark surface (Figure 14–24). The result is a substantial increase in the solar energy absorbed at the surface. This in turn feeds back to the atmosphere and magnifies the initial temperature increase created by higher levels of greenhouse gases.

So far the climate-feedback mechanisms discussed have magnified the temperature rise caused by the buildup of carbon dioxide. Because these effects reinforce the initial change, they are called **positive-feedback mechanisms.** However, other effects must be classified as **negative-feedback mechanisms** because they produce results that are just the opposite of the initial change and tend to offset it.

One probable result of a global temperature rise would be an accompanying increase in cloud cover due to the higher moisture content of the atmosphere. Most clouds are good reflectors of solar radiation. At the same time, however, they are also good absorbers and emitters of radiation emitted by Earth. Consequently, clouds produce two opposite effects. They are a negative-feedback mechanism because they increase albedo and thus diminish the amount of solar energy available to heat the atmosphere. On the other hand, clouds act as a positive-feedback mechanism by absorbing and emitting radiation that would otherwise be lost from the troposphere.

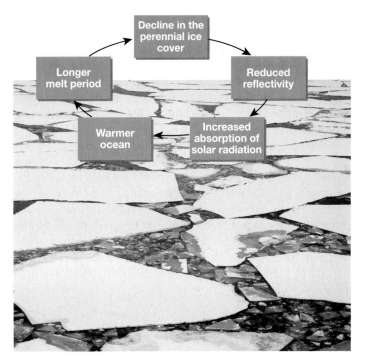

Figure 14–24 This image shows the springtime breakup of sea ice near Antarctica. The inset shows a likely feedback loop. A reduction in sea ice acts as a positive-feedback mechanism because surface albedo would decrease and the amount of energy absorbed at the surface would increase. (Photo by Radius Images/ Alamy)

Which effect, if either, is stronger? Scientists still are not sure whether clouds will produce a net positive or negative feedback. Although recent studies have not settled the question, they seem to lean toward the idea that clouds do not dampen global warming but rather produce a small positive feedback overall.*

The problem of global warming caused by human-induced changes in atmospheric composition continues to be one of the most studied aspects of climate change. Although no models yet incorporate the full range of potential factors and feedbacks, the scientific consensus is that the increasing levels of atmospheric carbon dioxide and trace gases will lead to a warmer planet with a different distribution of climate regimes.

Computer Models of Climate: Important yet Imperfect Tools

Earth's climate system is amazingly complex. Comprehensive state-of-the-science climate simulation models are among the basic tools used to develop possible climate-change scenarios. Called *general circulation models (GCMs),* they are based on fundamental laws of physics and chemistry and incorporate human and biological interactions. GCMs are used to simulate many variables, including temperature, rainfall, snow cover, soil moisture, winds, clouds, sea ice, and ocean circulation over the entire globe through the seasons and over spans of decades.

In many other fields of study, hypotheses can be tested by direct experimentation in a laboratory or by observations and measurements in the field. However, this is often not possible in the study of climate. Rather, scientists must construct computer models of how our planet's climate system works. If we understand the climate system correctly and construct the model appropriately, then the behavior of the model climate system should mimic the behavior of Earth's climate system (Figure 14–25).

What factors influence the accuracy of climate models? Clearly, mathematical models are *simplified* versions of the real Earth and cannot capture its full complexity, especially at smaller geographic scales. Moreover, when computer models are used to simulate future climate change, many assumptions have to be made that significantly influence the outcome. They must consider a wide range of possibilities for future changes in population, economic growth, consumption of fossil fuels, technological development, improvements in energy efficiency, and more.

Despite many obstacles, our ability to use supercomputers to simulate climate continues to improve. Although today's models are far from infallible, they are powerful tools for understanding what Earth's future climate might be like.

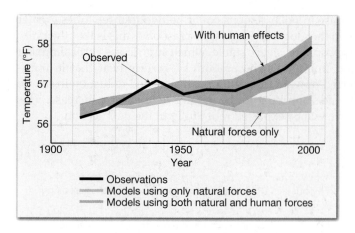

Figure 14–25 The blue band shows how global average temperatures would have changed due to natural forces only, as simulated by climate models. The red band shows model projections of the effects of human and natural forces combined. The black line shows actual observed global average temperatures. As the blue band indicates, without human influences, temperature over the past century would actually have first warmed and then cooled slightly over recent decades. Bands of color are used to express the range of uncertainty. (U.S. Global Change Research Program)

Concept Check 14.6

1 Distinguish between positive and negative climate-feedback mechanisms.

2 Provide at least one example of each type of feedback mechanism.

3 What factors influence the accuracy of computer models of climate?

How Aerosols Influence Climate

Increasing the levels of carbon dioxide and other greenhouse gases in the atmosphere is the most direct human influence on global climate. But it is not the only impact. Global climate is also affected by human activities that contribute to the atmosphere's aerosol content. Recall that *aerosols* are the tiny, often microscopic, liquid and solid particles that are suspended in the air. Unlike cloud droplets, aerosols are present even in relatively dry air. Atmospheric aerosols are composed of many different materials, including soil, smoke, sea salt, and sulfuric acid. Natural sources are numerous and include such phenomena as wildfires, dust storms, breaking waves, and volcanoes.

Most human-generated aerosols come from the sulfur dioxide emitted during the combustion of fossil fuels and from burning vegetation to clear agricultural land. Chemical reactions in the atmosphere convert the sulfur dioxide into sulfate aerosols, the same material that produces acid precipitation.

---

*A. E. Dessler, "A Determination of the Cloud Feedback from Climate Variations over the Past Decade," *Science,* 330 (December 10, 2010), 1523–1526.

How do aerosols affect climate? Most aerosols act directly by reflecting sunlight back to space and indirectly by making clouds "brighter" reflectors. The second effect relates to the fact that many aerosols (such as those composed of salt or sulfuric acid) attract water and thus are especially effective as cloud condensation nuclei. The large quantity of aerosols produced by human activities (especially industrial emissions) trigger an increase in the number of cloud droplets that form within a cloud. A greater number of small droplets increases the cloud's brightness causing more sunlight to be reflected back to space.

One category of aerosols, called *black carbon,* is soot generated by combustion processes and fires. Unlike other aerosols, black carbon warms the atmosphere because it is an effective absorber of incoming solar radiation. In addition, when deposited on snow and ice, black carbon reduces surface albedo, thus increasing the amount of light absorbed. Nevertheless, despite the warming effect of black carbon, the overall effect of atmospheric aerosols is to cool Earth.

Studies indicate that the cooling effect of human-generated aerosols offsets a portion of the global warming caused by the growing quantities of greenhouse gases in the atmosphere. The magnitude and extent of the cooling effect of aerosols is uncertain. This uncertainty is a hurdle in advancing our understanding of how humans alter Earth's climate.

It is important to point out some significant differences between global warming by greenhouse gases and aerosol cooling. After being emitted, greenhouse gases such as carbon dioxide remain in the atmosphere for many decades. By contrast, aerosols released into the troposphere remain there for only a few days or, at most, a few weeks before they are "washed out" by precipitation. Because of their short lifetime in the troposphere, aerosols are distributed unevenly over the globe. As expected, human-generated aerosols are concentrated near the areas that produce them, namely industrialized regions that burn fossil fuels and land areas where vegetation is burned (Figure 14–26).

Because their lifetime in the atmosphere is short, the effect of aerosols on today's climate is determined by the amount emitted during the preceding couple weeks. By contrast, the carbon dioxide and trace gases released into the atmosphere remain for much longer spans and thus influence climate for many decades.

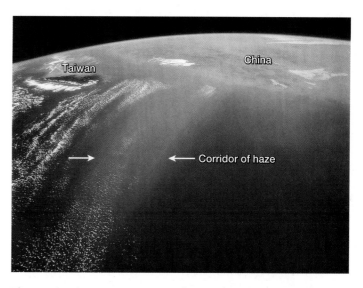

Figure 14–26 Human-generated aerosols are concentrated near the areas that produce them. Because aerosols reduce the amount of solar energy available to the climate system, they have a net cooling effect. This satellite image shows a dense blanket of pollution moving away from the coast of China. The plume is about 200 kilometers wide and more than 600 kilometers long. (NASA Image)

Some Possible Consequences of Global Warming

What consequences can be expected as the carbon dioxide content of the atmosphere reaches a level that is twice what it was early in the twentieth century? Because the climate system is so complex, predicting the occurrence of specific effects in particular places is speculative. It is not yet possible to pinpoint such changes. Nevertheless, there are plausible scenarios for larger scales of space and time.

As noted, the magnitude of the temperature increase will not be the same everywhere. The temperature rise will probably be smallest in the tropics and increase toward the poles. As for precipitation, the models indicate that some regions will experience significantly more precipitation and runoff. However, others will experience a decrease in runoff due to reduced precipitation or greater evaporation caused by higher temperatures.

Table 14–1 summarizes some of the most likely effects and their possible consequences. The table also provides the IPCC's estimate of the probability of each effect. Levels of confidence for these projections vary from *likely* (67 to 90 percent probability) to *very likely* (90 to 99 percent probability) to *virtually certain* (greater than 99 percent probability).

Sea-Level Rise

A significant impact of human-induced global warming is a rise in sea level. As this occurs, coastal cities, wetlands, and low-lying islands could be threatened with more frequent

Concept Check 14.7

1 What are the main sources of human-generated aerosols?

2 What effect does black carbon have on atmospheric temperatures?

3 What is the net effect of aerosols on temperatures in the troposphere?

4 How long do aerosols remain in the atmosphere before they are removed? How does this residence time compare to CO_2?

TABLE 14–1 Projected Changes and Effects of Global Warming in the Twenty-First Century

Projected Changes and Estimated Probability*	Examples of Projected Impacts
Higher maximum temperatures; more hot days and heat waves over nearly all land areas (*virtually certain*)	Increased incidence of death and serious illness in older age groups and urban poor Increased heat stress in livestock and wildlife Shift in tourist destinations Increased risk of damage to a number of crops Increased electric cooling demand and reduced energy supply reliability
Higher minimum temperatures; fewer cold days, frost days, and cold waves over nearly all land areas (*virtually certain*)	Decreased cold-related human morbidity and mortality Decreased risk of damage to a number of crops and increased risk to others Extended range and activity of some pest and disease vectors Reduced heating energy demand
Frequency of heavy precipitation events increasing over most areas (*very likely*)	Increased flood, landslide, avalanche, and mudflow damage Increased soil erosion Increased recharge of some floodplain aquifers due to increased runoff Increased pressure on government and private flood insurance systems and disaster relief
Increases in area affected by drought (*likely*)	Decreased crop yields Increased damage to building foundations caused by ground shrinkage Decreased water-resource quantity and quality Increased risk of wildfires
Increases in intense tropical cyclone activity (*likely*)	Increased risks to human life, risk of infectious-disease epidemics, and many other risks Increased coastal erosion and damage to coastal buildings and infrastructure Increased damage to coastal ecosystems, such as coral reefs and mangroves

Virtually certain indicates a probability greater than 99 percent, *very likely* indicates a probability of 90–99 percent, and *likely* indicates a probability of 67–90 percent. *Source:* Adapted from Intergovernmental Panel on Climate Change, *Climate Change 2007: Synthesis Report*, "Summary for Policy Makers." Table SPM-3, p. 13.

flooding, increased shoreline erosion, and saltwater encroachment into coastal rivers and aquifers.

Research indicates that sea level has risen between 10 and 25 centimeters (4 and 8 inches) over the past century and that the trend will continue at an accelerated rate. Some models indicate that the rise may approach or even exceed 50 centimeters (20 inches) by the end of the twenty-first century. Such a change may seem modest, but scientists realize that any rise in sea level along a *gently* sloping shoreline, such as the Atlantic and Gulf coasts of the United States, will lead to significant erosion and severe permanent inland flooding (Figure 14–27). If this happens, many beaches and wetlands will be eliminated, and coastal civilization will be severely disrupted.

How is a warmer atmosphere related to a rise in sea level? One significant factor is thermal expansion. Higher air temperatures warm the adjacent upper layers of the ocean, which in turn causes the water to expand and sea level to rise.

Perhaps a more easily visualized contributor to global sea-level rise is melting glaciers. With few exceptions, glaciers around the world have been retreating at unprecedented rates over the past century. Some mountain glaciers have disappeared altogether (Figure 14–28). A recent nearly 20-year satellite study showed that the mass of the Greenland and Antarctic ice sheets dropped an average of 475 gigatons per year. (A gigaton is 1 billion metric tons.) That is enough water to raise sea level 1.5 millimeters

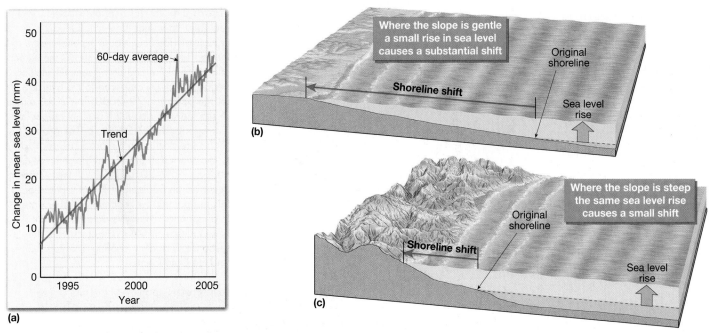

1941

2004

Figure 14–27 (a) Using data from satellites and floats, it was determined that sea level rose an average of 3 millimeters (0.1 inch) per year between 1993 and 2005. Researchers attributed about half the rise to melting glacial ice and the other half to thermal expansion. Rising sea level can affect some of the most densely populated areas on Earth. (NASA/Jet Propulsion Laboratory) (b) The slope of a shoreline is critical to determining the degree to which sea-level changes will affect it. When the slope is gentle, small changes in the sea level cause a substantial shift. (c) The same sea-level rise along a steep coast results in only a small shoreline shift.

Figure 14–28 Two images taken 63 years apart from the same spot in Alaska's Glacier Bay National Park. Muir Glacier, which is prominent in the 1941 photo, has retreated out of the field of view in the 2004 image. Also, Riggs Glacier (upper right) has thinned and retreated significantly. (Photos courtesy of National Snow and Ice Data Center)

(0.05 inch) per year. The loss of ice was not steady but was occurring at an accelerating rate during the study period. During the same span, mountain glaciers and ice caps lost an average of slightly more than 400 gigatons per year.

Because rising sea level is a gradual phenomenon, it may be overlooked by coastal residents as an important contributor to shoreline erosion problems. Rather, the blame may be assigned to other forces, especially storm activity. Although a given storm may be the immediate cause, the magnitude of its destruction may result from the relatively small sea-level rise that allowed the storm's power to cross a much greater land area.

Students Sometimes Ask...

What are scenarios, and why are they used?

A scenario is an example of what might happen under a particular set of assumptions. Using scenarios is a way of examining questions about an uncertain future. For example, future trends in fossil-fuel use and other human activities are uncertain. Therefore, scientists have developed a set of scenarios for how climate may change based on a wide range of possibilities for these variables.

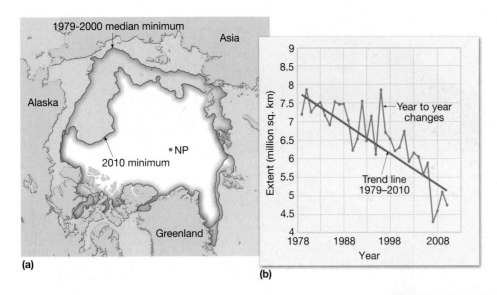

(a)

(b)

Figure 14–29 (a) Sea ice is frozen seawater. In winter the Arctic Ocean is completely ice covered. In summer, some of the sea ice melts. This map shows the extent of sea ice at the end of the summer melting period in 2010 compared to the average for the period 1979–2000. The 2010 extent of sea ice was about 31 percent below the average for 1979–2000. (b) The graph shows the decline in the area covered by Arctic sea ice at the end of the summer melt period, 1979–2010. In 2010, the minimum occurred on September 19. In other years this point might be reached on other dates in September. After this date, the extent of sea ice begins its cycle of growth in response to autumn cooling. (National Snow and Ice Data Center)

The Changing Arctic

A 2005 study of climate change in the Arctic began with the following statement:

> For nearly 30 years, Arctic sea ice extent and thickness have been falling dramatically. Permafrost temperatures are rising and coverage is decreasing. Mountain glaciers and the Greenland ice sheet are shrinking. Evidence suggests we are witnessing the early stage of an anthropogenically induced global warming superimposed on natural cycles, reinforced by reductions in Arctic ice.*

Arctic Sea Ice Climate models are in general agreement that one of the strongest signals of global warming should be a loss of sea ice in the Arctic. This is indeed occurring. The map in Figure 14–29a compares the average sea ice extent for September 2010 to the long-term average for the period 1979–2000. September represents the end of the melt period, when the area covered by sea ice is at a minimum. The September 2010 Arctic sea ice extent was the third lowest in the satellite record (since 1979). The trend is also clear when you examine the graph in Figure 14–29b. Is it possible that this trend may be part of a natural cycle? Yes, but it is more likely that the sea-ice decline represents a combination of natural variability and human-induced global warming, with the latter being increasingly evident in coming decades. As was noted in the section "Climate-Feedback Mechanisms," a reduction in sea ice represents a positive-feedback mechanism that reinforces global warming.

Permafrost During the past decade, evidence has mounted to indicate that the extent of permafrost in the Northern Hemisphere has decreased, as would be expected under long-term warming conditions. Figure 14–30 presents one example of this decline.

*J. T. Overpeck, et al., "Arctic System on Trajectory to New, Seasonally Ice-Free States," *EOS, Transactions, American Geophysical Union*, 86, no. 34 (August 23, 2005), 309.

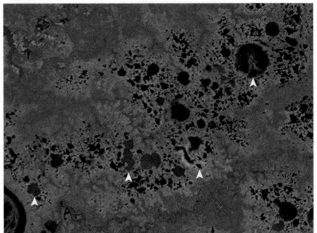

(a) June 27, 1973

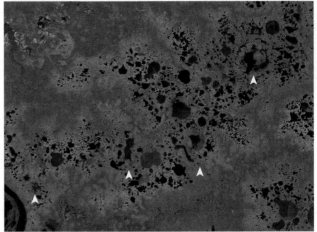

(b) July 2, 2002

Figure 14–30 This false-color image pair shows lakes dotting the tundra in northern Siberia in 1973 and 2002. The tundra vegetation is colored a faded red, whereas lakes appear blue or blue-green. Many lakes disappeared or shrunk considerably between 1973 and 2002. After studying satellite imagery of about 10,000 large lakes in a 500,000-square-kilometer area in northern Siberia, scientists documented an 11 percent decline in the number of lakes, at least 125 of which disappeared completely. (NASA)

EYE ON THE ATMOSPHERE

This ice breaker is plowing through sea ice in the Arctic Ocean. Sea ice is frozen seawater. The area covered by sea ice expands during the winter and shrinks in the summer, reaching a minimum sometime in September. (Photo by Wolfgang Bechtold/ Photolibrary)

Question 1 What parts (spheres) of the climate system are represented in this photograph?

Question 2 How has the summer sea ice minimum in the Arctic Ocean been changing since 1979 (see Figure 14–29)?

Question 3 How does the trend you identified in Question 2 relate to climate change in the Arctic?

In the Arctic, short summers thaw only the top layer of frozen ground. The permafrost beneath this *active layer* is like the cement bottom of a swimming pool. In summer, water cannot percolate downward, so it saturates the soil above the permafrost and collects on the surface in thousands of lakes. However, as Arctic temperatures climb, the bottom of the "pool" seems to be "cracking." Satellite imagery shows that over a 20-year span, a significant number of lakes have shrunk or disappeared altogether. As the permafrost thaws, lake water drains deeper into the ground.

Thawing permafrost represents a potentially significant positive-feedback mechanism that may reinforce global warming. When vegetation dies in the Arctic, cold temperatures inhibit its total decomposition. As a consequence, over thousands of years a great deal of organic matter has become stored in the permafrost. When the permafrost thaws, organic matter that may have been frozen for millennia comes out of "cold storage" and decomposes. The result is the release of carbon dioxide and methane—greenhouse gases that contribute to global warming.

Increasing Ocean Acidity

The human-induced increase in the amount of carbon dioxide in the atmosphere has some serious implications for ocean chemistry and marine life. Nearly half of the human-generated carbon dioxide ends up dissolved in the oceans. When carbon dioxide (CO_2) from the atmosphere dissolves in seawater (H_2O), it forms carbonic acid (H_2CO_3), which lowers the ocean's pH and changes the balance of certain chemicals found naturally in seawater.* In fact, the oceans have already absorbed enough carbon dioxide for surface waters to have experienced a pH decrease of 0.1 since preindustrial times, with an additional pH decrease likely in the future. Moreover, if the current trend in carbon dioxide emissions continues, by the year 2100 the ocean will have experienced a pH decrease of at least 0.3, which represents a change in ocean chemistry that has not occurred for millions of years. This shift toward acidity and the changes in ocean chemistry that result make it more difficult for certain marine creatures to build hard parts out of calcium carbonate. The decline in pH thus threatens a variety of calcite-secreting organisms as diverse as microbes and corals. This concerns marine scientists because of the potential consequences for other sea life that depend on the health and availability of these organisms.

The Potential for "Surprises"

You have seen that climate in the twenty-first century, unlike the preceding thousand years, is not expected to be stable. Rather, a change is very likely. The amount and rate of future climate change depends primarily on current and future human-caused emissions of heat-trapping gases and airborne particles. Many of the changes will probably be gradual envi-

ronmental shifts, imperceptible from year to year. Nevertheless, the effects, accumulated over decades, will have powerful economic, social, and political consequences.

Despite our best efforts to understand future climate shifts, there is also the potential for "surprises." This simply means that due to the complexity of Earth's climate system, we might experience relatively sudden, unexpected changes or see some aspects of climate shift in an unexpected manner. The report *Climate Change Impacts on the United States* describes the situation like this:

> Surprises challenge humans' ability to adapt, because of how quickly and unexpectedly they occur. For example, what if the Pacific Ocean warms in such a way that El Niño events become much more extreme? This could reduce the frequency, but perhaps not the strength, of hurricanes along the East Coast, while on the West Coast, more severe winter storms, extreme precipitation events, and damaging winds could become common. What if large quantities of methane, a potent greenhouse gas currently frozen in icy Arctic tundra and sediments, began to be released to the atmosphere by warming, potentially creating an amplifying "feedback loop" that would cause even more warming? We simply do not know how far the climate system or other systems it affects can be pushed before they respond in unexpected ways.
>
> There are many examples of potential surprises, each of which would have large consequences. Most of these potential outcomes are rarely reported, in this study or elsewhere. Even if the chance of any particular surprise happening is small, the chance that at least one such surprise will occur is much greater. In other words, while we can't know which of these events will occur, it is likely that one or more will eventually occur.**

The impact on climate of an increase in atmospheric CO_2 and trace gases is obscured by some uncertainties. Yet climate scientists continue to improve our understanding of the climate system and the potential impacts and effects of global climate change. Policymakers are confronted with responding to the risks posed by emissions of greenhouse gases, knowing that our understanding is imperfect. However, they are also faced with the fact that climate-induced environmental changes cannot be reversed quickly, if at all, due to the lengthy time scales associated with the climate system.

Concept Check 14.8

1 List and describe the factors that are causing sea level to rise.

2 Is global warming greater near the equator or near the poles? Explain.

3 Based on Table 14–1, what projected changes relate to something other than temperature?

---

*See Figure 13–18, page 372, to review the pH scale.

**National Assessment Synthesis Team, *Climate Change Impacts on the United States: The Potential Consequences of Climate Variability and Change* (Washington, D.C.: U.S. Global Change Research Program, 2000), p. 19.

Michael Mann: Climate Change Scientist

In the last few decades of the twentieth century, evidence of global warming was growing too obvious to ignore. Vast ice shelves in the Antarctic were breaking into floating islands, glaciers worldwide were in full retreat, and each turn of the calendar seemed to bring a new record for the hottest year yet measured. But whether this warming was due to humans or nature, no one was sure. Researchers needed to study a longer history of Earth's climate than just the twentieth century.

Michael Mann, of Penn State University, knew where to find the answers. Ice cores, sediments, coral skeletons, and tree rings all contain natural records of past temperature fluctuations. However, Mann, a meteorology professor, points out that "it's a fuzzy record, not as precise as thermometer readings." For example, while individual annual layers of ice in ice cores provide a direct reading of ancient atmospheric conditions, deciphering which layers belonged to which years can be difficult. And snow might have fallen only in winter in some areas, versus year round in others, making it difficult to compare cores from different areas.

These variables can make interpreting such proxy data a complex affair. Even so, with the help of modern statistical methods and climate models, Mann was able to piece together a reconstruction of the average temperature of the Northern Hemisphere going back 1000 years. Mann was most interested in the surface temperature patterns he uncovered. But colleagues suggested that he also plot out the average temperature of the Northern Hemisphere.

> It was that single result—the item we thought was least interesting in our work—that got the most attention.

"It was that single result—the item we thought was least interesting in our work—that got the most attention. It spoke to the question of how anomalous recent warming was," Mann says. The graph revealed an alarming trend. Temperatures stayed relatively warm during medieval times, cooled after that, but zoomed sharply higher starting around 1900—the same period when humans began burning large quantities of fossil fuels.

> I realized I had a responsibility to use that spotlight to educate the public and policymakers.

Known as the "hockey stick graph" for its distinctive shape, Mann's 1998 paper caused a sensation. Such vivid evidence of human contributions to global warming helped catapult both Mann and the issue of climate change onto the public stage. Meanwhile, some commentators encouraged their peers to "break the hockey stick" and discredit the idea of a warming Earth. The attacks convinced Mann to spread the word about climate change research. "I realized I had a responsibility to use that spotlight to educate the public and policymakers, to try to do some good with that opportunity," he says.

In 2001, Mann's research became a centerpiece of the *Third Report of the Intergovernmental Panel on Climate Change (IPCC)*. These reports, written by scientists, are summaries of the most current and important climate research. Mann served as a lead author of the *Third Report* and received a singular honor for his labors: For their work to raise awareness of the threat of human-caused climate change, the IPCC and former Vice President Al Gore were awarded the 2007 Nobel Peace Prize.

Mann continues to spread the word about global warming today. In addition to his scientific research, he has helped found the website www.RealClimate.org, featuring commentary on climate news by climate researchers. Mann also co-authored the book *Dire Predictions*, which explains the complexities of climate change science to the public.

Mann says that humans can still steer a course that could evade the worst consequences of a warming globe. "I like to think that together, those of us who are engaging the public on this issue are making some positive steps forward," he says. His main message is that "we hold the future in our own hands."

DR. MICHAEL MANN is a member of the Penn State University faculty, holding joint appointments in the Departments of Meteorology and Geosciences, as well as the Earth and Environmental Systems Institute. He is also director of the Penn State Earth System Science Center. (Photo courtesy of Jon Golden)

Give It Some Thought

1. The five major parts (spheres) of the climate system are represented in Figure 14–1. List an example of each that is apparent in the image.

2. Refer to Figure 14–2, which illustrates various components of Earth's climate system. Boxes represent interactions or changes that occur in the climate system. Select three boxes and provide an example of an interaction or change associated with each. Explain how these interactions may influence temperature.

3. Describe one way in which changes in the biosphere can cause changes in the climate system. Next, suggest one way in which the biosphere is affected by changes in some other part of the climate system. Finally, indicate one way in which the biosphere records changes in the climate system.

4. Recent volcanic events, such as the eruptions of El Chichón and Mount Pinatubo, were associated with a drop in global temperatures. During the Cretaceous Period, volcanic activity was associated with global warming. Explain this apparent paradox.

5. The accompanying photo is a 2005 view of Athabasca Glacier in the Canadian Rockies. A line of boulders in the foreground marks the outer limit of the glacier in 1992. Is the behavior of Athabasca Glacier shown in this image typical of other glaciers around the world? Describe a significant impact of such behavior.

(Photo by Hughrocks)

6. It has been suggested that global warming over the past 40 years might have been even greater were it not for the effect of certain types of air pollution. Explain how this could be true.

7. During a conversation, an acquaintance indicates that he is skeptical about global warming. When you ask him why he feels that way, he says, "The past couple of years in this area have been among the coolest I can remember." While you assure this person that it is useful to question scientific findings, you suggest to him that his reasoning in this case may be flawed. Use your understanding of the definition of *climate* along with one or more graphs in the chapter to persuade this person to reevaluate his reasoning.

THE CHANGING CLIMATE IN REVIEW

- The *climate system* includes the atmosphere, hydrosphere, solid Earth, biosphere, and cryosphere (the ice and snow that exists at Earth's surface). The system involves the exchanges of energy and moisture that occur among the five spheres.

- Techniques for analyzing Earth's climate history on a scale of hundreds to thousands of years include evidence from *seafloor sediments* and *oxygen-isotope analysis*. Seafloor sediments are useful recorders of worldwide climate change because the numbers and types of organic remains included in the sediment are indicative of past sea-surface temperatures. Using oxygen-isotope analysis, scientists can use the $^{18}O/^{16}O$ ratio found in the shells of microorganisms in sediment and layers of ice and snow to detect past temperatures. Other sources of data used for the study of past climates include the growth rings of trees, pollen contained in sediments, coral reefs, and information contained in historical documents.

- Several explanations have been formulated to explain climate change. Current hypotheses for the "natural" mechanisms (causes unrelated to human activities) of climate change include (1) plate tectonics, rearranging Earth's continents closer to or farther from the equator; (2) volcanic activity, reducing the solar radiation that reaches the surface; (3) variations in Earth's orbit, involving changes in the shape of the orbit (*eccentricity*), the angle that Earth's axis makes with the plane of its orbit (*obliquity*), and/or the wobbling of the axis (*precession*); and (4) changes in the Sun's output associated with *sunspots*.

- Humans have been modifying the environment for thousands of years. By altering ground cover with the use of fire and the overgrazing of land, people have modified important climatological factors such as surface albedo, evaporation rates, and surface winds. Along with water vapor, carbon dioxide is largely responsible for the *greenhouse effect* of the atmosphere. Therefore, by adding carbon dioxide and other trace gases (methane, nitrous oxide, and chlorofluorocarbons) to the atmosphere, humans are likely contributing to global warming in a significant way.

- When any component of the climate system is altered, scientists must consider the many possible outcomes, called *climate-feedback mechanisms*. Changes that reinforce the initial change are called *positive-feedback mechanisms*. On the other hand, *negative-feedback mechanisms* produce results that are the opposite of the initial change and tend to offset it.

- Global climate is affected by human activities that contribute to the atmosphere's *aerosol* (tiny liquid and solid particles that are suspended in air) content. Black carbon aerosols absorb incoming solar radiation and warm the atmosphere. However, most aerosols reflect sunlight back to space. Overall, aerosols have a cooling effect. The effect of aerosols on today's climate is determined by the amount emitted during the preceding couple weeks, while carbon dioxide remains for much longer spans and influences climate for many decades.

- Because the climate system is complex, predicting specific regional changes that may occur as a result of increased levels of carbon dioxide in the atmosphere is speculative. However, some possible consequences of greenhouse warming include (1) altering the distribution of the world's water resources, (2) a significant rise in sea level, (3) a greater intensity of tropical cyclones, and (4) changes in the extent of Arctic sea ice and permafrost.

- Due to the complexity of the climate system, not all future shifts can be foreseen. Thus, "surprises" (relatively sudden unexpected changes in climate) are possible.

VOCABULARY REVIEW

climate system (p. 380)
climate-feedback mechanism (p. 397)
eccentricity (p. 389)
negative-feedback mechanism (p.397)

obliquity (p. 389)
oxygen-isotope analysis (p. 382)
paleoclimatology (p. 381)
plate-tectonics theory (p. 386)

positive-feedback mechanism (p. 397)
precession (p. 389)
proxy data (p. 381)
sunspot (p. 390)

MyMeteorologyLab™

Log in to www.mymeteorologylab.com for animations, videos, MapMaster interactive maps, GEODe media, *In the News* RSS feeds, web links, glossary flashcards, self-study quizzes and a Pearson eText version of this book to enhance your study of *The Changing Climate.*

World Climates

The varied nature of Earth's surface (oceans, mountains, plains, ice sheets) and the many interactions that occur among atmospheric processes give every location on our planet a distinctive (sometimes unique) climate. However, we cannot describe the climatic character of countless locales; that would require many volumes. Our purpose is to introduce you to the *major climate regions* of the world. We will examine large areas and zoom in on particular places to illustrate the characteristics of these major climate regions. In addition, for regions that are probably unfamiliar to you (the tropical, desert, and polar realms), we briefly describe the natural landscape.

A British-American research station in Antarctica. Antarctica has an ice cap climate--the most extreme cold on the planet. (Photo by David Vaughan/ Photo Researchers, Inc.)

Focus On Concepts

After completing this chapter, you should be able to:

- Explain why classification is a necessary process when studying world climates.

- Discuss the criteria used in the Köppen system of climate classification and the nature of climate boundaries.

- List and briefly discuss the major controls of climate.

- Summarize the characteristics associated with the five major climate groups in the Köppen classification.

- Identify the subtypes in each of the five major climate groups.

- Analyze and classify climate data.

- Explain the distribution of climates as shown on a world map in terms of the major controls involved.

In Chapter 1 we mentioned the common misconception that climate is only "the average state of the atmosphere." Although averages are certainly important to climate descriptions, variations and extremes must also be included to accurately portray the character of an area.

Temperature and precipitation are the most important elements in climate descriptions because they have the greatest influence on people and their activities and also have an important impact on the distribution of vegetation and the development of soils. Nevertheless, other factors are also important for a complete climate description. When possible, some of these factors are introduced into our discussion of world climates.

Climate Classification

The worldwide distribution of temperature, precipitation, pressure, and wind is, to say the least, complex. Because of the many differences from place to place and time to time, it is unlikely that any two places that are more than a very short distance apart can experience identical weather. The virtually infinite variety of places on Earth makes it apparent that the number of different climates must be extremely large. Having such a diversity of information to investigate is not unique to the study of the atmosphere. It is a problem basic to all science. (Consider astronomy, which deals with billions of stars, and biology, which studies millions of complex organisms.) To cope with such variety, we must devise some means of *classifying* the vast array of data to be studied. By establishing groups of items that have common characteristics, order and manageability are introduced. Bringing order to large quantities of information not only aids comprehension and understanding but also facilitates analysis and explanation.

One of the first attempts at climate classification was made by the ancient Greeks, who divided each hemisphere into three zones: *torrid, temperate,* and *frigid* (Figure 15–1). The basis of this simple scheme was Earth–Sun relationships. The boundaries were the four astronomically important parallels of latitude: the Tropic of Cancer (23.5° north), the Tropic of Capricorn (23.5° south), the Arctic Circle (66.5° north), and the Antarctic Circle (66.5° south). Thus, the globe was divided into winterless climates and summerless climates and an intermediate type that had features of the other two.

Few other attempts were made until the beginning of the twentieth century. Since then, many climate-classification schemes have been devised. Remember that the classification of climates (or of anything else) is not a natural phenomenon but the product of human ingenuity. The value of any particular classification system is determined largely by its *intended use.* A system designed for one purpose may not work well for another.

In this chapter we use a classification devised by the German climatologist Wladimir Köppen (1846–1940). As a tool for presenting the general world pattern of climates, the **Köppen classification** has been the best-known and most-used system for decades. It is widely accepted for

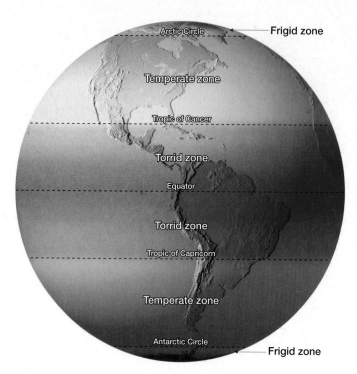

Figure 15–1 Probably the first attempt at climate classification was made by the ancient Greeks. They divided each hemisphere into three zones. The winterless *torrid* zone was separated from the summerless *frigid* zone by the *temperate* zone, which had features of the other two.

many reasons. For one, it uses easily obtained data: mean monthly and annual values of temperature and precipitation. Furthermore, the criteria are unambiguous, relatively simple to apply, and divide the world into climate regions in a realistic way.

Köppen believed that the distribution of natural vegetation was the best expression of overall climate. Consequently, the boundaries he chose were largely based on the limits of certain plant associations. He recognized five principal climate groups, each designated by a capital letter:

A *Humid tropical.* Winterless climates, with all months having a mean temperature above 18°C (64°F).

B *Dry.* Climates where evaporation exceeds precipitation, and there is a constant water deficiency.

C *Humid middle-latitude, mild winters.* The average temperature of the coldest month is below 18°C (64°F) but above −3°C (27°F).

D *Humid middle-latitude, severe winters.* The average temperature of the coldest month is below −3°C (27°F), and the warmest monthly mean exceeds 10°C (50°F).

E *Polar.* Summerless climates, where the average temperature of the warmest month is below 10°C (50°F).

Notice that four of these major groups (A, C, D, and E) are defined on the basis of *temperature*. The fifth, the B group, has *precipitation* as its primary criterion.

Each of the five groups is further subdivided by using the criteria and symbols presented in Figure 15–2.

Letter Symbol 1st	2nd	3rd		
A			Average temperature of the coldest month is 18°C or higher.	
	f		Every month has 6 cm of precipitation or more.	
	m		Short dry season; precipitation in driest month less than 6 cm but equal to or greater than 10 − R/25 (R is annual rainfall in cm).	
	w		Well-defined winter dry season; precipitation in driest month less than 10 − R/25.	
	s		Well-defined summer dry season (rare).	
B			Potential evaporation exceeds precipitation. The dry–humid boundary is defined by the following formulas: (Note: R is the average annual precipitation in cm, and T is the average annual temperature in °C.) $R < 2T + 28$ when 70% or more of rain falls in warmer 6 months. $R < 2T$ when 70% or more of rain falls in cooler 6 months. $R < 2T + 14$ when neither half year has 70% or more of rain.	
	S		Steppe The BS–BW boundary is 1/2 the dry–humid boundary.	
	W		Desert	
		h	Average annual temperature is 18°C or greater.	
		k	Average annual temperature is less than 18°C.	
C			Average temperature of the coldest month is under 18°C and above −3° C.	
	w		At least 10 times as much precipitation in a summer month as in the driest winter month.	
	s		At least three times as much precipitation in a winter month as in the driest summer month; precipitation in driest summer month less than 4 cm.	
	f		Criteria for w and s cannot be met.	
		a	Warmest month is over 22° C; at least 4 months over 10° C.	
		b	No month above 22° C; at least 4 months over 10° C.	
		c	One to 3 months above 10° C.	
D			Average temperature of coldest month is −3° C or below; average temperature of warmest month is greater than 10° C.	
	w		Same as under C.	
	s		Same as under C.	
	f		Same as under C.	
		a	Same as under C.	
		b	Same as under C.	
		c	Same as under C.	
		d	Average temperature of the coldest month is −3° C or below.	
E			Average temperature of the warmest month is below 10° C.	
	T		Average temperature of the warmest month is greater than 0° C and less than 10° C.	
	F		Average temperature of the warmest month is 0° C or below.	

Figure 15–2 The Köppen system of climate classification. This system uses easily obtained data: mean monthly and annual values of temperature and precipitation. When using this figure to classify climate data, first determine whether the data meet the criteria for the *E* climates. If the station is not a polar climate, proceed to the criteria for *B* climates. If the data do not fit into either the *E* or *B* groups, check the data against the criteria for *A*, *C*, and *D* climates, in that order. You may find Figure G-1 in Appendix G helpful when using this chart. (Photos, in order from A-E are Michael Collier, Marek Zak/ Alamy, Ed Reschke/Photolibrary, Michael Collier, J.G. Paren/Photo Researchers, Inc.)

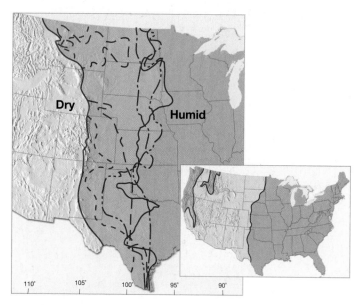

Figure 15–3 Yearly fluctuations in the dry-humid boundary during a five-year period using the Köppen classification. The small inset shows the average position of the dry-humid boundary for the five-year period.

A strength of the Köppen system is the relative ease with which boundaries are determined. However, these boundaries cannot be viewed as fixed. On the contrary, all climate boundaries shift their positions from one year to the next (Figure 15–3). The boundaries shown on climate maps are simply average locations based on data collected over many years. Thus, a climate boundary should be regarded as a broad transition zone and not a sharp line (Box 15–1).

The world distribution of climates, according to the Köppen classification, is shown in Figure 15–4. We will refer you to this map several times as we examine Earth's climates in the following pages.

Concept Check 15.1

1 Why is classification often a necessary task in science?

2 What climate data are needed to classify a climate using the Köppen scheme?

3 Should climate boundaries, such as those shown on the world map in Figure 15–4, be regarded as fixed? Explain.

Climate Controls: A Summary

If Earth's surface were completely homogeneous, the map of world climates would be simple. It would look much like the ancient Greeks must have pictured it—a series of latitudinal bands girdling the globe in a symmetrical pattern on each side of the equator (see Figure 15–1). This is not the case, of course. Earth is not a homogeneous sphere, and many factors disrupt the symmetry just described.

At first glance, the world climate map (see Figure 15–4) reveals what appears to be a scrambled or even haphazard pattern, with similar climates located in widely sepa-

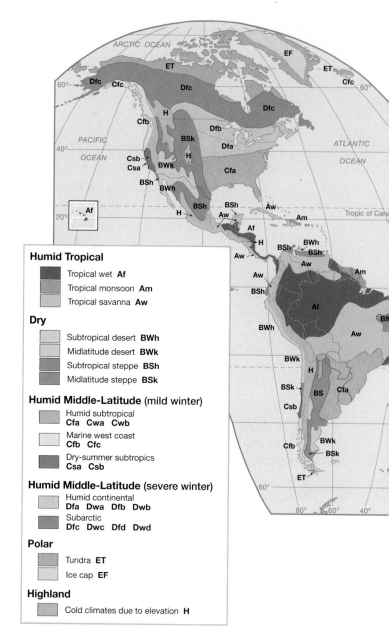

Humid Tropical

- Tropical wet **Af**
- Tropical monsoon **Am**
- Tropical savanna **Aw**

Dry

- Subtropical desert **BWh**
- Midlatitude desert **BWk**
- Subtropical steppe **BSh**
- Midlatitude steppe **BSk**

Humid Middle-Latitude (mild winter)

- Humid subtropical **Cfa Cwa Cwb**
- Marine west coast **Cfb Cfc**
- Dry-summer subtropics **Csa Csb**

Humid Middle-Latitude (severe winter)

- Humid continental **Dfa Dwa Dfb Dwb**
- Subarctic **Dfc Dwc Dfd Dwd**

Polar

- Tundra **ET**
- Ice cap **EF**

Highland

- Cold climates due to elevation **H**

rated parts of the world. A closer examination shows that although they may be far apart, similar climates generally have similar latitudinal and continental positions. This consistency suggests an order in the distribution of climate elements and that the pattern of climates is not by chance. Indeed, the climate pattern reflects a regular and dependable operation of the major climate controls. So before we examine Earth's major climates, it will be worthwhile to review each of the major controls: latitude, land and water, geographic position and prevailing winds, mountains and highlands, ocean currents, and pressure and wind systems.

Latitude

Fluctuations in the amount of solar radiation received at Earth's surface represent the single greatest cause of temperature differences. Although variations in such factors as cloud coverage and the amount of dust in the air may be

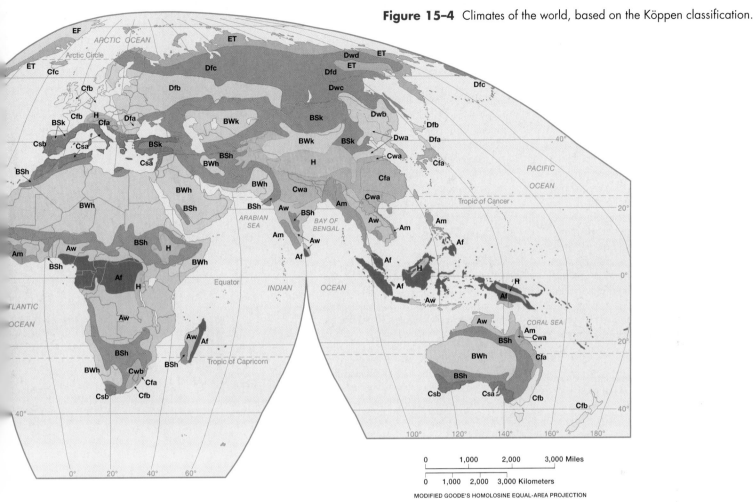

Figure 15–4 Climates of the world, based on the Köppen classification.

locally influential, seasonal changes in Sun angle and length of daylight are the most important factors controlling the global temperature distribution. Because all places situated along the same parallel of latitude have identical Sun angles and lengths of daylight, variations in the receipt of solar energy are largely a function of latitude. Moreover, because the vertical rays of the Sun migrate annually between the Tropic of Cancer and the Tropic of Capricorn, there is a regular latitudinal shifting of temperatures. Temperatures in the tropical realm are consistently high because the vertical rays of the Sun are never far away. As one moves farther poleward, however, greater seasonal fluctuations in the receipt of solar energy are reflected in larger annual temperature ranges.

Land and Water

The distribution of land and water must be considered second in importance only to latitude as a control of temperature. Recall that water has a greater *heat capacity* than rock and soil. Thus, land heats more rapidly and to higher temperatures than water, and it cools more rapidly and to lower temperatures than water. Consequently, variations in air temperatures are much greater over land than over water. This differential heating of Earth's surface has led to

climates being divided into two broad classes—marine and continental.

Marine climates are considered relatively mild for their latitude because the moderating effect of water produces summers that are warm but not hot and winters that are cool but not cold. In contrast, **continental climates** tend to be much more extreme. Although a marine station and a continental station along the same parallel in the middle latitudes may have similar annual mean temperatures, the annual temperature *range* will be far greater at the continental station.

The differential heating and cooling of land and water can also have a significant effect on pressure and wind systems and hence on seasonal precipitation distribution. High summer temperatures over the continents can produce low-pressure areas that allow the inflow of moisture-laden maritime air. Conversely, the high pressure that forms over the chilled continental interiors in winter causes a reverse flow of dry air toward the oceans.

Geographic Position and Prevailing Winds

To understand fully the influence of land and water on the climate of an area, the position of that area on the continent

Box 15–1 Climate Diagrams

Throughout this chapter climate diagrams accompany climate descriptions. They are useful tools for presenting the basic data needed in the study of world climates. Figure 15–A shows a typical climate diagram for Portland, Oregon. It has 12 columns, one for each month of the year. A temperature scale is on the left side, and a red line connects monthly mean temperatures. A precipitation scale is on the right side, and blue bars show average monthly precipitation totals.

Just a glance at a climate diagram reveals whether the annual temperature range is great or small and clearly shows the seasonal distribution of precipitation. Is the station in the Northern or Southern Hemisphere? Is it near the equator? Does it experience a monsoon precipitation regime or one more characteristic of a Mediterranean climate? Such information is basic in any discussion or comparison of world climates.

The diagrams presented in this chapter include a location map in the background

and other information, including the latitude, longitude, and climate classification of the station. However, only temperature and precipitation data are essential to construct a climate diagram.

Answer the following questions about Figure 15–A, the climate diagram for Portland. What are the warmest and coldest months? What are the approximate monthly mean temperatures for the warmest and coldest months? What are the four driest months? What is the rainfall for the wettest month?

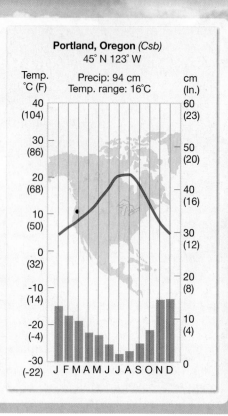

FIGURE 15–A Climate diagrams, such as this one for Portland, Oregon, are useful tools because they show important details at a glance. The plot of temperature immediately indicates whether the station is in the Northern Hemisphere or the Southern Hemisphere and whether it is near the equator. The combination line and bar graph provides a visual summary that allows one to readily recognize annual patterns of temperature and precipitation.

and its relationship to the prevailing winds must be considered. The moderating influence of water is much more pronounced along the windward side of a continent, for here the prevailing winds may carry the maritime air masses far inland. On the other hand, places on the lee side of a continent, where the prevailing winds blow from the land toward the ocean, are likely to have a more continental temperature regime.

Mountains and Highlands

Mountains and highlands play an important part in the distribution of climates. This impact may be illustrated by examining western North America. Because prevailing winds are from the west, the mountain chains that trend north–south are major barriers. They prevent the moderating influence of maritime air masses from reaching far inland. Consequently, although stations may lie within a few hundred kilometers of the Pacific Ocean, their temperature regime is essentially continental.

Also, these topographic barriers trigger orographic rainfall on their windward slopes, often leaving a dry rain shadow on the leeward side. Similar effects may be seen in South America and Asia, where the towering Andes and the massive Himalayan system are major barriers. In compari-

son, Western Europe lacks a mountain barrier to obstruct the free movement of maritime air masses from the North Atlantic. As a result, moderate temperatures and sufficient precipitation mark the entire region.

Extensive highlands create their own climatic regions. Because of the drop in temperature with increasing altitude, areas such as the Tibetan Plateau, the Altiplano of Bolivia, and the uplands of East Africa are cooler and drier than their latitudinal locations alone would indicate.

Ocean Currents

The effect of ocean currents on the temperatures of adjacent land areas can be significant. Poleward-moving currents, such as the Gulf Stream and Kurashio Currents in the Northern Hemisphere and the Brazil and East Australian Currents in the Southern Hemisphere, cause air temperatures to be warmer than would be expected for their latitudes. This influence is especially pronounced in winter. Conversely, the Canary and California Currents in the Northern Hemisphere and the Peru and Benguela Currents south of the equator reduce the temperatures of bordering coastal zones. In addition, the chilling effects of these cold currents act to stabilize the air masses moving across them. The result is marked aridity and often considerable advection fog.

Pressure and Wind Systems

The world distribution of precipitation shows a close relationship to the distribution of Earth's major pressure and wind systems. Although the latitudinal distribution of these systems does not generally take the form of simple "belts," it is still possible to identify a zonal arrangement of precipitation from the equator to the poles (see Figure 7–27, page 214).

In the realm of the equatorial low, the convergence of warm, moist, and unstable air makes this zone one of heavy rainfall. In the regions dominated by subtropical highs, general aridity prevails, creating major deserts. Farther poleward, in the middle-latitude zone dominated by the irregular subpolar low, the influence of the many traveling cyclonic disturbances again increases precipitation. Finally, in polar regions, where temperatures are low and the air can hold only small quantities of moisture, precipitation totals decline.

The seasonal shifting of the pressure and wind belts, which follows the movement of the Sun's vertical rays, significantly affects areas in intermediate positions. Such regions are alternately influenced by two different pressure and wind systems. A station located poleward of the equatorial low and equatorward of the subtropical high, for example, will experience a summer rainy period as the low migrates poleward and a wintertime dry season as the high moves equatorward. This latitudinal shifting of pressure belts is largely responsible for the seasonality of precipitation in many regions.

Concept Check 15.2

1 List the major climate controls and briefly describe their influence.

2 Which control has the greatest influence on global temperature differences?

3 Contrast continental and marine climates.

4 What is the connection between pressure systems and the world distribution of precipitation?

World Climates— An Overview

The remainder of this chapter is a tour of world climates. Our tour is organized like this:

- Beginning along the equator, we visit the *wet tropics* (the *A* climates), studying their temperature and precipitation characteristics, which foster the great tropical rain forests.
- North and south of the wet tropics is the *tropical wet and dry* climates (still *A* climates), including the monsoon areas.

- North and south of the tropical wet and dry areas are the *dry* climates (the *B* climates). Dominating large parts of the subtropics and extending into the interiors of continents in the middle latitudes, deserts and steppes cover nearly one-third of Earth's land surface.
- Moving poleward from the dry subtropical realm, we visit regions exhibiting the *humid middle-latitude* climate (one of the *C* climates). These prevail on the eastern sides of continents between 25° and 40° latitude; the southeastern United States is an example.
- On the windward coasts of continents, we next visit *marine west coast* climates (also *C* climates), such as that of Western Europe.
- We complete our look at *C* climates with a visit to *dry-summer subtropical* or *Mediterranean* climates, such as those of Italy, Spain, and parts of California.
- In the Northern Hemisphere, where the continents extend into the middle and high latitudes, *C* climates give way to *D* climates called *humid continental*. These are "breadbasket" areas hospitable to growing grains and meat that feed much of the world.
- Verging on the polar climates are the *subarctic* climates. These vast zones of coniferous forest in Canada and Siberia are known for their long and bitter winters.
- Around the poles we visit the *polar* climates (the *E* climates). These are summerless areas of tundra and permafrost or permanent ice sheets.
- Finally, we visit some cool places that are not necessarily near the poles but are chilled by their high elevation. The *highland* climates even occur on mountaintops near the equator and are characteristic of the Rockies, Andes, Himalayas, and other mountain regions.

We begin our tour with the *wet tropics*.

Concept Check 15.3

1 Briefly describe the sequence of climates you would encounter going from equatorial Africa through central Europe and Scandinavia to the North Pole.

The Wet Tropics (*Af, Am*)

In the wet tropics, constant high temperatures and year-round rainfall combine to produce the most luxuriant vegetation in any climatic realm: the **tropical rain forest** (Figure 15–5). Unlike the forests that we North Americans are accustomed to, the tropical rain forest is made up of broadleaf trees that remain green throughout the year. In addition, instead of being dominated by a few species, these forests are characterized by many. It is not uncommon for hundreds of different species to inhabit a single square kilometer of the forest. As a consequence, the individuals of a single species are often widely spaced.

Standing on the shaded floor of the forest looking upward, one sees tall, smooth-barked, vine-entangled trees, the trunks branchless in their lower two-thirds, with an almost continuous canopy of foliage above. A closer look reveals a three-level structure. Nearest the ground, perhaps 5 to 15 meters (16 to 50 feet) above, the narrow crowns of rather slender trees are visible. Rising above these relatively short components of the forest, a more continuous canopy of foliage occupies the height range of 20 to 30 meters (65 to 100 feet). Finally, visible through an occasional opening in the second level, a third level may be seen at the very top of the forest. Here the crowns of the trees tower 40 meters (130 feet) or more above the forest floor.

The environment of the wet tropics just described covers almost 10 percent of Earth's land area (Box 15–2). Figure 15–4 shows that *Af* and *Am* climates form a discontinuous belt astride the equator that typically extends 5° to 10° into each hemisphere. The poleward margins are most often marked by diminishing rainfall, but occasionally decreasing temperatures mark the boundary. Because of the general decrease in temperature with height in the troposphere, this climate region is restricted to elevations below 1000 meters (3300 feet). Consequently, the major interruptions near the equator are mostly cooler highland areas.

Also note that the rainy tropics tend to have greater north–south extent along the eastern side of continents (especially South America) and along some tropical coasts. The greater span on the eastern side of a continent is due primarily to its windward position on the weak western side of the subtropical high, a zone dominated by neutral or unstable air. In other cases, as along the eastern side of Central America, the coast, backed by interior highlands, intercepts the flow of trade winds. Orographic uplift thus greatly enhances the rainfall total.

Data for some representative stations in the wet tropics are shown in Table 15–1 and Figure 15–6a and b. A brief examination of the numbers reveals the most obvious features that characterize the climate in these areas:

1. Temperatures usually average 25°C (77°F) or higher each month. Not only is the annual mean high, but the annual range is very small (note the flat temperature curves in the graphs in Figure 15–6a and b).

2. The total precipitation for the year is high, often exceeding 200 centimeters (80 inches).

3. Although rainfall is not evenly distributed throughout the year, tropical rain forest stations are generally wet in all months. If a dry season exists, it is very short.

Figure 15–5 Unexcelled in luxuriance and characterized by hundreds of different species per square kilometer, the tropical rain forest is a broadleaf evergreen forest that dominates the wet tropics. This image shows Borneo's Segama River passing through virgin tropical rain forest. (Photo by Peter Lilja/agefotostock)

TABLE 15-1 Data for Wet Tropical Stations

	J	F	M	A	M	J	J	A	S	O	N	D	YR
Singapore 1° 21' N; 10 m													
Temp. (°C)	26.1	26.7	27.2	27.6	27.8	28.0	27.4	27.3	27.3	27.2	26.7	26.3	27.1
Precip. (mm)	285	164	154	160	131	177	163	200	122	184	236	306	2282
Belém, Brazil, 1° 18' S; 10 m													
Temp. (°C)	25.2	25.0	25.1	25.5	25.7	25.7	25.7	25.9	25.7	26.1	26.3	25.9	25.7
Precip. (mm)	340	406	437	343	287	175	145	127	119	91	86	175	2731
Douala, Cameroon, 4° N; 13 m													
Temp. (°C)	27.1	27.4	27.4	27.3	26.9	26.1	24.8	24.7	25.4	25.9	26.5	27.0	26.4
Precip. (mm)	61	88	226	240	353	472	710	726	628	399	146	60	4109

Temperature Characteristics

Because places with an *Af* or *Am* designation lie near the equator, the reason for their uniform temperature rhythm is clear: Insolation is consistently intense. The Sun's rays are always near to vertical, and changes in day length are slight throughout the year. Therefore, seasonal temperature variations are minimal. The small difference that exists between the warmest and coolest months often reflects changes in cloud cover rather than in the position of the Sun. In the case of Belém, Brazil, for example, you can see that the highest temperatures occur during the months when rainfall (and hence cloud cover) is least.

Students Sometimes Ask...

Isn't jungle just another word for tropical rain forest?

No. Although both refer to vegetation in the wet tropics, they are not the same. In the tropical rain forest, there is a high canopy of foliage that does not allow much light to penetrate to the ground. As a result, plant foliage is relatively sparse on the dimly lit forest floor. By contrast, anywhere considerable light makes its way to the ground, as along riverbanks or in human-made clearings, an almost impenetrable growth of tangled vines, shrubs, and short trees exists. The familiar term *jungle* is used to describe such areas.

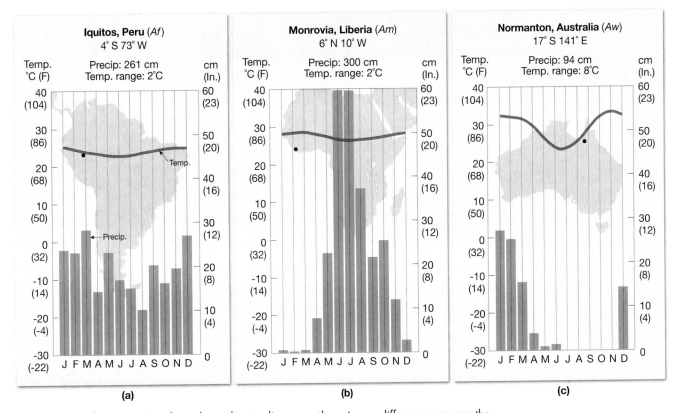

Figure 15-6 By comparing these three climate diagrams, the primary differences among the *A* climates can be seen. (a) Iquitos, the *Af* station, is wet throughout the year. (b) Monrovia, the *Am* station, has a short dry season. (c) As is true for all *Aw* stations, Normanton has an extended dry season and a greater annual temperature range than the others.

Box 15–2   Clearing the Tropical Rain Forest—The Impact on Its Soils

Thick red soils are common in the wet tropics and subtropics. They are the end product of extreme chemical weathering. Because lush tropical rain forests are associated with these soils, we might assume that they are fertile and have great potential for agriculture. However, just the opposite is true: They are among the poorest soils for farming. How can this be?

Because rain forest soils develop under conditions of high temperature and heavy rainfall, they are severely leached. Leaching removes soluble materials such as calcium carbonate, and the great quantities of percolating water also remove much of the silica; the result is that insoluble oxides of iron and aluminum become concentrated in the soil. Iron oxides give the soil its distinctive red color. Because bacterial activity is very high in the tropics, rain forest soils contain practically no humus. Moreover, leaching destroys fertility because the large volume of downward-percolating water removes most plant nutrients. Therefore, even though the vegetation may be dense and luxuriant, the soil itself contains few available nutrients.

Most nutrients that support the rain forest are locked up in the trees themselves. As vegetation dies and decomposes, the roots of the rain forest trees quickly absorb the nutrients before they are leached from the soil. The nutrients are continuously recycled as trees die and decompose.

FIGURE 15–B Clearing the Amazon rain forest in Surinam. The thick orange soil is highly leached. (Photo by Wesley Bocxe/Photo Researchers, Inc.)

FIGURE 15–C This ancient temple at Angkor Wat, Cambodia, was built of bricks made of laterite. (Photo by Nestor Noci/Shutterstock)

Therefore, when forests are cleared to provide land for farming or to harvest the timber, most of the nutrients are removed as well (Figure 15–B). What remains is a soil that contains little to nourish planted crops.

The clearing of rain forests not only removes plant nutrients but also accelerates erosion. When vegetation is present, its roots anchor the soil, and its leaves and branches provide a canopy that protects the ground by deflecting the full force of the frequent heavy rains.

The removal of the vegetation also exposes the ground to strong direct sunlight. When baked by the Sun, these tropical soils can harden to a bricklike consistency and become practically impenetrable by water and crop roots. In only a few years, soils in a freshly cleared area may no longer be cultivable.

The term *laterite*, which is often applied to these soils, is derived from the Latin word *latere*, meaning "brick," and was first applied to the use of this material for brick-making in India and Cambodia. Laborers simply excavated the soil, shaped it, and allowed it to harden in the Sun. Ancient but still well-preserved structures built of laterite remain standing today in the wet tropics (Figure 15–C). Such structures have withstood centuries of weathering because all of the original soluble materials were already removed from the soil by chemical weathering. Laterites are therefore virtually insoluble and very stable.

In summary, we have seen that some rain forest soils are highly leached products of extreme chemical weathering in the warm, wet tropics. Although they may be associated with lush tropical rain forests, these soils are unproductive when vegetation is removed. Moreover, when cleared of plants, these soils are subject to accelerated erosion and can be baked to bricklike hardness by the Sun.

A striking characteristic of temperatures in the wet tropics is that daily temperature variations greatly exceed seasonal differences. Whereas annual temperature ranges in the wet tropics rarely exceed 3°C (6°F), daily temperature ranges are between two and five times greater. Thus, there is a greater variation between day and night than there is seasonally. It is interesting that monthly and daily mean temperatures in the tropics are no greater than those in many U.S. cities during the summer. For example, the highest temperature recorded at Jakarta, Indonesia, over a 78-year period and at Belém, Brazil, over a 20-year period has been only 36.6°C (98°F), compared with extremes of 40.5°C (105°F) at Chicago and 41.1°C (106°F) at New York City.

The uniqueness of the wet tropical temperature regime is its day-in and day-out, month-in and month-out regularity. Although the thermometer may not indicate abnormal or extreme conditions, the warm temperatures combined with the high humidity and meager winds make apparent temperatures particularly high. The reputation of the wet tropics as being oppressive and monotonous is mostly well deserved.

Precipitation Characteristics

The regions dominated by *Af* and *Am* climate normally receive from 175 to 250 centimeters (68 to 98 inches) of rain each year. But a glance at the data in Table 15–1 reveals more variability in rainfall than in temperature, both seasonally and from place to place. The rainy nature of the equatorial realm is partly related to the extensive heating of the region and the consequent thermal convection. In addition, this is the zone of the converging trade winds, often referred to as the **intertropical convergence zone,** or simply the **ITCZ.** The thermally induced convection, coupled with convergence, leads to widespread ascent of the warm, humid, unstable air. Conditions near the equator are thus ideal for precipitation.

Rain typically falls on more than half of the days each year. In fact, at some stations, three-quarters of the days experience rain. There is a marked daily regularity to the rainfall at many places. Cumulus clouds begin forming in late morning or early afternoon. The buildup continues until about 3 or 4 P.M., the time when temperatures are highest and thermal convection is at a maximum; then the cumulonimbus towers yield showers. Figure 15–7, showing the hourly distribution of rainfall at Kuala Lumpur, Malaysia, exemplifies this pattern.

The cycle is different at many marine stations, with the rainfall maximum occurring at night. Here the environmental lapse rate is steepest and hence instability is greatest during the dark hours instead of in the afternoon. The environmental lapse rate steepens at night because the radiation heat loss from the air at heights of 600 to 1500 meters (2000 to 5000 feet) is greater than near the surface, where the air continues to be warmed by conduction and low-level turbulence created when air is heated by the warm water.

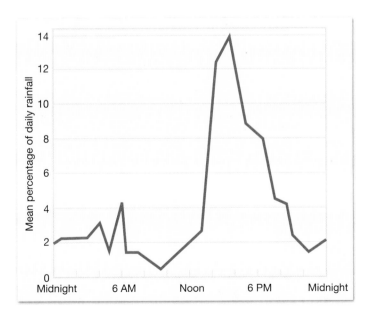

Figure 15–7 The distribution of rainfall by hour of the day at Kuala Lumpur, Malaysia, with its midafternoon maximum, illustrates the typical pattern at many wet tropical stations. (Data from Ooi Jin-bee, "Rural Development in Tropical Areas," *Journal of Tropical Geography,* 12, no. 1 [1959])

Portions of the rainy tropics are wet throughout the year. According to the Köppen scheme, at least 6 centimeters (2.3 inches) of rain falls each month. Yet extensive areas (those having the *Am* designation) are characterized by a brief dry season of one or two months. Despite the short dry season, the annual precipitation total in *Am* regions closely corresponds to the total in areas that are wet year-round (*Af*). Because the dry period is too brief to deplete the supply of soil moisture, the rain forest is maintained.

The seasonal pattern of precipitation in wet tropical climates is complex and not yet fully understood. Month-to-month variations are, at least in part, caused by the seasonal migration of the ITCZ, which follows the migration of the vertical rays of the Sun.

Concept Check 15.4

1 How does the tropical rain forest differ from a typical middle-latitude forest?

2 Explain each of the following characteristics of the wet tropics:

   a The climate is restricted to elevations below 1000 meters.
   b Mean monthly and annual temperatures are high, and the annual temperature range is low.
   c The climate is rainy throughout the year, or nearly so.

3 What is the difference between *Af* areas and *Am* areas?

Tropical Wet and Dry (Aw)

Between the rainy tropics and the subtropical deserts lies a transitional climatic region referred to as **tropical wet and dry.** Along its equatorward margin, the dry season is short, and the boundary between *Aw* and the rainy tropics is difficult to define. Along the poleward side, however, the dry season is prolonged, and conditions merge into those of the semiarid realm.

Here in the tropical wet and dry climate the rain forest gives way to the **savanna,** a tropical grassland with scattered deciduous trees (Figure 15–8). In fact, *Aw* is often called the *savanna climate.* This name may not be appropriate technically, because some ecologists doubt that these grasslands are climatically induced. It is believed that woodlands once dominated this zone and that the savanna grasslands developed in response to seasonal burnings by native populations.

Temperature Characteristics

The temperature data in Table 15–2 show only modest differences between the wet tropics and the tropical wet and dry regions. Because of the somewhat higher latitude of most *Aw* stations, annual mean temperatures are slightly lower. In addition, the annual temperature range is a bit greater, varying from 3°C (5°F) to perhaps 10°C (20°F). The daily temperature range, however, still exceeds the annual variation.

Figure 15–8 The tropical savanna, with its stunted, drought-resistant trees scattered amid a grassland may have resulted from seasonal burnings by native human populations. Serengeti National Park, Tanzania. (Photo by Kondrachov Vladimir/Shutterstock)

Because seasonal fluctuations in humidity and cloudiness are more pronounced in *Aw* areas, daily temperature ranges vary noticeably during the year. Generally, they are small during the rainy season, when humidity and cloud cover are at a maximum, and large during droughts, when clear skies and dry air prevail. Furthermore, because of the more persistent summertime cloudiness, many *Aw* stations experience their warmest temperatures at the end of the dry season, just prior to the summer solstice. Thus, in tropical wet and dry regions of the Northern Hemisphere, March, April, and May are often warmer than June and July.

Precipitation Characteristics

Because temperature regimes among the *A* climates are similar, the primary factor that distinguishes the *Aw* climate from *Af* and *Am* is precipitation. *Aw* stations typically receive 100 to 150 centimeters (40 to 60 inches) of rainfall each year, an amount often appreciably less than in the wet tropics. The most distinctive characteristic of this climate, however, is the *markedly seasonal character of the rainfall—wet summers followed by dry winters.* This is clearly shown by the climate diagram in Figure 15–6c.

TABLE 15–2 Data for Tropical Wet and Dry Stations

	J	F	M	A	M	J	J	A	S	O	N	D	YR
Calcutta, India, 22° 32' N; 6 m													
Temp. (°C)	20.2	23.0	27.9	30.1	31.1	30.4	29.1	29.1	29.9	27.9	24.0	20.6	26.94
Precip. (mm)	13	24	27	43	121	259	301	306	290	160	35	3	1582
Cuiaba, Brazil, 15° 30' S; 165 m													
Temp. (°C)	27.2	27.2	27.2	26.6	25.5	23.8	24.4	25.5	27.7	27.7	27.7	27.2	26.5
Precip. (mm)	216	198	232	116	52	13	9	12	37	130	165	195	1375

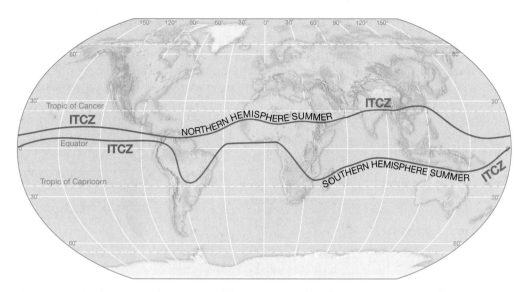

Figure 15–9 The seasonal migration of the ITCZ strongly influences precipitation distribution in the tropics.

The alternating wet and dry periods are due to the latitudinal position of the *Aw* climate region. It lies between the intertropical convergence zone, with its sultry weather and convective thundershowers, and the stable, subsiding air of the subtropical highs. Following the spring equinox, the ITCZ and the other wind and pressure belts all shift poleward as they migrate with the vertical rays of the Sun (Figure 15–9). With the advance of the ITCZ into a region, the summer rainy season commences with weather patterns typical of the wet tropics. Later, with the retreat of the ITCZ back toward the equator, the subtropical high advances into the region and brings with it intense aridity.

During the dry season, the landscape grows parched and nature seems to become dormant as water-stressed trees shed their leaves and the abundant tall grasses turn brown and wither. The dry season's duration depends primarily on distance from the ITCZ. Typically, the farther an *Aw* station is from the equator, the shorter the period of ITCZ control and the longer the locale will be influenced by the stable subtropical high. Consequently, the higher the latitude, the longer the dry season and the shorter the wet period.

Understanding the movement of the ITCZ is essential to understanding rainfall distribution in the tropics. This is clear from Table 15–3, which shows precipitation data for six African stations. Notice that at Malduguri (farthest north) and Francistown (farthest south), there are single rainfall maxima (bold print) that occur when the ITCZ reaches its most poleward positions. Between these stations, double maxima (bold print) represent the passage of the ITCZ on its way to and from these extreme locations. It is important to remember that these statistics represent long-term averages and that on a year-to-year basis the movement of the ITCZ is far from regular. There is, nevertheless, no doubt that in the tropics, rainfall follows the Sun.

The Monsoon

In much of India, Southeast Asia, and portions of Australia, the alternating periods of rainfall and dryness characteristic of the *Aw* precipitation regime are associated with the **monsoon.** The term is derived from the Arabic word *mausim,*

TABLE 15–3 Rainfall Regimes and the Movement of the ITCZ in Africa

	J	F	M	A	M	J	J	A	S	O	N	D
Malduguri, Nigeria, 11° 51' N												
Precip. (mm)	0	0	0	7.6	40.6	68.6	180.3	**220.9**	106.6	17.7	0	0
Yaounde, Cameroon, 3° 53' N												
Precip. (mm)	22.8	66.0	147.3	170.1	**195.6**	152.4	73.7	78.7	213.4	**294.6**	116.8	22.9
Kisangani, Dem. Rep. of the Congo, 0° 26' N												
Precip. (mm)	53.3	83.8	**177.8**	157.5	137.2	114.3	132.0	165.1	182.9	**218.4**	198.1	83.8
Kananga, Dem. Rep. of the Congo, 5° 54' S												
Precip. (mm)	137.2	142.2	**195.6**	193.0	83.8	20.3	12.7	58.4	116.8	165.1	**231.1**	226.0
Zomba, Malawi, 15° 23' S												
Precip. (mm)	274.3	**289.6**	198.1	76.2	27.9	12.7	5.1	7.6	17.8	17.8	134.6	**279.4**
Francistown, Botswana, 21° 13' S												
Precip. (mm)	**106.7**	78.7	71.1	17.8	5.1	2.5	0	0	0	22.9	58.4	86.4

which means "season" and typically refers to wind systems that have a pronounced seasonal reversal of direction. During the summer, conditions are conducive to rainfall because humid, unstable air moves from the oceans toward the land. In winter this reverses, and a dry wind, which originates over the continent, blows toward the sea.

This monsoonal circulation system develops partly in response to the differences in annual temperature variations between continents and oceans. In principle, the processes associated with the monsoon are similar to those described in connection with the land and sea breeze (Chapter 7) except that the scales, in both time and space, are much larger.

During spring in the Northern Hemisphere, an irregular area of thermally induced low pressure gradually develops over the interior of southern Asia. It is further strengthened by the poleward advance of the ITCZ (Figure 15–9). Thus, the *summertime* circulation is from the higher pressure over the ocean toward the lower pressure over the continent. As winter approaches, winds reverse direction as the ITCZ migrates southward, and a deep anticyclone develops over the chilled continent. By *midwinter* dry winds blow from the continent southward to converge on Australia and southern Africa.

The *Cw* Variant

Adjacent to the wet and dry tropics in southern Africa, South America, northeastern India, and China are areas that are sometimes designated *Cw*. Although *C* has been substituted for *A*, indicating that these regions are subtropical instead of tropical, the *Cw* climate is nevertheless a variant of *Aw*, for the only major difference is somewhat lower temperatures. In Africa and South America, *Cw* climates are highland extensions of *Aw*. Because they occupy elevated

sites, they have lower temperatures than the adjacent wet and dry tropics. In India and China, *Cw* areas are middle-latitude extensions of the tropical monsoon realm. In some cases, especially in India, the *Cw* areas are barely poleward enough to have winter temperatures below those of the *A* climates.

The Dry Climates (*B*)

The dry regions of the world cover some 42 million square kilometers (nearly 16.5 million square miles), or about 30 percent, of Earth's land surface. No other climatic group covers so large a land area (Figure 15–10).

The characteristic features of dry climates are their meager yearly rainfall and the fact that their precipitation is very unreliable. Generally, the smaller the mean annual rainfall, the greater its variability. As a result, yearly rainfall averages are often misleading.

For example, during one seven-year period, Trujillo, Peru, had an average rainfall of 6.1 centimeters per year (2.4 inches). Yet a closer look reveals that during the first 6 years and 11 months of the period, the station received a scant 3.5 centimeters (1.4 inch) (an annual average of slightly more

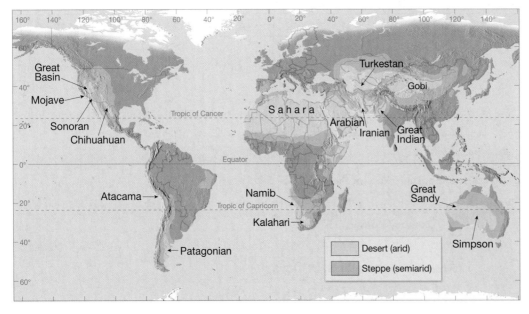

Figure 15–10 Arid and semiarid climates cover about 30 percent of Earth's land area. These dry (*B*) climates constitute the single largest climatic group.

than 0.5 centimeter). Then during the twelfth month of the seventh year, 39 centimeters (15.2 inches) of rain fell, 23 centimeters (9 inches) of it during a 3-day span.

This extreme case illustrates the irregularity of rainfall in most dry regions. Also, there are usually more years when rainfall totals are below the average than years when they are above. As the foregoing example shows, the occasional wet period tends to lift the average.

What Is Meant by "Dry"?

It is important to realize that dryness is relative and simply refers to any *water deficiency*. Thus, climatologists define a **dry climate** as one in which the *yearly precipitation is less than the potential water loss by evaporation*. Dryness, then, is not only related to annual rainfall. It is also a function of evaporation, which in turn depends closely on temperature. As temperatures climb, potential evaporation also increases.

For example, 25 centimeters (10 inches) of precipitation may be sufficient to support forests in northern Scandinavia, where evaporation is slight into the cool, humid air and a surplus of water remains in the soil. However, the same amount of rain falling on Nevada or Iran supports only a sparse vegetative cover because evaporation into the hot, dry air is great. Clearly, no specific amount of precipitation can define a dry climate.

To establish a meaningful boundary between dry and humid climates, the Köppen classification uses formulas that involve three variables: (1) average annual precipitation, (2) average annual temperature, and (3) seasonal distribution of precipitation.

The use of average annual precipitation is obvious. Average annual temperature is used because it is an index of evaporation; the amount of rainfall defining the humid–dry boundary is greater where mean annual temperatures are high and less where temperatures are low. The third variable, seasonal distribution of precipitation, is also related to this idea. If rain is concentrated in the warmest months,

TABLE 15–4 Average Annual Precipitation at *BS*–Humid Boundary

Average Annual Temp. (°C)	Average Annual Precipitation (mm)		
	Summer Dry Season	Even Distribution	Winter Dry Season
5	100	240	380
10	200	340	480
15	300	440	580
20	400	540	680
25	500	640	780
30	600	740	880

loss to evaporation is greater than if it is concentrated in the cooler months. Thus, considerable differences exist in the precipitation amounts received at various stations in the *B* climates.

Table 15–4 summarizes these differences. For example, if a station with an annual mean of 20°C (68°F) and a summer wet season ("winter dry") receives less than 680 millimeters (26.5 inches) of precipitation per year, it is classified as *dry*. If the rain falls primarily in winter ("summer dry"), however, the station must receive only 400 millimeters (15.6 inches) or more to be considered humid. If the precipitation is more evenly distributed, the figure defining the humid–dry boundary is between the other two.

Within the regions defined by a general water deficiency, there are two climatic types: **arid, or desert (*BW*)**, and **semiarid, or steppe (*BS*)**. Figure 15–11 presents climatic diagrams for both types. Stations (a) and (b) are in the subtropics, and (c) and (d) are in the middle latitudes. Deserts and steppes have many features in common; their differences are primarily a matter of degree. The semiarid is a marginal and more humid variant of the arid and represents a transition zone that surrounds the desert and separates it from the bordering humid climates (Box 15–3). The arid–semiarid boundary is commonly set at one-half the annual precipitation separating dry regions from humid. Thus, if the humid–dry boundary happens to be 40 centimeters, the steppe–desert boundary will be 20 centimeters.

Subtropical Desert (*BWh*) and Steppe (*BSh*)

The heart of low-latitude dry climates lies along the vicinity of the Tropic of Cancer and the Tropic of Capricorn. A glance at Figure 15–10 reveals a virtually unbroken desert environment stretching for more than 9300 kilometers (nearly 6000 miles) from the Atlantic coast of North Africa to the dry lands of northwestern India. In addition to this single great expanse, the Northern Hemisphere contains another much smaller area of subtropical desert (*BWh*) and steppe (*BSh*) in northern Mexico and the southwestern United States. In the Southern Hemisphere, dry climates dominate Australia. Almost 40 percent of that continent is desert, and much of the remainder is steppe.

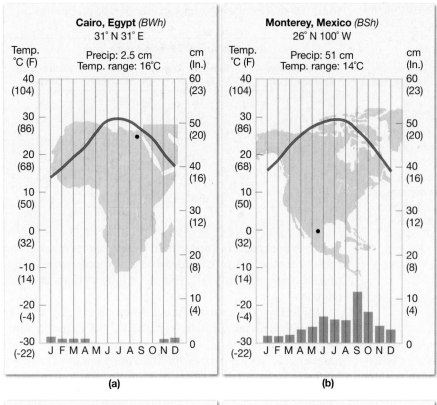

(a)

(b)

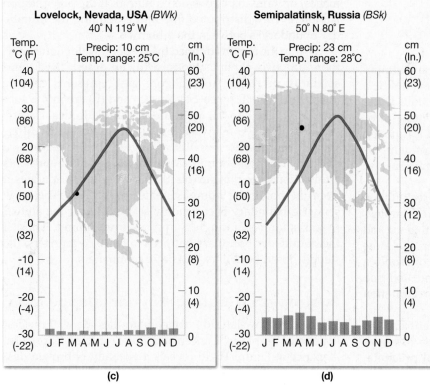

(c)

(d)

Figure 15–11 Climate diagrams for representative arid and semiarid stations. Cairo and Lovelock are classified as deserts; Monterey and Semipalatinsk are steppes. Stations (a) and (b) are in the subtropics, whereas (c) and (d) are middle-latitude sites.

Precipitation Within subtropical deserts, the scant precipitation is both infrequent and erratic. Indeed, no well-defined seasonal precipitation pattern exists in subtropical deserts. The reason is that these areas are too far poleward to be influenced by the ITCZ and too far equatorward to benefit from the frontal and cyclonic precipitation of the middle latitudes. Even during summer, when daytime heating produces a steep environmental lapse rate and considerable convection, clear skies still rule. In this case, subsidence aloft prevents the lower air, with its modest moisture content, from rising high enough to penetrate the condensation level.

The situation is different in the semiarid transitional belts surrounding the desert. Here a seasonal rainfall pattern becomes better defined. As shown by the data for Dakar in Table 15–5 (see page 426), stations located on the equatorward side of low-latitude deserts have a brief period of relatively heavy rainfall during the summer, when the ITCZ is farthest poleward. The rainfall regime should look familiar, for it is similar to that found in the adjacent wet and dry tropics, except that the amount is less and the dry period lasts longer.

For steppe areas on the poleward margins of the tropical deserts, the precipitation regime is reversed. As the data for Marrakech illustrate (Table 15–5), the cool season is the period when nearly all precipitation falls. At this time of year, middle-latitude cyclones often take more equatorward routes and so bring occasional periods of rain.

Temperature The keys to understanding temperatures in the desert environment are humidity and cloud cover. The cloudless sky and low humidity allow an abundance of solar radiation to reach the ground during the day and permit the rapid exit of terrestrial radiation at night. As would be expected, relative humidities are low throughout the year. Relative humidities between 10 and 30 percent are typical at midday for interior locations.

In addition, arid and semiarid areas exist in southern Africa and make a rather limited appearance in coastal Chile and Peru. The distribution of this dry subtropical realm is primarily a consequence of the subsidence and marked stability of the subtropical highs.

Box 15–3   The Disappearing Aral Sea—A Large Lake Becomes a Barren Wasteland

The Aral Sea lies on the border between Uzbekistan and Kazakhstan in central Asia (Figure 15–D). The setting is the Turkestan Desert, a middle-latitude desert in the rain shadow of Afghanistan's high mountains. In this region of interior drainage, two large rivers, the Amu Darya and the Syr Darya, carry water from the mountains of northern Afghanistan across the desert to the Aral Sea. Water leaves the sea by evaporation. Thus, the size of the water body depends upon the balance between river inflow and evaporation.

In 1960 the Aral Sea was one of the world's largest inland water bodies, with an area of about 67,000 square kilometers (26,000 square miles). Only the Caspian Sea, Lake Superior, and Lake Victoria were larger. By 2010 the area of the Aral Sea was about 10 percent of its 1960 size. Its volume was reduced by nearly 90 percent, from 708 cubic kilometers to 75 cubic kilometers. The shrinking Aral Sea is depicted in Figure 15–E. All that remains are three shallow remnants.

What caused the Aral Sea to dry up? The answer is that the flow of water from the mountains that supplied the sea was significantly reduced and then all but eliminated. The waters of the Amu Darya and Syr Darya were diverted to supply a major expansion of irrigated agriculture in this dry realm.

The intensive irrigation greatly increased agricultural productivity, but not without significant costs. The deltas of the two major rivers have lost their wetlands, and wildlife has disappeared. The once-thriving fishing industry is dead, and the 24 species of fish that once lived in the Aral Sea are no longer there. The shoreline is now tens of kilometers from the towns that were once fishing centers.

The shrinking sea has exposed millions of acres of former seabed to sun and wind. The surface is encrusted with salt and with agricultural chemicals brought by the rivers. Strong winds routinely pick up and deposit thousands of tons of newly exposed material every year. This process has not only contributed to a significant reduction in air quality for people living in the region but has also appreciably affected crop yields due to the deposition of salt-rich sediments on arable land.

The shrinking Aral Sea has had a noticeable impact on the region's climate. Without the moderating effect of a large water body, there are greater extremes of temperature, a shorter growing season, and reduced local precipitation. These changes have caused many farms to switch from growing cotton to growing rice, which demands even more diverted water.

Could this crisis be reversed if enough freshwater were to once again flow into the Aral Sea? Prospects appear grim. Experts estimate that restoring the Aral Sea to about twice its present size would require stopping all irrigation from the two major rivers for decades. This could not be done without ruining the economies of the countries that rely on that water. One effort has improved the situation in the northern portion of the Aral. In November 2005, a large earthen dam was completed that blocked the southward outflow of water. Prior to the construction of this structure, water contributed by the Syr Darya was lost as it flowed southward and evaporated. The dam allowed water from the Syr Darya to recharge and partially restore this portion of the water body. The North Aral Sea is now larger and less salty than before.

The decline of the Aral Sea is a major environmental disaster that, sadly, is of human making.

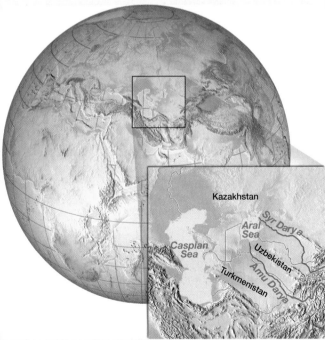

FIGURE 15–D The Aral Sea lies east of the Caspian Sea in the Turkestan Desert. Two rivers, the Amu Darya and Syr Darya, bring water from the mountains to the south. This map shows the outline of the water body prior to 1960. Today it is much smaller.

FIGURE 15–E The shrinking Aral Sea. Small remnants are all that remain. (NASA)

TABLE 15-5 Data for Subtropical Steppe and Desert Stations

	J	F	M	A	M	J	J	A	S	O	N	D	YR
Marrakech, Morocco, 31° 37' N; 458 m													
Temp. (°C)	11.5	13.4	16.1	18.6	21.3	24.8	28.7	28.7	25.4	21.2	16.5	12.5	19.9
Precip. (mm)	28	28	33	30	18	8	3	3	10	20	28	33	242
Dakar, Senegal, 14° 44' N; 23 m													
Temp. (°C)	21.1	20.4	20.9	21.7	23.0	26.0	27.3	27.3	27.5	27.5	26.0	25.2	24.49
Precip. (mm)	0	2	0	0	1	15	88	249	163	49	5	6	578
Alice Springs, Australia, 23° 38' S; 570 m													
Temp. (°C)	28.6	27.8	24.7	19.7	15.3	12.2	11.7	14.4	18.3	22.8	25.8	27.8	20.8
Precip. (mm)	43	33	28	10	15	13	8	8	8	18	30	38	252

Regarding cloud cover, desert skies are almost always clear (Figure 15–12). In the Sonoran Desert region of Mexico and the United States, for example, most stations receive nearly 85 percent of the possible sunshine. Yuma, Arizona, averages 91 percent for the year, with a low of 83 percent in January and a high of 98 percent in June. The Sahara has an average winter cloud cover of about 10 percent, which in summer drops to a mere 3 percent.

During summer, the desert surface heats rapidly after sunrise because the clear skies permit almost all the solar energy to reach the surface, as we saw in preceding examples. By midafternoon ground-surface temperatures may approach 90°C (nearly 200°F)! Under such circumstances, it is not surprising that the world's highest temperatures are recorded in the subtropical deserts. Nor is it unexpected that the daily maximums at many stations during the hot season are consistently close to the absolute maximum (the highest temperature ever recorded at a station). At Abadán, Iran, the average daily maximum in July is a scorching 44.7°C (112°F), or only 8.3°C (15°F) lower than the record high for that station. Phoenix, Arizona, is not much better,

recording an average July maximum of 40.5°C (105°F) compared with an absolute maximum of 47.7°C (118°F).

A contributing factor to the high ground and air temperatures is that little energy from insolation goes to power evaporation. Thus, almost all the energy goes to heating the surface. In contrast, humid regions are less likely to have such extreme ground and air temperatures, for more energy from solar radiation goes to evaporate water and less remains to heat the ground.

At night, temperatures typically drop rapidly, partly because the water vapor content of the air is fairly low. Ground-surface temperature is also a factor, however. Recall from the discussion of radiation in Chapter 2 that the higher the temperature of a radiating body, the faster it loses heat. Thus, applied to a desert setting, such environments not only heat up quickly by day but also cool rapidly at night.

Consequently, low-latitude deserts in the interiors of continents have the greatest daily temperature ranges on Earth. Daily ranges from 15 to 25°C (27 to 45°F) are common, and they occasionally reach even higher values. The highest daily temperature range ever recorded was at In Salah, Algeria, in the Sahara. On October 13, 1927, this station experienced a 24-hour range of 55.5°C (100°F), from 52.2 to −3.3°C (126 to 26°F).

Because most areas of *BWh* and *BSh* are poleward of the *A* climates, annual temperature ranges are the highest among the tropical climates. During the low-Sun period, averages are below those in other parts of the tropics, with monthly means of 16 to 24°C (60 to 75°F) being typical. Still, temperatures during the summer are higher than those in the humid tropics. Consequently, annual means at many subtropical desert and steppe stations are similar to those in the *A* climates.

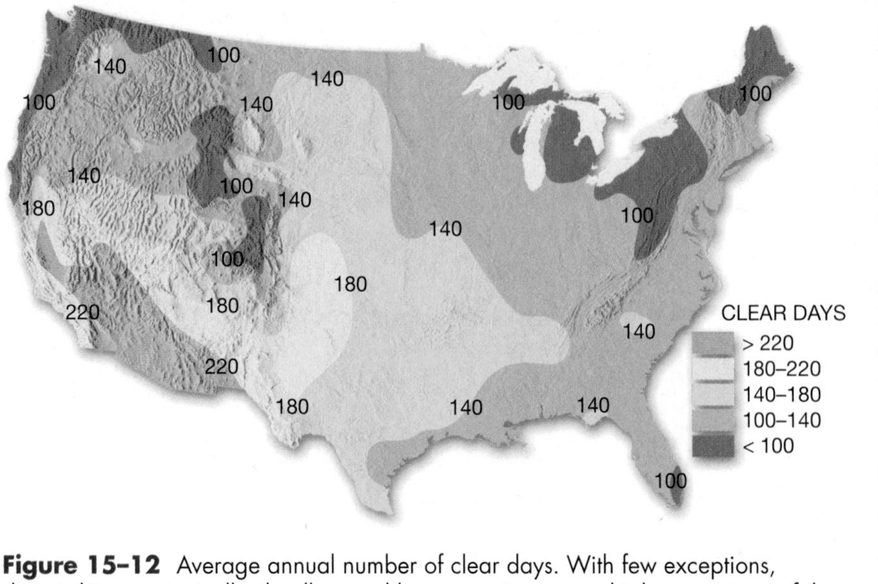

West Coast Subtropical Deserts

Where subtropical deserts are found along the west coasts of continents, cold ocean currents have a dramatic influence on the climate. The principal west coast deserts are the

Figure 15-12 Average annual number of clear days. With few exceptions, desert skies are typically cloudless and hence receive a very high percentage of the possible sunlight. This is strikingly illustrated in the U.S. Southwest desert.

CLEAR DAYS
> 220
180–220
140–180
100–140
< 100

TABLE 15–6 Data for West Coast Tropical Desert Stations

	J	F	M	A	M	J	J	A	S	O	N	D	YR
Port Nolloth, South Africa, 29° 14' S; 7 m													
Temp. (°C)	15	16	15	14	14	13	12	12	13	13	15	15	14
Precip. (mm)	2.5	2.5	5.1	5.1	10.2	7.6	10.2	7.6	5.1	2.5	2.5	2.5	63.4
Lima, Peru, 12° 02' S; 155 m													
Temp. (°C)	22	23	23	21	19	17	16	16	16	17	19	21	19
Precip. (mm)	2.5	T	T	T	5.1	5.1	7.6	7.6	7.6	2.5	2.5	T	40.5

Atacama in South America and the Namib in southern and southwestern Africa. Other areas include portions of the Sonoran Desert in Baja, California, and coastal areas of the Sahara in northwestern Africa.

A Different Kind of Desert West coast subtropical deserts deviate considerably from the general image we have of subtropical deserts. The most obvious effect of the cold current is reduced temperatures, as exemplified by the data for Lima, Peru, and Port Nolloth, South Africa (Table 15–6). Compared with other stations at similar latitudes, these places have lower annual mean temperatures and subdued annual and daily ranges. Port Nolloth, for example, has an annual mean of only 14°C (57°F) and an annual range of just 4°C (7.2°F). Contrast this with Durban, on the opposite side of South Africa, which has a yearly mean of 20°C (68°F) and an annual range twice that at Port Nolloth.

Although these stations are adjacent to the oceans, their yearly rainfall totals are among the lowest in the world. The aridity of these coasts is intensified because the lower air is chilled by the cold offshore waters and hence further stabilized. In addition, the cold currents cause temperatures often to reach the dew point. As a result, these areas are characterized by high relative humidity and much advection fog and dense stratus cloud cover.

The point is that not all subtropical deserts are sunny, hot places with low humidity and cloudless skies. Indeed, the presence of cold currents causes west coast subtropical deserts to be relatively cool, humid places, often shrouded by low clouds or fog.

Chile's Atacama Desert: Driest of the Dry Stretching for nearly 1000 kilometers (600 miles) in northern Chile, the Atacama Desert is situated between the Pacific Ocean on the west and the towering Andes Mountains on the east (see Figure 15–10). This slender arid zone extends inland an average distance of just 50 to 80 kilometers (30 to 50 miles). At its widest, the Atacama spans only 160 kilometers (100 miles).

The Atacama has the distinction of being the world's driest desert (Figure 15–13). In many places, measurable rain occurs only at intervals of several years. The average rainfall at the Atacama's wettest locations is not more than 3 millimeters (0.12 inch) per year. At Arica, a coastal town near Chile's border with Peru, the average annual rainfall is a mere 0.5 millimeter (0.02 inch). Further inland, some stations have *never* recorded rainfall.

Why is this narrow band of land so dry? First, the region is dominated by the dry, subsiding air associated with the semi-permanent cell of high pressure that dominates the eastern South Pacific (see Figure 7–10, page 200). Second, the cold Peru Current flows northward along the coast and contributes to aridity because it chills and stabilizes the lower atmosphere (see Figure 7–21, page 209). A third factor that adds to the Atacama's extraordinary aridity is the fact that the Andes Mountains shield the Pacific coast from incursions of humid air from the east.

As is the case for other west coast subtropical deserts, the Atacama is a relatively cool place where advection fogs are common. Both phenomena are related to the cold Peru Current. The fogs, called *camanchacas*, form as

Figure 15–13 Chile's Atacama Desert is classified as a west coast subtropical desert. It is the driest desert on Earth. The wettest places here receive an average of about 3 millimeters (0.12 inch) of rain per year. (Photo by Jaques Jangoux/Photo Researchers, Inc.)

Figure 15-14 Mountains frequently contribute to the aridity of middle-latitude deserts and steppes by creating a rain shadow. Orographic lifting leads to precipitation on the windward slopes. By the time air reaches the leeward side of the mountains, much of the moisture has been lost. The Great Basin Desert is a rain shadow desert that covers nearly all of Nevada and portions of adjacent states. (Photo on left by Dean Pennala/Shutterstock; photo on right by Dennis Tasa)

moist air moves over the cold current and is chilled below its dew point.

Middle-Latitude Desert (BWk) and Steppe (BSk)

Unlike their low-latitude counterparts, middle-latitude deserts and steppes are not controlled by the subsiding air masses of the subtropical highs. Instead, these dry lands exist principally because of their position in the deep interiors of large landmasses, far removed from the main moisture source—the oceans. In addition, the presence of high mountains across the paths of the prevailing winds further separates these areas from water-bearing maritime air masses (Figure 15–14). In North America the Coast ranges, Sierra Nevada, and Cascades are the foremost barriers; in Asia the great Himalayan chain prevents the summertime monsoon flow of moist air off the Indian Ocean from reaching far into the interior.

A glance at Figure 15–10 reveals that middle-latitude desert and steppe climates are most widespread in North America and Eurasia. The Southern Hemisphere lacks extensive land areas in the middle latitudes, so it has a much smaller area of *BWk* and *BSk*; it is found only at the southern tip of South America, in the rain shadow of the towering Andes.

Like subtropical deserts and steppes, the dry regions of the middle latitudes have meager and unreliable precipitation. Unlike the dry lands of the low latitudes, however, these more poleward regions have much lower winter temperatures and hence lower annual means and higher annual ranges of temperature.

The data in Table 15–7 illustrate this point nicely. The data also reveal that rainfall is most abundant during the warm months. Although not all *BWk* and *BSk* stations have a summer precipitation maximum, most do because in winter, high pressure and cold temperatures tend to dominate the continents. Both factors oppose precipitation. In summer, however, conditions are somewhat more conducive to cloud formation and precipitation because the anticyclone disappears over the heated continent, and higher surface temperatures and greater mixing ratios prevail.

TABLE 15-7 Data for Middle-Latitude Steppe and Desert Stations

	J	F	M	A	M	J	J	A	S	O	N	D	YR
Ulan Bator, Mongolia, 47° 55' N; 1311 m													
Temp. (°C)	−26	−21	−13	−1	6	14	16	14	9	−1	−13	−22	−3
Precip. (mm)	1	2	3	5	10	28	76	51	23	7	4	3	213
Denver, Colorado, 39° 32' N; 1588 m													
Temp. (°C)	0	1	4	9	14	20	24	23	18	12	5	2	11
Precip. (mm)	12	16	27	47	61	32	31	28	23	24	16	10	327

situated areas: all of Uruguay and portions of Argentina and southern Brazil in South America, eastern China and southern Japan in Asia, and the eastern coast of Australia.

Humid Middle-Latitude Climates with Mild Winters (*C*)

The Köppen classification recognizes two groups of humid middle-latitude climates. One group has mild winters (the *C* climates) and the other experiences severe winters (the *D* climates). The following three sections pertain to the *C*-type mild winter group. Figure 15–15 presents climate diagrams of the three types of *C* climates.

Humid Subtropical Climate (*Cfa*)

Humid subtropical climates are found on the eastern sides of the continents, in the 25° to 40° latitude range. They dominate the southeastern United States and other similarly

In the summer a visitor to the humid subtropics would experience hot, sultry weather of the type expected in the rainy tropics. Daytime temperatures are generally in the lower 30s (°C) (high 80s°F), but it is not uncommon for the thermometer to reach into the upper 30s (90s°F) or even 40 (°C) (more than 100°F) on many afternoons. Because the mixing ratio and relative humidity are high, the night brings little relief. An afternoon or evening thunderstorm is possible because these areas experience them between 40 and 100 days each year, the majority during summer.

The primary reason for the tropical summer weather in *Cfa* regions is the dominating influence of maritime tropical air masses. During the summer months, this warm, moist,

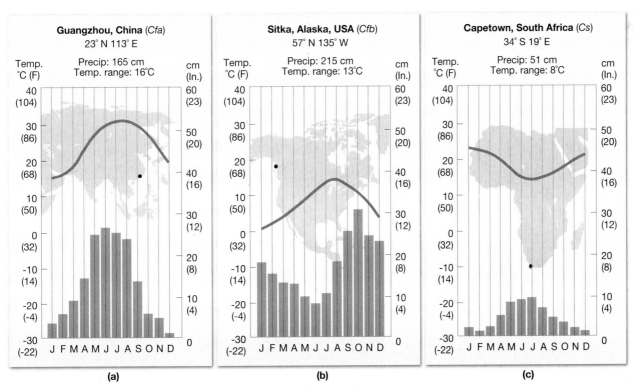

Figure 15–15 Each of these climate diagrams represents a type of *C* climate: (a) humid subtropical, (b) marine west coast, and (c) dry-summer subtropical.

unstable air moves inland from the western portions of the oceanic subtropical anticyclone. As the maritime tropical (*mT*) air passes over the heated continent, it becomes increasingly unstable, giving rise to the common convectional showers and thunderstorms.

As summer turns to autumn, the humid subtropics lose their similarity to the rainy tropics. Although winters are mild, frosts are common in higher-latitude *Cfa* areas and occasionally plague the tropical margins. The winter precipitation is also different in character. Some is in the form of snow, and most is generated along fronts of the frequent middle-latitude cyclones that sweep over these regions.

Because the land surface is colder than the maritime air, the air becomes chilled in its lower layers as it moves poleward. Consequently, convectional showers are rare, for the stabilized *mT* air masses produce clouds and precipitation only when forced to rise.

The data for two humid subtropical stations in Table 15–8 serve to summarize the general characteristics of the *Cfa* climate. Yearly precipitation usually exceeds 100 centimeters (39 inches), and the rainfall is well distributed throughout the year. Summer normally brings the most precipitation, but there is considerable variation. In the United States, for example, precipitation in the Gulf states is very evenly distributed. But as one moves poleward or toward the drier western margins, much more falls in summer. Some coastal stations have rainfall maximums in late summer or autumn, when tropical cyclones or their remnants visit the area. In Asia the well-developed monsoon circulation favors a summer precipitation maximum (see the Chinese station in Figure 15–15a).

Temperature figures show that summer temperatures are comparable to temperatures in the tropics and that winter values are markedly lower. This is to be expected because the higher-latitude position of the subtropics experiences a wider variation in Sun angle and day length, plus occasional (even frequent) invasions of continental polar (*cP*) air masses during winter.

The Marine West Coast Climate (*Cfb*)

On the western (windward) side of continents from about 40° to 65° north and south latitude is a climate region dominated by the onshore flow of oceanic air. The prevalence of maritime air masses means mild winters, cool summers, and ample rainfall throughout the year. In North America this **marine west coast climate** (*Cfb*) extends from near the

EYE ON THE ATMOSPHERE

This classic view of Earth from space was taken in December 1968 by an *Apollo 8* astronaut. The image shows the Western Hemisphere. Dense clouds cover much of North America. A large portion of South America is also cloud covered. However, the western margin of South America is cloud free.

(NASA)

Question 1 What is the name of the narrow cloudless desert along the West Coast of South America?

Question 2 An ocean current flows just offshore of this desert area. What is the name of the current? Is it warm or cold?

Question 3 How does this desert's close proximity to the Pacific Ocean influence its aridity?

Question 4 In what ways is this desert different from subtropical deserts such as the Sahara and Arabian? Name a desert that has characteristics that are similar to this one.

U.S.–Canada border northward as a narrow belt into southern Alaska (Figure 15–16). A similar slender strip occurs in South America along the coast of Chile. In both instances, high mountains parallel the coast and prevent the marine climate from penetrating far inland. The largest area of *Cfb* climate is in Europe, where there is no mountain barrier blocking the movement of cool maritime air from the North Atlantic. Other locations include most of New Zealand as well as tiny slivers of South Africa and Australia.

TABLE 15–8 Data for Humid Subtropical Stations

	J	**F**	**M**	**A**	**M**	**J**	**J**	**A**	**S**	**O**	**N**	**D**	**YR**
New Orleans, Louisiana, 29° 59' N; 1 m													
Temp. (°C)	12	13	16	19	23	26	27	27	25	21	15	13	20
Precip. (mm)	98	101	136	116	111	113	171	136	128	72	85	104	1371
Buenos Aires, Argentina, 34° 35' S; 27 m													
Temp. (°C)	24	23	21	17	14	11	10	12	14	16	20	22	17
Precip. (mm)	104	82	122	90	79	68	61	68	80	100	90	83	1027

Figure 15–16 Fog is common along the rocky Pacific coastline at Olympic National Park, Washington. This area is classified as a *marine west coast climate.* As the name implies, the ocean exerts a strong influence. (Photo by Richard J. Green/ Photo Researchers, Inc.)

The data for representative marine west coast stations in Table 15–9 and Figure 15–15b reveal no pronounced dry period, although monthly precipitation drops during the summer. The reduced summer rainfall is due to the poleward migration of the oceanic subtropical highs. Although the areas of marine west coast climate are too far poleward to be dominated by these dry anticyclones, their influence is sufficient to cause a decrease in warm-season rainfall. This is illustrated nicely by two precipitation graphs in Figure 4–D (page 123)—for Quinault Ranger Station and Rainier Paradise Station.

A comparison of precipitation data for London and Vancouver (Table 15–9) also demonstrates that coastal mountains have a significant influence on yearly rainfall. Vancouver has about two and a half times that of London. In settings like Vancouver's, the precipitation totals are higher not only because of orographic uplift but also because the mountains slow the passage of cyclonic storms, allowing them to linger and drop a greater quantity of water.

The ocean is near, so winters are mild and summers relatively cool. Therefore, a low annual temperature range is characteristic of the marine west coast climate. Because *cP*

TABLE 15–9 Data for Marine West Coast Stations

	J	F	M	A	M	J	J	A	S	O	N	D	YR
Vancouver, British Columbia, 49° 11' N; 0 m													
Temp. (°C)	2	4	6	9	13	15	18	17	14	10	6	4	10
Precip. (mm)	139	121	96	60	48	51	26	36	56	117	142	156	1048
London, United Kingdom, 51° 28' N; 5 m													
Temp. (°C)	4	4	7	9	12	16	18	17	15	11	7	5	10
Precip. (mm)	54	40	37	38	46	46	56	59	50	57	64	48	595

air masses generally drift eastward in the zone of the westerlies, periods of severe winter cold are rare. The western edge of North America is especially sheltered from incursions of frigid continental air by the high mountains that intervene between the coast and the source regions for *cP* air masses. Because no such mountain barrier exists in Europe, cold waves there are somewhat more frequent.

The ocean's control of temperatures can be further demonstrated by a look at temperature gradients (that is, changes in temperature per unit distance). Although this climate encompasses a wide latitudinal span, temperatures change much more abruptly moving inland from the coast than they do in a north–south direction. The transport of heat from the oceans more than offsets the latitudinal variation in the receipt of solar energy. For example, in both January and July, the temperature change from coastal Seattle to inland Spokane, a distance of about 375 kilometers (230 miles), is equal to the variation between Seattle and Juneau, Alaska. Juneau is at about 11° of latitude, or roughly 1200 kilometers (750 miles), north of Seattle.

The Dry-Summer Subtropical (Mediterranean) Climate (*Csa, Csb*)

The **dry-summer subtropical climate** is typically located along the west sides of continents between latitudes 30° and 45°. Situated between the marine west coast climate on the poleward side and the subtropical steppes on the equatorward side, this climate region is transitional in character. It is the only humid climate that has a pronounced winter rainfall maximum, a feature that reflects its intermediate position (see Figure 15–15c).

In summer the region is dominated by the stable eastern side of the oceanic subtropical highs. In winter, as the wind and pressure systems follow the Sun equatorward, it is within range of the cyclonic storms of the polar front. Thus, during the course of a year, these areas alternate between being a part of the dry tropics and being an extension of the humid middle latitudes. Although middle-latitude changeability characterizes the winter, tropical constancy describes the summer.

As is the case for the marine west coast climate, mountain ranges limit the dry-summer subtropics to a relatively narrow coastal zone in both North and South America. Because Australia and southern Africa barely extend to the latitudes where dry-summer climates exist, the development of this climatic type is limited on these continents as well.

Because of the arrangement of the continents and their mountain ranges, inland development occurs only in the Mediterranean basin (Figure 15–17). Here the zone of subsidence extends far to the east in summer; in winter the sea is a major route of cyclonic disturbances. Because the dry-summer climate is particularly extensive in this region, the name **Mediterranean climate** is often used as a synonym.

Temperature Two types of Mediterranean climate are recognized and are based primarily on summertime temperatures. The cool summer type (*Csb*), as exemplified by San Francisco and Santiago, Chile (Table 15–10), is limited to coastal areas. Here the cooler summer temperatures one expects on a windward coast are further intensified by cold ocean currents.

Figure 15–17 The dry-summer subtropical climate is especially well developed in the Mediterranean region. Italy's Tuscany region is an example. (Photo by Stefano Caporali/agefotostock)

TABLE 15–10 Data for Dry-Summer Subtropical Stations

	J	F	M	A	M	J	J	A	S	O	N	D	YR
San Francisco, California, 37° 37' N; 5 m													
Temp. (°C)	9	11	12	13	15	16	17	17	18	16	13	10	14
Precip. (mm)	102	88	68	33	12	3	0	1	5	19	40	104	475
Sacramento, California, 38° 35' N; 13 m													
Temp. (°C)	8	10	12	16	19	22	25	24	23	18	12	9	17
Precip. (mm)	81	76	60	36	15	3	0	1	5	20	37	82	416
Izmir, Turkey, 38° 26' N; 25 m													
Temp. (°C)	9	9	11	15	20	25	28	27	23	19	14	10	18
Precip. (mm)	141	100	72	43	39	8	3	3	11	41	93	141	695
Santiago, Chile, 33° 27' S; 512 m													
Temp. (°C)	19	19	17	13	11	8	8	9	11	13	16	19	14
Precip. (mm)	3	3	5	13	64	84	76	56	30	13	8	5	360

The data for Izmir, Turkey, and Sacramento illustrate the features of the warm summer type (*Csa*). At both places winter temperatures are not very different from those in the *Csb* type. But Sacramento, in the Central Valley of California, is removed from the coast, and Izmir is bordered by the warm waters of the Mediterranean; consequently, summer temperatures are noticeably higher. As a result, annual temperature ranges are also higher in *Csa* areas.

Precipitation Yearly precipitation within the dry-summer subtropics ranges between about 40 and 80 centimeters (16 and 31 inches). In many areas such amounts mean that a station barely escapes being classified as semiarid. As a result, some climatologists refer to the dry-summer climate as *subhumid* instead of humid. This is especially true along the equatorward margins because rainfall totals increase in the poleward direction. Los Angeles, for example, receives 38 centimeters (15 inches) of precipitation annually, whereas San Francisco, 400 kilometers (250 miles) to the north, receives 51 centimeters (20 inches) per year. Still farther north, at Portland, Oregon, the yearly rainfall average is over 90 centimeters (35 inches).

Concept Check 15.7

1 Describe and explain the differences between summertime and wintertime precipitation in the humid subtropics (*Cfa*).

2 Why is the marine west coast climate (*Cfb*) represented by only slender strips of land in North and South America but is extensive in Western Europe?

3 How do temperature gradients (north–south versus east–west) reveal the strong oceanic influence along the west coast of North America?

4 What other name is given to the dry-summer subtropical climate?

5 Why are summer temperatures cooler at San Francisco than at Sacramento (Table 15–10)?

Humid Continental Climates with Severe Winters (*D*)

The *C* climates just described have mild winters. By contrast, *D* climates experience severe winters. In this and the following section, two types of *D* climates are discussed—the humid continental and the subarctic. Climate diagrams of representative locations are shown in Figure 15–18.

Students Sometimes Ask...

Do soils differ from one climate to another?

The answer is a definite yes. Climate is one of the most influential soil-forming factors. Variations in temperature and precipitation determine the type of weathering that occurs and also greatly influence the rate and depth of soil formation. For example, a hot, wet climate may produce a thick layer of chemically weathered soil in the same amount of time that a cold, dry climate produces a thin mantle of mechanically weathered debris. Also, the amount of precipitation influences the degree to which various materials are leached from the soil, thereby affecting soil fertility. Finally, climate is an important control on the type of plant and animal life present, which in turn influences the nature of the soil that forms.

This climate region is dominated by the polar front and thus is a battleground for tropical and polar air masses. No other climate experiences such rapid, nonperiodic changes in the weather. Cold waves, heat waves, droughts, blizzards, and heavy downpours are all yearly events in the humid continental realm. For more about droughts, see *Severe and Hazardous Weather*, page 440.

Humid Continental Climate (*Dfa*)

The **humid continental climate** (*Dfa*), as its name implies, is a land-controlled climate. It is the product of broad conti-

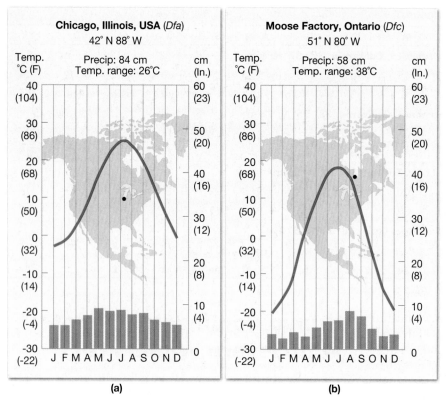

Figure 15–18 *D* climates are associated with the interiors of large landmasses in the mid- to high latitudes of the Northern Hemisphere. Although winters can be harsh in Chicago's humid continental (*Dfa*) climate (a), the subarctic environment (*Dfc*) of Moose Factory is more extreme (b).

nents located in the middle latitudes. Because continentality is a basic feature, this climate does not occur in the Southern Hemisphere, where the middle-latitude zone is dominated by the ocean. Instead, it is confined to central and eastern North America and Eurasia in the latitude range 40° to 50° north.

It may at first seem unusual that a continental climate should extend eastward to the margins of the ocean. How-

ever, the prevailing atmospheric circulation is from the west, so deep and persistent incursions of maritime air from the east are unlikely.

Temperature Both winter and summer temperatures in the *Dfa* climate are relatively severe. Consequently, annual temperature ranges are great. A comparison of the stations in Table 15–11 illustrates this. July means are generally near and often above 20°C (68°F), so summertime temperatures, although lower in the north than in the south, are not markedly different.

This is illustrated by the temperature distribution map of the eastern United States in Figure 15–19a. The summer map has only a few widely spaced isotherms, indicating a weak summer temperature gradient. The winter map, however, shows a stronger temperature gradient (Figure 15–19b). The decrease in midwinter values with increasing latitude is appreciable. Confirming this, Table 15–11 reveals that the temperature change between Omaha and Winnipeg is more than twice as great in winter as in summer.

Because of the steeper winter temperature gradient, shifts in wind direction during the cold season often result in sudden large temperature changes. Such is not the case in summer, when temperatures throughout the region are more uniform.

Annual temperature ranges also vary within this climate, generally increasing from south to north and from the coast toward the interior. Comparing the data for Omaha and Winnipeg illustrates the first situation, and comparing New York and Omaha illustrates the second.

TABLE 15–11 Data for Humid Continental Stations

	J	F	M	A	M	J	J	A	S	O	N	D	YR
Omaha, Nebraska, 41° 18' N; 330 m													
Temp. (°C)	–6	–4	3	11	17	22	25	24	19	12	4	–3	10
Precip. (mm)	20	23	30	51	76	102	79	81	86	48	33	23	652
New York, New York, 40° 47' N; 40 m													
Temp. (°C)	–1	–1	3	9	15	21	23	22	19	13	7	1	11
Precip. (mm)	84	84	86	84	86	86	104	109	86	86	86	84	1065
Winnipeg, Canada, 49° 54' N; 240 m													
Temp. (°C)	–18	–16	–8	3	11	17	20	19	13	6	–5	–13	3
Precip. (mm)	26	21	27	30	50	81	69	70	55	37	29	22	517
Harbin, Manchuria, 45° 45' N; 143 m													
Temp. (°C)	–20	–16	–6	6	14	20	23	22	14	6	–7	–17	3
Precip. (mm)	4	6	17	23	44	92	167	119	52	36	12	5	577

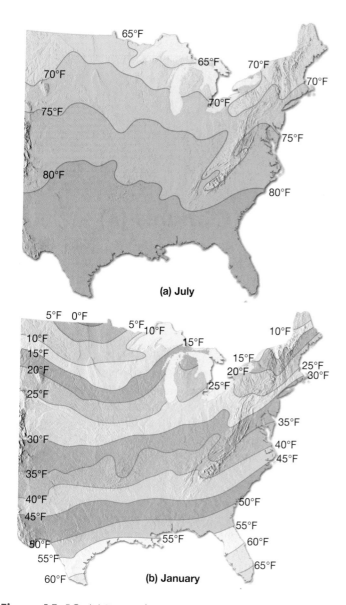

Figure 15–19 (a) During the summer months, north–south temperature variations in the eastern United States are small—that is, the temperature gradient is weak. (b) In winter, however, north–south temperature contrasts are sharp.

northerly stations are also influenced for a greater part of the year by drier polar air masses.

Wintertime precipitation is chiefly associated with the passage of fronts connected with traveling middle-latitude cyclones. Part of this precipitation is snow, the proportion increasing with latitude. Although precipitation is often considerably less during the cold season, it is usually more conspicuous than the greater amounts that fall during summer. An obvious reason is that snow remains on the ground, often for extended periods, and rain, of course, does not. Moreover, summer rains are often in the form of relatively short convective showers, whereas winter snows are more prolonged.

The Subarctic Climate (*Dfc, Dfd*)

North of the humid continental climate and south of the polar tundra is an extensive **subarctic climate.** It covers broad, uninterrupted expanses in North America (western Alaska to Newfoundland) and in Eurasia (Norway to the Pacific coast of Russia). It is often called the **taiga climate,** for it closely corresponds to the northern coniferous forest region of the same name (Figure 15–20). Although scrawny, the spruce, fir, larch, and birch trees in the taiga represent the largest stretch of continuous forest on Earth.

Temperature   The subarctic is well illustrated by the climate diagram in Figure 15–18b and by the data for Yakutsk, Russia, and Dawson, Yukon Territory, in Table 15–12. Here in the source regions of continental polar (*cP*) air masses, the outstanding feature is the dominance of winter. Not only is it long, it is bitterly cold. Winter minimum temperatures are

Precipitation   Records of the four stations in Table 15–11 reveal the general precipitation pattern for *Dfa* climates. A summer maximum occurs at each station. But it is weakly defined at New York because the east coast is more accessible to maritime air masses throughout the year. For the same reason, New York also has the highest total of the four stations. Harbin, Manchuria, on the other hand, shows the most pronounced summer maximum, followed by a winter dry season. This is characteristic of most eastern Asian stations in the middle latitudes and reflects the powerful control of the monsoon.

Another pattern revealed by the data is that precipitation generally decreases toward the continental interior and from south to north, primarily because of increasing distance from the sources of *mT* air. Furthermore, the more

Figure 15–20 The northern coniferous forest is also called the *taiga.* Denali National Park, Alaska. (Photo by Bill Brooks/Alamy)

TABLE 15–12 Data for Subarctic Stations

	J	F	M	A	M	J	J	A	S	O	N	D	YR
Yakutsk, Russia, 62° 05' N; 103 m													
Temp. (°C)	−43	−37	−23	−7	7	16	20	16	6	−8	−28	−40	−10
Precip. (mm)	7	6	5	7	16	31	43	38	22	16	13	9	213
Dawson, Yukon Territory, Canada, 64° 03' N; 315 m													
Temp. (°C)	−30	−24	−16	−2	8	14	15	12	6	−4	−17	−25	−5
Precip. (mm)	20	20	13	18	23	33	41	41	43	33	33	28	346

among the lowest recorded outside the ice caps of Greenland and Antarctica. In fact, for many years the world's coldest temperature was attributed to Verkhoyansk in east-central Siberia, where the temperature dropped to −68°C (−90°F) on February 5 and 7, 1892. Over a 23-year period this same station had an average monthly minimum of −62°C (−80°F) during January. Although exceptional, these temperatures illustrate the extreme cold that envelops the taiga in winter.

In contrast, subarctic summers are remarkably warm, despite their short duration. When compared to regions farther south, however, this short season must be characterized as cool; for despite the many hours of daylight, the Sun never rises very high in the sky, so solar radiation is not intense. The extremely cold winters and the relatively warm summers of the taiga combine to produce the greatest annual temperature ranges on Earth. Yakutsk holds the distinction of having the greatest average temperature range in the world, 63°C (113°F). As the data for Dawson show, the North American subarctic is less severe.

Precipitation Because these far northerly continental interiors are the source regions of cP air masses, only limited moisture is available throughout the year. Precipitation totals are therefore small, seldom exceeding 50 centimeters (20 inches). By far the greatest precipitation comes as rain from scattered summer convectional showers. Less snow falls than in the humid continental climate to the south, yet there is the illusion of more. The reason is simple: No melting occurs for months at a time, so the entire winter accumulation (up to 1 meter) is visible all at once. Furthermore, during blizzards, high winds swirl the dry, powdery snow into high drifts, giving the false impression that more snow is falling than is actually the case. So although snowfall is not excessive, a visitor to this region could leave with that impression.

Concept Check 15.8

1 Why is the humid continental climate confined to the Northern Hemisphere?

2 Why do coastal stations such as New York City experience primarily continental climatic conditions?

3 Using the four stations in Table 15–11, describe the general pattern of precipitation in humid continental climates.

4 Describe and explain the annual temperature range one should expect in the realm of the taiga.

The Polar Climates (E)

According to the Köppen classification, **polar climates** are those in which the mean temperature of the warmest month is below 10°C (50°F). Two types are recognized: the tundra climate (ET) and the ice cap climate (EF). Climate diagrams of representative stations are presented in Figure 15–21.

Just as the tropics are defined by their year-round warmth, so the polar realm is known for its enduring cold, with the lowest annual means on the planet. Because polar winters are periods of perpetual night, or nearly so, temperatures are understandably bitter. During the summer, temperatures remain cool despite the long days, because the Sun is so low in the sky that its oblique rays produce little warming. In addition, much solar radiation is reflected by the ice and snow or used in melting the snow cover. In either case, energy that could have warmed the land is lost. Although summers are cool, temperatures are still much higher than those experienced during the severe winter months. Consequently, annual temperature ranges are extreme.

Although polar climates are classified as humid, precipitation is generally meager, with many nonmarine stations receiving less than 25 centimeters (10 inches) annually. Evaporation, of course, is also limited. The scanty precipitation is easily understood in view of the temperature characteristics of the region. The amount of water vapor in the air is always small because low mixing ratios must accompany low temperatures. In addition, steep lapse rates are not possible. Usually precipitation is most abundant during the warmer summer months, when the air's moisture content is highest.

The Tundra Climate (ET)

The **tundra climate** on land is found almost exclusively in the Northern Hemisphere. It occupies the coastal fringes of the Arctic Ocean, many Arctic islands, and the ice-free shores of northern Iceland and southern Greenland. In the Southern Hemisphere no extensive land areas exist in the latitudes where tundra climates prevail. Consequently, except for some small islands in the southern oceans, the ET climate occupies only the southwestern tip of South America and the northern portion of the Palmer Peninsula in Antarctica.

The 10°C (50°F) summer isotherm that marks the equatorward limit of the tundra also marks the poleward limit of

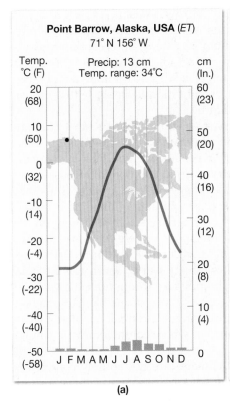

Point Barrow, Alaska, USA (*ET*)
71° N 156° W

Temp.
°C (F)   Precip: 13 cm   cm
                Temp. range: 34°C   (In.)

20 (68)   60 (23)
10 (50)   50 (20)
0 (32)   40 (16)
-10 (14)   30 (12)
-20 (-4)   20 (8)
-30 (-22)   10 (4)
-40 (-40)
-50 (-58)   J F M A M J J A S O N D   0

(a)

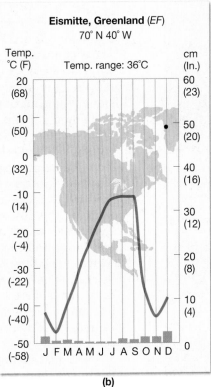

Eismitte, Greenland (*EF*)
70° N 40° W

Temp.
°C (F)   Temp. range: 36°C   cm
                                       (In.)

20 (68)   60 (23)
10 (50)   50 (20)
0 (32)   40 (16)
-10 (14)   30 (12)
-20 (-4)   20 (8)
-30 (-22)   10 (4)
-40 (-40)
-50 (-58)   J F M A M J J A S O N D   0

(b)

Figure 15–21 These climate diagrams represent the two basic polar climates. (a) Point Barrow, Alaska, exhibits a tundra (*ET*) climate. (b) Eismitte, Greenland, a station located on a massive ice sheet, is classified as an ice cap (*EF*) climate.

Large portions of the tundra are characterized by **permafrost** in which deep ground remains frozen year-round (Figure 15–23).* Strictly speaking, permafrost is defined only on the basis of temperature; that is, it is ground with temperatures that have remained below 0°C (32°F) continuously for two years or more. The degree to which ice is present in the ground strongly affects the behavior of the surface material. Knowing how much subsurface ice is present and where it is located is very important in constructing roads, buildings, and other projects in areas underlain by permafrost.

When the activities of people disturb the surface, such as by removing the insulating vegetation mat or by constructing roads and buildings, the delicate thermal balance is disturbed, and the permafrost can thaw. Thawing produces unstable ground that may slide, slump, subside, and undergo severe frost heaving. When a heated structure is built directly on ice-rich permafrost, thawing creates soggy material into which the structure can sink.

tree growth. Thus, the tundra is a treeless region of grasses, sedges, mosses, and lichens (Figure 15–22). During the long cold season, plant life is dormant, but once the short, cool summer commences, these plants mature and produce seeds very rapidly.

Because summers are cool and short, the frozen soils of the tundra generally thaw to depths of less than 1 meter.

*Permafrost is not confined to the tundra. It is also common in subarctic taiga regions.

Figure 15–22 The tundra is a region almost completely devoid of trees. Bogs and marshes are common, and plant life frequently consists of mosses, low shrubs, and flowering herbs. (Photo by Arcticphoto/Alamy)

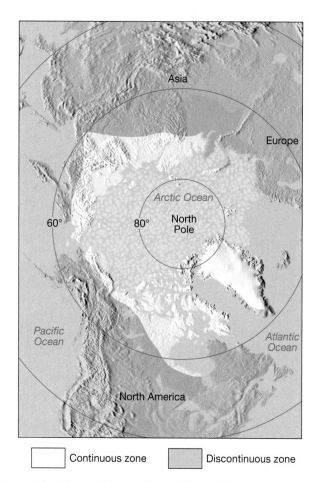

Figure 15–23 Distribution of permafrost in the Northern Hemisphere. More than 80 percent of Alaska and about 50 percent of Canada are underlain by permafrost. Two zones are recognized. In the continuous zone, the only ice-free areas are beneath deep lakes or rivers. In the higher-latitude portions of the discontinuous zones, there are only scattered islands of thawed ground. Moving southward, the percentage of unfrozen ground increases until all the ground is unfrozen. (After the U.S. Geological Survey)

The data for Point Barrow, Alaska (Table 15–13), on the shores of the frozen Arctic Ocean, exemplify the most common type of *ET* station, where continentality prevails. The combination of high latitude and continentality makes winters severe, summers cool, and annual temperature ranges great. Yearly precipitation is small, with a modest summertime maximum.

Although Point Barrow represents the most common type of tundra setting, the data for Angmagssalik, Greenland (see Table 15–13), reveal that some *ET* stations are different. Summer temperatures at the two stations are equivalent, yet winters at Angmagssalik are much warmer, and the annual precipitation is eight times greater than at Point Barrow. The reason is Angmagssalik's location on the southeastern coast of Greenland, where there is considerable marine influence. The warm North Atlantic Drift keeps winter temperatures relatively warm, and maritime polar (*mP*) air masses supply moisture throughout the year. Because winters are less severe at stations like Angmagssalik, annual temperature ranges are much smaller than at stations like Point Barrow, where continentality is a major control.

Note that tundra climates are not entirely confined to the high latitudes. The summer coolness of this climate is also found at higher elevations as one moves equatorward. Even in the tropics, you can find *ET* climates if you go high enough. When compared with the Arctic tundra, however, winter temperatures in these lower-latitude counterparts become milder and less distinct from summer, as the data for Cruz Loma, Ecuador (see Table 15–13), illustrate.

The Ice-Cap Climate (EF)

The **ice-cap climate**, designated by Köppen as *EF*, has no monthly mean above 0°C (32°F). Because the average temperature for all months is below freezing, the growth of vegetation is prohibited, and the landscape is one of per-

TABLE 15–13 Data for Polar Stations

	J	F	M	A	M	J	J	A	S	O	N	D	YR
Eismitte, Greenland, 70° 53' N; 2953 m													
Temp. (°C)	−42	−47	−40	−32	−24	−17	−12	−11	−11	−36	−43	−38	−29
Precip. (mm)	15	5	8	5	3	3	3	10	8	13	13	25	111
Angmagssalik, Greenland, 65° 36' N; 29 m													
Temp. (°C)	−7	−7	−6	−3	2	6	7	7	4	0	−3	−5	0
Precip. (mm)	57	81	57	55	52	45	28	70	72	96	87	75	775
Point Barrow, Alaska, 71° 18' N; 9 m													
Temp. (°C)	−28	−28	−26	−17	−8	0	4	3	−1	−9	−18	−24	−12
Precip. (mm)	5	5	3	3	3	10	20	23	15	13	5	5	110
Cruz Loma, Ecuador, 0° 08' S; 3888 m													
Temp. (°C)	6.1	6.6	6.6	6.6	6.6	6.1	6.1	6.1	6.1	6.1	6.6	6.6	6.4
Precip. (mm)	198	185	241	236	221	122	36	23	86	147	124	160	1779

Continuous zone   Discontinuous zone

manent ice and snow. This climate of perpetual frost covers a surprisingly large area of more than 15.5 million square kilometers (6 million square miles), or about 9 percent of Earth's land area. Aside from scattered occurrences in high mountain areas, it is largely confined to the ice sheets of Greenland and Antarctica.

Average annual temperatures are extremely low. For example, the annual mean at Eismitte, Greenland (see Table 15–13), is −29°C (−20°F); at Byrd Station, Antarctica, −21°C (−6°F); and at Vostok, the Russian Antarctic Meteorological Station, −57°C (−71°F). Vostok also experienced the lowest temperature ever recorded, −88.3°C (−127°F), on August 24, 1960.

In addition to latitude, the primary reason for such temperatures is the presence of permanent ice. Ice has a very high albedo, reflecting up to 80 percent of the meager sunlight that strikes it. The energy that is not reflected is used largely to melt the ice and so is not available for raising the temperature of air.

Another factor at many *EF* stations is elevation. Eismitte, at the center of the Greenland ice sheet, is almost 3000 meters above sea level (10,000 feet), and much of Antarctica is even higher. Thus, the permanent ice and high elevations further reduce the already low temperatures of the polar realm.

The intense chilling of the air close to the ice sheet means that strong surface-temperature inversions are common. Near-surface temperatures may be as much as 30°C (54°F) colder than air just a few hundred meters above. Gravity pulls this cold, dense air downslope, often producing strong winds and blizzard conditions. Such air movements, called *katabatic winds,* are an important aspect of ice cap weather at many locations. Where the slope is sufficient, these gravity-induced air movements can be strong enough to flow in a direction that is opposite the pressure gradient.

Concept Check 15.9

1 Although polar regions experience extended periods of sunlight in the summer, temperatures remain cool. Explain.

2 What is the significance of the 10°C (50°F) summer isotherm?

3 Why is the tundra landscape characterized by poorly drained, boggy soils?

4 The tundra climate is not confined solely to high latitudes. Under what circumstances might the *ET* climate be found in lower-latitude locations?

5 Where are *EF* climates developed most extensively?

Highland Climates

It is well known that mountain climates are distinctly different from those in adjacent lowlands. Sites with **highland climates** are cooler and usually wetter. The world climate types already discussed consist of large, relatively homogeneous regions. But highland climates are characterized by a great diversity of climatic conditions over small areas. Because large differences occur over short distances, the pattern of climates in mountainous areas is a complex mosaic, too complicated to depict on a world map.

In North America highland climates characterize the Rockies, Sierra Nevada, Cascades, and mountains and interior plateaus of Mexico. In South America the Andes create a continuous band of highland climate that extends for nearly 8000 kilometers (5000 miles). The greatest span of highland climates stretches from western China, across southern Eurasia, to northern Spain, from the Himalayas to the Pyrenees. Highland climates in Africa occur in the Atlas Mountains in the north and in the Ethiopian Highlands in the east.

The best-known climate effect of increased altitude is lower temperatures. Greater precipitation due to orographic lifting is also common at higher elevations. The precipitation map for Nevada in Figure 15–24 illustrates this nicely. The long, slender zones of highest precipitation coincide with areas of mountainous topography. Despite the fact that mountain stations are colder and often wetter than locations at lower elevations, highland climates are often very similar to those in adjacent lowlands in terms of seasonal

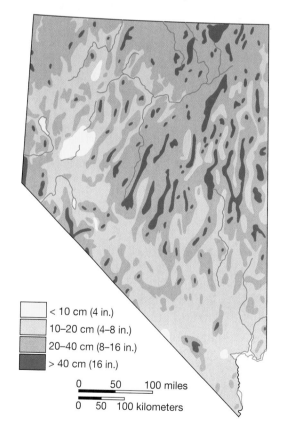

	< 10 cm (4 in.)
	10–20 cm (4–8 in.)
	20–40 cm (8–16 in.)
	> 40 cm (16 in.)

0   50   100 miles

0   50   100 kilometers

Figure 15–24 Precipitation map for the state of Nevada. Nevada is part of a region known as the Basin and Range that is characterized by many small mountain ranges that rise 900 to 1500 meters (3000–5000 feet) above the basins that separate them. Where there are mountains, there is more precipitation; where there are basins, there is less precipitation. The relationship is simple and straightforward.

SEVERE AND HAZARDOUS WEATHER

Drought—A Costly Atmospheric Hazard*

Drought is a period of abnormally dry weather that persists long enough to produce a significant hydrologic imbalance such as crop damage or water supply shortages. Drought severity depends upon the degree of moisture deficiency, its duration, and the size of the affected area.

> **Drought severity depends upon the degree of moisture deficiency, its duration, and the size of the affected area.**

During the summer of 2011, extreme or exceptional drought conditions extended from southern Arizona across New Mexico and Texas and into Georgia and Florida. The drought status map in Figure 15–F shows the extent and severity of the drought on July 12. The drought was a result of months of meager precipitation accompanied by near-record summer temperatures and desiccating winds. In the hardest-hit areas, crops failed and rangeland grasses were in such poor condition that many ranchers were forced to sell their herds.

Although natural disasters such as floods and hurricanes usually generate more attention, droughts can be just as devastating and carry a bigger price tag. On average, droughts cost the United States $6 to $8 billion annually compared to $2.4 billion for floods and $1.2 to $4.8 billion for hurricanes. Direct economic losses from a severe drought in 1988 were estimated at $40 billion.

Many people consider drought a rare and random event, yet it is actually a normal, recurring feature of climate. It occurs in virtually all climate zones, although its characteristics vary from region to region. The concept of drought differs from that of aridity. Drought is a temporary happening, whereas aridity describes regions where low rainfall is a permanent feature of the climate.

Drought is different from other natural hazards in several ways. First, it occurs in a gradual, "creeping" way, making its onset and end difficult to determine. The effects of drought accumulate slowly over an extended time span and sometimes linger for years after the drought has ended. Second, there is not a precise and universally accepted definition of drought. This adds to the confusion about whether a drought is actually occurring and, if it is, its severity. Third, drought seldom produces structural damages, so its social and economic effects are less obvious than are damages from other natural disasters.

> **Drought is a temporary happening, whereas aridity describes regions where low rainfall is a permanent feature of the climate.**

No single definition of drought works in all circumstances. Most definitions pertain to a specific region because of differences in climate characteristics. Thus, it is usually

*Based in part on material prepared by the National Drought Mitigation Center (http://drought.unl.edu).

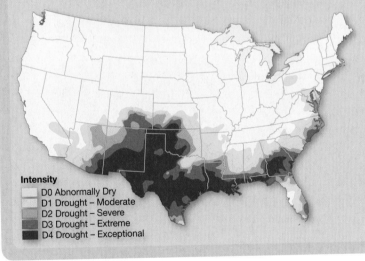

Intensity
- D0 Abnormally Dry
- D1 Drought – Moderate
- D2 Drought – Severe
- D3 Drought – Extreme
- D4 Drought – Exceptional

FIGURE 15–F Drought-status map for July 12, 2011. The percent of the contiguous U.S. experiencing "exceptional" drought in July 2011 reached the highest levels in the 12-year history of the U.S. Drought Monitor. Eighteen percent of the country was classified as under either "extreme" or "exceptional" drought. The entire state of Texas was experiencing drought—three-quarters of it "exceptional." Maps such as this are based on a blend of several criteria, including precipitation, soil moisture, water levels in reservoirs, observers reports, and satellite-based maps of vegetation health. To examine current and archived national, regional, and state-by-state drought maps and conditions, go to http://droughtmonitor.unl.edu. (NOAA)

---

temperature cycles and precipitation distribution. Figure 15–25 (see page 442) illustrates this relationship.

Phoenix, at an elevation of 338 meters (1109 feet), lies in the desert lowlands of southern Arizona. By contrast, Flagstaff is located at an altitude of 2100 meters (about 7000 feet) on the Colorado Plateau in northern Arizona. When summer averages climb to 34°C (93°F) in Phoenix, Flagstaff is experiencing a pleasant 19°C (66°F), a full 15°C (27°F) cooler.

Although the temperatures at the two cities are quite different, the annual march of temperature for both is similar. Both experience minimum and maximum monthly means in the same months. When precipitation data are examined, both places have a similar seasonal pattern, but the amounts at Flagstaff are higher in every month. In addition, much of Flagstaff's winter precipitation is snow, whereas it only rains in Phoenix.

difficult to apply to one region a definition that was developed for another.

Definitions reflect different approaches to measuring drought: meteorological, agricultural, and hydrological. *Meteorological drought* deals with the degree of dryness based on the departure of precipitation from normal values and the duration of the dry period. *Agricultural drought* is usually linked to a deficit of soil moisture. A plant's need for water depends on prevailing weather conditions, biological characteristics of the particular plant, the plant's stage of growth, and various soil properties. *Hydrological drought* refers to deficiencies in surface and subsurface water supplies. It is measured as streamflow and as lake, reservoir, and groundwater levels (Figure 15–G). There is a time lag between the onset of dry conditions and a drop in streamflow or the lowering of

lakes, reservoirs, or groundwater levels. So, hydrological measurements are not the earliest indicators of drought.

There is a sequence of impacts associated with meteorological, agricultural, and hydrological drought (Figure 15–H). When meteorological drought begins, the agricultural sector is usually the first to be affected because of its heavy dependence on soil moisture. Soil moisture is rapidly depleted during extended dry periods. If precipitation deficiencies continue, those dependent on rivers, reservoirs, lakes, and groundwater may be affected.

When precipitation returns to normal, meteorological drought comes to an end. Soil moisture is replenished first, followed by streamflow, reservoirs, and lakes, and finally groundwater. Thus, drought impacts may diminish rapidly in the agricultural sector because of its reliance on soil moisture,

but they may linger for months or years in other sectors that depend on stored surface or subsurface water supplies. Groundwater users, who are often the last to be affected following the onset of meteorological drought, may also be the last to experience a return to normal water levels. The length of the recovery period depends upon the intensity of the meteorological drought, its duration, and the quantity of precipitation received when the drought ends.

The impacts suffered because of drought are the product of both the meteorological event and the vulnerability of society to periods of precipitation deficiency. As demand for water increases as a result of population growth and regional population shifts, future droughts can be expected to produce greater impacts whether or not there is any increase in the frequency or intensity of meteorological drought.

FIGURE 15–G One indicator of hydrological drought is significantly diminished streamflow. Continuous records of streamflow are collected by the U.S. Geological Survey at more than 7000 gaging stations in the United States. This station is on the Rio Grande, south Taos, New Mexico. (Photo by E. J. Tarbuck)

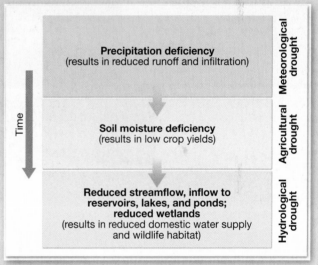

FIGURE 15–H Sequence of drought impacts. After the onset of meteorological drought, agriculture is affected first, followed by reductions in streamflow and water levels in lakes, reservoirs, and underground. When meteorological drought ends, agricultural drought ends as soil moisture is replenished. It takes a considerably longer time span for hydrological drought to end.

Because topographic variations are pronounced in mountains, every change in slope with respect to the Sun's rays produces a different microclimate. In the Northern Hemisphere, south-facing slopes are warmer and dryer because they receive more direct sunlight than do north-facing slopes and deep valleys. Wind direction and speed in mountains can be highly variable and quite different from the movement of air aloft or over adjacent plains. Moun-

tains create various obstacles to winds. Locally, winds may be funneled through valleys or forced over ridges and around mountain peaks. When weather conditions are fair, mountain and valley breezes are created by the topography itself.

We know that climate strongly influences vegetation, which is the basis for the Köppen system. Thus, where there are vertical differences in climate, we should expect a

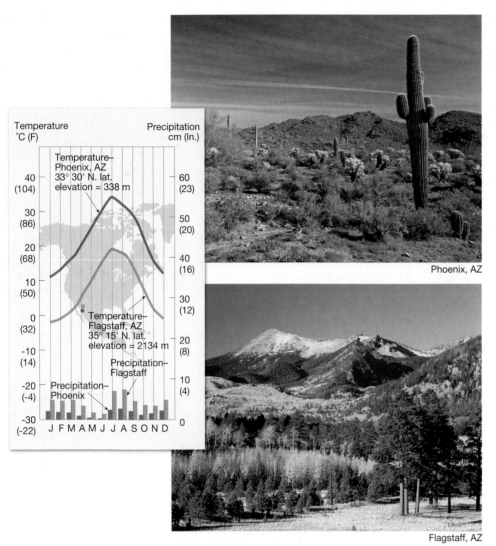

Figure 15-25 Climate diagrams for two Arizona stations illustrate the general influence of elevation on climate. Flagstaff is cooler and wetter because of its position on the Colorado Plateau, nearly 1800 meters (6000 feet) higher than Phoenix. Only scanty drought-tolerant natural vegetation can survive in the hot, dry climate of southern Arizona, near Phoenix. (Photo by Design Pics/Shutterstock) The natural vegetation associated with the cooler, wetter highlands near Flagstaff, Arizona, is much different from that in the desert lowlands. (Photo by Jim Cole/Alamy)

Phoenix, AZ

Flagstaff, AZ

EYE ON THE ATMOSPHERE

These images show an aerial view and close-up of a portion of the Trans-Alaska Pipeline. This 1300 kilometer- (800 mile-) long structure was constructed in the 1970s to transport oil from Prudhoe Bay on Alaska's North Slope, southward to the ice-free port of Valdez on the Gulf of Alaska. Because temperatures in this region range from cool to frigid, the oil must be heated to flow properly. Pumping stations along the way keep the oil moving through the pipeline. (Photos by Michael Collier)

Question 1 The pipeline passes through sub-arctic (*Dfc*) and tundra (*ET*) climate zones. In which climate zone were these photos taken? What clue in the photos helped you figure this out?

Question 2 Notice that in these views, the pipeline is suspended above ground. Based on your knowledge of the climate and landscape of the region, suggest a reason why the pipeline is elevated above ground level.

vertical zonation of vegetation as well. Ascending a mountain can let us view dramatic vegetation changes that otherwise might require a poleward journey of thousands of kilometers. This occurs because altitude duplicates, in some respects, the influence of latitude on temperature and hence on vegetation types. However, we know other factors, such as slope orientation, exposure, winds, and orographic effects also play a role in controlling the climate of highlands. Consequently, although the concept of vertical life zones applies on a broad regional scale, the details within an area vary considerably. Some of the most obvious variations result from differences in rainfall or the receipt of solar radiation on opposite sides of a mountain.

To summarize, *variety* and *changeability* best describe highland climates. Because atmospheric conditions fluctuate with altitude and exposure, a nearly limitless variety of local climates occurs in mountainous regions. The climate in a protected valley is very different from that on an exposed peak. Conditions on windward slopes contrast sharply with those on the leeward sides. Slopes facing the Sun are unlike those that lie mainly in the shadows.

Concept Check 15.10

1 The Arizona cities of Flagstaff and Phoenix are relatively close to one another yet have contrasting climates.
 a In what ways do they differ?
 b Briefly explain why the differences occur.

2 Refer to the precipitation map of Nevada in Figure 15–24. The rainiest areas (>40 centimeters per year) appear as isolated and disconnected zones. Explain this pattern.

Give It Some Thought

1. The tropical soils described in Box 15–2 are considered to have low fertility. Nevertheless, these soils support luxuriant rain forest vegetation like that shown in the accompanying photo. Explain how lush plant growth can occur in poor soils.

(Photo by I.K. Lee/Alamy)

2. In which of the following climates is annual rainfall likely to be most consistent—that is, which would probably experience the smallest percentage change in rainfall from year to year: *BSh, Aw, BWh,* or *Af*? In which of these climates is the annual rainfall most variable from year to year? Explain your answers.

3. Albuquerque, New Mexico, receives an average of 20.7 centimeters (8.07 inches) of rainfall annually.

Albuquerque is considered a desert when the Köppen classification is applied. The Russian city of Verkhoyansk is located near the Arctic Circle in Siberia. The yearly precipitation at Verkhoyansk averages 15.5 centimeters (6.05 inches), about 5 centimeters (2 inches) less than Albuquerque, yet it is classified as a humid climate. Explain how this can occur.

4. This problem examines two places in North Africa that are adjacent to the Sahara Desert. Both are classified as having a subtropical steppe (*BSh*) climate. One place is on the southern margin of the Sahara and the other is north of the Sahara near the Mediterranean Sea.
 a. In which season, summer or winter, does each place receive its maximum precipitation? Explain.
 b. If both stations barely meet the requirements for steppe climates (that is, with only a little more rainfall, both would be considered humid), which station would probably have the lower rainfall total? Explain.

5. Refer to Figure 15–4, which shows climates of the world. Humid continental (*Dfb* and *Dwb*) and subarctic (*Dfc*) climates are usually described as being "land controlled"—that is, they lack marine influence. Nevertheless, these climates are found along the margins of the North Atlantic and North Pacific oceans. Explain why this occurs.

6. In the dry-summer subtropics precipitation totals increase with an increase in latitude, but in the humid continental climate the reverse is true. Explain.

7. Although snowfall in the subarctic climate is relatively scant, a wintertime visitor might leave with the impression that snowfall is great. Explain.

8. Refer to the monthly rainfall data (in millimeters) for three cities in Africa. Their locations are shown on the accompanying map. Match the data for each city to the correct location (1, 2, or 3) on the map. How were you able to figure this out? *Bonus:* Select a figure in Chapter 7 that would be especially useful in explaining or illustrating why these places have rainfall maximums and minimums when they do.

	J	F	M	A	M	J	J	A	S	O	N	D
CITY A	81	102	155	140	133	119	99	109	206	213	196	122
CITY B	0	2	0	0	1	15	88	249	163	49	5	6
CITY C	236	168	86	46	13	8	0	3	8	38	94	201

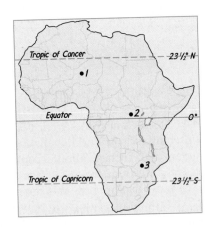

9. Identify three deserts that are visible on this classic view of Earth from space. Briefly explain why these regions are so dry. Identify two other prominent climates in this image. Can you also explain the cause of the band of clouds in the region of the equator?

(NASA)

10. When the rail line in the accompanying photo was built in rural Alaska in the 1940s, the terrain was relatively level. Not long after the railroad was completed, a great deal of subsidence and shifting of the ground occurred, turning the tracks into the "roller coaster" shown here. As a result, the rail line had to be abandoned. Suggest a reason why the ground became unstable and shifted.

(Photo by U.S. Geological Survey)

WORLD CLIMATES IN REVIEW

- Climate is more than "the average state of the atmosphere" because a complete climate description should also include variations and extremes to accurately portray the total character of an area. The most important elements in climate descriptions are *temperature* and *precipitation* inasmuch as they have the greatest influence on people and their activities and also have an important impact on the distribution of vegetation and the development of soils.

- Perhaps the first attempt at climate classification was made by the ancient Greeks, who divided each hemisphere into three zones: *torrid, temperate,* and *frigid.* Many climate-classification schemes have been devised. The classification of climates is the product of human ingenuity, and its value is determined largely by its intended use.

- For decades, a climate classification devised by Wladimir Köppen (1846–1940) has been the best-known and most-used tool for presenting the world pattern of climates. The *Köppen classification* uses easily obtained data: mean monthly and annual values of temperature and precipitation. Furthermore, the criteria are unambiguous, are simple to apply, and divide the world into climate regions in a realistic way. Köppen believed that the distribution of natural vegetation was the best expression of an overall climate. Consequently, the boundaries he chose were largely based on the limits of certain plant associations.

- Köppen recognized five principal climate groups, each designated with a capital letter: A (humid tropical), B (dry), C (humid middle-latitude, mild winters), D (humid middle-latitude, severe winters), and E (polar). Four groups (A, C, D, and E) are defined by temperature. The fifth, the B group, has precipitation as its primary criterion.

- Order exists in the distribution of climate elements, and the pattern of climates is not by chance. The world's climate pattern reflects a regular and dependable operation of the major *climate controls*. The major controls of climate are (1) *latitude* (variations in the receipt of solar energy and temperature differences are largely a function of latitude), (2) *land/water influence* (marine climates are generally mild, whereas *continental climates* are typically more extreme), (3) *geographic position and prevailing winds* (the moderating effect of water is more pronounced along the windward side of a continent), (4) *mountains and highlands* (mountain barriers prevent maritime air masses from reaching far inland, trigger orographic rainfall, and where they are extensive, create their own climatic regions), (5) *ocean currents* (poleward-moving currents cause air temperatures to be warmer than otherwise would be expected), and (6) *pressure and wind systems* (the world distribution of precipitation is closely related to the distribution of Earth's major pressure and wind systems).

- Situated astride the equator, the *wet tropics (Af, Am)* exhibit constant high temperatures and year-round rainfall that combine to produce the most luxuriant vegetation in any climatic realm—the *tropical rain forest*. Temperatures in these regions usually average 25°C (77°F) or more each month. Precipitation in *Af* and *Am* climates is normally from 175 to 250 centimeters (68 to 98 inches) per year and is more variable than temperature, both seasonally and from place to place. Thermally induced convection coupled with convergence along the *intertropical convergence zone (ITCZ)* leads to widespread ascent of the warm, humid, unstable air and ideal conditions for cloud formation and precipitation.

- The *tropical wet and dry (Aw)* climate region is a transitional zone between the rainy tropics and the subtropical steppes. Here, the rain forest gives way to the *savanna*, a tropical grassland with scattered deciduous trees. Only modest temperature differences exist between the wet tropics and the tropical wet and dry regions. The primary factor that distinguishes the *Aw* climate from *Af* and *Am* is precipitation. In *Aw* regions the precipitation is typically between 100 and 150 centimeters (40 to 60 inches) per year and exhibits some seasonal character—wet summers followed by dry winters. In much of India, southeast Asia, and portions of Australia, these alternating periods of rainfall and drought are associated with the *monsoon*, wind systems with a pronounced seasonal reversal of direction. The *Cw* climate, which is subtropical instead of tropical, is a variant of *Aw*.

- Dry regions of the world cover about 30 percent of Earth's land area. Other than their meager yearly rainfall, the most characteristic feature of dry climates is that precipitation is unreliable. Climatologists define a *dry climate* as one in which the yearly precipitation is less than the potential water loss by evaporation. To define the boundary between dry and humid climates, the Köppen classification uses formulas that involve three variables: (1) average annual precipitation, (2) average annual temperature, and (3) seasonal distribution of precipitation.

- The two climates defined by a general water deficiency are *arid* or *desert (BW)* and *semiarid* or *steppe (BS)*. The differences between deserts and steppes are primarily a matter of degree. Semiarid climates are a marginal and more humid variant of arid climates that represent transitional zones that surround deserts and separate them from the bordering humid climates.

- Under the strong influence of the subtropical highs, the heart of the *subtropical desert (BWh)* and *steppe (BSh)* climates lies in the vicinity of the Tropic of Cancer and the Tropic of Capricorn. Within subtropical deserts, the scant precipitation is both infrequent and erratic. In the semiarid transitional belts surrounding the desert, a seasonal rainfall pattern becomes better defined. Due to cloudless skies and low humidities, low-latitude deserts in the interiors of continents have the greatest daily temperature ranges on Earth. Where subtropical deserts are found along the west coasts of continents, cold ocean currents produce cool, humid conditions, often shrouded by low clouds or fog. Unlike their low-latitude counterparts, *middle-latitude deserts (BWk)* and *steppes (BSk)* are not controlled by the subsiding air masses of the subtropical highs. Instead, these lands exist principally because of their position in the deep interiors of large landmasses.

- *Humid middle-latitude climates with mild winters (C climates)* occur where the average temperature of the coldest month is less than 18°C (64°F) but above −3°C (27°F). Several C climate subgroups exist. *Humid subtropical climates (Cfa)* are on the eastern sides of the continents, in the 25° to 40° latitude range. Because of the dominating influence of maritime tropical air masses, summer weather in these regions is hot and sultry, and winters are mild. In North America, the *marine west coast climate (Cfb)* extends from near the U.S.–Canada border northward as a narrow belt into southern Alaska. The prevalence of maritime air masses means mild winters, cool summers, and ample rainfall throughout the year. *Dry-summer subtropical (Mediterranean) climates (Csa and Csb)* are typically found along the west sides of continents between latitudes 30° and 45°. In summer, the regions are dominated by the stable eastern sides of the oceanic subtropical highs. In winter, as the wind and pressure systems follow the Sun equatorward, they are within range of the cyclonic storms of the polar front.

- The average temperature of the coldest month in *humid continental climates with severe winters (D climates)* is −3°C (27°F) or below, and the average temperature of the warmest month exceeds 10°C (50°F). *Humid continental climates (Dfa)* are land controlled and do not occur in the Southern Hemisphere. They are confined to central and eastern North America and Eurasia in the latitude range 40° to 50°N. Both winter and summer temperatures in *Dfa* climates can be characterized as severe, and annual temperature ranges are large. Precipitation is generally greater in summer and generally decreases toward the continental interior and from south to north. Wintertime precipitation is chiefly associated with the passage of fronts connected with traveling midlatitude cyclones. *Subarctic climates* (*Dfc* and *Dfd*), often called *taiga climates* because they correspond to the northern coniferous forests of the same name, are situated north of the humid continental climates and south of the polar tundras. The outstanding feature of subarctic climates is the dominance of winter. By contrast, summers in the subarctic are remarkably warm, despite their short duration. The greatest annual temperature ranges on Earth occur here.

- *Polar climates* (*ET* and *EF*) are those in which the mean temperature of the warmest month is below 10°C (50°F). Polar climates have the lowest annual means on the planet. Although polar climates are classified as humid, precipitation is generally meager, with many nonmarine stations receiving less than 25 centimeters (10 inches) annually. Two types of polar climates are recognized. Found almost exclusively in the Northern Hemisphere, the *tundra climate* (*ET*), marked by the 10°C (50°F) summer isotherm at its equatorward limit, is a treeless region of grasses, sedges, mosses, and lichens with permanently frozen subsoil, called *permafrost*. The *ice-cap climate* (*EF*) does not have a single monthly mean above 0°C (32°F). Consequently, the growth of vegetation is prohibited, and the landscape is one of permanent ice and snow.

- *Highland climates* are characterized by a great diversity of climatic conditions over a small area. In North America highland climates characterize the Rockies, Sierra Nevada, Cascades, and mountains and interior plateaus of Mexico. Although the best-known climatic effect of increased altitude is lower temperatures, greater precipitation due to orographic lifting is also common. Variety and changeability best describe highland climates. Because atmospheric conditions fluctuate with altitude and exposure to the Sun's rays, a nearly limitless variety of local climates occur in mountainous regions.

VOCABULARY REVIEW

arid, or desert (p. 423)
continental climate (p. 413)
dry climate (p. 423)
dry-summer subtropical climate (p. 432)
highland climate (p. 439)
humid continental climate (p. 433)
humid subtropical climate (p. 429)
ice-cap climate (p. 438)

intertropical convergence zone (ITCZ) (p. 419)
Köppen classification (p. 410)
marine climate (p. 413)
marine west coast climate (p. 430)
Mediterranean climate (p. 432)
monsoon (p. 421)
permafrost (p. 437)

polar climate (p. 436)
savanna (p. 419)
semiarid, or steppe (p. 423)
subarctic climate (p. 435)
taiga climate (p. 435)
tropical rain forest (p. 415)
tropical wet and dry (p. 419)
tundra climate (p. 436)

PROBLEMS

1. Use Figure 15–2 to determine the appropriate classification for stations a, b, and c in the table below. You may find the flow chart (Figure G-1) in Appendix G a helpful tool.

2. Using the maps in Figure 15–19, determine the approximate January and July temperature gradients between the southern tip of mainland Florida and the point where the Minnesota–North Dakota border touches Canada. Assume the distance to be 3100 kilometers (1900 miles). Express your answers in °C per 100 kilometers or °F per 100 miles.

	J	F	M	A	M	J	J	A	S	O	N	D	YR
Station a.													
Temp. (°C)	−18.7	−18.1	−16.7	−11.7	−5.0	0.6	5.3	5.8	1.4	−4.2	−12.3	−15.8	−7.5
Precip. (mm)	8	8	8	8	15	20	36	43	43	33	13	12	247
Station b.													
Temp. (°C)	24.6	24.9	25.0	24.9	25.0	24.2	23.7	23.8	23.9	24.2	24.2	24.7	24.4
Precip. (mm)	81	102	155	140	133	119	99	109	206	213	196	122	1675
Station c.													
Temp. (°C)	12.8	13.9	15.0	16.1	17.2	18.8	19.4	22.2	21.1	18.8	16.1	13.9	15.9
Precip. (mm)	53	56	41	20	5	0	0	2	5	13	23	51	269

MyMeteorologyLab™

Log in to www.mymeteorologylab.com for animations, videos, MapMaster interactive maps, GEODe media, *In the News* RSS feeds, web links, glossary flashcards, self-study quizzes and a Pearson eText version of this book to enhance your study of *World Climates*.

Optical Phenomena of the Atmosphere

One of Earth's most spectacular and intriguing natural phenomena is the rainbow. Its "surprise" appearance and splash of colors make it the subject of both poets and artists, not to mention every casual photographer within reach of a camera. In addition to rainbows, many other stunning optical phenomena occur in our atmosphere. In this chapter, we consider how the most familiar of these displays occur. Knowing when and where to look for these phenomena will allow you to identify each type and, ultimately, witness and appreciate them more frequently.

Rainbows are among the most common and spectacular of the atmosphere's optical phenomena. (Photo by Michael Routh/Ambient Images/Glow Images)

Focus On Concepts

After completing this chapter, you should be able to:

- Explain the principle called the *law of reflection*.
- Discuss how refraction causes white light to separate into the spectrum of colors.
- Define *internal refraction*.
- Describe how a mirage is created.
- Sketch a raindrop and how sunlight travels through it to form a primary rainbow.

- Distinguish among three different optical phenomena that are produced as a result of the interaction between sunlight and hexagonal ice crystals.
- Describe how a glory forms.
- Compare and contrast glories and rainbows.
- Distinguish between coronas and halos.

Interactions of Light and Matter

The interaction of white (visible) sunlight with our atmosphere creates the numerous optical phenomena we observe in the sky. In Chapter 2 we considered some properties of light and how they contribute to occurrences such as the blue color of the sky and the red color of sunset. This chapter examines other interactions between sunlight and the gases, ice crystals, and water droplets found in the atmosphere that generate still other optical phenomena. We begin with two fundamental ways light and matter interact: reflection, and refraction.

Reflection

Light traveling from the Sun toward Earth travels at a uniform speed and in a straight line. However, when light encounters a transparent material, such as a water body, a portion bounces off the surface, and some is transmitted through the material at a slower velocity. The light that bounces back from the surface is *reflected*. Reflected light allows you to see yourself in a mirror. The image that you see in a mirror originates as light that first reflected off you toward the mirror and then was bounced from the silver- or aluminum-coated surface of the mirror back to your eyes. When light is reflected, the rays bounce off the reflecting surface at the same angle at which they met that surface (Figure 16–1). This principle is called the **law of reflection.** It states that the angle that the incidence (incoming) ray makes with a line perpendicular to the reflective surface is equal to the angle that the reflected (outgoing) ray makes to that same line.

On a smooth, highly reflective surface like a mirror, about 90 percent of the parallel incoming rays leave the surface by following outgoing parallel paths. However, when light encounters a slightly irregular surface, the rays are re-

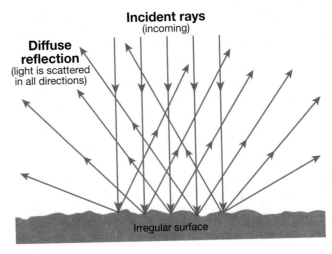

Figure 16–2 Sunlight striking a rough surface is dispersed (scattered) in various directions.

flected in many directions; this is referred to as *diffuse reflection* (Figure 16–2). Even something that appears as smooth as a page in this book is sufficiently rough to scatter light in all directions. This type of reflection makes it possible to read the print from almost any angle. Most of the objects in our surroundings are seen by diffuse reflection.

When light is reflected from a very rough surface, the image you see is generally distorted, or it may appear as multiple images. For example, the reflected image of the Sun when viewed on a wavy ocean surface is not a circular disk but a long narrow band, as shown in Figure 16–3. What is seen is not one image of the Sun but many distorted images of a single light source. This bright band of light is produced because only a portion of the sunlight that strikes each wave is reflected toward the viewer's eyes.

Another type of reflection that is important to our discussion of optical phenomena is called **internal reflection.** Internal reflection occurs when light that travels through a transparent material, such as water, reaches the opposite surface and is reflected back into the same transparent material. You can easily demonstrate this phenomenon by holding a glass of water directly overhead and looking up through the water. You should be able to see clearly through the water because very little internal reflection results when light strikes perpendicular to the surface of a transparent material. While keeping the glass overhead, move it sideways so that you look through the glass at an angle. The underside of the surface of the water takes on the appearance of a silvered mirror. What you are observing here is the internal reflection that occurs when light strikes the surface at an angle greater than 48° from the vertical. Internal reflection plays an important role in the formation of optical phenomena such as rainbows.

Refraction

When light strikes a transparent material, the portions that are not reflected are transmitted through the material and un-

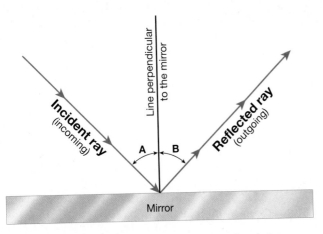

Figure 16–1 Reflection of light by a mirror. When light rays are reflected, they bounce off the reflecting surface at the same angle at which they met the surface.

Figure 16–3 This long, narrow band of sunlight is really multiple distorted images of the Sun reflected from the water surface. Deception Pass Bridge, Washington. (Photo by Peter Skinner/Photo Researchers, Inc.)

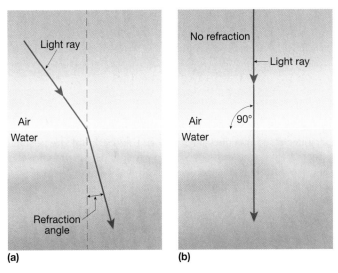

(a)

(b)

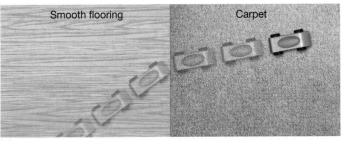

(c)

dergo another well-known effect, called *refraction*. **Refraction** is the change of direction of light as it passes obliquely from one transparent medium to another. Refraction occurs because the velocity of light is dependent on the material that transmits it. In a vacuum, radiation travels at the speed of light (3.0×10^{10} centimeters per second); and when it travels through air, its speed is slowed slightly. However, in substances such as water, ice, or glass, its speed is slowed considerably.*

When light encounters a transparent medium at some angle other than 90°, the light bends, as shown in Figure 16–4a, b. Refraction can be demonstrated by seeing how

---

*The speed of light is a constant; however, as it passes through various substances, the light is delayed because of interaction with the electrons in the material. The speed of light as it travels between these intervening electrons is the same as in a vacuum.

Figure 16–4 Refraction. (a) Light waves change direction (bend) as they pass from one transparent material into another. The speed of light is slower in water than in air, which causes light rays to bend toward a line drawn perpendicular to the water surface. (b) When light rays enter water at a 90° angle, they are not refracted. (c) A toy car traveling from a smooth surface to a carpeted floor illustrates the concept of refraction.

a toy car responds when it is rolled from smooth flooring onto a carpeted surface. If the toy car meets the carpet at a 90° angle, it slows down but does not change direction. However, if the front wheels of the car hit the carpet at an angle different from 90°, the car not only slows but changes direction, as shown in Figure 16–4c. In this analogy, the directional change occurs because the front wheel that hit the carpet first begins to slow while the other front wheel, which is still on the smooth flooring, continues at the same speed, resulting in a sudden turn of the toy car. Now imagine that the path of the toy car represents the path of light rays, and the smooth flooring and carpet represent air and water, respectively. As light enters the water and is slowed, its path is diverted toward a line extending perpendicularly from the water's surface (Figure 16–4a). If the light passes from water into air, the bending occurs in the opposite direction—away from the perpendicular.

Refraction can be demonstrated by immersing a pencil halfway into a glass of water and viewing the pencil obliquely—an angle other than 90° to the surface (Figure 16–5a). How refraction produces the apparent bending of a pencil is illustrated in Figure 16–5b. The solid lines in this sketch show the actual path taken by the light, whereas the dashed lines indicate the straight path a human brain perceives. As we look down at the pencil, the point appears closer to the surface than it actually is. This occurs because our brain perceives the light as coming along the straight path indicated by the dashed line rather than along the actual bent path. Because all the light coming from the submerged portion of the pencil is bent similarly, this portion of the pencil appears nearer the surface. Therefore, where the pencil enters the water, it appears to be bent upward toward the surface.

In addition to the abrupt bending of light as it passes obliquely from one transparent substance to another, light also gradually bends as it traverses a material of varying density. As the density of a material changes, the velocity of light changes as well. Within Earth's atmosphere, for example, the density of air usually increases Earthward. The result of this gradual density change is an equally gradual slowing and bending of light rays. Rays that travel from areas of lower air density to areas of higher air density curve in a direction that has the same orientation as Earth's curvature.

The bending of light caused by refraction is responsible for a number of common optical phenomena. These phenomena result because our brain perceives bent light as if it has traveled to our eyes along a straight path. Try to imagine "looking down" a bent light ray to view a light source located around a corner. If you could see the object, your brain would place that object away from the corner in "plain sight." On occasion, we see things that are "around the corner." One example is our view of the setting Sun. Several minutes after the Sun has actually slipped below the horizon, it still appears to us as a full disk. Figure 16–6 illustrates this situation. It is our *inability to perceive light bending* that causes the position of the Sun to appear above the horizon after it has set.

(a)

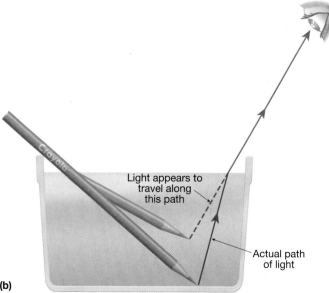
(b)

Light appears to travel along this path

Actual path of light

Figure 16–5 Refraction. (a) Image illustrating refraction of a pencil immersed in water. (Photo by E. J. Tarbuck) (b) The pencil appears bent because the eye perceives light as if it were traveling along the dashed line rather than along the solid line, which represents the path of the refracted light.

Concept Check 16.1

1 State the *law of reflection*.

2 Briefly describe the optical phenomenon known as *internal reflection*.

3 What happens to light as it passes obliquely from one transparent medium into another?

4 Provide one example of how *refraction* changes the way we perceive an object.

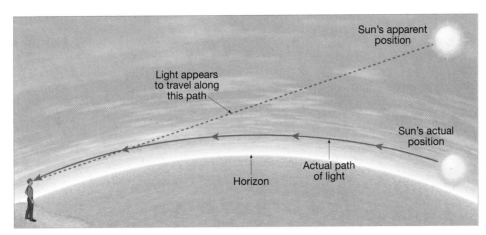

Figure 16–6 Light is refracted as it passes through layers of the atmosphere having different densities. The result is an apparent displacement of the position of the Sun.

Mirages

One of the most interesting optical events common to our atmosphere is the **mirage.** Although this phenomenon is most often associated with desert regions, it can be experienced anywhere (Box 16–1). One type of mirage occurs on very hot days when the air near the ground is much less dense than the air aloft. As noted earlier, a change in the density of air is accompanied by a gradual bending of the light rays. When light travels through air that is less dense near the surface, the rays bend in a direction opposite to Earth's curvature. As we can see in Figure 16–7, this direction of bending will cause the light reflected from a distant object to approach the observer from below eye level. Because the brain perceives the light as following a straight path, the object appears below its original position and is often inverted, as shown

with the palm tree in Figure 16–7. The palm tree appears inverted because the rays that originate near the top of the tree are bent more than those that originate near the base of the tree.

In the classic desert mirage, a lost and thirsty wanderer encounters an image consisting of an oasis of palm trees and a shimmering water surface on which he can see a reflection of the palms. Although the trees are real, the water and reflected palms are part of the mirage. Light traveling to the observer through the cooler air above produces the image of the actual tree. The reflected image of the palms is produced from the light that traveled downward from the trees and was gradually bent upward as it traveled through the hot (less dense) air near the ground. The image of water is produced by light that travels downward from the sky and is bent upward. Thus the water mirage is actually in the sky. Such desert mirages are called **inferior mirages** because the images appear *below* the true location of the observed object.

In addition to the "desert mirage," another common type of mirage occurs when the air near the ground is much cooler than the air aloft. Consequently, this effect is observed most frequently in polar regions or over cool ocean surfaces. When the air near the ground is substantially colder than the air aloft, light bends in the same direction as Earth's curvature. As shown in Figure 16–8, this effect allows ships to be seen where ordinarily Earth's curvature would block them from view. This phenomenon, referred to as **looming,** occurs when the refraction of light is significant enough to make the object appear suspended above the horizon. In contrast to a desert mirage, looming is considered a **superior mirage** because the image is seen *above* its true position.

In addition to the easily explained inferior and superior mirages, several more complex mirages have been observed. They occur when the atmosphere develops a temperature profile in which the temperature changes rapidly with height, causing a similar change in the air's density. Under these conditions, each thermal layer acts like a glass lens. Because each layer bends the light rays somewhat differently, the size and shape of the objects observed through these thermal layers are greatly distorted. You may have observed an analogous sight

Figure 16–7 In this classic desert mirage, light travels more rapidly in the hot air near the surface. Thus, as downward-directed rays enter this warm zone, they are bent (refracted) upward so that they reach the observer from below eye level.

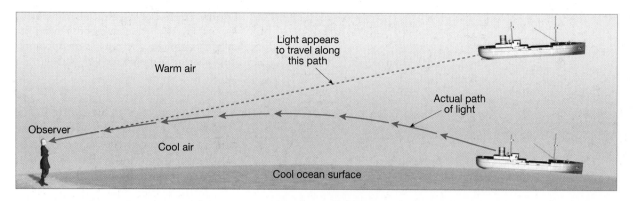

Figure 16-8 As light enters a cool layer of air, it will slow and bend downward. This results in objects appearing to loom above their true position.

if you have ever entered a "house of mirrors" at a carnival. While one of the mirrors makes you look taller, others either stretch or compress the image of your body. Mirages are capable of similarly distorting objects and occasionally form a mountain-like image over a barren ice cap or an open ocean.

A mirage that changes the apparent size of an object is called **towering.** As the name implies, towering creates a much larger image of the original object. This optical phenomenon is most frequently observed in coastal areas, where sharp temperature contrasts are common. An interesting type of towering is called the *Fata Morgana*, named for the legendary sister of King Arthur, who was credited with the magical power of being able to create towering castles out of thin air. In addition to generating magical castles, the Fata Morgana probably explains the towering mountains that early explorers of the north polar region observed but that never materialized.

Box 16–1 Are Highway Mirages Real?

You have undoubtedly seen a mirage while traveling along a highway on a hot summer afternoon. The most common highway mirages are in the form of "wet areas" that appear on the pavement ahead, only to disappear as you approach (Figure 16–A). Because these "wet areas" always disappear as a person gets closer, many people believe they are optical illusions. This is not the case. Highway mirages, as well as all other types of mirages, are as real as the images observed in a mirror. As can be seen in Figure 16–A, highway mirages can be photographed and are not tricks played by the mind.

What causes the "wet areas" that appear on dry pavement? On hot summer days, the layer of air near Earth's surface is much warmer than the air aloft. Sunlight traveling from a region of colder (more dense) air into the warmer (less dense) air near the surface bends in a direction opposite to Earth's curvature (see Figure 16–7). As a consequence, light rays that began traveling downward from the sky are refracted upward and appear to the observer to have originated on the pavement ahead. What appears to the traveler as water is really just an inverted image of the sky. This can be verified by careful observation. The next time you view a highway mirage, look closely at any vehicle ahead of you at about the same distance as the "wet" area. Below the vehicle you should be able to see an inverted image of it. Such an image is produced in the same manner as the inverted image of the sky.

FIGURE 16–A A classic highway mirage. (Photo by Joe Orman)

EYE ON THE ATMOSPHERE

The accompanying image is an optical phenomenon known as a *green flash* that appears as a green cap above the Sun at sunset or sunrise. Green flashes result from atmospheric changes in density that cause sunlight to bend as it travels from the thinner air in the upper atmosphere into the denser air near Earth's surface, where the observer is located. Green flashes are produced because the light at the blue/green end of the spectrum is bent more than the light in the red/orange range. As a result, the blue rays from the upper limb of the setting Sun remain visible to an Earthbound observer after the red rays are obstructed by the curvature of Earth. We should, therefore, expect to see a glimpse of blue/green light at each sunset. Clear conditions and an unobstructed view of the sunset produce the most vivid emerald green flashes. Green flashes are rare, however, because almost all of the blue light and much of the green light is removed from a sunbeam by scattering (see Chapter 2). (Photo by Pekka Parviainen/Photo Researchers, Inc.)

Question 1 What term is used for the bending of light as it travels through a transparent material that exhibits various densities?

Question 2 Green flashes are most commonly photographed over the ocean at sunset. Suggest a reason for this.

Concept Check 16.2

1 What happens to light if the density of the air through which it travels changes with height?

2 When light travels from warm (less dense) air into a region of colder (denser) air, its path will curve. Does it bend in the same direction as Earth's curvature or in the opposite direction?

3 What is meant by the term *inferior mirage*?

4 How is an inferior mirage different from a superior mirage?

Students Sometimes Ask . . .

Why is the Moon much larger when it is low on the horizon than when it is overhead?

This phenomenon is an optical illusion that has nothing to do with the refraction of light. The size of the Moon does not actually change as it rises above the horizon. As meteorologist Craig Bohren stated, "The Moon illusion results from refraction by the mind, mirages from refraction by the atmosphere."

Rainbows

Probably the most spectacular and best known of all optical phenomena in our atmosphere is the **rainbow** (Figure 16–9). An observer on the ground sees the rainbow as an arch-shaped array of colors that trails across a large segment of the sky. Although the clarity of the colors varies with each rainbow, the observer can usually discern six rather distinct bands of color. The outermost band is red and blends gradually to orange, yellow, green, blue, and eventually violet light. These

Figure 16–9 Rainbow over spruce trees in the taiga north of the Alaska Range. (Photo by Michael Giannechini/Photo Researchers, Inc.)

spectacular splashes of color are seen when the Sun is behind the observer and a rain shower is in front. A fine mist of water droplets generated by a waterfall or lawn sprinkler can also generate miniature rainbows.

To understand how raindrops disperse sunlight to generate a rainbow, recall our discussion of refraction. Remember that as light passes obliquely from air to water, its speed slows, which causes it to be refracted (change direction). In addition, each color of light travels at a different velocity in water; consequently, each color will be bent at a slightly different angle. Violet-colored light, which travels at the slowest rate, is refracted the most, whereas red light, which travels most rapidly, is bent the least. Thus, when sunlight (which consists of all colors) enters water, refraction separates it into colors according to their velocity.

Seventeenth-century scientist Sir Isaac Newton used a prism to demonstrate the concept of color separation. Light transmitted through a prism is refracted twice—once as it passes from the air into the glass and again as it leaves the prism and reenters the air. Newton noted that when light is refracted twice, as by a prism, the separation of sunlight into its component colors is quite noticeable (Figure 16–10). We refer to this separation of colors by refraction as **dispersion.**

Rainbows are generated because water droplets act like prisms, dispersing sunlight into the spectrum of colors. Upon entering a droplet, the sunlight is refracted, with violet light bent the most and red the least. Then, when the sunlight reaches the opposite side of the droplet, the rays are internally reflected backward and exit the droplet on the same side they entered. As they leave the droplet, further refraction increases the amount of dispersion and accounts for the nearly complete color separation.

The angle between the incoming (incident) rays and the dispersed colors that constitute the rainbow is 42° for red light and 40° for violet. The other colors—orange, yellow, green, and blue—are dispersed at intermediate angles. Although each droplet disperses the full spectrum of colors, an observer sees only one color from any single raindrop. For example, if red light from a particular droplet reaches an observer's eye, the violet light from that droplet is visible only to an observer in a different location (Figure 16–11a). Consequently, each observer sees his or her "own" rainbow, which is generated by a different set of droplets and different rays of light. In effect, a rainbow is produced through the interaction of sunlight with millions of raindrops, each of which acts as a miniature prism.

The curved shape of the rainbow results because the rainbow rays always travel toward the observer at an angle of about 42° from the path of the sunlight. Thus, we experience a 42° semicircle of color across the sky that we identify as the arch shape of the rainbow. If the Sun is higher than 42° above the horizon, an Earthbound observer will not see a rainbow. Under certain conditions, an observer in an airplane will see the rainbow as a full circle.

When a spectacular rainbow is visible, an observer will sometimes see a dimmer secondary rainbow. The secondary bow will be visible about 8° above the primary bow and will suspend a larger arc across the sky.* The secondary bow also has a slightly narrower band of colors than the primary rainbow, and the colors are in reverse order. Red makes up the innermost band of the secondary rainbow and violet the outermost.

The secondary rainbow is generated in much the same way as the primary bow. The main difference is that the dispersed light that constitutes the secondary bow is reflected twice within a raindrop before it exits, as shown in Figure 16–12. The extra internal reflection results in a 50° angle for the dispersion of the color red (about 8° greater than the primary rainbow) and a reverse order of the colors.

The extra reflection also accounts for the fact that the secondary bow is less frequently observed than the primary

---

*For reference, the diameter of the Sun is equal to about 0.5° of arc.

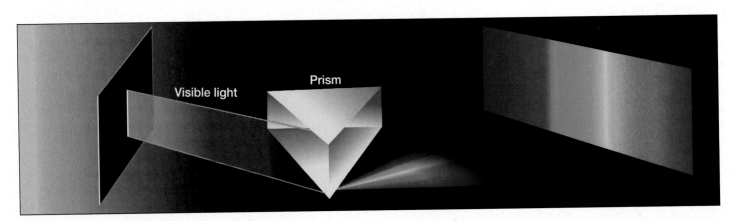

Figure 16–10 The spectrum of colors is produced when sunlight is passed through a prism. Notice that each wavelength (color) of light is bent differently.

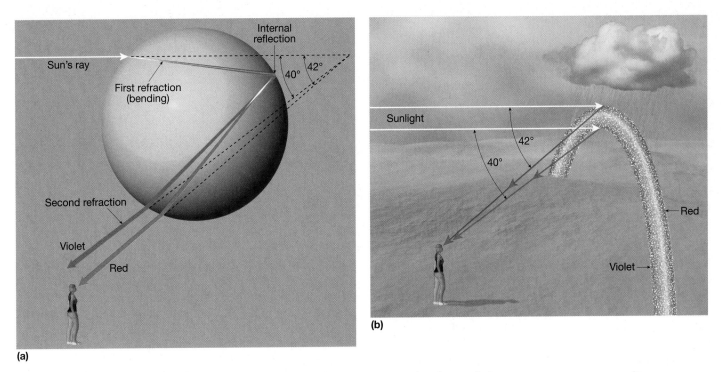

(a)

(b)

Figure 16–11 The formation of a primary rainbow. (a) Color separation results when sunlight is refracted as it enters a raindrop and again when it leaves the raindrop. This results in the colors of the rainbow. (b) It takes millions of raindrops to produce a rainbow. Each observer sees only one color from each of these raindrops.

bow (Figure 16–13). Each time light strikes the inner surface of the droplet, some of the light is transmitted through the reflecting surface. Because this light is "lost," it does not contribute to the brightness of the rainbow. Although secondary rainbows always form, they are rarely discernible.

Humans use rainbows, like other optical phenomena, as a means of predicting the weather. A well-known weather proverb illustrates this point:

Rainbow in the morning gives you fair warning,
Rainbow in the afternoon, good weather coming soon.

This bit of weather lore relies on the fact that midlatitude weather systems usually move from west to east. Remember that observers must be positioned with their backs to the Sun and facing the rain in order to see a rainbow. When a rainbow is seen in the morning, the Sun is located to the east of the observer, and the clouds and raindrops that are responsible for its formation must therefore be located to the west. Thus, we predict the advance of foul weather when the rainbow is seen in the morning because the rain is located to the west and is traveling toward the observer. In the afternoon, the opposite situation exists: the rain clouds

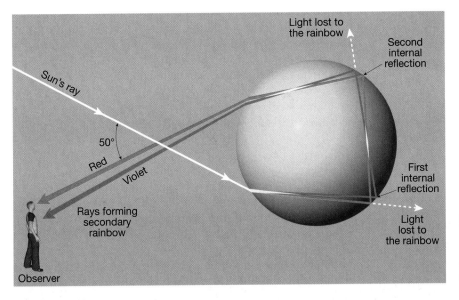

Figure 16–12 Formation of a secondary rainbow. By comparing this figure with Figure 16–11, you will see that the positions of the red and violet rays are reversed, which accounts for the order of the colors.

Figure 16–13 Double rainbow. The secondary rainbow is dimmer than the primary bow and the order of colors is reversed. (Photo by Sean Davey/Aurora Open/Glow Images)

are located to the east of the observer. Therefore, when the rainbow is seen in the afternoon, the rain has already passed. Although this famous proverb has a scientific basis, a small break in the clouds, which lets the sunshine through, can generate a late-afternoon rainbow. In this situation, a rainbow may certainly be followed shortly by more rainfall.

Concept Check 16.3

1 List the colors of a primary rainbow, in order, from the outer edge to the inner edge.

2 If you were looking for a rainbow in the morning, which direction would you look? Explain.

3 Why is a secondary rainbow dimmer than a primary rainbow?

4 What term is used to describe the separation of colors that occurs when light is refracted as it passes from one transparent material into another?

Halos, Sun Dogs, and Solar Pillars

Although halos are a fairly common occurrence, they are rarely seen by casual observers. When noticed, a **halo** appears as a narrow whitish ring centered on the Sun, or less often, the Moon (Figure 16–14). Halos appear on days when the sky is covered with a thin layer of cirrus or cirrostratus clouds. In addition, halos are seen more often in the morning or late afternoon, when the Sun is near the horizon. Residents of polar regions, where a low Sun and cirrus clouds are common, are frequently treated to views of halos and related phenomena.

The most common halo is called the *22° halo,* named because it has a radius of 22 degrees. Less frequently observed is the larger *46° halo.* Like a rainbow, a halo is produced by dispersion of sunlight. In the case of a halo, however, ice crystals rather than raindrops refract the light. Thus, the clouds most often associated with halo formation are high clouds. Because cirrus clouds often form as a result of frontal lifting, halos have been accurately described as harbingers of foul weather, as the following weather proverb attests:

The moon with a circle brings water in her beak.

Four basic types of hexagonal (six-sided) ice crystals contribute to the formation of halos: plates, columns, capped columns, and bullets (Figure 16–15). Because ice crystals typically have a random orientation, the diffused light produces a nearly circular halo centered on the illuminating object (Sun or Moon).

The primary difference between 22° and 46° halos is the path that light takes through the ice crystals. The scat-

tered sunlight responsible for the 22° halo strikes one face and exits from an alternate face, as shown in Figure 16–16a. The angle of separation between the alternating faces of ice crystals is 60°, the same as in a common glass prism. Consequently, ice crystals disperse light in a manner similar to a prism to produce a 22° halo. A 46° halo, by contrast, is formed from light that passes through one face of the crystal and exits either the top or base of the crystal (Figure 16–16b). The angle separating these two surfaces is 90°. Light that passes through ice crystal faces separated by 90° is refracted so it is concentrated 46° from the light source, which accounts for its name. Many other types of halos and partial halos have been observed, all of which owe their formation to the relative abundance of ice crystals of particular shapes and orientations.

Although ice crystals disperse light in the same manner as raindrops (or prisms), halos are generally whitish and do not display the colors of the rainbow. This difference is primarily attributed to the fact that raindrops tend to be uniform in both size and shape, whereas ice crystals vary considerably in size and have imperfect shapes. Although individual ice crystals produce the rainbow of colors in a manner similar to raindrops, the colors tend to overlap and wash each other out. Occasionally, halos display some coloration; in such cases, a reddish band surrounds a whitish ring.

One of the most spectacular features associated with halos is **sun dogs.** These two bright regions, or "mock suns," as they are often called, can be seen adjacent to the 22° halo and on opposite sides of the Sun (Figure 16–17). Sun dogs form under the same conditions as, and in conjunction with, halos except that their existence depends on the presence of numerous vertically oriented ice crystals. This particular orientation results when elongated ice crystals are slowly descending. Vertically oriented ice crystals cause the Sun's rays to be concentrated (much as with a magnifying glass) in two distinct areas at a distance of about 22° on opposite sides of the Sun.

Another optical phenomenon caused by falling ice crystals is **sun pillars.** These vertical shafts of light are

Figure 16–15 Common hexagonal ice crystals that contribute to the formation of certain optical phenomena: (a) plate, (b) column, (c) capped column, and (d) bullet.

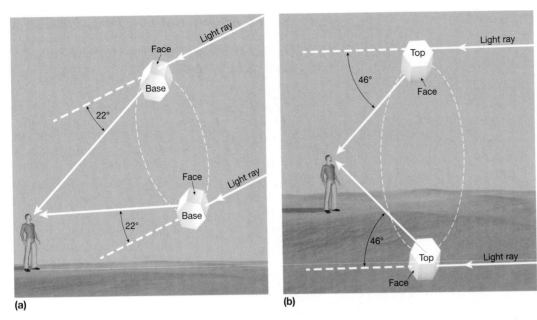

Figure 16–16 A halo forms when light is refracted as it passes through hexagonal ice crystals. (a) When the majority of sunlight or moonlight enters one face of an ice crystal, where it is refracted, and exits one of the opposite faces, a 22° halo is produced. (b) A 46° halo is caused by refraction that occurs when light enters one of the crystal faces of a hexagonal crystal and exits either the top or base of the crystal.

most often viewed shortly before sunset or shortly after sunrise, when they appear to extend upward from the Sun (Figure 16–18). These bright pillars of light are created when sunlight is reflected toward the observer from the base (bottom) of descending ice crystals that have plate-like shapes oriented like slowly falling leaves. Because direct sunlight is often reddish when the Sun is low in the sky, pillars appear similarly colored. Occasionally, pillars that extend below the Sun can also be viewed.

Concept Check 16.4

1 How are halos and rainbows similar?

2 In what two ways are halos and rainbows different?

3 What is the orientation of the ice crystals that produce a halo? Sun dogs?

4 If a halo is present around the Sun, where would you look to find sun dogs?

Figure 16–17 Sun dogs are produced by the dispersion of sunlight by the ice crystals of cirrus clouds. (Photo by Stefan Wackerhagen/agephotostock)

through one internal reflection—and the remaining distance along the surface of the droplet. This path causes the rays to be backscattered directly toward the Sun. Because glories always form opposite the Sun's position, the observer's shadow will always be found within a glory.

Concept Check 16.5

1 Explain the origin of the term *glory*.

2 What material interacts with sunlight to produce a glory?

Other Optical Phenomena

Thus far, we have considered optical phenomena produced when light is reflected and/or refracted (bent) as it travels through a medium. Optical phenomena also occur when light passes close to water droplets—a process called *diffraction*.

Like refraction, diffraction also separates white light into the colors of the rainbow, but by a more complex process. Two related optical phenomena produced by diffraction are coronas and iridescent clouds.

Coronas

A **corona** most often appears as a bright whitish disk centered on the Moon or Sun. When colors are discernible, the central white disk of the corona is surrounded by one or

Figure 16–18 A Sun pillar is a shaft of sunlight reflected off the bases of ice crystals in high clouds. (Photo by Masaaki Tanaka/Aflo/Glow Images)

Glories

A **glory** is a spectacle that is rarely witnessed by Earthbound observers. However, the next time you are in an airplane and have a window seat, look for the shadow of your aircraft projected on the clouds below. The airplane shadow will often be surrounded by one or more colored rings that constitute the glory. Each ring will be colored in a manner similar to a rainbow, with red being the outermost band and violet the innermost. Generally, however, the colors are not as discernible as those of a rainbow. When two or more sets of rings are seen, the inner one will be the brightest and thinnest.

Although most commonly seen by airplane pilots, glories are named for their appearance when viewed by an observer on the ground. Hikers have reported seeing glories when they have climbed above a cloud or fog layer and had the Sun at their backs. When the hiker's shadow is cast on the cloud or fogbank, the glory will enshroud the observer's head (Figure 16–19). If two or more persons experience a glory simultaneously, they see only their own shadow surrounded by their own glory.

Glories have been observed for centuries and appear much like the iconic halos often seen in religious paintings. In China, this phenomenon is commonly called *Buddha's light*.

Glories form by backscattered light in a manner similar to rainbows. However, the cloud droplets that are responsible for the glory are much smaller and more uniform in size than the raindrops that produce rainbows. The light that becomes the glory strikes the very edge of the droplet and then travels to the opposite side of the droplet—partly

Figure 16–19 The colored rings of a glory surround the shadow of an observer that is projected on the fog below. (Photo by Mila Zinkova)

(a)

(b)

Figure 16–20 A corona. (a) Most often, a corona appears as a bright whitish disk centered on the Moon or Sun. (Photo by E. J. Tarbuck) (b) Those that display the colors of the rainbow are rare. (Photo by Mike Hollingshead/Photo Researchers, Inc.)

more concentric rings that exhibit the colors of the rainbow (Figure 16–20). Unique features of coronas include the possible repetition of the corona color sequence and the fact that they are one of the few optical phenomena observed around the Moon more frequently than around the Sun.

A corona is produced when a thin layer of clouds, often altostratus or cirrostratus, veils the illuminating body (Sun or Moon). When the droplets (or sometimes ice crystals) that produce coronas are small and nearly the same size, the rings are easily discernable and have the most distinct colors. The colors of the rings produced by large cloud droplets tend to appear washed out or whitish.

Coronas can be easily distinguished from 22° halos because their color sequence is bluish-white on the inside and reddish on the outside—opposite the halo color sequence. Coronas also form closer to the illuminating body than halos.

Iridescent Clouds

Iridescent clouds are among the more spectacular and elusive of optical phenomena. Cloud iridescence, most often associated with altostratus, cirrocumulus, or lenticular clouds, appears as areas of bright colors—violet, pink, and green—generally seen near the edges of a cloud (Figure 16–21). Common examples of iridescence are the spectrum of colors reflected from soap bubbles and from a thin layer of gasoline spilled on a wet surface.

As with the corona shown in Figure 16–20b, the display of colors associated with iridescent clouds is produced by

EYE ON THE ATMOSPHERE

The accompanying image of noctilucent clouds was taken over Bilund, Denmark, on July 15, 2010. *Noctilucent clouds* are thin wavy clouds that form high in the atmosphere (50–53 miles above Earth) and are observed only for a short time following sunset. Because they are observed only at high latitudes and lie within the mesosphere, they are also known as *polar mesospheric clouds*. (Photo by NASA)

Question 1 Explain how noctilucent clouds might be illuminated so they are visible to an Earthbound observer after sunset.

Question 2 Do you think noctilucent clouds are composed of water droplets or ice crystals?

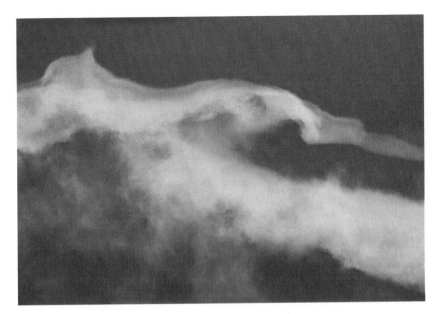

Figure 16–21 Iridescent cloud. (Photo by Tom Schlatter, National Oceanic and Atmospheric Administration/Seattle)

the diffraction of sunlight or moonlight by small, uniform cloud droplets or occasionally small ice crystals. The best time to view iridescent clouds is when the Sun is behind a cloud or just after the Sun has set behind a building or topographic barrier.

Concept Check 16.6

1 What process is responsible for the colors exhibited by coronas and iridescent clouds?

2 How can coronas be distinguished from 22° halos?

Give It Some Thought

1. Explain why a mirage disappears when the observer gets near.

2. What particles (cloud droplets, rain drops, or ice crystals) are usually associated with each of these optical phenomena: rainbows, halos, coronas, glories, and sun dogs?

3. The accompanying drawing illustrates how white sunlight is separated into the colors of the rainbow. Match the features numbered 1 through 5 with the following terms: internal reflection, red light, violet light, incident ray, and refracted ray.

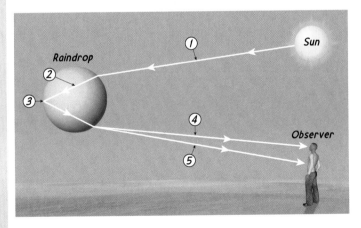

4. If you have ever been close to a large movie theater screen, you might have noticed that the screen is made of small glassy particles oriented at slightly different angles rather than one very smooth surface. Based on what you know about how light is reflected from various surfaces, why do you think this type of screen is used?

5. Explain why the refraction of sunlight causes the length of daylight to be longer than if the Earth had no atmosphere.

6. Explain why a person in Lincoln, Nebraska (41° north latitude), will never see a rainbow at noon on June 21.

7. Name each of the optical phenomena shown in the accompanying photos.

(a)

(b)

(c)

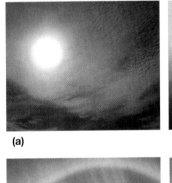

(d)

(Photos by (a) blickwinkle/ Alamy (b) Mila Zinkova (c) Peter Marenfeld/ Photo Researchers, Inc. (d)Exactostock/SuperStock)

OPTICAL PHENOMENA OF THE ATMOSPHERE IN REVIEW

- The two basic interactions of light with matter are *reflection* and *refraction*. The *law of reflection* states that when light rays are reflected, they bounce off the reflecting surface at the same angle (the angle of reflection) at which they meet that surface (the angle of incidence). Refraction is the bending of light due to a change in velocity as it passes obliquely from one transparent medium to another. Furthermore, light will gradually bend as it traverses a layer of air that varies in density.

- A *mirage* is an optical effect of the atmosphere caused by refraction when light passes through air with different densities. The result is that objects appear to be displaced from their true position. An *inferior mirage* produces an image that appears below the true location of the observed object.

- Perhaps the most spectacular and most commonly known atmospheric optical phenomenon is the *rainbow*. When a rainbow forms, rain drops act as prisms and disperse the sunlight into the spectrum of colors in a process called *dispersion*. In a *primary rainbow*, sunlight is reflected once within a raindrop. In a dimmer, less frequently observed *secondary rainbow*, light is reflected twice within a raindrop before it is backscattered.

- A *halo* is a narrow whitish ring centered on the Sun. Halos occur most often when the sky is covered with a thin layer of cirrus clouds. The most common halo is the 22° halo, so named because its radius subtends an angle of 22° from the observer. Halos are produced by dispersion of sunlight by ice crystals. A spectacular effect associated with a halo is called *sun dogs*. These two bright regions, or mock suns, can be seen adjacent to a 22° halo.

- *Glories,* most commonly seen by pilots, consist of one or more colored rings that surround the observer's (or airplane's) shadow projected on the clouds below. They form in a manner similar to rainbows. The phenomenon can also be viewed by an observer on the ground when he or she is above a layer of fog, with the Sun at his or her back.

- *Coronas* are bright whitish disks centered on the Moon or Sun and are the only optical phenomenon more commonly witnessed in association with the Moon than with the Sun.

VOCABULARY REVIEW

corona (p. 461)
dispersion (p. 456)
glory (p. 461)
halo (p. 458)
inferior mirage (p. 453)
internal reflection (p. 450)

iridescent cloud (p. 462)
law of reflection (p. 450)
looming (p. 453)
mirage (p. 453)
rainbow (p. 455)
refraction (p. 451)

sun dogs (p. 459)
sun pillar (p. 459)
superior mirage (p. 453)
towering (p. 454)

MyMeteorologyLab™

Log in to www.mymeteorologylab.com for animations, videos, MapMaster interactive maps, GEODe media, *In the News* RSS feeds, web links, glossary flashcards, self-study quizzes and a Pearson eText version of this book to enhance your study of *Optical Phenomena of the Atmosphere.*

Appendix A Metric Units

TABLE A–1 The International System of Units (SI)

I. Basic units

Quantity	Unit	SI symbol
Length	meter	m
Mass	kilogram	kg
Time	second	s
Electric current	ampere	A
Thermodynamic temperature	kelvin	K
Amount of substance	mole	mol
Luminous intensity	candela	cd

II. Prefixes

Prefix	Factor by which unit is multiplied	Symbol
tera	10^{12}	T
giga	10^{9}	G
mega	10^{6}	M
kilo	10^{3}	k
hecto	10^{2}	h
deka	10	da
deci	10^{-1}	d
centi	10^{-2}	c
milli	10^{-3}	m
micro	10^{-6}	μ
nano	10^{-9}	n
pico	10^{-12}	p
femto	10^{-15}	f
atto	10^{-18}	a

III. Derived units

Quantity	Units	Expression
Area	square meter	m^2
Volume	cubic meter	m^3
Frequency	hertz (Hz)	s^{-1}
Density	kilograms per cubic meter	kg/m^3
Velocity	meters per second	m/s
Angular velocity	radians per second	rad/s
Acceleration	meters per second squared	m/s^2
Angular acceleration	radians per second squared	rad/s^2
Force	newton (N)	$kg \cdot ms^2$
Pressure	newtons per square meter	N/m^2
Work, energy, quantity of heat	joule (J)	$N \cdot m$
Power	watt (W)	J/s
Electric charge	coulomb (C)	$A \cdot s$
Voltage, potential difference, electromotive force	volt (V)	W/A
Luminance	candelas per square meter	cd/m^2

TABLE A–2 Metric–English Conversion

When you want to convert to	Multiply by:	To find:
Length		
inches	2.54	centimeters
centimeters	0.39	inches
feet	0.30	meters
meters	3.28	feet
yards	0.91	meters
meters	1.09	yards
miles	1.61	kilometers
kilometers	0.62	miles
Area		
square inches	6.45	square centimeters
square centimeters	0.15	square inches
square feet	0.09	square meters
square meters	10.76	square feet
square miles	2.59	square kilometers
square kilometers	0.39	square miles
Volume		
cubic inches	16.38	cubic centimeters
cubic centimeters	0.06	cubic inches
cubic feet	0.028	cubic meters
cubic meters	35.3	cubic feet
cubic miles	4.17	cubic kilometers
cubic kilometers	0.24	cubic miles
liters	1.06	quarts
liters	0.26	gallons
gallons	3.78	liters
Masses and Weights		
ounces	28.33	grams
grams	0.035	ounces
pounds	0.45	kilograms
kilograms	2.205	pounds
Temperature		

When you want to convert degrees Fahrenheit (°F) to degrees Celsius (°C), subtract 32 degrees and divide by 1.8 (also see Table A–3).

When you want to convert degrees Celsius (°C) to degrees Fahrenheit (°F), multiply by 1.8 and add 32 degrees (also see Table A–3).

When you want to convert degrees Celsius (°C) to kelvins (K), delete the degree symbol and add 273.

When you want to convert kelvins (K) to degrees Celsius (°C), add the degree symbol and subtract 273.

TABLE A–3   Temperature Conversion Table. (To find either the Celsius or the Fahrenheit equivalent, locate the known temperature in the center column. Then read the desired equivalent value from the appropriate column.)

°C		°F	°C		°F	°C		°F	°C		°F
−40.0	−40	−40	−17.2	+1	33.8	5.0	41	105.8	27.2	81	177.8
−39.4	−39	−38.2	−16.7	2	35.6	5.6	42	107.6	27.8	82	179.6
−38.9	−38	−36.4	−16.1	3	37.4	6.1	43	109.4	28.3	83	181.4
−38.3	−37	−34.6	−15.4	4	39.2	6.7	44	111.2	28.9	84	183.2
−37.8	−36	−32.8	−15.0	5	41.0	7.2	45	113.0	29.4	85	185.0
−37.2	−35	−31.0	−14.4	6	42.8	7.8	46	114.8	30.0	86	186.8
−36.7	−34	−29.2	−13.9	7	44.6	8.3	47	116.6	30.6	87	188.6
−36.1	−33	−27.4	−13.3	8	46.4	8.9	48	118.4	31.1	88	190.4
−35.6	−32	−25.6	−12.8	9	48.2	9.4	49	120.2	31.7	89	192.2
−35.0	−31	−23.8	−12.2	10	50.0	10.0	50	122.0	32.2	90	194.0
−34.4	−30	−22.0	−11.7	11	51.8	10.6	51	123.8	32.8	91	195.8
−33.9	−29	−20.2	−11.1	12	53.6	11.1	52	125.6	33.3	92	197.6
−33.3	−28	−18.4	−10.6	13	55.4	11.7	53	127.4	33.9	93	199.4
−32.8	−27	−16.6	−10.0	14	57.2	12.2	54	129.2	34.4	94	201.2
−32.2	−26	−14.8	−9.4	15	59.0	12.8	55	131.0	35.0	95	203.0
−31.7	−25	−13.0	−8.9	16	60.8	13.3	56	132.8	35.6	96	204.8
−31.1	−24	−11.2	−8.3	17	62.6	13.9	57	134.6	36.1	97	206.6
−30.6	−23	−9.4	−7.8	18	64.4	14.4	58	136.4	36.7	98	208.4
−30.0	−22	−7.6	−7.2	19	66.2	15.0	59	138.2	37.2	99	210.2
−29.4	−21	−5.8	−6.7	20	68.0	15.6	60	140.0	37.8	100	212.0
−28.9	−20	−4.0	−6.1	21	69.8	16.1	61	141.8	38.3	101	213.8
−28.3	−19	−2.2	−5.6	22	71.6	16.7	62	143.6	38.9	102	215.6
−27.8	−18	−0.4	−5.0	23	73.4	17.2	63	145.4	39.4	103	217.4
−27.2	−17	+1.4	−4.4	24	75.2	17.8	64	147.2	40.0	104	219.2
−26.7	−16	3.2	−3.9	25	77.0	18.3	65	149.0	40.6	105	221.0
−26.1	−15	5.0	−3.3	26	78.8	18.9	66	150.8	41.1	106	222.8
−25.6	−14	6.8	−2.8	27	80.6	19.4	67	152.6	41.7	107	224.6
−25.0	−13	8.6	−2.2	28	82.4	20.0	68	154.4	42.2	108	226.4
−24.4	−12	10.4	−1.7	29	84.2	20.6	69	156.2	42.8	109	228.2
−23.9	−11	12.2	−1.1	30	86.0	21.1	70	158.0	43.3	110	230.0
−23.3	−10	14.0	−0.6	31	87.8	21.7	71	159.8	43.9	111	231.8
−22.8	−9	15.8	0.0	32	89.6	22.2	72	161.6	44.4	112	233.6
−22.2	−8	17.6	+0.6	33	91.4	22.8	73	163.4	45.0	113	235.4
−21.7	−7	19.4	1.1	34	93.2	23.3	74	165.2	45.6	114	237.2
−21.1	−6	21.2	1.7	35	95.0	23.9	75	167.0	46.1	115	239.0
−20.6	−5	23.0	2.2	36	96.8	24.4	76	168.8	46.7	116	240.8
−20.0	−4	24.8	2.8	37	98.6	25.0	77	170.6	47.2	117	242.6
−19.4	−3	26.6	3.3	38	100.4	25.6	78	172.4	47.8	118	244.4
−18.9	−2	28.4	3.9	39	102.2	26.1	79	174.2	48.3	119	246.2
−18.3	−1	30.2	4.4	40	104.0	26.7	80	176.0	48.9	120	248.0
−17.8	0	32.0									

TABLE A-4 Wind-Conversion Table (Wind-Speed Units: 1 Mile Per Hour = 0.868391 Knot = 1.609344 km/h = 0.44704 m/s)

Miles per Hour	Knots	Meters per Second	Kilometers per Hour	Miles per Hour	Knots	Meters per Second	Kilometers per Hour
1	0.9	0.4	1.6	51	44.3	22.8	82.1
2	1.7	0.9	3.2	52	45.2	23.2	83.7
3	2.6	1.3	4.8	53	46.0	23.7	85.3
4	3.5	1.8	6.4	54	46.9	24.1	86.9
5	4.3	2.2	8.0	55	47.8	24.6	88.5
6	5.2	2.7	9.7	56	48.6	25.0	90.1
7	6.1	3.1	11.3	57	49.5	25.5	91.7
8	6.9	3.6	12.9	58	50.4	25.9	93.3
9	7.8	4.0	14.5	59	51.2	26.4	95.0
10	8.7	4.5	16.1	60	52.1	26.8	96.6
11	9.6	4.9	17.7	61	53.0	27.3	98.2
12	10.4	5.4	19.3	62	53.8	27.7	99.8
13	11.3	5.8	20.9	63	54.7	28.2	101.4
14	12.2	6.3	22.5	64	55.6	28.6	103.0
15	13.0	6.7	24.1	65	56.4	29.1	104.6
16	13.9	7.2	25.7	66	57.3	29.5	106.2
17	14.8	7.6	27.4	67	58.2	30.0	107.8
18	15.6	8.0	29.0	68	59.1	30.4	109.4
19	16.5	8.5	30.6	69	59.9	30.8	111.0
20	17.4	8.9	32.2	70	60.8	31.3	112.7
21	18.2	9.4	33.8	71	61.7	31.7	114.3
22	19.1	9.8	35.4	72	62.5	32.2	115.9
23	20.0	10.3	37.0	73	63.4	32.6	117.5
24	20.8	10.7	38.6	74	64.3	33.1	119.1
25	21.7	11.2	40.2	75	65.1	33.5	120.7
26	22.6	11.6	41.8	76	66.0	34.0	122.3
27	23.4	12.1	43.5	77	66.9	34.4	123.9
28	24.3	12.5	45.1	78	67.7	34.9	125.5
29	25.2	13.0	46.7	79	68.6	35.3	127.1
30	26.1	13.4	48.3	80	69.5	35.8	128.7
31	26.9	13.9	49.9	81	70.3	36.2	130.4
32	27.8	14.3	51.5	82	71.2	36.7	132.0
33	28.7	14.8	53.1	83	72.1	37.1	133.6
34	29.5	15.2	54.7	84	72.9	37.6	135.2
35	30.4	15.6	56.3	85	73.8	38.0	136.8
36	31.3	16.1	57.9	86	74.7	38.4	138.4
37	32.1	16.5	59.5	87	75.5	38.9	140.0
38	33.0	17.0	61.2	88	76.4	39.3	141.6
39	33.9	17.4	62.8	89	77.3	39.8	143.2
40	34.7	17.9	64.4	90	78.2	40.2	144.8
41	35.6	18.3	66.0	91	79.0	40.7	146.5

(Continued)

TABLE A–4   (Continued)

Miles per Hour	Knots	Meters per Second	Kilometers per Hour	Miles per Hour	Knots	Meters per Second	Kilometers per Hour
42	36.5	18.8	67.6	92	79.9	41.1	148.1
43	37.3	19.2	69.2	93	80.8	41.6	149.7
44	38.2	19.7	70.8	94	81.6	42.0	151.3
45	39.1	20.1	72.4	95	82.5	42.5	152.9
46	39.9	20.6	74.0	96	83.4	42.9	154.5
47	40.8	21.0	75.6	97	84.2	43.4	156.1
48	41.7	21.5	77.2	98	85.1	43.8	157.7
49	42.6	21.9	78.9	99	86.0	44.3	159.3
50	43.4	22.4	80.5	100	86.8	44.7	160.9

Appendix B Explanation and Decoding of the Daily Weather Map

 Introduction to the Atmosphere
▶ In the Lab: Reading Weather Maps

Weather maps showing the development and movement of weather systems are among the most important tools used by the meterologist. Some maps portray conditions near the surface of Earth and others depict conditions at various heights in the atmosphere. Some cover the entire Northern Hemisphere and others cover only local areas as required for special purposes.

Principal Surface Weather Map

To prepare the surface map and present the information quickly and pictorially, two actions are necessary: (1) Weather observers and automated observing stations must send data to the offices where the maps are prepared; (2) the information must be quickly transcribed to the maps. In order for the necessary speed and economy of space and transmission time to be realized, codes have been devised for sending the information and for plotting it on the maps.

Codes and Map Plotting

A great deal of information is contained in a brief coded weather message. If each item were named and described in plain language, a very lengthy message would be required, one confusing to read and difficult to transfer to a map. A code permits the message to be condensed to a few five-figure numeral groups, each figure of which has a meaning, depending on its position in the message. People trained in the use of the code can read the message as easily as plain language (see Figure B–1).

The location of the reporting station is printed on the map as a small circle (the station circle). A definite arrangement of the data around the station circle, called the *station model*, is used. When the report is plotted in these fixed positions around the station circle on the weather map, many code figures are transcribed exactly as sent. Entries in the station model that are not made in code figures or actual values found in the message are usually in the form of symbols that graphically represent the element concerned. In some cases, certain of the data may or may not be reported by the observer, depending on local weather conditions. Precipitation and clouds are examples. In such cases, the absence of an entry on the map is interpreted as nonoccurrence or nonobservance of the phenomenon. The letter *M* is entered where data are normally observed but not received.

Both the code and the station model are based on international agreements. These standardized numerals and symbols enable a meteorologist of one country to use the weather reports and weather maps of another country even though that person does not understand the language. Weather codes are, in effect, an international language that permits complete interchange and use of worldwide weather reports so essential in present-day activities.

The boundary between two different air masses is called a *front*. Important changes in weather, temperature, wind direction, and clouds often occur with the passage of a front. Half circles or triangular symbols or both are placed on the lines representing fronts to indicate the kind of front. The side on which the symbols are placed indicates the direction of frontal movement. The boundary of relatively cold air of polar origin advancing into an area occupied by warmer air, often of tropical origin, is called a *cold front*. The boundary of relatively warm air advancing into an area occupied by colder air is called a *warm front*. The line along which a cold front has overtaken a warm front at the ground is called an *occluded front*. A boundary between two air masses, which shows at the time of observation little tendency to advance into either the warm or cold areas, is called a *stationary front*. Air-mass boundaries are known as *surface fronts* when they intersect the ground and as *upper-air fronts* when they do not. Surface fronts are drawn in solid black; fronts aloft are drawn in outline only. Front symbols are given in Table B–1.

A front that is disappearing or weak and decreasing in intensity is labeled *frontolysis*. A front that is forming is labeled *frontogenesis*. A *squall line* is a line of thunderstorms or squalls usually accompanied by heavy showers and shifting winds (Table B–1).

The paths followed by individual disturbances are called *storm tracks* and are shown by arrows (Table B–1). A symbol (a box containing an X) indicates past positions of a low-pressure center at six-hour intervals. HIGH (H) and LOW (L) indicate the centers of high and low barometric pressure. Solid lines are isobars and connect points of equal sea-level barometric pressure. The spacing and orientation of these lines on weather maps are indications of speed and direction of windflow. In general, wind direction is parallel to these lines with low pressure to the left of an observer looking downwind. Speed is directly proportional to the closeness of the lines (called *pressure gradient*). Isobars are labeled in millibars.

Isotherms are lines connecting points of equal temperature. Two isotherms are frequently drawn on large

FIGURE B-1 Explanation of Station Symbols and Map Entries

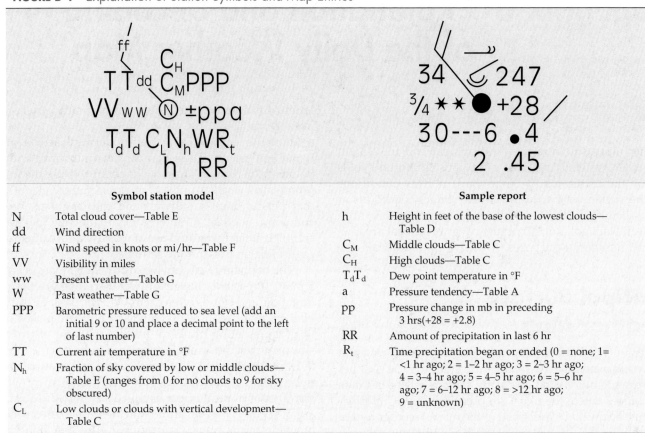

Symbol station model		
N	Total cloud cover—Table E	
dd	Wind direction	
ff	Wind speed in knots or mi/hr—Table F	
VV	Visibility in miles	
ww	Present weather—Table G	
W	Past weather—Table G	
PPP	Barometric pressure reduced to sea level (add an initial 9 or 10 and place a decimal point to the left of last number)	
TT	Current air temperature in °F	
N_h	Fraction of sky covered by low or middle clouds—Table E (ranges from 0 for no clouds to 9 for sky obscured)	
C_L	Low clouds or clouds with vertical development—Table C	

Sample report		
h	Height in feet of the base of the lowest clouds—Table D	
C_M	Middle clouds—Table C	
C_H	High clouds—Table C	
T_dT_d	Dew point temperature in °F	
a	Pressure tendency—Table A	
pp	Pressure change in mb in preceding 3 hrs(+28 = +2.8)	
RR	Amount of precipitation in last 6 hr	
R_t	Time precipitation began or ended (0 = none; 1= <1 hr ago; 2 = 1–2 hr ago; 3 = 2–3 hr ago; 4 = 3–4 hr ago; 5 = 4–5 hr ago; 6 = 5–6 hr ago; 7 = 6–12 hr ago; 8 = >12 hr ago; 9 = unknown)	

surface weather maps when applicable. The freezing, or 32°F, isotherm is drawn as a dashed line, and the 0°F isotherm is drawn as a dash–dot line (Table B–1). Areas where precipitation is occurring at the time of observation are shaded.

TABLE B-1 Weather Map Symbols

Symbol	Explanation
▲▲▲	Cold front (surface)
●●●	Warm front (surface)
▲●▲●	Occluded front (surface)
▲●▲●	Stationary front (surface)
●●●	Dryline
–··–··–··–	Squall line
⟶ ⟶	Path of low-pressure center
⊠	Location of low pressure at 6-hour intervals
– – – – –	32° F isotherm
–·–··–··–	0° F isotherm

Auxiliary Maps

500-Millibar Map

Contour lines, isotherms, and wind arrows are shown on the 500-millibar contour level. Solid lines are drawn to show height above sea level and are labeled in feet. Dashed lines are drawn at 5° intervals of temperature and are labeled in degrees Celsius. True wind direction is shown by "arrows" that are plotted as flying with the wind. The wind speed is shown by flags and feathers. Each flag represents 50 knots, each full feather represents 10 knots, and each half feather represents 5 knots.

Temperature Map (Highest and Lowest)

Temperature data are entered from selected weather stations in the United States. The figure entered above the station dot shows the maximum temperature for the 12-hour period ending 7:00 P.M. EST of the previous day. The figure entered below the station dot shows the minimum temperature during the 12 hours ending at 7:00 A.M. EST. The letter *M* denotes missing data.

Precipitation Map

Precipitation data are entered from selected weather stations in the United States. When precipitation has occurred at any of these stations in the 24-hour period ending at 7:00 A.M. EST, the total amount, in inches and hundredths, is entered above the station dot. When the figures for total precipita-tion have been compiled from incomplete data and entered on the map, the amount is underlined. *T* indicates a trace of precipitation (less than 0.01 inch) and the letter *M* denotes missing data. The geographical areas where precipitation has fallen during the 24 hours ending at 7:00 A.M. EST are shaded. Dashed lines show depth of snow on ground in inches as of 7:00 A.M. EST.

TABLE A Air Pressure Tendency

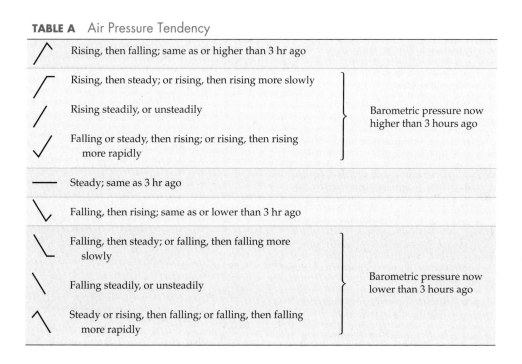

⟋	Rising, then falling; same as or higher than 3 hr ago	
⌐⟋	Rising, then steady; or rising, then rising more slowly	Barometric pressure now higher than 3 hours ago
⟋	Rising steadily, or unsteadily	
✓	Falling or steady, then rising; or rising, then rising more rapidly	
—	Steady; same as 3 hr ago	
⟍	Falling, then rising; same as or lower than 3 hr ago	
⟍_	Falling, then steady; or falling, then falling more slowly	Barometric pressure now lower than 3 hours ago
⟍	Falling steadily, or unsteadily	
⋀	Steady or rising, then falling; or falling, then falling more rapidly	

TABLE B Cloud Abbreviations

St	stratus
Fra	fractus
Sc	stratocumulus
Ns	nimbostratus
As	altostratus
Ac	altocumulus
Ci	cirrus
Cs	cirrostratus
Cc	cirrocumulus
Cu	cumulus
Cb	cumulonimbus

TABLE C Cloud Types

Low Clouds and Clouds of Vertical Development

Cu of fair weather, little vertical development and seemingly flattened

Cu of considerable development, generally towering, with or without other Cu or Sc, bases all at same level

Cb with tops lacking clear-cut outlines, but distinctly not cirriform or anvil shaped; with or without Cu, Sc, or St

Sc formed by spreading out of Cu; Cu often present also

Sc not formed by spreading out of Cu

St or StFra, but no StFra of bad weather

StFra and/or CuFra of bad weather (scud)

Cu and Sc (not formed by spreading out of Cu) with bases at different levels

Cb having a clearly fibrous (cirriform) top, often anvil shaped, with or without Cu, Sc, St, or scud

Middle Clouds

Thin As (most of cloud layer semitransparent)

Thick As, greater part sufficiently dense to hide Sun (or Moon), or Ns

Thin Ac, mostly semitransparent; cloud elements not changing much and at a single level

Thin Ac in patches; cloud elements continually changing and/or occurring at more than one level

Thin Ac in bands or in a layer gradually spreading over sky and usually thickening as a whole

Ac formed by the spreading out of Cu or Cb

Double-layered Ac, or a thick layer of Ac, not increasing; or Ac with As and/or Ns

Ac in the form of Cu-shaped tufts or Ac with turrets

Ac of a chaotic sky, usually at different levels; patches of dense Ci usually present also

High Clouds

Filaments of Ci, or "mares' tails," scattered and not increasing

Dense Ci in patches or twisted sheaves, usually not increasing, sometimes like remains of Cb; or towers or tufts

Dense Ci, often anvil shaped, derived from or associated with Cb

Ci, often hook shaped, gradually spreading over the sky and usually thickening as a whole

Ci and Cs, often in converging bands, or Cs alone; generally overspreading and growing denser; the continuous layer not reaching 45° altitude

Ci and Cs, often in converging bands, or Cs alone; generally overspreading and growing denser; the continuous layer exceeding 45° altitude

Veil of Cs covering the entire sky

Cs not increasing and not covering entire sky

Cc alone or Cc with some Ci or Cs, but the Cc being the main cirriform cloud

TABLE D Height of Base of Lowest Cloud

Code	Feet	Meters
0	0–149	0–49
1	150–299	50–99
2	300–599	100–199
3	600–999	200–299
4	1000–1999	300–599
5	2000–3499	600–999
6	3500–4999	1000–1499
7	5000–6499	1500–1999
8	6500–7999	2000–2499
9	8000 or above or no clouds	2500 or above or no clouds

TABLE E Cloud Cover

Symbol	Description
 ○	No clouds
 ◔ (one line)	One-tenth or less
 ◔	Two-tenths or three-tenths
 ◔	Four-tenths
 ◑	Five-tenths
 ◒	Six-tenths
 ◕	Seven-tenths or eight-tenths
 ◍	Nine-tenths or overcast with openings
●	Completely overcast (ten-tenths)
 ⊗	Sky obscured

TABLE F Wind Speed

Symbol	Miles per hour	Knots	Kilometers per hour
 ◎	calm	calm	calm
	1–2	1–2	1–3
	3–8	3–7	3–13
	9–14	8–12	14–19
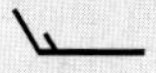	15–20	8–12	14–19
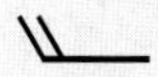	21–25	18–22	14–19
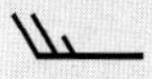	26–31	23–27	41–50
	32–37	28–32	51–60
	38–43	33–37	61–69
	44–49	38–42	70–79
	50–54	43–47	80–87
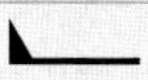	55–60	48–52	88–96
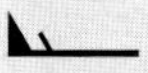	61–66	53–57	97–106
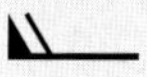	67–71	58–62	107–114
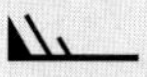	72–77	63–67	115–124
	78–83	68–72	125–134
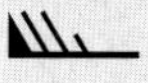	84–89	73–77	135–143
	119–123	103–107	192–198

TABLE G Weather Conditions

Cloud development NOT observed or NOT observable during past hour	Clouds generally dissolving or becoming less developed during past hour	State of sky on the whole unchanged during past hour	Clouds generally forming or developing during past hour	Visibility reduced by smoke
Light fog (mist)	Patches of shallow fog at station, NOT deeper than 6 feet on land	More or less continuous shallow fog at station, NOT deeper than 6 feet on land	Lightning visible, no thunder heard	Precipitation within sight, but NOT reaching the ground
Drizzle (NOT freezing) or snow grains (NOT falling as showers) during past hour, but NOT at time of observation	Rain (NOT freezing and NOT falling as showers) during past hour, but NOT at time of observation	Snow (NOT falling as showers) during past hour, but NOT at time of observation	Rain and snow or ice pellets (NOT falling as showers) during past hour, but NOT at time of observation	Freezing drizzle or freezing rain (NOT falling as showers) during past hour, but NOT at time of observation
Slight or moderate dust storm or sandstorm has decreased during past hour	Slight or moderate dust storm or sandstorm, no appreciable change during past hour	Slight or moderate dust storm or sandstorm has begun or increased during past hour	Severe dust storm or sandstorm, has decreased during past hour	Severe dust storm or sandstorm, no appreciable change during past hour
Fog or ice fog at distance at time of observation, but NOT at station during past hour	Fog or ice fog in patches	Fog or ice fog, sky discernible, has become thinner during past hour	Fog or ice fog, sky NOT discernible, has become thinner during past hour	Fog or ice fog, sky discernible, no appreciable change during past hour
Intermittent drizzle (NOT freezing), slight at time of observation	Continuous drizzle (NOT freezing), slight at time of observation	Intermittent drizzle (NOT freezing), moderate at time of observation	Continuous drizzle (NOT freezing), moderate at time of observation	Intermittent drizzle (NOT freezing), heavy at time of observation
Intermittent rain (NOT freezing), slight at time of observation	Continuous rain (NOT freezing), slight at time of observation	Intermittent rain (NOT freezing), moderate at time of observation	Continuous rain (NOT freezing), moderate at time of observation	Intermittent rain (NOT freezing), heavy at time of observation
Intermittent fall of snowflakes, slight at time of observation	Continuous fall of snowflakes, slight at time of observation	Intermittent fall of snowflakes, moderate at time of observation	Continuous fall of snowflakes, moderate at time of observation	Intermittent fall of snowflakes, heavy at time of observation
Slight rain shower(s)	Moderate or heavy rain shower(s)	Violent rain shower(s)	Slight shower(s) of rain and snow mixed	Moderate or heavy shower(s) of rain and snow mixed
Moderate or heavy shower(s) of hail, with or without rain, or rain and snow mixed, not associated with thunder	Slight rain at time of observation; thunderstorm during past hour, but NOT at time of observation	Moderate or heavy rain at time of observation; thunderstorm during past hour, but NOT at time of observation	Slight snow, or rain and snow mixed, or hail at time of observation; thunderstorm during past hour, but NOT at time of observation	Moderate or heavy snow, or rain and snow mixed, or hail at time of observation; thunderstorm during past hour, but NOT at time of observation

(Continued)

TABLE G   Continued

Symbol	Description	Symbol	Description	Symbol	Description	Symbol	Description	Symbol	Description
∞	Haze	S	Widespread dust in suspension in the air, NOT raised by wind, at time of observation	$	Dust or sand raised by wind at time of observation	⚡	Well developed dust whirl(s) within past hour	(S)	Dust storm or sand-storm within sight of or at station during past hour
)•(	Precipitation within sight, reaching the ground but distant from station	(•)	Precipitation within sight, reaching the ground, near to but NOT at station	⟋	Thunderstorm, but no precipitation at the station	∨	Squall(s) within sight during past hour or at time of observa-tion	)(	Funnel cloud(s) within sight of station at time of observation
•∇	Showers of rain during past hour, but NOT at time of observation	✷∇	Showers of snow, or of rain and snow, during past hour, but NOT at time of observation	▲∇	Showers of hail, or of hail and rain, during past hour, but NOT at time of observation	=	Fog during past hour, but NOT at time of observation	⟋	Thunderstorm (with or without precipi-tation) during past hour, but NOT at time of observation
S	Severe dust storm or sandstorm has begun or in-creased during past hour	↓	Slight or moderate drifting snow, generally low (less than 6 ft)	�	Heavy drifting snow, generally low	↑	Slight or moderate blowing snow, generally high (more than 6 ft)	⇑	Heavy blowing snow, generally high
≡	Fog or ice fog, sky NOT discernible, no appreciable change during past hour	|– –	Fog or ice fog, sky discernible, has begun or become thicker during past hour	|≡	Fog or ice fog, sky NOT discernible, has begun or become thicker during past hour	⩔	Fog depositing rime, sky discernible	⩔	Fog depositing rime, sky NOT discernible
,',	Continuous drizzle (NOT freezing), heavy at time of observation	⌇	Slight freezing drizzle	⌇	Moderate or heavy freezing drizzle	• ;	Drizzle and rain, slight	• ;	Drizzle and rain, moderate or heavy
•∴•	Continuous rain (NOT freezing), heavy at time of observation	∿	Slight freezing rain	∿•	Moderate or heavy freezing rain	•✱	Rain or drizzle and snow, slight	✱✱	Rain or drizzle and snow, moderate or heavy
✷✷✷	Continuous fall of snowflakes, heavy at time of observation	↔	Ice prisms (with or without fog)	⟶△	Snow grains (with or without fog)	⟶✕	Isolated starlike snow crystals (with or without fog)	△	Ice pellets or snow pellets
✱∇	Slight snow shower(s)	✱∇	Moderate or heavy snow shower(s)	△∇	Slight shower(s) of snow pellets, or ice pellets with or without rain, or rain and snow mixed	△∇	Moderate or heavy shower(s) of snow pellets, or ice pellets, or ice pellets with or without rain or rain and snow mixed	△∇	Slight shower(s) of hail, with or without rain or rain and snow mixed, not associated with thunder
•/✱⟋	Slight or moderate thunderstorm without hail, but with rain, and/or snow at time of observation	△⟋	Slight or moderate thunderstorm, with hail at time of observation	•/✱⟋	Heavy thundestorm, without hail, but with rain and/or snow at time of observation	S⟋	Thunderstorm combined with dust storm or sandstorm at time of observation	△⟋	Heavy thunder-storm with hail at time of observa-tion

Appendix C Relative Humidity and Dew-Point Tables

TABLE C–1 Relative Humidity (percent)*

Dry Bulb (°C)	Depression of Web-Bulb Temperature (Dry-bulb temperature − Wet-bulb temperature = Depression of the wet bulb)																					
	1	2	3	4	5	6	7	8	9	10	11	12	13	14	15	16	17	18	19	20	21	22
−20	28																					
−18	40																					
−16	48	0																				
−14	55	11																				
−12	61	23																				
−10	66	33	0																			
−8	71	41	13																			
−6	73	48	20	0																		
−4	77	54	32	11																		
−2	79	58	37	20	1																	
0	81	63	45	28	11																	
2	83	67	51	36	20	6																
4	85	70	56	42	27	14																
6	86	72	59	46	35	22	10	0														
8	87	74	62	51	39	28	17	6														
10	88	76	65	54	43	33	24	13	4													
12	88	78	67	57	48	38	28	19	10	2												
14	89	79	69	60	50	41	33	25	16	8	1											
16	90	80	71	62	54	45	37	29	21	14	7	1										
18	91	81	72	64	56	48	40	33	26	19	12	6	0									
20	91	82	74	66	58	51	44	36	30	23	17	11	5									
22	92	83	75	68	60	53	46	40	33	27	21	15	10	4	0							
24	92	84	76	69	62	55	49	42	36	30	25	20	14	9	4	0						
26	92	85	77	70	64	57	51	45	39	34	28	23	18	13	9	5						
28	93	86	78	71	65	59	53	45	42	36	31	26	21	17	12	8	4					
30	93	86	79	72	66	61	55	49	44	39	34	29	25	20	16	12	8	4				
32	93	86	80	73	68	62	56	51	46	41	36	32	27	22	19	14	11	8	4			
34	93	86	81	74	69	63	58	52	48	43	38	34	30	26	22	18	14	11	8	5		
36	94	87	81	75	69	64	59	54	50	44	40	36	32	28	24	21	17	13	10	7	4	
38	94	87	82	76	70	66	60	55	51	46	42	38	34	30	26	23	20	16	13	10	7	5
40	94	89	82	76	71	67	61	57	52	48	44	40	36	33	29	25	22	19	16	13	10	7

Relative humidity values

*To determine the relative humidity, find the air (dry-bulb) temperature on the vertical axis (far left) and the depression of the wet bulb on the horizontal axis (top). Where the two meet, the relative humidity is found. For example, when the dry-bulb temperature is 20°C and a wet-bulb temperature is 14°C, then the depression of the wet-bulb is 6°C (20°C − 14°C). From Table C-1, the relative humidity is 51 percent and from Table C-2, the dew point is 10°C.

TABLE C–2 Dew-Point Temperature (°C)

Dry-bulb temperature − Wet-bulb temperature = Depression of the wet bulb

Dry Bulb (°C)	1	2	3	4	5	6	7	8	9	10	11	12	13	14	15	16	17	18	19	20	21	22
−20	−33																					
−18	−28																					
−16	−24																					
−14	−21	−36																				
−12	−18	−28																				
−10	−14	−22																				
−8	−12	−18	−29																			
−6	−10	−14	−22																			
−4	−7	−12	−17	−29																		
−2	−5	−8	−13	−20																		
0	−3	−6	−9	−15	−24																	
2	−1	−3	−6	−11	−17																	
4	1	−1	−4	−7	−11	−19																
6	4	1	−1	−4	−7	−13	−21															
8	6	3	1	−2	−5	−9	−14															
10	8	6	4	1	−2	−5	−9	−14	−18													
12	10	8	6	4	1	−2	−5	−9	−16													
14	12	11	9	6	4	1	−2	−5	−10	−17												
16	14	13	11	9	7	4	1	−1	−6	−10	−17											
18	16	15	13	11	9	7	4	2	−2	−5	−10	−19										
20	19	17	15	14	12	10	7	4	2	−2	−5	−10	−19									
22	21	19	17	16	14	12	10	8	5	3	−1	−5	−10	−19								
24	23	21	20	18	16	14	12	10	8	6	2	−1	−5	−10	−18							
26	25	23	22	20	18	17	15	13	11	9	6	3	0	−4	−9	−18						
28	27	25	24	22	21	19	17	16	14	11	9	7	4	1	−3	−9	−16					
30	29	27	26	24	23	21	19	18	16	14	12	10	8	5	1	−2	−8	−15				
32	31	29	28	27	25	24	22	21	19	17	15	13	11	8	5	2	−2	−7	−14			
34	33	31	30	29	27	26	24	23	21	20	18	16	14	12	9	6	3	−1	−5	−12	−29	
36	35	33	32	31	29	28	27	25	24	22	20	19	17	15	13	10	7	4	0	−4	−10	
38	37	35	34	33	32	30	29	28	26	25	23	21	19	17	15	13	11	8	5	1	−3	−9
40	39	37	36	35	34	32	31	30	28	27	25	24	22	20	18	16	14	12	9	6	2	−2

Dry-bulb (air) temperature

Dew-point temperatures

Appendix D Laws Relating to Gases

Kinetic Energy

All moving objects, by virtue of their motion, are capable of doing work. We call this energy of motion, or *kinetic energy*. The kinetic energy of a moving object is equal to one-half its mass (M) multiplied by its velocity (v) squared. Stated mathematically:

$$\text{Kinetic energy} = \frac{1}{2} Mv^2$$

Therefore, by doubling the velocity of a moving object, the object's kinetic energy will increase four times.

First Law of Thermodynamics

The first law of thermodynamics is simply the thermal version of the law of conservation of energy, which states that energy cannot be created or destroyed, only transformed from one form to another. Meteorologists use the first law of thermodynamics along with the principles of kinetic energy extensively in analyzing atmospheric phenomena. According to the kinetic theory, the temperature of a gas is proportional to the kinetic energy of the moving molecules. When a gas is heated, its kinetic energy increases because of an increase in molecular motion. Further, when a gas is compressed, the kinetic energy will also be increased and the temperature of the gas will rise. These relationships are expressed in the first law of thermodynamics, as follows: The temperature of a gas may be changed by the addition or subtraction of heat, or by changing the pressure (compression or expansion), or by a combination of both. It is easy to understand how the atmosphere is heated or cooled by the gain or loss of heat. However, when we consider rising and sinking air, the relationships between temperature and pressure become more important. Here an increase in temperature is brought about by performing work on the gas and not by the addition of heat. This phenomenon is called the *adiabatic form* of the first law of thermodynamics.

Boyle's Law

About 1600, the Englishman Robert Boyle showed that if the temperature is kept constant when the pressure exerted on a gas is increased, the volume decreases. This principle, called *Boyle's law,* states: At a constant temperature, the volume of a given mass of gas varies inversely with the pressure. Stated mathematically:

$$P_1 V_1 = P_2 V_2$$

The symbols P_1 and V_1 refer to the original pressure and volume, respectively, and P_2 and V_2 indicate the new pressure and volume, respectively, after a change occurs. Boyle's law shows that if a given volume of gas is compressed so that the volume is reduced by one-half, the pressure exerted by the gas is doubled. This increase in pressure can be explained by the kinetic theory, which predicts that when the volume of the gas is reduced by one-half, the molecules collide with the walls of the container twice as often. Because density is defined as the mass per unit volume, an increase in pressure results in increased density.

Charles's Law

The relationships between temperature and volume (hence, density) of a gas were recognized about 1787 by the French scientist Jacques Charles, and were stated formally by J. Gay-Lussac in 1802. *Charles's law* states: At a constant pressure, the volume of a given mass is directly proportional to the absolute temperature. In other words, when a quantity of gas is kept at a constant pressure, an increase in temperature results in an increase in volume and vice versa. Stated mathematically:

$$\frac{V_1}{V_2} = \frac{T_1}{T_2}$$

where V_1 and T_1 represent the original volume and temperature, respectively, and V_2 and T_2 represent the final volume and temperature, respectively. This law explains the fact that a gas expands when it is heated. According to the kinetic theory, when heated, particles move more rapidly and therefore collide more often.

The Ideal Gas Law or Equation of State

In describing the atmosphere, three variable quantities must be considered: pressure, temperature, and density (mass per unit volume). The relationships among these variables can be found by combining in a single statement the laws of Boyle and Charles as follows:

$$PV = RT \quad \text{or} \quad P = \rho RT$$

where P is pressure, V volume, R the constant of proportionality, T absolute temperature, and   density. This law, called the *ideal gas law,* states:

1. When the volume is kept constant, the pressure of a gas is directly proportional to its absolute temperature.
2. When the temperature is kept constant, the pressure of a gas is proportional to its density and inversely proportional to its volume.
3. When the pressure is kept constant, the absolute temperature of a gas is proportional to its volume and inversely proportional to its density.

Appendix E Newton's Laws, Pressure–Gradient Force, and Coriolis Force

Newton's Laws of Motion

Because air is composed of atoms and molecules, its motion is governed by the same natural laws that apply to all matter. Simply put, when a force is applied to air, it will be displaced from its original position. Depending on the direction from which the force is applied, the air may move horizontally to produce winds, or in some situations vertically to generate convective flow. To better understand the forces that produce global winds, it is helpful to become familiar with Newton's first two laws of motion.

Newton's first law of motion states that an object at rest will remain at rest, and an object in motion will continue moving at a uniform speed and in a straight line unless a force is exerted upon it. In simple terms this law states that objects at rest tend to stay at rest, and objects in motion tend to continue moving at the same rate in the same direction. The tendency of things to resist change in motion (including a change in direction) is known as inertia.

You have experienced Newton's first law if you have ever pushed a stalled auto along flat terrain. To start the automobile moving (accelerating) requires a force sufficient to overcome its inertia (resistance to change). However, once this vehicle is moving, a force equal to that of the frictional force between the tires and the pavement is enough to keep it moving.

Moving objects often deviate from straight paths or come to rest, whereas objects at rest begin to move. The changes in motion we observe in daily life are the result of one or more applied forces.

Newton's second law of motion describes the relationship between the forces that are exerted on objects and the observed accelerations that result. Newton's second law states that the acceleration of an object is directly proportional to the net force acting on that body and inversely proportional to the mass of the body. The first part of Newton's second law means that the acceleration of an object changes as the intensity of the applied force changes.

We define acceleration as the rate of change in velocity. Because velocity describes both the speed and direction of a moving body, the velocity of something can be changed by changing its speed or its direction or both. Further, the term acceleration refers both to decreases and increases in velocity.

For example, we know that when we push down on the gas pedal of an automobile, we experience a positive acceleration (increase in velocity). On the other hand, using the brakes retards acceleration (decreases velocity).

In the atmosphere, three forces are responsible for changing the state of motion of winds. These are the pressure–gradient force, the Coriolis force, and friction. From the preceding discussion, it should be clear that the relative strengths of these forces will determine to a large degree the role of each in establishing the flow of air. Further, these forces can be directed in such a way as to increase the speed of airflow, decrease the speed of airflow, or, in many instances, just change the direction of airflow.

Pressure–Gradient Force

The magnitude of the pressure–gradient force is a function of the pressure difference between two points and air density. It can be expressed as

$$F_{PG} = \frac{1}{d} \times \frac{\Delta_p}{\Delta_n}$$

where:

F_{PG} = pressure–gradient force per unit mass

d = density of air

p = pressure difference between two points

n = distance between two points

Let us consider an example where the pressure 5 kilometers above Little Rock, Arkansas, is 540 millibars, and at 5 kilometers above St. Louis, Missouri, it is 530 millibars. The distance between the two cities is 450 kilometers, and the air density at 5 kilometers is 0.75 kilogram per cubic meter. In order to use the pressure–gradient equation, we must use compatible units. We must first convert pressure from millibars to pascals, another measure of pressure that has units of (kilograms $\times$ meters^{-1} $\times$ second^2).

In our example, the pressure difference above the two cities is 10 millibars, or 1000 pascals (1000 kg/m·s^{-2}). Thus, we have:

$$F_{PG} = \frac{1}{0.75} \times \frac{1000}{450,000} = 0.0029 \, \frac{m}{s^2}$$

Newton's second law states that force equals mass times acceleration ($F = m \times a$). In our example, we have considered pressure–gradient force *per unit mass*; therefore, our result is an acceleration ($F/m = a$). Because of the small units shown, pressure–gradient *acceleration* is often expressed as centimeters per second squared. In this example, we have 0.296 cm/s^2.

Coriolis Force as a function of Wind Speed and Latitude

Figure 6–14 (page 171) shows how wind speed and latitude conspire to affect the Coriolis force. Consider a west wind at four different latitudes (0°, 20°, 40°, and 60°). After several hours Earth's rotation has changed the orientation of latitude and longitude of all locations except the equator such that the wind appears to be deflected to the right. The degree of deflection for a given wind speed increases with latitude because the orientation of latitude and longitude lines changes more at higher latitudes. The degree of deflection of a given latitude increases with wind speed because greater distances are covered in the period of time considered.

We can show mathematically the importance of latitude and wind speed on Coriolis force:

$$F_{CO} = 2v \, \Omega \sin \phi$$

where:

F_{CO} = Coriolis force *per unit mass of air*

v = wind speed

Ω = Earth's rate of rotation or angular velocity (which is 7.29×10^{-5} radians per second)

ϕ = latitude

Note that $\sin \phi$ is a trigonometric function equal to zero for an angle of 0° (equator) and 1 when $\phi = 90°$ (poles).

As an example, the Coriolis force per unit mass that must be considered for a 10-meters-per-second (m/s) wind at 40° is calculated as:

$$F_{CO} = 2\Omega \sin \phi \, v$$

$$F_{CO} = 2\Omega \sin 40° \times 10 \text{ m/s}$$

$$F_{CO} = 2\left(7.29 \times 10^{-5} \text{s}^{-1}\right) 0.64(10 \text{ m/s})$$

$$F_{CO} = 0.00094 \text{ meter per second squared}$$

$$= 0.094 \text{ cm s}^{-2}$$

The result (0.094 cm/s^{-2}) is expressed as an acceleration because we are considering force per unit mass and Force = Mass × Acceleration.

Using this equation, one could calculate the Coriolis force for any latitude or wind speed. Consider Table E–1, which shows the Coriolis force per unit mass for three specific wind speeds at various latitudes. All values are expressed in centimeters per second squared (cm/s^{-2}). Because pressure–gradient force and Coriolis force approximately balance under geostrophic conditions, we can see from our table that the pressure–gradient force (per unit mass) of 0.296 cm/s^{-2} illustrated in the preceding discussion of pressure–gradient force would produce relatively strong winds.

TABLE E–1 Coriolis Force for Three Wind Speeds at Various Latitudes

Wind Speed		Latitude (ϕ)			
		0°	20°	40°	60°
(m/s)	(kph)	Coriolis Force (cm/s^2)			
5	18	0	0.025	0.047	0.063
10	36	0	0.050	0.094	0.126
25	90	0	0.125	0.235	0.316

Appendix F Saffir–Simpson Hurricane Scale

Scale Number (category)	Central Pressure (millibars)	Wind Speed (kph)	Wind Speed (mph)	Storm Surge (meters)	Storm Surge (feet)	Damage
1	≥980	119–153	74–95	1.2–1.5	4–5	*Minimal.* No real damage to building structures. Damage primarily to unanchored mobile homes, shrubbery, and trees. Also, some coastal-road flooding and minor pier damage.
2	965–979	154–177	96–110	1.6–2.4	6–8	*Moderate.* Some roofing material, door, and window damage to buildings. Some trees blown down. Considerable damage to mobile homes. Coastal and low-lying escape routes flood 2 to 4 hours before arrival of the hurricane center. Small craft in unprotected anchorages break moorings.
3	945–964	178–209	111–130	2.5–3.6	9–12	*Extensive.* Some structural damage to small residences and utility buildings. Large trees blown down. Mobile homes are destroyed. Flooding near the coast destroys smaller structures with larger structures damaged due to battering by floating debris. Terrain lower than 2 meters above sea level may be flooded inland 13 km or more. Evacuation of low-lying residences within several blocks of the shoreline may be required.
4	920–944	210–250	131–155	3.7–5.4	13–18	*Extreme.* Some complete roof structure failures on small residences. Extensive damage to doors and windows. Low-lying escape routes may be cut by rising water 3 to 5 hours before arrival of the hurricane center. Major damage to lower floors of structures near the shore. Terrain lower than 3 meters above sea level may be flooded, requiring massive evacuation of residential areas as far inland as 10 km.
5	<920	>250	>155	>5.4	>18	*Catastrophic.* Complete roof failure on many residences and industrial buildings. Some complete building failures. Severe window and door damage. Low-lying escape routes are cut by rising water 3 to 5 hours before arrival of the hurricane center. Major damage to lower floors of all structures located less than 5 meters above sea level and within 500 meters of the shoreline. Massive evacuation of residential areas on low ground within 8 to 16 km of the shoreline may be required.

Appendix G Climate Data

Table G-1 includes data for 51 stations around the world that represent many different climate types. Temperatures are given in degrees Celsius and precipitation in millimeters. Names and locations are given in Table G-2, along with the elevation (in meters) of each station and its Köppen classification. This format was used so that you can use the data in exercises to reinforce your understanding of climate controls and climate classification.

Use Figure 15-2 (page 411) to determine the proper Köppen classification. The flowchart in Figure G-1 will help guide you through the classification process. Once you have classified a station, determine a likely location, based on factors such as mean annual temperature, annual temperature range, total precipitation, and seasonal precipitation distribution. Your location need not be a specific city; it could be a description of the station's setting, such as "middle-latitude continental" or "subtropical with a strong monsoon influence." It would also be a good idea to list the reasons for your selection. You may check your answer by examining the list of stations in Table G-2.

If you simply wish to examine the data for a specific place or data for a specific climate type, consult the list in Table G-2.

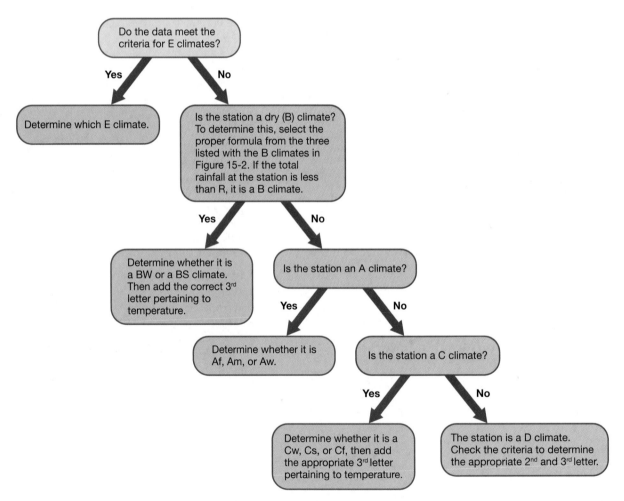

Figure G–1 Classifying Climates Using Figure 15-2, page 411

TABLE G–1 Selected Climate Data for the World

	J	F	M	A	M	J	J	A	S	O	N	D	YR.
1	1.7	4.4	7.9	13.2	18.4	23.8	25.8	24.8	21.4	14.7	6.7	2.8	13.8
	10	10	13	13	20	15	30	32	23	18	10	13	207
2	−10.4	−8.3	−4.6	3.4	9.4	12.8	16.6	14.9	10.8	5.5	−2.3	−6.4	3.5
	18	25	25	30	51	89	64	71	33	20	18	15	459
3	10.2	10.8	13.7	17.9	22.2	25.7	26.7	26.5	24.2	19.0	13.3	10.0	18.4
	66	84	99	74	91	127	196	168	147	71	53	71	1247
4	−23.9	−17.5	−12.5	−2.7	8.4	14.8	15.6	12.8	6.4	−3.1	−15.8	−21.9	−3.3
	23	13	18	8	15	33	48	53	33	20	18	15	297
5	−17.8	−15.3	−9.2	−4.4	4.7	10.9	16.4	14.4	10.3	3.3	−3.6	−12.8	−0.3
	58	58	61	48	53	61	81	71	58	61	64	64	738
6	−8.2	−7.1	−2.4	5.4	11.2	14.7	18.9	17.4	12.7	7.4	−0.4	−4.6	5.4
	21	23	29	35	52	74	39	40	35	29	26	21	424
7	12.8	13.9	15.0	15.0	17.8	20.0	21.1	22.8	22.2	18.3	17.2	15.0	17.6
	69	74	46	28	3	3	0	0	5	10	28	61	327
8	18.9	20.0	21.1	22.8	25.0	26.7	27.2	27.8	27.2	25.0	21.1	20.0	23.6
	51	48	58	99	163	188	172	178	241	208	71	43	1520
9	−4.4	−2.2	4.4	10.6	16.7	21.7	23.9	22.7	18.3	11.7	3.8	−2.2	10.4
	46	51	69	84	99	97	97	81	97	61	61	51	894
10	−5.6	−4.4	0.0	6.1	11.7	16.7	20.0	18.9	15.5	10.0	3.3	−2.2	7.5
	112	96	109	94	86	81	74	61	89	81	107	99	1089
11	−2.1	0.9	4.7	9.9	14.7	19.4	24.7	23.6	18.3	11.5	3.4	−0.2	10.7
	34	30	40	45	36	25	15	22	13	29	33	31	353
12	12.8	13.9	15.0	16.1	17.2	18.8	19.4	22.2	21.1	18.8	16.1	13.9	15.9
	53	56	41	20	5	0	0	2	5	13	23	51	269
13	−0.1	1.8	6.2	13.0	18.7	24.2	26.4	25.4	21.1	14.9	6.7	1.6	13.3
	50	52	78	94	95	109	84	77	70	73	65	50	897
14	2.7	3.2	7.1	13.2	18.8	23.4	25.7	24.7	20.9	15.0	8.7	3.4	13.9
	77	63	82	80	105	82	105	124	97	78	72	71	1036
15	12.8	15.0	18.9	21.1	26.1	31.1	32.7	33.9	31.1	22.2	17.7	13.9	23.0
	10	9	6	2	0	0	6	13	10	10	3	8	77
16	25.6	25.6	24.4	25.0	24.4	23.3	23.3	24.4	24.4	25.0	25.6	25.6	24.7
	259	249	310	165	254	188	168	117	221	183	213	292	2619
17	25.9	25.8	25.8	25.9	26.4	26.6	26.9	27.5	27.9	27.7	27.3	26.7	26.7
	365	326	383	404	185	132	68	43	96	99	189	143	2433
18	13.3	13.3	13.3	13.3	13.9	13.3	13.3	13.3	13.9	13.3	13.3	13.9	13.5
	99	112	142	175	137	43	20	30	69	112	97	79	1115
19	25.9	26.1	25.2	23.9	22.3	21.3	20.8	21.1	21.5	22.3	23.1	24.4	23.2
	137	137	143	116	73	43	43	43	53	74	97	127	1086
20	13.8	13.5	11.4	8.0	3.7	1.2	1.4	2.9	5.5	9.2	11.4	12.9	7.9
	21	16	18	13	25	15	15	17	12	7	15	18	171
21	1.5	1.3	3.1	5.8	10.2	12.6	15.0	14.7	12.0	8.3	5.5	3.3	7.8
	179	139	109	140	83	126	141	167	228	236	207	203	1958
22	−0.5	0.2	3.9	9.0	14.3	17.7	19.4	18.8	15.0	9.6	4.7	1.2	9.5
	41	37	30	39	44	60	67	65	45	45	44	39	556

(Continued)

TABLE G–1   *(Continued)*

	J	F	M	A	M	J	J	A	S	O	N	D	YR.
23	6.1	5.8	7.8	9.2	11.6	14.4	15.6	16.0	14.7	12.0	9.0	7.0	10.8
	133	96	83	69	68	56	62	80	87	104	138	150	1126
24	10.8	11.6	13.6	15.6	17.2	20.1	22.2	22.5	21.2	18.2	14.4	11.5	16.6
	111	76	109	54	44	16	3	4	33	62	93	103	708
25	−9.9	−9.5	−4.2	4.7	11.9	16.8	19.0	17.1	11.2	4.5	−1.9	−6.8	4.4
	31	28	33	35	52	67	74	74	58	51	36	36	575
26	8.0	9.0	10.9	13.7	17.5	21.6	24.4	24.2	21.5	17.2	12.7	9.5	15.9
	83	73	52	50	48	18	9	18	70	110	113	105	749
27	−9.0	−9.0	−6.6	−4.1	0.4	3.6	5.6	5.5	3.5	−0.6	−4.5	−7.6	−1.9
	202	180	164	166	197	249	302	278	208	183	190	169	2488
28	−2.9	−3.1	−0.7	4.4	10.1	14.9	17.8	16.6	12.2	7.1	2.8	0.1	6.6
	43	30	26	31	34	45	61	76	60	48	53	48	555
29	12.8	13.9	17.2	18.9	22.2	23.9	25.5	26.1	25.5	23.9	18.9	15.0	20.3
	66	41	20	5	3	0	0	0	3	18	46	66	268
30	24.6	24.9	25.0	24.9	25.0	24.2	23.7	23.8	23.9	24.2	24.2	24.7	24.4
	81	102	155	140	133	119	99	109	206	213	196	122	1675
31	21.1	20.4	20.9	21.7	23.0	26.0	27.3	27.3	27.5	27.5	26.0	25.2	24.3
	0	2	0	0	1	15	88	249	163	49	5	6	578
32	20.4	22.7	27.0	30.6	33.8	34.2	33.6	32.7	32.6	30.5	25.5	21.3	28.7
	0	0	0	0	0	2	1	11	2	0	0	0	16
33	17.8	18.1	18.8	18.8	17.8	16.2	14.9	15.5	16.8	18.6	18.3	17.8	17.5
	46	51	102	206	160	46	18	25	25	53	109	81	922
34	20.6	20.7	19.9	19.2	16.7	13.9	13.9	16.3	19.1	21.8	21.4	20.9	18.7
	236	168	86	46	13	8	0	3	8	38	94	201	901
35	11.7	13.3	16.7	18.6	19.2	20.0	20.3	20.5	20.5	19.1	15.9	12.9	17.4
	20	41	179	605	1705	2875	2455	1827	1231	447	47	5	11437
36	26.2	26.3	27.1	27.2	27.3	27.0	26.7	27.0	27.4	27.4	26.9	26.6	26.9
	335	241	201	141	116	97	61	50	78	91	151	193	1755
37	−0.8	2.6	5.3	8.5	13.1	17.0	17.2	17.3	15.3	11.5	5.7	0.3	9.4
	0	0	1	1	18	72	157	151	68	4	1	0	473
38	24.5	25.8	27.9	30.5	32.7	32.5	30.7	30.1	29.7	28.1	25.9	24.6	28.6
	24	7	15	25	52	53	83	124	118	267	308	157	1233
39	−18.7	−18.1	−16.7	−11.7	−5.0	0.6	5.3	5.8	1.4	−4.2	−12.3	−15.8	−7.5
	8	8	8	8	15	20	36	43	43	33	13	12	247
40	−21.9	−18.6	−12.5	−5.0	9.7	15.6	18.3	16.1	10.3	0.8	−10.6	−18.4	−1.4
	15	8	8	13	30	51	51	51	28	25	18	20	318
41	−4.7	−1.9	4.8	13.7	20.1	24.7	26.1	24.9	19.9	12.8	3.8	−2.7	11.8
	4	5	8	17	35	78	243	141	58	16	10	3	623
42	24.3	25.2	27.2	29.8	29.5	27.8	27.6	27.1	27.6	28.3	27.7	25.0	27.3
	8	5	6	17	260	524	492	574	398	208	34	3	2530
43	25.8	26.3	27.8	28.8	28.2	27.4	27.1	27.1	26.7	26.5	26.1	25.7	27.0
	6	13	12	65	196	285	242	277	292	259	122	37	1808
44	3.7	4.3	7.6	13.1	17.6	21.1	25.1	26.4	22.8	16.7	11.3	6.1	14.7
	48	73	101	135	131	182	146	147	217	220	101	60	1563

(Continued)

TABLE G–1 *(Continued)*

	J	F	M	A	M	J	J	A	S	O	N	D	YR.
45	−15.8	−13.6	−4.0	8.5	17.7	21.5	23.9	21.9	16.7	6.1	−6.2	−13.0	5.3
	8	15	15	33	25	33	16	35	15	47	22	11	276
46	−46.8	−43.1	−30.2	−13.5	2.7	12.9	15.7	11.4	2.7	−14.3	−35.7	−44.5	−15.2
	7	5	5	4	5	25	33	30	13	11	10	7	155
47	19.2	19.6	18.4	16.4	13.8	11.8	10.8	11.3	12.6	16.3	15.9	17.7	15.2
	84	104	71	109	122	140	140	109	97	106	81	79	1242
48	28.2	27.9	28.3	28.2	26.8	25.4	25.1	25.8	27.7	29.1	29.2	28.7	27.6
	341	338	274	121	9	1	2	5	17	66	156	233	1562
49	21.9	21.9	21.2	18.3	15.7	13.1	12.3	13.4	15.3	17.6	19.4	21.0	17.6
	104	125	129	101	115	141	94	83	72	80	77	86	1205
50	−7.2	−7.2	−4.4	−0.6	4.4	8.3	10.0	8.3	5.0	1.1	−3.3	−6.1	0.7
	84	66	86	62	89	81	79	94	150	145	117	79	1132
51	−4.4	−8.9	−15.5	−22.8	−23.9	−24.4	−26.1	−26.1	−24.4	−18.8	−10.0	−3.9	−17.4
	13	18	10	10	10	8	5	8	10	5	5	8	110

TABLE G–2 Locations and Climate Classifications for Table G–1

Station No.	City	Location	Elevation (m)	Köppen Classification
		North America		
1	Albuquerque, N.M.	lat. 35°05′N	1593	BWk
		long. 106°40′W		
2	Calgary, Canada	lat. 51°03′N	1062	Dfb
		long. 114°05′W		
3	Charleston, S.C.	lat. 32°47′N	18	Cfa
		long. 79°56′W		
4	Fairbanks, Alaska	lat. 64°50′N	134	Dfc
		long. 147°48′W		
5	Goose Bay, Canada	lat. 53°19′N	45	Dfb
		long. 60°33′W		
6	Lethbridge, Canada	lat. 49°40′N	920	Dfb
		long. 112°39′W		
7	Los Angeles, Calif.	lat. 34°00′N	29	BSk
		long. 118°15′W		
8	Miami, Fla.	lat. 25°45′N	2	Am
		long. 80°11′W		
9	Peoria, Ill.	lat. 40°45′N	180	Dfa
		long. 89°35′W		
10	Portland, Me.	lat. 43°40′N	14	Dfb
		long. 70°16′W		
11	Salt Lake City, Utah	lat. 40°46′N	1288	BSk
		long. 111°52′W		
12	San Diego, Calif.	lat. 32°43′N	26	BSk
		long. 117°10′W		
13	St. Louis, Mo.	lat. 38°39′N	172	Cfa
		long. 90°15′W		

(Continued)

TABLE G-2   *(Continued)*

Station No.	City	Location	Elevation (m)	Köppen Classification
14	Washington, D.C.	lat. 38°50′N	20	Cfa
		long. 77°00′W		
15	Yuma, Ariz.	lat. 32°40′N	62	BWh
		long. 114°40′W		
South America				
16	Iquitos, Peru	lat. 3°39′S	115	Af
		long. 73°18′W		
17	Manaus, Brazil	lat. 3°01′S	60	Am
		long. 60°00′W		
18	Quito, Ecuador	lat. 0°17′S	2766	Cfb
		long. 78°32′W		
19	Rio de Janeiro, Brazil	lat. 22°50′S	26	Aw
		long. 43°20′W		
20	Santa Cruz, Argentina	lat. 50°01′S	111	BSk
		long. 60°30′W		
Europe				
21	Bergen, Norway	lat. 60°24′N	44	Cfb
		long. 5°20′E		
22	Berlin, Germany	lat. 52°28′N	50	Cfb
		long. 13°26′E		
23	Brest, France	lat. 48°24′N	103	Cfb
		long. 4°30′W		
24	Lisbon, Portugal	lat. 38°43′N	93	Csa
		long. 9°05′W		
25	Moscow, Russia	lat. 55°45′N	156	Dfb
		long. 37°37′E		
26	Rome, Italy	lat. 41°52′N	3	Csa
		long. 12°37′E		
27	Santis, Switzerland	lat. 47°15′N	2496	ET
		long. 9°21′E		
28	Stockholm, Sweden	lat. 59°21′N	52	Dfb
		long. 18°00′E		
Africa				
29	Benghazi, Libya	lat. 32°06′N	25	BSh
		long. 20°06′E		
30	Coquilhatville, Zaire	lat. 0°01′N	21	Af
		long. 18°17′E		
31	Dakar, Senegal	lat. 14°40′N	23	BSh
		long. 17°28′W		
32	Faya, Chad	lat. 18°00′N	251	BWh
		long. 21°18′E		
33	Nairobi, Kenya	lat. 1°16′S	1791	Csb
		long. 36°47′E		
34	Harare, Zimbabwe	lat. 17°50′S	1449	Cwb
		long. 30°52′E		

(Continued)

TABLE G–2 *(Continued)*

Station No.	City	Location	Elevation (m)	Köppen Classification
Asia				
35	Cherrapunji, India	lat. 25°15′N	1313	Cwb
		long. 91°44′E		
36	Djakarta, Indonesia	lat. 6°11′S	8	Am
		long. 106°45′E		
37	Lhasa, Tibet	lat. 29°40′N	3685	Cwb
		long. 91°07′E		
38	Madras, India	lat. 13°00′N	16	Aw
		long. 80°11′E		
39	Novaya Zemlya, Russia	lat. 72°23′N	15	ET
		long. 54°46′E		
40	Omsk, Russia	lat. 54°48′N	85	Dfb
		long. 73°19′E		
41	Beijing, China	lat. 39°57′N	52	Dwa
		long. 116°23′E		
42	Rangoon, Myanmar	lat. 16°46′N	23	Am
		long. 96°10′E		
43	Ho Chi Minh City, Viet Nam	lat. 10°49′N	10	Aw
		long. 106°40′E		
44	Tokyo, Japan	lat. 35°41′N	6	Cfa
		long. 139°46′E		
45	Urumchi, China	lat. 43°47′N	912	Dfa
		long. 87°43′E		
46	Verkhoyansk, Russia	lat. 67°33′N	137	Dfd
		long. 133°23′E		
Australia and New Zealand				
47	Auckland, New Zealand	lat. 37°43′S	49	Csb
		long. 174°53′E		
48	Darwin, Australia	lat. 12°26′S	27	Aw
		long. 131°00′E		
49	Sydney, Australia	lat. 33°52′S	42	Cfb
		long. 151° 179E		
Greenland				
50	Ivigtut, Greenland	lat. 61°12′N	29	ET
		long. 48° 109W		
Antarctica				
51	McMurdo Station, Antarctica	lat. 77°53′S	2	EF
		long. 167°00′E		

Glossary

Absolute Humidity The mass of water vapor per volume of air (usually expressed as grams of water vapor per cubic meter of air).

Absolute Instability The condition of air that has an environmental lapse rate that is greater than the dry adiabatic rate (1°C per 100 meters).

Absolute Stability The condition of air that has an environmental lapse rate that is less than the wet adiabatic rate.

Absolute Zero The zero point on the Kelvin temperature scale, representing the temperature at which all molecular motion is presumed to cease.

Absorptivity A measure of the amount of radiant energy absorbed by a substance.

Acid Precipitation Rain or snow with a pH value that is less than the value for uncontaminated rain.

Adiabatic Temperature Change The cooling or warming of air caused when air is allowed to expand or is compressed, not because heat is added or subtracted.

Advection Horizontal convective motion, such as wind.

Advection Fog Fog formed when warm moist air is blown over a cool surface and chilled below the dew point.

Aerosols Tiny solid and liquid particles suspended in the atmosphere.

Aerovane A device that resembles a wind vane with a propeller at one end. Used to indicate wind speed and direction.

Air A mixture of many discrete gases, of which nitrogen and oxygen are most abundant, in which varying quantities of tiny solid and liquid particles are suspended.

Air Mass A large body of air, usually 1600 kilometers or more across, that is characterized by homogeneous physical properties at any given altitude.

Air-Mass Thunderstorm A localized thunderstorm that forms in a warm, moist, unstable air mass. This type of storm occurs most frequently in the afternoon in spring and summer.

Air-Mass Weather The conditions experienced in an area as an air mass passes over it. Because air masses are large and relatively homogeneous, air-mass weather will be fairly constant and may last for several days.

Air Pollutants Airborne particles and gases occurring in concentrations that endanger the health and well-being of organisms or disrupt the orderly functioning of the environment.

Air Pressure The force exerted by the weight of a column of air above a given point.

Air Quality Index (AQI) A standardized indicator for reporting daily air quality to the general public. It is calculated for five major pollutants regulated by the Clean Air Act.

Albedo The reflectivity of a substance, usually expressed as a percentage of the incident radiation reflected.

Aleutian Low A large cell of low pressure centered over the Aleutian Islands of the North Pacific during the winter.

Altimeter An aneroid barometer calibrated to indicate altitude instead of pressure.

Altitude (of the Sun) The angle of the Sun above the horizon.

Analog Method A statistical approach to weather forecasting in which current conditions are matched with records of similar past weather events with the idea that the succession of events in the past will be paralleled by current conditions.

Anemometer An instrument used to determine wind speed.

Aneroid Barometer An instrument for measuring air pressure; it consists of evacuated metal chambers that are very sensitive to variations in air pressure.

Annual Mean Temperature An average of the 12 monthly means.

Annual Temperature Range The difference between the warmest and coldest monthly means.

Anticyclone An area of high atmospheric pressure characterized by diverging and rotating winds and subsiding air aloft.

Anticyclonic Flow Winds blow out and flow clockwise about an anticyclone (high) in the Northern Hemisphere, and they blow out and flow counterclockwise about an anticyclone in the Southern Hemisphere.

Aphelion The point in the orbit of a planet that is farthest from the Sun.

Apparent Temperature The air temperature perceived by a person.

Arctic (A) Air Mass A bitterly cold air mass that forms over the frozen Arctic Ocean.

Arctic Sea Smoke A dense and often extensive steam fog occurring over high-latitude ocean areas in winter.

Arid *See* Desert.

Atmosphere The gaseous portion of a planet, the planet's envelope of air; one of the traditional subdivisions of Earth's physical environment.

Atmospheric Window Terrestrial radiation between 8 and 11 micrometers in length to which the troposphere is transparent.

Aurora A bright and ever-changing display of light caused by solar radiation interacting with the upper atmosphere in the region of the poles. It is called *aurora borealis* in the Northern Hemisphere and *aurora australis* in the Southern Hemisphere.

Automated Surface Observing System (ASOS) A widely used, standardized set of automated weather instruments that provide routine surface observations.

Autumnal Equinox *See* Equinox.

Azores High The name given to a subtropical anticyclone when it is situated over the eastern part of the North Atlantic Ocean.

Backdoor Cold Front A cold front moving toward the west or southwest along the Atlantic Seaboard.

Backing Wind Shift A wind shift in a counterclockwise direction, such as a shift from east to north.

Barograph A recording barometer.

Barometric Tendency *See* Pressure Tendency.

Beaufort Scale A scale that can be used for estimating wind speed when an anemometer is not available.

Bergeron Process A theory that relates the formation of precipitation to supercooled clouds, freezing nuclei, and the different saturation levels of ice and liquid water.

Bermuda High The name given to the subtropical high in the North Atlantic during the summer, when it is centered near the island of Bermuda.

Bimetal Strip A thermometer consisting of two thin strips of metal welded together, which have widely different coefficients of thermal expansion. When temperature changes, the two metals expand or contract unequally and cause changes in the curvature of the element. Commonly used in thermographs.

Biosphere The totality of life-forms on Earth.

Blizzard A violent and extremely cold wind, laden with dry snow picked up from the ground.

Bora In the region of the eastern shore of the Adriatic Sea, a cold, dry northeasterly wind that blows down from the mountains.

Buys Ballot's Law A law which states that with your back to the wind in the Northern Hemisphere, low pressure will be to your left and high pressure to your right. The reverse is true in the Southern Hemisphere.

Calorie The amount of heat required to raise the temperature of 1 gram of water 1°C.

Ceiling The height ascribed to the lowest layer of clouds or obscuring phenomena when the sky is reported as broken, overcast, or obscured and the clouds are not classified "thin" or "partial." The ceiling is termed *unlimited* when the foregoing conditions are not present.

Celsius Scale A temperature scale (at one time called the centigrade scale) devised by Anders Celsius in 1742 and used where the metric system is in use. For water at sea level, 0° is designated the ice point and 100° the steam point.

Chinook The name applied to a foehn wind in the Rocky Mountains.

Circle of Illumination The line (great circle) separating daylight from darkness on Earth.

Cirrus One of three basic cloud forms; also one of the three high cloud types. They are thin, delicate ice-crystal clouds often appearing as veil-like patches or thin, wispy fibers.

Climate A description of aggregate weather conditions; the sum of all statistical weather information that helps describe a place or region.

Climate Change A study dealing with variations in climate on many different time scales from decades to millions of years, and the possible causes of such variations.

Climate-Feedback Mechanisms Several different possible outcomes that may result when one of the atmospheric system's elements is altered.

Climate System The exchanges of energy and moisture occurring among the atmosphere, hydrosphere, lithosphere, biosphere, and cryosphere.

Cloud A form of condensation best described as a dense concentration of suspended water droplets or tiny ice crystals.

Cloud Condensation Nuclei Microscopic particles that serve as surfaces on which water vapor condenses.

Cloud Seeding The introduction into clouds of particles (most commonly dry ice or silver iodide) for the purpose of altering the cloud's natural development.

Clouds of Vertical Development A cloud that has its base in the low height range and extends upward into the middle or high altitudes.

Cold Front The discontinuity at the forward edge of an advancing cold air mass that is displacing warmer air in its path.

Cold-Type Occluded Front A front that forms when the air behind the cold front is colder than the air underlying the warm front it is overtaking.

Cold Wave A rapid and marked fall in temperature. The National Weather Service applies this term to a fall in temperature in 24 hours equaling or exceeding a specified number of degrees and reaching a specified minimum temperature or lower. These specifications vary for different parts of the country and for different periods of the year.

Collision–Coalescence Process A theory of raindrop formation in warm clouds (above 0°C) in which large cloud droplets ("giants") collide and join together with smaller droplets to form a raindrop. Opposite electrical charges may bind the cloud droplets together.

Condensation The change of state from a gas to a liquid.

Conditional Instability The condition of moist air with an environmental lapse rate between the dry and wet adiabatic rates.

Conduction The transfer of heat through matter by molecular activity. Energy is transferred during collisions among molecules.

Conservation of Angular Momentum, Law of The product of the velocity of an object around a center of rotation (axis) and the distance of the object from the axis is constant.

Constant-Pressure Surface A surface along which the atmospheric pressure is everywhere equal at any given moment.

Continental (c) Air Mass An air mass that forms over land; it is normally relatively dry.

Continental Climate A climate lacking marine influence and characterized by more extreme temperatures than in marine climates; therefore, it has a relatively high annual temperature range for its latitude.

Contrail A cloudlike streamer frequently observed behind aircraft flying in clear, cold, and humid air and caused by the addition to the atmosphere of water vapor from engine exhaust gases.

Controls of Temperature Factors that cause variations in temperature from place to place, such as latitude and altitude.

Convection The transfer of heat by the movement of a mass or substance. It can take place only in fluids.

Convection Cell Circulation that results from the uneven heating of a fluid; the warmer parts of the fluid expand and rise because of their buoyancy, and the cooler parts sink.

Convergence The condition that exists when the wind distribution within a given region results in a net horizontal inflow of air into the area. Because convergence at lower levels is associated with an upward movement of air, areas of convergent winds are regions favorable to cloud formation and precipitation.

Cooling Degree-Day Each degree of temperature of the daily mean above 65°F. The amount of energy required to maintain a certain temperature in a building is proportional to the cooling degree-days total.

Coriolis Effect The deflective effect of Earth's rotation on all free-moving objects, including the atmosphere and oceans. Deflection is to the right in the Northern Hemisphere and to the left in the Southern Hemisphere.

Corona A bright, whitish disk centered on the Moon or Sun that results from diffraction when the objects are veiled by a thin cloud layer.

Country Breeze A circulation pattern characterized by a light wind blowing into a city from the surrounding countryside. It is best developed on clear and otherwise calm nights when the urban heat island is most pronounced.

Cryosphere Collective term for the ice and snow that exist on Earth. One of the *spheres* of the climate system.

Cumulus One of three basic cloud forms; also the name of one of the clouds of vertical development. Cumulus are billowy, individual cloud masses that often have flat bases.

Cumulus Stage The initial stage in thunderstorm development, in which the growing cumulonimbus is dominated by strong updrafts.

Cup Anemometer *See* Anemometer.

Cyclogenesis The process that creates or develops a new cyclone; also the process that produces an intensification of a preexisting cyclone.

Cyclone An area of low atmospheric pressure characterized by rotating and converging winds and ascending air.

Cyclonic Flow Winds blowing in and counterclockwise about a cyclone (low) in the Northern Hemisphere and in and clockwise about a cyclone in the Southern Hemisphere.

Daily Mean Temperature The mean temperature for a day that is determined by averaging the hourly readings or, more commonly, by averaging the maximum and minimum temperatures for a day.

Daily Temperature Range The difference between the maximum and minimum temperatures for a day.

Dart Leader *See* Leader.

Deposition The process whereby water vapor changes directly to ice, without going through the liquid state.

Desert One of the two types of dry climate—the driest of the dry climates.

Dew A form of condensation consisting of small water drops on grass or other objects near the ground that forms when the surface temperature drops below the dew point. Usually associated with radiation cooling on clear, calm nights.

Dew Point The temperature to which air has to be cooled in order to reach saturation.

Diffused Light Solar energy is scattered and reflected in the atmosphere and reaches Earth's surface in the form of diffuse blue light from the sky.

Discontinuity A zone characterized by a comparatively rapid transition of meteorological elements.

Dispersion The separation of colors by refraction.

Dissipating Stage The final stage of a thunderstorm that is dominated by downdrafts and entrainment leading to the evaporation of the cloud structure.

Diurnal Daily, especially pertaining to actions that are completed within 24 hours and that recur every 24 hours.

Divergence The condition that exists when the distribution of winds within a given area results in a net horizontal outflow of air from the region. In divergence at lower levels, the resulting deficit is compensated for by a downward movement of air from aloft; hence, areas of divergent winds are unfavorable to cloud formation and precipitation.

Doldrums The equatorial belt of calms or light variable winds lying between the two trade wind belts.

Doppler Radar A type of radar that has the capacity of detecting motion directly.

Drizzle Precipitation from stratus clouds consisting of tiny droplets.

Dry Adiabatic Rate The rate of adiabatic cooling or warming in unsaturated air. The rate of temperature change is 1°C per 100 meters.

Dry Climate A climate in which yearly precipitation is not as great as the potential loss of water by evaporation.

Dryline A narrow zone in the atmosphere along which there is an abrupt change in moisture as when dry continental tropical air converges with humid maritime tropical air. The denser cT air acts to lift the less dense mT air producing clouds and storms.

Dry-Summer Subtropical Climate A climate located on the west sides of continents between latitudes 30° and 45°. It is the only humid climate with a strong winter precipitation maximum.

Dynamic Seeding A type of cloud seeding that uses massive seeding, a process resulting in an increase of the release of latent heat and causing the cloud to grow larger.

Easterly Wave A large migratory wavelike disturbance in the trade winds that sometimes triggers the formation of a hurricane.

Eccentricity The variation of an ellipse from a circle.

Electromagnetic Radiation *See* Radiation.

Elements (Atmospheric) Quantities or properties of the atmosphere that are measured regularly and that are used to express the nature of weather and climate.

El Niño The name given to the periodic warming of the ocean that occurs in the central and eastern Pacific. A major El Niño episode can cause extreme weather in many parts of the world. The opposite of La Niña.

Entrainment The infiltration of surrounding air into a vertically moving air column. For example, the influx of cool, dry air into the downdraft of a cumulonimbus cloud; a process that acts to intensify the downdraft.

Environmental Lapse Rate The rate of temperature decrease with height in the troposphere.

Equatorial Low A quasi-continuous belt of low pressure lying near the equator and between the subtropical highs.

Equinox The point in time when the vertical rays of the Sun are striking the equator. In the Northern Hemisphere, March 20 or 21 is the *vernal*, or *spring, equinox* and September 22 or 23 is the *autumnal equinox*. Lengths of daylight and darkness are equal at all latitudes at equinox.

Evaporation The process by which a liquid is transformed into gas.

Eye A roughly circular area of relatively light winds and fair weather at the center of a hurricane.

Eye Wall The doughnut-shaped area of intensive cumulonimbus development and very strong winds that surrounds the eye of a hurricane.

Fahrenheit Scale A temperature scale devised by Gabriel Daniel Fahrenheit in 1714 and used in the English system. For water at sea level, 32° is designated the ice point and 212° the steam point.

Fall Wind *See* Katabatic Wind.

Ferrel Cell The middle cell in the three-cell global circulation model; named for William Ferrel.

Fixed Points Reference points, such as the steam point and the ice point, used in the construction of temperature scales.

Flash The total discharge of lightning, which is usually perceived as a single flash of light but which actually consists of several flashes. *See also* Stroke.

Foehn A warm, dry wind on the lee side of a mountain range that owes its relatively high temperature largely to adiabatic heating during descent down mountain slopes.

Fog A cloud with its base at or very near Earth's surface.

Forecasting Skill *See* Skill.

Freezing The change of state from liquid to solid.

Freezing Nuclei Solid particles that have a crystal form resembling that of ice; they serve as cores for the formation of ice crystals.

Freezing Rain *See* Glaze.

Front A boundary (discontinuity) separating air masses of different densities, one warmer and often higher in moisture content than the other.

Frontal Fog Fog formed when rain evaporates as it falls through a layer of cool air.

Frontal Wedging The lifting of air resulting when cool air acts as a barrier over which warmer, lighter air will rise.

Frontogenesis The beginning or creation of a front.

Frontolysis The destruction and dying of a front.

Frost Ice crystals that occur when the temperature falls to 0°C or below. *See also* White Frost.

Fujita Intensity Scale (F-scale) A scale developed by T. Theodore Fujita for classifying the severity of a tornado, based on the correlation of wind speed with the degree of destruction.

Geosphere The solid Earth, the largest of Earth's four major spheres.

Geostationary Satellite A satellite that remains over a fixed point because its rate of travel corresponds to Earth's rate of rotation. Because the satellite must orbit at distances of about 35,000 kilometers, images from this type of satellite are not as detailed as those from polar satellites.

Geostrophic Wind A wind, usually above a height of 600 meters, that blows parallel to the isobars.

Glaze A coating of ice on objects formed when supercooled rain freezes on contact. A storm that produces glaze is termed an "icing storm."

Global Circulation The general circulation of the atmosphere; the average flow of air over the entire globe.

Glory A series of rings of colored light, most commonly appearing around the shadow of an airplane that is projected on clouds below.

Gradient Wind The curved airflow pattern around a pressure center resulting from a balance among pressure–gradient force, Coriolis force, and centrifugal force.

Greenhouse Effect The transmission of shortwave solar radiation by the atmosphere coupled with the selective absorption of longer-wavelength terrestrial radiation, especially by water vapor and carbon dioxide, resulting in warming of the atmosphere.

Growing Degree-Days A practical application of temperature data for determining the approximate date when crops will be ready for harvest.

Gust Front The boundary separating the cold downdraft from a thunderstorm and the relatively warm, moist surface air. Lifting along this boundary may initiate the development of thunderstorms.

Hadley Cell The thermally driven circulation system of equatorial and tropical latitudes consisting of two convection cells, one in each hemisphere. The existence of this circulation system was first proposed by George Hadley in 1735, as an explanation for the trade winds.

Hail Precipitation in the form of hard, round pellets or irregular lumps of ice that may have concentric shells formed by the successive freezing of layers of water.

Halo A narrow whitish ring of large diameter centered around the Sun. The commonly observed 22° *halo* subtends an angle of 22° from the observer.

Heat The kinetic energy of random molecular motion.

Heat Budget The balance of incoming and outgoing radiation.

Heating Degree-Day Each degree of temperature of the daily mean below 65°F is counted as one heating degree-day. The amount of heat required to maintain a certain temperature in a building is proportional to the heating degree-days total.

Heterosphere A zone of the atmosphere beyond about 80 kilometers, where the gases are arranged into four roughly spherical shells, each with a distinctive composition.

High Cloud A cloud that normally has its base above 6000 meters; the base may be lower in winter and at high-latitude locations.

Highland Climate Complex pattern of climate conditions associated with mountains. Highland climates are characterized by large differences that occur over short distances.

Homosphere A zone of atmosphere extending from Earth's surface to about 80 kilometers that is uniform in terms of the proportions of its component gases.

Horse Latitudes A belt of calms or light variable winds and subsiding air located near the center of the subtropical high.

Humid Continental Climate A relatively severe climate characteristic of broad continents in the middle latitudes between approximately 40° and 50° north latitude. This climate is not found in the Southern Hemisphere, where the middle latitudes are dominated by the oceans.

Humidity A general term referring to water vapor in the air.

Humid Subtropical Climate A climate generally located on the eastern side of a continent and characterized by hot, sultry summers and cool winters.

Hurricane A tropical cyclonic storm having minimum winds of 119 kilometers per hour; also known as *typhoon* (western Pacific) and *cyclone* (Indian Ocean).

Hurricane Warning A warning issued when sustained winds of 119 kilometers per hour or higher are expected within a specified coastal area in 24 hours or less.

Hurricane Watch An announcement aimed at specific coastal areas that a hurricane poses a possible threat, generally within 36 hours.

Hydrologic Cycle The continuous movement of water from the oceans to the atmosphere (by evaporation), from the atmosphere to the land (by condensation and precipitation), and from the land back to the sea (via stream flow).

Hydrophobic Nuclei Particles that are not efficient condensation nuclei. Small droplets will form on them whenever the relative humidity reaches 100 percent.

Hydrosphere The water portion of our planet; one of the traditional subdivisions of Earth's physical environment.

Hydrostatic Equilibrium The balance maintained between the force of gravity and the vertical pressure gradient that does not allow air to escape to space.

Hygrometer An instrument designed to measure relative humidity.

Hygroscopic Nuclei Condensation nuclei having a high affinity for water, such as salt particles.

Hypothesis A tentative explanation that is tested to determine whether it is valid.

Ice Cap Climate A climate that has no monthly means above freezing and supports no vegetative cover except in a few scattered high mountain areas. This climate, with its perpetual ice and snow, is confined largely to the ice sheets of Greenland and Antarctica.

Icelandic Low A large cell of low pressure centered over Iceland and southern Greenland in the North Atlantic during the winter.

Ice Point The temperature at which ice melts.

Ideal Gas Law The pressure exerted by a gas is proportional to its density and absolute temperature.

Inclination of the Axis The tilt of Earth's axis from the perpendicular to the plane of Earth's orbit (plane of the ecliptic). Currently, the inclination is about 23 1/2° away from the perpendicular.

Inferior Mirage A mirage in which the image appears below the true location of the object.

Infrared Radiation Radiation with a wavelength from 0.7 to 200 micrometers.

Interface A common boundary where different parts of a system interact.

Interference A phenomenon that occurs when light rays of different frequencies (i.e., colors) meet. Such interference results in the cancellation or subtraction of some frequencies, which is responsible for the colors associated with coronas.

Internal Reflection A reflection that occurs when light that is traveling through a transparent material, such as water, reaches the opposite surface and is reflected back into the material. This is an important factor in the formation of optical phenomena such as rainbows.

Intertropical Convergence Zone (ITCZ) The zone of general convergence between the Northern and Southern Hemisphere trade winds.

Ionosphere A complex atmospheric zone of ionized gases extending between 80 and 400 kilometers, thus coinciding with the lower thermosphere and heterosphere.

Isobar A line drawn on a map connecting points of equal barometric pressure, usually corrected to sea level.

Isohyet A line connecting places having equal rainfall.

Isotach A line connecting places having equal wind speed.

Isotherm A line connecting points of equal air temperature.

ITCZ *See* Intertropical Convergence Zone.

Jet Streak Areas of higher-velocity winds found within the jet stream.

Jet Stream Swift geostrophic airstreams in the upper troposphere that meander in relatively narrow belts.

Jungle An almost impenetrable growth of tangled vines, shrubs, and short trees characterizing areas where the tropical rain forest has been cleared.

Katabatic Wind The flow of cold, dense air downslope under the influence of gravity; the direction of this wind's flow is controlled largely by topography. Also called *fall winds*.

Kelvin Scale A temperature scale (also called the *absolute scale*) used primarily for scientific purposes and having intervals equivalent to those on the Celsius scale but beginning at absolute zero.

Köppen Classification Devised by Wladimir Köppen, a system for classifying climates that is based on mean monthly and annual values of temperature and precipitation.

Lake-Effect Snow Snow showers associated with a *cP* air mass to which moisture and heat are added from below as it traverses a large and relatively warm lake (such as one of the Great Lakes), rendering the air mass humid and unstable.

Land Breeze A local wind blowing from the land toward the sea during the night in coastal areas.

La Niña An episode of strong trade winds and unusually low sea-surface temperatures in the central and eastern Pacific. The opposite of *El Niño*.

Lapse Rate *See* Environmental Lapse Rate; Normal Lapse Rate.

Latent Heat The energy absorbed or released during a change of state.

Latent Heat of Condensation The energy released when water vapor changes to the liquid state. The amount of energy released is equivalent to the amount absorbed during evaporation.

Latent Heat of Vaporization The energy absorbed by water molecules during evaporation. It varies from about 600 calories per gram for water at 0°C to 540 calories per gram at 100°C.

Leader The conductive path of ionized air that forms near a cloud base prior to a lightning stroke. The initial conductive path is referred to as a *step leader* because it extends itself earthward in short, nearly invisible bursts. A *dart leader,* which is continuous and less branched than a step leader, precedes each subsequent stroke along the same path.

Lifting Condensation Level The height at which rising air that is cooling at the dry adiabatic rate becomes saturated and condensation begins.

Lightning A sudden flash of light generated by the flow of electrons between oppositely charged parts of a cumulonimbus cloud or between the cloud and the ground.

Liquid-in-Glass Thermometer A device for measuring temperature that consists of a tube with a liquid-filled bulb at one end. The expansion or contraction of the fluid indicates temperature.

Long-Range Forecasting Estimating rainfall and temperatures for a period beyond 3 to 5 days, usually for 30-day periods. Such forecasts are not as detailed or reliable as those for shorter periods.

Longwave Radiation A reference to radiation emitted by Earth. Wavelengths are roughly 20 times longer than those emitted by the Sun.

Looming A mirage that allows objects that are below the horizon to be seen.

Low Cloud A cloud that forms below a height of about 2000 meters.

Macroscale Winds Phenomena such as cyclones and anticyclones that persist for days or weeks and have a horizontal dimension of hundreds to several thousands of kilometers; also, features of the atmospheric circulation that persist for weeks or months and have horizontal dimensions of up to 10,000 kilometers.

Marine Climate A climate dominated by the ocean; because of the moderating effect of water, sites having this climate are considered relatively mild.

Marine West Coast Climate A climate found on windward coasts from latitudes 40° to 65° and dominated by maritime air masses. In this climate, winters are mild and summers are cool.

Maritime (m) Air Mass An air mass that originates over the ocean. These air masses are relatively humid.

Mature Stage The second of the three stages of a thunderstorm. This stage is characterized by violent weather as downdrafts exist side-by-side with updrafts.

Maximum Thermometer A thermometer that measures the maximum temperature for a given period in time, usually 24 hours. A constriction in the base of the glass tube allows mercury to rise but prevents it from returning to the bulb until the thermometer is shaken or whirled.

Mediterranean Climate A common name applied to the dry-summer subtropical climate.

Melting The change of state from solid to liquid.

Mercury Barometer A mercury-filled glass tube in which the height of the column of mercury is a measure of air pressure.

Mesocyclone A vertical cylinder of cyclonically rotating air (3 to 10 kilometers in diameter) that develops in the updraft of a severe thunderstorm and that often precedes the development of damaging hail or tornadoes.

Mesopause The boundary between the mesosphere and the thermosphere.

Mesoscale Convective Complex (MCC) A slow-moving roughly circular cluster of interacting thunderstorm cells covering an area of thousands of square kilometers that may persist for 12 hours or more.

Mesoscale Winds Small convective cells that exist for minutes or hours, such as thunderstorms, tornadoes, and land and sea breezes. Typical horizontal dimensions range from 1 to 100 kilometers.

Meteorology The scientific study of the atmosphere and atmospheric phenomena; the study of weather and climate.

Microscale Winds Phenomena such as turbulence, with life spans of less than a few minutes that affect small areas and are strongly influenced by local conditions of temperature and terrain.

Middle Cloud A cloud occupying the height range from 2000 to 6000 meters.

Midlatitude Cyclone A large low-pressure center with diameter often exceeding 1000 kilometers that moves from west to east and may last from a few days to more than a week and usually has a cold front and a warm front extending from the central area of low pressure.

Midlatitude Jet Stream A jet stream that migrates between latitudes 30° and 70°.

Millibar The standard unit of pressure measurement used by the National Weather Service. One millibar (mb) equals 100 newtons per square meter.

Minimum Thermometer A thermometer that measures the minimum temperature for a given period of time, usually 24 hours. By checking the small dumbbell-shaped index, the minimum temperature can be read.

Mirage An optical effect of the atmosphere caused by refraction in which the image of an object appears displaced from its true position.

Mistral A cold northwest wind that blows into the western Mediterranean basin from higher elevations to the north.

Mixing Depth The height to which convectional movements extend above Earth's surface. The greater the mixing depths, the better the air quality.

Mixing Ratio The mass of water vapor in a unit mass of dry air; commonly expressed as grams of water vapor per kilogram of dry air.

Monsoon The seasonal reversal of wind direction associated with large continents, especially Asia. In winter, the wind blows from land to sea; in summer, it blows from sea to land.

Monthly Mean Temperature The mean temperature for a month that is calculated by averaging the daily means.

Mountain Breeze The nightly downslope winds commonly encountered in mountain valleys.

Multiple-Vortex Tornado A tornado that contains several smaller intense whirls called suction vortices that orbit the center of the larger tornado circulation.

National Weather Service (NWS) The federal agency responsible for gathering and disseminating weather-related information.

Negative-Feedback Mechanism As used in climatic change, any effect that is opposite of the initial change and tends to offset it.

Newton A unit of force used in physics. One newton is the force necessary to accelerate 1 kilogram of mass 1 meter per second squared.

Nor'easter The term used to describe the weather associated with an incursion of *mP* air into the Northeast from the North Atlantic; strong northeast winds, freezing or near-freezing temperatures, and the possibility of precipitation make this an unwelcome weather event.

Normal Lapse Rate The average drop in temperature with increasing height in the troposphere; about 6.5°C per kilometer.

Nowcasting Short-term weather forecasting techniques that are generally applied to predicting severe weather.

Numerical Weather Prediction (NWP) Forecasting the behavior of atmospheric disturbances based upon the solution of the governing fundamental equations of hydrodynamics, subject to the observed initial conditions. Because of the vast number of calculations involved, high-speed computers are always used for NWP.

Obliquity The angle between the planes of Earth's equator and orbit.

Occluded Front A front formed when a cold front overtakes a warm front.

Occlusion The overtaking of one front by another.

Ocean Current The mass movement of ocean water that is either wind driven or initiated by temperature and salinity conditions that alter the density of seawater.

Orographic Lifting The process in which mountains or highlands act as barriers to the flow of air and force the air to ascend. The air cools adiabatically, and clouds and precipitation may result.

Outgassing The release of gases dissolved in molten rock.

Overrunning Warm air gliding up a retreating cold air mass.

Oxygen-Isotope Analysis A method of deciphering past temperatures based on the precise measurement of the ratio between two isotopes of oxygen, ^{16}O and ^{18}O. Analysis is commonly made of seafloor sediments and cores from ice sheets.

Ozone A molecule of oxygen containing three oxygen atoms.

Paleoclimatology The study of ancient climates; the study of climate and climate change prior to the period of instrumental records using proxy data.

Paleosol Old, buried soil that may furnish some evidence of the nature of past climates because climate is the most important factor in soil formation.

Parcel An imaginary volume of air enclosed in a thin elastic cover. Typically it is considered to be a few hundred cubic meters in volume and is assumed to act independently of the surrounding air.

Perihelion The point in the orbit of a planet closest to the Sun.

Permafrost The permanent freezing of the subsoil in tundra regions.

Persistence Forecast A forecast that assumes that the weather occurring upstream will persist and move on and will affect the areas in its path in much the same way. Persistence forecasts do not account for changes that might occur in the weather system.

pH Scale A 0-to-14 scale that is used for expressing the exact degree of acidity or alkalinity of a solution. A pH of 7 signifies a neutral solution. Values below 7 signify an acid solution, and values above 7 signify an alkaline solution.

Photochemical Reaction A chemical reaction in the atmosphere triggered by sunlight, often yielding a secondary pollutant.

Photosynthesis The production of sugars and starches by plants using air, water, sunlight, and chlorophyll. In the process, atmospheric carbon dioxide is changed to organic matter and oxygen is released.

Plane of the Ecliptic The plane of Earth's orbit around the Sun.

Plate Tectonics Theory A theory which states that the outer portion of Earth is made up of several individual pieces, called *plates*, which move in relation to one another upon a partially molten zone below. As plates move, so do continents, which explains some climatic changes in the geologic past.

Polar (P) Air Mass A cold air mass that forms in a high-latitude source region.

Polar Climate A climate in which the mean temperature of the warmest month is below 10°C; a climate that is too cold to support the growth of trees.

Polar Easterlies In the global pattern of prevailing winds, winds that blow from the polar high toward the subpolar low. These winds, however, should not be thought of as persistent winds, such as the trade winds.

Polar Front The stormy frontal zone separating air masses of polar origin from air masses of tropical origin.

Polar Front Theory A theory developed by J. Bjerknes and other Scandinavian meteorologists in which the polar front, separating polar and tropical air masses, gives rise to cyclonic disturbances that intensify and move along the front and pass through a succession of stages.

Polar High Anticyclones that are assumed to occupy the inner polar regions and are believed to be thermally induced, at least in part.

Polar-Orbiting Satellite Satellites that orbit the poles at rather low altitudes of a few hundred kilometers and require only 100 minutes per orbit.

Positive-Feedback Mechanism As used in climatic change, any effect that acts to reinforce an initial change.

Potential Energy Energy that exists by virtue of a body's position with respect to gravity.

Precession The slow migration of Earth's axis that traces a cone over a period of 26,000 years.

Precipitation Fog *See* Frontal Fog.

Pressure Gradient The amount of pressure change occurring over a given distance.

Pressure Tendency The nature of the change in atmospheric pressure over the past several hours. It can be a useful aid in short-range weather prediction.

Prevailing Westerlies The dominant west-to-east motion of the atmosphere that characterizes the regions on the poleward side of the subtropical highs.

Prevailing Wind A wind that consistently blows from one direction more than from any other.

Primary Pollutant A pollutant emitted directly from an identifiable source.

Prognostic Chart A computer-generated forecast showing the expected pressure pattern at a specified future time. Anticipated positions of fronts are also included. They usually represent the graphical output associated with a numerical weather prediction model.

Proxy Data Data gathered from natural recorders of climate variability, such as tree rings, ice cores, and ocean-floor sediments.

Psychrometer A device consisting of two thermometers (wet bulb and dry bulb) that is rapidly whirled and, with the use of tables, yields the relative humidity and dew point.

Radiation The wavelike energy emitted by any substance that possesses heat. This energy travels through space at 300,000 kilometers per second (the speed of light).

Radiation Fog Fog resulting from radiation cooling of the ground and adjacent air; primarily a nighttime and early morning phenomenon.

Radiosonde A lightweight package of weather instruments fitted with a radio transmitter and carried aloft by a balloon.

Rainbow A luminous arc formed by the refraction and reflection of light in drops of water.

Rain Shadow Desert A dry area on the lee side of a mountain range.

Rawinsonde A radiosonde that is tracked by radio-location devices in order to obtain data on upper-air winds.

Reflection, Law of A law that states that the angle of incidence (incoming ray) is equal to the angle of reflection (outgoing ray).

Refraction The bending of light as it passes obliquely from one transparent medium to another.

Relative Humidity The ratio of the air's water-vapor content to its water-vapor capacity.

Return Stroke The electric discharge resulting from the downward (earthward) movement of electrons from successively higher levels along the conductive path of lightning.

Revolution The motion of one body about another, as Earth about the Sun.

Ridge An elongate region of high atmospheric pressure.

Rime A delicate accumulation of ice crystals formed when supercooled fog or cloud droplets freeze on contact with objects.

Rossby Waves Upper-air waves in the middle and upper troposphere of the middle latitudes with wavelengths of from 4000 to 6000 kilometers; named for C. G. Rossby, the meteorologist who developed the equations for parameters governing the waves.

Rotation The spinning of a body, such as Earth, about its axis.

Saffir–Simpson Scale A scale, from 1 to 5, used to rank the relative intensities of hurricanes.

Santa Ana The local name given a foehn wind in southern California.

Saturation The maximum possible quantity of water vapor that the air can hold at any given temperature and pressure.

Saturation Vapor Pressure The vapor pressure, at a given temperature, wherein the water vapor is in equilibrium with a surface of pure water or ice.

Savanna A tropical grassland, usually with scattered trees and shrubs.

Sea Breeze A local wind that blows from the sea toward the land during the afternoon in coastal areas.

Secondary Pollutant A pollutant that is produced in the atmosphere by chemical reactions occurring among primary pollutants.

Semiarid *See* Steppe.

Sensible Heat The heat we can feel and measure with a thermometer.

Severe Thunderstorm A thunderstorm that produces frequent lightning, locally damaging wind, or hail that is 2 centimeters or more in diameter. In the middle latitudes, most thunderstorms form along or ahead of cold fronts.

Shortwave Radiation Radiation emitted by the Sun.

Siberian High The high-pressure center that forms over the Asian interior in January and produces the dry winter monsoon for much of the continent.

Skill An index of the degree of accuracy of a set of forecasts as compared to forecasts based on some standard, such as chance or climatic data.

Sleet Frozen or semifrozen rain formed when raindrops pass through a subfreezing layer of air.

Smog A word currently used as a synonym for general air pollution. It was originally created by combining the words "smoke" and "fog."

Snow Precipitation in the form of white or translucent ice crystals, chiefly in complex branched hexagonal form and often clustered into snowflakes.

Solstice The point in time when the vertical rays of the Sun are striking either the Tropic of Cancer (summer solstice in the Northern Hemisphere) or the Tropic of Capricorn (winter solstice in the Northern Hemisphere). Solstice represents the longest or shortest day (length of daylight) of the year.

Source Region The area where an air mass acquires its characteristic properties of temperature and moisture.

Southern Oscillation The seesaw pattern of atmospheric pressure change that occurs between the eastern and western Pacific. The interaction of this effect and that of El Niño can cause extreme weather events in many parts of the world.

Specific Heat The amount of heat needed to raise 1 gram of a substance 1°C at sea-level atmospheric pressure.

Specific Humidity The mass of water vapor per unit mass of air, including the water vapor (usually expressed as grams of water vapor per kilogram of air).

Squall Line Any nonfrontal line or narrow band of active thunderstorms.

Stable Air Air that resists vertical displacement. If it is lifted, adiabatic cooling will cause its temperature to be lower than the surrounding environment; if it is allowed, it will sink to its original position.

Standard Rain Gauge A gauge that has a diameter of about 20 centimeters and funnels rain into a cylinder that magnifies precipitation amounts by a factor of 10, allowing for accurate measurement of small amounts.

Static Seeding The most commonly used technique of cloud seeding, based on the assumption that cumulus clouds are deficient in freezing nuclei and that the addition of nuclei will spur additional precipitation formation.

Stationary Front A situation in which the surface position of a front does not move; the flow on either side of such a boundary is nearly parallel to the position of the front.

Statistical Methods (Forecasting) Methods in which tables or graphs are prepared from a long series of observations to show the probability of certain weather events under certain conditions of pressure, temperature, or wind direction.

Steam Fog Fog that has the appearance of steam and that is produced by evaporation from a warm water surface into the cool air above.

Steam Point The temperature at which water boils.

Step Leader *See* Leader.

Steppe One of the two types of dry climate; a marginal and more humid variant of the desert that separates it from bordering humid climates. Steppe also refers to the short-grass vegetation associated with this semiarid climate.

Storm Surge The abnormal rise of the sea along a shore as a result of strong winds.

Stratopause The boundary between the stratosphere and the mesosphere.

Stratosphere The zone of the atmosphere above the troposphere characterized at first by isothermal conditions and then a gradual temperature increase. Earth's ozone is concentrated here.

Stratus One of three basic cloud forms; also the name of a type of low cloud. Stratus clouds are sheets or layers that cover much or all of the sky.

Stroke One of the individual components that make up a flash of lightning. There are usually three to four strokes per flash, roughly 50 milliseconds apart.

Subarctic Climate A climate found north of the humid continental climate and south of the polar climate that is characterized by bitterly cold winters and short cool summers. Places within this climatic realm experience the highest annual temperature ranges on Earth.

Sublimation The process whereby a solid changes directly to a gas, without going through the liquid state.

Subpolar Low Low pressure located at about the latitudes of the Arctic and Antarctic Circles. In the Northern Hemisphere, the low takes the form of individual oceanic cells; in the Southern Hemisphere, there is a deep and continuous trough of low pressure.

Subsidence An extensive sinking motion of air, most frequently occurring in anticyclones. The subsiding air is warmed by compression and becomes more stable.

Subtropical High Several semipermanent anticyclonic centers characterized by subsidence and divergence located roughly between latitudes 25° and 35°.

Summer Solstice *See* Solstice.

Sun Dogs Two bright spots of light, sometimes called "mock suns," that sit at a distance of 22° on either side of the Sun.

Sun Pillar Shafts of light caused by reflection from ice crystals that extend upward or, less commonly, downward from the Sun when the Sun is near the horizon.

Sunspot A dark area on the Sun associated with powerful magnetic storms that extend from the Sun's surface deep into the interior.

Supercell A type of thunderstorm that consists of a single, persistent, and very powerful cell (updraft and downdraft) and that often produces severe weather, including hail and tornadoes.

Supercooled The condition of water droplets that remain in the liquid state at temperatures well below 0°C.

Superior Mirage A mirage in which the image appears above the true position of the object.

Synoptic Weather Forecasting A system of forecasting based on careful studies of synoptic weather charts over a period of years; from such studies a set of empirical rules is established to aid the forecaster in estimating the rate and direction of weather-system movements.

Synoptic Weather Map A weather map describing the state of the atmosphere over a large area at a given moment.

Taiga The northern coniferous forest; also a name applied to the subarctic climate.

Temperature A measure of the degree of hotness or coldness of a substance.

Temperature Gradient The amount of temperature change per unit of distance.

Temperature Inversion A layer in the atmosphere of limited depth where the temperature increases rather than decreases with height.

Theory A well-tested and widely accepted view that explains certain observable facts.

Thermal An example of convection that involves the upward movements of warm, less dense air. In this manner, heat is transported to greater heights.

Thermal Low An area of low atmospheric pressure created by abnormal surface heating.

Thermistor An electric thermometer consisting of a conductor whose resistance to the flow of current is temperature dependent; commonly used in radiosondes.

Thermocouple An electric thermometer that operates on the principle that differences in temperature between the junction of two unlike metal wires in a circuit will induce a current to flow.

Thermograph An instrument that continuously records temperature.

Thermometer An instrument for measuring temperature; in meteorology, a thermometer is generally used to measure the temperature of the air.

Thermosphere The zone of the atmosphere beyond the mesosphere in which there is a rapid rise in temperature with height.

Thunder The sound emitted by rapidly expanding gases along the channel of a lightning discharge.

Thunderstorm A storm produced by a cumulonimbus cloud and always accompanied by lightning and thunder. It is of relatively short duration and usually accompanied by strong wind gusts, heavy rain, and sometimes hail.

Tipping-Bucket Gauge A recording rain gauge consisting of two compartments ("buckets"), each capable of holding 0.025 centimeter of water. When one compartment fills, it tips, and the other compartment takes its place.

Tornado A violently rotating column of air attended by a funnel-shaped or tubular cloud extending downward from a cumulonimbus cloud.

Tornado Warning A warning issued when a tornado has actually been sighted in an area or is indicated by radar.

Tornado Watch A forecast issued for areas of about 65,000 square kilometers, indicating that conditions are such that tornadoes may develop; a tornado watch is intended to alert people to the possibility of tornadoes.

Towering A mirage in which the size of an object is magnified.

Trace of Precipitation An amount of precipitation less than 0.025 centimeter.

Trade Winds Two belts of winds that blow almost constantly from easterly directions and are located on the equatorward sides of the subtropical highs.

Transpiration The release of water vapor to the atmosphere by plants.

Trend Forecast A short-range forecasting technique which assumes that the weather occurring upstream will persist and move on to affect the area in its path.

Tropic of Cancer The parallel of latitude, 23 1/2° north latitude, marking the northern limit of the Sun's vertical rays.

Tropic of Capricorn The parallel of latitude, 23 1/2° south latitude, marking the southern limit of the Sun's vertical rays.

Tropical (T) Air Mass A warm-to-hot air mass that forms in the subtropics.

Tropical Depression By international agreement, a tropical cyclone with maximum winds that do not exceed 61 kilometers per hour.

Tropical Disturbance A term used by the National Weather Service for a cyclonic wind system in the tropics that is in its formative stages.

Tropical Rain Forest A luxuriant broadleaf evergreen forest; also the name given the climate associated with this vegetation.

Tropical Storm By international agreement, a tropical cyclone with maximum winds between 61 and 115 kilometers per hour.

Tropical Wet and Dry A climate that is transitional between the wet tropics and the subtropical steppes.

Tropopause The boundary between the troposphere and the stratosphere.

Troposphere The lowermost layer of the atmosphere, marked by considerable turbulence and, in general, a decrease in temperature with increasing height.

Trough An elongate region of low atmospheric pressure.

Tundra Climate A climate found almost exclusively in the Northern Hemisphere and at high altitudes in many mountainous regions. A treeless climatic realm of sedges, grasses, mosses, and lichens dominated by a long, bitterly cold winter.

Ultraviolet Radiation Radiation with a wavelength from 0.2 to 0.4 micrometer.

Unstable Air Air that does not resist vertical displacement. If it is lifted, its temperature will not cool as rapidly as the surrounding environment, and so it will continue to rise on its own.

Upslope Fog Fog created when air moves up a slope and cools adiabatically.

Upwelling The process by which deep, cold, nutrient-rich water is brought to the surface, usually by coastal currents that move water away from the coast.

Urban Heat Island A city area where temperatures are generally higher than in surrounding rural areas.

U.S. Standard Atmosphere The idealized vertical distribution of atmospheric pressure, temperature, and density that represents average conditions in the atmosphere.

Valley Breeze The daily upslope winds commonly encountered in a mountain valley.

Vapor Pressure The part of the total atmospheric pressure attributable to its water-vapor content.

Veering Wind Shift A wind shift in a clockwise direction, such as a shift from east to south.

Vernal Equinox *See* Equinox.

Virga Wisps or streaks of water or ice particles that fall out of a cloud and evaporate before reaching Earth's surface.

Visibility The greatest distance at which prominent objects can be seen and identified by unaided, normal eyes.

Visible Light Radiation with a wavelength from 0.4 to 0.7 micrometer.

Warm Front The discontinuity at the forward edge of an advancing warm air mass that displaces cooler air in its path.

Warm-Type Occluded Front A front that forms when the air behind the cold front is warmer than the air underlying the warm front it is overtaking.

Water Hemisphere A term used to refer to the Southern Hemisphere, where the oceans cover 81 percent of the surface (compared to 61 percent in the Northern Hemisphere).

Wavelength The horizontal distance separating successive crests or troughs.

Weather The state of the atmosphere at any given time.

Weather Analysis The stage prior to developing a weather forecast. This stage involves collecting, compiling, and transmitting observational data.

Weather Forecasting Predicting the future state of the atmosphere.

Weather Modification Deliberate human intervention to influence and improve atmospheric processes.

Weighting Gauge A recording precipitation gauge consisting of a cylinder that rests on a spring balance.

Westerlies *See* Prevailing Westerlies.

Wet Adiabatic Rate The rate of adiabatic temperature change in saturated air. The rate of temperature change is variable, but it is always less than the dry adiabatic rate.

White Frost Ice crystals that form on surfaces instead of dew when the dew point is below freezing.

Wind Air flowing horizontally with respect to Earth's surface.

Windchill A measure of apparent temperature that uses the effects of wind and temperature on the cooling rate of the human body. The windchill chart translates the cooling power of the atmosphere with the wind to a temperature under nearly calm conditions.

Wind Vane An instrument used to determine wind direction.

Winter Solstice *See* Solstice.

World Meteorological Organization (WMO) Established by the United Nations, an organization that consists of more than 130 nations and is responsible for gathering needed observational data and compiling some general prognostic charts.

Index

Cloud Guide

JIM LEE/NOAA

Cumulus These clouds often have flat bottoms, rounded tops, and a "cellular" structure made up of individual clouds. (The word "cumulus" comes from the Latin word for "heap.") Cumulus clouds tend to grow vertically.

JIM LEE/NOAA

Cumulus Humilis Often called fair-weather cumulus, these small white individual masses lack conspicuous vertical development and rarely produce precipitation.

JIM LEE/NOAA

Nimbostratus These low clouds are thick gray layers that contain sufficient water to yield light-to-moderate precipitation.

JIM LEE/NOAA

Stratocumulus These are low, layered clouds that have regions of some vertical development. Differences in thickness create varying degrees of darkness when seen from below.

Middle Clouds: cloud bases 2-6km (6,500–20,000 ft)

SHUTTERSTOCK

Altocumulus These midlevel clouds are horizontally layered but exhibit varying thicknesses across their bases. Thicker areas can be arranged as parallel linear bands or as a series of individual puffs.

SHUTTERSTOCK

Altocumulus (Lenticular) These clouds are marked by their lens-shaped appearance. They usually form downwind of mountain barriers as horizontal airflow is disrupted into a sequence of waves.

JIM LEE/NOAA

Altostratus These are midlevel, layered clouds that produce gray skies and obscure the Sun or Moon enough to make them appear as poorly defined bright spots. In this example, the setting sun brightens the clouds near the horizon but the gray appearance remains elsewhere.

JIM LEE/NOAA

Altostratus (Multilayer) These are midlevel layered clouds that are dense enough to completely hide the Sun or Moon.

Cloud Guide

SHUTTERSTOCK

Cirrus These clouds are made exclusively of ice crystals. They are not as horizontally extensive as cirrostratus clouds.

SHUTTERSTOCK

Cirrocumulus These high clouds can produce striking skies. Composed of ice crystals, they often contain linear bands, numerous patches of greater vertical development, or both.

© Jim W. Lee

JIM LEE/NOAA

Cirrostratus These are thin layered clouds composed of ice crystals. They are relatively indistinct and give the sky a whitish appearance.

DENNIS TASA

Contrails A contrail is a long, narrow cloud that is formed as exhaust from a jet aircraft condenses in cold air at high altitude. Upper level winds may gradually cause contrails to spread out.

Cumulus Congestus These clouds have considerably more vertical development than cumulus humilis. They may produce heavy precipitation, but not the severe weather associated with some cumulonimbus clouds.

JIM LEE/NOAA

SHUTTERSTOCK

Cumulonimbus These clouds result from very strong updrafts that may push the cloud tops up to several kilometers into the stratosphere. Their characteristic feature is the anvil, a zone of ice crystals extending outward from the main portion of the cloud.

PUBLIC DOMAIN

Cumulonimbus with Mammatus These are dramatic features associated with some cumulonimbus, resulting from strong downdrafts and turbulence along the bases or margins of the clouds.

NOAA

Cumulonimbus with Wall Cloud A feature associated with some cumulonimbus clouds. When wall clouds are present, heavy rain, hail, and sometimes tornadoes can be expected.

Temperature and Precipitation Extremes

Northice, Greenland
-87°F (-66°C)

Snag, Yukon
-81°F (-63°C)

Henderson Lake, British Columbia
256" (650 cm)

Death Valley, California
134°F (57°C)

Batagues, Mexico
1.2" (3.0 cm)

Mt. Waialeale, Hawaii
460" (1168 cm)

Puako, Hawaii
8.9" (23 cm)

Mauna Kea Observatory, Hawaii
12°F (-11°C)

NORTH AMERICA

Quibdó, Colombia
354" (899 cm)

OCEANIA

SOUTH AMERICA

Arica, Chile
0.03" (0.1 cm)

Rivadavia, Argentina
120°F (49°C)

Sarmiento, Argentina
-27°F (-33°C)